Chemostratigraphy

Chemostratigraphy
Concepts, Techniques, and Applications

Edited by

Mu. Ramkumar

Department of Geology, Periyar University, Salem, Tamilnadu, India;
South East Asia Carbonate Research Laboratory (Seacarl), Universiti Teknologi Petronas, Tronoh, Malaysia

ELSEVIER

AMSTERDAM • BOSTON • HEIDELBERG • LONDON • NEW YORK • OXFORD
PARIS • SAN DIEGO • SAN FRANCISCO • SINGAPORE • SYDNEY • TOKYO

Elsevier
Radarweg 29, PO Box 211, 1000 AE Amsterdam, Netherlands
The Boulevard, Langford Lane, Kidlington, Oxford OX5 1GB, UK
225 Wyman Street, Waltham, MA 02451, USA

ISBN: 978-0-12-419968-2

British Library Cataloguing in Publication Data
A catalogue record for this book is available from the British Library

Library of Congress Cataloguing-in-Publication Data
A catalog record for this book is available from the Library of Congress

For Information on all Elsevier publications
visit our website at http://store.elsevier.com/

Contents

Contributors

Leho Ainsaar
Department of Geology, University of Tartu, Tartu, Estonia

Matthias Alberti
GeoZentrum Nordbayern, Fachgruppe Paläoumwelt, Universität Erlangen-Nürnberg, Erlangen, Germany

V. Balaram
CSIR, National Geophysical Research Institute, Hyderabad, Telangana, India

Z. Berner
Institute of Mineralogy and Geochemistry, Karlsruhe Institute of Technology, Karlsruhe, Germany

Leah Calvert
Barker Lemar Companies, West Des Moines, IA, USA; Department of Earth & Environmental Sciences, The University of Iowa, Iowa City, IA, USA

Wilker S. Cezario
NEG-LABISE, Department of Geology, Federal University of Pernambuco, Recife, Pernambuco, Brazil

Leticia Chiglino
NEG-LABISE, Department of Geology, Federal University of Pernambuco, Recife, Pernambuco, Brazil; PDU Geología y Recursos Minerales, CURE, Universidad de la República, Treinta y Tres, Uruguay

María José Comas-Rengifo
Departamento de Paleontología, Facultad de Ciências Geológicas, Universidad Complutense de Madrid, Madrid, Spain

A.M. Dayal
CSIR-National Geophysical Research Institute, Hyderabad, India

Luís V. Duarte
Departamento de Ciências da Terra, IMAR-CMA/MARE, Faculdade de Ciências e Tecnologia, Universidade de Coimbra, Largo Marquês de Pombal, Coimbra, Portugal

Valderez P. Ferreira
NEG-LABISE, Department of Geology, Federal University of Pernambuco, Recife, Pernambuco, Brazil

Robert Frei
Department of Geosciences and Natural Resource Management, Geology Section, University of Copenhagen and Nordic Center for Earth Evolution (NordCEE), Copenhagen, Denmark

Franz T. Fürsich
GeoZentrum Nordbayern, Fachgruppe Paläoumwelt, Universität Erlangen-Nürnberg, Erlangen, Germany

Claudio Gaucher
Departamento de Geología, Facultad de Ciencias, Universidad de la República, Montevideo, Uruguay

Esmat Ebne Haddad
Department of Geology, Fars Science and Research Branch, Islamic Azad University, Shiraz, Iran

Hiromichi Hirano
Department of Earth Sciences, Waseda University, Tokyo, Japan

S.M. Hussain
Department of Geology, University of Madras, Chennai, India

Elena Letnikova
Sobolev Institute of Geology and Mineralogy, Siberian Branch of the Russian Academy of Sciences, Novosibirsk, Russia; Geology-Geophysics Department of Novosibirsk State University, Novosibirsk, Russia

Haroldo M. Lima
NEG-LABISE, Department of Geology, Federal University of Pernambuco, Recife, Pernambuco, Brazil

J. Madhavaraju
Estación Regional del Noroeste, Instituto de Geología, Universidad Nacional Autónoma de México, Hermosillo, Sonora, México

P. Mahalakshmi
Department of Geology, University of Madras, Chennai, India

Devleena Mani
CSIR-National Geophysical Research Institute, Hyderabad, India

Tõnu Meidla
Department of Geology, University of Tartu, Tartu, Estonia

Nasim Mohammadzadeh
Department of Geology, Fars Science and Research Branch, Islamic Azad University, Shiraz, Iran

Philip Morath
EQT Production, Pittsburgh, PA, USA; Department of Geosciences, The Pennsylvania State University, University Park, PA, USA

R. Nagarajan
Department of Applied Geology, School of Engineering and Science, Curtin University, Miri, Sarawak, Malaysia

D.J. Patil
CSIR-National Geophysical Research Institute, Hyderabad, India

Natan S. Pereira
NEG-LABISE, Department of Geology, Federal University of Pernambuco, Recife, Pernambuco, Brazil

Natalia Pisareva
Sobolev Institute of Geology and Mineralogy, Siberian Branch of the Russian Academy of Sciences, Novosibirsk, Russia; Geology-Geophysics Department of Novosibirsk State University, Novosibirsk, Russia

Artem Proshenkin
Sobolev Institute of Geology and Mineralogy, Siberian Branch of the Russian Academy of Sciences, Novosibirsk, Russia; Geology-Geophysics Department of Novosibirsk State University, Novosibirsk, Russia

Ritesh Purohit
Government College, Sirohi, Rajasthan, India

Jyotsana Rai
Birbal Sahni Institute of Paleobotany, Lucknow, Uttar Pradesh, India

S. Ramasamy
Department of Geology, University of Madras, Chennai, India

Mu. Ramkumar
Department of Geology, Periyar University, Salem, Tamilnadu, India;
South East Asia Carbonate Research Laboratory (Seacarl), Universiti Teknologi Petronas, Tronoh, Malaysia

Mohammad Sahraeyan
Department of Geology, Khorasgan (Isfahan) Branch, Islamic Azad University, Isfahan, Iran

Pratul Kumar Saraswati
Department of Earth Sciences, IIT Bombay, Mumbai, India

M. Satyanarayanan
CSIR, National Geophysical Research Institute, Hyderabad, Telangana, India

Hossein Seif
Department of Geology, North Tehran Branch, Islamic Azad University, Tehran, Iran

Alcides N. Sial
NEG-LABISE, Department of Geology, Federal University of Pernambuco, Recife, Pernambuco, Brazil

Ricardo L. Silva
Basin and Reservoir Lab, Department of Earth Sciences, Faculty of Sciences, Dalhousie University, Life Science Center, Dalhousie University Halifax, Canada

Ashutosh K. Singh
Department of Geology, Centre of Advanced Studies, University of Delhi, Delhi, India

Devesh K. Sinha
Department of Geology, Centre of Advanced Studies, University of Delhi, Delhi, India

T. Sugantha
Department of Geology, Periyar University, Salem, Tamilnadu, India

Rui Tahara
Redpath Museum, McGill University, Montreal, QC, Canada

Vahid Tavakoli
School of Geology, College of Science, University of Tehran, Tehran, Iran

Manish Tiwari
National Centre for Antarctic & Ocean Research, Vasco-da-Gama, Goa, India

Janika Truumees
Institute for Geology and Palaeontology, University of Hamburg, Bundesstrasse, Hamburg, Germany

J. Ugeswari
Department of Geology, University of Madras, Chennai, India

Go-Ichiro Uramoto
Kochi Institute for Core Sample Research, Japan Agency for Marine-Earth Science and Technology (JAMSTEC), Nankoku, Kochi, Japan

Irina Vishnevskaya
Sobolev Institute of Geology and Mineralogy, Siberian Branch of the Russian Academy of Sciences, Novosibirsk, Russia; Geology-Geophysics Department of Novosibirsk State University, Novosibirsk, Russia

Timothy White
EMS Earth and Environmental Systems Institute, The Pennsylvania State University, University Park, PA, USA

Foreword

The study of earth materials, such as minerals, rocks, and fossils; soils, sediment, and dust; and water for their chemical makeup, is an emerging subdiscipline in geological sciences, but it started with a delay in India. This subject is commonly known now as *geochemistry.* With advancements in various analytical techniques and computing, it is now possible to determine elemental and isotopic abundances precisely and accurately, even at ultratrace levels. Until about three decades ago, geochemical tools were largely used for magmatic and metamorphic rocks (hard rock geochemistry). Realizing the potential of geochemistry and the progress made in understanding processes, timing, and evolution of and sources to hard rocks, geochemical tools are being extensively applied to sedimentary systems. This has recently led to the emergence of the subject *chemostratigraphy*, in which chemical signatures (elemental, radioactive–radiogenic, stable, and cosmogenic isotopes) in sediments and sedimentary rocks are used to unravel environmental conditions of sediment production and deposition, their sources, and their evolutionary changes from sedimentary packages. This is a significant development in geology and the book edited by Dr. Mu. Ramkumar has come at an appropriate time. The various papers in the book illustrate how chemical signatures of different kinds of sedimentary rocks, formed from Paleoproterozoic to Paleogene, could be used to infer physical/biological conditions involved in sediment production and deposition and their sources. Future generations of geoscience students will find this book very useful to begin their research in soft rock geochemistry.

27/02/2015
New Delhi

V. Rajamani
Emeritus Professor, JNU

Acknowledgments

A work of this magnitude requires the efforts of many individuals, in their individual capacity and their inclination toward a common goal. This edited volume is the result of such efforts by many. I have been put in the center stage as the editor not by my work alone, but also by the kind-hearted and cooperative and understanding souls they are: the contributors, the reviewers, the editorial team at Elsevier, the reviewers who have recommended my proposal in the first place and those at the helm of administrative and technical approval committees at Elsevier.

I record a deep sense of gratitude to Professor V. Rajamani, School of Environmental Sciences, Jawaharlal Nehru University, New Delhi for "initiating" me into the sediment geochemistry and laboratory protocols. The classic articles of M.A. Pearce and T.J. Jarvis have introduced me to this new realm of geosciences and I owe a great deal of thankfulness to them. Understanding on this subject was made easy through intense and lively discussions with Dr. Zsolt Berner, Senior Scientist and Head of Laboratories, Institute of Mineralogy and Geochemistry, Karlsruhe Institute of Technology (erstwhile University of Karlsruhe), Germany. Scientific collaborations with German institutions and scientists at Institute of Mineralogy and Geochemistry, Karlsruhe, Institute of Palaeoenvironment, Erlangen, University of Freiburg, Marburg University, etc., were made possible by the sponsorships from Alexander von Humboldt Foundation since the year 2000, for which I express my thankfulness. My involvement on this subject and scientific collaboration with national and international academic and research institutions has been and is being supported by research grants from various organizations, namely, the Alexander von Humboldt Foundation, Germany, the University Grants Commission, the Council of Scientific and Industrial Research, the Department of Science and Technology, the Oil Industry Development Board, the Oil and Natural Gas Corporation Limited, India, and the German Research Foundation, Germany, etc., for which I am thankful. My interest in this subject was sustained by reading the publications of Ian Veizer, J.L. Banner, Susan Gaffey, Uwe Brand, M.E. Tucker, H. Strauss, M.M. Joachimski, Gerta Keller, G. Dromart, A. Rangel, G.A. Shields, A. Kampschulte, P.B. Wignall, A.D. Miall, L.A. Spalletti, J. McManus, and many more authors whose papers became my favorites. Thanks are due to all of them for driving me to think ahead and experiment with chemostratigraphy. Ms. Sharmila Vadivelan and her team at Elsevier's production unit have done excellent work and are thanked for their professional and flawless handling.

This acknowledgment section would not be complete without thanking my wife Smt. A. Shanthy, daughter, Ra. Krushnakeerthana and son Ra. Shreelakshminarasimhan for their support and also for forbearing my absence while editing this volume.

Above all, I submit my thankfulness unto the lotus feet of The Lord Shree Ranganatha, for his boundless mercy showered on me, and by whose ordinance those mentioned above have shouldered the responsibilities and helped me to complete this work.

Mu. Ramkumar

CHAPTER

TOWARD STANDARDIZATION OF TERMINOLOGIES AND RECOGNITION OF CHEMOSTRATIGRAPHY AS A FORMAL STRATIGRAPHIC METHOD

1

Mu. Ramkumar[1,*]

*[1]Department of Geology, Periyar University, Salem, Tamilnadu, India; *South East Asia Carbonate Research Laboratory (Seacarl), Universiti Teknologi Petronas, Tronoh, Malaysia*

1.1 INTRODUCTION

For a long time, sedimentary geochemistry has been in use to understand the conditions of deposition, climatic variations, tectonic setting, provenance, reservoir characteristics, etc. However, characterization of depositional units for distinction and correlation based on stratigraphic variation of geochemical traits and usage of the term "chemostratigraphy" have been more frequent only from the 1980s. A search for this term in popular scientific databases such as www.Sciencedirect.com, www.GeoscienceWorld.org, and www.Springerlink.com, etc., also returns articles only from 1980s. Thenceforth, the number of articles published every year shows a steady increase. For example, www.Sciencedirect.com returns a total of 2151 journal articles, 125 books, and 32 reference works published by Elsevier; out of which only 142 were published prior to the year 1995. Oldest article was published in the year 1986 (Renard, 1986). However, there are other previous and contemporaneous publications that used stratigraphic geochemical variation to infer paleoclimate and paleoenvironments and to define specific geochronological or lithostratigraphic boundaries (for example, Keith and Weber, 1964; Scholle and Arthur, 1980; Berger and Vincent, 1981; Romein and Smit, 1981; Odin et al., 1982; Williams et al., 1983; Renard et al., 1984; Shackleton and Hall, 1984; Jorgensen, 1986; and many others and references cited therein). Already 266 articles were published in the year 2014, 254 articles were published in the year 2013, and the year 2012 ranks third with publication of 218 articles. Similar trends are observable in other scientific databases too. These statistics indicate the growing popularity and expanding applications of this subdiscipline of geoscience.

During the initial years, the publications documented stratigraphic variation of selective elemental concentrations and isotopic compositions for relating the observed changes with known geological events, and/or chrono, litho, and biostratigraphic boundaries (*for example*, Kaminski and Malmgren, 1989; Nandy et al., 1995). It means, chemostratigraphy was utilized only as a supplement to other lines of geological evidence for analyzing and/or documenting geological phenomena. Currently, chemostratigraphy finds its place in every conceivable geoscientific problem (Weissert et al., 2008) as could be observed in the published literature (*for example*: Brasier and Shields, 2000; Hurst and Morton, 2001;

Chemostratigraphy. http://dx.doi.org/10.1016/B978-0-12-419968-2.00001-7

Jenkyns et al., 2002; Saltzman, 2002a; Mutti and Bernoulli, 2003; Korte et al., 2004; Schroeder et al., 2004; Ramkumar et al., 2005; Zachos et al., 2005; Bergström et al., 2006; Jarvis et al., 2006; Mutti et al., 2006; Nedelec et al., 2007; Kouchinsky et al., 2007; Marquillas et al., 2007; Schroeder and Grotzinger, 2007; Alvaro et al., 2008; Cramer et al., 2008; Racki et al., 2008; Elrick et al., 2009; Kakizaki and Kano, 2009; Robinson et al., 2009; Ruhl et al., 2009; Gouldey et al., 2010; Kiipli et al., 2010; Cui et al., 2011; Grotzinger et al., 2011; Salzman and Thomas, 2012; Aehnelt et al., 2013; Saltzman and Sedlacek, 2013; Uramoto et al., 2013).

Thus, chemostratigraphy *a la* chemical stratigraphy has evolved to current stage from a humble beginning of identification of "patterns in geochemical profile" of sedimentary records of event beds, barren sequences, and rocks deposited across specific chronostratigraphic and/or litho-biostratigraphic boundaries. While it would be beyond the scope of this chapter to review all the published literature on this subject, an attempt is made to present basic concepts involved in the chemostratigraphy of sedimentary deposits. They are then examined in the light of major criteria suggested by the International Stratigraphic Commission (ISC) for formal stratigraphy; based on which, a plea to formalize the chemostratigraphy as an independent stratigraphic classification method is made.

1.2 BASIS OF CHEMOSTRATIGRAPHY

It is an established fact that the sediments are faithful recorders of the changes in provenance, environment of deposition, and postdepositional history. Such changes mean that apparently uniform successions may show primary differences in the chemistry of their constituent minerals and also in the proportions of accessory phases such as heavy minerals and clays, many of which have distinct chemical compositions (Das, 1997). In a stratigraphic context, these traits led to the proposition that, stratigraphic record is a product of a geochemical system consisting of geological setting, climate, and processes of sediment production (Berger and Vincent, 1981) and preservation. The sedimentary record also shows changes of certain elements with time (Morante et al., 1994). An ability to infer the spatiotemporal distinctness of chemical compositions of sedimentary record enables apparently uniform thick successions to be subdivided and correlated with coeval strata located elsewhere (Ramkumar, 1999). It helps to recognize completeness of the stratigraphic record (for example, Saltzman and Sedlacek, 2013; Ramkumar et al. this volume) and found its usefulness to discern distinct lava flows, interlayers with volcanoclastic or even with "true" sedimentary deposits (Zsolt Berner, personal communication). Thus, chemostratigraphy is not restricted to sedimentary sequences alone, but also found its usefulness in other lithologies too. Realization of the potential of this ability and the developments in sophisticated equipment for precise and rapid determination of chemical and isotopic compositions of earth materials have contributed towards the birth of chemostratigraphy (Ramkumar, 2014).

As each sedimentary environment is characterized by unique physical, chemical, and biological milieu in a geomorphic setup (Reineck and Singh, 1980), the resultant sediments are subjected to varying spatiotemporal scales and intensities of these processes, chemostratigraphy is inherently interdisciplinary in nature and requires inputs from many branches of geosciences including, but not limited to paleontology, tectonics, petrology, mineralogy, geochemistry, geomorphology, geophysics, hydrogeology, taphonomy, oceanography, limnology, climatology, and litho, bio, chrono, magneto, and sequence stratigraphy, and other allied fields, such as physics, statistics, biology, etc.

1.3 DEVELOPMENT OF CHEMOSTRATIGRAPHY

Initially, chemostratigraphy was applied to recognize unique geochemical compositions for characterizing depositional units and correlate them with coeval strata elsewhere. It was also utilized as a subsidiary tool to supplement inferences drawn from other lines of geoscientific data. Later, chemostratigraphy found its utility for comparing and contrasting geochemical compositions of strata across major geochronological boundaries, such as Precambrian–Cambrian, Permian–Triassic, Triassic–Jurassic, Jurassic–Cretaceous, Cretaceous–Tertiary, Palaeocene–Eocene, etc. The documentation of iridium anomaly in the Late Cretaceous strata and cataclysmic origin assigned to it (Alvarez et al., 1980) might have prompted the geoscientists to examine coeval stratigraphic sections located elsewhere and also the strata deposited across other major geochronological boundaries. Thus, chemostratigraphy was initially applied to examine presence/absence of specific boundaries, and found its use in the stratigraphic location of the boundaries, and later expanded to examination of specific causes to the stratigraphic variations of geochemical compositions (*see* Keller, 1988; Kaminski and Malmgren, 1989; Glasby and Kunzendorf, 1996; Wallmann, 2001; Adatte et al., 2004; Keller et al., 2003, 2004; Ramkumar et al., 2004, 2005; Stüben et al., 2005; Ramkumar et al., 2010a for a detailed review) by integration of various types data from other branches of geosciences.

There were many contemporaneous works that documented systematic variations of certain elements and isotopic compositions over time (for example, Claypool et al., 1980; Veizer, 1985) and the list of standard curve for various time spans started growing (see Strauss, 1997; Veizer et al., 1999; Weissert et al., 2008; Saltzman and Thomas, 2012 for a detailed review). Almost simultaneously, the availability of continuous stratigraphic records in the form of undisturbed core samples, raised through DSDP (Deep Sea Drilling Project) and ODP (Ocean Drilling Project) programs, and facilitation of extensive scientific collaborations and access to analytical facilities have all contributed to prolific documentation of stratigraphic variations of elemental, isotopic, and other criteria of ancient sediments. It also permitted integration of a wide variety of data with stratigraphic variation of geochemistry, to understand geological processes of varying spatiotemporal scales.

Two other factors, namely, improvements in instrumentation for rapid and precise analyses of geochemical composition and the introduction of sequence stratigraphic concepts have contributed significantly to the development of chemostratigraphy and its application widely and frequently than ever. These have also provided impetus to test the applicability of chemostratigraphy at varying spatial and temporal scales from local to global and few tens of millions of years to few 1000 years and at instances, at tidal cycles. The third, but unrelated development has been that the economic liberalization in many countries following the end of Cold War has heightened the energy demand that in turn necessitated hydrocarbon exploration in areas hitherto remained unexplored and/or less explored. Hydrocarbon exploration industry has been the beneficiary of the developments in chemostratigraphy as it has provided a reliable tool for hydrocarbon exploration where new reservoirs have to be identified, applicable analogues have to be tracked in the vicinity of known reservoirs and when reservoirs are identified, characterization of them for designing better exploitation strategies. Soon, chemostratigraphy has occupied unassailable position among the tools and techniques used for hydrocarbon exploration and reservoir characterization as it has proved its mettle even where other stratigraphic methods have limitations. Finding utility in one of the daunting tasks of humankind has given rise to further sophistications in terms of generous funding for research and development, improvements in instrumentation, academic, and professional collaboration, etc., which, together contributed to the current status of chemostratigraphy.

1.4 PREVAILING TERMINOLOGIES, THEIR INTENDED MEANINGS AND DEFINITIONS

The ISC defines stratigraphy as *the description of all rock bodies forming the Earth's crust and their organization into distinctive, useful, mappable units based on their inherent properties or attributes in order to establish their distribution and relationship in space and their succession in time, and to interpret geologic history*. Wikipedia defines chemostratigraphy as *the study and dating of sedimentary strata*. Wiktionary defines chemostratigraphy as *the study and dating of sedimentary strata by the analysis of trace elements and isotopic ratios*. In the scientific literature also, no clear definition or consensus could be found. It is largely due to the absence of formalized terms, and radiogenic or other dates for correlation of chemically distinct units. However, it does not thwart chemostratigraphy being applied widely. Though, many terms are in common use. At this juncture, chemostratigraphy needs to prove itself fit enough to be included in the exclusive club of stratigraphic methods such as lithostratigraphy or biostratigraphy.

In this section, the terminologies that are available in the published literature and their widely followed meanings are collected and presented. In addition, definitions for these terminologies are added to provide clarity in their usage and to evolve consensus.

1.4.1 CHEMOSTRATIGRAPHY

Based on the definition of the ISC for the stratigraphic methods and the ways the stratigraphic geochemical variability is described in the literature, chemostratigraphy could be described as *the study of temporal and spatial geochemical variability of rocks with a view to establish mappable rock units, defined in terms of unique geochemical composition, isotopic ratios and elemental and/or elemental oxide ratios or a combination of any or all of these*.

1.4.2 FINGERPRINTING OR GEOCHEMICAL FINGERPRINTING

Fingerprinting or geochemical fingerprinting (Miller et al., 2009; Robinson et al., 2009) or recognition of signal (Wynn and Read, 2007) is the essential task of chemostratigraphy. This term could be defined as an *act of geochemical characterization of rock record affiliated to specific geological event, process, age, depositional unit, or a part thereof, and/or a combination of any or all of these*. The resulting fingerprint or geochemical fingerprint is a trait akin to geochemical signature based on which similar events, processes, depositional units ascribable to particular age and/or corresponding depositional units or parts thereof could be recognized elsewhere on a spatiotemporal scale. In order to utilize this term, emphasis has to be placed on discriminating the rock record with a specific elemental, oxide, isotopic, and/or elemental ratios from that of other rock records, so that, when a particular compositional value or range of values are determined, the rock record is also distinctly recognized.

1.4.3 GEOCHEMICAL SIGNATURE

Geochemical signature (Staron et al., 2001; Sluijs et al., 2007; Bergström et al., 2006, 2009, 2010; Krause et al., 2014) or geochemical fingerprint or geochemical marker (Kiipli et al., 2010; Grandpre et al., 2013) is the *distinct geochemical trait of a depositional unit defined in terms of point or range values of elemental composition, or a combination of many elements* (for example, Aldahan et al., 2000; Aehnelt et al., 2013) *or ratios or isotopic composition* with which chemostratigraphic classification or distinction of rock records based on geochemistry is made. The geochemical signature may also indicate specific nature and/or trends of compositional and other values; for example, deviation from normality, commonly referred as "anomaly." It could either be "positive" or "negative." The deviation may also be described in terms of be "abrupt" or "gradual." Here, the emphasis is made on the unique geochemical trait. Robustness of chemostratigraphic classification and correlation rests on the uniqueness of the geochemical fingerprint or the signature or the marker.

1.4.4 EXCURSION, SHIFT, FLUCTUATION, PERTURBATION, AND ANOMALY

These terms are normally used to denote the *change of geochemical compositional trend from normality either from a point value or a range of value that is normally observed in a depositional unit specific to a particular age, event, etc.* The term "normal" suggests background value of the strata under study and/or "normal" value as established by previous studies for a given time slice or depositional unit. Hence, it is variable in space and time and also with reference to the value of individual element or isotope or elemental ratio or a combination of any of these. *Increase of values from normality is referred as positive excursion* (Saltzman et al., 2000; Cramer et al., 2006; Zachos et al., 2007; Dzyuba et al., 2013) or shift (Immenhauser et al., 2002; Calner et al., 2014) or fluctuation (Scholle and Arthur, 1980; Payne and Kump, 2007) and *the decrease is referred as negative excursion* (Wang et al., 2012) or shift. The term perturbation (Payne et al., 2004; Racki et al., 2008; Littler et al., 2010; Takahashi et al., 2014) refers to a *significant change from normality regardless of positive or negative shift or excursion*. However, it is more often found its use to denote positive excursion. The term anomaly (Matsumoto, 1995; Thomas and Shackleton, 1996; Palfy et al., 2001) implies *significant positive or negative change of compositional values from normality*. Thus, it stands distinct/anomalous in a geochemical profile (*curve of geochemical compositional values drawn in stratigraphic order*). These are also termed as signals (Marshall, 1992) and/or events (Marshall et al., 1997; Melezhik et al., 2007). As geochemical compositions of sedimentary records are intricately linked with the changes in physical, chemical, biological, and atmospheric processes, fluctuation, perturbation, excursion, shift, and anomaly are always indicative of prevalence of such processes and thus extensively used as geochemical signatures associated with myriad varieties of processes and events (for example, Magioncalda et al., 2004; Gouldey et al., 2010; Grotzinger et al., 2011) and these terms are used interchangeably (for example, Calner et al., 2014). Timing of prevalence of such processes could also be identified depending on their episodic (Yin et al., 2007; Xie et al., 2008) or stepwise incremental and/or abrupt nature as has been recorded from classic boundaries such as Palaeocene–Eocene (Zachos et al., 2007; Cui et al., 2011), Maastrichtian–Danian (Ramkumar et al., 2005; Grandpre et al., 2013; Punekar et al., 2014), Cenomanian–Turonian (Sageman et al., 2006), Jurassic–Cretaceous (Zakharov et al., 2014), Triassic–Jurassic (Palfy et al., 2001; Korte et al., 2009; Williford et al., 2009), Permian–Triassic (Berner, 2002), Devonian–Carboniferous (Brand et al., 2004), Silurian–Devonian (Saltzman, 2002b), Ordovician–Silurian (Kaljo et al., 2003), Cambrian–Ordovician (Alvaro et al., 2008), Precambrian–Cambrian (Kouchinsky et al., 2007), and Meso–Neoproterozoic (Bartley et al., 2001). Depending on the nature and quantum of trends of shifts, distinctness of perturbations, and the ranges of spatiotemporal scale and the intensities of the processes and events that caused the changes in geochemical compositional values from normality, these excursions, shifts, perturbations, and anomalies are used for recognition of similar features and correlation of geographically widely separated strata, examination of the completeness of the stratigraphic records and also to locate stratigraphic positions of strata—all of which are the essential and three major functions of any stratigraphic method.

1.4.5 TREND

The term "trend" (Magaritz et al., 1992; Zachos et al., 2001; Kikumoto et al., 2014) denotes the *nature of geochemical compositional variations observable in a geochemical profile*. The nature could be featureless, secular, cyclic, and highly variable (irregular). The *systematic increase or decrease of geochemical composition is considered to be secular* (Knoll et al., 1986; Martin, 1995; Gao and Land, 1991; Chu et al., 2007) and/or linear trend. The cycles (Martin, 1995; Zachos et al., 2001, 2010) are *repetitive occurrences of combination of positive and negative excursions at regular intervals*, the durations of which may vary either systematically and/or otherwise.

The extent of excursions could also be variable, namely, progressively incremental or decreasing and/or a combination of all of these. Many a times, the cycles of short duration are found embedded within a longer cycles (cycle in cycle). Cyclic sedimentation has been documented in numerous sedimentary basins that led to the development of cyclostratigraphy (Weedon, 2005) and there are many lines of evidence that relate those cycles to short-term (Milankovitch band) glacio-eustatic pulses (Grammer et al., 1996). Documentation of cyclic nature of stable isotope compositions of sediment records of DSDP and ODP and other sites (for example, Renard et al., 1984; Shackleton and Hall, 1984; Srinivasan, 1989; Pearce and Jarvis, 1991; Bruckschen et al., 1999; Pearce et al., 1999; Korte et al., 2004; Bergström et al., 2006, 2009, 2010; Ramkumar et al., 2010, 2011) have helped establish the utilities of chemostratigraphy in various fields.

Spectral analysis of $\delta^{18}O$ and $\delta^{13}C$ of DSDP core samples shows typical cycles of 100, 43, 23, and 19 Ka spans (Oppo et al., 1990; Oppo and Fairbanks, 1989). While examining $\delta^{18}O$ of Phanerozoic seawater, Veizer et al. (1997) observed the presence of high-frequency cycles within first order cycle. Strauss (1997) recorded fourth order cycles of sulfur isotope that stacked up to form third order cycles that in turn were found to be accommodated within a second order cycle. Goldhammer et al. (1991) showed that the sequences of Paradox Basin exhibited a distinct cyclicity characterized by a hierarchical stacking pattern of fifth order shallowing upward cycles grouped into fourth order cycles, which in turn stacked vertically into part of a third order cycle. It is also possible to define the secular/linear and cyclic trends through statistical linear, polynomial, and exponential functions. Geographically separated stratigraphic sections are frequently correlated through the trends in terms of similarities, dissimilarities etc. Application of statistical procedures for chemostratigraphic correlation was suggested to be more reliable than visual inferences (Whitford et al., 1996). Statistical characterization of these similarities and dissimilarities through mismatch computation analysis or by applying dynamic time warping algorithm (Weedon, 2005; Hladil et al., 2011) could thwart human error and establish reliable correlation.

The utility of chemostratigraphy for age determination was demonstrated through documentation of chronological variations of isotopic trends (for example, Sr isotopic curve—Whitford et al., 1996; Denison et al., 1997; Veizer et al., 1997, 1999; Ebneth et al., 2001; Mearon et al., 2003; C and O isotopic curve—Williams et al., 1983; Renard et al., 1984; Shackleton and Hall, 1984; Mount et al., 1986; Koch et al., 1992; Friedman and Chakraborty, 1997; Saltzman and Thomas, 2012; sulfur isotopic curve—Strauss, 1997; Kampschulte et al., 2001; and references cited therein).

1.4.6 CHEMOSTRATIGRAPHIC INDICES

Chemostratigraphic indices *are those geochemical parameters or unique geochemical traits of strata under study defined by elements or elemental ratios or isotopic values or unique nature of geochemical profile (shift, signal, excursion, anomaly, and perturbation) or a combination of any or all of these that designate specifically a depositional unit/strata/bedset/facies sequence and differentiate it from others.* Chemostratigraphic indices could be point value or a range of values. Depending on the uniqueness and significance, the indices help correlate the depositional unit spatially and temporally (Aehnelt et al., 2013). While the geochemical signature expresses the trait of a depositional unit, the index refers to the element or elemental ratio or isotope itself.

A term "geochemical proxy" is also in use by those practice chemostratigraphy. It is considered to be the fundamental notion in chemostratigraphy (Zsolt Berner, personal communication). It denotes the *specific element or isotope or any elemental ratio or a combination of any of these, utilized as an indicator of particular geological or related event or a process prevalent during geological past* (for example, Suarez et al., 2013; Reuter et al., 2013; Li et al., 2013). However, there are usages of this term for denoting specific beds themselves (for example, Compton et al., 2000—who used phosphorite beds as "proxy" to short-term carbon burial) or specific fauna (for example, Dubicka et al., 2014—who used fossil foraminifera as "proxy" to past environmental conditions) or the geochemical excursions (for example, Lehnert et al. 2014—who used middle Darriwilian isotope carbon excursion as "proxy" for intercontinental correlation of Lower–Middle Ordovician strata of Sweden). Hence, caution and clarity are required for using this term.

1.4.7 CHEMOZONE

Chemozone is the *unique rock record defined by chemostratigraphic indices and recognizable through unique geochemical signature(s) which in turn helps distinction of a designated rock record from other rock records and also correlation with applicable analogues at appropriate/applicable spatiotemporal scale.* Recently, Montero-Serrano et al. (2010) used the term "chemofacies" which seems to be synonymous with "chemozone".

As the bounding surfaces of each chemozone are defined by geochemical indices/markers, chemozones may or may not coincide with litho, bio, chrono, magneto, and sequence stratigraphic boundaries (for example, Cramer et al., 2008; Bergström et al., 2010; Ramkumar et al., 2011). Similar to the recognition of sequences and sea level cycles, the chemozones are recognized hierarchically and are termed as level 1 chemozone or major chemozone, within which many level 2 or minor chemozones, and so on. The practice of recognizing hierarchical nature of chemozones is influenced by the sequence stratigraphic concepts, the common prevalence of cycle-in-cycle pattern of the depositional units, and the existence of intimate relationships between depositional conditions and geochemistry.

The relationship between the characteristics of organic matter and sea level changes have been well documented (for example, Middleberg et al., 1991; Pasley et al., 1993; Tu et al., 1999; Calver, 2000) and between the geochemical parameters and sequence cycles (Pelechaty, 1998; Rangel et al., 2000; McKirdy et al., 2001; Fanton and Holmden, 2007; Bergström et al., 2010; Ramkumar et al., 2011). The introduction of concept of global sea level changes and their control over stratal patterns and facies distribution and the fact that such changes are recorded in sedimentary geochemistry (Jarvis et al., 2001) give rise to the possibility of correlating stratigraphic records with their counterparts located anywhere in the world. The relative sea level cycles, first published by Vail et al. (1977) and revised by Haq et al. (1987) espoused that sedimentary sequences are produced principally under the influence of sea level cycles that vary between few tens of millions of years (first order cycle) to few Ma (third order cycle). Successive studies have shown that distinct sedimentary sequences could be traced to sea level cycles of up to infra seventh order (Nelson et al., 1985; Williams et al., 1988; Carter et al., 1991) i.e., few 1000 years.

Vail et al. (1977) stated that the sea level chart published by them is incomplete and cycles of varying order could be added as the studies on sedimentary sequences progress, so that, more complete chart could be produced. The aim behind this statement was to incorporate sea level cycles at Milankovitch scale, to which the response of the sedimentation system is proved beyond reasonable doubt (Carter et al., 1991). Hays et al. (1976) first convincingly demonstrated that climatic records were dominated by frequencies characteristic of variations in the Earth's tilt, precession and eccentricity relative to the Sun. In the years since, numerous studies have upheld the validity of the Milankovitch climatic cycles in terms of 100, 41, 23 Ka orbital periods that influence or control variations in global ice volume, thermohaline circulation, continental aridity and run off, sea surface temperature, deep ocean carbonate preservation and atmospheric CO_2, and methane concentrations (Raymo et al., 1997). There are ample evidences of facies successions representing infra seventh order spanning few 1000 years of duration (Carter et al., 1991). This information, together with the established fact of stacking pattern of sedimentary records, i.e., fifth and sixth order cycles that stack up to form fourth order cycles, which in turn form part of third order cycle, affirms the possibility and utility of recognizing chemozones at varying scales, in a stacked pattern.

1.4.8 CHEMOCHRON

The chemozones calibrated with absolute time are termed as chemochrons (Pelechaty et al., 1996). Thus, *unique rock record, defined to be distinct in terms of geochemical signature and calibrated with absolute time* is termed as a chemochron.

Establishing chemochrons with unique chemostratigraphic indices is still under developmental stage, although many standard curves for Phanerozoic and specific time slices are available in the published literature. Many of those curves have been validated through independent studies of geographically widely separated strata. In view of the widely differing geochemical compositions of sedimentary rocks as a result of inherent heterogeneities of depositional and diagenetic sedimentary processes, establishing unique markers/indices applicable to all the sedimentary basins would be a difficult and most daunting task. Nevertheless, unique signals of stable isotopes (Jarvis et al., 2006; Saltzman and Thomas, 2012), strontium isotope (Veizer et al., 1999), and sulfur isotope (Chu et al., 2007), at varying temporal resolutions are documented.

Global chemostratigraphic signals such as those carried by organic matter (Middleberg et al., 1991; Pasley et al., 1993; Meyers and Simoneit, 1989; Tu et al., 1999; Calver, 2000; Saltzman and Thomas, 2012) oxygen isotope (Anderson et al., 1996; Veizer et al., 1999), carbon isotope (Chu et al., 2007; Saltzman and Thomas, 2012), and strontium isotope (Veizer, 1985; Veizer et al., 1999; Mearon et al., 2003; Thomas et al., 2004) and their relationships with sea level changes (Jarvis et al., 2001) are well known and thus they may help establish unique chemochrons (Cramer et al., 2010).

1.4.9 RESOLUTION

Resolution (Xiao et al., 1997) refers to the *least measurable depositional unit defined by geochemical signature through a chemostratigraphic index or select set of chemostratigraphic indices in a chemostratigraphic model of a given stratigraphic section, on a spatiotemporal scale*. However, in the literature, terminologies abound in terms of high resolution (Stoll and Schrag, 2000; Hutchison et al., 2001; Brenchley et al., 2003; Magioncalda et al., 2004; Du Vivier et al., 2014), very high resolution (Boyer et al., 2011; Hermoso et al., 2012; Pellenard et al., 2014), ultra high resolution (Uramoto et al., 2013), etc., and there are no uniformities in the usage. It is suggested that the practice of classification of sequences may be applied. For example, the basic building blocks of sequences—the third order sequences and comparable chemozones may be considered as baseline sequences, the fourth and higher orders may be termed as "high resolution," fifth and higher orders may be termed as "very high resolution," infra seventh orders may be termed as "ultra high resolution." This suggestion need not necessarily be followed, but it is made only to emphasize the need to establish standard usage and to evolve a consensus. It is to be remembered that unlike any other stratigraphic method, resolution of chemostratigraphic model relies on the sampling interval and the ability of the interpreter to define chemozones based on a combination of chemostratigraphic indices. Hence, the need for establishing standard usage for resolution is more important.

1.4.10 SCALE OF CORRELATION

It refers to *the range of spatiotemporal geochemical homogeneity of chemozone, as defined by chosen chemostratigraphic index*. The spatial scale depends on the cause of the chemostratigraphic index/marker being local, regional, basinal, and global, based on which, the chemozone recognized could be correlated at an appropriate scale (Ramkumar, 1999; Samptleben et al., 2000; Panchuk et al., 2005; Young et al., 2005; Mazumdar and Strauss, 2006; Ratcliffe et al., 2006; Ramkumar and Sathish, 2007; Ramkumar et al., 2010, 2011; Albanesi et al., 2013). The temporal scale refers to the fineness or coarseness of the measurable units defined by chemostratigraphic indices and geochemical signature. Many a times, temporal scale is limited or enhanced by the sampling interval, affordability of geochemical analyses and the requirement for which chemostratigraphic modeling is attempted.

Theoretically, sedimentary cycles vary from few seconds (catastrophic events, mass wasting, avalanche, etc.), through few hours (tidal cycles), diurnal variations, seasonal variations, annual climatic cycles, decadal cycles, Milankovitch cycles, and finally first order cycles of few tens of millions of years, all of which might have unique geochemical compositions at corresponding scales. However, many causes, namely, the stacking pattern of these cycles (cycle-in-cycle pattern), omission surfaces, stratigraphic gaps, and diagenetic overprinting, etc., alter the pristine signatures significantly. An ability to characterize the depositional units in terms of unique geochemical signatures would help establish robust chemostratigraphic model and correlation of widely separated strata with measurable confidence levels.

Montero-Serrano et al. (2010) tested discrimination of chemofacies through multivariate statistics. Albertão et al. (2004) discriminated the Cretaceous and Tertiary deposits based on geochemical data with the help of factor and discriminant analyses. Applying a combination of multivariate statistical analyses on textural, petrographic, facies, geochemical, mineralogical, and other data, distinction of depositional units in terms of sequences, chemozones, lithostratigraphic formations, and members of

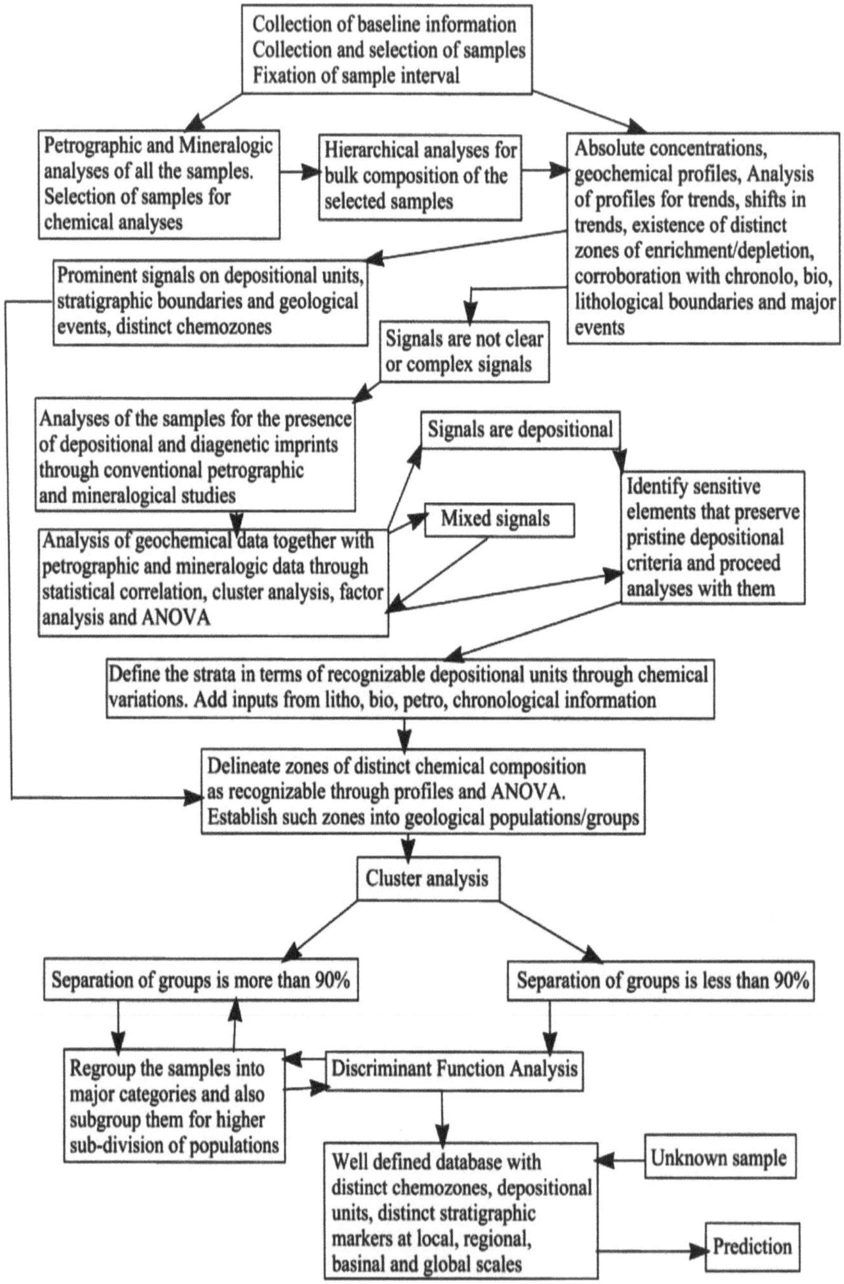

FIGURE 1

Characterization and discrimination of depositional units.

the Cauvery Basin, that spanned a temporal resolution of about 1–18 Ma (Ramkumar et al., 2010b), petrographic and standard microfacies types of Maastrichtian carbonates (Ramkumar et al., 2002), lithostratigraphic members, formations, petrographic types, standard microfacies types, and facies zones of Tertiary carbonates (Ramkumar and Guha, 2000), 21 types of subenvironments of modern delta (Ramkumar, 2001) and estuarine sediments deposited under flood and ebb tides (Ramkumar, 2004) were discriminated (Ramkumar and Sathish, 2007). Based on the results, a scheme of discrimination and prediction of depositional units (Fig. 1) has been developed (Ramkumar, 2014). The results have also demonstrated the plausibility of defining and discriminating the depositional units through unique geochemical traits at various temporal scales. It was also demonstrated statistically that the accuracy of the defined units could thwart human error or bias.

1.5 TERMINOLOGIES AND APPLICATIONS OF CHEMOSTRATIGRAPHY: A FIT CASE FOR FORMALIZATION

With these terminologies, definitions, and generally followed meanings, it is now possible to meet the criteria of formal stratigraphy as defined by the ISC.

> **Criterion No. 1**: *Stratigraphy should describe and organize rock bodies into distinctive, useful, and mappable units based on their inherent properties or attributes.* Geochemical signatures, fingerprints, and chemozones do this job perfectly as any other stratigraphic method do. In fact, geochemistry can do better than many other methods or when other methods have limitations.
> **Criterion No. 2**: Subject to the requirement as defined in criterion No. 1, *the units should help establish the distribution and relationship of mappable units in space and time.* Extensive literature is available in proving the capability and utility of chemostratigraphy in this respect.
> **Criterion No. 3**: Fulfilling the conditions set forth in criterion No. 1 and 2, *the stratigraphic method should help interpret geological history.* Again, chemostratigraphic research over the years has proved beyond any reasonable doubt, the capability of geochemical signatures to interpret precisely the geological history.

In the stratigraphic code itself, the use of informal stratigraphic terminologies is strongly discouraged. Nevertheless, despite being widely practiced and prolific publications by many hundreds of geoscientists every year, chemostratigraphic studies continue using informal terminologies only. For this reason also, it is strongly felt that it is time, geoscientific community takes cognizance of the reality and makes initiative to formalize chemostratigraphy and standardize the terminologies.

ACKNOWLEDGMENTS

I record a deep sense of gratitude to Professor V. Rajamani, School of Environmental Sciences, Jawaharlal Nehru University, New Delhi for "initiating" me into sediment geochemistry and laboratory protocols. The classic articles of Pearce and Jarvis have introduced me to this new realm of geosciences and I owe a great deal of thankfulness to them. Many of the authors whose articles are cited in this chapter are thanked for generous supply of their reprints during "hard copy reprint age." Understanding on this subject was made easy through intense and lively discussions with Dr. Zsolt Berner, Senior Scientist and Head of Laboratories, Institute of Mineralogy and Geochemistry,

Karlsruhe Institute of Technology (erstwhile University of Karlsruhe), Germany. Scientific collaborations with German institutions and scientists at Institute of Mineralogy and Geochemistry, Karlsruhe, Institute of Palaeoenvironment, Erlangen, University of Freiburg, Marburg University, etc., were made possible by sponsorships by Alexander von Humboldt Foundation since the year 2000, for which I express my thankfulness. My interest on this subject and scientific collaboration with national and international academic and research institutions have been sustained by research grants from various organizations such as Alexander von Humboldt Foundation, Germany, University Grants Commission, Council of Scientific and Industrial Research, Department of Science and Technology, Oil Industry Development Board, Oil and Natural Gas Corporation Limited, India, German Research Foundation, Germany, etc., for which I am thankful.

REFERENCES

Adatte, T., Keller, G., Stüben, D., Harting, M., Kramar, U., Stinnesbeck, W., Abramovitch, B.C., 2004. Late Maastrichtian and K/T paleoenvironment of the Eastern Tethys (Israel): mineralogy, trace element and platinum group elements, biostratigraphy and faunal turnovers. Bulletin de la Societe Geologique de France 176, 35–53.

Aehnelt, M., Worden, R.H., Canham, A.C., Hill, S.J., Hodgson, D.M., Flint, S.S., 2013. Geochemical correlation in an exhumed submarine channel complex (Tabernas Basin, Se Spain): comparison to sedimentological correlation at various length scales. Journal of Sedimentary Research 83, 669–690.

Albanesi, G.L., Bergström, S.M., Schmitz, B., Serra, F., Feltes, N.A., Voldman, G.G., Ortega, G., 2013. Darriwilian (Middle Ordovician) $\delta^{13}C_{carb}$ chemostratigraphy in the Precordillera of Argentina: documentation of the middle Darriwilian isotope carbon excursion (MDICE) and its use for intercontinental correlation. Palaeogeography, Palaeoclimatology, Palaeoecology 389, 48–63.

Albertão, G.A., Grassi, A.A., Marini, F., Martins Jr., P.P., de Ros, L.F., 2004. The K–T boundary in Brazilian marginal sedimentary basins and related spherules. Geochemical Journal 38, 121–128.

Aldahan, A., Possnert, G., Scherer, R., Shi, N., Backman, J., Boström, K., 2000. Trace-element and major-element stratigraphy in quaternary sediments from the Arctic Ocean and implications for glacial termination. Journal of Sedimentary Research 70, 1095–1106.

Alvarez, L.W., Alvarez, W., Asaro, F., Michael, H.V., 1980. Extraterrestrial cause for the Cretaceous–Tertiary extinction. Science 208, 1095–1098.

Alvaro, J.J., Bauluz, B., Subias, I., Pierre, C., Vizcaino, D., 2008. Carbon chemostratigraphy of the Cambrian–Ordovician transition in a midlatitude mixed platform, Montagne Noire, France. Geological Society of America Bulletin 120, 962–975.

Anderson, J.B., Abdulah, K., Sarzalejo, S., 1996. Late quaternary sedimentation and high-resolution sequence stratigraphy of the east Texas shelf. In: De Batist, M., Jacobs, P. (Eds.), Geology of Siliciclastic Shelf Seas. Geol. Soc. Spec. Publ., vol. 117, pp. 95–124.

Bartley, J.K., Semikhatov, M.A., Kaufman, A.J., Knoll, A.H., Pope, M.C., Jacobsen, S.B., 2001. Global events across the Mesoproterozoic–Neoproterozoic boundary: C and Sr isotopic evidence from Siberia. Precambrian Research 111, 165–202.

Berger, W.H., Vincent, E., 1981. Chemostratigraphy and biostratigraphic correlation: exercises in systematic stratigraphy. Oceanologica Acta 115–127 Suppl.

Bergström, S.M., Saltzman, M.M., Schmitz, B., 2006. First record of the Hirnantian (Upper Ordovician) $\delta^{13}C$ excursion in the North American Midcontinent and its regional implications. Geological Magazine 143, 657–678.

Bergström, S.M., Schmitz, B., Rong, J., Young, S., Saltzman, M.R., 2009. First documentation of the Ordovician Guttenberg $\delta^{13}C$ excursion (GICE) in Asia: chemostratigraphy of the Pagoda and Yanwashan formations in southeastern China. Geological Magazine 146, 1–11.

Bergström, S.M., Young, S.A., Schmitz, B., 2010. Katian (Upper Ordovician) $\delta^{13}C$ chemostratigraphy and sequence stratigraphy in the United States and Baltoscandia: a regional comparison. Palaeogeography, Palaeoclimatology, Palaeoecology 296, 217–234.

Berner, R.A., 2002. Examination of hypotheses for the Permo–Triassic boundary extinction by carbon cycle modeling. Proceedings of the National Academy of Sciences of the United States of America 99, 4172–4177.

Boyer, D.L., Owens, J.D., Lyons, T.W., Droser, M.L., 2011. Joining forces: combined biological and geochemical proxies reveal a complex but refined high-resolution palaeo-oxygen history in Devonian epeiric seas. Palaeogeography, Palaeoclimatology, Palaeoecology 306, 134–146.

Brand, U., Legrand-Blain, M., Streel, M., 2004. Biochemostratigraphy of the Devonian–Carboniferous boundary global stratotype section and Point, Griotte formation, La Serre, Montagne Noire, France. Palaeogeography, Palaeoclimatology, Palaeoecology 205, 337–357.

Brasier, M.D., Shields, G., 2000. Neoproterozoic chemostratigraphy and correlation of the Port Askaig glaciation, Dalradian Supergroup of Scotland. Journal of the Geological Society 157, 909–914.

Brenchley, P.J., Carden, G.A., Hints, L., Kaljo, D., Marshall, J.D., Martma, T., Meidla, T., Nolvak, J., 2003. High-resolution stable isotope stratigraphy of Upper Ordovician sequences: constraints on the timing of bioevents and environmental changes associated with mass extinction and glaciations. Geological Society of America Bulletin 115, 89–104.

Bruckschen, P., Oesmann, S., Veizer, J., 1999. Isotope stratigraphy of the European Carboniferous: proxy signals for ocean chemistry, climate and tectonics. Chemical Geology 161, 127–164.

Calner, M., Lehnert, O., Wu, R., Dahlqvist, P., Joachimski, M.M., 2014. $\delta^{13}C$ chemostratigraphy in the Lower–Middle Ordovician succession of Öland (Sweden) and the global significance of the MDICE. Geologiska Forenmgens i Stockholm Forhandlingar. http://dx.doi.org/10.1080/11035897.2014.901409.

Calver, C.R., 2000. Isotope stratigraphy of the Ediacarian (Neoproterozoic III) of the Adelaide rift complex, Australia, and the overprint of water column stratification. Precambrain Research 100, 121–150.

Carter, R.M., Abbott, S.T., Fulthorpe, C.S., Haywick, D.W., Henderson, R.A., 1991. Application of global sealevel and sequence stratigraphic models on Southern Hemisphere Neogene strata from New Zealand. International Association of Sedimentologists Special Publication 12, 41–65.

Chu, X.L., Zhang, T.G., Zhang, Q.R., Lyons, T.W., 2007. Sulfur and carbon isotope records from 1700 to 800 Ma carbonates of the Jixian section, northern China: implications for secular isotope variations in Proterozoic seawater and relationships to global supercontinental events. Geochimica et Cosmochimica Acta 71, 4668–4692.

Claypool, G.E., Holser, W.T., Kaplan, I.R., Sakai, H., Zak, I., 1980. The age curves of sulfur and oxygen isotopes in marine sulfate and their mutual interpretations. Chemical Geology 28, 199–260.

Compton, J., Mallinson, D., Glenn, C.R., Filippelli, G., Follmi, K., Shields, G., Zanin, Y., 2000. Variations in the global phosphorus cycle. In: Marine Authigenesis: From Global to Microbial. SEPM Spl. Publn. No. 66, pp. 21–33.

Cramer, B.D., Saltzman, M.R., Kleffner, M.A., 2006. Spatial and temporal variability in organic carbon burial during global positive carbon isotope excursions: new insight from high resolution carbon isotope stratigraphy from the type area of the Niagaran Provincial Series. Stratigraphy 2, 327–340.

Cramer, B.D., Saltzman, M.R., Day, J.E., Witzke, B.J., 2008. Lithological expression of global positive carbon isotope excursions in epeiric sea settings: carbonate production, organic carbon burial, and oceanography during the late Famennian. In: Pratt, B.R., Holmden, C. (Eds.), Geol. Assoc. Canada Spec. Publ., vol. 48, pp. 103–118.

Cramer, B.D., Loydell, D.K., Samtleben, C., Munnecke, A., Kaljo, D., Mannik, P., Martma, T., Jeppsson, L., Kleffner, M.A., Barrick, J.E., Johnson, C.A., Emsbo, P., Joachimski, M.M., Saltzman, M.R., 2010. Testing the limits of Paleozoic chronostratigraphic correlation via high resolution (<500 kyr) integrated conodont, graptolite, and carbon isotope ($\delta^{13}C_{carb}$) biochemostratigraphy across the Llandovery–Wenlock (Silurian) boundary: is a unified Phanerozoic timescale achievable? Geological Society of America Bulletin 122, 1700–1716.

Cui, Y., Kump, L.R., Ridgwell, A.J., Charles, A.J., Junium, C.K., Diefendorf, A.F., Freeman, K.H., Urban, N.M., Harding, I.C., 2011. Slow release of fossil carbon during the Palaeocene–Eocene Thermal Maximum. Nature Geoscience 4, 481–485.

Das, N., 1997. Chemostratigraphy of sedimentary sequences: a review of the state-of-the-art. Journal of the Geological Society of India 49, 621–628.

Denison, R.E., Koepnick, R.B., Burke, W.H., Hetherington, E.A., Fletcher, A., 1997. Construction of the Silurian and Devonian seawater $^{87}Sr/^{86}Sr$ curve. Chemical Geology 140, 109–121.

Dubicka, Z., Peryt, D., Szuszkiewicz, M., 2014. Foraminiferal evidence for paleogeographic and palaeoenvironmental changes across the Coniacian–Santonian boundary in western Ukraine. Palaeogeography, Palaeoclimatology, Palaeoecology 401, 43–56.

Du Vivier, A.D.C., Selby, D., Sageman, B.B., Jarvis, I., Grocke, D.R., Voigt, S., 2014. Marine $^{187}Os/^{188}Os$ isotope stratigraphy reveals the interaction of volcanism and ocean circulation during Oceanic Anoxic Event 2. Earth and Planetary Science Letters 389, 23–33.

Dzyuba, O.S., Izokh, O.P., Shurygin, B.N., 2013. Carbon isotope excursions in boreal Jurassic–Cretaceous boundary sections and their correlation potential. Palaeogeography, Palaeoclimatology, Palaeoecology 381–382, 33–46.

Ebneth, S., Shields, G.A., Veizer, J., Miller, J.F., Shergold, J.H., 2001. High-resolution strontium isotope stratigraphy across the Cambrian–Ordovician transition. Geochimica et Cosmochimica Acta 65, 2273–2292.

Elrick, M., Molina-Garza, R., Duncan, R., 2009. C-isotope stratigraphy and paleoenvironmental changes across OAE2 (mid-Cretaceous) from shallow-water platform carbonates of southern Mexico. Earth and Planetary Science Letters 277, 295–306.

Fanton, K.C., Holmden, C., 2007. Sea-level forcing of carbon isotope excursions in epeiric seas: implications for chemostratigraphy. Canadian Journal of Earth Sciences 44, 807–818.

Friedman, G.M., Chakraborty, C., 1997. Stable isotopes in marine carbonates: their implications for the palaeoenvironment with special reference to the Proterozoic Vindhyan carbonates (Central India). Journal of the Geological Society of India 50, 131–159.

Gao, G., Land, L.S., 1991. Geochemistry of Cambrian–Ordovician Arbuckle Limestone, Oklahoma: implications for diagenetic $\delta^{18}O$ alteration and secular $\delta^{13}C$ and $^{87}Sr/^{86}Sr$ variation. Geochimica et Cosmochimica Acta 55, 2911–2920.

Glasby, G.P., Kunzendorf, H., 1996. Multiple factors in the origin of the Cretaceous/Tertiary boundary: the role of environmental stress and Deccan Trap volcanism. Geologische Rundschau 85, 191–210.

Goldhammer, R.K., Oswald, E.J., Dunn, P.A., 1991. Hierarchy of stratigraphic forcing: example from Middle Pennsylvanian shelf carbonates of the Paradox basin. In: Franseen, E.K., Watney, W.L., Kendall, C.G.St.C. (Eds.). Sedimentary Modeling, vol. 233. Kansas Geol. Surv. Bull., pp. 361–413.

Gouldey, J., Saltzman, M.R., Young, S.A., Kaljo, D., 2010. Strontium and carbon isotope stratigraphy of the Llandovery (Early Silurian): implications for tectonics and weathering. Palaeogeography, Palaeoclimatology, Palaeoecology 296, 264–275.

Grandpre, R., Schauer, A., Samek, K., Veeger, A., Ward, P., Fastovsky, D., 2013. Testing the terrestrial $\delta^{13}C$ Cretaceous–Paleogene (K–Pg) chemostratigraphic marker. Palaeogeography, Palaeoclimatology, Palaeoecology 381–382, 67–75.

Grammer, G.M., Eberli, G.P., Van Buchem, F.S.P., 1996. Application of high resolution sequence stratigraphy to evaluate lateral variability in outcrop and subsurface—Desert Creek and Ismay intervals, Paradox basin. In: Longman, M.W., Sonnelfeld, M.D. (Eds.), Paleozoic Systems of the Rocky Mountain Region. Soc. Economic Palaeont. Mineral., Rocky Mountain Section, pp. 235–266.

Grotzinger, J.P., Fike, D.A., Fischer, W., 2011. Enigmatic origin of the largest-known carbon isotope excursion in Earth's history. Nature Geoscience 4, 285–292.

Haq, B.U., Hardenbol, J., Vail, P.R., 1987. Chronology of fluctuating sealevels since the Triassic. Science 235, 1156–1167.

Hays, J.D., Imbrie, J., Shackleton, N.J., 1976. Variations in the earth's orbit: pacemaker of the ice ages. Science 194, 1121–1132.

Hermoso, M., Minoletti, F., Rickaby, R.E.M., Hesselbo, S.P., Baudin, F., Jenkyns, H.C., 2012. Dynamics of a stepped carbon-isotope excursion: ultra high-resolution study of Early Toarcian environmental change. Earth and Planetary Science Letters 319–320, 45–54.

Hladil, J., Slavik, L., Vondra, M., Koptikova, L., Cejchan, P., Schnabl, P., Adamovic, J., Vacek, F., Vich, R., Lisa, L., Lisy, P., 2011. Pragian–Emsian successions in Uzbekistan and Bohemia: magnetic susceptibility logs and their dynamic time warping alignment. Stratigraphy 8, 217–235.

Hurst, A., Morton, A.C., 2001. Generic relationships in the mineral-chemical stratigraphy of turbidite sandstones. Journal of the Geological Society 158, 401–404.

Hutchison, A., Hartley, A., Hole, M., Whear, E., Preston, J., 2001. High resolution geochemical and petrographical correlation of deep marine Palaeocene sandstones from the Foinaven Field, West of Shetland. Petroleum Geoscience 7, 123–129.

Immenhauser, A., Kenter, J.A.M., Ganssen, G., Bahamonde, J.R., Van Vliet, A., Saher, M.H., 2002. Origin and significance of isotope shifts in Pennsylvanian carbonates (Asturias, northwest Spain). Journal of Sedimentary Research 72, 82–94.

Jarvis, I., Murphy, A.M., Gale, A.S., 2001. Geochemistry of pelagic and hemipelagic carbonates: criteria for identifying systems tracts and sea-level change. Journal of the Geological Society 158, 685–696.

Jarvis, I., Gale, A.S., Jenkyns, H.C., Pearce, M.A., 2006. Secular variation in Late Cretaceous carbon isotopes: a new $\delta^{13}C$ carbonate reference curve for the Cenomanian–Campanian (99.6–70.6 Ma). Geological Magazine 143, 561–608.

Jenkyns, H.C., Jones, C.E., Groecke, D.R., Hesselbo, S.P., Parkinson, D.N., 2002. Chemostratigraphy of the Jurassic System: applications, limitations and implications for palaeoceanography. Journal of the Geological Society 159, 351–378.

Jorgensen, N.O., 1986. Chemostratigraphy of Upper Cretaceous chalk in the Danish Sub-basin. Bulletin of American Association of Petroleum Geologists 70, 309–317.

Kakizaki, Y., Kano, A., 2009. Architecture and chemostratigraphy of Late Jurassic shallow marine carbonates in NE Japan, western Paleo-Pacific. Sedimentary Geology 214, 49–61.

Kaljo, D., Martma, T., Mannik, P., Viira, V., 2003. Implications of Gondwana glaciations in the Baltic late Ordovician and Silurian and a carbon isotopic test of environmental cyclicity. Bulletin de la Societe Geologique de France 174, 59–66.

Kaminski, M.A., Malmgren, B.A., 1989. Stable isotope and trace element stratigraphy across the Cretaceous/Tertiary boundary in Denmark. Geologiska Forenmgens i Stockholm Forhandlingar 111, 305–312.

Kampschulte, A., Bruckschen, P., Strauss, H., 2001. The sulphur isotopic composition of trace sulphates in Carboniferous brachiopods: implications for coeval seawater, correlation with other geochemical cycles and isotope stratigraphy. Chemical Geology 175, 149–173.

Keith, M.L., Weber, J.N., 1964. Carbon and oxygen isotopic composition of selected limestones and fossils. Geochimica et Cosmochimica Acta 28, 1787–1816.

Keller, G., 1988. Extinction, survivorship and evolution of planktonic foraminifera across the Cretaceous/Tertiary boundary at El Kef, Tunisia. Marine Micropalaeontology 13, 239–263.

Keller, G., Stinnesbeck, W., Adatte, T., Stüben, D., 2003. Multiple impacts across the Cretaceous–Tertiary boundary. Earth Science Reviews 62, 327–363.

Keller, G., Adatte, T., Stinnesbeck, W., Stüben, D., Berner, Z., Kramar, U., Harting, M., 2004. More evidence that the Chicxulub impact predates the K/T mass extinction. Meteoritics & Planetary Science 39, 1127–1144.

Kiipli, T., Kallaste, T., Nestor, V., 2010. Composition and correlation of volcanic ash beds of Silurian age from the eastern Baltic. Geological Magazine 147, 895–909.

Kikumoto, R., Tahata, M., Nishizawa, M., Sawaki, Y., Maruyama, S., Shu, D., Han, J., Komiya, T., Takai, K., Ueno, Y., 2014. Nitrogen isotope chemostratigraphy of the Ediacaran and Early Cambrian platform sequence at Three Gorges, South China. Gondwana Research 25, 1057–1069.

Knoll, A.H., Hayes, J.M., Kaufman, A.J., Lambert, I.B., Swett, K., 1986. Secular variation in carbon isotope ratios from Upper Proterozoic successions of Svalbard and East Greenland. Nature 321, 832–838.

Koch, P.L., Zachos, J.C., Gingerich, P.D., 1992. Correlation between isotope records in marine and continental carbon reservoirs near the Palaeocene/Eocene boundary. Nature 358, 319–322.

Korte, C., Kozur, H.W., Joachimski, M.M., Strauss, H., Veizer, J., Schwark, L., 2004. Carbon, sulfur, oxygen, and strontium isotope records, organic geochemistry and biostratigraphy across the Permian/Triassic boundary in Abadeh, Iran. International Journal of Earth Science 93, 565–581.

Korte, C., Hesselbo, S.P., Jenkyns, H.C., Rickaby, R.E.M., Spötl, C., 2009. Palaeoenvironmental significance of carbon- and oxygen-isotope stratigraphy of marine Triassic–Jurassic boundary sections in SW Britain. Journal of the Geological Society 166, 431–445.

Kouchinsky, A., Bengtson, S., Pavlov, V., 2007. Carbon isotope stratigraphy of the Precambrian–Cambrian Sukharikha River section, northwestern Siberian platform. Geological Magazine 144, 609–618.

Krause, S., Steeb, P., Hensen, C., Liebetrau, V., Dale, A.W., Nuzzo, M., Treude, T., 2014. Microbial activity and carbonate isotope signatures as a tool for identification of spatial differences in methane advection: a case study at the Pacific Costa Rican margin. Biogeosciences 11, 507–523.

Lehnert, O., Meinhold, G., Wu, R., Calner, M., Joachimski, M.M., 2014. $\delta^{13}C$ chemostratigraphy in the upper Tremadocian through lower Katian (Ordovician) carbonate succession of the Siljan district, Central Sweden. Estonian Journal of Earth Sciences 63, 277–286.

Li, Q., McArthur, J.M., Doyle, P., Janssen, N., Leng, N.J., Müller, W., Reboulet, S., 2013. Evaluating Mg/Ca in belemnite calcite as palaeo-proxy. Palaeogeography, Palaeoclimatology, Palaeoecology 388, 98–109.

Littler, K., Hesselbo, S.P., Jenkyns, H.C., 2010. A carbon-isotope perturbation at the Pliensbachian-Toarcian boundary: evidence from the Lias Group, NE England. Geological Magazine 147, 181–192.

Magioncalda, R., Dupuis, C., Smith, T., Steurbaut, E., Gingerich, P.D., 2004. Paleocene-Eocene carbon isotope excursion in organic carbon and pedogenic carbonate: direct comparison in a continental stratigraphic section. Geology 32, 553–556.

Magaritz, M., Krishnamurthy, R.V., Holser, W.T., 1992. Parallel trends in organic and inorganic carbon isotopes across the Permian/Triassic boundary. American Journal of Science 292, 727–739.

Marquillas, R., Sabino, I., Sial, A.N., Papa, C., Ferreira, V., Matthews, S., 2007. Carbon and oxygen isotopes of Maastrichtian-Danian shallow marine carbonates: Yacoraite Formation, Northwestern Argentina. Journal of South American Earth Sciences 23, 304–320.

Marshall, J.D., 1992. Climatic and oceanographic isotopic signals from the carbonate rock record and their preservation. Geological Magazine 129, 143–160.

Marshall, J.D., Brenchley, P.J., Mason, P., Wolff, G.A., Astini, R.A., Hints, L., Meidla, T., 1997. Global carbon isotopic events associated with mass extinction and glaciation in the late Ordovician. Palaeogeography, Palaeoclimatology, Palaeoecology 132, 195–210.

Martin, R.E., 1995. Cyclic and secular variation in microfossil biomineralization nucleus to the biogeochemical evolution of Phanerozoic oceans. Global and Planetary Change 11, 1–23.

Matsumoto, R., 1995. Causes of the $\delta^{13}C$ anomalies of carbonates and a new paradigm "gas hydrate hypothesis". Journal of the Geological Society of Japan 101, 902–924.

Mazumdar, A., Strauss, H., 2006. Sulfur and strontium isotopic compositions of carbonate and evaporite rocks from the late Neoproterozoic–early Cambrian Bilara Group (Nagaur-Ganganagar Basin, India): constraints on intrabasinal correlation and global sulfur cycle. Precambrian Research 149, 217–230.

McKirdy, D.M., Burgess, J.M., Lemon, N.M., Yu, X., Cooper, A.M., Gostin, V.A., Jenkins, R.J.F., Both, R.A., 2001. A chemostratigraphic overview of the Late Cryogenian interglacial sequence in the Adelaide fold-thrust belt, South Australia. Precambrian Research 106, 149–186.

Mearon, S., Payton, A., Bralower, T.J., 2003. Cretaceous strontium isotope stratigraphy using marine barite. Geology 31, 15–18.

Melezhik, V.A., Huhma, A., Condon, D.J., Fallick, A.E., Whitehouse, M.J., 2007. Temporal constraints on the Paleoproterozoic Lomagundi-Jatuli carbon isotopic event. Geology 35, 655–658.

Meyers, P.A., Simoneit, B.R.T., 1989. Global comparisons of organic matter in sediments across the Cretaceous/Tertiary boundary. Advances in Organic Geochemistry 16, 641–648.

Middleberg, J., Calvert, S., Karlin, R., 1991. Organic rich transitional facies in silled basins: response to sealevel change. Geology 19, 679–682.

Miller, J.R., Lechler, P.J., Hudson-Edwards, K.A., Macklin, M.G., 2009. Lead isotopic fingerprinting of heavy metal contamination, Rio Pilcomayo basin, Bolivia. Geochemistry: Exploration, Environment, Analysis 9, 115–123.

Montero-Serrano, J.C., Palarea-Albaladejo, J., Martin-Fernandez, J.A., Martinez-Santana, M., Gutierrez-Martin, J.V., 2010. Sedimentary chemofacies characterization by means of multivariate analysis. Sedimentary Geology 228, 218–228.

Morante, R., Veevers, J.J., Andrew, A.S., Hamilton, P.J., 1994. Determination of the Permian–Triassic boundary in Australia. APEA Journal 330–336.

Mount, J.F., Margolis, S.V., Showers, W., Ward, P., Doehne, E., 1986. Carbon and oxygen isotope stratigraphy of the upper Maastrichtian, Zumaya, Spain: a record of oceanographic and biologic changes at the end of the Cretaceous Period. Palaios 1, 87–92.

Mutti, M., Bernoulli, D., 2003. Early marine lithification and hardground development on a Miocene ramp (Maiella, Italy): key surfaces to track changes in trophic resources in non-tropical carbonate settings. Journal of Sedimentary Research 73, 296–308.

Mutti, M., John, C.M., Knoerich, A.C., 2006. Chemostratigraphy in Miocene heterozoan carbonate settings: applications, limitations and perspectives. Geological Society, London, Special Publications 255, 307–322.

Nandy, P., Saraswati, P.K., Kumar, A., 1995. Chemostratigraphy of the Eocene–Oligocene sequence of southwestern Kutch, Gujarat. In: Proc. Recent Researches in Geology of Western India, pp. 53–59.

Nedelec, A., Affaton, P., France-Lanord, C., 2007. Sedimentology and chemostratigraphy of the Bwipe Neoproterozoic cap dolostones (Ghana, Volta Basin): a record of microbial activity in a peritidal environment. Comptes Rendus Geoscience 339, 223–239.

Nelson, C.S., Hendry, C.H., Jarrett, G.R., 1985. Near-synchroneity of New Zealand Alpine glaciations during the past 750 Kyr. Nature 318, 361–363.

Odin, G.S., Renard, M., Grazzini, C.U., 1982. Geochemical events as a means of correlation. In: Odin, G.S. (Ed.), Numerical Dating in Stratigraphy. Part I. Wiley & Sons, Chichester, pp. 37–71.

Oppo, D.W., Fairbanks, R.G., 1989. Carbon isotope composition of tropical surface water during the past 22,000 years. Paleoceanography 4, 333–351.

Oppo, D.W., Fairbanks, R.G., Gordon, A.L., 1990. Late Pleistocene southern ocean $\delta^{13}C$ variability. Paleoceanography 5, 43–54.

Palfy, J., Demeny, A., Haas, J., Hetenyi, M., Orchard, M.J., Veto, I., 2001. Carbon isotope anomaly and other geochemical changes at the Triassic–Jurassic boundary from a marine section in Hungary. Geology 29, 1047–1050.

Panchuk, K.M., Holmden, C., Kump, L.R., 2005. Sensitivity of the epeiric sea carbon isotope record to local-scale carbon cycle processes: tales from the Mohawkian sea. Palaeogeography, Palaeoclimatology, Palaeoecology 228, 320–337.

Pasley, M., Riley, G., Nummedal, D., 1993. Sequence stratigraphic significance of organic matter variations: examples from the Upper Cretaceous Mancos shale of the San Juan Basin, New Mexico. In: Kabz, B., Prat, L. (Eds.), Source Rocks in a Sequence Stratigraphic Framework. Amer. Assoc. Petrol. Geol. Studies in Geology, vol .37, pp. 221–241.

Payne, J.L., Lehrmann, D.J., Wei, J.Y., Orchard, M.J., Schrag, D.P., Knoll, A.H., 2004. Large perturbations of the carbon cycle during recovery from the end-Permian extinction. Science 305, 506–509.

Payne, J.L., Kump, L.R., 2007. Evidence for recurrent Early Triassic massive volcanism from quantitative interpretation of carbon isotope fluctuations. Earth and Planetary Science Letters 256, 264–277.

Pearce, T.J., Jarvis, I., 1991. Applications of geochemical data to modeling sediment dispersal patterns in distal turbidites: late quaternary of the Madeira abyssal plain. Journal of Sedimentary Petrology 62, 1112–1129.

Pearce, T.J., Besly, B.M., Wray, D.S., 1999. Chemostratigraphy: a method to improve interwell correlation in barren sequences—a case study using onshore Duckmantian/Stephanian sequences (West Midlands, U.K.). Sedimentary Geology 124, 197–220.

Pelechaty, S.M., 1998. Integrated chronostratigraphy of the Vendian System of Siberia: implications for a global stratigraphy. Journal of the Geological Society 155, 957–973.

Pelechaty, S.M., Grotzinger, J.P., Kashirtsev, V.A., Zhernovsky, V.P., 1996. Chemostratigraphic and sequence stratigraphic constraints, Vendian–Cambrian basin dynamics, Northeast Siberian Craton. Journal of Geology 104, 543–563.

Pellenard, P., Tramoy, R., Pucéat, E., Huret, E., Martinez, M., Bruneau, L., Thierry, J., 2014. Carbon cycle and seawater palaeotemperature evolution at the Middle-Late Jurassic transition, eastern Paris Basin (France). Marine and Petroleum Geology 53, 30–43.

Punekar, J., Keller, G., Khozyemb, H., Hamming, C., Adatte, T., Tantawy, A.A., Spangenberg, J.E., 2014. Late Maastrichtian early Danian high-stress environments and delayed recovery linked to Deccan volcanism. Cretaceous Research 49, 63–82.

Racki, G., Joachimski, M.M., Morrow, J.R., 2008. A major perturbation of the global carbon budget in the Early-Middle Frasnian transition (Late Devonian). Palaeogeography, Palaeoclimatology, Palaeoecology 269, 127–129.

Ramkumar, M., 1999. Role of chemostratigraphic technique in reservoir characterization and global stratigraphic correlation. Indian Journal of Geochemistry 14, 33–45.

Ramkumar, M., Guha, A.K., 2000. Multivariate statistical verification of petrographic and standard microfacies types and lithostratigraphy of Tertiary carbonates of western Kutch, Gujarat, India: implications on global stratigraphic correlation and hydrocarbon exploration. Indian Journal of Petroleum Geology 9, 52–74.

Ramkumar, M., 2001. Sedimentary microenvironments of modern Godavari delta: characterization and statistical discrimination—towards computer assisted environment recognition scheme. Journal of the Geological Society of India 57, 49–63.

Ramkumar, M., 2004. Dynamics of moderately well mixed tropical estuarine system, Krishna estuary, India: part IV Zones of active exchange of physico-chemical properties. Indian Journal of Geochemistry 19, 245–269.

Ramkumar, M., 2014. Characterization of depositional units for stratigraphic correlation, petroleum exploration and reservoir characterization. In: Sinha, S., (Ed.), Advances in Petroleum Engineering. Studium Press L.L.C, U.S.A, pp. 1–13.

Ramkumar, M., Berner, Z., Stüben, D., 2002. Multivariate statistical discrimination of selected carbonate petrographic classifications: implications on applicability of classification systems and predictability of petrographic types. International Journal of Geochemistry 62, 145–159.

Ramkumar, M., Stüben, D., Berner, Z., Schneider, J., 2004. Isotopic and geochemical anomalies preceding K/T boundary in the Cauvery Basin, south India: timing of events in the context of global scenario. Current Science 87, 1738–1747.

Ramkumar, M., Harting, M., Stüben, D., 2005. Barium anomaly preceding K/T boundary: possible causes and implications on end Cretaceous events of K/T sections in Cauvery Basin (India), Israel, NE-Mexico and Guatemala. International Journal of Earth Sciences 94, 475–489.

Ramkumar, M., Sathish, G., 2007. Integrated sequence and chemostratigraphic modelling: a sure-fire technique for stratigraphic correlation, petroleum exploration and reservoir characterization. In: Rajendran, S., Srinivasamoorty, K., Aravindan, S. (Eds.), Mineral Exploration: Recent Strategies. New India Publishers, New Delhi, pp. 21–40.

Ramkumar, M., Stüben, D., Berner, Z., Schneider, J., 2010a. $^{87}Sr/^{86}Sr$ anomalies in Late Cretaceous–Early Tertiary strata of the Cauvery Basin, south India: constraints on nature and rate of environmental changes across K–T boundary. Journal of Earth System Science 119, 1–17.

Ramkumar, M., Stüben, D., Berner, Z., 2010b. Hierarchical delineation and multivariate statistical discrimination of chemozones of the Cauvery Basin, South India: implications on Spatio-temporal scales of stratigraphic correlation. Petroleum Science 7, 435–447.

Ramkumar, M., Stüben, D., Berner, Z., 2011. Barremian-Danian chemostratigraphic sequences of the Cauvery Basin, South India: implications on scales of stratigraphic correlation. Gondwana Research 19, 291–309.

Ramkumar, M., Alberti, M. and Fürsich, F.T., (this volume) Chemostratigraphy of the Dhosa Oolite Member (Oxfordian), Kachchh Basin, western India: Implications for completeness of the stratigraphic record and correlation with global oolite peak. In: Ramkumar, M., (Ed.). Chemostratigraphy: Concepts, techniques and applications. Elsevier.

Rangel, A., Parra, P., Nino, C., 2000. The La Luna Formation: chemostratigraphy and organic facies in the middle Magdalena Basin. Organic Geochemistry 31, 1267–1284.

Ratcliffe, K.T., Martin, J., Pearce, T.J., Hughes, A.D., Lawton, A.D., Lawton, D.E., Wray, D.S., Bessa, F., 2006. A regional chemostratigraphically-defined correlation framework for the late Triassic TAG-I Formation in blocks 402 and 405a, Algeria. Petroleum Geoscience 12, 3–12.

Raymo, E., Oppo, D.W., Curry, W., 1997. The mid-Pleistocene climate transition: a deepsea carbon isotopic perspective. Paleoceanography 12, 546–559.

Reineck, H.E., Singh, I.B., 1980. Depositional Sedimentary Environments. Springer-Verlag, New York. p. 439.

Renard, M., 1986. Pelagic carbonate chemostratigraphy (Sr, Mg, ^{18}O, ^{13}C). Marine Micropaleontology 10, 117–164.

Renard, M., Richebois, G., Letolle, R., 1984. Trace element and stable isotope geochemistry of Paleocene to Coniacian carbonate samples from hole 516F, comparison with North Atlantic and Tethys sites. In: Barker, P.F., Johnson, D.A., Carlson, R.L., Cepek, P., Coulbourn, W.T., Gamboa, L.A., Hamilton, N., de Melo, U., Pujol, C., Shor, A.N., Suzyumov, A.E., Tjalsma, L.R.C., Walton, W.H. (Eds.), Initial Reports of the Deep Sea Drilling Project, vol. 72, pp. 399–420.

Reuter, M., Piller, W.E., Brandano, M., Harzhauser, M., 2013. Correlating Mediterranean shallow water deposits with global Oligocene–Miocene stratigraphy and oceanic events. Global and Planetary Change 111, 226–236.

Robinson, N., Ravizza, G., Coccioni, R., 2009. High-resolution marine $^{187}Os/^{188}Os$ record for the late Maastrichtian: distinguishing the chemical fingerprints of Deccan volcanism and the K–P impact event. Earth and Planetary Science Letters 281, 159–168.

Romein, A.J.T., Smit, J., 1981. Carbon–Oxygen stable isotope stable isotope stratigraphy of the Cretaceous–Tertiary boundary interval: data from the Blarritz section (SW France). Geologie en Mijnbouw 60, 514–544.

Ruhl, M., Kueschner, W.M., Krystyn, L., 2009. Triassic–Jurassic organic carbon isotope stratigraphy of key sections in the western Tethys realm (Austria). Earth and Planetary Science Letters 281, 169–187.

Sageman, B.B., Meyers, S.R., Arthur, M.A., 2006. Orbital time scale and new C-isotope record for Cenomanian–Turonian boundary stratotype. Geology 34, 125–128.

Samtleben, C., Munnecke, A., Bickert, T., 2000. Development of facies and C/O-isotopes in transects through the Ludlow of Gotland: evidence for global and local influences on a shallow marine environment. Facies 43, 1–38.

Saltzman, M.R., 2002a. Carbon and oxygen isotope stratigraphy of the Lower Mississippian (Kinderhookian-Lower Osagean), western United States: implications for seawater chemistry and glaciation. Geological Society of America Bulletin 114, 96–108.

Saltzman, M.R., 2002b. Carbon isotope ($\delta^{13}C$) stratigraphy across the Silurian–Devonian transition in North America: evidence for a perturbation of the global carbon cycle. Palaeogeography, Palaeoclimatology, Palaeoecology 187, 83–100.

Saltzman, M.R., Thomas, E., 2012. Carbon isotope stratigraphy. In: Gradstein, F.M., Ogg, J.G., Schmitz, M., Ogg, G. (Eds.), The Geologic Time Scale. Elsevier, The Netherlands, pp. 207–232.

Saltzman, M.R., Brasier, M.D., Ripperdan, R.L., Ergaliev, G.K., Lohmann, K.C., Robison, R.A., Chang, W.T., Peng, S., Runnegar, B., 2000. A global carbon isotope excursion during the Late Cambrian: relation to trilobite extinctions, organic-matter burial and sea level. Palaeogeography, Palaeoclimatology, Palaeoecology 162, 211–223.

Saltzman, M.R., Sedlacek, A.R.C., 2013. Chemostratigraphy indicates a relatively complete Late Permian to Early Triassic sequence in the western United States. Geology 41, 399–440.

Scholle, P.A., Arthur, M.A., 1980. Carbon isotope fluctuations in Cretaceous pelagic limestones: potential stratigraphic and petroleum exploration tool. American Association of Petroleum Geologists Bulletin 64, 67–87.

Schroeder, S., Grotzinger, J.P., 2007. Evidence for anoxia at the Ediacaran–Cambrian boundary: the record of redox-sensitive trace elements and rare earth elements in Oman. Journal of the Geological Society 164, 175–187.

Schroeder, S., Schreiber, B.C., Amthor, J.A., 2004. Stratigraphy and environmental conditions of the terminal Neoproterozoic–Cambrian Period in Oman: evidence from sulphur isotopes. Journal of the Geological Society 161, 489–499.

Shackleton, N.J., Hall, M.A., 1984. Carbon isotope data from Leg 74 sediments. In: Moore Jr., T.C., Rabinowitz, P.D., Boersma, A., borella, P.E., Chave, A.D., Duee, G., Futterer, D.K., Jiang, M.J., Kleinert, K., Lever, A., Manivit, H., O'Connel, S., Richardson, S.H., Shackleton, N.J. (Eds.), Initial Reports of the Deep Sea Drilling Project, vol. 74, pp. 613–619.

Sluijs, A., Bowen, G.J., Brinkhuis, H., Lourens, L.J., Thomas, E., 2007. The Palaeocene–Eocene Thermal maximum super greenhouse: biotic and geochemical signatures, age models and mechanisms of global change. In: Williams, M., Haywood, A.M., Gregory, F.J., Schmidt, D.N. (Eds.), Deep Time Perspectives on Climate Change: Marrying the Signal from Computer Models and Biological Proxies. The Micropalaeontological Society Special Publications, The Geological Society, London, pp. 323–351.

Srinivasan, M.S., 1989. Recent advances in Neogene planktonic foraminiferal biostratigraphy, chemostratigraphy and paleoceanography: northern Indian Ocean. Journal of the Palaeontological Society of India 34, 1–18.

Staron, R.M., Grandstaff, B.S., Gallagher, W.B., Grandstaff, D.E., 2001. REE signatures in vertebrate fossils from Sewell, NJ: implications for location of the K–T boundary. Palaios 16, 255–265.

Stoll, H.M., Schrag, D.P., 2000. High-resolution stable isotope records from the Upper Cretaceous rocks of Italy and Spain: glacial episodes in a greenhouse planet? Geological Society of America Bulletin 112, 308–319.

Strauss, H., 1997. The isotopic composition of sedimentary sulphur through time. Palaeogeography, Palaeoclimatology, Palaeoecology 132, 97–118.

Stüben, D., Kramar, U., Harting, M., Stinnesbeck, W., Keller, G., 2005. High resolution geochemical record of Cretaceous–Tertiary boundary sections in Mexico: new constrains on the K/T and Chicxulub events. Geochimica et Cosmochimica Acta 69, 2559–2579.

Suarez, C.A., Ludvigson, G.A., Gonzalez, L.A., Fiorillo, A.R., Flaig, P.P., McCarthy, P.J., 2013. Use of multiple oxygen isotope proxies for elucidating Arctic Cretaceous palaeo-hydrology. In: Bojar, A.-V., Melinte-Dobrinescu, M.C., Smit, J. (Eds.), Isotopic Studies in Cretaceous Research. Geological Society, London. Special Publications, 382. http://dx.doi.org/10.1144/SP382.3.

Takahashi, S., Yamasaki, S., Ogawa, Y., Kimura, K., Kaiho, K., Yoshida, T., Tsuchiya, N., 2014. Bioessential element-depleted ocean following the euxinic maximum of the end-Permian mass extinction. Earth and Planetary Science Letters 393, 94–104.

Thomas, E., Shackleton, N.J., 1996. The Palaeocene–Eocene Benthic Foraminiferal Extinction and Stable Isotope Anomalies, vol. 101,Geol. Soc. London Spec. Publ. pp. 401–441.

Thomas, C.W., Graham, C.M., Ellam, R.M., Fallick, A.E., 2004. $^{87}Sr/^{86}Sr$ chemostratigraphy of Neoproterozoic Dalradian limestones of Scotland and Ireland: constraints on depositional ages and time scales. Journal of the Geological Society 161, 229–242.

Tu, T.T.N., Bocherens, H., Mariotti, A., Baudin, F., Pons, D., Broutin, J., Derenne, S., Largeau, C., 1999. Ecological distribution of Cenomanian terrestrial plants based on $^{13}C/^{12}C$ ratios. Palaeogeography, Palaeoclimatology, Palaeoecology 145, 79–93.

Uramoto, G., Tahara, R., Sekiya, T., Hirano, H., 2013. Carbon isotope stratigraphy of terrestrial organic matter for the Turonian (Upper Cretaceous) in northern Japan: implications for ocean-atmosphere $\delta^{13}C$ trends during the mid-Cretaceous climatic optimum. Geosphere 9, 355–366.

Vail, P.R., Mitchum, R.M., Thompson, S., 1977. Seismic stratigraphy and global changes of sealevel. Part 4 global cycles of relative changes of sealevel. In: Payton, C.E. (Ed.), Seismic Stratigraphy—Applications to Hydrocarbon Exploration. Mem. Amer. Assoc. Petrol. Geol. vol. 26, pp. 83–97.

Veizer, J., 1985. Carbonates and ancient oceans: isotopic and chemical record on timescales of 10^7—10^9 years. In: Sunquist, E.T., Broecker, W.S. (Eds.), The Carbon Cycle and Atmospheric CO_2: Natural Variations—Archaen to Present. Amer. Geophy. Union Monogr. vol. 32, pp. 595–601.

Veizer, J., Buhl, D., Diener, A., Ebneth, S., Podlaha, O.G., Bruckschen, P., Jasper, T., Korte, C., Schaaf, M., Ala, D., Azmy, K., 1997. Strontium isotope stratigraphy: potential resolution and event correlation. Palaeogeography, Palaeoclimatology, Palaeoecology 132, 65–77.

Veizer, J., Ala, D., Azmy, K., Bruckschen, P., Buhl, D., Bruhn, F., Carden, G.A.F., Diener, A., Ebneth, S., Godderis, Y., Jasper, T., Korte, C., Pawellek, F., Podlaha, O.G., Strauss, H., 1999. $^{87}Sr/^{86}Sr$ and $\delta^{18}O$ evolution of Phanerozoic seawater. Chemical Geology 161, 59–88.

Wallmann, K., 2001. Controls on the Cretaceous and Cenozoic evolution of seawater composition, atmospheric CO_2 and climate. Geochimica et Cosmochimica Acta 65, 3005–3025.

Wang, W., Zhou, C., Yuan, X., Chen, Z., Xiao, S., 2012. A pronounced negative $\delta^{13}C$ excursion in an Ediacaran succession of western Yangtze Platform: a possible equivalent to the Shuram event and its implication for chemostratigraphic correlation in South China. Gondwana Research 22, 1091–1101.

Weedon, G.P., 2005. Time-Series Analysis and Cyclostratigraphy: Examining Stratigraphic Records of Environmental Cycles. Cambridge University Press, Cambridge. p. 259.

Weissert, H., Joachimski, M., Sarnthein, M., 2008. Chemostratigraphy. Newsletters on Stratigraphy 42, 145–179.

Whitford, D.J., Allan, T.L., Andrew, A.S., Craven, S.J., Hamilton, P.J., Korsch, M.J., Trotter, J.A., Valenti, G.L., 1996. Strontium isotope chronostratigraphy and geochemistry of the Darai limestone: Juha 1X well, Papua New Guinea. In: Proc. III PNG Petroleum Convention, Port Moresby, pp. 369–380.

Williams, D.F., Healy-Williams, N., Thunell, R.C., Baruch, B.W., Leventer, A., 1983. Detailed stable isotope and carbonate records from the upper Maastrichtian-lower Paleogene section of hole 516F (Leg 72) including the Cretaceous/Tertiary boundary. In: Barker, P.F., Johnson, D.A., Carlson, R.L., Cepek, P., Coulbourn, W.T., Gamboa, L.A., Hamilton, N., de Melo, U., Pujol, C., Shor, A.N., Suzyumov, A.E., Tjalsma, L.R.C., Walton, W.H. (Eds.), Initial Reports of the Deep Sea Drilling Project, vol. 72, pp. 921–929.

Williams, D.F., Thunell, R.C., Tappa, E., 1988. Chronology of the Pleistocene oxygen isotope record 0–1.88 My BP. Palaeogeography, Palaeoclimatology, Palaeoecology 64, 221–240.

Williford, K.H., Foriel, J., Ward, P.D., Steig, E.J., 2009. Major perturbation in sulfur cycling at the Triassic–Jurassic boundary. Geology 37, 835–838.

Wynn, T.C., Read, J.F., 2007. Carbon–oxygen isotope signal of Mississippian slope carbonates, Appalachians, USA: a complex response to climate-driven fourth-order glacio-eustasy, Palaeoceanographic implications. Palaeogeography, Palaeoclimatology, Palaeoecology 256, 254–272.

Xiao, S.H., Knoll, A.H., Kaufman, A.J., Yin, L., Zhang, Y., 1997. Neoproterozoic fossils in Mesoproterozoic rocks? Chemostratigraphic resolution of a biostratigraphic conundrum from the north China platform. Precambrain Research 84, 197–220.

Xie, S., Pancost, R.D., Huang, J., Wignall, P.B., Yu, J., Tang, X., Chen, L., Huang, X., Lai, X., 2008. Changes in the global carbon cycle occurred as two episodes during the Permian–Triassic crisis. Geology 35, 1083–1086.

Yin, H.F., Feng, Q., Lai, X., Baud, A., Tong, J., 2007. The protracted Permo–Triassic crisis and multi-episode extinction around the Permian–Triassic boundary. Global and Planetary Change 55, 1–20.

Young, S.A., Saltzman, M.R., Bergström, S.M., 2005. Upper Ordovician (Mohawkian) carbon isotope ($\delta^{13}C$) stratigraphy in eastern and central North America: regional expression of a perturbation of the global carbon cycle. Palaeogeography, Palaeoclimatology, Palaeoecology 222, 53–76.

Zachos, J., Pagani, M., Sloan, L., Thomas, E., Billups, K., 2001. Trends, rhythms, and aberrations in global climate 65 Ma to present. Science 292, 686–693.

Zachos, J.C., Röhl, U., Schellenberg, S.A., Sluijs, A., Hodell, D.A., Kelly, D.C., Thomas, E., Nicolo, M., Raffi, I., Lourens, L., Dick Kroon, D., McCarren, H., 2005. Extreme acidification of the Atlantic Ocean at the Paleocene–Eocene boundary (~55 Mya). Science 308, 1611–1615.

Zachos, J.C., Bohaty, S.M., John, C.M., McCarren, H., Kelly, D.C., Nielsen, T., 2007. The Palaeocene–Eocene carbon isotope excursion: constraints from individual shell planktonic foraminifer records. Philosophical Transactions of the Royal Society A 365, 1829–1842.

Zachos, J.C., McCarren, H., Murphy, B., Roehl, U., Westerhold, T., 2010. Tempo and scale of late Paleocene and early Eocene carbon isotope cycles: Implications for the origin of hyperthermals. Earth and Planetary Science Letters 29, 242–249.

Zakharov, V.A., Rogov, M.A., Dzyuba, O.S., Zak, K., Kostak, M., Pruner, P., Skupien, P., Chadima, M., Mazuch, M., Nikitenko, B.L., 2014. Palaeoenvironments and palaeoceanography changes across the Jurassic/Cretaceous boundary in the Arctic realm: case study of the Nordvik section (north Siberia, Russia). Polar Research 33 19714. http://dx.doi.org/10.3402/polar.v33.19714.

CHAPTER

ISOTOPE AND ELEMENTAL CHEMOSTRATIGRAPHY

2

Alcides N. Sial[1], Claudio Gaucher[2], Valderez P. Ferreira[1], Natan S. Pereira[1], Wilker S. Cezario[1,†], Leticia Chiglino[1,3], Haroldo M. Lima[1]

[1]*NEG-LABISE, Department of Geology, Federal University of Pernambuco, Recife, Pernambuco, Brazil;*
[2]*Departamento de Geología, Facultad de Ciencias, Universidad de la República, Montevideo, Uruguay;*
[3]*PDU Geología y Recursos Minerales, CURE, Universidad de la República, Treinta y Tres, Uruguay*

2.1 INTRODUCTION

The use of chemostratigraphy as a tool in the correlation of global events in the Precambrian followed the pioneer research by William T. Holser on ancient ocean water chemistry (Kaufman et al., 2007). Several researchers focused on the carbon isotope variations across thick successions to investigate long-term fluctuations in the chemistry of the seawater (SW) (Veizer et al., 1980; Magaritz et al., 1986 and references therein). As a result, despite widespread effects of late diagenesis on the isotope record, important isotope events were demonstrated on a global scale (Knoll et al., 1986; Magaritz et al., 1986; Holser, 1997 among many others) as it became evident that contemporaneous, apparently well-preserved, geographically distant marine strata registered similar isotopic compositions. Thereafter, chemostratigraphy has served as one of the principal means of intra- and extra-basinal stratigraphic correlation to assemble Precambrian stratigraphic record from fragments preserved in scattered successions (Karhu et al., 2010). It has emerged as an important tool for providing a timeline compensating for poor biostratigraphic resolution of the Precambrian fossils (Veizer et al., 1980; Magaritz et al., 1986; Knoll et al., 1986; Knoll and Walter, 1992; Kaufman et al., 1997; Corsetti and Kaufman, 2003; Halverson et al., 2005). Correlations established through chemostratigraphy can also be used to comment on biogeochemical and climate changes through time. The determination of the base of the Ediacaran Period was largely based on chemostratigraphy of cap carbonates that overlie glacial diamictites (Knoll et al., 2006) that have been accumulated during the postglacial transgression, recording the most profound carbon, sulfur, and strontium isotope variations in Earth's history, and the "Snowball Earth" hypothesis (Kirschvink, 1992; Hoffman et al., 1998; Bekker et al., 2005) rests on them. The paucity of radiometric constraints on the absolute age of most of these extreme isotope excursions has led to debates on their temporal equivalence (Kaufman et al., 1997; Kennedy et al., 1998; Calver et al., 2004; Allen and Etienne, 2008).

Development of high-resolution deep-sea oxygen and carbon isotope records in the last four decades has served as the principal means of reconstructing global and regional climate change on a variety of

† In memoriam.

Chemostratigraphy. http://dx.doi.org/10.1016/B978-0-12-419968-2.00002-9

geologic timescales, from millennia to gigayears. These records are multidimensional in that they provide both climatic and stratigraphic information, and can be quickly generated with automated mass spectrometers (Zachos et al., 2001). As a consequence, the number of studies relying on isotope stratigraphy has grown substantially and it is worth noting that in the case of carbon isotope stratigraphy, it can be applied with care to metamorphic rocks up to the amphibolites facies (Melezhik et al., 2005b; Nascimento et al., 2007; Chiglino et al., 2010; Sial et al., 2010a) and may retain depositional $\delta^{13}C$ values in even diagenetically altered carbonates (Kaufman et al., 2007).

It would be impossible to discuss here the myriad types of isotope systems that have been successfully used in chemostratigraphy. Thus, we offer herein an overview of the most widely used isotope systems, as well as those recently developed methods that may potentially become important in chemostratigraphy in the near future.

2.2 ISOTOPE AND ELEMENTAL CHEMOSTRATIGRAPHY: USE AND LIMITATIONS

2.2.1 CARBON ISOTOPES

A priori knowledge on the secular variations of marine carbon isotope ratios is essential for application of carbon isotope chemostratigraphic correlation of sedimentary sequences. An attempt to compile carbon isotope data to obtain a secular variation curve of $\delta^{13}C$ (Karhu and Holland, 1996; Hoffman et al., 1998; Kah et al., 1999; Melezhik et al., 1999, 2007; Lindsay and Brasier, 2002; Bekker et al., 2006; Halverson et al., 2005, 2010b; Saltzman and Thomas, 2012 among others) has revealed remarkable $\delta^{13}C$ anomalies (Figs 1–3). $\delta^{13}C$ can be measured on carbonates and organic matter. However, due to a larger database and to a higher resilience against postdepositional alteration, the $\delta^{13}C$ of carbonates is more widely used in chemostratigraphy, except in carbonate-poor successions characterized by black shales (e.g., Johnston et al., 2010). Carbon isotope stratigraphy has also demonstrated that $\delta^{13}C$ minima follow the main extinction events (Magaritz, 1989). It is known that carbon isotopes speak with a robust voice (Knauth and Kennedy, 2009) and have tales to tell about Earth's history, although postdepositional alteration of the carbonate rocks under study may radically alter the story (Bristow and Kennedy, 2008). The necessity of careful examination of the diagenetic history of rocks cannot be underestimated. Petrographic, elemental (e.g., Mn/Sr, Sr concentration, and Rb/Sr vs $\delta^{13}C$), and isotopic ($\delta^{18}O$ vs $\delta^{13}C$) proxies are fundamental for the assessment of the nature of carbon isotopic signals (e.g., Marshall, 1992; Jacobsen and Kaufman, 1999; Melezhik et al., 2001). In doing so, dolostones and limestones have to be dealt with separately, due to their originally different elemental and isotopic compositions (e.g., Kah et al., 1999; Gaucher et al., 2007).

In the Paleoproterozoic, the Lomagundi–Jatuli carbon positive anomaly, between 2.3 and 2.1 Ga, shows a magnitude and a duration which are unique in Earth's history (see review by Melezhik et al., 1999, 2007; Karhu et al., 2008; Master et al., 2010; Maheshwari et al., 2010). In the Neoproterozoic, unprecedented carbon isotope fluctuations in the Cryogenian and Ediacaran—including the Shuram–Wonoka anomaly—reach values often below the canonical mantle composition though thick successions of sedimentary rocks were deposited during tens of millions of years in Oman, Australia, Brazil, southern China, Namibia, Norway, Siberia, and Uruguay (Brasier et al., 2000; Walter et al., 2000; Halverson et al., 2005; Melezhik et al., 2005b; Fike et al., 2006;

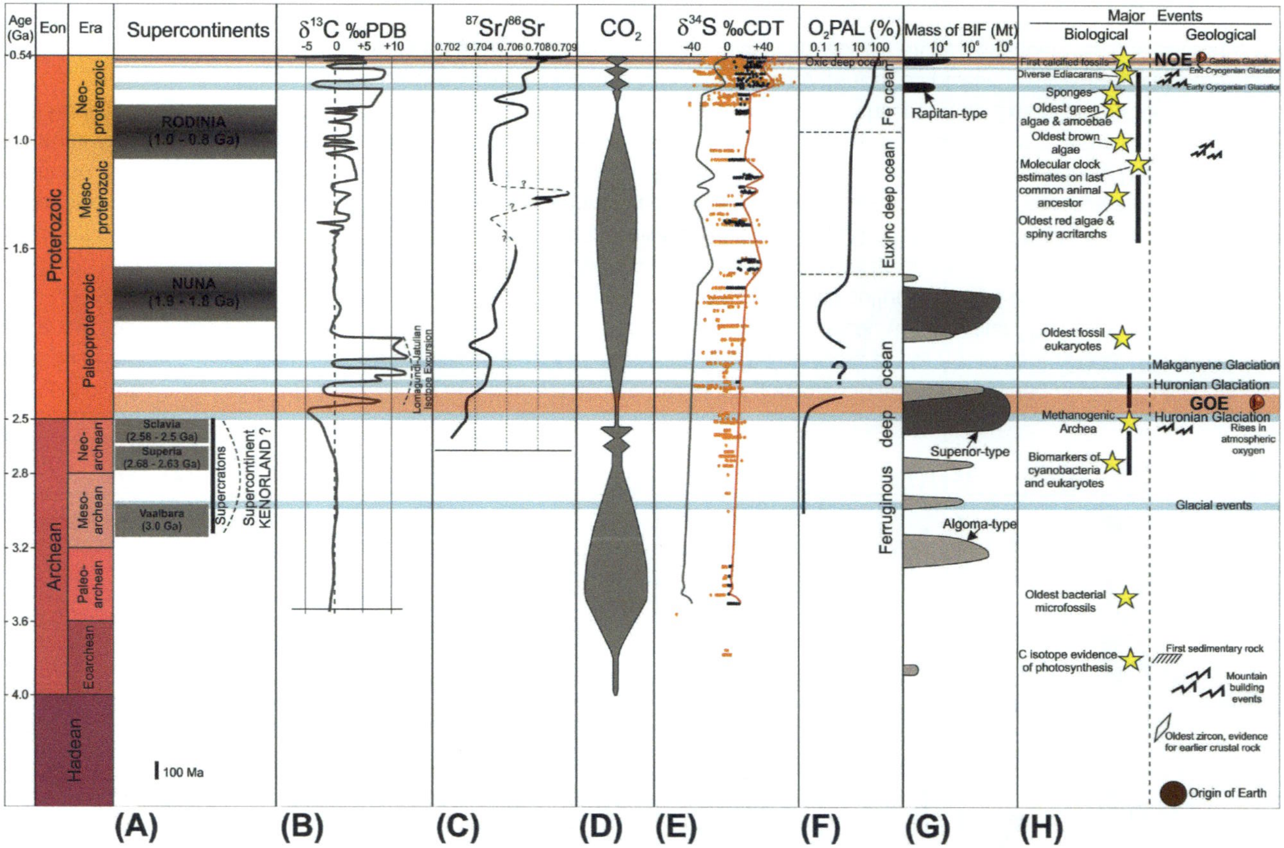

FIGURE 1

(A) Supercontinents (Bleeker, 2003; Reddy and Evans, 2009); (B) $\delta^{13}C$ record throughout the Precambrian *(Modified from Bartley and Kah (2004), Melezhik et al. (1999), Kah et al. (1999), and Halverson et al. (2005, 2010a,b).)*. (C) Seawater $^{87}Sr/^{86}Sr$ curve *(Modified from Veizer and Compston (1976), Kah et al. (2001).)*. (D) CO_2-estimated relative abundance of carbon dioxide in the atmosphere (Young, 2013). (E) Compilation of sulfur isotope data measured in sulphide- and sulphate-bearing minerals (Canfield and Farquhar, 2009; Shield-Zhou and Och, 2011). Upper red curve illustrates maximal $\delta^{34}S_{sulfate}$ values and its offset by −55‰ (lower line). (F) Reconstruction of atmospheric O_2 content through time expressed as percentage of present atmospheric level (PAL) of oxygen (Berner et al., 2003). (G) Variations in the abundance of BIF through time (Huston and Logan, 2004). (H) Great biological and geological events (Kaufman et al., 2007). Abbreviations: GOE, Great Oxidation Event; NOE, Neoproterozoic Oxidation Event.

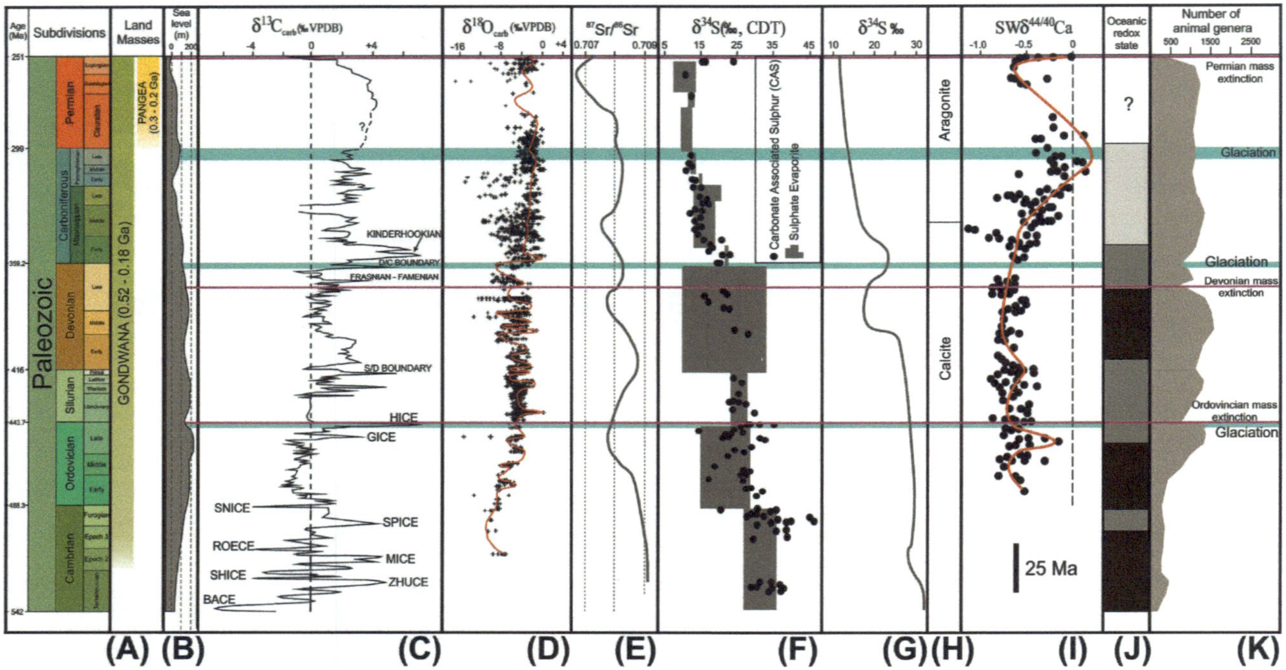

FIGURE 2

(A) Supercontinents (Reddy and Evans, 2009) (B) Sea-level variation curve for the Paleozoic (Haq and Schutter, 2008). (C) $\delta^{13}C$ record throughout the Paleozoic (Compiled from Joachimski et al., 2002; Saltzman, 2005; Zhu et al., 2006; Álvaro et al., 2008). (D) $\delta^{18}O$ variation throughout the Paleozoic (database from Veizer et al.,1999). The red line is from this work. (E) Seawater $^{87}Sr/^{86}Sr$ curve (Burke et al., 1982). (F) Compilation of sulfur isotope data: sulphate evaporite and carbonate- associated sulfur (modified from Kampschulte and Strauss, 2004). (G) $\delta^{34}S$ record for sulphates throughout the Paleozoic (Claypool et al., 1980). (H) Calcite and aragonite-sea intervals for the Paleozoic (Blättler et al., 2012). (I) Seawater $\delta^{44/40}Ca$, relative to modern seawater (Blätler et al., 2012). Red line is from this work. (J) Oceanic redox state: oxic ocean, icehouse (light gray), oscillating oxic, anoxic, cool climate (gray), anoxic-euxinic deep ocean, warm climate (dark gray) (modified from Saltzman, 2005); (K) Phanerozoic mass extinction events (Walochnik et al., 2010). Abbreviations: DICE, Drumian carbon isotope excursion; BACE, Basal Cambrian Carbon Isotope Excursion; ZHUCE, Zhujiaqing Carbon Isotope Excursion; CARE, Cambrian Arthropod Radiation Isotope Excursion; MICE, Mingxinsi Carbon Isotope Excursion; AECE, Archaeocyathid Extinction Carbon Isotope Excursion; ROECE, Redlchiid–Olenellid Extinction Carbon Isotope Excursion (Zhu et al., 2006); SPICE, Steptoean Positive Carbon Isotope Excursion (Saltzman et al., 1998); SNICE, Steptoean Negative Carbon Isotope Excursion (Sial et al., 2008); GICE, Guttenberg Isotope Carbon Excursion (Hatch et al., 1987); HICE (Orth et al., 1986).

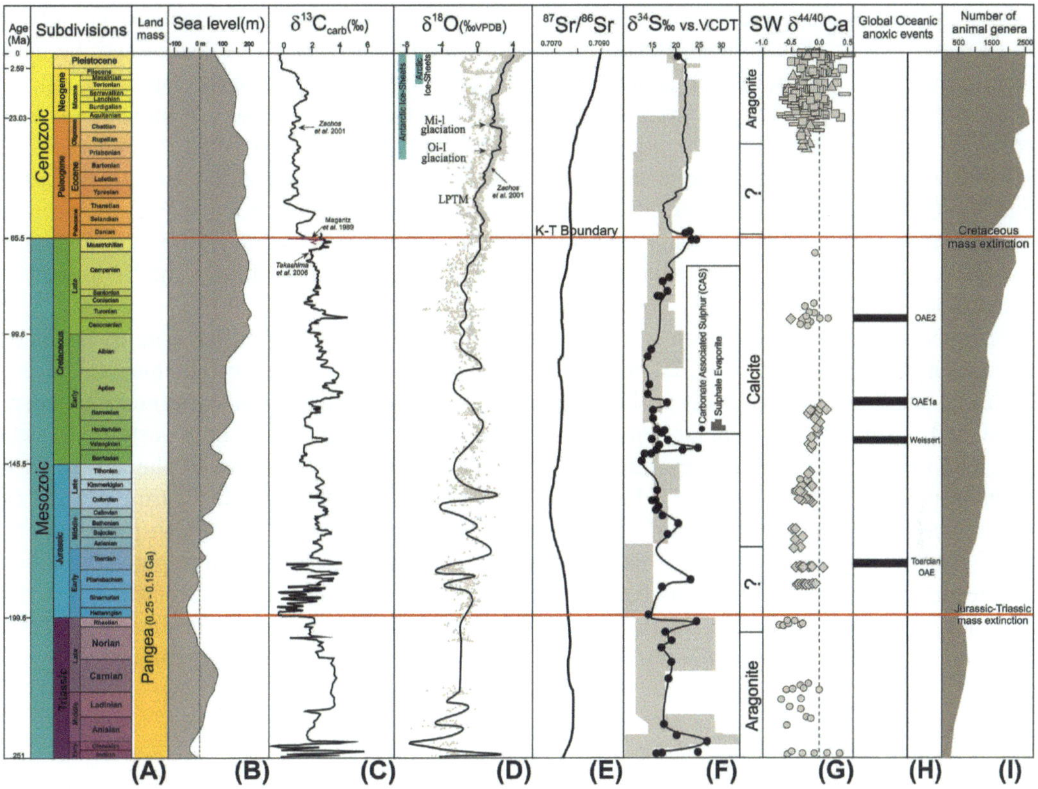

FIGURE 3

(A) Pangea supercontinent (Reddy and Evans, 2009). (B) Sea-level variation curve for the Mesozoic and Paleogene-Neogene (Haq et al., 1987). (C) $\delta^{13}C$ record throughout the Mesozoic (modified from Takashima et al., 2006; Magaritz et al., 1989) and Paleogene-Neogene that corresponds to a single global deep-sea isotope record (Zachos et al., 2001). (D) $\delta^{18}O$ record throughout the Mesozoic, Paleogene and Neogene (for the Mesozoic, modified from Veizer et al., 1999, and for the Cenozoic, from Zachos et al., 2001). LPTM (Late Paleocene Thermal Maximum) is a climatic aberration by the Paleocene-Eocene boundary; Oi-1(Eocene-Oligocene boundary) and Mi-1 (Oligocene-Miocene boundary) are climatic aberrations. (E) Seawater $^{87}Sr/^{86}Sr$ curve (modified from Burke et al., 1982). (F) $\delta^{34}S$ vs. VCDT record for sulphates throughout the Mesozoic and Cenozoic (modified from Kampschulte and Strauss, 2004; Paytan et al., 1998, Riccardi, 2007). Black dots represent whole-rock carbonates, and gray areas, structurally substituted sulfate in carbonates (SSS). Evaporite-based $\delta^{34}S$ is from a compilation by Strauss (1997). (G) Seawater $\delta^{44/40}Ca$, relative to modern seawater (Blätler et al., 2012). (H) Oceanic anoxic events (Takashima et al., 2006). (I) Main mass extinction events in the Mesozoic and Paleogene (Walochnik et al., 2010). Dots: brachiopods and belemnites; squares: foraminifera; bar: barite; triangles: bulk carbonate.

Le Guerroué et al., 2006a,b,c, 2010; Figueiredo, 2006; McFadden et al., 2008; Melezhik et al., 2005a,b; Alvarenga et al., 2009; Gaucher et al., 2009; Le Guerroué and Cozzi, 2010; Derry, 2010; Boggiani et al., 2010; Frei et al., 2011). Besides, in the Cambrian, the Steptoean positive C-isotope excursion or SPICE (Saltzman et al., 1998, 2000) and Sunwaptan negative C-isotope excursion or SNICE (Sial et al., 2008, 2013) are important tools in refining the Upper Cambrian stratigraphy.

Many studies have reported carbon isotope variations in Neoproterozoic carbonate rocks and linked them to perturbations of the global carbon cycle. Knauth and Kennedy (2009) examined the data by concentrating on substantial oxygen isotope measurements obtained as part of the carbon isotope analysis but often overlooked. They came to the conclusion that the combined oxygen and carbon isotope systematics are identical to those Phanerozoic examples that lithified in coastal pore fluids receiving a groundwater influx of photosynthetic carbon from terrestrial phytomass. According to Knauth and Kennedy (2009), instead of being perturbations to the carbon cycle, decrease in $^{13}C/^{12}C$ in Neoproterozoic carbonates, would be more easily interpreted by an extensive greening in the late Precambrian times. However, no direct evidence exists of a widespread terrestrial microflora in the Neoproterozoic, as pointed out by Arthur (2009).

The indiscriminate use of carbon isotope stratigraphy to correlate Neoproterozoic carbonates has been cautioned by Frimmel (2008, 2009, 2010) who has studied the rare-earth elements (REE) plus yttrium (Y) distribution in a variety of Neoproterozoic (Cryogenian and Ediacaran) carbonates from different settings in the Saldania, Gariep, Damara, and West Congo belts in southwestern and central Africa. His studies have revealed systematic differences in REE+Y patterns, explained by varying paleoenvironmental factors and have led Frimmel to consider that a short residence time of REE and a comparatively short residence time of carbon and a strong influence of river-born particles on the REE+Y patterns pose some doubt on the usefulness of cap carbonates for stratigraphic correlation of Neoproterozoic sediment successions based on carbon isotopes. One can argue that REEs and dissolved inorganic carbon behave differently in SW and are affected in a complete different way during diagenesis. Nevertheless, Frimmel's observations seem to deserve further investigation.

2.2.2 NITROGEN ISOTOPES

The utility of nitrogen isotopes for stratigraphic correlation or study of the evolution of atmospheric chemistry remains ambiguous, although some results have suggested they are a promising tool (Beaumont and Robert, 1999; Papineau et al., 2005; Algeo et al., 2008; Cremonese et al., 2009). Nitrogen isotope values for bulk samples from sections in South China that straddle the Ediacaran–Cambrian boundary show secular variations (positive values in upper Ediacaran strata and strong negative shift in Cambrian strata, especially in black shales) and testify to the changes in the biogeochemical cycle of the ancient ocean (Cremonese et al., 2009). The use of $\delta^{15}N$ profiles is perhaps a promising chemostratigraphic correlation tool, especially where biostratigraphy is of limited usefulness. Denitrification, the process by which nitrate and nitrite are reduced to nitrogen gas, is an important part of the global nitrogen cycle in modern oceans (Algeo et al., 2008) and its variations over Quaternary glacial–interglacial time scales may have affected the global climate. It is unclear whether denitrification rates varied turning pre-Quaternary glacial cycles. Algeo et al. (2008) discussed ratios of C_{org}/N and nitrogen isotope data from the Upper Carboniferous black shales and found evidence of variations in the intensity of denitrification associated with glacially driven sea-level changes. According to these authors, sedimentary $\delta^{15}N$ increases during the interval of rapid sea-level rise in each cycle, indicative of intensified denitrification, before returning to background levels when sea level remains stabilized during the interglacial phase. Therefore, ice ages have produced similar oceanographic conditions and nitrogen cycle dynamics over the past 300 Ma. Whether Cryogenian–Ediacaran ice ages had a similar behavior in relation to the nitrogen cycle, however, is something that remains to be investigated.

2.2.3 OXYGEN ISOTOPES

Oxygen isotope chemostratigraphy forms the backbone of Meso- and Cenozoic stratigraphic correlation of marine sediments (e.g., Friedrich et al., 2012). $\delta^{18}O$ is usually measured on benthic foraminifera, to avoid isotopic gradient effects. In older successions, however, the demonstration of the primary nature of oxygen isotope signals is often difficult. Based on a large database, Veizer et al. (1999) showed a general increase from −8‰ to 0‰ V-PDB (Vienna-PeeDee Belemnite) in the Phanerozoic, punctuated by positive excursions coincident with cold intervals. These authors suggested that $\delta^{18}O$ analyses of carefully screened, well-preserved brachiopods and mollusks can still retain primary SW values even for Paleozoic samples. The absence of calcified metazoans in the Precambrian, except for the Ediacaran, makes this kind of studies effectively impossible. $\delta^{18}O$ analyses of whole-rock samples of Precambrian successions are usually interpreted as reflecting diagenetic conditions, although in exceptional occasions primary trends have been reported (Tahata et al., 2013).

A different approach in the discrimination of early-Cryogenian from end-Cryogenian cap carbonates is the use of triple oxygen isotope evidence as suggested by Bao et al. (2008, 2009). They demonstrated that triple oxygen isotope composition of sulfate from ancient evaporites and barite shows variable negative ^{17}O isotope anomalies over the past 750 Ma. An important difference in ^{17}O isotope anomalies of barite at the top of dolostones from Marinoan cap carbonates (negative spike of approximately −0.70‰) suggest that by the time this mineral was precipitated, P_{CO_2} was at the highest level in the past 750 Ma (CO_2 levels reached 0.01–0.08 bar during and just after ~635 Ma glacial event; Bao et al., 2008, 2009).

2.2.4 SULFUR ISOTOPES

Sulfur isotopes are largely used in Neoproterozoic and Phanerozoic isotope chemostratigraphy, especially after Claypool et al. (1980) reported a secular variation curve of $\delta^{34}S$ values of evaporites (1.0 Ga to present). A critical evaluation of the isotopic composition of sedimentary sulfur through time and the rather limited potential of sulfur isotopic investigations of sedimentary sulfides from time boundaries have been reviewed by Strauss (1997). Sedimentary sulfur isotope data are subdivided into those recording SW sulfate ($\delta^{34}S_{sulf}$) and epi- or authigenic pyrite ($\delta^{34}S_{pyr}$) (Halverson et al., 2010 a,b). Sulfur isotope data on sulfate are recovered from evaporites, barite, phosphorites, and carbonates (as carbonate-associated sulfate, or CAS). Pyrite data record the fractionation that occurs during bacterial sulfate reduction, plus additional fractionation effects of the reactions during oxidative recycling of sulfides (Canfield and Teske, 1996). A detailed discussion on the use of sulfur isotopes on Neoproterozoic stratigraphy is found in Halverson et al. (2010a,b).

A recent development is the use of mass-independent fractionation (MIF) of sulfur isotopes as a tracer of atmospheric oxygenation. The method implies measurements of multiple sulfur isotopes ($\delta^{33}S$, $\delta^{34}S$, and $\delta^{36}S$) on CAS and sulfides. Present-day sulfur isotopic composition of marine rocks is dominated by fractionation processes (kinetic, biological, and thermodynamic) that depend on atomic mass of the different sulfur isotopes. MIF, linked to photochemical reactions in the atmosphere, is not recorded in the last 2 Gyr (Farquhar et al., 2000). However, a significant MIF signal of sulfur isotopes has been reported for rocks older than 2.45 Ga which is interpreted as indicating very low atmospheric oxygen concentrations in the Archean and earliest Paleoproterozoic (Farquhar et al., 2000; Guo et al., 2009).

2.2.5 CALCIUM AND BORON ISOTOPES

Recent technological advances in the analyses of some nontraditional stable isotopes (Johnson et al., 2004) with MC-ICP-MS, mainly Li, Mg, Ca, Mo, Cr, Fe, Cu, Zn, and Se, have opened new avenues, some still to be explored in terms of isotope stratigraphy. Among these elements, particularly Ca, Mo, and Fe have received more attention in the Precambrian isotope stratigraphy (Kasemann et al., 2005; Arnold et al., 2004; Siebert et al., 2003; Johnson and Beard, 2006; Staubwasser et al., 2006 among others) and Cr isotopes may become an important tool in this regard (Frei et al., 2009, 2011).

The global Ca-isotope signal displayed by end-Cryogenian (Marinoan) carbonate successions suggests the use of Ca-isotope stratigraphy as an additional tool to correlate postglacial Neoproterozoic carbonate successions (Kasemann et al., 2005; Higgins and Schrag, 2003; Silva Tamayo et al., 2007, 2010a,b). Although it is not well understood how Ca isotopes work in modern carbonate rocks, or the extension on how diagenesis affects them, these authors have claimed that the Neoproterozoic Ca isotopic record, still far from complete, is perhaps an archive of changes in the oceanic Ca isotopic composition. Calcium isotopes suggest rapid glacier melting and important increase in the Ca input to the ocean immediately after deglaciation, followed by progressive increase in carbonate precipitation and burial compensating for the large initial Ca input. Post-Sturtian global $\delta^{44/40}$Ca patterns are, perhaps, different from the post-Marinoan ones, implying a difference in Ca mass balance evolution among these two deglaciation events due to the contrasting glacier melting regimes (Silva Tamayo et al., 2010a,b). The different Ca isotopic evolution makes Ca-isotope stratigraphy a promise, perhaps, to discriminate and correlate Neoproterozoic postglacial carbonate successions. Recently, Blättler et al. (2012) have demonstrated a close connection between Ca isotopic cycling in the Phanerozoic carbonate sedimentation, SW chemistry, and evolutionary trends.

Boron and calcium isotopes have been used to understand the paleoenvironmental conditions in the aftermath of major Neoproterozoic glaciations in Namibia (Kasemann et al., 2005). Negative δ^{11}B excursions were observed in postglacial carbonates and interpreted as an indication of temporary decrease in SW pH.

2.2.6 CHROMIUM AND MOLYBDENUM ISOTOPES

It is known that the Neoproterozoic is transitional between an earlier Proterozoic world with widespread deep-water anoxia and a Phanerozoic world with largely oxygenated oceans. Whereas, in the Mesoproterozoic sulfidic (euxinic) marine conditions prevailed, ferruginous oceans characterized the Archean–Paleoproterozoic and Neoproterozoic oceans, and oxygenated conditions dominated the Phanerozoic ocean chemistry (Canfield et al., 2008). The study of Fe, Cr, and Mo isotopes has provided further insight into the question of surface ocean oxygenation (Frei et al., 2009; Scott et al., 2008).

Chromium is highly sensitive to the redox state of the surface environment, and the oxidative weathering processes producing the oxidized hexavalent Cr. Oxidation of the reduced Cr on land is accompanied by positive isotopic fractionation (Frei et al., 2009 and references therein). The isotopic systematic of the chromium cycle including incorporation into banded iron formation (BIF) has been discussed in detail by Lyons and Reinhard (2009) and Døssing et al. (2011). The δ^{53}Cr from BIFs can track the presence of hexavalent Cr in the Precambrian oceans providing a picture of the oxygenation history of the Earth's atmosphere–hydrosphere system as demonstrated by Frei et al. (2009, 2011). According to Frei et al. (2011), the Cr isotopes in the carbonates may be useful as a tracer for climate change and for reconstructing the redox state of the ancient SW and atmosphere. Furthermore, because

Cr- and C-isotope curves in carbonates are virtually parallel (Frei et al., 2011), coupled $\delta^{13}C$–$\delta^{53}Cr$ chemostratigraphy of mixed BIF/carbonate/chert successions may provide more continuous curves than C isotopes alone. The method can be applied to chemical sediments (BIF, chert, and carbonates), provided that they contain low amounts of terrigenous material, which would otherwise dominate the Cr isotopic composition of the rock (Frei et al., 2013).

Molybdenum concentrations in black shales (Scott et al., 2008) and its isotopic composition (Arnold et al., 2004) have been used to assess the evolution of the redox state of the oceans. The method allows to differentiate between euxinic (i.e., sulfidic) and oxygenated environments. Three oxygenation events at 2.65 Ga, c. 2.5 Ga, and 550 Ma were recognized, with the late Paleoproterozoic and Mesoproterozoic (1.8–1.0 Ga) being characterized by euxinic conditions ("Canfield Ocean": Canfield, 1998; Arnold et al., 2004; Scott et al., 2008). This is consistent with other proxies, such as MIF of sulfur isotopes and chromium isotopes (see above).

2.2.7 STRONTIUM AND NEODYMIUM ISOTOPES

Fluctuations of the $^{87}Sr/^{86}Sr$ ratios are independent of those of $\delta^{13}C$ and the simultaneous utilization of the two systems has been proven to be ideal in high-resolution stratigraphic studies. Combined, these systems are an even more powerful tool in the resolution of geological problems. The secular $^{87}Sr/^{86}Sr$ variation curve is better known in the Phanerozoic (McArthur et al., 2001, Fig. 2) showing a cycling of about 500–550 Ma from the Upper Cambrian (0.709; Montañez et al., 2000) gradually decreasing with a nadir of 0.7068 at 250 Ma and reassuming values of 0.709 again in the present-day ocean (Macdougall, 1991; McArthur et al., 2001). Despite some efforts to compile Sr-isotope data to determine the secular $^{87}Sr/^{86}Sr$ variation curve for the entire Proterozoic or just for the Neoproterozoic (e.g., Jacobsen and Kaufman, 1999; Halverson et al., 2007), the use of strontium isotopes in chemostratigraphy is limited by the paucity of limestone in many successions, and by the likelihood of alteration in samples with low strontium contents through the incorporation of ^{87}Sr from the decay of ^{87}Rb in coexisting clay minerals (Kaufman et al., 2009). Furthermore, the variety of leaching techniques, currently being used by different laboratories, hampers interlaboratory comparison of results (e.g., Bailey et al., 2000). Geochemical screens ($^{87}Rb/^{86}Sr$, Mn/Sr, and $\delta^{18}O$) have been developed to evaluate the degree of alteration of strontium isotopes (Veizer et al., 1983; Kaufman et al., 1992, 1993; Marshall, 1992; Jacobsen and Kaufman, 1999; Melezhik et al., 2001) and have been widely adopted. Unfortunately, because of differences in the tectonic and diagenetic history of individual basins, the empirical limits delimiting altered and unaltered samples has varied—leading to some confusion in the literature (see Melezhik et al., 2001 for a discussion). Dolostones are usually not suitable for Sr-isotope studies, due to the lower Sr concentrations (usually a few tens of ppm—Kah et al., 1999; Gaucher et al., 2007), although a few exceptions have been reported (Sawaki et al., 2010).

Apart from its use in geochronology, Sr isotopes provide valuable information about the predominant tectonic regimes and their evolution in time. The main source of ^{87}Sr is the weathering of Rb-rich granites. The hydrothermal vents near mid-ocean ridges are enriched in nonradiogenic ^{86}Sr (Shields, 2007a). Thus high $^{87}Sr/^{86}Sr$ ratios were usually interpreted as reflecting periods of enhanced orogenesis, and low ratios characterize periods of continental breakup and enhanced hydrothermal activity. Recently, Flament et al. (2013) argued that $^{87}Sr/^{86}Sr$ is influenced by the area of emerged land rather than orogenic processes alone. This is especially important for calculations of continental growth in the Archean, when maybe less than 4% of Earth's area was emerged (Flament et al., 2013).

Neodymium isotopes have been little used as stratigraphic tool although a "global average" εNd curve for the oceans since 800 Ma has been constructed (Keto and Jacobsen, 1988; Macdougall, 1991). This curve, similar to the Sr-isotope secular curve, indicates indifferent εNd values at the end of the Precambrian oceans and those at present oceans. Average εNd values decrease remarkably in the time interval from 700 to 550 Ma (−5 to −15). Even with the precision of modern instruments, the scatter in measured values is substantial and this has limited the use of εNd in chemostratigraphic studies. Nevertheless, secular εNd variations coupled to $\delta^{13}C$ and $\delta^{53}Cr$ have been reported from Ediacaran rocks in Uruguay (Frei et al., 2011, 2013), which yield valuable information regarding the tectonic evolution of the shelf.

2.2.8 ELEMENTAL CHEMOSTRATIGRAPHY

Hg, Mo, and V seem to be elements of particular interest. The first one tends to concentrate in sediments deposited right after major glacial events as a result from leaching of volcanogenic Hg from land surface and accumulation along argillaceous sediments (Santos et al., 2001). The usually low geological background concentrations of Hg makes this trace element suitable for identifying accumulation pulses in sediments that can be tentatively related to weathering processes and thus to climatic changes.

Volcanism is assumed to be the main source for CO_2 buildup in the atmosphere during Neoproterozoic glacial events, eventually leading to enhanced greenhouse effect, catastrophic ice melting, and cap carbonate deposition (e.g., Hoffman et al., 1998; Hoffman, 2011). Moreover, intense volcanism may have witnessed the Cretaceous–Paleogene transition (KTB) and was, perhaps, co-responsible for dramatic climatic changes responsible for the decrease in biodiversity and mass extinction (Keller, 2005; Archibald et al., 2010). Sial et al. (2010b) used Hg as a proxy for volcanic activity and CO_2 buildup during snowball events in Neoproterozoic cap carbonates of northeastern Brazil. Nascimento-Silva et al. (2011, 2013) and Sial et al. (2013, 2014) applied it to investigate the role of volcanism in the KTB. Combined Hg and C-isotope chemostratigraphies may become a powerful tool for the eventual quantification of climatic changes as a proxy for volcanic activity during extreme environmental turnover as witnessed during the Neoproterozoic Snowball Earth and the Permian–Triassic and the Cretaceous–Paleogene boundaries (KTB).

Molybdenum concentrations increase to Phanerozoic values in black shales in the interval of 665–530 Ma (Scott et al., 2008), a pattern also recognized in vanadium concentrations, usually enriched in similar conditions (Shields-Zhou and Och, 2011). Molybdenum and vanadium chemostratigraphies may be useful in the investigation of the redox state of deep-ocean water. Moreover, some elemental ratios can be useful as proxies in chemostratigraphy for investigation of salinity variation (Rb/K), redox state (V/Cr), or provenance (Zr/Ti) according to Scheffler et al. (2003).

Poulton and Canfield (2005) developed a method of measuring the ratio of highly reactive iron to total iron (Fe_{HR}/Fe_T) to unravel the redox evolution of sedimentary successions. Highly reactive iron includes iron oxide, carbonate, and sulfide minerals and represents the iron geochemically and biologically active during early sediment diagenesis. The Fe_{HR}/Fe_T ratio is lower than 0.38 for sediments deposited in oxygenated environments, and exceeds 0.38 in anoxic conditions (Poulton and Canfield, 2005). The amount of iron stored in sulfides can be quantified and the ratio of pyrite iron to highly reactive iron (Fe_{Py}/Fe_{HR}) allows to distinguish euxinic from ferruginous environments (e.g., Johnston et al., 2010 and references therein). The method has been widely and successfully applied to Neoproterozoic (e.g., Canfield et al., 2008; Johnston et al., 2010; Frei et al., 2013) and older successions (Shen et al., 2002), but also to recent lake and marine sediments (Zegeye et al., 2012).

Iodine has a well-known redox behavior (Broecker et al., 1982) and it is also known that the concentration of carbonate-associated iodine increases linearly with those of IO_3^- during calcite precipitation, but I^- is completely excluded (Lu et al., 2010). This behavior in association with the residence time of iodine in SW (300 k.y.; Broecker et al., 1982) and concentration near 450 nM in modern ocean, makes I/Ca (or I/Ca+Mg) ratios in carbonates, a robust indicator of the presence of IO_3^-, and hence oxygen in the water column. Therefore, I/Ca ratios have been used as a paleoredox indicator to verify the surface ocean oxygenation (e.g., Hardisty et al., 2014).

Ambiguity in interpreting the oxygen isotope signal has necessitated independent estimates of ocean temperatures (e.g., Lear et al., 2000) and for that a useful paleothermometer is the Mg/Ca ratio of foraminiferal shells (e.g., Lea et al., 2000; Lear et al., 2000). When calibrated for the vital effects of the organism, the difference in the Mg/Ca ratio from a baseline value (defined by the global ocean Mg and Ca concentration) is a function of temperature. Inferring the baseline composition of SW is relatively straightforward because both Mg and Ca are major components of ocean salts and both have long residence times in the oceans (>1 Ma).

2.3 TEMPORAL TRENDS AND SIGNATURES

2.3.1 ARCHEAN

The Precambrian comprises 82% of the geological evolution of the Earth and represents a time in which sedimentary rocks contain only the small, mostly microscopic, and relatively scarce fossils which have been studied systematically only in the last decades. Isotope chemostratigraphy has helped unravel these radically different conditions that prevailed in the Archean. First recognized from sedimentological evidence, such as the occurrence of detrital pyrite and uraninite in the fluvial conglomerates (Cloud, 1976), the reducing nature of the Archean surface environments has been confirmed by several chemostratigraphic methods, including: (1) the occurrence of MIF of sulfur isotopes (Farquhar et al., 2000); (2) the concentration of Mo in shales (Scott et al., 2008), and (3) chromium isotope chemostratigraphy of BIF and cherts (Frei et al., 2009).

The oldest Archean microfossils occur in 3465 Ma cherts (Schopf and Kudryavtsev, 2012). This information, together with the low $\delta^{13}C$ ratios (−25‰) of reduced carbon in shale, strongly suggests that photosynthetic organisms were extant already by 3.7 Ga (Rosing and Frei, 2004; Shields, 2007b). Although $\delta^{13}C$ data of the Archean carbonates are scarce, they yield values around 0‰ V-PDB, suggesting a biomass in the oceans similar to the modern-day biosphere (Shields, 2007b). All the lines of evidence point to an early origin of life and a rapid development of a relatively large biomass in the early oceans.

It has been widely assumed that the atmospheric oxygen concentrations remained persistently low (less than 10^{-5} times present levels) for about the first 2 gigayears of Earth's history. Relevant chemostratigraphic studies for this Eon have been mostly oriented to the investigation of distribution of redox-sensitive elements and their isotope variations in marine sediments to track a possible oxygenation of atmosphere at, and before, the Great Oxidation Event (GOE) (2.4–2.2 Ga). Among the redox-sensitive elements, Cr and Mo isotopes and Cr, Mo, and U elemental signatures have been extensively used. Based on various geochemical tracers, contrasting results have advocated for and against the rise of free oxygen in the atmosphere prior to 2.4 Ga, and evidence for a partially oxidized environment remains vague. Wille et al. (2013) used Mo and Cr isotopes and trace element signatures from four

Archean low-grade metamorphic black shale successions from Pilbara, Australia, to investigate environmental redox conditions in the 3.46–2.76 Ga interval. Their study was aimed at determining the source and mobility of Mo, Cr, and U to decipher the redox state of Archean oceanic basins and to answer the question of existence of free oxygen prior to 2.7 Ga. The results of their investigation have confirmed previous works that have shown an increase in Mo and/or Cr mobility due to the partial oxidation of the atmosphere since 2.7 Ga (Anbar et al., 2007; Wille et al., 2007; Frei et al., 2009; Duan et al., 2010; Kendall et al., 2010; Voegelin et al., 2010). Wille et al. (2013) have concluded that Mo, Cr and, U elemental signatures, within all the studied sedimentary units, were largely dependent on the continental input. Both Mo isotopic values and Mo and U elemental signatures have suggested anoxic conditions with no indication of redox cycling of these elements. This contention contradicts the interpretations of an oxidized atmosphere based on published sulfur isotopic data (Ohmoto et al., 2006) and oxidized hematite occurrences at ~3.45 Ga (Hoashi et al., 2009).

Crowe et al. (2013) examined the distribution of Cr isotopes and Cr and U elemental signatures in the approximately 3-Ga-old Nsuze paleosol and in the near-contemporaneous Ijzermyn iron formation from the Pongola Supergroup, South Africa. They observed extensive mobilization of redox-sensitive elements through oxidative weathering and Cr isotopic fractionations. Whereas the iron formations yielded positive δ^{53}Cr values, the paleosols are characterized by negative δ^{53}Cr values below those typical for high-temperature Cr. This is consistent with the interpretation of oxidative weathering that removed isotopically heavy, soluble Cr (VI) from the host rocks and soils and later deposited this positively fractionated Cr in the iron formations. Crowe et al. (2013) have also computed a best minimum estimate for atmospheric oxygen concentrations at that time to be of 3×10^{-4} times than the present levels. Their findings suggest that there were appreciable levels of atmospheric oxygen about 3 billion years ago (>600 Ma before the GOE and some 300–400 Ma earlier than previous indications for Earth surface oxygenation).

Zerkle et al. (2012) used multiproxy geochemical analyses (S and C isotopes) of sediments from the 2.65 to 2.5-Ga-old Ghaap Group, South Africa, to reconstruct the oceanic and atmospheric chemistry during this time interval. They have found evidence for oxygen production in microbial mats and localized oxygenation of surface waters. Carbon and sulfur isotopes have indicated that this oxygen production occurred under a reduced atmosphere, periodically rich in methane, and consistent with the prediction of a hydrocarbon haze. From the isotopic signatures observed, Zerkle et al. (2012) deduced that methane was an important component of the atmosphere throughout the Archean.

2.3.2 PALEOPROTEROZOIC

Marine carbonates are a good proxy of ancient SW carbon isotope composition. Measurements of well-preserved, and even metamorphosed, carbonates have provided valuable insights on the biogeochemical cycling of carbon in the geological past (e.g., Schidlowski et al., 1976a,b; Knoll et al., 1986; Veizer and Hoefs, 1976; Derry et al., 1992; Kah et al., 1999; Maheshwari et al., 1999; Sreenivas et al., 2001; Bekker et al., 2003, 2008). Carbon isotope chemostratigraphy has become an important tool in the Precambrian geochronology (e.g., Knoll et al., 1986). However, the oscillatory nature due to a comparatively short residence time of carbon makes dating by means of carbon isotopes alone impossible. Instead, the use of ^{87}Sr/^{86}Sr ratios in combination with carbon isotopes on marine carbonates has been successful in many cases, especially in the Neoproterozoic (Halverson et al., 2007, 2010a,b; Kaufman et al., 2009; Frei et al., 2011) and Paleoproterozoic (e.g., Kuznetsov et al., 2010).

Studies of Proterozoic rocks reveal time intervals, at the beginning and at the end of the Proterozoic, with highly positive carbon isotope values (e.g., Schidlowski et al., 1976b; Knoll et al., 1986; Veizer and Hoefs, 1976; Derry et al., 1992; Kah et al., 1999; Maheshwari et al., 1999; Sreenivas et al., 2001; Bekker et al., 2003). The $\delta^{13}C$ chemostratigraphic curve for marine carbonates starting at 2.5 Ga (Fig. 1) shows a widespread positive $\delta^{13}C$ anomaly in the age interval of 2.33–2.06 Ga age (Baker and Fallick, 1989; Melezhik and Fallick, 1996; Melezhik et al., 1997; Hoffman et al., 1998; Maheshwari et al., 2010) and the remainder of the Paleoproterozoic (2.0–1.6 Ga) shows $\delta^{13}C$ near 0‰. This unprecedented positive $\delta^{13}C$ anomaly known as the Lomagundi or Jatulian event is the most striking phenomenon recorded by C isotopes in the Paleoproterozoic, originally reported by Schidlowski et al. (1976b) from carbonates of the Lomagundi Group, Zimbabwe ($\delta^{13}C$ up to 17‰). According to Melezhik et al. (1997), this phenomenon is linked with the explosion of stromatolites, expressed as a peak in the abundance and taxonomic diversity of stromatolites.

Similar positive anomalies were reported from the Francevillian Series, Gabon, West Africa (Gauthier-Lafaye and Weber, 1989), Pretoria and Elim groups, South Africa (Buick et al., 1998; Bekker et al., 2001, 2008; Schröder et al., 2008), Fennoscandian Shield (Karhu, 1993; Melezhik and Fallick, 1996), Scotland (Baker and Fallick, 1989) and North America (Schidlowski et al., 1983; Mirota and Veizer, 1994; Bekker et al., 2003, 2006). The magnitude and extended duration of this carbon isotope anomaly suggest that vast quantities of reduced carbon were buried in the sediments, thereby releasing oxidants to the oceans and atmosphere. This, conceivably, led to an increase in atmospheric oxygen levels, which might have resulted in profound consequences for surface environments and biological processes worldwide (Karhu and Holland, 1996). The increase in oxygen levels in the atmosphere may also explain the development of "red beds" associated with ^{13}C-enriched carbonates.

The duration of the Lomagundi phenomenon is still a matter of debate. Melezhik et al. (1999), based on the C-isotope data available at the time, assumed that the Paleoproterozoic positive excursion of $\delta^{13}C$ values actually represent three positive shifts separated by returns to 0‰ values that occurred during the 2.40–2.06 Ga time interval (Fig. 1). This interpretation was followed by Lindsay and Brasier (2002). However, very few reliable isotopic ages directly constrain the age of the Lomagundi carbon isotope excursion. New ages from the Pechenga greenstone Belt and Perapohja Belt in the Fennoscandia Shield provide more accurate constraints for the shift from high to low values of $\delta^{13}C$ at the end of the Paleoproterozoic carbon isotope excursion (Melezhik et al., 2007; Karhu et al., 2008). U–Pb single zircon (SHRIMP) age for the pyroclastic Hirsimaa Formation, underlain by dolostones of the Poikkimaa Formation and overlain by dolostones of the Rantamaa Formation, is 2108 ± 8 Ma, therefore constraining depositional age of the Hirsimaa Formation (Karhu et al., 2008). $\delta^{13}C$ values for the underlying and overlying dolostones show a generally decreasing trend from +10‰ to +1‰. It was therefore assumed that the age of the Hirsimaa Formation constrains the end of the carbon isotope excursion, when $\delta^{13}C$ values of SW were in the 4–6‰ range. U–Pb single zircon (TIMS and SHRIMP) ages for detrital zircons in volcaniclastic conglomerate of the Kuetsjärvi Volcanic Formation overlying carbonates with the positive carbon isotope excursion in the Kuetsjärvi sedimentary Formation provide the minimum age for the Lomagundi anomaly at 2056.6 ± 0.8 Ma (Martin et al., 2013). These studies suggest a single hump rather than three-spike excursions between 2.22 and 2.06 Ga. Bekker et al. (2006), however, suggested that the early stage of the Lomagundi excursion is marked by a return to 0‰ values in the Lake Superior area.

The Lomagundi excursion was unknown from South America and India until the last decade. Maheshwari et al. (1999, 2002), Sreenivas et al. (2001), Sial et al. (2000), and Bekker et al. (2003)

documented this biogeochemical anomaly in Paleoproterozoic sedimentary rocks from these areas for the first time. Based on the previously published chemostratigraphic and geochronologic data (Dix et al., 1995; Fortes and Jost, 1996; Fortes et al., 2003), Bekker et al. (2003) proposed that carbon isotope values as high as +14.5‰ in dolomites of the Ribeirão das Antas Formation of the Crixás greenstone belt, Brazil, also reflect SW composition during the Lomagundi event. Recent detailed chemostratigraphic and geochronologic studies (Jost et al., 2008a,b; Tassinari et al., 2006; Santos et al., 2008) confirmed this signal in the Crixás and other correlative greenstone belts in the State of Goiás (Brazil), and delimited it in age between c. 2.22 and 2.15 Ga.

Positive carbon isotope excursion has been reported by Maheshwari et al. (2010) from Paleoproterozoic carbonates of the Aravalli Supergroup (northwestern India), the Minas Supergroup (Brazil), and sections of the Paso Severino Formation (Uruguay). The 2.42 Ga Gandarela Formation, State of Minas Gerais, Brazil, contains red carbonate-facies BIF grading into dolostones and limestones and yielding $\delta^{13}C$ values ranging from −1.6‰ to +0.4‰. The positive C-isotope excursion (up to +11‰) in marine shallow-water carbonates in India and Brazil (Jhamarkotra Formation in northwestern India, and Cercadinho and Fecho do Funil formations in the State of Minas Gerais, Brazil) is comparable to that observed in 2.22–2.1 Ga carbonate successions worldwide that were deposited during the Lomagundi excursion (Maheshwari et al., 2010). In Uruguay, $\delta^{13}C$ values up to +11.6‰ in the deep-water Paso Severino Formation of the Piedra Alta Terrane are compatible with U–Pb SHRIMP ages of dacites of 2146 ± 7 Ma occurring at the top of the unit (Maheshwari et al., 2010, and references therein). Negative $\delta^{13}C$ values are also present in carbonates of the Paso Severino Formation, but an origin related to organic-matter remineralization cannot be ruled out. Thin carbonate beds in the Rio Itapicuru greenstone belt, Bahia State, Brazil, are associated, as in the Paso Severino Formation, with deep-water black shales and yielded carbon isotope values up to +9‰. High-grade marbles of the Jacurici terrane in the Medrado-Ipueira area, Bahia (Brazil) yielded carbon isotope values up to +6.9‰, and the same is consistent with their minimum age of 2085 ± 5 Ma inferred from datings of the intrusive Medrado norite.

No evidence was found in India, Brazil, or Uruguay for Paleoproterozoic glacial events recognized in the 2.45–2.22 Ga sedimentary successions worldwide (Maheshwari et al., 2010). Unconformities between the Gandarela and Cercadinho formations in Brazil, and between the banded gneissic complex and the Lower Aravalli Supergroup in India might explain the absence of glacial record. Compositional and isotopic data presented here for the studied Paleoproterozoic carbonate successions allow their integration into the global record of the Paleoproterozoic evolution as well as correlation with other successions of similar age.

The study by Maheshwari et al. (2010) indicated that the Lomagundi excursion is recorded in both shallow-water (Aravalli and Minas supergroups) and deep-water carbonates (Paso Severino Formation and Rio Itapicuru greenstone belt) and questioned the roles of stromatolite productivity and hypersaline conditions on carbon isotope values of the shallow-water carbonates. Cap carbonates associated to Neoproterozoic glacial diamictites are the record of postglacial deposition and abrupt climatic change (Hoffman and Schrag, 2002). The possibility that cap carbonates similar to Neoproterozoic examples were deposited during the Paleoproterozoic has been raised (e.g., Kirschvink et al., 2000; Bekker et al., 2005; Kopp et al., 2005) but this is a matter which demands further insight.

Paleoproterozoic ice ages (2.45–2.22 Ga) are supposedly associated with the breakup and assembly of the Archean Kenorland supercontinent (Williams et al., 1991) probably positioned at low latitudes (Evans et al., 1997). In North America, the Early Paleoproterozoic Huronian and Snowy Pass supergroups, Ontario and Wyoming, preserve lithologic and geochemical evidence of three glaciations with intervening warm climate periods (Young, 1991). Although these three episodes of glaciations have been recorded, no true iron formation was found in this interval and just scarce carbonate rocks have been observed. Based on paleomagnetic, litho- and chemostratigraphic data, at least one of these ice ages has been interpreted as a Snowball event, potentially related to the transition from an anoxic methane-rich atmosphere to an oxygenated atmosphere with high CO_2 levels (Evans et al., 1997; Kirschvink et al., 2000; Pavlov et al., 2000; Bekker et al., 2001). Due to the paucity of thick carbonates, little is known on the temporal variation of the contemporaneous SW composition during the Early Paleoproterozoic glaciation.

A compilation of sulfur isotope data by Shields-Zhou and Och (2011) have revealed that $\delta^{34}S$ values in the Archean do not exceed 20‰ and maximum Precambrian values do not exceed 45‰. The maximum $\delta^{34}S$ values are in the Paleo-Mesoproterozoic boundary, in the late Mesoproterozoic, and in the late Cryogenian (Fig. 1).

2.3.3 MESOPROTEROZOIC

Carbon isotope chemostratigraphic studies on the Late Paleoproterozoic (1.8–1.6 Ga) showed that $\delta^{13}C$ values were mainly negative (−3 to 0‰) and increased slowly, reaching a level of zero, with abrupt oscillation at the beginning of 1.6 Ga (Zhao et al., 1997; Xiao et al., 1997; Chu et al., 2004; Li et al., 2002 in Hongwei et al., 2011). $\delta^{13}C$ values of Mid-Riphean carbonates from the southern Urals of Russia (1.4–1.2 Ga) show a low positive range of 0–3‰ and a negative anomaly below −2.0‰ (Bartley et al., 2007) in the black shale. The $\delta^{13}C$ in the Mesoproterozoic Belt Supergroup (1.45–1.20 Ga), North America and Canada, consisting mainly of carbonate-bearing molar-tooth structure, varies from −0.2‰ to 1.4‰ from base to top (Frank et al., 1997), while $\delta^{13}C$ in carbonates of 1.70–1.60 Ga in the McArthur and Mount Isa basins in northern Australia displays a variation of −0.6 to 1‰ (Lindsay and Brasier, 2000). Kha et al. (1999) studied the carbonates of the Bylot Supergroup, Baffin Islands, Canada (1.27–0.72 Ga) and have improved the global carbon isotopic curve (Fig. 1) for the Mesoproterozoic to earliest Neoproterozoic interval. The carbon isotope curve for c. 1.6–0.6 Ga reported by Kah et al. (2001) presents a gradual increase in $\delta^{13}C$ values above 0‰, comparable to the $\delta^{13}C$ profiles reported by Hongwey et al. (2011) in China. However, the Chinese sections already show positive $\delta^{13}C$ plateaus reaching +1.5‰ between c. 1.5–1.4 Ga. This is confirmed by data from Mesoproterozoic carbonates in Uruguay, which yielded $\delta^{13}C$ values between −3‰ and +1.6‰ (Chiglino et al., 2010), increasing up section to +2.4‰ (Gaucher et al., 2009). The age of the Uruguayan sections is well constrained by U–Pb zircon ages of synsedimentary volcanism between 1.49 and <1.43 Ga (Gaucher et al., 2011). Therefore, after a prolonged interval between 2.0 Ga and 1.5 Ga characterized by invariant $\delta^{13}C$ values around 0‰, positive $\delta^{13}C$ plateaus around 1–2.5‰, punctuated by short-lived negative excursions, are typical of the 1.5–1.3 Ga interval (Fig. 1). Afterward, the amplitude and frequency of $\delta^{13}C$ excursions increase, a trend that continues into the Neoproterozoic (Kah et al., 1999).

Late Paleoproterozoic and Mesoproterozoic oceans are also interesting because euxinic conditions dominated for almost 1 Ga. This is demonstrated by sulfur isotope chemostratigraphy, iron speciation data, and molybdenum isotopes and elemental concentrations (Shen et al., 2002; Arnold et al., 2004; Scott et al., 2008). The oxidation of sulfides on land after the GOE may have provided the mass of sulfate necessary to shift ocean chemistry from ferruginous to sulfidic (Canfield, 1998).

Based on the study of global average $\delta^{18}O$ and $\delta^{30}Si$ isotopes in carbonates, it has been suggested that during the interval of 1.4–1.2 Ga, the ocean temperature reached about 70 °C (Knauth and Lowe, 2003; De La Rocha, 2006; Robert and Chaussidon, 2006).

2.3.4 NEOPROTEROZOIC

In the Neoproterozoic, the $\delta^{13}C$ secular variation curve shows strong positive and negative excursions (Fig. 1), four of the major negative ones being interpreted as the result of ice ages which led to the proposal of the Snowball Earth theory (Hoffman et al., 1998; Hoffman and Schrag, 2002; Hoffman, 2011). In this era, the fragmentation of the ancient Rodinia supercontinent was accompanied by intermittent, but widespread glaciation in the 0.75–0.55 Ga interval. The isotopic excursions are enormous in this time interval (Fig. 1) in comparison to the excursions in the preceding 1.2 billion years or in the Phanerozoic (Hoffman et al., 1998b). In the Cryogenian and Ediacaran, it is common to observe post-glacial carbonate rocks even in terrigenous-dominated sections. This era is characterized by interrelated environmental, climatic, and biological events, probably associated with fluctuations in atmospheric concentrations of oxygen, carbon dioxide, and methane. Major orogenic and hydrothermal events influenced significantly the ocean chemistry and remarkably reorganized the carbon and sulfur cycles, as revealed by isotopic studies across this era. Stratigraphic studies, in particular, point to a close association of the biogeochemical and climatic events with the rise and fall of atmospheric oxygen (Kaufman, 2008a).

Neoproterozoic glacial deposits of the 750–635 Ma interval overlain by carbonates (cap carbonates) exhibit some features that are apparently contradictory to Phanerozoic glaciations. The most striking of these features are paleomagnetic data that point to glaciers at sea level in low latitudes and marked depletion in ^{13}C in carbonates above, and if preserved from erosion also below the glacial successions across the globe (Hoffman et al., 1998; Schrag et al., 2002; Halverson et al., 2002; Etienne et al., 2008 and references therein). Among the proposed models on such an intriguing observation, the "Snowball Earth" hypothesis (Kirschvink, 1992; Hoffman et al., 1998; Hoffman and Schrag, 2002), although seriously questioned (e.g., Eyles and Januszczak, 2004; Moczydlowska, 2008; Sansjofre et al., 2011), still fascinates geologists with the possibility that in the Cryogenian Period a totally ice-entombed Earth might have flown through space. An alternative "Slushball Earth" model proposed by Hyde et al. (2000) and Peltier et al. (2007) predicts that open glacial oceans would have constrained runaway refrigeration by allowing sunlight to warm the Earth's surface, driving an active hydrological cycle and permitting photosynthesis and the observed survival of acritarch species (Moczydlowska, 2008) in exposed seas. It has left unanswered several questions, however, as pointed out by Kaufman (2008b) in an overview of this alternative hypothesis.

The "Zipper Rift" model of Eyles and Januszczak (2004) and the "Snowball Earth" hypothesis advocate a genetic nature of diamictite facies and their paleoclimatic significance. They agree on at least local glacial activity, temporally subdivided into the Sturtian (c. 710 Ma), Marinoan (c. 635 Ma), and Gaskiers (c. 580 Ma) glacial epochs.

The possibility of the marine biosphere being a driving factor of the Neoproterozoic climate change has been suggested by Brasier (1992), Kaufman et al. (1997), Gaucher (2000), Gaucher et al. (2004), and Frei et al. (2013). Gaucher et al. (2004), Nagy et al. (2009), Gaucher and Sprechmann (2009), and Frei et al. (2013) observed that the low diversity and high abundance of preglacial assemblages are good indicators of eutrophication of large areas of the Neoproterozoic oceans. According to this view, the Neoproterozoic phytoplankton blooms exported huge amounts of ^{12}C-enriched organic matter to the sediments, which resulted in positive $\delta^{13}C$ values of coeval carbonates. The proportion of carbon deposited as organic matter was two to three times larger than in the present oceans (e.g., Kump and Arthur, 1999). The net effect of the cyanobacterial blooms plus concomitant carbonate deposition was the drawdown of atmospheric CO_2, leading to glaciation (Kaufman et al., 1997; Gaucher et al., 2004). Because ferruginous oceans prevailed during most of the Neoproterozoic (Canfield et al., 2008), iron fertilization is a mechanism which should be considered as a trigger of ocean eutrophication (Gaucher et al., 2004; Frei et al., 2013).

Evidence of a glacial event around the Mesoproterozoic–Neoproterozoic transition has been reported from units of the Macaúbas Group (1.05–0.95 Ga) that overlies the Espinhaço Supergroup rocks in Minas Gerais and central Bahia, Brazil (D'Agrella-Filho et al., 1990; Almeida-Abreu and Renger, 2002). This glacial event, if confirmed, would constitute the oldest among the glacial records in the Neoproterozoic of southwestern Gondwana. The possibility of the occurrence of an early-Cryogenian glaciation (c. 800 Ma) has been raised by Jacobsen and Kaufman (1999) although a c. 800 Ma negative carbon isotope interval in the northeastern Svalbard has received a different explanation by Halverson et al. (2007). Evidence from Ediacaran successions in Uruguay (Gaucher et al., 2004, 2009), Argentina (Barrio et al., 1991), Namibia (Praekelt et al., 2008), and South Africa (Germs and Gaucher, 2012) strongly suggests the occurrence of nonglobal post-Gaskiers, Phanerozoic-type glacial events in southwestern Gondwana.

Broadly equivalent carbonate successions that postdate known end-Cryogenian glacial deposits and predate the Precambrian–Cambrian boundary in Oman, India, South China, Australia, Namibia, Mexico, and the western USA, exhibit a remarkable negative carbon isotope anomaly (Le Guerroué et al., 2006a,b,c; Kaufman, 2008a; Boggiani et al., 2010; Grotzinger et al., 2011). This $\delta^{13}C$ negative anomaly (Shuram–Wonoka), which is the largest known in Earth history, according to Kaufman (2008a), accompanies a rise in the oceanic sulfate abundance and fall of $\delta^{34}S$ values in CAS and sulfides. Several authors have challenged the primary nature of the Shuram–Wonoka anomaly (Bristow and Kennedy, 2008; Derry, 2010), mainly on the basis of the absence of the anomaly in $\delta^{13}C$ of co-occurring organic matter and on problems arising from the redox budget of the oceans. The matter is still far from resolved.

Since the publication of the "Snowball Earth" hypothesis (Kirschvink, 1992; Hoffman et al., 1998) there has been an exponential growth in the number of publications focusing on the oscillating record of extreme climate change in the Cryogenian (generally constrained between 750 and 635 Ma) and through the Ediacaran and Cambrian periods. The paucity of a rigorous biostratigraphic framework through much of this time interval (e.g., Gaucher and Sprechmann, 2009 and references therein) and the general absence of reliable radiometric age constraints, led to the use of temporal carbon isotope trends (often constructed from single and likely incomplete sections) as a means of dating Neoproterozoic sedimentary successions through comparison with composite curves (e.g., Jacobsen and Kaufman, 1999). This is problematic insofar as carbon isotope variations, especially across glacial divides, are inherently oscillatory, the magnitude of $\delta^{13}C$ anomalies can vary between facies, and the age models used for the composites require periodic tuning (cf. Melezhik et al., 1999). Furthermore, chemostratigraphic correlations are complicated by the discovery of multiple diamictite/cap carbonate couplets in individual sections, the incomplete nature of the stratigraphic record in glaciated marginal marine environments and tectonic factors that either omit or duplicate strata (Kaufman et al., 2009). Further complicating the interpretation of Cryogenian to Cambrian paleoclimatic variations are suggestions that the negative $\delta^{13}C$ excursions are diagenetic artifacts, or are the result of strong surface-to-deep carbon isotope gradients in SW (Jiang et al., 2007; Swart, 2008; Kaufman et al., 2009; Frimmel, 2010). If the latter is true, the profound isotopic variations documented through the strata in any of the marginal marine successions may have more spatial than temporal significance (Kaufman et al., 2009).

2.3.5 PALEOZOIC

According to an international agreement, the end of the Ediacaran Period is defined by the first appearance datum of the trace fossil *Treptichnus pedum* (Narbonne et al., 1987). While this biological first appearance follows the typical pattern of Phanerozoic boundaries, the Precambrian–Cambrian transition is characterized by a profound negative $\delta^{13}C$ anomaly, more readily recognizable in carbonate-dominated successions worldwide (Magaritz, 1989; Brasier et al., 1990; Kirschvink et al., 1991; Narbonne et al., 1994; Knoll et al., 1995; Kaufman et al., 1996; Pelechaty et al., 1996; Bartley et al., 1998; Shen and Schidlowski, 2000; Kimura and Watanabe, 2001; Corsetti and Hagadorn, 2003; Tewari and Sial, 2007). The magnitude of this biogeochemical anomaly (down to −10‰ in many key sections) rivals that of the Shuram $\delta^{13}C$ excursion—notably this defines the interval of most Ediacaran fossils, although Lazarus specimens of *Swartpuntia* may also appear in Lower Cambrian strata of the Death Valley region (Hagadorn et al., 2000; Kaufman et al., 2009). Between these isotope extremes lies a unique window to climatic and environmental change during the nexus of Ediacaran organismal diversification. In Namibia, the boundary $\delta^{13}C$ anomaly is not recorded in 543 Ma carbonates of the Nama Group (Spitzkopf Member), which notably also contain Swartpuntia (Grotzinger et al., 1995; Narbonne, 1998). This member, however, is truncated by U-shaped valleys, which are filled with coarse clastic debris of the Nomtsas Formation dated around 539 Ma (Grotzinger et al., 1995; Blanco et al., 2009). Insofar as these have been interpreted as glacial in origin (Germs, 1972; Germs and Gaucher, 2012), the biogeochemical anomaly at the boundary might have been eroded in the Nama Group (Saylor et al., 1998), as was the positive Hüttenberg anomaly in northwestern Namibia earlier in Neoproterozoic time (Kaufman et al., 2009).

Oceanic $\delta^{13}C_{carb}$ fluctuations during the Paleozoic appear transitional between vigorous fluctuations observed in the Neoproterozoic (±10‰) and the increasingly stable Mesozoic and Cenozoic Eras (Saltzman, 2005). Thirteen globally recognized $\delta^{13}C$ excursions (e.g., Veizer et al., 1999; Joachimski et al., 2002; Saltzman, 2005; Zhu et al., 2006; Álvaro et al., 2008) have been documented in the Paleozoic (Fig. 2), with $\delta^{13}C$ shifts of ≥+3‰ and that lasted just for a few Ma or less, probably representing significant paleoceanographic events (Saltzman, 2005). Intervals characterized by $\delta^{13}C$ volatility in the Late Cambrian (Steptoean), Late Ordovician–Silurian, and Late Devonian–Early Mississippian are separated by relatively stable periods with values ranging between −2‰ and +2‰ (Fig. 2). The $\delta^{13}C$ excursions are commonly interpreted as episodes of enhanced organic carbon burial resulting from increments in oceanic nutrient availability (Kump and Arthur, 1999; Godderis and Joachimski, 2004). Phosphate probably limited the marine bioproductivity and controlled the $\delta^{13}C$ excursions by increasing P fluxes to the surface oceans, either from the continents or deep ocean (Saltzman, 2005). The control of primary production in the oceans by either nitrogen or phosphorous is, however, still a matter of debate in paleoceanography (e.g., Falkowski, 1997; Tyrrell, 1999).

The Paleozoic composite $\delta^{13}C_{carb}$ variation curve (Fig. 2) shows several positive excursions, some with peaks of about +7‰ (e.g., Kump et al., 1999; Saltzman, 2005), comparable in amplitude to the Neoproterozoic peaks but shorter in duration. These positive excursions lack associated negative excursions which are known only at the beginning and end of the Paleozoic (Knoll et al., 1996; Montañez et al.,

2000; Saltzman, 2005). However, a Sunwaptan (late Cambrian) negative $\delta^{13}C_{carb}$ anomaly (−4‰) associated with the positive Steptoean excursion in the Argentine Precordillera, the global nature of which needs further confirmation, has been reported by Sial et al. (2008, 2013). A composite high-resolution $\delta^{13}C$ curve for the Middle Cambrian—Pennsylvanian in North America (Saltzman, 2005), reveals that the large positive excursions were common during cool periods (e.g., Upper Ordovician–Silurian and Late Devonian–Early Mississippian). This author proposed that periods of stability in $\delta^{13}C$ reflect negative feedbacks on bioproductivity in a nitrogen-limited ocean where anoxia led to increased denitrification.

Ten Cambrian large-scale $\delta^{13}C$ excursions have been defined (Zhu et al., 2006; Álvaro et al., 2008) from the BACE (Basal Cambrian Isotope Excursion) to the TOCE (Top of Cambrian Excursion). None of these biogeochemical events, however, have been directly related to glaciation; some have been associated to eustatic sea-level fall (Ripperdan, 2002). Additionally, it seems likely that these excursions are related to faunal turnover and diversification, as well as to the progressive increase in bioturbation and ventilation of organic-rich sediments through Cambrian time (Droser and Bottjer, 1988).

A large and global positive $\delta^{13}C_{carb}$ excursion for Upper Cambrian (Steptoean) carbonates has been reported from North America, Kazakhstan, South China, and Australia and marks the beginning of a global extinction of trilobites (Saltzman et al., 1998, 2000). This Steptoean positive C-isotope excursion (~+5‰), described by the acronym SPICE (Saltzman et al., 1998) represents a major perturbation of the Cambrian carbon cycle at ~500 Ma and, as a peculiarity, a worldwide mass extinction (mainly trilobites among other faunas) coincides with the onset of this positive shift, rather than with the rising limb of the excursion. The peak of faunal diversity correlates in North America with the peak of SPICE (Saltzman et al., 2000). The SPICE is believed to have had a duration of about 4 Ma (Saltzman et al., 1998), although 3 Ma is probably a better estimate, according to Peng and Babcock (2008). One interpretation of the SPICE ascribes it to the increased burial of organic carbon, which is enriched in ^{12}C, thus increasing the ^{13}C pool in the oceans. However, terrestrial weathering impacts related to sea level and/or climate change have also been considered as explanations (Saltzman et al., 2000, 2004; Gill et al., 2007).

The $\delta^{13}C_{carb}$ variation curves for the Cambrian–Ordovician boundary at the Black Mountain, Australia (Ripperdan et al., 1992), and at the Lawson Cove, Ibex area, Utah (Ripperdan and Miller, 1995), reveal a pronounced negative $\delta^{13}C_{carb}$ excursion (−4‰) in the uppermost Cambrian (Sunwaptan Stage). The Cambrian–Ordovician boundary is characterized by synchronous changes in sea level and $\delta^{13}C$ of marine carbonates (Ripperdan et al., 1992). A Sunwaptan negative C-isotope excursion first observed in the La Flecha Formation of the Argentine Precordillera, has been named SNICE by Sial et al. (2008) and, if demonstrated to be a global event, when combined with SPICE, it can be a valuable tool for high-resolution correlation and can be used to locate primary subdivisions of the Cambrian system in an otherwise unfossiliferous carbonate sequences globally.

In the Ordovician, three major carbon isotope excursions are known. Two of them are recognized in the Upper Ordovician, one in the Early Chatfieldian North American Stage (Sandbian to early Katian), and another in the Gamachian Stage (Hirnantian) (Saltzman et al., 2003; Kaljo et al., 2007; Brenchley et al., 2003; Bergström et al., 2006 and references therein). The latter excursion is known as Hirnantian Isotope Carbon Excursion (HICE) (see Orth et al., 1986; Finney et al., 1999; Brenchley et al., 2003; Melchin et al., 2003; Fan et al., 2009; Jones et al., 2011) and the former as Guttenberg Isotope Carbon Excursion (GICE) and is recorded in Sweden, Estonia, Baltoscandia, and China (see Ainsaar et al., 1999, 2010; Kaljo et al., 2007; Saltzman et al., 2003; Saltzman and Young, 2005; Bergstrom et al., 2010; Brenchley et al., 2003; Young et al., 2008).

The GICE anomaly was first recognized in the Guttenberg Member of the Decorah Formation in Iowa (Hatch et al., 1987; Ludvigson et al., 1996, 2000). According to Saltzman et al. (2003), in this region, the GICE is expressed by a relatively abrupt change in the $\delta^{13}C$ curve from −2‰ to −1‰ to about +2 to +3‰ followed upward by a drop to previous values. In North American sections containing the widespread Millbrig K-bentonite, the base of the GICE starts in a transgressive interval above this ash bed (Bergström et al., 2010). This excursion ends at a regressive interval and its thickness varies from 1 to 100 m. Below the GICE interval, Ludvigson et al. (2004) recognized four minor pre-GICE $\delta^{13}C$ positive excursions in North America, regarded as regional, namely in descending order: Spechts Ferry, Quimbys Mill, Grand Detour, and Mifflin, in the Turinian (early Sandbian).

According to Schmitz et al. (2010), the relationship between the GICE and potential climatic and water temperature indicators (lithofacies, faunas, and geochemistry) does not suggest a close correlation to specific environmental conditions. However, Saltzman and Young (2005) interpreted this $\delta^{13}C$ excursion as a record of enhanced organic carbon burial, which lowered the atmospheric pCO_2 to levels near the threshold for ice buildup in the Ordovician greenhouse climate.

Toward the end of the Hirnantian, cold temperature, major glaciation, and severe drop in sea level, were replaced by rise in temperature, melting of glaciers, and return of sea level to the same or slightly higher level than it had been prior to glaciations (Munnecke et al., 2010). According to these authors, the Hirnantian mass extinction was the second largest extinction event in geologic time. HICE, the most prominent of the three major global $\delta^{13}C$ excursions in the Ordovician, has been recorded from the uppermost Ordovician in Nevada, Quebec, Arctic Canada, Baltoscandia, Scotland, China, and North American Midcontinent (Ripperdan et al., 1998; Finney et al., 1999; Bergstrom et al., 2006; Schmitz et al., 2010). HICE peak values of up to +5‰ have been reported from North America, and up to +6‰ from Estonia and Latvia (e.g., Bergstrom et al., 2010; Brenchley et al., 2003).

There is a consensus that HICE is of global distribution and associated with a major glaciation (Bergstrom et al., 2006; Jones et al., 2011). It occupies a biostratigraphically well-defined interval in the graptolite and chitinozoan zone successions (Finney et al., 1999; Soufiane and Achab, 2000; Melchin et al., 2003; Melchin and Holmden 2006; Kaljo et al., 2004). Sea-level change–HICE relationships in the Anticosti Island, Canada, indicate that they were not perfectly coupled (Jones et al., 2011). The commencement of the isotope excursion and the initial sea-level drawdown were coincident, but the excursion persisted through the deglacial sea-level rise, only reaching its peak after the sea level had begun to rise.

The stratigraphically oldest of the named Ordovician $\delta^{13}C$ excursions, the Middle Ordovician (Darriwilian) $\delta^{13}C$ excursion (MDICE), is the least known among the three major Ordovician positive $\delta^{13}C$ excursions. This excursion is known from the Middle Ordovician of Baltoscandia and recently has been identified from the Guniutan Formation at two localities in the Yangtze platform, China (Schmitz et al., 2010). According to these authors, such occurrences in strata with striking lithological and conodont faunal similarity to the Swedish Holen Limestone and some coeval units in Estonia, suggest that MDICE likely has a worldwide distribution, with potential for local and long-range chemostratigraphic correlations.

The $\delta^{44/40}Ca$ record of SW from fossil carbonate and barite for the Paleozoic is shown in Fig. 2. It is corrected for species-dependent fractionation for reconstruction of SW values but without the correction for temperature-dependent fractionation. The data shown are from diagenetically screened, primary calcite samples of fossil brachiopods from Farkaš et al. (2007a,b) and Blättler et al. (2012).

The main feature of this record is a shift toward higher Ca-isotope ratios in the Carboniferous, with no shift of comparable magnitude observed in the Mesozoic or Cenozoic (Fig. 3). The differing ratio of aragonite to calcite precipitated during aragonite seas and calcite seas as a mechanism for generating oscillations in SW Ca-isotope ratios has been proposed by Farkaš et al. (2007a). However, the isotope data do not show a strong oscillatory response to changing SW chemistry, particularly during the Cenozoic, as has been pointed by Blättler et al. (2012).

The composition of the carbonate sink, global temperature and dolomitization rate may contribute significantly to the SW $\delta^{44/40}Ca$ oscillation. Lowered global temperature may have been the cause of high $\delta^{44/40}Ca$ in the late Carboniferous icehouse relative to the lower values of the Permian–Triassic greenhouse, when there was no obvious change in proportions of aragonite to calcite sediment producers over this interval (Blättler et al., 2012).

2.3.6 MESOZOIC

It has been hypothesized that anomalous Hg levels caused by catastrophic Siberian Trap volcanic eruptions were the major reason for the mass extinction of the Permian–Triassic boundary (PTB; e.g., Renne et al., 1995; Sanei et al., 2012), the most important mass extinction in Earth's history and informally known as the "Great Dying" (Morrison, 2004). An asteroid impact as an alternative explanation for PTB mass extinction (e.g., Becker et al., 2001 and references therein) has never assembled as much evidence as for the KTB although a meteor crater dated at 250.7 ± 4.3 Ma ($^{39}Ar/^{40}Ar$ dating from a single plagioclase crystal; Morrison, 2004) was identified near the coast of Australia. Alternatively, global warming, anoxia, and the physiological effects of elevated CO_2 concentrations related to volcanism have been proposed as causes of the mass extinction (e.g., Knoll et al., 1996, 2007). No detailed Hg chemostratigraphic study has been performed across sedimentary rocks that record the PTB and that could contribute to the impact versus volcanism debate on this time boundary, as in the case of the KTB (Sial et al., 2013, 2014; Nascimento-Silva et al., 2011, 2013).

The Mesozoic–early Paleogene is a typical greenhouse period caused largely by increased CO_2 from high global igneous activity (Takashima et al., 2006). The mid-Cretaceous is marked by a major warming that reached globally averaged surface temperatures more than 14 °C higher than that of today (Tarduno et al., 1998). It was associated with a lack of permanent ice sheets (Frakes et al., 1992), and ~100–200 m higher sea level (Haq et al., 1987; Miller et al., 2005; Fig. 3). Studies using Deep Sea Drilling Project (DSDP) and Ocean Drilling Program (ODP) cores helped demonstrating that the Mesozoic ocean circulation and marine ecosystems differed greatly from those of today (Takashima et al., 2006).

Cretaceous marine sediments in Europe are mainly comprised of white limestone and chalk. However, distinct black, laminated organic-rich layers, termed "black shales," are occasionally intercalated within these sequences. Because organic carbon is preferentially preserved under anoxic conditions, earlier workers suggested that these black shales had accumulated locally in a weakly ventilated, restricted basin under regional anoxic conditions (Takashima et al., 2006). However, the discovery of black shales at many DSDP drill sites from the Atlantic, Indian, and Pacific Oceans led to recognition of widespread anoxic conditions in the global ocean spanning limited stratigraphic horizons. These widespread depositional black shale intervals have been called "Oceanic Anoxic Events" (OAE) by Schlanger and Jenkyns (1976). Burial of organic carbon, which preferentially retains isotopically light carbon during OAEs, resulted in a positive $\delta^{13}C$ excursion of 2–3‰ in the geologic record (Fig. 3).

Carbon isotope excursions are a useful marker for recognizing OAEs (e.g., Gröke et al., 1999; Takashima et al., 2004). Biostratigraphic studies and correlation using carbon isotopes have revealed that OAEs occurred at least eight times in the Cretaceous and at least one to four times in the Jurassic (Takashima et al., 2006). These authors regard that the Toarcian OAE, Weissert OAE, OAE 1a, and OAE 2 are global-scale anoxic events associated with prominent positive excursions of $\delta^{13}C$ and worldwide distribution of black shales (Fig. 3). Interestingly, Jones and Jenkyns (2001) demonstrated that three Jurassic and Cretaceous OAEs coincide with significant drops in $^{87}Sr/^{86}Sr$ ratios, strongly suggesting that increased ocean-crust production and hydrothermal activity may have provided the nutrients necessary to fuel-enhanced bioproductivity during OAEs. The ultimate trigger of anoxic events would be sea-level rise (Jones and Jenkyns, 2001). This scenario is very similar to the proposals of a link between Neoproterozoic glaciation, bioproductivity, and Rodinia breakup (Kaufman et al., 1997; Gaucher, 2000; Gaucher et al., 2004; Frei et al., 2013).

A $\delta^{34}S$ evolution curve for Phanerozoic SW, based on the analysis of structurally substituted sulfate (SSS) in carbonates, has been prepared by Kampschulte and Strauss (2004). In Fig. 3, we show the portion of their curve for the Mesozoic and Cenozoic Eras. These authors have documented large $\delta^{34}S$ variations in the Triassic, particularly in the Induan to Anisian Stages, with a maximum $\delta^{34}S$ value of 29.7‰. Values between 17.3‰ and 19.7‰ characterize the remainder of the Triassic, followed by peculiar short-term fluctuations from 11.1‰ to 24.3‰ in the uppermost Triassic (Rhaetian Stage). The Jurassic SW sulfate sulfur isotope record is represented by measurements of SSS in belemnites. The temporal trend shows that the bulk of $\delta^{34}S$ values fall within the 14.2–18.0‰ interval, with two maxima of 23.4‰ in the lower middle (Toarcian) and 20.7‰ in the upper middle Jurassic (Bathonian). The $\delta^{34}S$ values measured on belemnites, define a pronounced oscillations in the lower part of the early Cretaceous (Valanginian and Hauterivian) followed by well-defined succession of sulfur isotope values (~14‰) in the upper part of the early Cretaceous.

No data are available for the upper middle Cretaceous (Cenomanian, Turonian, and Coniacian). According to Kampschulte and Strauss (2004), the uppermost part of the late Cretaceous (Santonian to Campanian) displays $\delta^{34}S$ values around 18‰.

2.3.7 CENOZOIC

The Cenozoic era is divided into two periods, the Paleogene (Paleocene, Eocene, and Oligocene epochs) which extends from 65.5 to 23.03 Ma and the Neogene (Miocene and Pliocene epochs) from 23.03 to 2.59 Ma (base of the Gelasian stage) or to 1.81 Ma (base of Calabrian stage) (Ogg et al., 2008). The redefinition of the Quaternary and revision of the Pleistocene are under discussion and it is expected that the ages of important events for all the Cenozoic be well constrained in the near future.

The mass extinction recorded in the KTB is generally regarded as a consequence of single (e.g., Alvarez et al., 1980; Claeys et al., 2002; Schulte et al., 2010) and/or of multiple meteorite impacts (e.g., Keller et al., 2003; Stüben et al., 2005) and/or of intense volcanic eruptions (e.g., McLean, 1978; Courtillot, 1999; Hoffman et al., 2000; Keller, 2005; Archibald et al., 2010).

Alvarez et al. (1980) hypothesis assumes that a single meteorite impact led to a dust cloud that severely affected photosynthetic organisms, reduced global temperature, a phenomenon that has been called an "impact winter," which was derived from the term "nuclear winter" coined by Turco et al.

(1983) to explain the possible effect of a large thermonuclear war on the global climate. Their hypothesis rests on anomalous amounts of iridium (3000 ppt) in a 1-centimeter-thick clay layer within a sequence of pelagic limestones at Gubbio in Italy, which was later found at several localities worldwide (e.g., Stüben et al., 2002, 2005; Frei and Frei, 2002).

Stüben et al. (2005) reported a high-resolution geochemical data of eight KTB sequences from northeastern, east-central, and southern Mexico, around the Chicxulub meteor impact crater in the Yucatán peninsula, considered critical in the reconstruction of the nature of the KTB events and their environmental consequences. According to these authors, the C- and O-isotope patterns for these sections indicate a gradual climatic change during the latest Maastrichtian, an abrupt change at the KTB and a slight recovery during the lowermost Paleocene.

The hypotheses of one or multiple meteorite impacts as the main cause of mass extinction during the KTB, however, have never reached a consensus. It is known that the largest mass extinction, the PTB, coincides with basaltic flows in Siberia (e.g., Sanei et al., 2012). Likewise, it is feasible to suggest/assume that the perturbation in the carbon cycle and the iridium anomalies at the KTB resulted from volcanism of a magnitude equivalent to that of the Deccan traps of west-central India whose gigantic eruptions produced huge volumes of flood basalt (up to 2 km in thickness) between 68 and 60 Ma ago (Sheth, 2005). The discovery of rapid and voluminous Deccan basalt eruptions at the KTB led to the hypothesis that part of the iridium and other platinum-group elements (PGE) concentrations in KTB sections may well be attributable directly to a volcanic origin rather than to a meteoritic one. However, osmium ($^{186}Os/^{188}Os$ and $^{187}Os/^{188}Os$) and chromium isotope ratios strongly favor a cosmochemical, chondritic origin for this anomaly (Frei and Frei, 2002).

An Hg stratigraphic investigation across the KTB in the Yacoraite Formation, northwestern Argentina, revealed Hg contents as high as to 250 ng.g^{-1} (Sial et al., 2013). In three drill cores across the KTB in the Paraiba Basin, northeastern Brazil, Hg contents increase from the late Maastrichtian to early Danian (Nascimento-Silva et al., 2011, 2013; Sial et al., 2013). Hg spikes that predate the KTB are according to those authors, perhaps, the record of volcanism before (but very close to) this transition. At Stevns Klint, a classical locality of the KTB study in Denmark, Hg contents reach almost 260 ng.g^{-1} within a 5-centimeter-thick clay layer (Fish Clay) that registers the transition, and from where marked $^{87}Sr/^{86}Sr$ positive excursion and $^{206}Pb/^{204}Pb$ (T = 65 Ma) and $^{187}Os/^{188}Os$ (T = 65 Ma) negative excursions have been reported by Frei and Frei (2002). A covariation between Hg and Al_2O_3 contents is observed in all of the sections (Argentina, Brazil, and Denmark) across the KTB studied by Sial et al. (2013), suggesting that the Hg is adsorbed onto the clays. It is possible that $\delta^{202}Hg$ (mass-dependent fractionation) from volcanic sources differs from that of meteoritic sources (Berquist and Blum, 2009). Therefore, it is possible that Hg may become a proxy to determine the role of volcanism versus meteor impact as main trigger of past catastrophic climatic changes.

A good outline of the Cenozoic history is given by Zachos et al. (2001), especially concerning the paleoclimate. According to these authors, through the study of deep-sea sedimentary archives, it has become increasingly apparent that during much of the Cenozoic Era, Earth's climate system has experienced continuous change, drifting from extremes of expansive warmth with ice-free poles, to extremes of cold with massive continental ice sheets and polar ice caps. A rapid growth in the inventory of high-resolution isotope records across the Cenozoic, with a greater availability of high-quality sediment cores recovered by the DSDP and ODP has led to important scientific developments in the last two decades, including the discovery of geologically abrupt shifts in climate, as well as brief but extreme excursions often associated with profound impacts on global environments and the biosphere (Zachos et al., 2001).

The way the climate has evolved during the Cenozoic is revealed by a variety of geochemical records (e.g., O and C isotopes from Zachos et al., 2001; S isotopes from Paytan et al., 1998; Sr isotopes from McArthur et al., 2001). The compiled basic geochemical records of Cenozoic global changes (Fig. 3) lack unambiguous interpretations because all isotopic and major element data that record the paleocomposition of SW are affected by more than one process or physical state (Lyle et al., 2008). Oxygen isotopes record both temperature and ice volume, strontium isotopes record both the rate of weathering and the balance between hydrothermal and continental sources (e.g., Flament et al., 2013), and sulfur isotopes change because of continental weathering, marine pyrite burial, and submarine hydrothermal activity.

From the beginning of the Cenozoic until just after the end of the Eocene (Eocene–Oligocene boundary), the Earth experienced a warm climate, the "greenhouse world" (Miller et al., 1987). In this time period, abyssal water temperatures were about 10 °C warmer than modern bottom water in the early Eocene (Zachos et al., 2001), while mean annual temperatures in northern Greenland were above freezing (Fricke and Wing, 2004). The most prominent climatic aberration is known as Late Paleocene Thermal Maximum (LPTM; Fig. 3) when deep-sea temperatures rose 5–6 °C (~1‰ negative $\delta^{18}O$ excursion; Zachos et al., 2001). At the Eocene–Oligocene boundary, SW became significantly enriched in ^{18}O relative to ^{16}O. This marine $\delta^{18}O$ excursion, known as Oi-1 (Fig. 3), coincides with the major expansion of the Antarctic ice cap and the beginning of the "icehouse world" that has characterized much of the rest of the Cenozoic (Miller et al., 1987; Zachos et al., 2001; Coxall et al., 2005). Ice caps preferentially trap ^{16}O, so these events mark major changes in cryospheric mass. The Oi-1 glacial interval reached its maximum at 33.6 Ma (Coxall et al., 2005) as global climate cooled abruptly. At the Oligocene–Miocene boundary, another climatic cooling took place (Mi-1; Fig. 3) after a Late Oligocene warming. Both, Oi-1 and Mi-1, are characterized by positive $\delta^{18}O$ excursions. Following roughly 10 Ma of generally cool conditions, the climate gradually warmed and reached another warm maximum around 15 Ma (beginning of the middle Miocene). Cooling that eventually led to the Pleistocene ice ages began with another major Antarctic cooling event at 14 Ma, marked by another abrupt change in marine oxygen isotopes (Flower and Kennett, 1994). Between the onset of large continental ice sheets on Antarctica at 33.6 Ma and the development of Northern Hemisphere ice sheets at 2.5 Ma, global temperatures were generally warm compared to the Holocene but cooler than the early Cenozoic (Lyle et al., 2008).

The global $\delta^{13}C$ stratigraphic curve in Fig. 3 for Cenozoic pelagic sedimentary sections with predominantly fine-grained, carbonate-rich oozes or chalks (Zachos et al., 2001) displays a positive excursion (~+2.5‰) in the late Paleocene and a shift to zero or slightly negative values in the early Eocene. This picture changes to a pathway with values around 1‰ for the rest of the Eocene, Oligocene, and up to the Mid-Miocene when another positive excursion (~+2‰) is displayed. In the Pliocene, $\delta^{13}C$ values change abruptly to 0‰ or to slightly negative values and a small positive excursion is seen at the Pliocene–Pleistocene boundary.

Changes in the $^{87}Sr/^{86}Sr$ ratio of SW measure a mixed signal of continental exposure and weathering versus ocean spreading rates as well as a regional signal relying on the extent and rates that different terranes were weathered (Veizer, 1989; Lyle et al., 2008; Flament et al., 2013). The distinctive shape of the SW $^{87}Sr/^{86}Sr$ curve (Fig. 3) is recorded by all the marine carbonates and has been used as a coarse chronostratigraphy for the Phanerozoic and, especially for the Paleogene–Neogene (McArthur et al., 2001) it has been used as a high-resolution stratigraphic tool (DePaolo and Ingram, 1985).

Carbon and sulfur isotopes are related to the organic carbon burial, the redox state of the burial region, and the geochemistry of pore fluids (Paytan et al., 1998; Kurtz et al., 2003). A continuous SW $\delta^{34}S_{sulfate}$ for the Cenozoic (Fig. 3) with a resolution of about 1 Ma was generated using marine barite samples by Paytan et al. (1998). The sulfur isotopic composition decreased from 19‰ to 17‰ CDT in the 65–55 Ma ago interval, increased abruptly from 17‰ to 22‰ CDT from 55 to 45 Ma ago, remained nearly constant from 35 to about 2 Ma ago, and has decreased by 0.8‰ during the past 2 million. The SW $\delta^{34}S_{sulfate}$ and bulk deep-sea marine carbonate $\delta^{13}C_{carb}$ records show no clear systematic coupling between the sulfur and carbon cycles over one to several millions of years (Paytan et al., 1998).

According to Paytan et al. (1998), the low values of $\delta^{34}S_{sulfate}$ occurring between 65 and 55 Ma ago indicate that during this time interval, either pyrite deposition rates were lower or there was increased input of ^{32}S to the oceans. The minimum at the Paleocene–Eocene boundary (55 Ma) coincides with a large, negative, short-term $\delta^{13}C$ excursion (Fig. 3), an important benthic foraminiferal global extinction, a maximum in deep-ocean temperature (18 °C), and intense volcanism (Paytan et al., 1998).

The largest shift in the $\delta^{34}S_{sulfate}$ during the Cenozoic happened in the 55 to 45 Ma interval. The pronounced shift from 17‰ to 22‰ over a period of 10 Ma, according to Paytan et al. (1998) indicates that large changes in the forcing mechanisms (volcanism, tectonic uplift, and weathering, or oceanic vertical mixing and productivity) influenced the fluxes and isotopic compositions of S input or burial, or both.

The $\delta^{44/40}Ca$ record of SW for the Mesozoic–Cenozoic from fossil carbonate and barite for the Paleozoic is shown in Fig. 3. These data are from the diagenetically screened primary calcite of fossil brachiopods (fractionated from SW by 0.85‰), foraminifera, belemnite (fractionated by 1.40‰), and barite (from Farkaš et al., 2007a,b; Blättler et al., 2012). The important shift toward higher Ca-isotope ratios seen in the Carboniferous (Fig. 2) finds no comparable shift in the Mesozoic or Cenozoic (Fig. 3). The differing ratio of aragonite to calcite precipitated during aragonite seas and calcite seas as a mechanism for generating oscillations in SW Ca-isotope ratios caused no strong oscillatory response in the Cenozoic as already pointed out by Blättler et al. (2012).

2.4 CONCLUSIONS

The use of isotope stratigraphy is growing rapidly and has greatly benefitted from recent technological development on analytical facilities, opening new venues to explore. It is possible that in years to come, important information will be drawn from the use of nontraditional stable isotopes (e.g., Li, Mg, Ca, Cr, Mo, Fe, Cu, Zn, Se, and Cl) will be added to our geological knowledge.

Among the traditional stable isotopes, C, O, and S have been by far the most used with chemostratigraphical purposes. The Sr-isotope data so far generated have led to invaluable geological information as discussed in this review. In the near future, the amount of information derived from Ca, Mo, Cr, and other nontraditional isotopes will strengthen and broaden conclusions obtained from the traditional isotopes in this field.

ACKNOWLEDGMENTS

We thank Prof. Mu. Ramkumar, editor of this Special volume on Chemostratigraphy (Elsevier Publications), for the invitation to contribute this manuscript. This is the contribution n. 261 of the Nucleus for Geochemical Studies-Stable Isotope Laboratory (NEG-LABISE), Department of Geology, Federal University of Pernambuco, Brazil.

REFERENCES

Ainsaar, L., Meidla, T., Martma, T., 1999. Evidence for a widespread carbon isotopic event associated with late Middle Ordovician sedimentological and faunal changes in Estonia. Geological Magazine 136, 49–62.

Ainsaar, L., Kaljo, D., Martma, T., Meidla, T., Männik, P., Nõlvak, J., Tinn, T., 2010. Middle and Upper Ordovician carbon isotope chemostratigraphy in Baltoscandia: a correlation standard and clues to environmental history. Palaeogeography, Palaeoclimatology, Palaeoecology 294, 189–201.

Allen, P.A., Etienne, J.L., 2008. Sedimentary challenge to Snowball Earth. Nature Geoscience 1, 817–825.

Almeida-Abreu, P.A., Renger, F.E., 2002. Serra do Espinhaço Meridional: um orógeno de colisão do Mesoproterozóico. Revista Brasileira de Geociencias 32, 1–14.

Algeo, T., Rowe, H., Hower, J.C., Schwark, L., Herrmann, A., Heckel, P., 2008. Changes in ocean denitrification during late carboniferous glacial interglacial cycles. Nature Geoscience 1, 709–714.

Alvarenga, C.J.S., Boggiani, P.C., Babinski, M., Dardenne, M.A., Figueiredo, M., Santos, R.V., Dantas, E.L., 2009. The Amazonian Palaeocontinent. In: Gaucher, C., Sial, A.N., Halverson, G.P., Frimmel, H.E. (Eds.), Neoproterozoic–Cambrian Tectonics, Global Change and Evolution: A Focus on southwestern Gondwana. Developments in Precambrian Geology, vol. 16. Elsevier, pp. 15–28.

Alvarez, L.W., Alvarez, W., Asaro, F., Michael, H.V., 1980. Extraterrestrial cause for the Cretaceous-Tertiary extinction. Science 208, 1095–1108.

Álvaro, J.J., Bauluz, B., Subías, I., Pierre, C., Vizcaïno, D., 2008. Carbon chemostratigraphy of the Cambrian-Ordovician transition in a midlatitude mixed platform, Montagne Noire, France. Geological Society of America Bulletin 120, 962–975.

Anbar, A.D., Duan, Y., Lyons, T.W., Arnold, G.L., Kendall, B., Creaser, R.A., 2007. A whiff of oxygen before the Great Oxidation Event? Science 317, 1903–1906.

Archibald, J.D., 28 additional authors, 2010. Cretaceous extinctions: multiple causes. Science 328, 973–976.

Arnold, G.L., Anbar, A.D., Barling, J., Lyons, T.W., 2004. Molybdenum isotope evidence for widespread anoxia in mid-Proterozoic oceans. Science 304, 87–90.

Arthur, M.A., 2009. Carbonate rocks deconstructed. Nature 460, 698–699.

Bailey, T.R., McArthur, J.M., Prince, H., Thirlwall, M.F., 2000. Dissolution methods for strontium isotope stratigraphy: whole rock analysis. Chemical Geology 167, 313–319.

Baker, A.J., Fallick, A.E., 1989. Evidence from Lewisian limestone for isotopically heavy carbon in two thousand-million-old sea water. Nature 337, 352–354.

Bao, H., Lyons, J.R., Zhou, C., 2008. Triple oxygen isotope evidence for elevated CO_2 levels after a Neoproterozoic glaciations. Nature 453, 504–506.

Bao, H., Fairchild, I.J., Wynn, P.M., Spötl, C., 2009. Stretching the envelope of past surface environments: Neoproterozoic glacial lakes from Svalbard. Science 323, 119–122.

Barrio, C., Poiré, D.G., Iñiguez, A.M., 1991. El contacto entre la Formación Loma Negra (Grupo Sierras Bayas) y la Formación Cerro Negro, un ejemplo de paleokarst, Olavarría, provincia de Buenos Aires. Asociación Geológica Argentina, Revista 46, 69–76.

Bartley, J.K., Kah, L.C., 2004. Marine carbon reservoir, C_{org}–C_{carb} coupling, and the evolution of the Proterozoic carbon cycle. Geology 32, 129–132.

Bartley, J.K., Pope, M., Knoll, A.H., Semikhatov, M.A., Petrov, P.Yu., 1998. A Vendian-Cambrian boundary succession from the northwestern margin of the Siberian Platform; stratigraphy, palaeontology, chemostratigraphy and correlation. Geological Magazine 135, 473–494.

Bartley, J.K., Kah, L.C., McWilliams, J.L., Stagner, A.F., 2007. Carbon isotope chemostratigraphy of the Middle Riphean type section (Avzyan Formation, Southern Urals, Russia): signal recovery in a fold-and-thrust belt. Chemical Geology 237, 211–232.

Beaumont, V., Robert, F., 1999. Nitrogen isotope ratios of kerogens in Precambrian cherts: a record of the evolution of atmosphere chemistry? Precambrian Research 96, 63–82.

Becker, L., Poreda, R.J., Hunt, A.G., Bunch, T.E., Rampino, M., 2001. Impact event at the Permian–Triassic boundary: evidence from extraterrestrial noble gases in fullerenes. Science 291, 1530–1533.

Bekker, A., Kaufman, A.J., Karhu, J.A., Beukes, N.J., Swart, Q.D., Coetzee, L.L., Eriksson, K.A., 2001. Chemostratigraphy of the Paleoproterozoic Duitschland Formation. South Africa: implications for coupled climate change and carbon cycling. American Journal of Science 301, 261–285.

Bekker, A., Kaufman, A.J., Karhu, J.A., Eriksson, K.A., 2005. Evidence for Paleoproterozoic cap carbonates in North America. Precambrian Research 137, 167–206.

Bekker, A., Sial, A.N., Karhu, J.A., Ferreira, V.P., Noce, C.M., Kaufman, A.J., Romano, A.W., Pimentel, M.M., 2003. Chemostratigraphy of carbonates from the Minas Supergroup, Quadrilátero Ferrifero, Brazil: a stratigraphic record of early Proterozoic atmosphere, biogeochemical and climatic change. American Journal of Science 303, 865–904.

Bekker, A., Karhu, J.A., Kaufman, A.J., 2006. Carbon isotope record for the onset of the Lomagundi carbon isotope excursion in the Great Lakes area, North America. Precambrian Research 148, 145–180.

Bekker, A., Holmden, C., Beukes, N.J., Kenig, F., Eglington, B., Patterson, W.P., 2008. Fractionation between inorganic and organic carbon during the Lomagundi (2.22–2.1 Ga) carbon isotope excursion. Earth and Planetary Science Letters 271, 278–291.

Bergström, S.M., Saltzman, M.R., Schmitz, B., 2006. First record of the Hirnantian (Upper Ordovician) $\delta^{13}C$ excursion in the North American Midcontinent and its regional implications. Geological Magazine 143, 657–678.

Bergström, S.M., Schmitz, B., Saltzman, M.R., Huff, W.D., 2010. The Upper Ordovician Guttenberg $\delta^{13}C$ Excursion (GICE) in North America and Baltosdcandia: Occurrence, Chronostratigraphic Significance, and Paleoenvironmental Relationships, vol. 466, The Geological Society of America, Special Paper. 37–67.

Bergquist, B.A., Blum, J.D., 2009. The odds and evens of mercury isotopes: applications of mass-dependent and mass-independent isotope fractionation. Elements 5, 353–357.

Berner, R.A., Beerling, D.J., Dudley, R., Robinson, J.M., Wildman, R.A., 2003. Phanerozoic atmospheric oxygen. Annual Review of Earth and Planetary Sciences 31, 105–134.

Blanco, G., Germs, G.J.B., Rajesh, H.M., Chemale Jr., F., Dussin, I.A., Justino, D., 2009. Provenance and paleogeography of the Nama Group (Ediacaran to early Palaeozoic, Namibia): petrography, geochemistry and U–Pb detrital zircon geochronology. Precambrian Research 187, 15–32.

Blättler, C.L., Henderson, G.M., Jenkys, H.C., 2012. Explaining the Phanerozoic Ca isotope history of seawater. Geology 40, 843–846.

Bleeker, W., 2003. The Late Archean record: a puzzle in ca. 35 pieces. Lithos 71, 99–134.

Boggiani, P.C., Gaucher, C., Sial, A.N., Babinski, M., Simon, C.M., Riccomini, C., Ferreira, V.P., Fairchild, T., 2010. Chemostratigraphy of the Tamengo Formation (Corumbá Group, Brazil): a contribution to the calibration of the Ediacaran carbon-isotope curve. Precambrian Research 182, 382–401.

Brasier, M.D., 1992. Nutrient-enriched waters and the early skeletal fossil Record. Journal of the Geological Society of London 149, 621–629.

Brasier, M.D., Corfield, R., Luo, H., Wu, X., Ouyang, L., Jiang, Z., Hamdi, B., 1990. The carbon- and oxygen-isotope record of the Precambrian–Cambrian boundary interval in China and Iran and their correlation. Geological Magazine 127, 319–332.

Brasier, M., McCarron, G., Tucker, R., Leather, J., Allen, P., Shields, G., 2000. New U–Pb zircon dates for the Neoproterozoic Ghubrah glaciations and for the top of the Huqf Supergroup, Oman. Geology 28, 175–178.

Brenchley, P.J., Garden, G.A., Hints, L., Kaljo, D., Marshall, J.D., Martma, T., Meidla, T., Nõlvak, J., 2003. High-resolution stable isotope stratigraphy of Upper Ordovician sequences: constraints on the timing of bioevents and environmental changes associated with mass extinction and glaciations. Geological Society of America Bulletin 111, 89–104.

Bristow, T.F., Kennedy, M.J., 2008. Carbon isotope excursions and the oxidant budget of the Ediacaran atmosphere and ocean. Geology 36, 863–866.

Broecker, W.S., Peng, T.H., Beng, Z., 1982. Tracers in the Sea. Lamont-Doherty Geological Observatory, Columbia University, New York. 690 p.

Buick, I.S., Uken, R., Gibson, R.L., Wallmach, T., 1998. High $\delta^{13}C$ Paleoproterozoic carbonates from the Transvaal Supergroup, South Africa. Geology 26, 875–878.

Burke, W.H., Denison, R.E., Hetherington, E.A., Koepnick, R.B., Nelson, H.F., Otto, J.B., 1982. Variation of seawater $^{87}Sr/^{86}Sr$ throughout Phanerozoic time. Geology 10, 516–519.

Calver, C.R., Black, L.P., Everard, J.L., Seymour, D.B., 2004. U–Pb zircon age constraints on late Neoproterozoic glaciation in Tasmania. Geology 32, 892–896.

Canfield, D.E., 1998. A new model for Proterozoic ocean chemistry. Nature 396, 450–453.

Canfield, D.E., Farquhar, J., 2009. Animal evolution, bioturbation, and the sulfate concentration of the oceans. Proceedings of the National Academy of Sciences of the USA 106, 8123–8127.

Canfield, D.E., Teske, A., 1996. Late Proterozoic rise in atmospheric oxygen concentration inferred from phylogenetic and sulphur-isotope studies. Nature 381, 127–132.

Canfield, D.E., Poulton, S.P., Knoll, H., Narbonne, G.M., Ross, G., Goldberg, T., Strauss, H., 2008. Ferruginous conditions dominated later Neoproterozoic deep-water chemistry. Science 321, 949–952.

Chiglino, L., Gaucher, C., Sial, A.N., Bossi, J., Ferreira, V.P., Pimentel, M.M., 2010. Chemostratigraphy of Mesoproterozoic and Neoproterozoic carbonates of the Nico Pérez Terrane, Río de la Plata Craton, Uruguay. Precambrian Research 182, 313–336.

Chu, X., Zhang, T., Zhang, Q., Feng, L., Zhang, F., 2004. Carbon istopic variations of Proterozoic carbonates in Jixian, Tianjin, China. Science in China Series D 47, 160–170.

Claeys, P., Kiessling, W., Alvarez, W., 2002. Distribution of Chicxulub ejecta at the Cretaceous–Tertiary boundary. In: Koeberl, C., MacLeod, K.G. (Eds.), Catastrophic Events and Mass Extinctions: Impacts and Beyond, vol. 356. Geological Society of America Special Paper, Boulder, Colorado, pp. 55–68.

Claypool, G.E., Holser, W.T., Kaplan, I.R., Sakai, H., Zak, I., 1980. The ages curves of sulfur and oxygen isotopes in marine sulfate and their mutual interpretation. Chemical Geology 28, 199–260.

Cloud, P., 1976. Beginnings of biospheric evolution and their biogeochemical consequences. Paleobiology 2, 351–387.

Corsetti, F.A., Hagadorn, J.W., 2003. The Precambrian–Cambrian transition in the southern Great Basin, USA. The Sedimentary Record 1, 428.

Corsetti, F.A., Kaufman, A.J., 2003. Stratigraphic investigations of carbon isotope anomalies and Neoproterozoic ice ages in Death Valley, California. Geological Society of America Bulletin 115, 916–932.

Courtillot, V., 1999. Evolutionary Catastrophes: The Science of Mass Extinction. Cambridge University Press, Cambridge. 174 p.

Coxall, H.K., Wilson, P.A., Pälike, H., Lear, C.H., Backman, J., 2005. Rapid stepwise onset of Antarctic glaciation and deeper calcite compensation in the Pacific Ocean. Nature 433, 53–57.

Cremonese, L., Struck, U., Shields, Guo, G., Ling, H., Och, L., 2009. $\delta^{15}N$ chemostratigraphy of Ediacaran-Cambrian sections of South China. Supplement Geochimica et Cosmochimica Acta 73 (13S), A-251.

Crowel, S.A., Døssing, L.D., Beukes, N.J., Bau, M., Kruger, S.J., Frei, R., Canfield, D.E., 2013. Atmospheric oxygenation three billion years ago. Nature 501, 535–538.

D'Agrella Filho, M.S., Pacca, I.I.G., Teixeira, W., Onsttot, T.C., Renne, P.R., 1990. Paleomagnetic evidence for the evolution of Meso- to Neo-Proterozoic glaciogenic rocks in Central-Eastern Brazil. Palaeogeography, Palaeoclimatology, Palaeoecology 80, 255–265.

De La Rocha, C.L., 2006. Palaeoceanography: in hot water. Nature 443, 920–921.

DePaolo, D.J., Ingram, B.L., 1985. High resolution stratigraphy with strontium isotopes. Science 227, 938–941.

Derry, L.A., 2010. A burial diagenesis origin for the Ediacaran Shuram–Wonoka carbon isotope anomaly. Earth and Planetary Science Letters 294, 152–162.

Derry, L.A., Kaufman, A.J., Jacobsen, S.B., 1992. Sedimentary cycling and environmental change in the Late Proterozoic: evidence from stable and radiogenic isotopes. Geochimica et Cosmochimica Acta 56, 1317–1329.

Dix, G.R., Thomson, M.L., Longstaffe, F.J., McNutt, R.H., 1995. Systematic decrease of high $\delta^{13}C$ values with burial in late Archaean (2.8 Ga) diagenetic dolomite: evidence for methanogenesis from the Crixás Greenstone Belt, Brazil. Precambrian Research 70, 253–268.

Døssing, L.N., Dideriksen, K., Stipp, S.L.S., Frei, R., 2011. Reduction of hexavalent chromium by ferrous iron: a process of chromium isotope fractionation and its relevance to natural environments. Chemical Geology 285, 157–166.

Droser, M.L., Bottjer, D.J., 1988. Trends in depth and extent of bioturbation in Cambrian carbonate marine environments, Western United States. Geology 16, 233–236.

Duan, Y., Anbar, A.D., Arnold, G.L., Lyons, T.W., Gordon, G.W., Kendal, B., 2010. Molybdenum isotope evidence for mild environmental oxygenation before the Great Oxidation Event. Geochimica et Cosmochimica Acta 74, 6655–6668.

Evans, D.A., Beukes, N.J., Kirschvink, J.L., 1997. Low-latitude glaciation in the Paleoproterozoic era. Nature 386, 262–266.

Etienne, J.L., Allen, P.A., Rieu, R., Le Guerroué, E., 2008. Neoproterozoic glaciated basins: a critical review of the Snowball Earth hypothesis by comparison with Phanerozoic glaciations. In: Hamphrey, M.J., Christoffersen, P., Glasser, N.F., Hubbard, B. (Eds.), Glacial Sedimentary Processes and Products International Association of Sedimentologists, vol. 39, pp. 343–399.

Eyles, N., Januszczak, N., 2004. 'Zipper-Rift': a tectonic model for Neoproterozoic glaciations during the breakup up of Rodinia after 750 Ma. Earth Science Reviews 65, 1–73.

Falkowski, P.G., 1997. Evolution of the nitrogen cycle and its influence on the biological sequestration of CO_2 in the ocean. Nature 387, 272–275.

Fan, J., Peng, P., Melchin, M.J., 2009. Carbon isotopes and event stratigraphy near the Ordovician–Silurian boundary, Yichang, South China. Palaeogeography, Palaeoclimatology, Palaeoecology 276, 160–169.

Farkaš, J., Böhm, F., Wallmann, K., Blenkinsop, J., Eisenhauer, A., van Geldern, R., Munnecke, A., Voigt, S., Veizer, J., 2007a. Calcium isotope record of Phanerozoic oceans: Implications for chemical evolution of seawater and its causative mechanisms. Geochimica et Cosmochimica Acta 71, 5117–5134.

Farkaš, J., Buhl, D., Blenkinsop, J., Veizer, J., 2007b. Evolution of the oceanic calcium cycle during the late Mesozoic: evidence from $\delta^{44/40}Ca$ of marine skeletal carbonates. Earth and Planetary Science Letters 253, 96–111.

Farquhar, J., Bao, H., Thiemens, M., 2000. Atmospheric influence of Earth's earliest sulfur cycle. Science 289, 756–758.

Figueiredo, M.F., 2006. Quimioestratigrafia das rochas ediacarianas do extremo norte da Faixa Paraguai, Mato Grosso. (M.Sc. thesis), Universidade de São Paulo, Brazil.

Fike, D.A., Grotzinger, J.P., Pratt, L.M., Summons, R.E., 2006. Oxidation of the Ediacaran ocean. Nature 444, 744–747.

Finney, S.C., Berry, W.B.N., Cooper, J.D., Ripperdan, R.L., Sweet, W.C., Jacobson, S.R., Soufiane, A., Achab, A., Noble, P.J., 1999. Late Ordovician mass extinction: a new perspective from stratigraphic sections in central Nevada. Geology 27, 215–218.

Flament, N., Coltice, N., Rey, P.F., 2013. The evolution of $^{87}Sr/^{86}Sr$ of marine carbonates does not constrain continental growth. Precambrian Research 229, 177–188.

Flower, B.P., Kennett, J.P., 1994. The middle Miocene climatic transition: East Antarctic ice sheet development, deep ocean circulation and global carbon cycling. Palaeogeography, Palaeoclimatology, Palaeoecology 108, 537–555.

Fortes, P.T.F.O., Jost, H., 1996. Metallogeny of Mina III, Mina Nova and Mina Inglesa gold deposits, Crixás Greenstone Belt, Central Brazil. In: Symposium: Archaean Terranes of the South American Platform, Brasília, Brazil. Sociedade Brasileira de Geologia, pp. 18–19.

Fortes, P.T.F.O., Pimental, M.M., Santos, R.V., Junges, S.L., 2003. Sm–Nd studies at Mina III gold deposit, Crixás Greenstone Belt, Central Brazil: implications for the depositional age of the upper metasedimentary rocks and associated Au mineralization. Journal of South American Earth Sciences 16, 503–516.

Frakes, L.A., Francis, J.E., Syktus, J.I., 1992. Climate Modes of the Phanerozoic. Cambridge University Press, Cambridge, United Kingdom. 274 p.

Frank, T.D., Lyons, T.W., Lohmann, K.C., 1997. Isotopic evidence for the paleoenvironmental evolution of the mesoproterozoic Helena formation, belt supergroup, Montana, USA. Geochimica et Cosmochimica Acta 61, 5023–5041.

Frei, R., Frei, K.M., 2002. A multi-isotopic and trace element investigation of the Cretaceous–Tertiary boundary layer at Stevns Klint, Denmark: inferences for the origin and nature of siderophile and lithophile element geochemical anomalies. Earth and Planetary Science Letters 203, 691–708.

Frei, R., Gaucher, C., Poulton, S.W., Canfield, D.E., 2009. Fluctuations in Precambrian atmospheric oxygenation recorded by chromium isotopes. Nature 46, 250–254.

Frei, R., Gaucher, C., Døssing, L.N., Sial, A.N., 2011. Chromium isotopes in carbonates – a tracer for climate change and for reconstructing the redox state of ancient seawater. Earth and Planetary Science Letters 312, 114–125.

Frei, R., Gaucher, C., Stolper, D., Canfield, D.E., 2013. Fluctuations in late Neoproterozoic atmospheric oxidation – Cr isotope chemostratigraphy and iron speciation of the late Ediacaran lower Arroyo del Soldado Group (Uruguay). Gondwana Research 23, 797–811.

Fricke, H.C., Wing, S.L., 2004. Oxygen isotope and paleobotanical estimates of temperature and $\delta^{18}O$-latitude gradients over North America during the early Eocene. American Journal of Science 304, 612–635.

Friedrich, O., Norris, R.D., Erbacher, J., 2012. Evolution of middle to Late Cretaceous oceans—a 55 m.y. record of Earth's temperature and carbon cycle. Geology 40, 107–110.

Frimmel, H.E., 2008. REE Geochemistry of Neoproterozoic Carbonates: Deviations from Normal Marine Signatures. Abstract, 33 International Geological Congress, Oslo, Norway.

Frimmel, H.E., 2009. Trace element distribution in Neoproterozoic carbonates as palaeoenvironmental indicator. Chemical Geology 258, 338–353.

Frimmel, H.E., 2010. On the reliability of stable carbon isotopes for Neoproterozoic chemostratigraphic correlation. Precambrian Research 182, 239–252.

Gaucher, C., 2000. Sedimentology, palaeontology and stratigraphy of the Arroyo del Soldado Group (Vendian to Cambrian, Uruguay). Beringeria 26, 1–120.

Gaucher, C., Sial, A.N., Blanco, G., Sprechmann, P., 2004. Chemostratigraphy of the Lower Arroyo del Soldado Group (Vendian, Uruguay) and palaeoclimatic implications. Gondwana Research 7, 715–730.

Gaucher, C., Sial, A.N., Ferreira, V.P., Pimentel, M.M., Chiglino, L., Sprechmann, P., 2007. Chemostratigraphy of the Cerro Victoria Formation (Lower Cambrian, Uruguay): evidence for progressive climate stabilization across the Precambrian–Cambrian boundary. Chemical Geology 237, 28–46.

Gaucher, C., Sial, A.N., Poiré, D.G., Gómez Peral, L., Ferreira, V.P., Pimentel, M.M., 2009. Chemostratigraphy. Neoproterozoic–Cambrian evolution of the Río de la Plata Palaeocontinent. In: Gaucher, C., Sial, A.N., Halverson, G.P., Frimmel, H.E. (Eds.), Neoproterozoic–Cambrian Tectonics, Global Change and Evolution: A Focus on southwestern Gondwana. Developments in Precambrian Geology, vol. 16. Elsevier, pp. 115–122.

Gaucher, C., Sprechmann, P., 2009. Neoproterozoic acritarch evolution. In: Gaucher, C., Sial, A.N., Halverson, G.P., Frimmel, H.E. (Eds.), Neoproterozoic–Cambrian Tectonics, Global Change and Evolution: A Focus on southwestern Gondwana. Developments in Precambrian Geology, vol. 16. Elsevier, pp. 319–326.

Gaucher, C., Frei, R., Chemale Jr., F., Frei, D., Bossi, J., Martínez, G., Chiglino, L., Cernuschi, F., 2011. Mesoproterozoic evolution of the Río de la Plata Craton in Uruguay: at the heart of Rodinia? International Journal of Earth Sciences 100, 273–288.

Gauthier-Lafaye, F., Weber, F., 1989. The Francevillian (Lower Proterozoioc) uranium ore deposits of Gabon. Economic Geology 84, 2267–2285.

Germs, G.J.B., 1972. The Stratigraphy and Paleontology of the Lower Nama Group, South West Africa, vol. 12, Precambrian Research Unit, University of Cape Town, Bulletin. 1–250.

Germs, G.J.B., Gaucher, C., 2012. Nature and extent of a late Ediacaran (ca. 547 Ma) glacigenic erosion surface in southern Africa. South African Journal of Geology 115, 91–102.

Gill, B.C., Lyons, T.W., Saltzman, M.R., 2007. Parallel, high-resolution carbon and sulfur isotope records of the evolving Palaeozoic marine sulphur reservoir. Palaeogeography, Palaeoclimatology, Palaeoecology 256, 156–173.

Godderis, Y., Joachimski, M.M., 2004. Global change in the Late Devonian: modelling the Frasnian–Famennian short-term carbon isotope excursions. Palaeogeography, Palaeoclimatology, Palaeoecology 202, 309–329.

Gröcke, D.R., Hesselbo, S.P., Jenkyns, H.C., 1999. Carbon-isotope composition of Lower Cretaceous fossil wood: ocean-atmosphere chemistry and relation to sea-level change. Geology 27, 155–158.

Grotzinger, J.P., Bowring, S.A., Saylor, B.Z., Kaufman, A.J., 1995. Biostratigraphic and geochronologic constraints on early animal evolution. Science 270, 598–604.

Grotzinger, J.P., Fike, D.A., Fischer, W.W., 2011. Enigmatic origin of the largest-known carbon isotope excursion in Earth's history. Nature Geoscience 4, 285–291.

Guo, Q., Strauss, H., Kaufman, A.J., Schröder, S., Gutzmer, J., Wing, B., Baker, M.A., Bekker, A., Jin, Q., Kim, S.-T., Farquhar, J., 2009. Reconstructing Earth's surface oxidation across the Archean–Proterozoic transition. Geology 3, 399–402.

Hagadorn, J.W., Fedo, C.W., Waggoner, B.M., 2000. Lower cambrian ediacaran fossils from the Great Basin, USA. Journal of Paleontology 74, 731–740.

Halverson, G.P., Hoffman, P.F., Schrag, D.P., Kaufman, A.J., 2002. A major perturbation of the carbon cycle before the Ghaub glaciation (Neoproterozoic) in Namibia: prelude to Snowball Earth? Geochemistry, Geophysics, Geosystems 3, 1–24.

Halverson, G.P., Hoffman, P.F., Schrag, D.P., Maloof, A.C., Hugh, A., Rice, N., 2005. Towards a Neoproterozoic composite carbon-isotope record. Geological Society of America Bulletin 117, 1181–1207.

Halverson, G.P., Dudas, F.O., Maloof, A.C., Bowring, S.A., 2007. Evolution of the $^{87}Sr/^{86}Sr$ composition of the Neoproterozoic seawater. Palaeogeography, Palaeoclimatology, Palaeoecology 256, 103–129.

Halverson, G.P., Hurtgen, M.T., Porter, S.M., Collins, A.S., 2010a. Neoproterozoic–Cambrian biogeochemical evolution. In: Gaucher, C., Sial, A.N., Halverson, G.P., Frimmel, H. (Eds.), Neoproterozoic–CambrianTectonics, Global Change and Evolution: A Focus on South Western Gondwana Developments in Precambrian Geology, vol. 16, pp. 351–365.

Halverson, G.P., Wade, B.P., Hurtgen, M.T., Barovich, K., 2010b. Neoproterozoic chemostratigraphy. In: Karhu, J., Sial, A.N., Ferreira, V.P. (Eds.), Precambrian Isotope Stratigraphy, Special Issue Precambrian Research, vol. 182, pp. 337–350.

Haq, B.U., Hardenbol, J., Vail, P.R., 1987. Chronology of fluctuating sea levels since the Triassic. Science 235, 1156–1167.

Haq, B.U., Schutter, S.R., 2008. A chronology of paleozoic sea-level changes. Science 322, 64–68.

Hardisty, D.S., Lu, Z., Planavsky, N.J., Bekker, A., Philippot, P., Zhou, Z., Lyons, T.W., 2014. An iodine record of Paleoproterozoic surface ocean oxygenation. Geology 42, 619–622.

Hatch, J.R., Jacobson, S.R., Witzke, B.J., Risatti, J.B., Anders, D.E., Watney, W.L., Newell, K.D., Vuletich, A.K., 1987. Possible late Middle Ordovician carbon isotope excursion: evidence from Ordovician oils and hydrocarbon source rocks, mid-continent and east-central United States. American Association of Petroleum Geologists Bulletin 71, 1342–1354.

Higgins, J.A., Schrag, D.P., 2003. Aftermath of a Snowball Earth. Geophysics, Geochemistry, Geosystems 4. http://dx.doi.org/10.1029/2002GC000403.

Hoashi, M., Bevacqua, D.C., Otake, T., Watanabe, Y., Hickman, A.H., Utsunomiya, S., Ohmoto, H., 2009. Primary haematite formation in an oxygenated sea 3.46 billion years ago. Nature Geoscience 2, 301–306.

Hoffman, C., Feraud, G., Courtillot, V., 2000. $^{40}Ar/^{39}Ar$ dating of mineral separates and whole rocks from the Western Ghats lava pile: further constraints on duration and age of Deccan traps. Earth and Planetary Science Letters 180, 13–27.

Hoffman, P.F., 2011. Strange bedfellows: glacial diamictite and cap carbonate from the Marinoan (635 Ma) glaciation in Namibia. Sedimentology 58, 57–119.

Hoffman, P.F., Kaufman, A.J., Halverson, G.P., Schrag, D.P., 1998a. A neoproterozoic Snowball Earth. Science 281, 1342–1346.

Hoffman, P.F., Kaufman, A.J., Halverson, G.P., 1998b. Comings and goings of global glaciations on a Neoproterozoic tropical platform in Namibia. GSA Today 8, 129.

Hoffman, P.F., Schrag, D.P., 2002. The Snowball Earth hypothesis: testing the limits of global change. Terra Nova 14, 129–155.

Holser, W.T., 1997. Geochemical events documented in inorganic carbon isotopes. Palaeogeography, Palaeoclimatology, Palaeoecology 132, 173–182.

Hongwei, K., Yongqing, L., Jiahua, L., Nan, P., Shunshe, L., Chao, C., 2011. Carbon and oxygen isotopic stratigraphy of mesoproterozoic carbonate sequences (1.6–1.4 Ga) fromYanshan in North China. International Journal of Oceanography 2011, 1–11. http://dx.doi.org/10.1155/2011/410621.

Huston, D.L., Logan, G.A., 2004. Barite, BIFs and bugs: evidence for the evolution of the Earth's early hydrosphere. Earth and Planetary Science Letters 220, 41–55.

Hyde, W.T., Crowley, T.J., Baum, S.K., Peltier, W.R., 2000. Neoproterozoic "Snowball Earth" simulations with a coupled climate/ice sheet model. Nature 405, 425–429.

Jacobsen, S.B., Kaufman, A.J., 1999. The Sr, C and O isotopic evolution of Neoproterozoic seawater. Chemical Geology 161, 37–57.

Jiang, G., Christie-Blick, N., Kaufman, A.J., Zhang, S., Wu, H., 2007. Carbon isotope variability across the Ediacaran Yangtze in platform in South China: implications for a large surface-to-deep ocean $\delta^{13}C$ gradient. Earth and Planetary Science Letters 361, 303–320.

Joachimski, M.M., Pancost, R.D., Freeman, K.H., Ostertag-Henning, C., Buggisch, W., 2002. Carbon isotope geochemistry of the Frasnian–Famennian transition. Palaeogeography, Palaeoclimatology, Palaeoecology 181, 91–109.

Johnson, C., Leard, B., Albarede, F., 2004. Geochemistry of non-traditional stable isotopes. Reviews in Mineralogy and Geochemistry 55, 454.

Johnson, C., Beard, L., 2006. Fe isotopes: an emerging technique for understanding modern and ancient biogeochemical cycles. GSA Today 16, 4–10.

Johnston, D.T., Poulton, S.W., Dehler, C., Porter, S., Husson, J., Canfield, D.E., Knoll, A.H., 2010. An emerging picture of Neoproterozoic ocean chemistry: insights from the Chuar Group, Grand Canyon, USA. Earth and Planetary Science Letters 290, 64–73.

Jones, C.E., Jenkyns, H.C., 2001. Seawater strontium isotopes, oceanic anoxic events and sea-floor hydrothermal activity in the Jurassic and Cretaceous. American Journal of Science 301, 112–149.

Jones, D.S., Fike, D.A., Finnegan, S., Fischer, W.W., Schrag, D.P., McCay, D., 2011. Terminal Ordovician carbon isotope stratigraphy and glacioeustatic sea-level change across Anticosti Island (Québec, Canada). Geological Society of America Bulletin 123, 1645–1664.

Jost, H., Dussin, I.A., Chemale, F.J., Tassinari, C.C.G., Junges, S., 2008a. U–Pb and Sm–Nd constraints for the Paleoproterozoic age of the metasedimentary sequences of the Goiás Archean greenstone belts. In: VIth South American Symposium on Isotope Geology, San Carlos de Bariloche, Argentina, pp. 1–4.

Jost, H., Sial, A.N., Bennel, M.R., Ferreira, V.P., 2008b. Carbon isotopes in dolostones of the Goiás state greenstone Belt, Central Brazil. In: 44° Congresso Brasileiro de Geologia, Curitiba. Sociedade Brasileira de Geologia (CD-ROM).

Kah, L.C., Sherman, A.G., Narbonne, G.M., Knoll, A.H., Kaufman, A.J., 1999. $\delta^{13}C$ stratigraphy of the Proterozoic by lot supergroup, Baffin Islands, Canada: implications for regional lithostratigraphy correlations. Canadian Journal of Earth Sciences 36, 313–332.

Kah, L.C., Lyons, T.W., Chesley, J.T., 2001. Geochemistry of a 1.2 Ga carbonate-evaporite succession, northern Baffin and Bylot Islands: implications for Mesoproterozoic marine evolution. Precambrian Research 111, 203–234.

Kaljo, D., Hints, L., Martma, T., Nolvak, J., Oraspold, A., 2004. Late Ordovician carbon isotope trend in Estonia, its significance in stratigraphy and environmental analysis. Palaeogeography, Palaeoclimatology, Palaeoecology 210, 165–185.

Kaljo, D., Martma, T., Saadre, T., 2007. Post-Hunnebergian Ordovician carbon isotope trend in Baltoscandia, its environmental implications and some similarities with that of Nevada. Palaeogeography, Palaeoclimatology, Palaeoecology 245, 138–155.

Kampschulte, A., Strauss, H., 2004. The sulfur isotopic evolution of Phanerozoic seawater based on the analysis of structurally substituted sulfate in carbonates. Chemical Geology 204, 255–286.

Karhu, J.A., 1993. Paleoproterozoic evolution of the carbon isotope ratios of sedimentary carbonates in the Fennoscandian Shield: Espoo. Geological Survey of Finland Bulletin 371, 87.

Karhu, J.A., Holland, H.D., 1996. Carbon isotopes and the rise of atmospheric oxygen. Geology 24, 867–870.

Karhu, J., Kortelainen, N.M., Huhma, H., Perttunen, V., Sergeev, S., 2008. The end of the paleoproterozoic carbon isotope excursion: new time constraints. In: International Geological Congress. IUGS, Oslo.

Karhu, J., Sial, A.N., Ferreira, V.P., 2010. Insights from precambrian isotope stratigraphy. Precambrian Research 182, 239–412.

Kasemann, S.A., Hawkesworth, C.J., Prave, A.R., Fallick, A.E., Pearson, P.N., 2005. Boron and calcium isotope composition in Neoproterozoic carbonate rocks from Namibia: evidence for extreme environmental change. Earth Planetary Science Letters 23, 73–86.

Kaufman, A.J., 2008a. Atmospheric, climatic, and biological evolution at both ends of the Proterozoic Eon. In: 33rd International Geology. Congress Abstracts, Oslo, Norway (CD-ROM).

Kaufman, A.J., 2008b. Slush find. Nature 450, 807–808.

Kaufman, A.J., Knoll, A.H., Awramik, A.H.S., 1992. Biostratigraphic and chemostratigraphic correlation of Neoproterozoic sedimentary successions: upper Tindir Group, northwestern Canada, as a test case. Geology 20, 181–185.

Kaufman, A.J., Jacobsen, S.B., Knoll, A.H., 1993. The Vendian record of C- and Sr-isotopic variations: implications for tectonics and paleoclimate. Earth Planetary Science Letters 120, 409–430.

Kaufman, A.J., Knoll, A.H., Semikhatov, M.A., Grotzinger, J.P., Jacobsen, S.B., Adams, W., 1996. Integrated chronostratigraphy of Proterozoic-Cambrian boundary beds in the western Anabar region, northern Siberia. Geological Magazine 133, 509–533.

Kaufman, A.J., Knoll, A.H., Narbonne, G.M., 1997. Isotopes, ice ages, and terminal Proterozoic earth history. Proceedings of the National Academy of Sciences of the USA 94, 6600–6605.

Kaufman, A.J., Sial, A.N., Ferreira, V.P., 2007. Preface to special issue of chemical geology on precambrian chemostratigraphy in honor of the late William T. Holser. Chemical Geology 237, 1–4.

Kaufman, A.J., Sial, A.N., Frimmel, H.E., Misi, A., 2009. Neoproterozoic to cambrian palaeoclimatic events in southwestern gondwana. In: Gaucher, C., Sial, A.N., Halverson, G.P., Frimmel, H. (Eds.), Neoproterozoic–Cambrian Tectonics, Global Change and Evolution: A Focus on Southwestern Gondwana. Developments in Precambrian Geology, vol. 16. Elsevier, pp. 369–388.

Keller, G., 2005. Impacts, volcanism and mass extinction: random coincidence or cause and effect? Australian Journal of Earth Science 52, 725–757.
Keller, G., Stinnesbeck, W., Adatte, T., Stüben, D., 2003. Multiple impacts across the Cretaceous–Tertiary boundary. Earth-Science Reviews 62, 327–363.
Kendall, B., Reinhard, C.T., Lyons, T.W., Kaufman, A.J., Poulton, S.W., Anbar, A.D., 2010. Pervasive oxygenation along late Archaean ocean margins. Nature Geoscience 3, 647–652.
Kennedy, M.J., Runnegar, B., Prave, A.R., Hoffmann, K.H., Arthur, M.A., 1998. Two or four Neoproterozoic glaciations? Geology 26, 1059–1063.
Keto, L.S., Jacobsen, S.B., 1988. Nd isotopic variations of early Paleozoic oceans. Earth Planetary Science Letters 84, 27–41.
Kimura, H., Watanabe, Y., 2001. Ocean anoxia at the Precambrian Cambrian boundary. Geology 29, 995–998.
Kirschvink, J.L., 1992. Late Proterozoic low-latitude global glaciation: the Snowball Earth. In: Schopf, J.W., Klein, C. (Eds.), The Proterozoic Biosphere. Cambridge University Press, Cambridge, pp. 51–52.
Kirschvink, J.L., Magaritz, M., Ripperdan, R.L., Zhuravlev, A.Yu., Rozanov, A.Yu., 1991. The Precambrian/Cambrian boundary; magnetostratigraphy and carbon isotopes resolve correlation problems between Siberia, Morocco, and South China. GSA Today 1, 87–91.
Kirschvink, J.L., Gaidos, E.J., Bertani, L.E., Beukes, N.J., Gutzmer, J., Maepa, L.N., Steinberger, R.E., 2000. The Paleoproterozoic Snowball Earth: extreme climatic and geochemical global change and its biological consequences. Proceedings of the National Academy of Sciences of the USA 97, 1400–1405.
Knauth, L.P., Kennedy, M.J., 2009. The late Precambrian greening of the Earth. Nature 460, 728–732.
Knauth, L.P., Lowe, D.R., 2003. High Archean climatic temperature inferred from oxygen isotope geochemistry of cherts in the 3.5 Ga Swaziland Supergroup, South Africa. Bulletin of the Geological Society of America 115, 566–580.
Knoll, A.H., Hayes, J.M., Kaufman, A.J., Swett, K., Lambert, I.B., 1986. Secular variation in carbon isotope ratios from Upper Proterozoic successions of Svalbard and East Greenland. Nature 321, 832–838.
Knoll, A.H., Walter, M.R., 1992. Latest Proterozoic stratigraphy and Earth history. Nature 356, 673–678.
Knoll, A.H., Kaufman, A.J., Semikhatov, M.A., Grotzinger, J.P., Adams, W.R., 1995. Sizing up the sub-Tommotian unconformity in Siberia. Geology 23, 1139–1143.
Knoll, A.H., Bambach, R.K., Canfield, D.E., Grotzinger, J.P., 1996. Comparative earth history and late permian mass extinction. Science 273, 452–457.
Knoll, A.H., Walter, M.R., Narbonne, G.M., Christie-Blick, N., 2006. The Ediacaran period: a new addition to the geologic time scale. Lethaia 39, 13–30.
Knoll, A.H., Bambach, R.K., Payne, J.L., Pruss, S., Fischer, W.W., 2007. Paleophysiology and end-Permian mass extinction. Earth and Planetary Science Letters 256, 295–313.
Kopp, R.E., Kirschvink, J.L., Hilburn, I.A., Nash, C.Z., 2005. The Paleoproterozoic Snowball Earth: a climate disaster triggered by the evolution of oxygenic photosynthesis. Proceedings of the National Academy of Sciences of the USA 102, 11131–11136.
Kump, L.R., Arthur, M.A., 1999. Interpreting carbon-isotope excursions: carbonates and organic matter. Chemical Geology 161, 181–198.
Kump, L.R., Arthur, M.A., Patzkowsky, M.E., Gibbs, M.T., Pinkus, D.S., Shehan, P.M., 1999. A weathering hypothesis for glaciation at high atmospheric pCO_2 during the Late Ordovician. Palaeogeography, Palaeoclimatology, Palaeoecology 152, 173–187.
Kurtz, A.C., Kump, L.R., Arthur, M.A., Zachos, J.C., Paytan, A., 2003. Early Cenozoic decoupling of the global carbon and sulfur cycles. Paleoceanography 18, 1090.
Kuznetsov, A.B., Melezhik, V.A., Gorokhov, I.M., Melnikov, N.N., Konstantinova, G.V., Kutyavin, E.P., Turchenko, T.L., 2010. Sr isotopic composition of paleoproterozoic ^{13}C-rich carbonate rocks: the Tulomozero formation, SE fennoscandian shield. Precambrian Research 182, 300–312.
Lea, D.W., Pak, D.K., Spero, H.J., 2000. Climate impact of late quaternary equatorial Pacific sea surface temperature variations. Science 289, 1719–1724.

Lear, C.H., Elder¢eld, H., Wilson, P.A., 2000. Cenozoic deep-sea temperatures and global ice volumes from Mg/Ca in benthic foraminiferal calcite. Science 287, 269–272.

Le Guerroué, E., Cozzi, A., 2010. Veracity of Neoproterozoic negative C-isotope values: the termination of the Shuram negative excursion. Gondwana Research 17, 653–661.

Le Guerroué, E., Allen, P.A., Cozzi, A., 2006a. Chemostratigraphic and sedimentological framework of the largest negative carbon isotopic excursion in earth history: the Neoproterozoic Shuram Formation (Nafun Group, Oman). Precambrian Research 146, 68–92.

Le Guerroué, E., Allen, P.A., Cozzi, A., 2006b. Parasequence development in the Ediacaran Shuram Formation (Nafun Group, Oman): high resolution stratigraphic test for primary origin of negative carbon isotopic ratios. Basin Research 18, 205–220.

Le Guerroué, E., Allen, P.A., Cozzi, A., Etienne, J.L., Fanning, C.M., 2006c. 50 Myr recovery from the largest negative $\delta^{13}C$ excursion in the Ediacaran ocean. Terra Nova 18, 147–153.

Li, C., Peng, P., Sheng, G., Fu, J., Yan, Y., 2002. A carbon isotopic biogeochemical study of Meso- to Neoproterozoic sediments from the Jixian section, North China. Acta Geologica Sinica 76 (4), 433–440 2002 (Chinese).

Lindsay, J.F., Brasier, M.D., 2000. A carbon isotope reference curve for ca. 1700–1575 Ma, McArthur and Mount Isa Basins, Northern Australia. Precambrian Research 99, 271–308.

Lindsay, J.F., Brasier, M.D., 2002. Did global tectonics drive early biosphere evolution? Carbon isotope record from 2.6 to 1.9 Ga carbonates of Western Australian basins. Precambrian Research 114, 1–34.

Lu, Z., Jenkyns, H.C., Rickaby, R.E.M., 2010. Iodine to calcium ratios in marine carbonates as a paleo-redox proxy during oceanic anoxic events. Geology 38, 1107–1110.

Ludvigson, G.A., Jacobsen, S.R., Witzke, B.J., González, L.A., 1996. Carbonate Component Chemostratigraphy and Depositional History of the Ordovician Decorah Formation, Upper Mississipi Valley, vol. 306, Geological Society of America Special Paper. 67–86.

Ludvigson, G.A., Witzke, B.J., Schneider, C.L., Smith, E.A., Emerson, S.J., Gonzalez, L.A., 2000. A profile of the mid-Caradoc (Ordovician) carbon isotope excursion at the MacGregor Quarry, Clayton County, Iowa. Geological Society of Iowa Guidebook 70, 25–31.

Ludvingson, G.A., Witzke, B.J., Schneider, C.L., Smith, E.A., Emerson, S.J., Carpenter, S.J., Gonzalez, L.A., 2004. Late Ordovician (Turinian–Chatfieldian) carbon isotope excursions and their stratigraphic and paleoceanic significance. Palaeogeography, Palaeoclimatology, Palaeoecology 210, 187–214.

Lyle, M., Barron, J., Bralowe, T.J., Huber, M., Lyle, A.O., Ravelo, A.C., Rea, D.K., Wilson, P.A., 2008. Pacific ocean and Cenozoic evolution of climate. Reviews of Geophysics 46, 1–47.

Lyons, T.W., Reinhard, C.T., 2009. Oxygen for heavy-metal fans. Nature 461, 179–181.

Macdougall, J.D., 1991. Radiogenic isotopes in seawater and sedimentary systems. In: Heaman, L., Ludden, J.N. (Eds.), Applications of Radiogenic Isotope Systems to Problems in Geology, NMCA Short Course Handbook, pp. 337–364 (Chapter10).

Magaritz, M., 1989. $\delta^{13}C$ minima follow extinction events: a clue to faunal radiation. Geology 17, 337–340.

Magaritz, M., Holser, W.T., Kirschvink, J.L., 1986. Carbon-isotope events across the Precambrian/Cambrian boundary on the Siberian platform. Nature 320, 258–259.

Marshall, J.D., 1992. Climatic and oceanographic isotopic signals from the carbonate rock record and their preservation. Geological Magazine 129, 143–160.

Martin, A.P., Condon, D.J., Prave, A.R., Melezhik, V.A., Lepland, A., Fallick, A.E., 2013. Dating the termination of the Palaeoproterozoic Lomagundi–Jatuli carbon isotopic event in the North Transfennoscandian Greenstone Belt. Precambrian Research 224, 160–168.

Master, S., Bekker, A., Hoffmann, A., 2010. A review of the stratigraphy and geological setting of the Palaeoproterozoic Magondi Supergroup, Zimbabwe – type locality for the Lomagundi carbon isotope excursion. In: Karhu, J., Sial, A.N., Ferreira, V.P. (Eds.), Precambrian Isotope Stratigraphy, Special Issue Precambrian Research, vol. 182, pp. 254–273.

Maheshwari, A., Sial, A.N., Chittora, V.K., 1999. High-$\delta^{13}C$ paleoproterozoic carbonates from the Aravalli Supergroup, Western India. International Geology Review 41, 949–954.

Maheshwari, A., Sial, A.N., Chittora, V.K., Harsh, B., 2002. A positive $\delta^{13}C_{carb}$ anomaly in Paleoproterozoic carbonates of Aravalli Craton, Western India; "Support for a global isotopic excursion". Journal of Asian Earth Sciences 21, 59–67.

Maheshwari, A., Sial, A.N., Gaucher, C., Bossi, J., Bekker, A., Ferreira, V.P., Romano, A.W., 2010. Global nature of the Paleoproterozoic Lomagundi carbon isotope excursion: a review of occurrences in Brazil, India, and Uruguay. In: Karhu, J., Sial, A.N., Ferreira, V.P. (Eds.), Precambrian Isotope Stratigraphy, Special Issue Precambrian Research, vol. 182, pp. 274–299.

McArthur, J.M., Howarth, R.J., Bailey, T.R., 2001. Strontium isotope stratigraphy: Lowess version 3: best fit to the marine Sr-isotope curve for 0–509 Ma and accompanying look-up table for deriving numerical age. Journal of Geology 109, 155–170.

McFadden, K.A., Huang, J., Chu, X., Jiang, G., Kaufman, A.J., Zhou, C., Yuan, X., Xiao, S., 2008. Pulsed oxidation and biological evolution in Ediacaran Doushantuo Formation. Proceedings of the National Academy of Sciences of the USA 105, 3197–3202.

McLean, D.M., 1978. A terminal Mesozoic "greenhouse": lessons from the past. Science 201, 401–406.

Melchin, M.J., Holmden, C., Williams, S.H., 2003. Correlation of graptolite biozones, chitinozoan biozones and carbon isotope curves through the Hirnantian. In: Albanesi, G.L., Beresi, S., Peralta, S.H. (Eds.), Ordovician from the AndesINSUGEO, Serie Correlación Geológica, vol. 17, pp. 101–104.

Melchin, M.J., Holmden, C.E., 2006. Carbon isotope chemostratigraphy in Arctic Canada: sea-level forcing of carbonate platform weathering and implications for Hirnantian global correlation. Palaeogeography, Palaeoclimatology, Palaeoecology 234, 186–200.

Melezhik, V.A., Fallick, A.E., 1996. A widespread positive $\delta^{13}C_{carb}$ anomaly at around 2.33–2.06 Ga on the Fennoscandian Shield: a paradox? Terra Nova 8, 141–157.

Melezhik, V.A., Fallick, A.E., Clark, T., 1997. Two billion year old isotopically heavy carbon: evidence from the Labrador Trough, Canada. Canadian Journal of Earth Sciences 34, 271–287.

Melezhik, V.A., Fallick, A.E., Medvedev, P.V., Marakarikhin, V.V., 1999. Extreme $^{13}C_{carb}$ enrichment in ca. 2.0 Ga magnesite–stromatolite–dolomite-"redbeds" association in a global context: a case for the world-wide signal enhanced by a local environment. Earth Science Reviews 48, 71–120.

Melezhik, V.A., Gorokov, I.M., Kuznetsov, A.B., Fallick, A.E., 2001. Chemostratigraphy of Neoproterozoic carbonates: implications for "blind dating". Terra Nova 13, 1–11.

Melezhik, V.A., Fallick, A.E., Pokrovsky, B.G., 2005a. Enigmatic nature of thick sedimentary carbonates depleted in ^{13}C beyond the canonical mantle value: the challenges to our understanding of the terrestrial carbon cycle. Precambrian Research 137, 131–165.

Melezhik, V.A., Roberts, D., Fallick, A.E., Gorokhov, I.M., Kusnetzov, A.B., 2005b. Geochemical preservation potential of high-grade calcite marble versus dolomite marble: implication for isotope chemostratigraphy. Chemical Geology 216, 203–224.

Melezhik, V.A., Huhma, H., Fallick, A.E., Whitehouse, M.J., 2007. Temporal constraints on the Palaeoproterozoic Lomagundi-Jatuli carbon isotope event. Geology 35, 655–658.

Mirota, M.D., Veizer, J., 1994. Geochemistry of precambrian Carbonates: IV. Aphebian Albanel formations, Quebec, Canada. Geochimica et Cosmochimica Acta 58, 1735–1745.

Miller, K.G., Fairbanks, R.G., Mountain, G.S., 1987. Tertiary oxygen isotope synthesis, sea level history, and continental margin erosion. Paleoceanography 2, 1–21.

Miller, K.G., Kominz, M.A., Browning, J.V., Wright, J.D., Mountain, G.S., Katz, M.E., Sugarman, P.J., Cramer, B.S., Christie-Blick, N., Pekar, S.F., 2005. The Phanerozoic record of global sea-level change. Science 310, 293–298.

Moczydlowska, M., 2008. The Ediacaran microbiota and the survival of Snowball Earth conditions. Precambrian Research 167, 1–15.

Montañez, I.P., Osleger, D.A., Banner, J., Mack, L.E., Musgrove, M., 2000. Evolution of the Sr and C isotope composition of Cambrian Oceans. GSA Today 10, 1–7.

Morrison, D., May 17, 2004. Did an Impact Trigger the Permian–Triassic Extinction? NASA Astrobiology Institute.

Munnecke, A., Calner, M., Harper, D.A.T., Servais, T., 2010. Ordovician and Silurian sea-water chemistry, sea level, and climate: a synopsis. Palaeogeography, Palaeoclimatology, Palaeoecology 296, 389.

Nagy, R.M., Porter, S.M., Dehler, C.M., Shen, Y., 2009. Biotic turnover driven by eutrophication before the Sturtian low-latitude glaciation. Nature Geoscience 2, 415–418.

Narbonne, G.M., 1998. The Ediacara biota: a terminal Neoproterozoic experiment in the evolution of life. GSA Today 8, 126.

Narbonne, G.M., Myrow, P.M., Landing, E., Anderson, M.M., 1987. A candidate stratotype for the Precambrian–Cambrian boundary, Fortune Head, Burin Peninsula, southeastern Newfoundland. Canadian Journal of Earth Sciences 24, 1277–1293.

Narbonne, G.M., Kaufman, A.J., Knoll, A.H., 1994. Integrated carbon isotope and biostratigraphy of the upper Windermere group, MacKenzie Mountains, N.W. Territories, Canada. The Bulletin of the Geological Society of America 106, 1281–1292.

Nascimento, R.S.C., Sial, A.N., Pimentel, M.M., 2007. C- and Sr-isotope systematics applied to Neoproterozoic marbles of the Seridó Belt, northeastern Brazil. Chemical Geology 237, 209–228.

Nascimento-Silva, V.M., Sial, A.N., Ferreira, V.P., Neumann, V.H., Barbosa, J.A., Pimentel, M.M., Lacerda, L.D., 2011. Cretaceous–Paleogene transition at the Paraíba Basin, northeastern, Brazil: carbon-isotope and mercury subsurface stratigraphies. Journal of South American Earth Sciences 32, 379–392.

Nascimento-Silva, M.V., Sial, A.N., Ferreira, V.P., Barbosa, J.A., Neumann, V.H., Pimentel, M.M., Lacerda, L.D., 2013. Carbon Isotopes, rare-earth elements and mercury behavior of Maastrichtian–Danian carbonate succession of the Paraíba Basin, Northeastern Brazil. In: Bojar, A.V., Melinte-Dobrinescu, M.C., Smit, J. (Eds.), Isotopic Studies in Cretaceous Research, vol. 382. Geological Society, London, Special Publications, pp. 85–104.

Ogg, J.G., Ogg, G., Godstein, F.M., 2008. The Concise Geologic Time Scale. Cambridge University Press, Cambridge. 177 p.

Ohmoto, H., Watanabe, Y., Ikemi, H., Poulson, S.R., Taylor, B.E., 2006. Sulphur isotope evidence for an oxic Archaean atmosphere. Nature 442, 908–911.

Orth, C.J., Gilmore, J.S., Quintana, L.R., Sheehan, P.M., 1986. Terminal Ordovician extinction: geochemical analysis of the Ordovician/Silurian boundary, Anticosti island, Quebec. Geology 13, 433–436.

Papineau, D., Mojzsis, S.J., Karhu, J.A., Marty, B., 2005. Nitrogen isotopic composition of ammoniated phyllosilicates: case studies from Precambrian metamorphosed sedimentary rocks. Chemical Geology 216, 37–58.

Paytan, A., Kastner, M., Campbell, D., Thiemens, M.H., 1998. Sulfur isotopic composition of cenozoic seawater sulfate. Science 282, 1459–1462.

Pavlov, A.A., Kasting, J.F., Brown, L.L., 2000. Greenhouse warming by CH_4 in the atmosphere of early earth. Journal of Geophysical Research 105, 11981–11990.

Pelechaty, S., Kaufman, A.J., Grotzinger, J.P., 1996. Evaluation of $\delta^{13}C$ isotope stratigraphy for intrabasinal correlation: data from Vendian strata of the Olenek uplift and Kharaulakh Mountains, Siberian platform, Russia. The Bulletin of the Geological Society of America 108, 992–1003.

Peltier, W.R., Liu, Y., Crowley, J.W., 2007. Snowball Earth prevention by dissolved organic carbon remineralization. Nature 450, 813–818.

Peng, S., Babcock, L., 2008. Cambrian period. In: Ogg, J.G., Ogg, G., Gradstein, F.M. (Eds.), The Concise Geologic Time Scale. Cambridge University Press, Cambridge, pp. 37–46.

Poulton, S.W., Canfield, D.E., 2005. Development of a sequential extraction procedure for iron: implications for iron partitioning in continentally derived particulates. Chemical Geology 214, 209–221.

Praekelt, H.E., Germs, G.J.B., Kennedy, J.H., 2008. A distinct unconformity in the Cango Caves Group of the Neoproterozoic to early Paleozoic Saldania Belt in South Africa: its regional significance. South African Journal of Geology 111, 357–368.

Reddy, S.M., Evans, D.A.D., 2009. Palaeoproterozoic supercontinents and global evolution: correlations from core to atmosphere. Geological Society of London Special Publication 323, 1–26.

Renne, P.R., Black, M.T., Zichao, Z., Richards, M.A., Basu, A.R., 1995. Synchrony and causal relations between Permian–Triassic boundary crises and Siberian flood Volcanism. Science 269, 1413–1416.

Riccardi, A.L., 2007. Carbonate-associated sulfate: assessment of and use as an isotopic proxy for global sulfur cycling during end-Permian mass extinction (Ph.D. dissertation), Department of Geosciences, The Pennsylvania State University, 128 p.

Ripperdan, R., 2002. The HERB event: end of the Cambrian carbon cycle paradigm? Abstracts with Programs. Geological Society of America Annual Meeting, vol. 34, No. (6), Denver, Colorado, 413 p.

Ripperdan, R.L., Magaritz, M., Nicoll, R.S., Shergold, J.H., 1992. Simultaneous changes in carbon, sea level, and conodont biozones within Cambrian–Ordovician boundary interval at Black Mountain, Australia. Geology 20, 1039–1042.

Ripperdan, R.L., Miller, J., 1995. Carbon isotope ratios from the Cambrian–Ordovician boundary section at Lawson Cove, Ibex area, Utah. In: Cooper, J.D., Droser, M.L., Finney, S.C. (Eds.), Ordovician OdysseyPacific Section SEPM, vol. 77, pp. 129–132.

Ripperdan, R.L., Cooper, J.D., Finney, S.C., 1998. High resolution $\delta^{13}C$ and lithostratigraphic profiles Copenhagen Canyon, Nevada: clues to the behavior of the ocean carbon cycle during the Late Ordovician global crisis. Mineralogical Magazine 62, 1279–1280.

Robert, F., Chaussidon, M., 2006. A palaeotemperature curve for the Precambrian oceans based on silicon isotopes in cherts. Nature 443, 969–972.

Rosing, M.T., Frei, R., 2004. U-rich Archaean sea-floor sediments from Greenland-indications of >3700 Ma oxygenic photosynthesis. Earth and Planetary Science Letters 217, 237–244.

Saltzman, M.R., 2005. Phosphorus, nitrogen, and the redox evolution of the Paleozoic oceans. Geology 33, 573–576.

Saltzman, M.R., Thomas, E., 2012. Carbon isotope stratigraphy. In: Gradstein, F.M., Ogg, J.G., Schmitz, M., Ogg, G. (Eds.), The Geologic Time Scale. Elsevier. http://dx.doi.org/10.1016/B978-0-444-59425-9.00011-1.

Saltzman, M.R., Young, S.A., 2005. Long-lived glaciation in the Late Ordovician? Isotopic and sequence-stratigraphic evidence from western Laurentia. Geology 33, 109–112.

Saltzman, M.R., Runnegar, B., Lohmann, K.C., 1998. Carbon isotope stratigraphy of the Upper Cambrian (Steptoean Stage) sequence of the eastern Great basin: record of a global oceanographic event. Geological Society of America Bulletin 110, 285–297.

Saltzman, M.R., Ripperdan, R.L., Brasier, M.D., Lohmann, K.G., Robison, R.A., Chang, W.T., Pemf, S., Ergaliev, E.K., Runnegar, B., 2000. A global carbon isotope excursion (SPICE) during the Late Cambrian: relation to trilobite extinctions, organic-matter burial and sea level. Palaeogeography, Palaeoclimatology, Palaeoecology 162, 211–223.

Saltzman, M.R., Bergström, W.H., Huff, W.D., Kolata, D.R., 2003. Conodonts and graptolite Biostratigraphy and the Ordovician (Early Chatfieldian, Middle Caradocian) $\delta^{13}C$ excursion in North America and Baltoscandia: implications for the interpretation of the relations between the Millbrig and Kinnekulle K-bentonites. In: Albanesi, G.L., Beresi, M.S., Peralta, S.H. (Eds.), Ordovician from the AndesINSUGEO, Serie Correlación Geológica, vol. 17, pp. 137–142.

Saltzman, M.R., Cowan, C.A., Runkel, A.C., Runnegar, B., Stewart, M.C., Palmer, A.R., 2004. The Late Cambrian SPICE ($\delta^{13}C$) event and the Sauk II-Sauk III regression: new evidence from Laurentian basins in Utah, Iowa and Newfoundland. Journal of Sedimentary Research 74, 366–377.

Sanei, H., Grassby, S.E., Beauchamp, B., 2012. Latest Permian mercury anomalies. Geology 40, 63–66.

Sansjofre, P., Ader, M., Trindade, R.I.F., Elie, M., Lyons, J., Cartigny, P., Nogueira, A.C.R., 2011. A carbon isotope challenge to Snowball Earth. Nature 478, 93–97.

Santos, G.M., Cordeiro, R.C., Silva Filho, E.V., Turcq, B., Lacerda, L.D., Fifield, L.K., Gomes, P.R.S., Hauscaden, P.A., Sifeddine, A., Albuquerque, A.L.S., 2001. Chronology of the atmospheric mercury in Lagoa da Pata Basin, Upper Rio Negro of Brazilian Amazon. Radiocarbon 43, 801–808.

Santos, R.V., Oliveira, C.G., Sousa, V.H.V., Carvalho, M.J., Andrade, T.V., Azevedo, Souza, H.G., 2008. Correlação isotópica baseada em isótopos de carbono entre os greenstone belts de Goiás. In: 44° Congresso Brasileiro de Geologia. Sociedade Brasileira de Geologia, Curitiba (CD-ROM).

Sawaki, Y., Ohno, T., Tahata, M., Komiya, T., Hirata, T., Maruyama, S., Windley, B.F., Han, J., Shud, D., Li, Y., 2010. The ediacaran radiogenic Sr isotope excursion in the Doushantuo formation in the three Gorges area, South China. Precambrian Research 176, 46–64.

Saylor, B.Z., Kaufman, A.J., Grotzinger, J.P., Urban, F., 1998. A composite reference section for Terminal Proterozoic strata of Southern Namibia. Journal of Sedimentary Research 68, 1223–1235.

Scheffler, K., Hoernes, S., Schwark, L., 2003. Global changes during Carboniferous–Permian glaciation of Gondwana: linking polar and equatorial climate evolution by geochemical proxies. Geology 31, 505–608.

Schidlowski, M., Eichmann, R., Fiebiger, W., 1976a. Isotopic fractionation between organic carbon and carbonate carbon in Precambrian banded ironstones series from Brazil. Neues Jarhburch fur Mineralogie, Monatschefte 8, 344–353.

Schidlowski, M., Eichmann, R., Junge, C.E., 1976b. Carbon isotope geochemistry of the Precambrian Lomagundi carbonate province, Rhodesia. Geochimica et Cosmochimica Acta 40, 449–455.

Schidlowski, M., Hayes, J.M., Kaplan, I.R., 1983. Isotopic inferences of ancient biochemistries: carbon, sulfur, hydrogen and nitrogen. In: Schopf, J.W. (Ed.), Earth's Earliest Biosphere—Its Origin and Evolution. Princeton University Press, Princeton, New Jersey, pp. 149–186.

Schlanger, S.O., Jenkyns, H.C., 1976. Cretaceous oceanic anoxic events: causes and consequences. Geologie en Mijnbouw 55, 179–184.

Schmitz, B., Bergström, S., Xiaofeng, W., 2010. The middle Darriwilian (Ordovician) $\delta^{13}C$ excursion (MDICE) discovered in the Yangtze Platform succession in China: implications of its first recorded occurrences outside Baltoscandia. Journal of the Geological Society 167, 249–259.

Schopf, J.W., Kudryavtsev, A.B., 2012. Biogenicity of Earth's earliest fossils: a resolution of the controversy. Gondwana Research 22, 761–771.

Schröder, S., Bekker, A., Beukes, N.J., Strauss, H., van Niekerk, H.S., 2008. Rise in seawater sulphate concentration associated with the Paleoproterozoic positive carbon isotope excursion: evidence from sulphate evaporites in the ~2.2–2.1 Gyr shallow-marine Lucknow Formation, South Africa. Terra Nova 20, 108–117.

Schrag, D.P., Berner, R.A., Hoffman, P.F., Halverson, G.P., 2002. On the initiation of a Snowball Earth. Geochemistry, Geophysics, Geosystems 3 (6).

Schulte, P., 40 additional authors, 2010. The chicxulub asteroid impact and mass extinction at the Cretaceous–Paleogene boundary. Science 327, 1214–1218.

Scott, C., Lyons, T.W., Bekker, A., Shen, Y., Pultron, S.W., Chu, X., Anbar, A.D., 2008. Tracing the stepwise oxygenation of the Proterozoic ocean. Nature 452, 456–459.

Shen, Y., Schidlowski, M., 2000. New C isotope stratigraphy from Southwest China: implications for the placement of the Precambrian–Cambrian boundary on the Yangtze Platform and global correlations. Geology 28, 623–626.

Shen, Y., Canfield, D.E., Knoll, A.H., 2002. Middle proterozoic ocean chemistry: evidence from the McArthur Basin, northern Australia. American Journal of Science 302, 81–109.

Sheth, H.C., 2005. Were the Deccan flood basalts derived in part from ancient oceanic crust within the Indian continental lithosphere? Gondwana Research 8, 109–127.

Shields, G.A., 2007a. A normalised seawater strontium isotope curve and the Neoproterozoic-Cambrian chemical weathering event. Earth Discussions 2, 69–84.

Shields, G.A., 2007b. The marine carbonate and chert isotope records and their implications for tectonics, life and climate on the early Earth. In: van Kranendonk, M.J., Smithies, R.H., Bennett, V.C. (Eds.), Earth's Oldest Rocks. Developments in Precambrian Geology, vol. 15. Elsevier, pp. 971–983.

Shields-Zhou, G., Och, L., 2011. The case for a Neoproterozoic Oxygenation Event: geochemical evidence and biological consequences. GSA Today 21, 4–11.

Sial, A.N., Ferreira, V.P., Almeida, A.R., Romano, A.W., Parente, C.V., Costa, M.L., Santos, V.H., 2000. Carbon isotope fluctuations in Precambrian carbonate sequences of several localities in Brazil. Anais da Academia Brasileira de Ciencias 72, 539–558.

Sial, A.N., Karhu, J., Ferreira, V.P., 2010a. Insights from isotope stratigraphy. Preface to the special issue on precambrian isotope stratigraphy. Precambrian Research 182 (4), v–viii.

Sial, A.N., Gaucher, C., Silva Filho, M.A., Ferreira, V.P., Pimentel, M.M., Lacerda, L.D., Silva Filho, E.V., Cezario, W., 2010b. C-, Sr-isotope and Hg chemostratigraphy of Neoproterozoic cap carbonates of the Sergipano Belt, Northeastern Brazil. In: Karhu, J., Sial, A.N., Ferreira, V.P. (Eds.), Precambrian Isotope Stratigraphy Precambrian Research, vol. 182, pp. 351–372.

Sial, A.N., Peralta, S., Ferreira, V.P., Toselli, A.J., Aceñolaza, F.G., Parada, M.A., Gaucher, C., Alonso, R.N., Pimentel, M.M., 2008. Upper Cambrian carbonate sequences of the Argentine Precordillera and the Steptoean C-isotope positive excursion (SPICE). Gondwana Research 13, 437–452.

Sial, A.N., Lacerda, L.D., Ferreira, V.P., Frei, R., Marquillas, R.A., Barbosa, J.A., Gaucher, C., Windmöller, C.C., Pereira, N.S., 2013a. Mercury as a proxy for volcanic activity during extreme environmental turnover: the Cretaceous–Paleogene transition. Palaeogeography, Palaeoclimatology, Palaeoecology 387, 153–164.

Sial, A.N., Peralta, S., Gaucher, C., Toselli, A.J., Ferreira, V.P., Frei, R., Parada, M.A., Pimentel, M.M., Pereira, N.S., 2013b. High-resolution stable isotope stratigraphy of the Upper Cambrian and Ordovician of the Argentine Precordillera: carbon-isotope excursions and correlations. Gondwana Research 24, 330–348.

Sial, A.N., Chen, J.-B., Lacerda, L.D., Peralta, S., Gaucher, C., Frei, R., Cirilli, S., Ferreira, V.P., Marquillas, R.A., Barbosa, J.A., Pereira, N.S., Belmino, I.K.C., 2014. High-resolution Hg Chemostratigraphy: a contribution to the distinction of chemical fingerprints of the Deccan volcanism and Cretaceous–Paleogene Boundary impact event. Palaeogeography, Palaeoclimatology, Palaeoecology 414, 98–115.

Siebert, C., Nägler, T.F., von Blanckenburg, F., Kramers, J.D., 2003. Molybdenum isotope records as a potential new proxy for paleoceanography. Earth Planetary Science Letters 211, 159–171.

Silva Tamayo, J.C., Nägler, T.F., Villa, I., Nogueira, A., Kyser, K., Riccomini, C., Sial, A.N., Narbonne, G.M., James, N.P., 2010a. Global perturbation of the marine Ca-isotopic composition in aftermath of the Marinoan global glaciations. Precambrian Research 182, 373–381.

Silva Tamayo, J.C., Nägler, T., Villa, I.M., Kyser, K., Vieira, L.C., Sial, A.N., Narbonne, G.M., James, N.P., 2010b. Global Ca isotope variations in Post–Sturtian carbonate successions. Terra Nova 22, 188–194.

Silva Tamayo, J.C., Sial, A.N., Nagler, T.F., Villa, I.M., Silva Filho, M.A., 2007. Ca-versus C-isotope stratigraphy of two post-Snowball Earth cap-carbonate successions in NE Brazil. In: XI Congresso Brasileiro de Geoquímica, Ataibaia, São Paulo Extended Abstracts (CD-ROM).

Soufiane, A., Achab, A., 2000. Upper ordovician and lower silurian chitinozoans from central Nevada and arctic Canada. Review of Palaeobotany and Palynology 113, 165–187.

Sreenivas, B., Das Sharma, S., Kumar, B., Patil, D.J., Roy, A.B., Srinivasan, R., 2001. Positive $\delta^{13}C$ excursion in carbonate and organic fractions from the Paleoproterozoic Aravalli Supergroup, Northwestern India. Precambrian Research 106, 277–290.

Staubwasser, M., Von Blackenburg, F., Shoenberg, R., 2006. Iron isotopes in the early marine diagenetic iron cycle. Geology 34, 629–632.

Strauss, H., 1997. The isotopic composition of sedimentary sulfur through time. Paleogeography, Palaeoclimatology, Palaeoecology 132, 97–118.

Stüben, D., Kramar, U., Berner, Z., Eckhardt, J., Stinnesbeck, W., Keller, G., Adatte, T., Heide, K., 2002. Two Anomalies of Platinum Group Elements above the Cretaceous–Tertiary Boundary at Beloc, Haiti: Geochemical Context and Consequences for the Impact Scenario, vol. 356, Geological Society of America Special paper. 163–188.

Stüben, D., Kramar, U., Harting, M., Stinnesbeck, W., Keller, G., 2005. High-resolution geochemical record of Cretaceous–Tertiary boundary sections in Mexico: new constraints on the K/T and Chicxulub events. Geochimica et Cosmochimica Acta 69, 2559–2579.

Swart, P.K., 2008. Global synchronous changes in the carbon isotope composition of carbonate sediments unrelated to changes in the global carbon cycle. Proceedings of the National Academy of Sciences of the USA 105, 13741–13745.

Tahata, M., Ueno, Y., Ishikawa, T., Sawaki, Y., Murakami, K., Han, J., Shu, D., Li, Y., Guo, J., Yoshida, N., Komiya, T., 2013. Carbon and oxygen isotope chemostratigraphies of the Yangtze platform, South China: Decoding temperature and environmental changes through the Ediacaran. Gondwana Research 23 (1), 333–353.

Takashima, R., Kawabe, F., Nishi, H., Moriya, K., Wani, R., Ando, H., 2004. Geology and stratigraphy of forearc basin sediments in Hokkaido, Japan: Cretaceous environmental events on the Northwest Pacific margin. Cretaceous Research 25, 365–390. http://dx.doi.org/10.1016/j.gr.2012.04.005.

Takashima, R., Nishi, H., Huber, B.T., Leckie, M., 2006. Greenhouse world and the mesozoic ocean. Oceanography 19, 82–92.

Tarduno, J.A., Brinkman, D.B., Renne, P.R., Cottrell, R.D., Scher, H., Castillo, P., 1998. Evidence for extreme climatic warmth from the Late Cretaceous Arctic vertebrates. Science 282, 2241–2244.

Tassinari, C.C.G., Jost, H., Santos, J.C., Nutman, A.P., Bennell, M.R., 2006. Pb and Nd isotope signatures and SHRIMP U–Pb geochronological evidence of Paleoproterozoic age for Mina III gold mineralization, Crixás District, Central Brazil. In: 5th South American Symposium on Isotope Geology, Punta Del Este, Uruguay, pp. 527–529 Short Papers Volume.

Tewari, V.C., Sial, A.N., 2007. Neoproterozoic–Early cambrian isotopic variation and chemostratigraphy of the Lesser Himalaya, India, Eastern Gondwana. Chemical Geology 237, 82–106.

Turco, R.P., Toon, O.B., Ackerman, T.P., Pollack, J.B., Sagan, C., 1983. Nuclear Winter: global consequences of multiple nuclear explosions. Science 222, 1283–1292.

Tyrrell, T., 1999. The relative influences of nitrogen and phosphorus on oceanic primary production. Nature 400, 525–531.

Veizer, J., 1989. Strontium isotopes in seawater through time. Annual Review of Earth and Planetary Sciences 17, 141–167.

Veizer, J., Compston, W., 1976. $^{87}Sr/^{86}Sr$ in Precambrian carbonates as an index of crustal evolution. Geochimica et Cosmochimica Acta 40, 905–914.

Veizer, J., Hoefs, J., 1976. The nature of O^{18}/O^{16} and C^{13}/C^{12} secular trends in sedimentary carbonate rocks. Geochimica et Cosmochimica Acta 40, 1387–1395.

Veizer, J., Holser, W.T., Wilgus, C.K., 1980. Correlation of $^{13}C/^{12}C$ and $^{34}S/^{32}S$ secular variations. Geochimica et Cosmochimica Acta 44, 579–588.

Veizer, J., Compston, W., Clauer, N., Schidlowski, M., 1983. $^{87}Sr/^{86}Sr$ in late Proterozoic carbonates: evidence for a "mantle"event at approximately 900 Ma ago. Geochimica et Cosmochimica Acta 47, 295–302.

Veizer, J., Ala, D., Azmy, K., Bruckschen, P., Buhl, P., Bruhn, F., Carden, G.A.F., Diener, A., Ebneth, S., Godderis, Y., Jasper, T., Korte, C., Pawellek, F., Podlaha, O.G., Strauss, H., 1999. $^{87}Sr/^{86}Sr$, $\delta^{13}C$ and $\delta^{18}O$ evolution of Phanerozoic seawater. Chemical Geology 161, 59–88.

Voegelin, A.R., Nägler, T.F., Beukes, N.J., Lacassie, J.P., 2010. Molybdenum isotopes in late Archean carbonate rocks: implications for early Earth oxygenation. Precambrian Research 182, 70–82.

Walochnik, J., Harzhauser, M., Aspöck, H., 2010. Climate change as a driving force for evolution. Nova Acta Leopoldina NF 111 (38), 21–31.

Walter, M.R., Veevers, J.J., Calver, C.R., Gorjan, P., Hill, A.C., 2000. Dating the 840–544 Ma Neoproterozoic interval by isotopes of strontium, carbon and sulfur in seawater, and some interpretative models. Precambrian Research 100, 371–433.

Wille, M., Kramers, J.D., Nägler, T.F., Beukes, N.J., Schröder, S., Meisel, Th, Lacassie, J.P., Voegelin, A.R., 2007. Evidence for a gradual rise of oxygen between 2.6 and 2.5 Ga from Mo isotopes and Re-PGE signatures in shales. Geochimica et Cosmochimica Acta 71, 2417–2435.

Wille, M., Nebel, O., Kranendonk, M.J.V., Schoenberg, R., Kleinhanns, I.C., Ellwood, M.J., 2013. Mo–Cr isotope evidence for a reducing Archean atmosphere in 3.46–2.76 Ga black shales from the Pilbara, Western Australia. Chemical Geology 340, 68–76.

Williams, H., Hoffman, P.F., Lewry, J.F., Monger, J.W.H., Rivers, T., 1991. Anatomy of North America: thematic geologic portrayals of the continent. Tectonophysics 187, 117–134.

Xiao, S., Knoll, A.H., Kaufman, A.J., Yin, L., Zhang, Y., 1997. Neoproterozoic fossils in Mesoproterozoic rocks? Chemostratigraphic resolution of a biostratigraphic conundrum from the North China Platform. Precambrian Research 84, 197–220.

Young, G.M., 1991. Stratigraphy, sedimentology and tectonic setting of the Huronian Supergroup. Geological Association of Canada, Mineralogical Association of Canada, Society of Economic Geologists. In: Joint Annual Meeting, Toronto 91, Field Trip B5: Guidebook, p. 34.

Young, G.M., 2013. Secular changes at the Earth´s surface; evidence from paloesols, some sedimentary rocks, and plaeoclimatic pertubations of the Proterozoic Eon. Gondwana Research 24 (2), 453–467.

Young, S.A., Saltzman, M.R., Bergström, S.M., Leslie, S.A., Xu, C., 2008. Paired $\delta^{13}C_{carb}$ and $\delta^{13}C_{org}$ records of Upper Ordovician (Sandbian–Katian) carbonates in North America and China: implications for paleoceanographic change. Palaeogeography, Palaeoclimatology, Palaeoecology 270, 166–178.

Zachos, J., Pagani, M., Sloan, L., Thomas, E., Billups, K., 2001. Trends, rhythms, and aberrations in global climate 65 Ma to present. Science 292, 686–693.

Zegeye, A., 11 additional authors, 2012. Green rust formation controls nutrient availability in a ferruginous water column. Geology 40, 599–602.

Zerkle, A.L., Claire, M.W., Domagal-Goldman, S.D., Farquhar, J., Poulton, S.W., 2012. A bistable organic-rich atmosphere on the Neoarchaean Earth. Nature Geoscience 5, 359–363.

Zhao, C., Li, R., Zhou, J., 1997. Sedimentology and Petroleum Geology of the Meso- and Neo-Proterozoic in North China. Geological Publishing House, Beijing, China.

Zhu, M.Y., Babcock, L.E., Peng, S.C., 2006. Advances in Cambrian stratigraphy and paleontology: integrating correlation techniques, paleobiology, taphonomy and paleoenvironmental reconstruction. Palaeoworld 15, 217–222.

CHAPTER

STABLE ISOTOPES: TOOLS FOR UNDERSTANDING PAST CLIMATIC CONDITIONS AND THEIR APPLICATIONS IN CHEMOSTRATIGRAPHY

3

Manish Tiwari[1], Ashutosh K. Singh[2], Devesh K. Sinha[2]

[1]*National Centre for Antarctic & Ocean Research, Vasco-da-Gama, Goa, India;* [2]*Department of Geology, Centre of Advanced Studies, University of Delhi, Delhi, India*

3.1 INTRODUCTION

The climate of the earth can change drastically and has changed so in the past. This was first proposed by the European geologists in the late eighteenth and early nineteenth century including the Scottish geologist James Hutton who found the signature of glacial activity in presently unglaciated regions. The Swiss geologist Louis Agassiz and his colleagues proposed the theory of "Ice Age" in 1837 to explain the observations that hypothesized a succession of glaciation. Meanwhile, the French mathematician and physicist Joseph Fourier postulated in 1824 that the earth should have been much colder but for an insulating effect of its atmosphere (Fourier, 1824), which is widely regarded as the first reference toward the "greenhouse effect" of the atmosphere. The English physicist John Tyndall in mid-nineteenth century, intrigued by the Fourier's discovery and a need to explain the changing climate of the earth, found that water vapor, methane, and carbon dioxide strongly block infrared radiation and act as greenhouse gases (Tyndall, 1861). The celebrated Swedish scientist Svante August Arrhenius in 1896 quantified the effect of CO_2 increase on temperature rise and was first to predict that human produced CO_2 can cause warming (Arrhenius, 1896). Among these developments, the self educated Scottish scientist James Croll postulated in 1870s the link between the ice ages and the changes in earth's orbit. The Serbian climatologist, Milutin Milankovitch in 1920s, built upon the work of James Croll, and established that slight changes in earth's orbital parameters viz eccentricity, obliquity, and precession can cause the climate of the earth to change in a cyclic pattern (Milankovitch, 1930), which are known as Milankovitch Cycles. The theory, although mathematically elegant, was not accepted as it did not match with the timing of four ice ages (based on the records of glacial activities) widely recognized by geologists at that time. During that period, several theories including greenhouse effect, solar irradiance variability, orbital changes and so on were proposed but were not believed as they were not able to match the observations.

Chemostratigraphy. http://dx.doi.org/10.1016/B978-0-12-419968-2.00003-0

In such a puzzling scenario, chemostratigraphic techniques emerged in the forefront and helped to explain the past climatic changes. One of such advancements was the development of the "Radiocarbon Dating" method by Willard Libby in 1949 for which he was awarded the Nobel Prize in Chemistry. This revolutionary technique allowed the age determination of the organic matter till ~40,000 years using the radioactive isotope of carbon, which helped to precisely establish the chronology of the past climate change. But still, most of the records were land based and hence were disturbed due to erosion and other geological factors. In the year 1947, Harold Urey had found that the ratio of oxygen isotopes in carbonates depends on the ambient temperature using the newly developed technique of the isotope ratio mass spectrometry (Urey, 1947). This led the way for establishing the paleotemperature scale using marine carbonates (McCrea, 1950; Epstein et al., 1953) that avoided the limitations of the terrestrial records. Cesare Emiliani used oxygen isotope ratio of the calcium carbonate shells of microscopic organisms called foraminifera from deep-sea sediment cores to reconstruct the records of temperature changes for the past 300,000 years (Emiliani, 1955). These precisely dated, high-resolution records of the cold and warm periods stretching back continuously for thousands of years were instrumental in understanding the causes of climate change. Emiliani was able to correlate his record of climate change with the varying insolation according to Milankovitch's astronomical calculations for which he had to factor a lag of 5000 years considering the response time of the ice sheets. It was later proved that the oxygen isotopes of the foraminiferal shells not only reflected temperature but also the removal of large amounts of isotopically lighter water from the ocean and its storage over land during glacial periods—the so-called ice-volume effect (Shackleton, 1967). Still, the work of Emiliani and others provided a substantial credence to the orbital theory of climate change and to the fact that a little change in the forcing factors can have a major impact on the climate system of the earth (Broecker et al., 1968).

The next major step, when the chemostratigraphic techniques helped to decipher the climate change at much higher resolution than previously possible was the determination of isotopic ratios and trapped gases in the long ice cores obtained from Greenland and Antarctica. Willi Dansgaard and his team drilled several long cores in Greenland and measured oxygen isotope ratio that led to the discovery of the rapid climatic changes such as "Younger Dryas" when the earth cooled down abruptly within a few decades (Johnsen et al., 1992; Dansgaard et al., 1993; Alley, 2000). Such pattern of rapid climate change was also observed by Hans Oeschger in glacial-era lake deposits in Switzerland that came to be known as "Dansgaard-Oeschger events." Long ice cores from the Vostok Station from Antarctica revealed that the CO_2 also covaried with the temperature change implying a very strong link between them (Petit et al., 1999). Oceans are one of the largest reservoirs of the CO_2 that can exchange with the atmosphere on very short timescales. This led to the development of the hypothesis that reorganization of the thermohaline circulation can cause the abrupt climatic changes (e.g., Broecker and Denton, 1990). Although the magnitude of feedback from different components of the climate system was not clear, a consensus among the scientific community was established by 1980s that increasing anthropogenic CO_2 will result in adverse climatic change.

United Nations, in 1988, created an autonomous body—Intergovernmental Panel on Climate Change (IPCC)—that was entrusted with the task of producing reports that serve as the definitive word on the state of the climate change and its future impact. The first IPCC report was published in 1990 and since then it has produced periodic reports that serve a great purpose by putting before the general public the magnitude of anthropogenic climate change for which it was awarded the Nobel Peace Prize in 2007. The fifth and the most recent report of the IPCC shows that the earth's climate is warming rapidly with an estimated 0.85 °C increase in globally averaged combined land and ocean surface temperatures from 1880 to 2012 of which the anthropogenic factors are the major causes. This warming of the climate system is beginning to show an adverse effect on the sea level, ocean acidification, ice melting, water cycle, and so on (IPCC, 2013).

This discussion leads us to the fact that the earth's climate change consists of a diversity of events that have occurred in its various spheres, i.e., atmosphere, ocean circulation change (hydrosphere), organic evolution (biosphere), ice-volume changes related to glacial and interglacial stages (cryosphere), large scale movements of the continents (asthenosphere and lithosphere), convection currents and volcanism (mesosphere), and paleomagnetic reversals (barysphere). A comprehensive understanding of the "Earth System Science" reveals that none of these spheres work in isolation; the changes in one sphere can cause perturbations in other spheres. The interplay among various spheres of the earth can be seen in operation in modern times and have been found to have occurred in geological past as well. In order to study the geological past, the geoscientists make use of what is known as "proxy." The past events are preserved in the geological material and the geologists use various kinds of proxies to infer the various parameters of the geological past. This constitutes the science of "paleoclimatology," which is defined as the study of past climate and its variation in space and time during the period before the instrumental documentation began. Stable isotopes are one of such powerful proxies that help us to infer the past geological events that have occurred in the various spheres of the earth. Like all other proxies, the application of stable isotopes also has some limitations. Nevertheless, they have been found to be of immense use for stratigraphic, paleoceanographic, and paleoclimatic studies. We present here the basics of the stable isotope methodology along with case studies of their application including high resolution paleoceanographic and paleoclimatic studies. The purpose is to make a new researcher acquainted with the potential of stable isotopes for application in paleoclimatological studies. The discussion is in no way an exhaustive summary of the available literature and only basic facts largely agreed upon, based on published work, have been highlighted.

3.2 STABLE ISOTOPE SYSTEMATICS

Atoms of an element having the same number of protons but different number of neutrons are called *isotopes*, i.e., they possess same atomic number but different mass number. The word "isotopes" in Greek means "equal places" indicating that they occupy the same position in the periodic table by virtue of their same atomic numbers. Isotopes, which spontaneously disintegrate with a measurable half-life, are called *radioactive* while those that do not disintegrate or have extremely long half-lives (comparable to the age of the earth) are considered *stable*. Such isotopes have existed in their current form before the formation of the Earth and therefore are also known as the "primordial isotopes." An isotope becomes unstable or radioactive when the electrostatic coulomb repulsion of the positively charged proton within the nucleus results in its spontaneous disintegration. To make an isotope stable, the repulsive force of the protons needs to be counterbalanced by neutrons. Till around mass number 40, same number of neutrons is sufficient to balance the repulsion generated by protons, i.e., neutron to proton ration (N/Z) is 1. This is known as the "Symmetry Rule." With increasing numbers of protons, the repulsion increases, thus requiring more neutrons than protons causing the N/Z ratio to exceed unity. The abundance of stable isotopes is explained by the "Oddo–Harkins Rule," which states that the cosmic abundance of elements with even number of protons is much more than the adjacent elements with odd numbers of proton. It applies to the stability of isotopes as well. The elements with even number of protons and even number of neutrons have the maximum number of stable isotopes. Next in line are the elements with the combination of an even number of protons and odd number of neutrons followed by elements with odd number of protons and even number of neutrons. The elements with the least number of stable isotopes are those that possess an odd number of protons and odd number of neutrons (Hoefs, 2009). *Isotopologues* (shortened from "isotopic analogues") are molecules of the same compound that differ in isotopic composition and relative molecular mass, e.g., $H_2{}^{16}O$ and $H_2{}^{18}O$. Isotopomers (shortened from "isotopic isomers") differ only in the position of the isotopes and not in the composition, i.e., they possess same relative molecular mass, for e.g., $^{14}N^{15}N^{16}O$ and $^{15}N^{14}N^{16}O$.

The variations in the physical and chemical properties due to the presence of different isotopes (the so-called "Isotope Effect") arise due to very small, yet significant mass differences among them. For example, properties like density, melting point, boiling point, viscosity, etc., of the compounds with heavier isotopes are more than those composed of lighter isotopes. These differences are most significant for light elements (less than mass number 50) such as hydrogen, carbon, nitrogen, oxygen, etc., as the relative mass difference of their isotopes ($\Delta m/m$) is higher than heavier elements. In general, lighter isotopes are more mobile as the ratio of the velocity of the lighter molecule to that of the heavier molecule is inversely proportional to the square root of their masses. The mass of the constituent isotope also influences the vibrational motion of the molecule. The molecule with lighter isotope possesses a higher vibrational frequency, and hence higher vibrational energy than a molecule containing the heavier isotope. Therefore, lesser dissociation energy is required to break a bond formed by a lighter isotope than that formed by a heavier isotope. It implies that a molecule with lighter isotope will participate in a chemical process more readily than a molecule possessing heavier isotope. This leads to *isotopic fractionation.*

3.2.1 ISOTOPIC FRACTIONATION

Isotopic fractionation is defined as the relative partitioning of the heavier and lighter isotopes between two coexisting phases in a natural system. There are two types of isotopic fractionation processes, namely, "equilibrium" and "kinetic."

Equilibrium fractionation is a special case of chemical equilibrium reaction in which there is no net reaction but an exchange of isotopes takes place, which affects the equilibrium constant. In this case, isotopes can move to and fro and equilibrium is attained when the isotopic ratios do not change any more with time. For example, foraminifers (unicellular microorganism) secrete their calcitic shells in equilibrium with ambient seawater and the following isotopic exchange process takes place:

$$1/3CaC^{16}O_3 + H_2{}^{18}O \rightleftharpoons 1/3CaC^{18}O_3 + H_2{}^{16}O$$

The equilibrium constant "K" for this reaction can be written as:

$$K = \frac{[CaC^{18}O_3]^{1/3}\,[H_2{}^{16}O]}{[CaC^{16}O_3]^{1/3}\,[H_2{}^{18}O]}$$

$$\text{or, } K = \frac{([CaC^{18}O_3]\,/\,[CaC^{16}O_3])^{1/3}}{([H_2{}^{18}O]\,/\,[H_2{}^{16}O])}$$

Thus, equilibrium constant can be expressed as the ratio of $^{18}O/^{16}O$ in the carbonate phase to that in the water. This leads to the concept of *isotopic fractionation factor* "α" that represents the partitioning of isotopes between two phases. Fractionation factor is defined as the ratio of isotopes in one phase to the other coexisting phase. In the $CaCO_3$–H_2O system, the fractionation factor is defined as:

$$\alpha_{CaCO_3\text{-}H_2O} = \frac{\left(^{18}O/^{16}O\right)_{CaCO_3}}{\left(^{18}O/^{16}O\right)_{H_2O}}$$

The fractionation factor is related to the equilibrium constant by:

$$\alpha = K^{1/n}$$

where "n" is the number of atoms exchanged.

Kinetic fractionation is associated with incomplete and unidirectional processes and it affects the rate constant of the reaction. No isotopic equilibrium is attained in the case of kinetic fractionation. For example, in a reaction, where the k_L and k_H respectively depict different rate constants involving the lighter and the heavier isotopes. In such a case, the effect of isotopic substitution in the reactant can be expressed as the ratio of the rate constants, k_L/k_H. If this ratio is greater than 1, it means that the reactant with lighter isotope reacts more readily and is termed "normal." It is called "inverse" when the ratio is less than 1 signifying that the reactant with heavier isotope reacts more rapidly. Note that, if in an irreversible reaction, all the reactants are consumed to form a product, then no isotope effects and no fractionation will be observed in the final product.

The *isotopic fractionation*, ε, is defined as: $\varepsilon = \alpha - 1$. This is usually a small number, hence is multiplied by 10^3 and is then expressed in per mil (‰). The isotopic fractionation is temperature dependent (Urey, 1947). In this example, the $\alpha_{CaCO_3\text{-}H_2O}$ is 1.031 at 25 °C, i.e., the ε is $(1.031-1)10^3‰ = 31‰$. The $\alpha_{CaCO_3-H_2O}$ is 1.028 at 33.7 °C, i.e., the ε is 28‰. It indicates that the isotopic fractionation decreases with the increasing temperature. Theoretical studies demonstrate that α is inversely proportional to the square of the temperature (Hoefs, 2009).

Additionally, the nature of the chemical bonds also dictates the isotopic distribution. The bonds with higher ionic potential involving low atomic masses have higher vibrational frequencies and hence tend to accumulate heavier isotopes to reduce it. For example, the inorganically precipitated calcite is always enriched in the heavier isotope of the oxygen (^{18}O) as compared to the other minerals due to the presence of the small, highly charged C^{4+} ion to which the oxygen is bonded. Crystal structure also plays a role: the more orderly and closely packed structure tend to accumulate the heavier isotope. This effect is evident in the enrichment of ^{18}O in ice as compared to water (Beck and Münnich, 1988).

Rayleigh isotopic fractionation is an important type of fractional distillation or condensation under the *equilibrium* condition. It is used to calculate the changing isotopic ratios as the material is removed from one coexisting phase to the other during the evaporation–condensation processes. Considering an example of a cloud mass from which condensation (precipitation) is taking place the Rayleigh isotopic fractionation could be defined as: If, "R_o" is the original isotopic ratio of the vapor before the rain out, "R" is the new isotopic ratio of the remaining vapor, "α" is the vapor to liquid fractionation factor, and "f" is the fraction of the vapor remaining, i.e., ratio of the present amount to the initial total amount, then:

$$R = R_o(f)^{\alpha - 1}$$

In terms of δ-value (explained below):

$$\delta = \delta_o + \varepsilon \ln(f) \; \{\text{in ‰}\}$$

3.2.2 DELTA (δ) VALUE

It is very difficult to precisely measure the absolute abundance of isotopes owing to the very low abundance of heavier isotopes. Therefore, *isotope ratio* is determined, which is the ratio of the number of heavier isotopes to that of the lighter isotopes. The term "isotopic abundance" refers to the fraction or percentage of a particular isotope with respect to all other isotopes of that element in the same system (synonymous with the terms *isotope-amount fraction* or *atom fraction*). Isotope Ratio Mass Spectrometers (IRMS) are used for isotopic analysis, which are based on the principle of the separation of charged molecules and atoms under electromagnetic field according to their masses. The IRMS cannot measure the absolute isotopic ratio of a compound with sufficient precision. But it can measure the difference between the isotopic ratios of two compounds with very high precision and accuracy as both of them experience the same conditions during the analysis. Therefore, the isotopic abundances are reported in "δ" value, which is the relative difference of isotopic ratio of a sample from that of an international standard:

$$\delta A = \left[\left(\frac{R_A}{R_{st}}\right) - 1\right]$$

where, R_A and R_{st} are the ratios of the abundance of the less abundant (heavier) to more abundant (lighter) isotope (e.g., $^{18}O/^{16}O$) in the sample and standard, respectively. δ-value is a dimensionless quantity as it is the ratio of the two quantities of the same kind. It is multiplied by extraneous numerical factors like 10^3 and expressed in "per mil" (‰) units for the sake of readability and ease of comprehension. When it is multiplied by 10^4, it shall be expressed in "pptt" (parts per ten thousand) and shall not be expressed as "ε" that has been erroneously reported so in several places in literature earlier (see Coplen, 2011 for detailed discussion). While expressing isotopic ratios, the heavier isotope is kept in numerator so that the δ-value is positive. A positive "δ" means that the sample comprises more number of heavier isotopes than the standard or, in other words, the atomic weight of the particular element is more in the sample than the standard.

3.2.3 ISOTOPIC STANDARDS

To be used as a standard or reference material for isotopic measurements, the substance should be sufficiently stable, homogeneous, widely available, and its isotope ratio shall be near the mean of the natural variation. Any such substance prepared in an individual laboratory is called as a "lab standard" or "working standard" but it needs to be calibrated with respect to an international standard so that the δ-values of the samples from all over the world are finally reported with respect to a common scale. The *primary reference materials* (international standards) for a few of the light elements are Vienna Standard Mean Ocean Water (VSMOW) for $\delta^{18}O$ and $\delta^{2}H$, Vienna Pee Dee Belemnite (VPDB) for $\delta^{18}O$ and $\delta^{13}C$, air-N_2 for $\delta^{15}N$, Vienna Canyon Diablo Troilite (an iron meteorite) for $\delta^{34}S$, SRM 951 (Boric Acid; formerly, NBS-951) for $\delta^{11}B$, and LSVEC (a lithium carbonate) for $\delta^{7}Li$ among others. Reputed institutions like International Atomic Energy Agency (IAEA, Vienna), National Institute of Standards and Technology (NIST, USA), and Institute for Reference Materials and Measurements (Belgium) prepare and distribute such international standards and several other "certified isotopic reference materials" like VSMOW2, NBS-19, IAEA-N-1, etc., with an agreed upon δ-value, which are used to calibrate to the international standard's scale.

Two international standards viz PDB and SMOW are of historical importance in the development of the application of stable isotope techniques for climate related studies. For oxygen and hydrogen isotopic studies of water, the international standard SMOW holds special significance. SMOW is a hypothetical standard with $\delta^{18}O$ value close to the modern mean seawater value. The value of SMOW was defined by Craig (1961) with respect to an existing water standard NBS-1 (distilled Potomac River water) with $\delta^{18}O$ value of −7.94‰ versus SMOW. As SMOW does not actually exist as a real water sample so it cannot be used for calibrating laboratory measurements. Therefore a water sample with values identical to SMOW was distributed by IAEA that was called VSMOW where "V" stands for Vienna—the city where the IAEA is based. Prof. Harmon Craig of the University of California prepared it by mixing distilled ocean water (collected from the Pacific Ocean, 0° latitude 180° longitude, in July 1967) with small amounts of other waters to adjust the isotopic ratios to the required values (Gonfiantini, 1981). It has $\delta^{18}O$ and $\delta^{2}H$ value of 0‰ with respect to SMOW. This scale has been used for oxygen isotope analyses involving waters, silicates, phosphates, high temperature carbonates, etc. As VSMOW was getting exhausted, it was replaced by VSMOW2, which was prepared by the IAEA Isotope Hydrology Laboratory in 2006 with values similar to the original VSMOW. The other water standards currently in use are Standard Light Antarctic Precipitation 2 (SLAP2) and Greenland Ice Sheet Precipitation (GISP). The $\delta^{18}O$ and $\delta^{2}H$ value of SLAP2 are −55.50‰ and −427.5‰, respectively, with respect to VSMOW (Gröning, 2004). The $\delta^{18}O$ and $\delta^{2}H$ value of GISP are −24.76‰ and −189.5‰, respectively, with respect to VSMOW (Gröning, 2004). All the δ-values have to be reported on the VSMOW/SLAP scale, using VSMOW2 and SLAP2 as the calibration materials.

The PDB standard scale is used in the low temperature carbonate studies. The original PDB standard was a carbonate obtained from the fossilized rostrum (cigar shaped internal shell) of *Belemnitella americana*, a belemnite (an extinct cephalopod) from the Pee Dee Formation of the Cretaceous Period in South Carolina, USA. It was the laboratory working standard used by the Urey's group at the University of Chicago in the 1950s for developing the initial paleotemperature scale (McCrea, 1950). Over the decades, the original PDB standard has been exhausted and is no longer available. To overcome this problem, IAEA introduced a hypothetical standard VPDB with values nearly identical to PDB and is defined by its relationship to the carbonate reference material NBS-19 (NBS stands for the National Bureau of Standards—the earlier name of NIST). NBS-19 is calcium carbonate from a marble of geologically unidentified origin with $\delta^{18}O$ and $\delta^{13}C$ values of −2.20‰ and +1.95‰ respectively with respect to VPDB (Gröning, 2004).

3.2.4 A FEW IMPORTANT EQUATIONS USED IN STABLE ISOTOPE ANALYSIS

The conversion equations of $\delta^{18}O_{(VPDB)}$ to $\delta^{18}O_{(VSMOW)}$ and vice versa are (Coplen et al., 1983)

$$\delta^{18}O_{(VSMOW)} = 1.03091\delta^{18}O_{(VPDB)} + 30.91$$

$$\delta^{18}O_{(VPDB)} = 0.97002\delta^{18}O_{(VSMOW)} - 29.98$$

Usually in the mass spectrometry labs, a "Lab Standard" is used for measuring δ-values that need to be converted to the international standard scale. For example, δ-value of a sample "Smp" with respect to an international standard, say VPDB, can be determined if its value with respect to the lab standard, say "Lb," is measured and the δ-value of the lab standard with respect to the international standard is known after careful calibration. The conversion equation is

$$_{VPDB}\delta^{Smp} = {}_{Lb}\delta^{Smp} + {}_{VPDB}\delta^{Lb} + {}_{Lb}\delta^{Smp}\ {}_{VPDB}\delta^{Lb}\ 10^{-3}$$

The difference in the δ-values (*isotopic difference*, Δ) of the two coexisting phases, say $CaCO_3$ and H_2O, is approximately related to their fractionation factor (α) by the equation:

$$\delta_{CaCO_3} - \delta_{H_2O} = \Delta_{CaCO_3-H_2O} \approx 10^3 \ln\alpha_{CaCO_3-H_2O} \text{ (expressed in ‰)}$$

This approximation is especially good for δ-values where the difference ($\delta_A - \delta_B$) is less than 10.

Isotope ratio and δ-value of a mixture (say, M) is proportional to the mole fractions of its constituents (say, A and B):

$$R_M = X_A R_A + X_B R_B$$

$$\delta_M = X_A \delta_A + X_B \delta_B$$

where, R_M, R_A, R_B and δ_M, δ_A, R_B depict the *isotopic ratios* and the *δ-values* of the mixture, constituent A and constituent B, respectively; X_A, and X_B depict *mole fractions* of the constituent A and constituent B, respectively.

3.3 STABLES ISOTOPES OF A FEW ELEMENTS AND THEIR APPLICATIONS

Stable isotopes of several elements namely carbon and oxygen are used in paleoclimatic research. These are measured on the available geological archives such as foraminifera, speleothems, corals, and calcretes. In addition, carbon and nitrogen isotopes of organic matter preserved in sediments; oxygen and hydrogen isotopes of water including ice; carbon, hydrogen, nitrogen, and oxygen isotopes of greenhouse gases (CH_4, N_2O, CO_2) trapped in ice cores; oxygen and hydrogen isotopes of cellulose from trees; silicon and oxygen isotopes of silica from diatoms are also studied routinely. It is out of the scope of this paper to discuss the application of every isotope from different materials and environments. Therefore, as an example, we present a discussion on a few important isotopes from the marine realm used in paleoclimatological studies.

3.3.1 STABLE ISOTOPES OF OXYGEN AND CARBON IN CENOZOIC PALEOCEANOGRAPHY AND STRATIGRAPHY

The Cenozoic (past ~65 Ma) stable isotopic record of the oxygen and carbon has not only provided us the insight into global paleoceanographic signals but has been very useful for stratigraphic correlation of a large number of wells drilled for oil exploration. We shall first consider the basic premises of using oxygen and carbon isotopic record for long timescale, the major trends in the Cenozoic stable isotopic record, the limitations and lastly the future potential.

3.3.1.1 Oxygen Isotopes

Oxygen has three stable isotopes viz ^{16}O, ^{17}O, and ^{18}O with 99.763%, 0.0375%, and 0.1995% abundances, respectively. For paleoclimatological studies, the ratio $^{18}O/^{16}O$ is determined as ^{18}O has a higher abundance than ^{17}O and greater mass difference with ^{16}O. It was the Nobel laureate Harold Urey who in his classic paper published in the Journal of Chemical Society introduced the field of stable isotopic geochemistry (Urey, 1947). Urey was able to quantify the effect of temperature on oxygen isotope fractionation between calcium carbonate and water and came with the extraordinary idea that the isotopic composition of carbonate material could be used as a proxy for paleotemperature. Thereafter, McCrea (1950) studied the low temperature calcite–water oxygen isotopic fractionation and came out with the first empirical relationship between temperature and $\delta^{18}O$. However, these studies were based on inorganically precipitated calcites. Epstein et al. (1953) studied the biogenically precipitated calcite for establishing similar fractionation relationships. These workers gave an empirical relationship between the isotopic composition of water and temperature by growing mollusks in water at different temperatures. The oxygen isotope record of marine carbonates is a function of the isotopic composition of the ambient seawater besides temperature at which the calcification takes place. Emiliani (1955) used these studies to reconstruct the Pleistocene climatic cycles from the deep sea sediment cores. Another important study in this context was that of Craig (1965) who modified Epstein's equation and suggested corrections. Shackleton (1974) did an initial study of the T–$\delta^{18}O$ relationship in benthic foraminifera *Uvigerina*. Erez and Luz (1983) were among the first to derive temperature–$\delta^{18}O$ relationship in the tests of planktonic foraminifera *Globigerinoides sacculifer*. Bemis et al. (1998) did culture experiments on planktonic foraminiferal species *Orbulina universa* (with symbiotic algae) and *Globigerina bulloides* (without symbiotic algae) and gave revised paleotemperature equations. They demonstrated the influence over oxygen isotope of the *G. bulloides* by test size and offset in oxygen isotopic equilibration of *O. universa* introduced by the algal symbionts. Such studies have helped refine the paleoceanographic models based on isotopic analyses of the tests of various species. The researchers all over the world realized the importance of these empirical relationships as they provided a very powerful tool to infer the paleotemperature of the ambient water in which the ancient geological materials were formed. In most of the cases, the geological materials were the shells of macro- or microorganisms which were supposed to have grown in a medium whose paleotemperatures were to be estimated. The approximate temperature coefficient is −0.25‰ per °C, i.e., $\delta^{18}O$ of carbonate decreases by 0.25‰ for every 1 °C rise in temperature (Erez and Luz, 1983). These authors determined the following empirical temperature equation by comparing the isotopic composition of the planktonic foraminifera *G. sacculifer* with the actual growth temperature:

$$T\ {}^\circ C = 17.0 - 4.52\ (\delta^{18}O_c - \delta^{18}O_w)\ + 0.03\ (\delta^{18}O_c + \delta^{18}O_w)^2$$

where T, $\delta^{18}O_c$, $\delta^{18}O_w$, are the estimated temperature (in °C), the isotopic composition of the shell carbonate, and the seawater, respectively. But there are few limitations that inhibit the widespread application of paleotemperature scale. To apply the paleotemperature equation, we must know the $\delta^{18}O$ value of the seawater with which a given specimen of $CaCO_3$ has equilibrated. This is not easy to establish with certainty; because, the isotopic composition of seawater as a whole depends on the amount of ice stored on the continents, which gives rise to the so-called "ice-volume effect." When water evaporates from the ocean surface, the water vapor gets enriched in the lighter isotopes, as vapor pressure of $H_2{}^{16}O$ is more than $H_2{}^{18}O$. For example, at 20 °C, the $\delta^{18}O$ value of the vapor is lower by 10‰ than the liquid phase. Kinetic effects also take place that further deplete the vapor phase in the heavier isotope during molecular diffusion in the boundary layer at the water–air interface. This kinetic fractionation is directly proportional to the relative air humidity (Gonfiantini, 1986) and inversely proportional to the wind speed (Merlivat and Jouzel, 1979). With progressive evaporation, isotopically lighter water gets locked in the form of continental ice sheets, and the remaining ocean water gets increasingly enriched with the heavier isotope. It is believed that during the Last Glacial Maximum, the average seawater $\delta^{18}O$ was 1.3‰ more than the present (Labeyrie et al., 1987; Fairbanks, 1989). Simultaneously, the sea level also fell by 120 m that resulted in a scale of 0.011‰ increase in $\delta^{18}O$ of marine carbonate for every meter fall in sea level, which can be used to reconstruct the sea level curve (Fairbanks and Matthews, 1978; Fairbanks, 1989). Presently, the temperature of sea water is independently determined using techniques like trace element ratio (Mg/Ca, Sr/Ca) of carbonates, alkenone unsaturation index, Tex_{86} (membrane lipid composition of single celled microorganisms), clumped isotope method, etc., which is then used to calculate the $\delta^{18}O$ of seawater using the paleotemperature equations. The $\delta^{18}O$ of seawater thus obtained is used to calculate the past salinity values using the $\delta^{18}O$–salinity relationships for specific oceanic regions (Tiwari et al., 2013 and references cited therein).

3.3.1.2 Carbon Isotopes

Carbon has two stable isotopes, namely, ^{12}C and ^{13}C with an abundance of 98.89% and 1.11%, respectively (Craig, 1953). Photosynthesis causes preferential uptake of ^{12}C due to the kinetic fractionation in the organic matter. Within the inorganic carbon system, "atmospheric CO_2—dissolved bicarbonate—solid carbonate," equilibrium fractionation causes enrichment in the heavier isotope in the carbonates. The carbonate equilibria in the marine system can be represented by the following equations (Faure, 1986):

$$CO_2\ (g) \leftrightarrow CO_2\ (aq.)$$

$$CO_2\ (aq.) + H_2O \leftrightarrow H^+ + HCO_3^-\ (aq.)$$

$$CaCO_3\ (s) + H^+ \leftrightarrow Ca^{2+}\ (aq.) + HCO_3^-\ (aq.)$$

Total dissolved inorganic carbon (DIC) comprises HCO_3^- (aq.), CO_3^{2-} and CO_2 (aq.). The equilibrium fractionation factors for carbon isotope in this system at 20 °C are (Emrich et al., 1970)

$$\alpha_{CaCO_3(s) - HCO_3^-\ (aq.)} = 1.00185$$

$$\alpha_{HCO_3^-\ (aq.) - CO_2\ (g)} = 1.00838$$

$$\alpha_{CaCO_3(s) - CO_2\ (aq.)} = 1.01017$$

At low temperatures, the largest fractionation occurs between the dissolved CO_2 and the bicarbonate ion. The bicarbonate ion is the dominant species at the prevailing pH of seawater and DIC can be conveniently approximated by HCO_3^- (aq.). The temperature coefficient shows 0.035‰ increase of $\delta^{13}C$ per degree Celsius increase in temperature (Grossman, 1984; Emrich et al., 1970). Thus temperature has a negligible effect on $\delta^{13}C$ of calcite.

The $\delta^{13}C$ of the planktonic foraminiferal tests are also very valuable in ranking them according to the depth. This is possible because of the presence of a vertical gradient in the total dissolved CO_2 of the oceans (Kroopnick, 1985; Berger and Vincent, 1986). This gradient is a function of the biological activity and productivity at various depths in the water column. In the photic zone, the organic productivity is very high due to the phytoplanktons and thus the isotope of carbon used in this process is mostly the lighter isotope ^{12}C. Thus, the surface waters become enriched in ^{13}C and accordingly the planktonic foraminifera calcifying their test in isotopic equilibrium with sea water has higher $\delta^{13}C$. As one goes into the thermocline and deeper ocean, the biological productivity decreases and ^{12}C is released due to the decay of the organic matter and respiration thereby depleting the waters with ^{13}C. Thus the planktonic foraminifera living in the mixed layer, thermocline and below thermocline have progressively lower values of $\delta^{13}C$ in a region with well stratified water column and thus isotopic depth ranking can be made with the help of carbon isotope also. In summary, the planktonic foraminiferal oxygen and carbon isotope values are functions of their depth habitat and thus can be used as proxies for their isotopic depth ranking (Gasperi and Kennett, 1992). Such studies have been used to provide an isotopic depth ranking of a number of extinct foraminifera. There are many publications that successfully utilized this trait to infer the depth, thickness of the thermocline, and mixed layer during the Cenozoic. For example, Srinivasan and Sinha (1998, 2000), based on the Neogene planktonic foraminiferal biostratigraphy and isotopic depth ranking of planktonic foraminiferal species *Pulleniatina spectabilis*, an extinct species, inferred the closing of Indonesian seaway to thermocline water during the early Pliocene.

3.3.1.3 Oxygen Isotope Stratigraphy

The advent of the Deep Sea Drilling Project (DSDP) in 1968 was a major step in the development of Cenozoic stable isotope record and later resulted into a robust chemostratigraphic application on marine strata for worldwide correlation. An important aspect of the Cenozoic oxygen isotope record is its dependence on the global ice volume, which is termed "ice-volume effect" as explained in preceding sections. During glacial intervals, when the global ice volume increases, a large portion of the ocean water is locked in the form of polar ice sheets which are depleted in the heavier isotope of oxygen (^{18}O). Thus, the resultant effect during glacial interval is enriched ocean waters and depleted glacial ice with respect to the heavier isotope ^{18}O. The foraminifera calcifying their test in isotopic equilibrium with sea water during glacial times are thus enriched in ^{18}O as compared to those calcifying during interglacial intervals. Usually, benthic foraminifera are used as they, unlike planktonic, are not affected by short term temperature and salinity fluctuations (Shackleton and Opdyke, 1973). This leads to an oxygen isotope curve whose trends parallel the glacial–interglacial fluctuation, i.e., the changing global ice volumes. This phenomenon has resulted in the establishment of the marine oxygen isotope stages (MIS) for the later part of the Cenozoic while efforts are being made to extend the record back in time with more and more advancements of techniques. Several workers have stacked together $\delta^{18}O$ values of foraminifera from several well dated deep sea records using the orbitally forced ice-volume tuning to obtain global, high-resolution isotope chronostratigraphic curves (Imbrie et al., 1984; Pisias et al., 1984; Prell et al., 1986; Martinson et al., 1987; Imbrie et al., 1992; Lisiecki and Raymo, 2005). Such curves are not only used for global correlation but also as a relative dating method. The latest stack LR04 based on 57 deep sea records of $\delta^{18}O$ of benthic foraminifera is by Lisiecki and Raymo (2005) that spans for the past 5.3 Ma (Fig. 1).

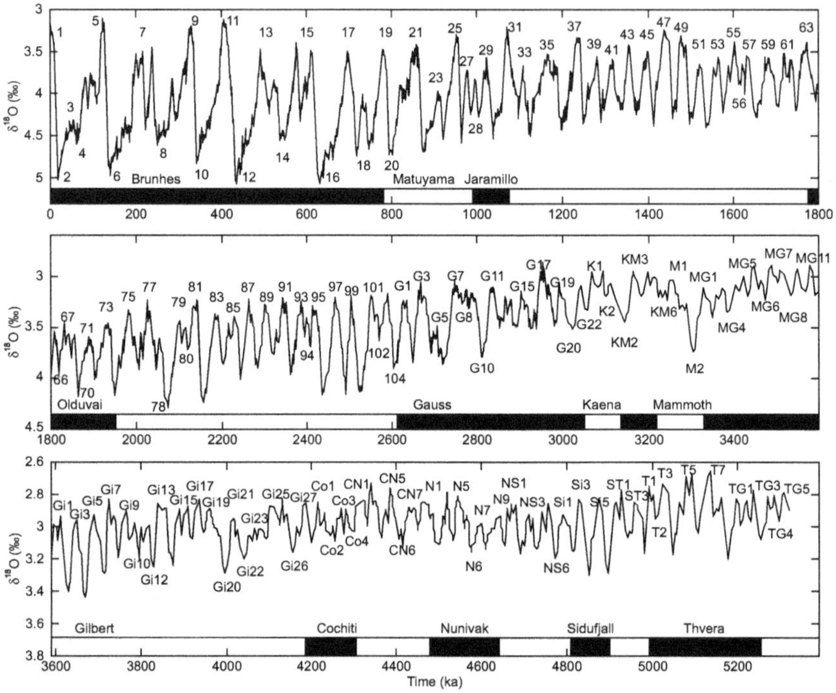

FIGURE 1

The oxygen isotope stratigraphy curve (LR04 stack) based on 57 stacked records of $\delta^{18}O$ values of benthic foraminifera spanning the past ~5.3 Ma.

Reprinted Lisiecki and Raymo (2005); Copyright 2005, with permission from John Wiley and Sons.

MIS 1 represents the Holocene and higher numbers indicate successively older periods with odd numbers depicting warmer stages and even numbers depicting colder stages. In addition, the later interglacial stages have been subdivided into separate warmer and colder episodes. For example, the MIS 5 is subdivided into five substages with 5a, 5c, and 5e as warmer conditions while 5b and 5d are colder intervals. Recently, decimal notation has been introduced to denote the isotopic substages so that the warmer episodes have odd numbers while the colder one possess even numbers. So, the MIS 5a, 5c, and 5e becomes 5.1, 5.3, and 5.5 respectively while the MIS 5b and 5d become 5.2 and 5.4 respectively. This scheme is followed till MIS 104 (~2.6 Ma). Thereafter, the isotopic stages are identified using the magnetic subchrons in which they appear. For example, the MIS G1 refers to the first warm episode in the Gauss subchron while the MIS G2 denotes the next colder episode; odd and even numbers again referring to the warmer and colder periods respectively. A distinctive feature of the oxygen isotope curve is the presence of the sawtooth pattern with the most rapid changes occurring at the termination of the glacial stages. These rapid changes from glacial to interglacial periods are called as "Terminations" and constitute major "events" in the oxygen isotope stratigraphy. For example, Termination 1 at ~14,000 years ago (Lisiecki and Raymo, 2005) is the rapid decline in $\delta^{18}O$ values at the end of the last glacial period, i.e., MIS 2.

3.3.1.4 Oxygen Isotopic Variation of Planktonic Foraminifera in Ocean Water Column

Though there is no strict rule to demonstrate the variation of $\delta^{18}O$ and $\delta^{13}C$ in the tests of planktonic foraminifera in the upper ocean water columns in the world's oceans, there is some agreement about the pattern of stable isotope composition if the water column is well stratified. The upper layer of the oceans receives kinetic energy from winds and heat energy from the sun. Due to this, the upper layer is well mixed and attains a uniform temperature. In the low latitudes this temperature is higher due to high amount of radiation received. This "mixed layer" can range in thickness from a few meters to few tens of meters and is characterized by a uniform temperature, i.e., the temperature does not decrease with depth within this layer. Below this mixed layer is the thermocline, a zone where the temperature declines sharply and reaches low values. Below the thermocline, the temperature declines very slowly till the bottom of the ocean is reached. This temperature distribution has implications on the oxygen isotopic composition of planktonic foraminifera. It is now well established that planktonic foraminifera are depth stratified in the upper ocean water columns. Several attempts were made to understand the distribution pattern with the help of plankton tows (Berger, 1969; Bé et al., 1973), sediment trap (Kahn and Williams, 1981; Fairbanks et al., 1982; Curry et al., 1983; Sautter and Thunell, 1991), and core top (Hemleben et al., 1989; Bijma et al., 1990). As described in the previous section, the oxygen isotope fractionation between sea water and calcium carbonate of the foraminifera is dependent on the ambient temperature in a way that lower the temperature of ambient water, more enrichment of the heavier isotope (^{18}O) takes place, i.e., $\delta^{18}O$ will be higher in $CaCO_3$. In a nutshell, the $\delta^{18}O$ of the planktonic foraminiferal test calcified in isotopic equilibrium indicates the temperature and isotopic composition of the sea water in which it lived (Gasperi and Kennett, 1992). Due to this relationship, the planktonic foraminifera residing at various depths in the ocean water column have different $\delta^{18}O$ values. Those residing in the warm mixed layer have lower values of $\delta^{18}O$ than those living in or below the thermocline. This property of depth (temperature) relationship with $\delta^{18}O$ values has been used to rank the extinct species also according to the depth habitat (Gasperi and Kennett, 1992; Douglas and Savin, 1978; Keller, 1985; Savin et al., 1985; Barrera et al., 1985).

3.3.1.5 Use of Oxygen and Carbon Isotopes in Upwelling Systems

The isotopic depth ranking as described in the preceding section can be used to detect the upwelling intervals in regions affected by seasonal upwelling. The planktonic foraminifera are depth stratified and their isotopic composition can be used to rank them according to depth. In the case of upwelling, the stratification of the water column is disturbed and surface waters are invaded by the cold, nutrient-rich waters. The invasion introduces two events, namely, the temperature gradient is reduced to minimum and the isotopic composition of the thermocline becomes uniform. Under such conditions, the foraminifera calcifying their tests in isotopic equilibrium with ambient water will have similar values of $\delta^{18}O$ and $\delta^{13}C$. Thus, if a time series data are obtained for the oxygen and carbon isotopic composition of shallow (mixed layer) and deep dwelling (thermocline and below) planktonic foraminifera, then the maximum divergence in the isotopic values indicate well-stratified water column while the minimum divergence or convergence indicate the upwelling conditions. Steens et al. (1992) studied the $\delta^{18}O$ and $\delta^{13}C$ of planktonic foraminiferal species including surface dwelling *G. bulloides* and the thermocline dwelling *Neogloboquadrina dutertrei* from the Hole 728A (ODP 117) located at a typical seasonal upwelling region of the Arabian Sea close to the Oman margin. These authors interpreted the mixing of the water column during upwelling situations when the isotopic values of the two species converge or have minimum difference and diverge during well-stratified conditions. The difference in $\delta^{13}C$ isotope distribution between the two species was found to be smallest within the upwelling areas (Steens et al., 1992). These authors also showed that from isotopic stage 2 till stage 11, the upwelling was stronger during interglacial periods, and that upwelling intensified downcore for the same period. In older sediments (stages 12 and 13), the signal from the upwelling indicators were found to be reversed, showing stronger upwelling during glacial (Steens et al., 1992) periods.

3.3.1.6 Major Trends in Cenozoic Oxygen Stable Isotopic Record

Perhaps one of the best curves to show the overall major trends in the Cenozoic oxygen isotopic record is by Miller et al. (1987) as shown in Fig. 2. It shows the composite $\delta^{18}O$ record of the last 70 Ma. As discussed previously, the curve shows the pattern of global ice volume and deep-sea temperature variations (benthic $\delta^{18}O$ value) since the beginning of the Cenozoic.

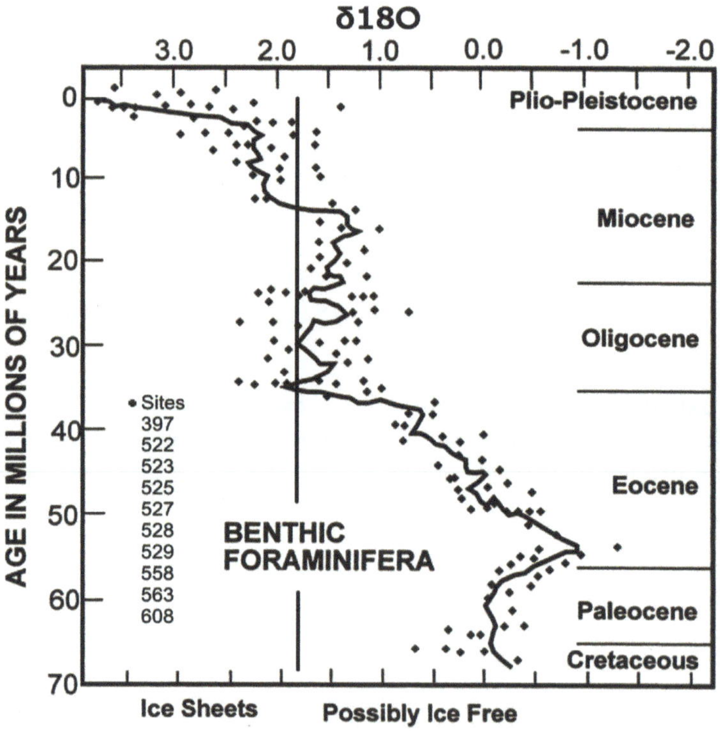

FIGURE 2

Cenozoic $\delta^{18}O$ record from the Atlantic Deep Sea Drilling Project sites. The vertical line is at 1.8‰ indicating the periods when the earth was either ice free or had major ice sheets.

Reprinted from Miller et al. (1987); Copyright 1987, with permission from John Wiley and Sons.

The minimum $\delta^{18}O$ value is observed at around ~55 Ma (million years ago)—a period known as the Paleocene-Eocene Thermal Maxima. Then onward, a long-term cooling trend is indicated by a steady increase in $\delta^{18}O$ values interrupted by times of relatively rapid change. This trend continues till the base of the Oligocene; after which, the cooling trend is not enhanced but remains steady till the Middle Miocene. After the Middle Miocene, the cooling trend is more significant and reaches the maximum values in the late Miocene. From the late Miocene to early Pliocene, again the steady state is maintained followed by a sharp increase in cooling trend during late Pliocene and Pleistocene. The steep increase in the isotopic values of the early Eocene–Oligocene boundary probably reflect the major ice cap expansion (Antarctic glaciation) event in the Cenozoic (Miller et al., 1987; Barrett et al., 1987). Later increases in the Middle Miocene and Late Pliocene indicate the subsequent growth of the ice volume with Arctic ice sheet coming into existence during the late Pliocene (Haug and Tiedemann, 1998). The high-resolution data superimposed over the long-term trends indicate Milankovitch forcing on the ice volume and all the three cycles of eccentricity, obliquity, and precession have been detected on a large number of deep sea cores.

3.3.1.7 Cenozoic Carbon Isotopic Record

Compared to the Cenozoic oxygen isotopic record, the interpretation of the carbon isotopic record has not been simple owing to its intrinsic interaction with the biological activity. The early work of Shackleton and Kennett (1975) provided an overview of the Cenozoic carbon isotopic record generated from planktonic and benthic foraminifera from DSDP sites 277, 279, and 281 from the southwest Pacific Ocean. While there is a general positive $\delta^{18}O$ shift from Paleocene to Pleistocene, no long term trend exists in the $\delta^{13}C$ record (Williams et al., 1988). While the oxygen isotope trend in the Neogene has been explained as the result of changing ice volume, initiation of major glaciations, waxing and waning of ice sheets, etc., the exact reasons for the shift in the carbon isotope record are not explained (Williams et al., 1988).

Kennett (1986) interpreted the Neogene $\delta^{13}C$ changes to have been caused by average oceanic compositional changes resulting from fractionation between continental and oceanic organic carbon reservoirs. Loutit et al. (1983) and Woodruff and Savin (1985) showed the existence of correlation between changes in Neogene carbon isotopic values and sea level as inferred from the onlap-offlap curve of Vail and Hardenbol (1979). During large-scale marine transgression corresponding to eustatic sea level rise during interglacials, the exposed land areas are reduced world over and continental shelves increase in size, which supports a diversity of organisms. Such organisms, their shell growth, and subsequent changes in food chain result in the removal of organic carbon from the ocean waters, which gets accumulated on the continental shelves and ocean waters became more enriched in ^{13}C. On the other hand, during a eustatic fall of sea level leading to large scale marine regression during glacial periods, large areas of the continental shelves are exposed and subjected to erosion. This results in the supply of organic carbon to the ocean basins and as a result, the oceans became more depleted in ^{13}C. Here, the scale of transgression and regression is of prime importance and should be large enough to cause the changes in organic carbon exchange at global level. Small rise and fall of sea level is reflected in the oxygen isotope values due to ice-volume effect but the same is not paralleled by the carbon isotopic fluctuation. The magnitude of the $\delta^{13}C$ change depends upon the areas of shallow shelves covered or exposed by a given change in sea level. This in turn, is directly related to the hypsometry (elevation relative to the sea level) of the continental margins and the magnitude of the sea-level change (Kennett, 1986).

3.3.1.7.1 Late Early to Early Middle Miocene Positive $\delta^{13}C$ Shift

In the early part of the Neogene, a positive shift in the $\delta^{13}C$ of 1‰ was demonstrated by Vincent and Berger (1985), which they termed as the "Monterey Carbon Isotope Excursion." Kennett (1986) explained this and agreed with several other workers regarding the explanation for this positive shift due to a global marine transgression (Loutit et al., 1983; Woodruff and Savin, 1985). Vincent and Berger (1985) stated that during this interval, large-scale transgression occurred throughout the circum-Pacific region, and, as explained above, huge amount of organic carbon was extracted from the oceans. This had resulted in the deposition of organic and phosphate-rich sediments within the Monterey Formation of California (John et al., 2002) throughout the Pacific margin. Vail and Hardenbol (1979) documented a period of coastal onlap that began at about 19 Ma (million years ago) and terminated about 14 Ma, and was contemporaneous with the interval containing the especially positive $\delta^{13}C$ values (Kennett, 1986). This positive carbon shift also corresponds to Early Neogene climatic optimum or warmth as reflected from the negative oxygen isotope excursion of benthic and planktonic foraminifera. Due to this climatic optimum, large amount of organic biomass in the form of forests could have grown on the continents and have contributed to the positive $\delta^{13}C$ values during this interval (Kennett, 1986). This climatic optimum was followed by middle Miocene global cooling as explained in the previous section corresponding with a negative carbon isotope excursion.

3.3.1.7.2 Late Miocene Negative Carbon Isotope Shift

The latest Miocene at 6.2 Ma is marked by a distinct, rapid shift of $\delta^{13}C$ up to–0.75‰ in all of the southwest Pacific sites, both in the benthic and planktonic records (Kennett, 1986). This depletion in $\delta^{13}C$ was followed markedly by low values during much of the remainder of the late Neogene, which led to the suggestion that it was "permanent" (Kennett, 1986; Keigwin, 1979). This well-known late Miocene carbon isotope shift is recorded throughout the Indo-Pacific and South Atlantic (Kennett, 1986; Keigwin, 1979; Bender and Keigwin, 1979; Loutit and Kennett, 1979; Keigwin and Shackleton, 1980; Vincent et al., 1980; Haq et al., 1980; Hodell and Kennett, 1984). This shift was of a very short duration and was assigned an age of ~6 Ma by Haq et al. (1980) and of 6.2 Ma by Loutit and Kennett (1979). Loutit and Keigwin (1982) summarized a number of global events associated with this negative carbon shift which included a global lowering of the sea level.

3.3.1.7.3 Messinian Salinity Crisis and Negative Carbon Shift

The world wide lowering of the global sea levels causing the negative carbon isotopic shift was also associated with isolation of the Mediterranean due to the closure of the Gibraltar Strait and resulted into the "Messinian Salinity Crisis" (Cita, 1982). The expansion of the ice sheets resulted in intensive circulation of bottom water and ocean ventilation (Kennett, 1986). Estimates of the magnitude of the marine regression are similar, ranging from 40 to 80 m (Loutit and Kennett, 1979; Berggren and Haq, 1976; Cita and Ryan, 1979).

3.3.1.8 Limitations

3.3.1.8.1 Vital Effects

The biological activities of foraminifera give rise to several effects due to which its calcite is not precipitated in isotopic equilibrium with the ambient water. The consequent offsets of isotopic values of the tests are collectively referred to as the "*Vital Effects*," which can be either metabolic or kinetic (McConnaughey, 1989a,b). Metabolic effects are due to the inclusion of ambient dissolved carbon and oxygen, which are produced during the respiration or photosynthesis into the calcite shell. Preferential uptake of lighter isotopes during the hydration and hydroxylation of CO_2 that are generally associated with rapid calcification constitute the kinetic effects. Vital effects, thus, are a result of the processes like respiration, ontogeny, and secretion of gametogenic calcite, etc. Lower $\delta^{18}O$ values result when respiration products depleted in the heavier isotopes (Lane and Doyle, 1956) are utilized for shell formation. This is the reason for the progressive $\delta^{18}O$ enrichment from juvenile to mature chambers as higher metabolic rates in the juvenile specimen as compared to an adult would cause the stronger depletions (Wefer and Berger, 1991). Planktonic foraminifera inhabit different depths at various stages of their ontogeny. Foraminifera secrete shells at deeper and cooler waters toward the end of their life cycles resulting in higher $\delta^{18}O$ values (Bouvier-Soumagnac and Duplessy, 1985). Toward the end of their life cycle, foraminifera also secrete gametogenic calcite. Foraminifera move to deeper, cooler waters before gamete release and secrete a layer of calcite over its shell. This calcite layer with higher $\delta^{18}O$ values can comprise 18–28% of the shell mass of foraminifera (Bé, 1980). Furthermore, the $\delta^{18}O$ decreases with increasing irradiance, which causes enhanced photosynthesis that leads to a higher rate of shell formation and hence stronger kinetic fractionation (Spero and Lea, 1993; Wefer and Berger, 1991). The $\delta^{18}O$ in foraminiferal carbonate decreases with increasing carbonate ion concentration that was also ascribed to biological, kinetic fractionation effect (Spero et al., 1997) as this effect was also observed in rapidly precipitating inorganic $CaCO_3$ (McCrea, 1950). The disequilibria due to the vital effects can be avoided by choosing those species, which are known to precipitate their shells in equilibrium with seawater and by picking mature shells from a particular size range.

3.3.1.8.2 Effect of Diagenesis

This section addresses the studies involving long term stable isotope record of benthic and planktonic foraminifera from stratigraphic columns where foraminiferal carbonate is supposed to have undergone diagenesis. Our entire premise of using stable isotopes of foraminifera for paleoceanographic interpretation is based on the belief that the long term burial and diagenesis will not affect the primary isotopic values. Although the oxygen isotopic values may be affected by diagenesis involving recrystallization and significant dissolution (Baker et al., 1982; Elderfield and Gieskes, 1982; Killingley, 1983), it has been observed that diagenesis is usually absent or negligible in sediments of Pleistocene–Pliocene age (Williams et al., 1988). Anderson and Arthur (1983) stated that the $\delta^{13}C$ values of marine carbonates are thought to be less affected by diagenesis than the oxygen isotope values because of the differences in respective reservoirs for potential isotopic changes. The controlling factor for the diagenetic effect on isotopic composition is the $\delta^{18}O$ of the pore water. Williams et al. (1988) found that unless the diagenesis is severe, the $\delta^{18}O$ pattern or signal is not obliterated in many sections. Presence of significant amounts of organic matter can affect the $\delta^{13}C$ values of marine carbonates.

In general, the $\delta^{13}C$ values of the marine carbonates can be affected only when the recrystallization occurs in the presence of pore water bicarbonate in which $\delta^{13}C$ has been considerably altered due to the addition of ^{12}C from organic matter oxidation and microbial activity. Recently, Edgar et al. (2013) concluded that even extensive recrystallization of benthic foraminiferal calcite results in minimal shifts from primary $\delta^{18}O$ and $\delta^{13}C$ values.

Thus, the Cenozoic oxygen and carbon isotopic records of foraminifera have a tremendous scope of paleoceanographic and stratigraphic studies and also in the hydrocarbon exploration where at times biostratigraphic datum do not serve the purpose or have limitations.

3.3.2 NITROGEN ISOTOPES OF SEDIMENTARY ORGANIC MATTER AS DENITRIFICATION INDICATOR

Nitrogen has two isotopes viz ^{14}N and ^{15}N with abundances of 99.64% and 0.36%, respectively. Nitrogen gets into the biological systems via "Nitrogen Fixation," which involves the reduction of N_2 to ammonia (NH_3), NH_4^+, or any other nitrogen compound by the microorganisms (e.g., *Rhizobium*) under anaerobic conditions. Nitrogen fixing bacteria are called as diazotrophs. "Nitrification" is the conversion of the NH_3, NH_4^+ to nitrate (NO_3^-) or nitrite (NO_2^-) by the bacteria of *Nitrosamonas* and *Nitrobacter*. If ammonia is directly assimilated, then it is called as "ammonia assimilation" but free ammonium ions cannot exist for long in the aerobic soils as they are oxidized to nitrate or nitrite and hence nitrification is the dominant step. Dentrification is the process by which nitrate is reduced to any gaseous nitrogen species, generally N_2 or nitrous oxide (N_2O) under the anaerobic conditions (Deuser et al., 1978; Naqvi, 1987).

In the world's oceans, there are three regions that support perennial anaerobic conditions in the water depths of ~250–1250 m, also called as oxygen minima zone (OMZ) where oxygen concentration fall below 0.05 ml/l (Wyrtki, 1971; Deuser et al., 1978; Naqvi, 1987). This is due to the occurrence of high overhead productivity that consumes the available oxygen during their descent/degradation. These regions are the eastern tropical North Pacific, eastern tropical South Pacific, and the Arabian Sea (Christensen et al., 1987; Naqvi, 1987, 1994). Due to the lack of oxygen in the OMZ, the NO_3^- acts as an electron acceptor and is reduced by the anaerobic bacteria for the decomposition of organic matter. During this process, they preferentially consume NO_3^- with lighter isotope (^{14}N), thus enriching the residual nitrate in the heavier isotope, which gets upwelled to the sea surface and is taken by the organisms as a nutrient. The $\delta^{15}N$ of the organic matter is governed by the isotopic composition of the source NO_3^- that is upwelled from below and fractionation experienced by NO_3^- during its uptake by the phytoplanktons (Altabet and Francois, 1994). In the oxic waters of the euphotic zone, the $\delta^{15}N$ of the nitrate is 5–6‰ (Liu and Kaplan, 1989) whereas, in the regions receiving waters upwelled from the OMZ the $\delta^{15}N$ is in the excess of 18‰ (Cline and Kaplan, 1975) that explains the highly enriched $\delta^{15}N$ values observed in the particulate organic matter. This enriched $\delta^{15}N$ signature is preserved even when the organic matter settles down and gets preserved in sea sediments (Saino and Hattori, 1987). Thus, a high $\delta^{15}N$ can be related to increased dentrification, which in turn is controlled by the climate induced productivity increase (Ganeshram et al., 1995).

In the Arabian Sea, the OMZ is due to the high oxygen consumption below the thermocline for the oxidation of organic matter supplied by the high overhead surface productivity. The sluggish flow of the oxygen-poor intermediate water (Olson et al., 1993; You and Tomczak, 1993) along with a strong tropical thermocline (due to relatively high sea surface temperature that prevents mixing of the oxygen rich surface waters with the deeper waters) maintains the OMZ (Qasim, 1982). As an example, a dentrification record from the western Arabian Sea that reflects overhead surface productivity related to the intensity of the Southwest (SW) Monsoon winds is presented herein. The surface circulation of the western Arabian Sea is controlled by the seasonal reversal of monsoon winds. During the boreal summer, the summer or the SW Monsoon winds blowing along the Somali and Omani coast induce intense upwelling with upwelling velocities close to 3×10^{-3} cm/s and an upwelling transport of 1.5–2 Sv in the upper 50 m (Smith and Bottero, 1977; Shi et al., 2000). The reasons attributed for such intense coastal upwelling is the Ekman divergence due to the flow of strong winds parallel to the coast. As the surface water moves away from the coast, water from below upwells to occupy its place. The offshore upwelling takes place due to the strong positive wind stress curl to the NW of the axis of the Somali jet, which is a low level cross-equatorial jet (Smith and Bottero, 1977). The upwelling injects nutrients into the euphotic zone leading to the increase in the productivity. The typical productivity values for the western Arabian Sea are 2.0 gC/m^2/day (gram carbon per meter square per day) during the SW monsoon period and 0.5 gC/m^2/day during the intermonsoon period (Codispoti, 1991; Barber et al., 2001). The Fig. 3 shows the $\delta^{15}N$ variations for the past 20,000 years from the western Arabian Sea (Altabet et al., 2002; Tiwari et al., 2010).

Altabet et al. (2002) analyzed the nitrogen isotope ratio ($\delta^{15}N$), nitrogen content, and chlorin (a pigment found in algae and plants, e.g., chlorophyll is a magnesium containing chlorin) abundance of sediment cores from Oman continental margin at millennial scale. Nitrogen and chlorin abundances are manifestations of the overhead surface productivity, which is controlled by the SW monsoon wind intensity. Altabet et al. (2002) found that all the three proxies varied in unison leading them to propose that $\delta^{15}N$ is a productivity indicator at least on century to millennial timescales. Their record also found a close correlation with the Greenland Ice Sheet Project 2 (GISP2) temperature record (Grootes and Stuiver, 1997) signifying strong teleconnection between the SW monsoon intensity and high latitude climate change. As evident from Fig. 3, during the Holocene (past 11,700 years), the $\delta^{15}N$ values show a maximum at around 10,000 years after which they decline slightly and then increase uniformly since 9000 years till the core top. It implies that an episode of monsoon intensification took place during early Holocene coinciding with the maximum summer insolation. Thereafter, the SW monsoon intensity did not decrease during the rest of Holocene unlike insolation that declined monotonically. Tiwari et al. (2010) also determined $\delta^{15}N$ values from a nearby sediment core (Fig. 3). The trend exhibited by the

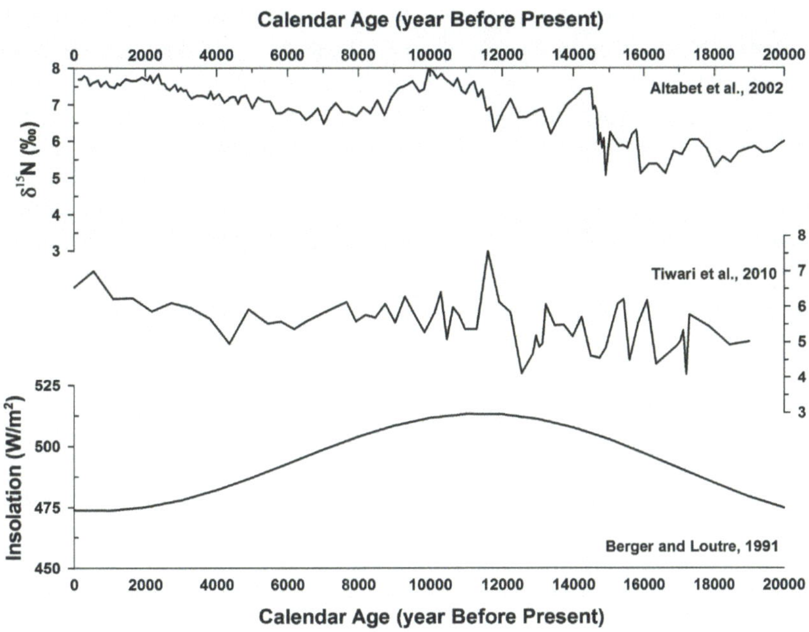

FIGURE 3

Denitrification records from the western Arabian Sea for the past 20,000 years.

Data taken from Berger and Loutre (1991), Berger (1992), Altabet et al. (2002), and Tiwari et al. (2010).

$\delta^{15}N$ values of the Tiwari et al. (2010) matches with that exhibited by Altabet et al. (2002) corroborating the fact that the SW monsoon did not decline during the Holocene. Thus, during the Holocene, SW monsoon did not follow insolation, signifying the importance of understanding internal feedbacks on nitrogen budgets.

3.3.3 CARBON ISOTOPES OF SEDIMENTARY ORGANIC MATTER AS PROVENANCE INDICATOR

The concentration of the organic carbon is a widely used proxy in the field of paleoclimatology to reconstruct the past productivity variability that can be related to various climatic factors. But this organic matter can be from marine as well as terrestrial sources. Therefore, it is important to ascertain the source of the organic matter for precise interpretation. Currently, the $\delta^{13}C$ value of the atmospheric CO_2 is around −8‰ (Hoefs, 2009). Plants preferentially incorporate ^{12}C due to the kinetic isotope effect during the photosynthesis. A two step model is proposed to explain the kinetic fractionation of carbon isotopes during photosynthesis that involves (1) fractionation during the uptake of gaseous CO_2 by the leaves through stomata and (2) then the fractionation during the carbon fixation using a suitable enzyme (Farquhar et al., 1989). In different categories of plants, the carbon fixation reaction involves different enzymes following diverse pathways. The first category comprising 90% of present day plants is of "C3 plants" that use an enzyme called as "ribulose bisphosphate carboxylase oxygenase (RuBisCO)." Such plants use what is known as Calvin Cycle wherein three molecules of CO_2 finally produce one molecule of a carbohydrate which is a three-carbon sugar phosphate molecule. The $\delta^{13}C$ values of C3 plants range from −20‰ to −36‰ with a mean value of −27‰ (Farquhar et al., 1989). The other important pathway is the Hatch–Slack Cycle which occurs in plants called as the C4 plants. The first stable compound of this pathway is a four-carbon compound namely oxaloacetic acid, hence such plants are called as C4 plants. This pathway utilizes water much more efficiently. Hence the C4 plants readily colonize the arid environments. The main enzyme used in this cycle is the PEP carboxylase that fractionates carbon isotopes much less than the RuBisCO leading to much higher $\delta^{13}C$ values than the C3 plants. The $\delta^{13}C$ values of the C4 plants range from −9‰ to −17‰ with a mean value of −13‰ (Farquhar et al., 1989). The marine organic matter plants/phytoplanktons are enriched in the heavier isotope (higher $\delta^{13}C$ values) as compared to the C3 plants as equilibrium fractionation leads to the enrichment of ^{13}C in the dissolved bicarbonate ion (HCO_3^-) as compared to the gaseous CO_2 ($\alpha_{HCO_3^-\,(aq.) - CO_2\,(g)} = 1.00838$); at 20 °C; (Faure, 1986). This enriched HCO_3^- is used by phytoplanktons for photosynthesis due to its higher concentration than the dissolved CO_2 at the prevailing pH of the ocean leading to the higher $\delta^{13}C$ values of the marine organic matter. The $\delta^{13}C$ value of marine organic matter ranges from −11 to −39‰ and the mean value is −21‰ (Farquhar et al., 1989).

It is clear from this discussion that the $\delta^{13}C$ value of the organic materials from different sources overlap. Therefore, it is advisable to use $\delta^{13}C$ in combination with C/N ratio, which is the carbon to nitrogen ratio by weight of the organic matter. The elemental composition of phytoplanktons and zooplanktons were given by Redfield (1934, 1958) and are called as "Redfield ratios." The Redfield ratios for marine organic matter are carbon:nitrogen:phosphorus = 106:16:1 that yields a C/N value of 6.6. This estimate was revised by Takahasi et al. (1985) who gave the ratio as C:N:P = 122 (±18):16:1 that provide a range of 6.5–8.7 as the C/N ratio. The C/N ratio of recent sediments comprising marine organic matter has a value of 8–10 whereas the ancient sediments have a value of 12–15 (Müller, 1977). In contrast, terrestrial sources exclusively possess some compounds such as cellulose, lignin, etc., that are rich in carbon and poor in nitrogen (protein) as compared to aquatic algae resulting in much higher C/N ratios. Organic matter derived from land plants has C/N values between 20 and 100 (Premuzic et al., 1982; Meyers, 1994). Diageneis of organic matter tends to lower the C/N ratio as nitrogenous compounds break down to produce ammonia, which is retained by clay minerals (Müller, 1977) whereas carbon in the form of CO_2 diffuses out.

As an example, new data from a sediment core—SK 274/4G—from the eastern Arabian Sea is presented here. The core—SK 274/4G—has a total length of 359 cm and was collected off Goa (13°59′45″ N; 72°0′24″ E) from a water depth of 1290 m. For the present study, the samples have been selected at every 5 cm. The Fig. 4 shows the $\delta^{13}C$ values of the sedimentary organic matter versus its C/N ratio that would help to identify the origin of the organic matter in the core.

Eastern Arabian Sea receives abundant surface runoff from the adjacent Western Ghats (*Sahyadri* Hills) that results in intense orographic precipitation during the SW monsoon. The monsoon winds also induce moderate upwelling in the coastal regions of the eastern Arabian Sea. The transfer of nutrients via upwelling and the surface runoff lead to the enhanced marine productivity. For the eastern Arabian Sea, the typical productivity values are 0.6 and 0.2 $gC/m^2/day$ during the SW monsoon and the inter-monsoon periods, respectively (Bhattathiri et al., 1996). However, there is a possibility of considerable contribution of the terrestrial organic matter as well through the surface runoff.

The $\delta^{13}C$ values in this core vary from −19‰ to −21‰ indicative of predominantly marine phytoplankton and some contribution from the terrestrial organic matter. The C/N ratio of the present study varies from 8 to 32, which is indicative of the presence of both terrestrial as well as the marine organic matter due to its coastal location. This example also stresses the fact that instead of relying solely on $\delta^{13}C$ values or C/N ratios to determine the provenance of organic matter, it is helpful to use them in conjunction.

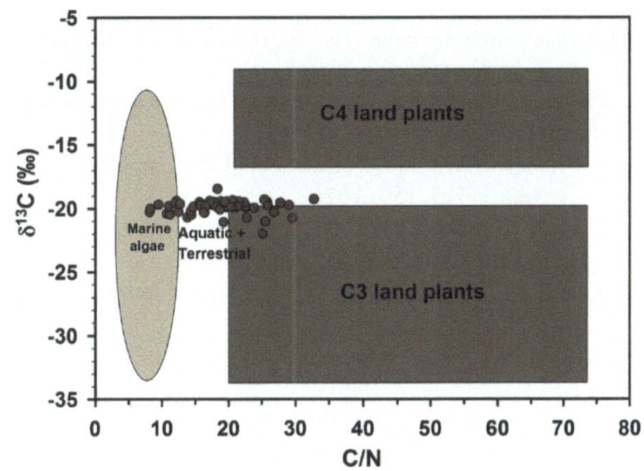

FIGURE 4

$\delta^{13}C$ values versus C/N ratios as indicator of the provenance of the organic matter; shaded areas approximately show the range of the $\delta^{13}C$ values for that particular group.

3.4 CONCLUSION

The stable isotopes of oxygen, carbon, and nitrogen are useful to reconstruct the past climatic and oceanographic changes. Besides, the oxygen isotopic stratigraphy of marine sediments provides a robust time frame work to correlate widely separated marine sections and also to correlate the paleoceanographic and paleoclimatic events. This helps us to interpret the cause and effect relationship among various events recorded in diversely located stratigraphic columns. The stable isotopic studies have become thus one of the most important tools in paleoceanographic and paleoclimatic studies. The aim of the paleoclimatology is far beyond the plain documentation of past climate changes. It will improve our understanding of the *mechanism* of climate variability including abrupt climate changes, especially under the greenhouse gases forcing expected toward the end of the century. It is necessary to determine the natural component to delineate the anthropogenic forcing to create viable models of future climate scenarios. Additionally, the trends/periodicities exhibited by the climatic phenomenon may indicate their future course. It would aid in gauging the severity, likely long-term effects, and possible consequences of changes.

ACKNOWLEDGMENT

MT is grateful to the Secretary, Ministry of Earth Sciences and Director NCAOR for their constant encouragement and support under the project Past Climate and Oceanic Variability (NCAOR Contribution no. 43/2014). DKS and MT thank ISRO-GBP Project for the financial support. AKS acknowledges DST Fast Track Project (SR/FTP/ES-85/2009) for the financial assistance. DKS and AKS thank Department of Geology, CAS in Geology, Delhi University for necessary facilities.

REFERENCES

Alley, R.B., 2000. The Younger Dryas cold interval as viewed from central Greenland. Quaternary Science Reviews 19, 213–226.

Altabet, M.A., Francois, R., 1994. Sedimentary nitrogen isotopic ratios as a recorder for surface ocean nitrate utilization. Global Biogeochemical Cycles 8, 103–116.

Altabet, M.A., Higginson, M.J., Murray, D.W., 2002. The effect of millennial-scale changes in Arabian Sea denitrification on atmospheric CO_2. Nature 415, 159–162.

Anderson, T.F., Arthur, M.A., 1983. Stable isotopes of oxygen and carbon and their application to sedimentologic and paleoenvironmental problems. In: Stable Isotopes in Sedimentary Geology M.A.

Arrhenius, S., 1896. On the influence of carbonic acid in the air upon the temperature of the ground. Philosophical Magazine 41, 237–276.

Baker, P.A., Gieskes, J.M., Elderfield, H., 1982. Diagenesis of carbonates in deep-sea sediments: evidence from Sr/Ca ratios and interstitial dissolved Sr^{2+} data. Journal of Sedimentary Petrology 52, 71–82.

Barber, R.T., Marra, J., Bidigare, R.C., Codispoti, L.A., Halpern, D., Johnson, Z., Latara, M., Goericke, R., Smith, S.L., 2001. Primary productivity and its regulation in the Arabian Sea during 1995. Deep-Sea Research Part II: Topical Studies in Oceanography 48, 1127–1172.

Barrera, E., Keller, G., Savin, S.M., 1985. Evolution of the Miocene ocean in the eastern North pacific as inferred from oxygen and carbon isotopic ratios of foraminifera. Geological Society of America, Memoir 163, 83–102.

Barrett, P.J., Elston, D.P., Harwood, D.M., McKelvey, B.C., Webb, P.N., 1987. Mid Cenozoic record of glaciation and sea level change on the margin of the Victorian land basin, Antarctica. Geology 15, 634–637.

Bé, A.W.H., 1980. Gametogenic calcification in a spinose planktonic foraminifer, *Globigerinoides sacculifer* (Brady). Marine Micropaleontology 5, 283–310.

Bé, A.W.H., Harrison, S.M., Lott, L., 1973. *Orbulina universa* in the Indian Ocean. Micropaleontology 19, 150–192.

Beck, N., Münnich, K.O., 1988. Freezing of water: isotopic fractionation. Chemical Geology 70, 168.

Bemis, B.E., Howard, J.S., Jelle, B., David, W.L., 1998. Reevaluation of the oxygen isotopic composition of planktonic foraminifera: experimental results and revised paleotemperature equations. Paleoceanography 13, 150–160.

Bender, M.L., Keigwin, L.D., 1979. Speculations about the Upper Miocene change in abyssal Pacific dissolved bicarbonate $\delta^{13}C$. Earth and Planetary Science Letters 45, 383–393.

Berger, A., 1992. Orbital Variations and Insolation Database. IGBP PAGES/World Data Center for Paleoclimatology Data Contribution Series # 92-007. NOAA/NGDC Paleoclimatology Program, Boulder CO.

Berger, A., Loutre, M.F., 1991. Insolation values for the climate of the last 10 million years. Quaternary Science Reviews 10, 297–317.

Berger, W.H., 1969. Ecological patterns of living foraminifera. Deep-Sea Research 16, 1–24.

Berger, W.H., Vincent, E., 1986. Deep-sea carbonates: reading the carbon isotope signal. Geologische Rundschau 75, 249–269.

Berggren, W.A., Haq, B.U., 1976. The Andalusian Stage (late Miocene): biostratigraphy, biochronology, and paleoecology. Palaeogeography, Palaeoclimatology, Palaeoecology 20, 67–129.

Bhattathiri, P.M.A., Pant, A., Sawant, S., Gauns, M., Matondakar, S.G.P., Mohanraju, R., 1996. Phytoplankton production and chlorophyll distribution in the eastern and central Arabian Sea in 1994–1995. Current Science 71, 857–862.

Bijma, J., Faber Jr., W.W., Hembleben, C., 1990. Temperature and salinity limits for the growth and survival of some planktonic foraminifera in laboratory cultures. Journal of Foraminiferal Research 20 (2), 95–116.

Bouvier-Soumagnac, Y., Duplessy, J.C., 1985. Carbon and isotopic composition of planktonic foraminifera from the laboratory culture, plankton tows and recent sediment: implications for the reconstruction of paleoclimatic conditions and of the global carbon cycle. Journal of Foraminiferal Research 15, 302–320.

Broecker, W.S., Thurber, D.L., Goddard, J., Ku, T.-L., Matthews, R.K., Mesolella, K.J., 1968. Milankovitch hypothesis supported by precise dating of coral reef and deep-sea sediments. Science 159, 297–300.

Broecker, W.S., Denton, G.H., 1990. The role of ocean-atmosphere reorganizations in glacial cycles. Quaternary Science Reviews 9, 305–341.

Cita, M.B., 1982. The Messinian salinity crisis in the mediterranean: a review. Alpine mediterranean Geodynamics series, vol. 7, American Geophysical Union, Washington. pp. 113–140.

Cita, M.B., Ryan, W.B.R., 1979. Late Neogene environmental evolution. In: von Rad, U., Ryan, W.B.R., et al. (Eds.), Init. Repts. DSDP, vol. 47 (Pt.1). U.S. Government Printing Office, Washington, pp. 447–459.

Christensen, J.P., Murray, J.W., Devol, A.H., Codispoti, L.A., 1987. Denitrification in continental shelf sediments has major impact on the oceanic nitrogen budget. Global Biogeochemical Cycles 1, 97–116.

Cline, J.D., Kaplan, I.R., 1975. Isotopic fractionation of dissolved nitrate during denitrification in the eastern tropical North Pacific Ocean. Marine Chemistry 3, 271–299.

Codispoti, L.A., 1991. Primary productivity and carbon and nitrogen cycling in the Arabian Sea. In: Smith, S.L., Banse, K., Cochran, J.K., Codispoti, L.A., Ducklow, H.W. (Eds.), Arabian Sea Process Study (M. E).

Coplen, T.B., 2011. Guidelines and recommended terms for expression of stable isotope-ratio and gas-ratio measurement results. Rapid Communications in Mass Spectrometry 25, 2538–2560.

Coplen, T.B., Kendall, C., Hopple, J., 1983. Comparison of stable isotope reference samples. Nature 302, 236–238.

Craig, H., 1953. The geochemistry of the stable carbon isotopes. Geochimica et Cosmochimica Acta 3, 53–92.

Craig, H., 1961. Standard for reporting concentrations of deuterium and oxygen-18 in natural waters. Science 133, 1833–1834.

Craig, H., 1965. Measurement of oxygen isotope paleotemperatures. In: Tongiorgi, E. (Ed.), Stable Isotopes in Oceanographic Studies and Paleotemperatures. Consiglio Nazionale delle Ricerche, Spoleto, Italy, pp. 161–182.

Curry, W.B., Thunnel, R.C., Honjo, S., 1983. Seasonal changes in the isotopic composition of planktonic foraminifera collected in Panama basin sediment traps. Earth and Planetary Science Letters 64, 33–43.
Dansgaard, W., Johnsen, S.J., Clausen, H.B., Dahl-Jensen, D., Gundestrup, N.S., Hammer, C.U., Hvidberg, C.S., Steffensen, J.P., Sveinbjörnsdottir, A.E., Jouzel, J., Bond, G., 1993. Evidence for general instability of past climate from a 250-kyr ice-core record. Nature 364, 218–220.
Deuser, W.G., Ross, E.H., Mlodzinska, Z.J., 1978. Evidence for and the rate of denitrification in the Arabian Sea. Deep-Sea Research 25, 431–445.
Douglas, R.G., Savin, S.M., 1978. Oxygen isotopic evidence for the depth stratification of Tertiary and Cretaceous planktonic foraminifera. Marine Micropaleontology 3, 175–196.
Edgar, K.M., Pälike, H., Wilson, P.A., 2013. Testing the impact of diagenesis on the $\delta^{18}O$ and $\delta^{13}C$ of benthic foraminiferal calcite from a sediment burial depth transect in the equatorial Pacific. Paleoceanography 28, 468–480.
Elderfield, H., Gieskes, J.M., 1982. Sr isotopes in interstitial waters of marine sediments from Deep Sea Drilling Project cores. Nature 300, 493–497.
Emiliani, C., 1955. Pleistocene temperatures. Journal of Geology 63, 538–578.
Emrich, K., Ehhalt, D.H., Vogel, J.C., 1970. Carbon isotope fractionation during the precipitation of calcium carbonate. Earth and Planetary Science Letters 8, 363–371.
Epstein, S., Buchsbaum, R., Lowenstam, H., Urey, H., 1953. Revised carbonate-water isotopic temperature scale. Geological Society of America Bulletin 64, 1315–1326.
Erez, J., Luz, B., 1983. Experimental paleotemperature equation for planktonic foraminifera. Geochimica et Cosmochimica Acta 47, 1025–1031.
Fairbanks, R.G., 1989. A 17,000-year glacio-eustatic sea level record: influence of glacial melting rates on the Younger Dryas event and deep ocean circulation. Nature 342, 637–642.
Fairbanks, R.G., Matthews, R.K., 1978. The marine oxygen isotope record in Pleistocene coral, Barbados, West Indies. Quaternary Research 10, 181–196.
Fairbanks, R.G., Sverdlove, M.S., Free, R., Wiebe, P.H., Bé, A.W.H., 1982. Vertical distribution and isotopic fractionation of living planktonic foraminifera from the Panama Basin. Nature 298, 841–844.
Farquhar, G.D., Ehleringer, J.R., Hubick, K.T., 1989. Carbon isotope discrimination and photosynthesis. Annual Review of Plant Physiology and Plant Molecular Biology 40, 503–537.
Faure, G., 1986. Principles of Isotope Geology, second ed. John Wiley and Sons. p. 589.
Fourier, J., 1824. Remarques générales sur les températures du globe terrestre et des espaces planétaires. Annales de chimie et de physique 27, 136–167.
Ganeshram, R.S., Pederson, T.F., Calvert, S.E., Murray, J.W., 1995. Large changes in oceanic inventories from glacial to interglacial periods. Nature 376, 755–758.
Gasperi, J.T., Kennett, J.P., 1992. Isotopic evidence for depth stratification and paleoecology of Miocene planktonic foraminifera: western equatorial Pacific DSDP site 289. In: Tsuchi, R., Ingle Jr., J. (Eds.), Pacific Neogene Events. University of Tokyo, press, pp. 117–147.
Gonfiantini, R., 1981. The δ-notation and the mass spectrometric measurement techniques. Technical Reports Series no. 210. In: Gat, J.R., Gonfiantini, R. (Eds.), Stable Isotope Hydrology: Deuterium and Oxygen-18 in the Water Cycle. IAEA, Vienna, pp. 35–84.
Gonfiantini, R., 1986. Environmental isotopes in lake studies. In: Fritz, P., Fontes, J.C. (Eds.), Handbook of Environmental Isotope Geochemistry, vol. 2. Elsevier, pp. 113–168.
Gröning, M., 2004. International stable isotope reference materials. In: first ed. In: de Groot, P.A. (Ed.), Handbook of Stable Isotope Analytical Techniques, vol. 1. Elsevier, pp. 874–906.
Grootes, P.M., Stuiver, M., 1997. Oxygen 18/16 variability in Greenland snow and ice with 103 to 105 year time resolution. Journal of Geophysical Research 102, 26455–26470.
Grossman, E.L., 1984. Carbon isotopic fractionation in live benthic foraminifera: comparison with inorganic precipitate studies. Geochimica et Cosmochimica Acta 48, 1505–1512.

Haq, B.U., Worsley, T.R., Burckle, L.H., Douglas, R.G., Keigwin Jr., L.D., Opdyke, N.D., Savin, S.M., Sommer II, M.A., Vincent, E., Woodruff, R., 1980. Late Miocene marine carbon isotopic shift and synchroneity of some phytoplanktonic biostratigraphic events. Geology 8, 427–431.

Haug, G.H., Tiedemann, R., 1998. Effect of the formation of the Isthmus of Panama on Atlantic Ocean thermohaline circulation. Nature 393, 673–676.

Hemleben, C., Spindler, M., Anderson, O.R., 1989. Modern Planktonic Foraminifera. Springer. p. 363.

Hodell, D., Kennett, J.P., 1984. Late Miocene carbon shift in DSDP Site 516A, western south Atlantic. Geological Society of America 16, 540.

Hoefs, J., 2009. Stable Isotope Geochemistry, sixth ed. Springer-Verlag. p. 285.

Imbrie, J., Boyle, E.A., Clemens, S.C., Duffy, A., Howard, W.R., Kukla, G., Kutzbach, J., Martinson, D.G., McIntyre, A., Mix, A.C., Molfino, B., Morley, J.J., Peterson, L.C., Pisias, N.G., Prell, W.L., Raymo, M.E., Shackleton, N.J., Toggweiler, J.R., 1992. On the structure and origin of major glaciation cycles. 1. Linear responses to Milankovitch forcing. Paleoceanography 7, 701–738.

Imbrie, J., Shackleton, N.J., Pisias, N.G., Morley, J.J., Prell, W.L., Martinson, D.G., Hays, J.D., McIntyre, A., Mix, A.C., 1984. The orbital theory of Pleistocene climate: support from a revised chronology of the marine $\delta^{18}O$ record. In: Berger, A., Imbrie, J., Hays, J., Kukla, G., Saltzman, B. (Eds.), Milankovitch and Climate: Understanding the Response to Astronomical Forcing. D. Reidel, Hingham, Massachusetts, USA, pp. 269–305.

IPCC, 2013. Summary for Policymakers. In: Stocker, T.F., Qin, D., Plattner, G.-K., Tignor, M., Allen, S.K., Boschung, J., Nauels, A., Xia, Y., Bex, V., Midgley, P.M. (Eds.), Climate Change 2013: The Physical Science Basis. Contribution of Working Group I to the Fifth Assessment Report of the Intergovernmental Panel on Climate Change. Cambridge University Press, Cambridge, UK and New York, NY, USA, pp. 3–29.

John, C.M., Follmi, K.B., Kaenel, E.D., Adatte, T., Steinman, P., Badertscher, C., 2002. Carbonaceous and phosphate rich sediments of the Miocene Monterey Formation at El Captain State Beach California, USA. Journal of Sedimentary Research 72, 252–267.

Johnsen, S.J., Clausen, H.B., Dansgaard, W., Fuhrer, K., Gundestrup, N., Hammer, C.U., Iversen, P., Jouzel, J., Stauffer, B., Steffensen, J.P., 1992. Irregular glacial interstadials recorded in a new Greenland ice core. Nature 359, 311–313.

Kahn, M.I., Williams, D.F., 1981. Oxygen and carbon isotopic composition of living planktonic foraminifera from the northeast Pacific Ocean. Palaeogeography, Palaeoclimatology, Palaeoecology 33, 47–69.

Keigwin Jr., L.D., 1979. Late Cenozoic stable isotope stratigraphy and paleoceanography of DSDP sites from the east equatorial and central North Pacific Ocean. Earth and Planetary Science Letters 45, 361–382.

Keigwin Jr., L.D., Shackleton, N.J., 1980. Uppermost Miocene carbon isotope stratigraphy of a piston core in the equatorial Pacific. Nature 284, 613–614.

Keller, G., 1985. Depth stratification of planktonic foraminifera in the Miocene ocean. Geological Society of America, Memoir 163, 177–195.

Kennett, J.P., 1986. Miocene to early Pliocene oxygen and carbon isotope stratigraphy in the Southwest Pacific. Initial reports. Deep Sea Drilling Project, vol. 90, pp. 1383–1412.

Killingley, J.S., 1983. Effect of diagenetic recrystallization on the $^{18}O/^{16}O$ values of the deep sea sediments. Nature 301, 594–597.

Kroopnick, P.M., 1985. The distribution of $\delta^{13}C$ of CO_2 in the world oceans. Deep-Sea Research 32, 57–84.

Labeyrie, L.D., Duplessy, J.C., Blanc, P.L., 1987. Deep water formation and temperature variations over the last 125,000 years. Nature 327, 477–482.

Lane, G.A., Doyle, M., 1956. Fractionation of oxygen isotopes during respiration. Science 123, 574–576.

Lisiecki, L.E., Raymo, M.E., 2005. A Pliocene-Pleistocene stack of 57 globally distributed benthic $\delta^{18}O$ records. Paleoceanography 20, PA1003.

Liu, K., Kaplan, I.R., 1989. The eastern tropical Pacific as a source of 15N-enriched nitrate in seawater off southern California. Limnology and Oceanography 34, 820–830.

Loutit, T.S., Keigwin Jr, L.D., 1982. Stable isotopic evidence for latest Miocene sea-level fall in the Mediterranean region. Nature 300 (5888), 163–166.

Loutit, T.S., Kennett, J.P., 1979. Application of carbon isotope stratigraphy to late Miocene shallow marine sediments. Science 204, 1196–1199.

Loutit, T.S., Kennett, J.P., Savin, S.M., 1983. Miocene equatorial and southwest Pacific paleoceanography from stable isotope evidence. Marine Micropaleontology 8, 215–233.

Martinson, D.G., Pisias, N.G., Hayes, J.D., Imbrie, J., Moore, T.C., Shackleton, N.J., 1987. Age dating and the orbital theory of the ice-ages: development of a high-resolution 0 to 300,000 year chronostratigraphy. Quaternary Research 27, 1–29.

McConnaughey, T., 1989a. ^{13}C and ^{18}O disequilibrium in biological carbonates.1. Patterns. Geochimica et Cosmochimica Acta 53, 151–162.

McConnaughey, T., 1989b. ^{13}C and ^{18}O disequilibrium in biological carbonates. 2. In vitro simulation of kinetic isotope effects. Geochimica et Cosmochimica Acta 53, 163–171.

McCrea, J.M., 1950. The isotopic chemistry of carbonates and a paleotemperature scale. Journal of Chemical Physics 18, 849–857.

Merlivat, L., Jouzel, J., 1979. Global climatic interpretation of the deuterium-oxygen 18 relationship for precipitation. Journal of Geophysical Research 84, 5029–5033.

Meyers, P.A., 1994. Preservation of elemental and isotopic source identification of sedimentary organic matter. Chemical Geology 114, 289–302.

Miller, K.G., Fairbanks, R.G., Mountain, G.S., 1987. Tertiary oxygen isotope synthesis, sea level history and continental margin erosion. Paleoceanography 2, 1–19.

Milankovitch, M., 1930. Mathematische Klimalehre und Astronomische Theorie der Kilimaschwankungen. In: Köppen, W., Geiger, R. (Eds.), Handbuch der Klimatologie, vol. 1 (Pt. A). Borntraeger, Berlin, pp. 1–176.

Müller, P.J., 1977. CN ratios in Pacific deep-sea sediments: effect of inorganic ammonium and organic nitrogen compounds sorbed by clays. Geochimica et Cosmochimica Acta 41, 765–776.

Naqvi, S.W.A., 1987. Some aspects of the oxygen deficient conditions and denitrification in the Arabian Sea. Journal of Marine Research 29, 459–469.

Naqvi, S.W.A., 1994. Denitrification process in the Arabian Sea. Proceedings of the Indian Academy of Sciences 103, 181–202.

Olson, D.B., Hitchcock, G.L., Fine, R.A., Warren, B.A., 1993. Maintenance of the low oxygen layer in the central Arabian Sea. Deep-Sea Research Part II: Topical Studies in Oceanography 40, 673–685.

Petit, et al., 1999. Climate and atmospheric history of the past 420,000 years from the Vostok ice core, Antarctica. Nature 399, 429–436.

Pisias, N.G., Martinson, D.G., Moore, T.C., Shackleton, N.J., Prell, W., Hays, J.D., Boden, G., 1984. High resolution stratigraphic correlation of benthic oxygen isotope records spanning the last 300,000 years. Marine Geology 56, 119–136.

Prell, W.L., Imbrie, J., Martinson, D.G., Morley, J.J., Pisias, N.G., Shackleton, N.J., Streeter, H.F., 1986. Graphic correlation of oxygen isotope stratigraphy application to the late Quaternary. Paleoceanography 1, 137–162.

Premuzic, E.T., Benkovitz, C.M., Gaffney, J.S., Walsh, J.J., 1982. The nature and distribution of organic matter in the surface sediments of the world oceans and seas. Organic Geochemistry 4, 63–77.

Qasim, S.Z., 1982. Oceanography of the northern Arabian Sea. Deep-Sea Research 29, 1041–1068.

Redfield, A.C., 1934. On the proportion of organic derivatives in seawater and their relation to the composition of plankton. In: James Johnstone Memorial Volume. Liverpool University Press, pp. 177–192.

Redfield, A.C., 1958. The biological controls of chemical factors in the environment. American Journal of Science 46, 205–221.

Saino, T., Hattori, A., 1987. Geographical variation of the water column distribution of suspended particulate organic nitrogen and its 15N natural abundance in the Pacific and its marginal seas. Deep-Sea Research 34, 807–827.

Savin, S.M., Abel, L., Barrera, E., Hodell, D., Keller, G., Kennett, J.P., Killingley, J., Murphy, M., Vincent, E., 1985. The evolution of Miocene surface and near surface marine temperatures: oxygen isotopic evidence. Geological Society of America, Memoir 163, 40–82.

Shackleton, N.J., 1967. Oxygen isotope analyses and Pleistocene temperatures re-assessed. Nature 215, 15–17.

Shackleton, N.J., 1974. Attainment of isotopic equilibrium between ocean water and benthonic foraminifera genus Uvigerina. Isotopic changes in the ocean during the last glacial. CNRS, Colloques Internationals 219, 203–209.

Shackleton, N.J., Kennett, J.P., 1975. Paleotemperature history of the Cenozoic and the initiation of the Antarctic glaciation:Oxygen and carbon isotope analyses in DSDP sites 277,279 and 281. Initial Reports. Deep Sea Drilling Project, vol. 29, pp. 743–756.

Shackleton, N.J., Opdyke, N.D., 1973. Oxygen isotope and paleomagnetic stratigraphy of equatorial Pacific core V28-238: oxygen isotope temperatures and ice volumes on a 105 and 106 year scale. Quaternary Research 3, 39–55.

Shi, W., Morrison, J.M., Bohm, E., Manghnani, V., 2000. The Oman upwelling zone during 1993, 1994 and 1995. Deep-Sea Research Part II: Topical Studies in Oceanography 47, 1227–1247.

Smith, R.L., Bottero, J.S., 1977. On upwelling in the Arabian Sea. In: Angel, M. (Ed.), A Voyage of Discovery. Pergamon Press, pp. 291–304.

Spero, H.J., Bijma, J., Lea, D.W., Bemis, B.E., 1997. Effect of seawater carbonate concentration on foraminiferal carbon and oxygen isotopes. Nature 390, 497–500.

Spero, H.J., Lea, D.W., 1993. Intraspecific stable isotope variability in the planktic foraminifera Globigerinoides sacculifer: results from laboratory experiments. Marine Micropaleontology 22, 221–234.

Srinivasan, M.S., Sinha, D.K., 1998. Early Pliocene closing of the Indonesian Seaway: evidence from northeast Indian Ocean and southwest Pacific deep sea cores. Journal of Southeast Asian Earth Sciences 16, 29–44.

Srinivasan, M.S., Sinha, D.K., 2000. Ocean circulation changes in the Indo-Pacific during 5.6 to 4.2 Ma: planktic foraminiferal and isotopic evidences. Proceedings of the Indian Academy of Sciences 109, 315–328.

Sautter, L., Thunell, R., 1991. Seasonal variability in the $\delta^{18}O$ and $\delta^{13}C$ of planktonic foraminifera from an upwelling environment: sediment trap results from the San Pedro Basin, Southern California Bight. Paleoceanography 6, 307–334.

Steens, T.N.F., Ganssen, G., Kroon, D., 1992. Oxygen and carbon isotope in planktonic foraminifera as indicators of upwelling intensity and upwelling induced high productivity in sediments from the northwestern Arabian Sea. In: Summerhayes, C.P., Prell, W.L., Emeis, K.C. (Eds.), Upwelling Systems Evolution since the early Miocene.Special Publication, vol. 64. Geological Society, London, pp. 107–120.

Takahasi, T., Broecker, W.S., Langer, S., 1985. Redfield ratio based on chemical data from isopycnal surfaces. Journal of Geophysical Research 90, 6907–6924.

Tiwari, M., Nagoji, S.S., Kartik, T., Drishya, G., Parvathy, R.K., Rajan, S., 2013. Oxygen isotope-salinity relationships of discrete oceanic regions from India to Antarctica vis-à-vis surface hydrological processes. Journal of Marine Systems 113–114, 88–93.

Tiwari, M., Ramesh, R., Sheshshayee, M., Bhushan, R., Somayajulu, B.L.K., Jull, A.J.T., Burr, G.S., 2010. Did the Indo-Asian summer monsoon decrease during the Holocene following insolation? Journal of Quaternary Science 25, 1179–1188.

Tyndall, J., 1861. On the absorption and radiation of heat by gases and vapours, and on the physical connection of radiation, absorption, and conduction. Philosophical Magazine Series 4, 22 (169–194), 273–285.

Urey, H., 1947. The thermodynamic properties of isotopic substances. Journal of Chemical Society 562–581.

Vail, P.R., Hardenbol, J., 1979. Sea-level changes during the Tertiary. Oceanus 22, 71–79.

Vincent, E., Berger, W.H., 1985. Carbon dioxide and polar cooling in the Miocene: the Monterey hypothesis. In: Sundquist, E.T., Broecker, W.S. (Eds.), The Carbon Cycle and Atmospheric CO_2: Natural Variations Archean to Present. American Geophysical Union, vol. 32, pp. 455–468.

Vincent, E., Killingley, J.S., Berger, W.H., 1980. The magnetic Epoch-6 carbon shift: a change in the ocean's 13C/12C ratio 6.2 million years ago. Marine Micropaleontology 5, 185–203.

Wefer, G., Berger, W.H., 1991. Isotope paleontology: growth and composition of extant calcareous species. Marine Geology 100, 207–248.
Williams, D., Lerche, I., Full, W.E., 1988. Isotope Chronostratigraphy, Theory and Methods. Academia Press Inc. p. 345.
Woodruff, F., Savin, S.M., 1985. $\delta^{13}C$ values of Miocene Pacific benthic foraminifera: correlations with sea level and biological productivity. Geology 13, 119–122.
Wyrtki, K., 1971. Oceanographic Atlas of the International Indian Ocean Expedition. National Science Foundation Publications, OCE/NSF 86-00-001, Washington, DC. p. 531.
You, Y., Tomczak, M., 1993. Thermocline circulation and ventilation in the Indian Ocean derived from water mass analysis. Deep-Sea Research 40, 13–56.

CHAPTER

TIME AVERAGING AND COMPOSITIONAL AVERAGING IN BIOGENIC CARBONATES: IMPLICATIONS FOR CHEMOSTRATIGRAPHY

4

Pratul Kumar Saraswati
Department of Earth Sciences, IIT Bombay, Mumbai, India

4.1 INTRODUCTION

A stratigraphic record is the product of a geochemical system consisting of geologic setting, climate, and processes of sediment production (Berger and Vincent, 1981). Several of these processes occur on global scale and as a result, chemostratigraphy concerned with variation in elemental and isotopic composition of the sedimentary strata has emerged as an important subdiscipline of stratigraphy. Chemostratigraphic markers are biostratigraphically dated to formally define age boundaries. The Global Stratotype Section and Point (GSSP) of the Eocene base, for example, is at a prominent carbon isotope excursion (CIE). The chemostratigraphy of carbonate sequences involves both stable isotopes and trace elements. The analysis is usually on bulk samples or on the shells of microfossils such as foraminifera and calcareous nannofossils. The variability in chemical data of marine sediments is due to change in climate or ocean water chemistry or many other processes. There are, however, two important factors that potentially influence the chemical data. These are time averaging and compositional averaging of the shells. The oceanic microfossils can provide biostratigraphic resolution of <1 Ma for many parts of the Cenozoic. Biostratigraphers, however, know well that resolutions finer than 1–2 Ma are not practically possible due to several constraints including diachroneity in the evolutionary first and last appearances of species, time averaging of fossil assemblages, and stratigraphic disorder due to the bioturbation. The effects of these processes were overlooked and considered inconsequential in the past. It has assumed greater significance today because of the attempts to interpret geological processes at <1 Ma or even at millennial to centennial and decadal scales. In addition to the biostratigraphic resolution among strata, within-stratum depositional resolution (Kowalewski and Bambach, 2008) further limits the use of geochemical data in high-resolution chemostratigraphy. The depositional resolution deteriorates due to temporal and spatial mixing within individual strata. The stratigraphers emphasize that depositional resolution of records should be carefully evaluated if the research objective is to analyze isotope ratios or any other parameter of the physical materials (such as fossils) extracted from the strata (Kowalewski and Bambach, 2008). The geochemical data from the same bed may therefore be subject to (1) variation in time and (2) mixing of geochemical signatures of different environments.

Chemostratigraphy. http://dx.doi.org/10.1016/B978-0-12-419968-2.00004-2

The modern analytical techniques allow chemical analyses of recent and fossil shells at microscales such that oxygen and carbon isotopes or trace elements of single chambers of microfossils can be determined. These studies have unambiguously shown that the chemical composition of shells is not homogeneous because individuals of a species may incorporate different amounts of trace elements and stable isotopes at different growth stages of calcification (see Tiwari et al. in this volume for more information). As a result, there is compositional averaging when bulk analysis is carried out on a multiple or single shells. This chapter discusses how the time averaging and compositional averaging influence the chemical data and particularly the interpretation of patterns and shifts in chemical variables in chemostratigraphy at submillenial scale.

4.2 BIOGENIC CARBONATE CHEMOSTRATIGRAPHY

The chemical approach to stratigraphy became routine in deep-sea studies after the classic paper of Emiliani (1955) on stable isotopic composition of foraminifera. It has since found widespread use because of the fact that global climatic and oceanographic processes drive isotopic signals. The oxygen and carbon isotopic variation in Cenozoic marine sediments are therefore also of chronologic significance and, when integrated with biostratigraphy, enhances the stratigraphic resolution of sedimentary successions. Chemostratigraphy initially applied in pelagic sequences has since been successfully used for shallow water carbonates and in land-based sections. The following attributes of isotopic records are used in chemostratigraphy (Williams et al., 1988):

1. Overall pattern (long-term trends, events, and cycles),
2. Absolute values of $\delta^{18}O$ and $\delta^{13}C$ with respect to time or depth,
3. Occurrence of particular shifts in the isotopic record,
4. Magnitude of shifts, long-term trends, cycles, and events as a function of time,
5. Direction of shifts (positive or negative).

Similar attributes also characterize the element-based chemostratigraphy. The chemostratigraphic units, referred as "chemozones," are defined based on the above listed attributes in the geochemical profile and delineated visually. In the absence of a definite trend and shift, visual delineation may be subjective and therefore statistical discrimination of chemozones is advisable (Ramkumar et al., 2010, 2011).

A well-known application of stable isotope method in stratigraphy is recognition of marine isotope stages (MIS) in Quaternary (Shackleton and Opdyke, 1973). It followed Emiliani' (1955) conclusion that the cyclical variation in oxygen isotopic composition of foraminiferal carbonates corresponds to glacial–interglacial periods. Shackleton and Opdyke (1973) applied it to stratigraphy and defined the boundaries of isotope stages at the times of rapid isotopic change in the oceans. The odd-numbered stages are characterized by lighter oxygen isotope values (less continental ice) and even-numbered stages by heavier oxygen isotope values (more continental ice). The MIS 1 contains the recent sediments. Shackleton and Opdyke (1976) integrated MIS with magnetostratigraphy and defined 23 stages in the past ~1.0 Ma record (extending to the Jaramillo magnetic event). The International Commission on Stratigraphy has used a prominent shift of 3.0–5.0‰ in organic CIE to define the GSSP for the base of Eocene Series in Dababiya section in Egypt. This prominent geochemical marker is reported worldwide in shallow-marine, deep-marine, and terrestrial sections alike. The boundary is also substantiated biostratigraphically by foraminifera and dinoflagellate taxa (Aubry et al., 2007).

In further deep time, seven major excursions of $\delta^{13}C$ are recorded in the consistent stratigraphic positions of the Cretaceous limestones of widely distributed sections in circum Atlantic–western Tethyan regions (Scholle and Arthur, 1980). The positive excursions (shifts to higher $\delta^{13}C$ values) are noted near Aptian–Albian and Cenomanian–Turonian boundaries and negative excursions (shifts to lower $\delta^{13}C$ values) are noted near Jurassic–Cretaceous, Albian–Cenomanian, Turonian–Coniacian, and Cretaceous–Tertiary boundaries. The pronounced excursions of $\delta^{13}C$ in Early Cretaceous pelagic carbonates are also observed to occur in the platform carbonates (Ferreri et al., 1997), proving the efficacy of carbon isotope stratigraphy in correlating sections of contrasting paleoenvironmental settings. The synchronicity of several of the geochemical markers is validated by biostratigraphy and magnetostratigraphy. Carbon isotopic shift across Cenomanian–Turonian boundary occurs uniformly relative to major biostratigraphic markers in Europe and North America (Gale et al., 1993). Many workers have integrated chemostratigraphy with biostratigraphy to enhance stratigraphic resolution (Crespo de Cabrera et al., 1999; De Gea et al., 2003) or used it as an alternative tool in stratigraphic correlation where biostratigraphy is of little use (Valladeres et al., 1996). The isotopic trends are often used to infer stratigraphic hiatuses (Magaritz, 1991).

The major, minor, and trace elements are also used in chemostratigraphy. The absolute values of elemental concentrations, ratios of two elements or shifts, and anomalies in concentration of elements characterize the chemozones. Each element has its own significance as fingerprinting of the geological processes, events, and specific beds. Some of the elements, for example, Si, Al, Ti, and Zr, are indicative of detrital flux in the basin and therefore related to the provenance. The trace elements in carbonate shells of foraminifera respond to ocean chemistry, productivity, and climate. Mo, Fe, Mn, V, Ni, and U characterize the level of oxygenation and the content of C, N, P, Cd, and Ba are related to nutrient and biological productivity. Magnesium content of the foraminiferal shells is related to seawater temperatures and Sr concentration changes with sea level. A comprehensive study on Sr/Ca and Mg/Ca of Jurassic–Cenozoic pelagic carbonates has led to chemostratigraphic zonation for wider application, validated by their synchronicity and worldwide occurrence of the geochemical shifts. It has been suggested that variations in submarine hydrothermal activity and fluctuations in sea level are primary processes for the systemic change in the chemistry of the pelagic carbonates (Renard, 1986).

A stratigraphic section thus contains elemental signatures of local, regional, and global processes. There have been numerous studies using chemical characteristics of stratigraphic sections to demarcate geochemical events, correlate the sections, and integrate isotopic and elemental variation with biostratigraphy or chronostratigraphy (Berger and Vincent, 1981; Jorgensen, 1986; Renard, 1986; Kaminski and Malmgren, 1989; Saraswati and Ramesh, 1992; Nandy et al., 1995; Madhavaraju et al., 2004).

4.3 IMPLICATIONS OF TIME AVERAGING AND COMPOSITIONAL AVERAGING

"Time averaging" refers to pooling of successive populations and communities into a single assemblage of remains (Kidwell, 1998). All the fossil assemblages are time averaged at different orders of magnitude. It arises due to the faster rate of population turnover compared with the rate of sediment accumulation. The assemblage therefore comprises of noncontemporaneous remains. The assemblage of shells on a modern-day beach represents populations over several decades to hundreds of years until covered by sediments (Fig. 1). By a similar analogy, different shells on bedding plain of geologically

FIGURE 1

Accumulation of mollusk shells on a beach in Gulf of Cambay, India. The individual shells in the assemblage may be of different generations and the input of newer generations of shells will continue till covered by the sediments.

FIGURE 2

Fossil mollusks on the bedding plain of Miocene limestone in Kutch, India. It represents a time-averaged assemblage of shells such that two shells may be of different generations, tens to hundreds of years apart.

older strata (Fig. 2) may differ in age by hundreds to thousands of years. The AMS ^{14}C dates of Hawaian beach sands have indicated that the duration of time averaging in foraminifer *Amphistegina* on the beach is of the order of 2 ka (Table 1; Martin et al., 1996; Resig, 2004). Fresh-looking, uncemented, late Pleistocene foraminifer *Marginopora vertebralis* co-occurs with recent specimens, recording a time averaging of 125 ka (Murray-Wallace and Belperio, 1995). Other physical and biotic processes

Table 1 Time Averaging in Foraminiferal Shells in Shallow Marine and Beach Environments

Foraminifera	Time-Averaging	Dating Method	Authors
Archaias, Amphistegina	2 ka	AMS^{14}C	Martin et al. (1996)
Marginopora	125 ka	Amino acid racemization	Murray Wallace and Belperio (1995)
Amphistegina	>1.5 ka	AMS^{14}C	Resig (2004)

FIGURE 3

A crab burrow on tidal flats of Daman, India. Older sediments (white colored) several centimeters below are brought to the surface and mixed with the present-day sediments by crab burrowing into the sediments for shelter. It results into mixing of sediments and fossils of different ages and possibly of different environments.

also cause the mixing of fossils of different ages. Bioturbation on the present-day tidal-flats by crabs bring sediments several centimeters below the surface to the top (Fig. 3), mixing the sediments and shells of different ages. By analogy, the fossils in bioturbated beds may contain typically time-averaged assemblages, particularly the microfossils. The magnitude of time averaging may not be significant for routine stratigraphic studies and therefore overlooked by most workers. It becomes critically important, however, when interpretation demands higher resolution such as in paleoenvironment or paleoclimate reconstruction at millennial and centennial to decadal scales. The estimate of time averaging for different types of fossil assemblages is given in Fig. 4. The snapshots (or ecological census) contain contemporaneous fossils while time-averaged records contain fossils mixed over time intervals of hundreds to thousands of years. Snapshots are rare and most geological records are time averaged. The environmental and biostratigraphic condensations have much higher magnitudes of time averaging up to millions of years. It implies that two pristine shells in the same bed may differ in age by more than a hundred thousand years. These observations indicate that geochemical data collected on two different shells from the same bed may represent the values of the variables in two different times. It severely compromises with climatic or oceanographic interpretations at millennial and submillenial scales.

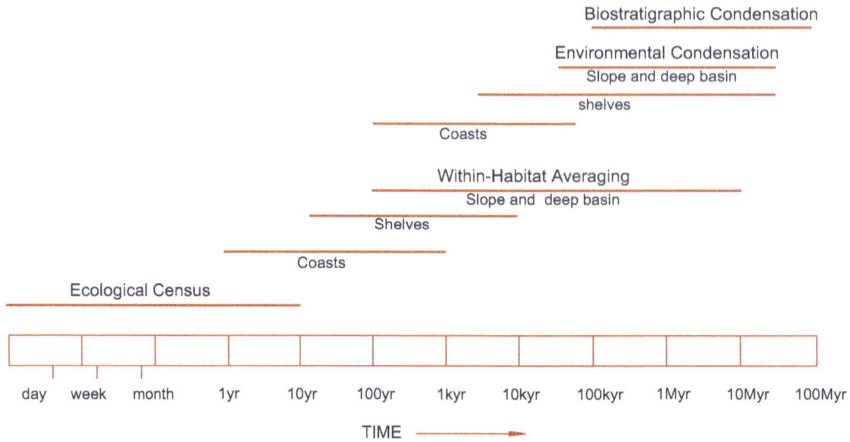

FIGURE 4

Magnitude of time averaging in different types of fossil assemblages.

After Kidwell (1998).

Advancements in analytical techniques have made it possible to analyze stable isotopic composition of single shells of microfossils. The laser-ablated, micron-sized samples of shells can be analyzed for trace elements by ICP-MS and EPMA and SIMS can map spatial variation in chemical composition of microfossils. While these techniques have greatly improved our understanding of biomineralization, chemostratigraphers need to carefully plan the analysis and objectively interpret the data. The compositional data of biogenic carbonate is highly heterogeneous, and most carbonates are biogenic in nature. It is important what is analyzed: bulk carbonate, multiple or single shells, and from which part of the shell in case of microanalyses. The result may differ substantially. The whole-shell oxygen isotopic composition of foraminifera corresponds to average temperature during the calcification that may range from a few days to weeks and more than a year depending on the life span of the foraminifer. It may be several years in the cases of mollusca and brachiopoda. The oxygen isotopic composition of dissected chambers at several growth stages of the foraminifer *Marginopora* shows seasonal variation corresponding to change in ambient water temperature (Fig. 5). It is noted in a study that $\delta^{18}O$ of the bulk carbonate of a single shell is −2.2‰ but it varies from −1.2‰ to −3.7‰ for the dissected chambers (*see* Saraswati, 2004; Saraswati et al., 2004 for more details). Inter- and intraspecies differences are also recorded in elemental composition of shells of foraminifera, ostracoda, and corals. A large variation in Mg content between the species of larger benthic foraminifera inhabiting the same temperature conditions (Raja et al., 2005, 2007) is attributed to physiological activities. The stable isotopes and trace elements in foraminifera are controlled by the life history including habitat temperature, reproduction, depth distribution, life span, calcification, and photosymbiosis (Saraswati, 2008).

The chemostratigraphic data is thus susceptible to large variation and compositional averaging of different magnitudes whether the analysis refers to bulk samples of several species, multiple specimens of the same species, or single individuals of the same species. The absolute values of the chemical variables and the magnitudes of shifts should be cautiously used in stratigraphic interpretation. The trends, events, and shifts in chemical variables may be more helpful. In view of the time-averaged fossil assemblage in

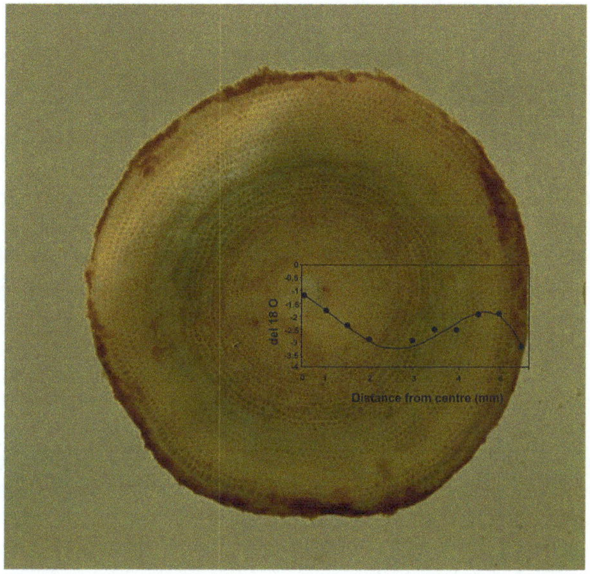

FIGURE 5

Variation in oxygen isotope values ($\delta^{18}O$) in a single shell of a modern foraminifer *Marginopora*. Depending on which part of the shell is analyzed, the difference in isotopic value may be of the order of ~2.5‰.

Data after Saraswati et al. (2004).

sedimentary strata and constraints of various types in the geochemical data of shells, higher resolution interpretations at millennial and submillennial scales need to be subjected to strict scrutiny.

4.4 CONCLUSIONS

- The sedimentary strata have temporal and spatial mixing of different orders in fossil assemblages. The geochemical data of the shells are therefore likely to be constrained by time averaging and mixing of geochemical signatures of different environments. The magnitude of time averaging is highly variable and for most geological records it is estimated to be hundreds to thousands of years.
- The stable isotopes and trace elements incorporated in the shells during calcification are clearly influenced by the life processes of organisms. The geochemical data are thus subject to compositional averaging.
- The sampling methodology and the generated data should be carefully scrutinized for high resolution chemostratigraphy, particularly when practiced at submillenial scale.

ACKNOWLEDGMENT

I thank the anonymous reviewer for his useful suggestions to improve the manuscript and to Dr. Mu. Ramkumar for inviting me to contribute to this volume.

REFERENCES

Aubry, M.P., Ouda, K., others, 2007. The global standard stratotype-section and point (GSSP) for the base of the Eocene Series in the Dababiya section (Egypt). Episodes 30 (4), 271–286.

Berger, W.H., Vincent, E., 1981. Chemostratigraphy and biostratigraphic correlation: exercises in systemic stratigraphy. In: Oceanologia Acta, Proceedings 26th IGC, Paris, pp. 115–127.

Crespo de Cabrera, S., Sliter, W.V., Jarvis, I., 1999. Integrated foraminiferal biostratigraphy and chemostratigraphy of the Querecual formation (Cretaceous), Eastern Venezuela. Journal of Foraminiferal Research 29, 487–499.

De Gea, G.A., Castro, J.M., Aguado, R., Ruiz-Ortiz, P.A., Company, M., 2003. Lower Aptian carbon isotope stratigraphy from a distal carbonate shelf setting: the Cau section, Prebetic Zone, SE Spain. Palaeogeography Palaeoclimatology Palaeoecology 200, 207–219.

Emiliani, C., 1955. Pleistocene temperatures. Journal of Geology 63, 538–578.

Ferreri, V., Weissert, H., D'Argenio, B., Buonocunto, F.P., 1997. Carbon isotope stratigraphy: a tool for basin to carbonate platform correlation. Terra Nova 9, 57–61.

Gale, A.S., Jenkyns, H.C., Kennedy, W.J., Corfield, R.M., 1993. Chemostratigraphy versus biostratigraphy: data from around the Cenomanian – Turonian boundary. Journal of the Geological Society 150, 29–32.

Jorgensen, N.O., 1986. Chemostratigraphy of Upper Cretaceous chalk in the danish sub-Basin. Bulletin American Association of Petroleum Geologists 70, 309–317.

Kaminski, M.A., Malmgren, B.A., 1989. Stable isotope and trace element stratigraphy across the Cretaceous/Tertiary boundary in Denmark. Geologiska Foreningens I Stockholm Forhandlingar 111 (4), 305–312.

Kidwell, S.M., 1998. Time-averaging in the marine fossil record: overview of strategies and uncertainties. Geobios 30, 977–995.

Kowalewski, M., Bambach, R.K., 2008. The limits of paleontological resolution. In: Harries, P.J. (Ed.), High-Resolution Approaches in Stratigraphic Paleontology. Springer Science, pp. 1–48.

Madhavaraju, J., Kolosov, I., Buhlak, D., Armstrong-Altrin, J.S., Ramasamy, S., Mohan, S.P., 2004. Carbon and oxygen isotopic signatures in Albian-Danian limestones of Cauvery Basin, southeastern India. Gondwana Research 7 (2), 519–529.

Magaritz, M., 1991. Carbon isotopes, time boundaries and evolution. Terra Nova 3, 251–256.

Martin, R.E., Wehmiller, J.F., Harris, M.S., Liddell, W.D., 1996. Comparative taphonomy of foraminifera and bivalves in holocene shallow water carbonate and siliciclastic regimes: taphonomic grades and temporal resolution. Paleobiology 22, 80–90.

Murray-Wallace, C.V., Belperio, A.P., 1995. Identification of remanie fossils using amino acid racemisation. Alcheringa 18, 219–227.

Nandy, P., Saraswati, P.K., Kumar, A., 1995. Chemostratigraphy of the Eocene – Oligocene sequence of southwestern Kutch. In: Desai, N., Ganapathi, S. (Eds.), National Seminar on Recent Researches in Geology of Western India, pp. 53–59.

Raja, R., Saraswati, P.K., Rogers, K.M., Iwao, K., 2005. Magnesium and strontium compositions of recent symbiont-bearing benthic foraminifera. Marine Micropaleontology 58, 31–44.

Raja, R., Saraswati, P.K., Iwao, K., 2007. A field based study on variation in Mg/Ca and Sr/Ca in larger benthic foraminifera. Geochemistry Geophysics Geosystems 8. http://dx.doi.org/10.1029/2006/GC001478.

Ramkumar, M., Berner, Z., Stüben, D., 2010. Hierarchical delineation and multivariate statistical discrimination of chemozones of the Cauvery Basin, South India: implications on spatio-temporal scales of stratigraphic correlation. Petroleum Science 7, 435–447.

Ramkumar, M., Stüben, D., Berner, Z., 2011. Barremian – Danian chemostratigraphic sequences of the Cauvery Basin, India: implications on scales of stratigraphic correlation. Gondwana Research 19, 291–309.

Renard, M., 1986. Pelagic carbonate chemostratigraphy. Marine Micropaleontology 10, 117–164.

Resig, J.M., 2004. Age and preservation of *Amphistegina* (Foraminifera) in Hawaiian beach sand: implications for sand turnover rate and resource renewal. Marine Micropaleontology 50 (3–4), 225–236.

Saraswati, P.K., 2004. Ontogenetic isotopic variation in foraminifera – implications for palaeo proxy. Current Science 86, 858–860.

Saraswati, P.K., 2008. Life history of foraminifera: stable isotopic and elemental proxies. Journal Palaeontological Society of India 53, 1–8.

Saraswati, P.K., Ramesh, R., 1992. Eocene – Oligocene stable isotope stratigraphy of Kutch. Journal Geological Society of India 39, 427–432.

Saraswati, P.K., Seto, K., Nomura, R., 2004. Oxygen and carbon isotopic variation in co-existing larger foraminifera from reef flat at Akajima, Okinawa, Japan. Marine Micropaleontology 50, 339–349.

Scholle, P.A., Arthur, M.A., 1980. Carbon isotope fluctuations in Cretaceous pelagic limestones - potential stratigraphic and petroleum exploration tool. American Association of Petroleum Geologists Bulletin 64 (1), 67–87.

Shackleton, N.J., Opdyke, N.D., 1973. Oxygen isotope and palaeomagnetic stratigraphy of equatorial pacific core V28-238: oxygen isotope temperatures and ice volumes on a 10^5 and 10^6 year scale. Quaternary Research 3, 39–55.

Shackleton, N.J., Opdyke, N.D., 1976. Oxygen isotope and paleomagnetic stratigraphy of Pacific core V28-239, late pliocene to latest pleistocene. In: Cline, R.M., Hays, J.D. (Eds.), Investigation of Late Quaternary Paleoceanography and Paleoclimatology Geological Society of America Memoir, 145, pp. 449–464.

Valladares, I., Recio, C., Lendinez, A., 1996. Sequence stratigraphy and stable isotopes ($\delta^{13}C$, $\delta^{18}O$) of the late Cretaceous carbonate ramp of western margin of the Iberian chain (Soria, Spain). Sedimentary Geology 105, 11–28.

Williams, D.F., Lerche, I., Full, W.E., 1988. Isotope Chronostratigraphy - Theory and Methods. Academic Press, Inc. 345 pp.

CHAPTER

SEDIMENTOLOGY AND GEOCHEMISTRY OF THE LATE MIOCENE–PLIOCENE SUCCESSION IN THE FARS INTERIOR (SW IRAN): IMPLICATIONS ON DEPOSITIONAL AND TECTONIC SETTING, PROVENANCE AND PALEOWEATHERING IN THE ZAGROS BASIN

5

Mohammad Sahraeyan[1], Hossein Seif[2], Esmat Ebne Haddad[3], Nasim Mohammadzadeh[3]

[1]Department of Geology, Khorasgan (Isfahan) Branch, Islamic Azad University, Isfahan, Iran; [2]Department of Geology, North Tehran Branch, Islamic Azad University, Tehran, Iran; [3]Department of Geology, Fars Science and Research Branch, Islamic Azad University, Shiraz, Iran

5.1 INTRODUCTION

The Aghajari Formation, was developed during last orogenic phase (Mitchell and Reading, 1986) and can be traced throughout the Zagros Basin. The Aghajari Formation is equivalent to Fat'ha and Injana formations in Iraq, Upper Fars and Lower Fars in Syria, Lower Fars in Kuwait, and Kial and Jabal Kibrit formations in Saudi Arabia (Al-Husseini, 2008). It is best developed in the Dezful Embayment (2966 m; Motiei, 1994). Stratigraphy of this formation at its type section (Aghajari oil field) was studied by James and Wynd (1965). It consists of alternate brown to gray calcareous sandstones and red marls with gypsum interlayers and red siltstones in its type section (N:30°48′42″/E:49°47′29″; Motiei, 1994) and is characterized by two lithofacies groups. The first one mostly consists of continental clastic sediments (mudstone, sandstone, and conglomerate) best developed in Fars Interior and Dezful Embayment. The second group has marine characteristics (marl with gypsum interlayers) and spread in the Coastal Fars (Motiei, 1994; Bahrami, 1997). Deposition of the Aghajari Formation took place during the Late Miocene–Pliocene (Stöcklin, 1977), but it does not have unity in age and becomes younger when studied from northwestern to southeastern (Motiei, 1994) exposures. Its lower contact with Mishan Formation is conformable and the upper contact with Bakhtiari Formation is sometimes conformable and occasionally unconformable. In the study area, the lower and upper contacts are transitional. The rocks of the Aghajari Formation, described as lithic arenite and sublitharenite (Bahrami, 2009; Khanehbad et al., 2012; Falahatgar et al., 2012) that were sourced from an active

Chemostratigraphy. http://dx.doi.org/10.1016/B978-0-12-419968-2.00005-4

continental margin (Afghah and Farhoudi, 2012; Falahatgar et al., 2012) were interpreted to be fluvial in origin (Junhui et al., 2012; Sahraeyan and Bahrami, 2012a; Falahatgar et al., 2012; Ning et al., 2012). However, studies on the sedimentology, petrography, and geochemistry of the Aghajari Formation in the Fars Interior are lacking, which prompted the authors to document in detail its lithofacies, architectural elements, paleocurrents, and geochemistry to constrain on depositional sedimentary environment, provenance, tectonic setting, paleoclimate, and source-area weathering for corroborating the interpretations with that of published data. As this formation spreads over a wide geographic area, it is believed that documentation of the lithic, architectural, and geochemical characteristics at the Fars Interior section would help in improving the existing understanding and also in regional correlation.

5.2 REGIONAL SETTING

The Zagros Basin was a part of the Arabian shield, which subsequently underwent rifting during the Permo-Triassic and collision with Iranian Plate during the Late Tertiary (Berberian and King, 1981; Stöcklin, 1974). The Zagros Basin sprawls over the northeastern (present coordinates) Afro-Arabian continental margin and is affected by the Zagros orogeny (Early Cretaceous to present). The Zagros Basin, as the external part of the Zagros orogen (Alavi, 1980, 1994, 2007), extends southeast for nearly 2000 km from southeastern Turkey through northern Syria and northeastern Iraq to western and southwestern Iran. The northwestern boundary of the Zagros Basin is delimited by the East Anatolian strike-slip fault in southeastern Turkey and the southeastern boundary is confined by the Oman Line (Falcon, 1969). It lies on the northeastern margin of the Arabian Shield and has NW–SE trending structural zones (imbricated and folded) aligned subparallel to the plate margin separated by major fault zones such as the High Zagros and the mountain front faults. It is further laterally divided into the Lurestan, Dezful Embayment, and Fars Salient (interior and coastal) from northwest to southeast (Motiei, 1994; Falcon, 1974; Sahraeyan and Bahrami, 2012b,c; Falahatgar et al., 2012; Sahraeyan et al., 2013; Figs 1 and 2).

The study area is located about 35 km southeast of Jahrom city in the Fars Interior, southwestern Iran (Fig. 2). Tectonically this area is a part of foreland basin filled dominantly with a thick sedimentary sequence of clastic compositions (Alavi, 1994, 2004, 2007). In the study area, Aghajari Formation siliciclastic succession overlies Mishan Formation and underlies Bakhtiari Formation, both are conformably and gradually (Figs 3 and 4).

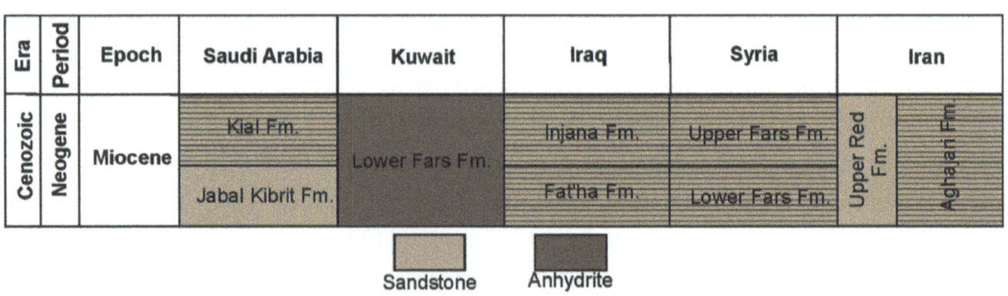

FIGURE 1

Equivalent of the Aghajari Formation in Saudi Arabia, Syria, Kuwait, Iraq, and Iran.

After Al-Husseini (2008).

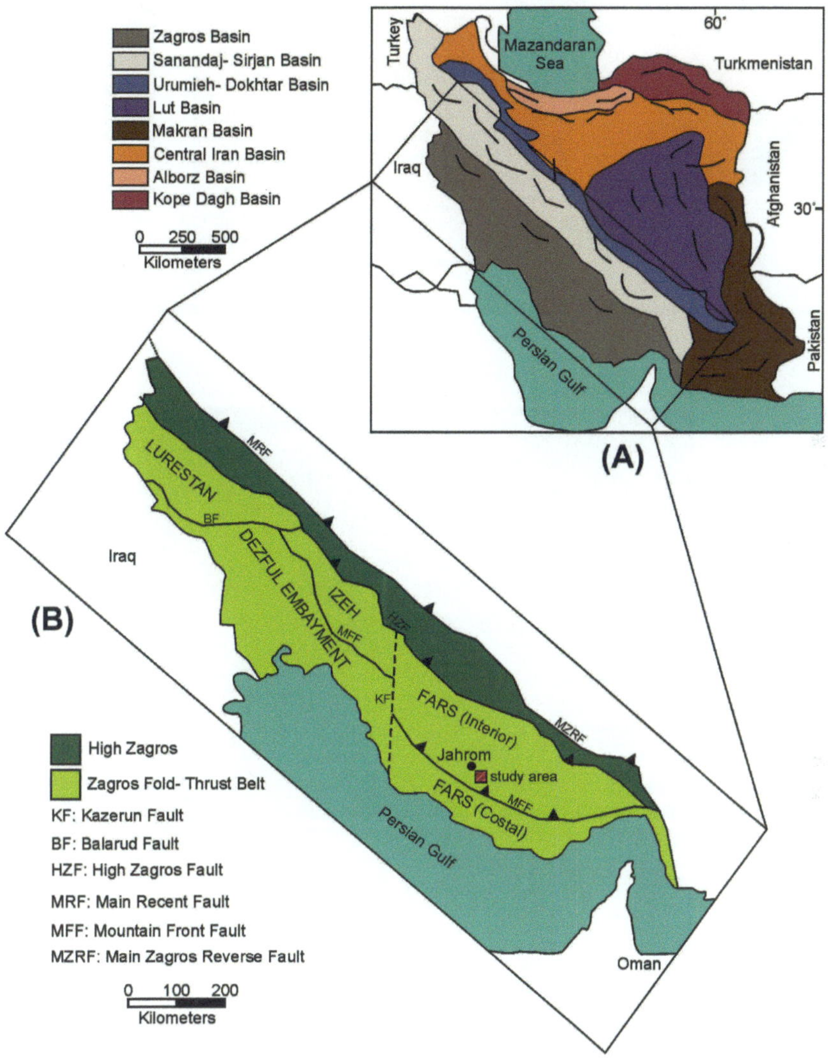

FIGURE 2

Location and geological map of the study area. (A) General map of Iran showing eight geologic provinces; the Barus section is located in the folded Zagros, *(adapted from Lacombe et al. (2006) and Mobasher and Babaie (2008))*, and (B) Subdivisions of the Zagros Mountains and Fars Subbasin, *(after Motiei (1994), with situation of the study sections in Interior Fars Subbasin)*.

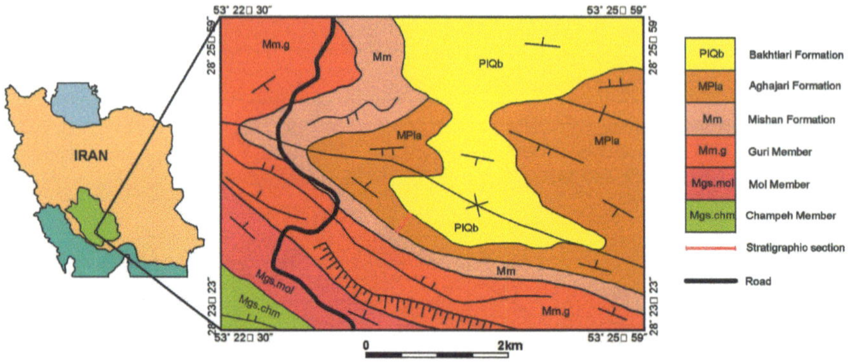

FIGURE 3

Location and geological map of the study area.

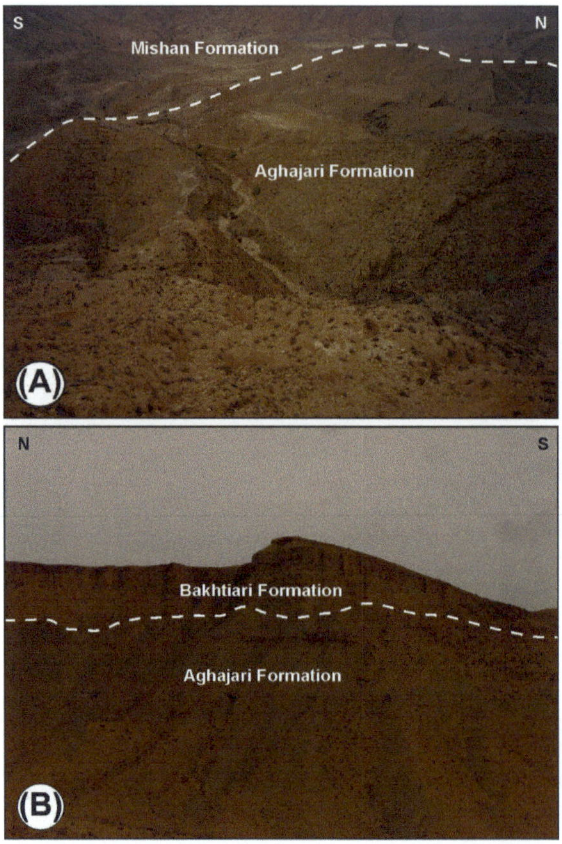

FIGURE 4

Field photographs showing (A) Lower boundary of Aghajari Formation with Mishan Formation, which is gradual and (B) Upper boundary of Aghajari Formation with Bakhtiari Formation is also gradual.

5.3 MATERIAL AND METHODS

Detailed facies analysis, petrographic and model analyses, paleocurrent measurements, definition of architectural elements, geochemical analysis of selected rock samples were attempted for the present study. It was followed by interpretation of the data toward depositional environments, determination of provenance and tectonic setting, and drawing inferences on source-area weathering.

For facies analysis, all the layers were identified in the 308 m thick studied section of the Aghajari Formation (Fig. 5) accurately and their individual thicknesses were measured and nominated following the procedures presented in Miall (1985, 1996). Nomenclatural assignment was based on particle size, sedimentary structures, and beds geometry (Miall, 1996), and the sedimentary structures were also utilized following the procedures enlisted by Tucker (2011). Paleocurrent analysis was performed given cognizance to the precepts of Potter and Pettijohn (1978) and Boggs (2006). Paleocurrent analysis is involved using sedimentary structures to determine the direction of flow like that of a river, a group of streams within a basin, the wind direction within a region, or the direction of oceanic currents

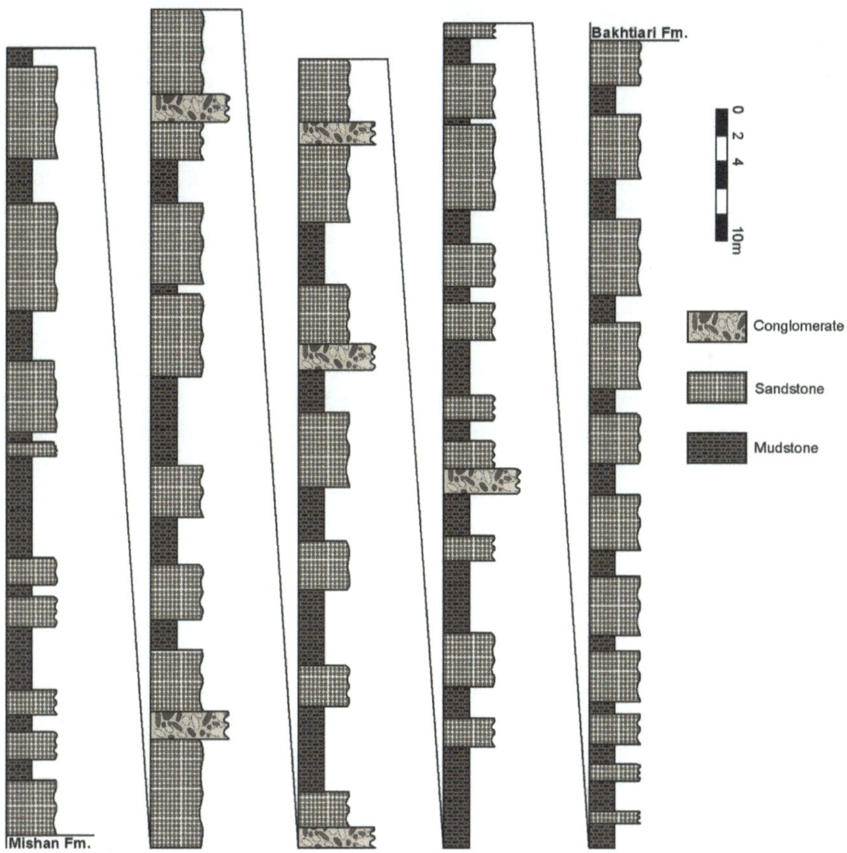

FIGURE 5

Lithostratigraphy of the Aghajari Formation in the study area.

(Potter and Pettijohn, 1978). However, as individual sedimentary structures indicate the flow direction at that geographic point and at that instant in time, collective average of the directions need to be considered for understanding regional current pattern (Boggs, 2006). Ten sites were measured for paleocurrent analysis. All the suitable structures (e.g., cross-bedding and imbrications) and clast orientations observable in the sandstone and conglomerate units were measured and then the data were plotted on the rose diagrams.

Thirty-five sandstone samples, collected from the siliciclastic deposits of Aghajari Formation (Late Miocene–Pliocene), were studied for model analysis. Framework mineral composition (modal analysis) was quantified using the point counting method of Gazzi-Dickinson (Ingersoll et al., 1984). For each thin section, 300 counts were made. To reconstruct the origin of detrital composition of the sandstone, the effects of diagenesis such as calcitization, zeolitization, and albitization of feldspars were taken into consideration during counting. Matrix-free recalculated data from point counting of the framework grains are listed in Table 1. These data were plotted in the QFR and QFLt diagrams of Folk (1974) and Pettijohn et al. (1987), respectively for classification of sandstones. Polycrystalline quartz, though not as durable as monocrystalline quartz, was placed at Q-pole to obviate the problems of distinction between plutonic polycrystalline quartz and metaquartzite fragments. In the triangular diagram constructed for delineating the provenance, polycrystalline quartz was placed at the rock fragment (RF) pole (in the Lt pole of QmFLt diagram of Dickinson and Suczek (1979)). Also, Chert was placed at the RF pole as its origin can be unequivocally traced to a sedimentary source, though it was plotted at the Q-pole by several earlier workers (Pettijohn et al., 1987; Blatt, 1967). Chert is less stable than quartz during transport and chert fragment placed into the RF-pole (Folk, 1974). The F-pole comprises all the types of feldspar grains (Folk, 1974; Pettijohn et al., 1987). For recognition of source rock lithology and tectonic setting of the Aghajari Formation, the modal analysis data were plotted in the diamond diagram of Basu et al. (1975) and Tortosa et al. (1991) as well as QtFL and QmFLt triangular diagrams of Dickinson and Suczek (1979).

Ten sandstone samples were analyzed by ICP-MS for the major and trace elements geochemistry, and gravimetric method for SiO_2 and volatile components. These analyses were carried out at the Geological Survey of Iran laboratory. The results are presented in Table 2. Sandstone geochemical data were plotted following the classification schemes of Herron (1988), Pettijohn et al. (1987). The major element oxides of the sandstone were used for provenance determination applying tectonic discriminatory plots (Roser and Korsch, 1986; Bhatia, 1983).

5.4 RESULTS AND INTERPRETATIONS

5.4.1 LITHOFACIES

Based on the classification scheme of Miall (1996), 11 lithofacies types, ascribable to three major lithofacies groups namely, the coarse-grained lithofacies consisting of horizontally stratified conglomerate (Gh), massive conglomerate (Gm), Planar cross-stratified conglomerate (Gp), the medium-grained lithofacies consisting of massive sandstone (Sm), planar cross-bedded sandstone (Sp), trough cross-bedded sandstone (St), horizontally stratified sandstone (Sh), ripple cross-laminated sandstone (Sr), low angle cross-bedded sandstone (Sl), and the fine-grained lithofacies consisting of laminated mudstone, siltstone and sandstone (Fl), massive mudstone and siltstone (Fm) are recognized from the studied section and are detailed herein.

Table 1 Recalculated Data of Modal Analysis of the Aghajari Formation Sandstones

Sample	Qt	Lt	F	Qm	Qmn	Qp	Ls	Lv	Lm	Ch
AJ-S1	54	44	2	33	15	5.9	33	0	1	10
AJ-S2	51	49	0	34	14	2.6	42	0	2	4.9
AJ-S3	53	47	0.5	34	12	5.9	36	0	0.5	11
AJ-S4	48	50	1.9	29	16	3.2	42	0	1.3	7
AJ-S5	47	52	0.9	24	19	4.3	42	0	0.9	9.5
AJ-S6	53	45	2.5	29	19	4.2	33	0	2.6	9.6
AJ-S7	53	45	1.4	30	17	6.1	36	0	1.5	7.7
AJ-S8	47	51	2.2	31	12	4.4	35	0	1.7	14
AJ-S9	49	51	0	22	21	6	37	0	3.8	10
AJ-S10	48	51	0.7	26	18	4.1	37	0	2	13
AJ-S11	56	43	0.7	35	17	3.9	31	0	1.2	11
AJ-S12	52	46	2.1	29	19	4.3	34	0	1.2	11
AJ-S13	49	48	2.4	33	13	3.6	30	0	0.4	18
AJ-S14	51	49	0.7	40	7.2	3.3	33	0	2.1	14
AJ-S15	55	42	2.4	32	19	4.8	28	0	1	14
AJ-S16	52	48	0.5	28	20	4.3	30	0	3	15
AJ-S17	52	47	1.5	33	14	5.2	30	0	1.4	15
AJ-S18	40	59	1.1	22	14	3.4	43	0	2.2	14
AJ-S19	50	47	2.5	28	19	3.1	33	0	1.7	13
AJ-S20	50	48	1.3	30	16	4.2	38	0	3.5	7.2
AJ-S21	44	55	1	27	14	3	46	0	3.2	6
AJ-S22	50	49	1.1	28	19	4	38	0	3	8.2
AJ-S23	52	46	1.6	26	22	4.1	33	0	0.5	13
AJ-S24	50	49	1.1	31	15	4.1	38	0	0.8	9.9
AJ-S25	52	48	0	27	21	4.7	35	0	0	13
AJ-S26	52	46	2.1	29	18	4.6	40	0	0	6
AJ-S27	49	48	3.9	25	21	2.7	40	0	2.3	4.9
AJ-S28	52	46	2.1	28	19	4.7	36	0	1.2	9.1
AJ-S29	51	49	0.8	28	19	3.9	37	0	2.1	9.6
AJ-S30	53	47	0	27	24	2.7	34	0	0	13
AJ-S31	51	45	4.1	25	23	3.4	31	0	2.4	12
AJ-S32	55	44	1.5	33	20	2	32	0	0.5	11
AJ-S33	41	59	0	19	16	5.5	44	0	1	14
AJ-S34	50	48	2.3	33	14	2.8	38	0	0.8	8.7
AJ-S35	54	44	1.5	31	18	5.3	32	0	2.2	9.8

Qt, total quartz; Lt, total lithic grains; F, feldspars; Qm, undulatory monocrystalline quartz; Qmn, non-undulatory monocrystalline quartz; Qp, polycrystalline quartz; Ls, sedimentary lithic grains; Lv, volcanic lithic grains; Lm, metamorphic lithic grains; Ch, chert.

Table 2 Major, Trace, and Rare Earth Element Geochemical Data of the Sandstones From the Aghajari Formation (Oxides in Percentage (%) and Trace and Rare Earth Elements in ppm)

Sample	AJ-S1	AJ-S6	AJ-S10	AJ-S13	AJ-S17	AJ-S20	AJ-S24	AJ-S28	AJ-S31	AJ-S35
Major elements										
SiO_2	55.68	48.17	57.82	57.35	53.07	50.65	51.13	54.1	55.3	57.63
TiO_2	0.25	0.16	0.28	0.22	0.3	0.27	0.3	0.38	0.29	0.32
Al_2O_3	6.28	5.07	6.6	6.32	7.16	6.45	6.18	7.43	5.25	6.42
Fe_2O_3	2.19	1.46	1.98	1.85	2.34	2	2.46	2.91	2.43	2.27
MnO	0.069	0.039	0.067	0.078	0.064	0.058	0.051	0.114	0.08	0.113
MgO	0.85	1.19	0.88	0.9	1.25	0.84	1.51	1.37	1.73	1.91
CaO	18.74	24.98	18.18	20.59	19.74	16.66	18.74	25.7	20.01	26.21
Na_2O	0.79	0.87	0.82	0.73	0.91	0.95	0.7	0.74	0.81	0.78
K_2O	1.5	1.37	1.64	1.45	1.41	1.22	1.14	1.32	1.64	1.21
P_2O_5	0.048	0.035	0.047	0.016	0.063	0.048	0.06	0.07	0.032	0.058
CIA	59.8	53.8	59.6	61.4	61.3	59.2	63.6	65.8	54.2	62.4
Trace elements										
U	2.63	2.45	3.43	2.51	3.63	2.34	3.49	4.11	3.27	3.17
W	1.47	1.95	1.19	1.29	1.31	1.28	1.4	1.59	1.37	1.56
Y	10.01	5.32	9.86	10.18	10.2	9.33	12.3	14.77	9.73	12.64
Mo	11.24	10.67	12.04	10.05	13.33	12.39	8.54	14.25	11.86	14.95
Ni	50.61	92.7	34.16	39.64	68.72	42.26	82.38	60.11	86	91.12
Nb	8.13	9.94	8.9	7.15	9.42	8.65	9.84	8.66	9.35	9.73
Th	2.98	3.3	3.92	3.8	2.43	2.97	3.28	4.85	3.01	4.91
Pb	4.5	8.41	0.5	14.06	4.69	4.05	13.52	1.01	5.06	10.72
Ga	8.11	5.33	6.82	7.82	6.03	5.31	12.96	7.55	4.92	11.98
Zn	13.01	14.68	10.52	10.76	15.86	8.6	18.4	19.65	11.27	18.75
Be	0.54	0.16	0.52	0.51	0.6	0.56	0.53	0.58	0.34	0.53
B	12.12	10.08	10.53	9.44	16.12	16.04	12.51	9.17	12.48	10.51
Sm	2.43	1.92	2.2	2.35	2.98	3.26	2.75	3.51	3.12	2.36
Eu	0.71	0.37	0.66	0.54	0.63	0.77	0.64	0.95	0.67	0.82
Sc	5.6	5.11	5.51	5.89	6.79	6.15	7.1	5.51	6.17	6.35
Ta	0.5	0.83	0.77	0.1	0.27	0.15	0.1	0.29	0.18	0.9

Table 2 Major, Trace, and Rare Earth Element Geochemical Data of the Sandstones From the Aghajari Formation (Oxides in Percentage (%) and Trace and Rare Earth Elements in ppm)—cont'd

Sample	AJ-S1	AJ-S6	AJ-S10	AJ-S13	AJ-S17	AJ-S20	AJ-S24	AJ-S28	AJ-S31	AJ-S35
Co	11.12	10.83	8.47	9.3	15.43	11.46	17.42	16.69	17.83	15.27
Li	0.3	0.3	0.3	0.3	2.48	0.3	1.03	1.93	0.3	1.59
Pr	1.69	1.37	1.77	1.9	1.68	1.76	2.12	2.66	2.57	1.71
Nd	6.24	7.14	8.81	7.3	7.89	8.68	6.48	8.54	7.22	9.94
Ba	297	245	220	189	245	267	519	192	221	218
Rb	150	132	146	148	146	150	151	126	120	137
Sr	212	211	215	197	222	180	346	288	277	436
Ho	0.74	0.7	0.59	0.68	0.75	0.65	0.77	0.94	0.78	0.79
Hf	6.09	4.24	7.63	7.29	3.83	2.8	4.34	4.82	3.48	5.66
Cs	3.12	3.43	2.29	2.21	3.09	2.78	3.44	3.71	3.76	3.32
Ce	25.65	20.75	18.99	17.79	19.25	17.78	19.88	21.15	13.47	21.98
Dy	1.18	0.7	0.91	1.14	0.96	0.61	1.08	1.32	1.3	1.42
Er	0.51	0.5	0.55	0.5	0.58	0.53	0.61	0.9	0.59	0.6
Tm	1.38	0.97	1.13	1.22	0.94	1.12	1.27	1.6	1.46	1.45
Yb	0.9	0.68	0.91	0.89	0.94	0.88	1.15	1.38	0.94	1.09
Sn	0.68	0.5	0.61	0.56	0.76	0.63	0.74	0.95	0.74	0.72
V	60.49	60.19	57.76	79.05	61.73	52.45	64.21	97.05	91.01	60.6
Cu	28.15	22.59	25.68	17.1	19.29	20.26	19.12	15.18	29.02	17.36
La	16.83	16.15	15.22	15.81	15.77	16.36	17.22	23.54	12.63	18.01
Zr	39.66	46.72	38.32	42.79	44.18	34.46	49	51.77	34.89	40.7
Gd	1.72	0.75	1.78	1.66	1.55	1.68	1.91	2.44	1.49	1.65

Present CIA, chemical index of alteration (after Nesbitt and Young, 1982).

The horizontally stratified conglomerate (Gh) is composed of coarse pebble to cobble-sized, clast-supported conglomerate with horizontal beds and imbrications (Fig. 6(A)). This lithofacies is interpreted as longitudinal bars and channel lag deposits in the rivers (Kim et al., 2009). The massive conglomerate (Gm) consists of massive, clast-supported conglomerate with a coarse sandstone or granule matrix filling the space between clasts. The basal contact of this facies is erosional. The internal sedimentary structural features and laterally extensive distribution of Gm suggest them to have been deposited as the longitudinal bars and channel lag deposits in a fluvial channel (Kim et al., 2009). The 15- to 100-cm thick planar cross-stratified conglomerate (Gp) is composed of clast-supported conglomerate with tabular cross-beds (Fig. 6(B)). The basal contacts of this lithofacies are erosional and sharp. The GP is interpreted as the product of a downstream movement of transverse bars in low sinuosity channels (Miall, 1996).

The massive sandstone (Sm) lithofacies comprises massive beds that consist of poor to well-sorted and sub- to well-rounded, fine- to very coarse-grained sandstones. This lithofacies is formed by rapid sedimentation (Tucker, 2011) as a result of a heavy sandy sediment flow (Simpson et al., 2002). The planar cross-bedded sandstone (Sp) is composed of solitary sets or as cosets (Fig. 7(A)) of planar cross-bedded fine- to coarse-grained sandstones. The contact boundaries of this lithofacies are sharp. This

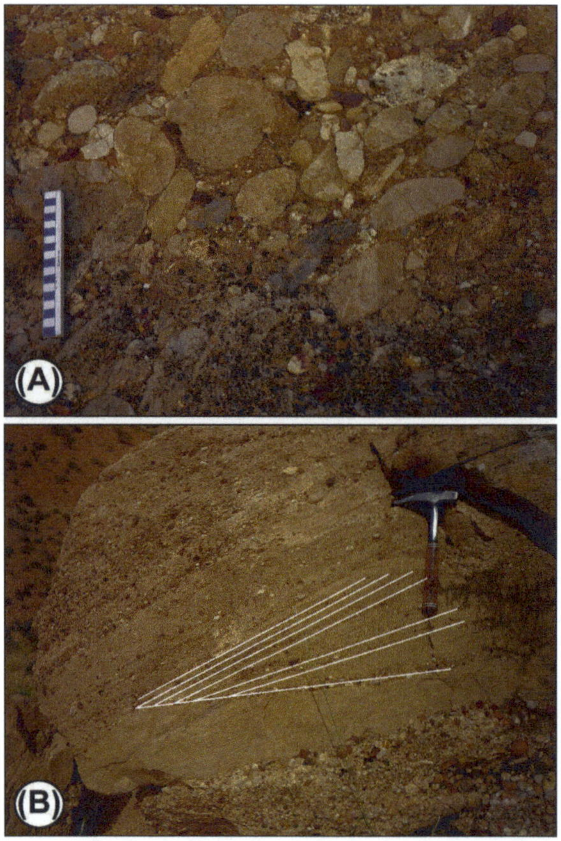

FIGURE 6

Field photographs showing (A) Lithofacies Gh and (B) Lithofacies Gp.

lithofacies is formed by the migration of straight crested dunes or bars deposited in the lower flow regime (Miall, 1996; Capuzzo and Wetzel, 2004). The trough cross-bedded sandstone (St) lithofacies is composed of medium- to very coarse-grained, moderately to poor-sorted sandstones. Sets of the trough cross-beds vary in thickness from 25 cm to 1.8 m (Fig. 7(B)). These are interpreted as the products of 3D dunes that migrated in the fluvial channels (Miall, 1996). The horizontally stratified sandstone (Sh) is composed of fine- to coarse-grained, moderately to well-sorted sandstone (Fig. 8(A)). This lithofacies might have been deposited by high-energy sheet floods that spilled into a lower energy environment from channels during flooding of the main fluvial channel system (Ghazi and Mountney, 2009). The ripple cross-laminated sandstone (Sr) lithofacies is composed of medium- to coarse-grained, moderately to well-sorted sandstones. The ripples are asymmetrical (Fig. 8(B)). This lithofacies is interpreted to have been deposited by alternate ripple migration and deposition from suspension (Miall, 1996). The low angle cross-bedded sandstone (Sl) is made up of low angle cross-beds that generally dip 5–10°. It contains moderately to well-sorted and fine- to coarse-grained sandstones. Low-angle cross-bedded sandstone is interpreted as the result of migration of either low amplitude bedform or upper flow regime plane bed (Hjellbakk, 1997).

FIGURE 7

Field photographs showing (A) Lithofacies Sp and (B) Lithofacies St.

FIGURE 8

Field photographs showing (A) Lithofacies Sh and (B) Lithofacies Sr.

The laminated mudstone, siltstone, and sandstone (Fl) is composed of parallel-laminated siltstone (mudstone) and very fine sandstone (Fig. 9(A)). The boundaries of this facies are sharp and planar. This lithofacies is interpreted to have been deposited from suspension (Hjellbakk, 1997). The massive mudstone and siltstone (Fm) is consists of massive siltstones and homogeneous mudstones (Fig. 9(B)). The lower boundary is commonly flat and irregular and the upper boundary is sharp. This lithofacies might have been deposited from suspension in overbank setting (Kim et al., 2009).

5.4.2 ARCHITECTURAL ELEMENTS

The reorganization of large morphological features of rivers such as bars and channels by architectural analysis is crucial for the interpretation of ancient fluvial sediments (Miall, 1996). Two major architectural elements namely (1) the channel consisting of channel fill element (CH), sandy bedform element (SB), laminated sand sheets element (LS), and lateral accretion element (LA) and (2) the overbank

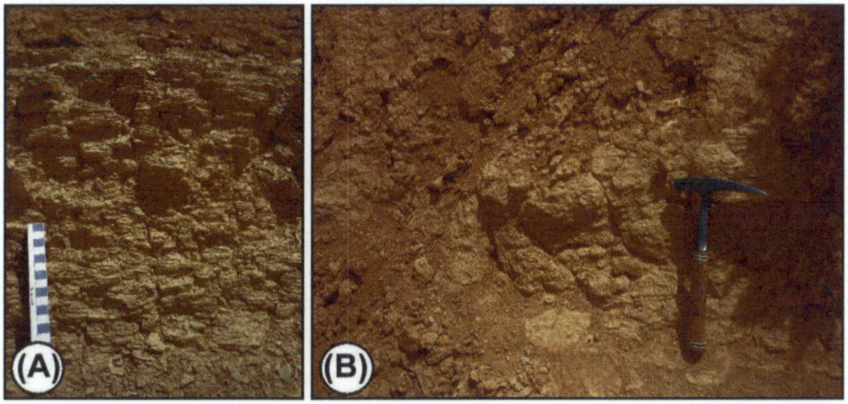

FIGURE 9

Field photographs showing (A) Lithofacies Fl and (B) Lithofacies Fm.

FIGURE 10

Field photographs showing (A) Channel fill element (CH), sandy bedform element (SB), and floodplain fines element (FF); and (B) Laminated sand element (LS).

consisting of levee element (LV) and floodplain fines element (FF) are recognized from the studied section and are detailed herein.

The CH is characterized by stacked, multistory channel bodies that cut down into the underlying sediments (Ghazi and Mountney, 2009; Kim et al., 2009). Lower boundary of this element is usually erosional and the upper boundary occasionally is sharp (Fig. 10(A)). The element often consists of lithofacies Gm, Gh, and Gp with subordinate sets of the facies St and Sp in the studied area. The SB is composed of trough or planar sandstone bodies (Fig. 10(A)) and represents the deposits of migratory dune scale bedform in either mid-channel bars or on the flanks of point bars (Miall, 1996). The LS forms discontinuous tabular sheets of fine- to medium-grained sandstone (Sh, Sl, and minor Sp) lithofacies (Fig. 10(B)). The depositional setting was related to the bar top or bar flank sand sheet under upper flow regime conditions (Miall, 1996; Ghazi and Mountney, 2009). The LA is characterized by low-angle inclined compound cosets of planar cross-bedded sandstones, and is interpreted as the components of point bars. This element occurs frequently all over the Aghajari Formation.

The deposits of the LV consist of tabular to wedge shaped, thick-bedded bedforms that show laterally thining and fining characteristics when observed from adjacently located fluvial channel deposits. The general geometry of the FF is a sheetlike shape, which occurs frequently in the upper part of the Aghajari Formation (Fig. 10(A)). Fine-grained nature of this element suggests deposition in the wide areas very far from the main channel (Kim et al., 2009).

5.4.3 DEPOSITIONAL ENVIRONMENTS

Analysis of the lithofacies and architectural elements on the basis of classification and nomenclatural scheme of Miall (1996), reveals the prevalence of deposition under fluvial regime. The Aghajari Formation deposits consist of stacked lenticular channels forming multistory channel-fill complexes. This fluvial system consists of CH, SB, LA, LV, LS, and FF elements. Fine-grained flood basin deposits are less abundant in these deposits. The occurrence of conglomerate facies with polymict character may indicate deposition from flash flood and sheet flood of large rivers (Luzón, 2005). The sedimentary environment of Aghajari Formation is interpreted as originating on large sandy meandering river (Miall, 1996). The occurrence of such river in an otherwise arid climatic zone was referred to the result of orogenic events (Miall, 1985, Fig. 11). Rose diagrams of the Aghajari Formation deposits showed that its source was from the NW and the paleoflow was into SE direction. The age of the Aghajari Formation is Late Miocene–Pliocene. At that time, Alpine orogeny was on its climax and the Iranian Plate was rotating in anticlockwise direction. The tectonically raised high relief and climatic change might have

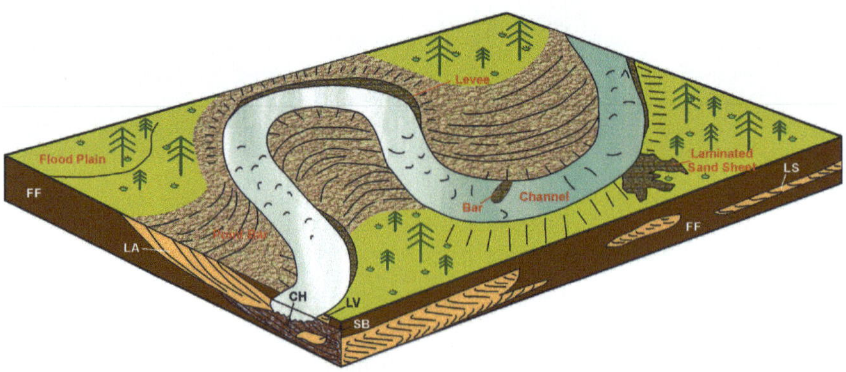

FIGURE 11

Depositional model of the Aghajari Formation in the Fars Interior. CH, channel fill element; SB, sandy bedform element; LS, laminated sand sheets element; LA, lateral accretion element; LV, levee element; and FF, floodplain fines element.

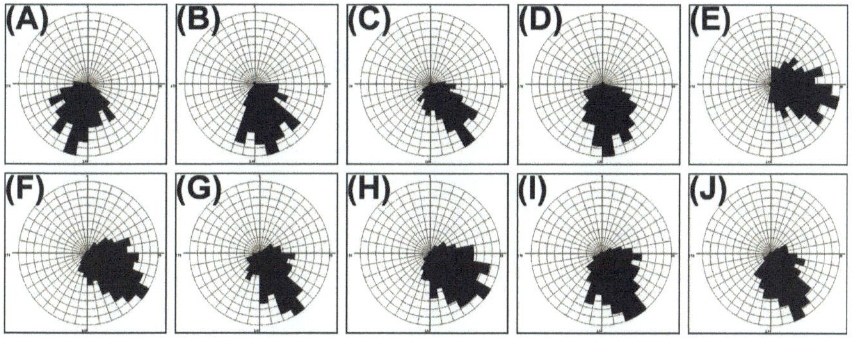

FIGURE 12

Rose diagrams of paleocurrent of Aghajari Formation deposits in the study area.

caused the great influx of sediments from the Iranian Plate and deposited in the foreland basin (Zagros fold–thrust belt). The paleoflow model of the Aghajari Formation is presented in Fig. 12.

5.4.4 DETRITAL FRAMEWORK COMPONENTS

The detrital framework grains of the Aghajari Formation sandstones consist of quartz, feldspars, RFs, and mica minerals such as biotite and muscovite. It is observed that fine grains are angular, and coarser ones are rounded to broken-rounded. These characteristics may have been the result of high-energy environment or through long-distance transport. The well-sorted, coarse-grained sands are found broken into fine-grained angular sands, which affect the sorting character. Thus, the heterogeneous roundness of grains of different grain sizes in the Aghajari Formation sandstones reflect the importance of mechanical factors for the final grain shapes.

Quartz grains include two groups: monocrystalline quartz (Qm) and polycrystalline quartz (Qp). Qm occurs in two variants: non-undulatory monocrystalline quartz (Qmn) (7.2–24, averaging 17%) and undulatory monocrystalline quartz (Qm) (19–40 with averaging 29%). The Qp (2–6.1, averaging 4.1%) was distinguished into two groups: polycrystalline quartz with two to three subgrains and polycrystalline quartz with more than three subgrains. Chert was considered as a monomineral (Folk, 1974) and was found in more samples (4.9–18, averaging 11%). Some of the non-undulatory monocrystalline quartz (Qmn) contains inclusions. Few of the studied samples contain small amounts of the feldspar (0–4.1, averaging 1.4%), including twinned microcline. Plagioclase occurs in few fine-grained sandstone samples.

Common occurrences of biotite mica and limited amounts of muscovite are observed. Considerable amounts of the lithic fragments, exceeding the quantum of the quartz and feldspar grains are noticed in the studied samples. Lithic fragments found in the studied samples include sedimentary fragments (Ls) (28–46, averaging 36%) (consist of siltstones, sandstones, and carbonate fragments), metamorphic fragments (Lm) (0–3.8, averaging 1.6%), and volcanic fragments (Lv) which are absent in these sandstones. Lithoclasts of metamorphic rocks are usually made of polycrystalline quartz fragments of high-grade metamorphic origin. Typical source rocks for these lithoclasts would be gneisses (Scholle, 1979). Calcite occurs in the well-cemented sandstone samples. Using Folk (1974) and Pettijohn et al. (1987) classifications (QFR and QFLt ternary plot, respectively; Fig. 13), the Aghajari Formation sandstone samples were classified as lithic arenite, thus reflecting their slightly mineralogical matured character (Fig. 14).

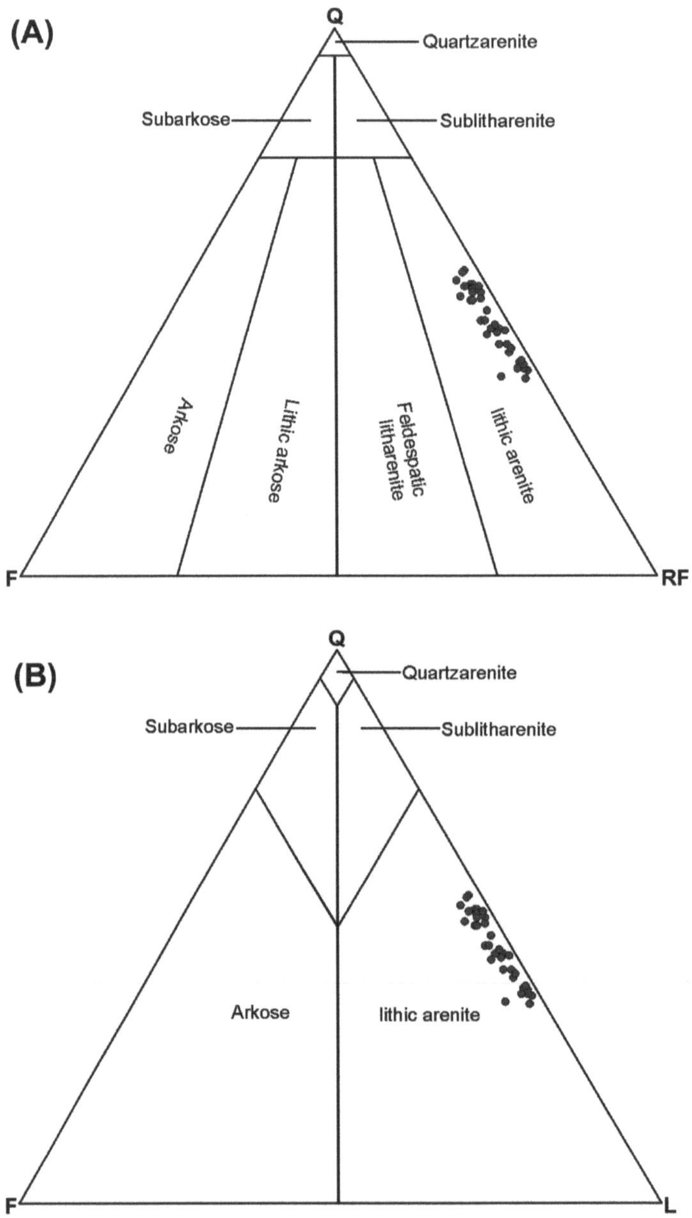

FIGURE 13

Triangular classification plot of different sandstone samples from the Aghajari Formation. (A) QFR triangular diagram *(after Folk (1974))* and (B) QFL triangular diagram *(after Pettijohn et al. (1987))*. Q, quartz; F, feldspar; RF, rock fragment; and L, lithic fragment.

FIGURE 14

Photomicrographs of lithic arenite in the sandstones from the Aghajari Formation (XPL). Qp, polycrystalline quartz; Qm, undulatory monocrystalline quartz; Qmn, non-undulatory monocrystalline quartz; Ch, chert; and L, lithic fragment.

5.4.5 SOURCE ROCK LITHOLOGY

Petrographic evidence such as incompatible roundness for different grains and resultant enforced sorting character of framework grains imply the role played by mechanical effects. The occurrences of coarse-grained feldspars indicate the prevalent low degree of chemical weathering (Osae et al., 2006) in the source area. The occurrence of rounded quartz overgrowths indicates recycling. Together, these evidences are interpreted as the result of scarcity of sediments (partly due to the necessity to transport sediments from high-relief Arabian Shield—Ziegler, 2001) at the receiving basin and recycling of sediments already deposited.

Petrographic analysis of the sandstones shows the occurrences of minor amounts of polycrystalline quartz grains (Table 1) exhibiting multiple (>5 crystals within a grain) crystal structure and irregular to crenulated intercrystal boundaries suggestive of their metamorphic source origin (Blatt et al., 1980;

Asiedu et al., 2000). In an earlier study, Asiedu et al. (2000) recorded the occurrences of monocrystalline quartz grains showing strong undulatory extinction and crypto-polycrystalline quartz grains in the Aghajari Formation sandstones. To evaluate the relative importance of metamorphic rocks as quartz grains sources, we plotted polycrystalline quartz versus non-undulatory and undulatory monocrystalline quartz in a double-triangular diagrams (Basu et al., 1975; Tortosa et al., 1991, Fig. 15). These diagrams suggested a low to high metamorphic origin, and a granitic/gneissic source rocks. These contrasting results can be partially explained by considering that the Aghajari Formation sandstones plot in a region of the diagram where only some of the granite sourced modern sands analyzed by Tortosa et al. (1991) plot.

The concentration of Zr is used for characterizing the nature and composition of source rocks (Hayashi et al., 1997; Paikaray et al., 2008). The high TiO_2/Zr ratios (34.2–83.1) of the sandstones indicate intermediate source rocks (Taylor and McLennan, 1985; Wronkiewicz and Condie, 1987; Condie, 1993; Cullers, 2000; Huntsman-Mapila et al., 2005; Spalletti et al., 2008). When plotted on the TiO_2–Zr diagram, these ratios help distinguish among three different source rock types (i.e., felsic, intermediate, and mafic). The TiO_2 versus Zr plot of the sandstones represents predominately intermediate source rocks (Fig. 16(A)). The variations in Th and La (indicative of felsic) and, Sc and Co (indicative of mafic) contents have been used to differentiate between felsic and mafic provenance by various authors (McLennan et al., 1980; Cullers, 2002; Kasanzu et al., 2008). Th/Co versus La/Sc bivariate diagram can provide information regarding the source rock characteristics (McLennan et al., 1980; McLennan and Taylor, 1991). The Th/Co versus La/Sc plot suggests a silicic nature of the source rocks (Cullers, 2002, Fig. 16(B)).

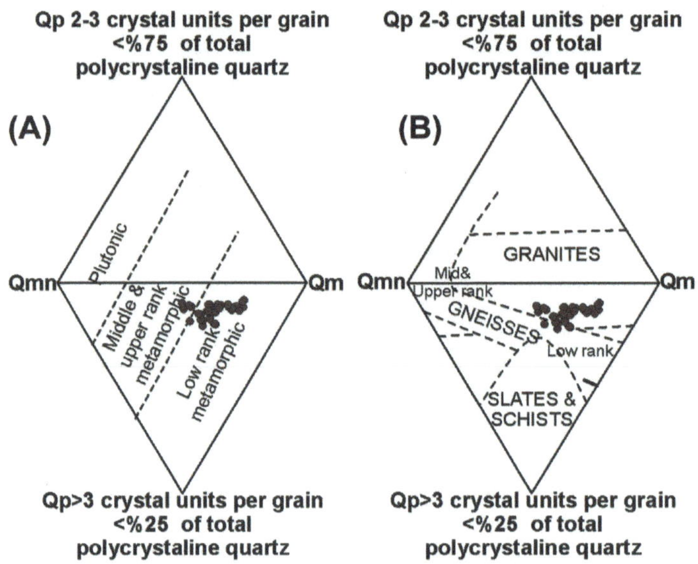

FIGURE 15

Varietal quartz diamond plot currently used to discriminate sands sourced by different types of crystalline rocks, on the basis of the extinction pattern and polycrystallinity of quartz grains. Qmn, non-undulatory monocrystalline quartz grains; Qm, undulatory monocrystalline quartz grains; Qp2-3, coarse-grained polycrystalline quartz grains; Qp > 3, fine-grained polycrystalline quartz grains. Aghajari Formation sandstones are compared with provenance fields after Basu et al. (1975) and Tortosa et al. (1991) (diagrams A and B, respectively).

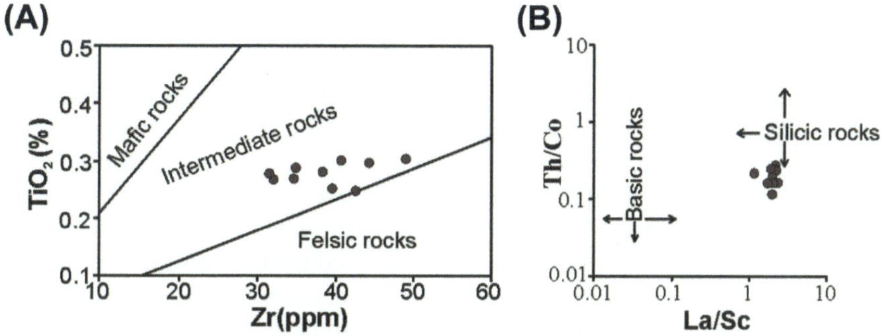

FIGURE 16

(A) TiO_2–Zr plot for the Aghajari Formation sandstones *(after Hayashi et al. (1997))*, and (B) Th/Co versus La/Sc plot showing source rock composition for the Aghajari Formation sandstones *(after Cullers (2002))*.

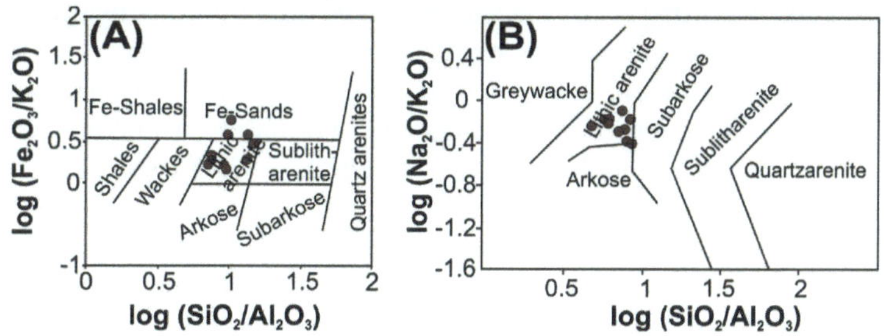

FIGURE 17

Chemical classification of samples from the Aghajari Formation sandstones based on binary diagrams. (A) log (SiO_2/Al_2O_3) versus log (Fe_2O_3/K_2O) diagram of Herron (1988), and (B) log (SiO_2/Al_2O_3) versus log (Na_2O/K_2O) diagram of Pettijohn et al. (1987).

5.4.6 SANDSTONE CLASSIFICATION

The composition of the terrigenous sedimentary rocks is affected by many factors including transportation (medium energy), diagenesis, and weathering rate of the source rock (Catuneanu, 2006). Table 2 presents the geochemical analytical results of the 10 sandstone samples of the Aghajari Formation. From this table, it is apparent that most of the samples contain moderate to high in SiO_2 (48.2–57.8%), and moderate Al_2O_3 (5.7–7.43%), and a low (1.14–1.64%). In contrast, low K_2O, TiO_2, MnO, CaO, Na_2O, Fe_2O_3, and MgO concentrations are observed in these rocks. The high K_2O/Na_2O ratios are attributed to the common presence of K-bearing minerals such as K-feldspar, illite, muscovite, and biotite (McLennan et al., 1983; Nath et al., 2000; Zhang, 2004; Osae et al., 2006). A positive correlation between K_2O and Al_2O_3 implies that the concentrations of the K-bearing minerals have significant influence on Al distribution and suggest that the abundances of these elements are primarily controlled by the content of clay minerals (McLennan et al., 1983). However, a contribution from the feldspars is indicated from their Al_2O_3/SiO_2 ratios. Based on these ratios, the Aghajari Formation sandstones can be classified chemically as lithic arenite and Fe-sands (Herron, 1988) (Fig. 17(A)) and as lithic arenite on diagram of Pettijohn et al. (1987) (Fig. 17(B)).

5.4.7 PROVENANCE AND TECTONIC SETTING

Based on the detrital sandstone compositions, major provenance types such as stable craton, basement uplift, magmatic arc, and recycled orogen (Dickinson et al., 1983; Dickinson and Suczek, 1979) can be discriminated as different tectonic settings contain characteristic rock types which, when eroded and produce sandstones with specific compositional ranges (Dickinson, 1985). In the QFL and QmFLt ternary diagrams of Dickinson et al. (1983), the analyzed samples plot exclusively in the recycled orogen (quartzose and transitional recycled) field (Fig. 18). As pointed out by Dickinson et al. (1983), sandstones plotting in this field are mature sandstones derived from relatively low-lying granitoid and gneissic sources, supplemented by recycled sands from associated platform or passive margin basins. Crook (1974) proposed the use of quartz content to infer tectonic setting, and linked quartz rich (>65% quartz) sandstones to passive continental margins (Atlantic type); intermediate quartz (15–65% quartz) sandstones to active continental margins (Andean type), and quartz poor (<15% quartz) sandstones to magmatic island arcs. The quartz rich (averaging 50%) sandstones studied here thus correspond to the Andean type sandstones.

The major element geochemistry of the sandstones can be used for drawing inferences related to the provenance type and the plate tectonic setting of ancient sedimentary basins (Armstrong-Altrin et al., 2004; Bhatia, 1983; Roser and Korsch, 1986). As reported by Bhatia (1983), the data generated by the present study had also indicated deposition of the Aghajari Formation sandstones, in the active margin setting (Bhatia, 1983) as evidenced by plots (Fig. 19) of $Fe_2O_3 + MgO$ versus TiO_2, Al_2O_3/SiO_2, K_2O/Na_2O, and $Al_2O_3/(CaO + Na_2O)$. Active continental margin sandstones are dominantly derived from the uplifted basement and reflect the composition of the upper continental crust (Bhatia, 1983). The K_2O/Na_2O versus SiO_2 diagram is widely used for discriminating between sediments deposited in the passive margin, active margin, and island arc settings. The ratios of K_2O/Na_2O variation with SiO_2 are plotted in a binary diagram (after Roser and Korsch, 1986) to understand the tectonic setting of the Aghajari Formation sandstones. It is found that all the samples are plotted in the active continental margin field (Fig. 20(A)). The same pattern is observed when the sandstone samples are plotted on the SiO_2/Al_2O_3 versus $Na_2O + K_2O$ diagram of Roser and Korsch (1986) (Fig. 20(B)). Trace elements (e.g., Ti, Co, La, Th, Sc, and Zr) in the clastic sedimentary rocks are considered to be immobile under conditions of weathering, diagenesis, and moderate levels of metamorphism, and are commonly preserved in sedimentary rocks. Therefore, the trace elements represent well-established provenance and tectonic setting indicators (McLennan and Taylor, 1991; Roser and Korsch, 1986; Spalletti et al., 2008). On the Th–Sc–Zr/10, La/Sc versus Ti/Zr, and La/Y versus Sc/Cr diagrams (Bhatia and Crook, 1986, Fig. 21), most of sandstone samples cluster within the active continental margin field.

5.4.8 SOURCE-AREA WEATHERING

Chemical weathering largely affects the composition of siliciclastic sediments. Through these processes, the large cations (e.g., Ba and Al) remain preserved in the weathering residue in contrast to the smaller cations (Na, Ca, and Sr) that can be selectively removed (Fedo et al., 1996; Nath et al., 2000). The degree of weathering may be assessed by the chemical index of alteration, commonly referred as CIA (Nesbitt and Young, 1982). According to the methods of McLennan (1993), we calculated the CIA for the samples under study. The CIA values of approximately 50 indicate simply an unweathered upper crust or weak weathering, whereas CIA values of 100 indicate intense weathering with a

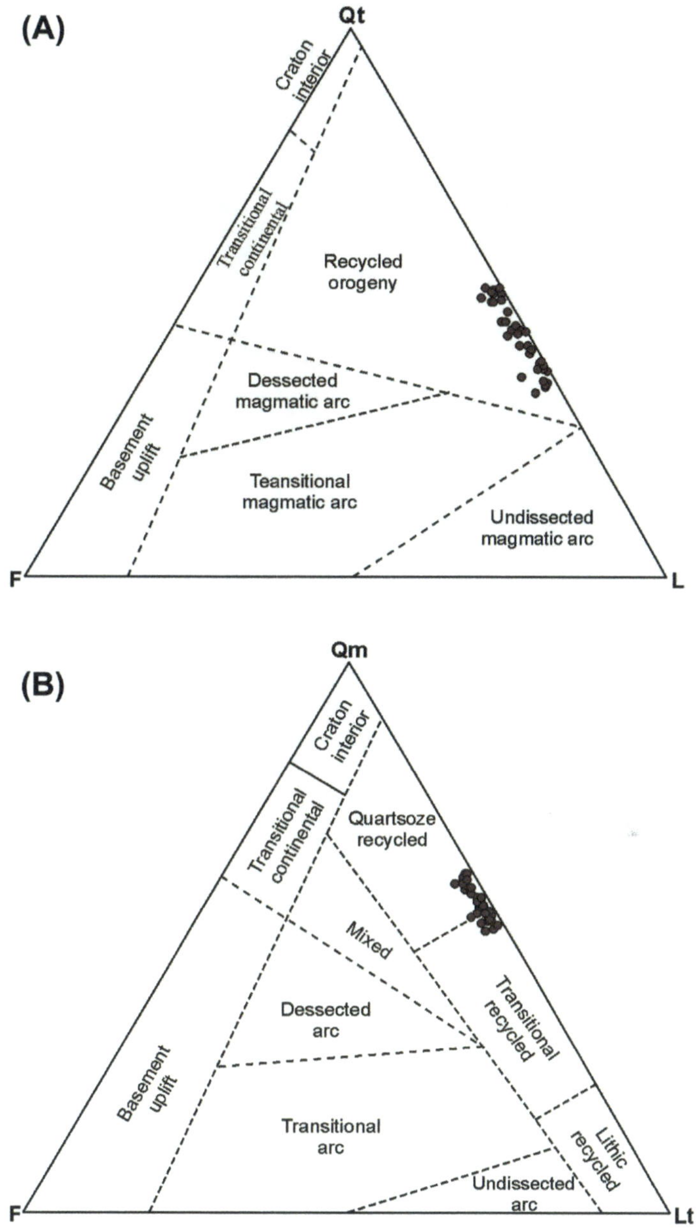

FIGURE 18

Ternary diagrams for the provenance of Aghajari Formation sandstones *(after Dickinson et al. (1983))*. (A) QtFL diagram and (B) QmFLt diagram. Qt, total quartz grain; Qm, monocrystalline quartz; F, feldspar; L, lithic fragment; Lt, total lithic fragment.

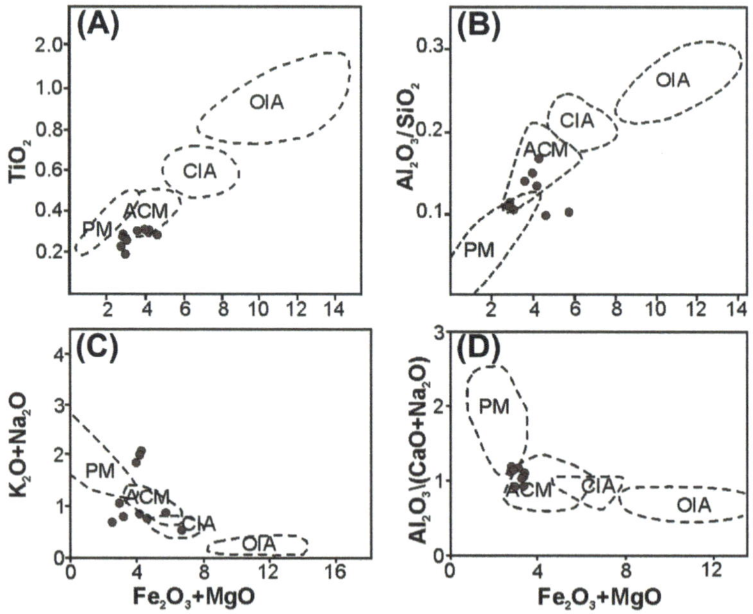

FIGURE 19

Plots of the major element composition of the Aghajari Formation sandstones on the tectonic setting discrimination diagrams of Bhatia (1983). OIA, oceanic island arc; CIA, continental island arc; ACM, active continental margin; and PM, passive continental margin.

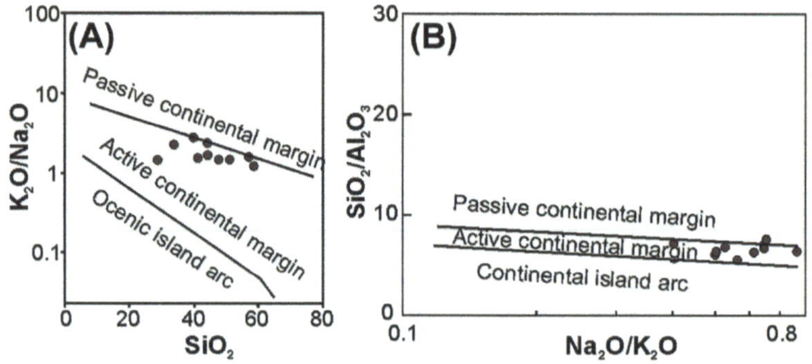

FIGURE 20

Tectonic setting discrimination diagrams based on major element compositions of Aghajari Formation sandstone samples. (A) K_2O/Na_2O versus SiO_2 and (B) SiO_2/Al_2O_3 versus K_2O/Na_2O *(after Roser and Korsch (1986)).*

complete removal of alkali and alkaline earth elements and an increase in Al_2O_3 (McLennan, 1993; Dupuis et al., 2006; Fedo et al., 1996). The calculated CIA values of sandstone samples range from 53.8 to 65.8, with an average of 60.11 (Table 2). However, all of the samples have values greater than 50, suggesting low to moderate weathering. In addition, occurrences of lowest CIA values in the lower part

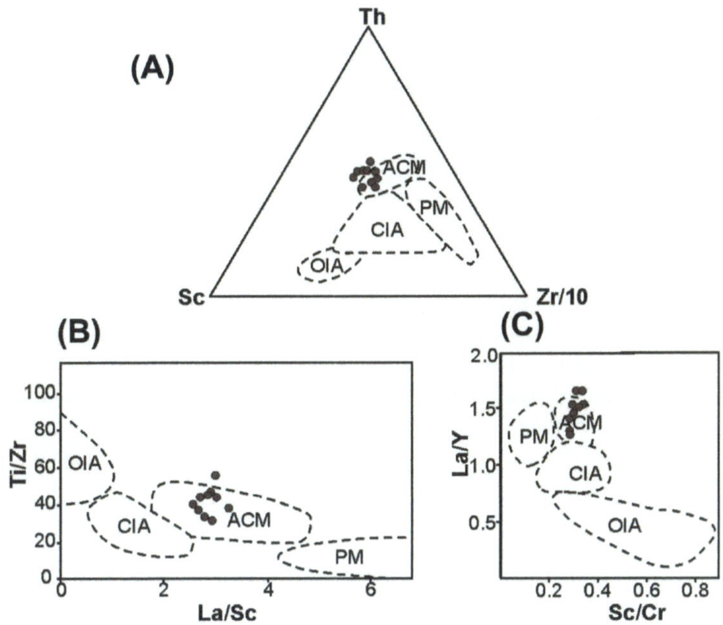

FIGURE 21

Plots of the trace element composition of the Aghajari Formation sandstone samples on the tectonic setting discrimination diagrams of Bhatia and Crook (1986). OIA, oceanic island arc; CIA, continental island arc; ACM, active continental margin; PM, passive continental margin.

of the formation and relatively higher CIA values in the upper part are observed, suggestive of initiation of the deposition of the Aghajari Formation immediately after major tectonic movement in the provenance as a result of which only mechanical weathering prevailed, which gradually changed to chemical weathering over time.

5.5 CONCLUSIONS

- The study had documented 11 lithofacies types that are grouped into three major facies types. These lithofacies types formed the basis of recognition of six architectural elements that are divisible into two major environmental settings, namely channel and bank environments of major fluvial system that flowed along the NW–SE trend in the study area. Deposition in meandering fluvial channel and adjoining regions is indicated by the vertical stacking pattern of these facies types and architectural elements.
- The rocks of the Aghajari Formation contain a wide variety of sediments sourced from granitic, gneissic, and sedimentary provenances. Prevalence of general sediment scarcity at the receiving basin and active tectonism in the source region during the deposition are also indicated by the petrographic model and geochemical data.

- The depositional environments were located on an active continental margin, as indicated by a variety of discrimination diagrams and elemental ratios.
- The study also indicates long transport distance for the sediments, that originated from the Zagros orogen, traversed variable climatic zones over the Iranian Plate and finally got accumulated in the Late Miocene–Pliocene foreland basin.

ACKNOWLEDGMENTS

We thank Mrs R. Jan Shekan, Mr H. Noroozi, and Mr A. Memar (Geological Survey of Iran) for their help in multivariate analysis; Professor S. Ghazi (Punjab University, Pakistan) and Professor A.I. Al-Juboury (Mosul University, Iraq) for their assistance in our research work. We are also grateful to Dr. B. Kowallis (Brigham Young University, USA) for correcting the language. We are especially indebted to the Dr. R. Sadeghi (Payam-E-Noor University of Jahrom, Iran) for providing the laboratory facilities.

REFERENCES

Afghah, M., Farhoudi, G., 2012. Boundary Between Upper Cretaceous and Lower Paleocene in the Zagros Mountain Ranges of Southwestern Iran. Acta Geologica Sinica - English Edition 86 (2), 325–338.

Al-Husseini, M.I., 2008. Middle east geological timescale. Journal of Middle East Petroleum and Geoscience 13 (4).

Alavi, M., 1980. Tectonostratigraphic evolution of the Zagrosides of Iran. Geology 8 (3), 144–149.

Alavi, M., 1994. Tectonics of the Zagros orogenic belt of Iran: new data and interpretations. Tectonophysics 229 (3–4), 211–238.

Alavi, M., 2004. Regional stratigraphy of the Zagros fold-thrust belt of Iran and its proforeland evolution. American Journal of Science 304 (1), 1–20.

Alavi, M., 2007. Structures of the Zagros fold-thrust belt in Iran. American Journal of Science 307 (9), 1064–1095.

Armstrong-Altrin, J.S., Lee, Y.I., Verma, S.P., Ramasamy, S., 2004. Geochemistry of sandstones from the upper miocene Kudankulam formation, Southern India: implications for provenance, weathering, and tectonic setting. Journal of Sedimentary Research 74 (2), 285–297.

Asiedu, D.K., Suzuki, S., Shibata, T., 2000. Provenance of sandstones from the Lower Cretaceous Sasayama group, inner zone of Southwest Japan. Sedimentary Geology 131 (1–2), 9–24.

Bahrami, M., 1997. Sedimentology and morphotectonical evolution of Aghajari and Bakhtiari formations in northwestern of Shiraz. Journal of Science, Islamic Azad University 8 (27–28), 1995–2010.

Bahrami, M., 2009. Lithofacies and sedimentary environments of Aghajari formation in Dehsheikh Mountain, west of Shiraz, Iran. World Applied Sciences Journal 6 (4), 464–473.

Basu, A., Young, S.W., Suttner, L.J., James, W.C., Mack, G.H., 1975. Re-evaluation of the use of undulatory extinction and polycrystallinity in detrital quartz for provenance interpretation. Journal of Sedimentary Research 45 (4), 873–882.

Berberian, M., King, G.C.P., 1981. Towards a paleogeography and tectonic evolution of Iran. Canadian Journal of Earth Sciences 18 (2), 210–265.

Bhatia, M., Crook, K.W., 1986. Trace element characteristics of graywackes and tectonic setting discrimination of sedimentary basins. Contributions to Mineralogy and Petrology 92 (2), 181–193.

Bhatia, M.R., 1983. Plate tectonics and geochemical composition of sandstones. The Journal of Geology 91 (6), 611–627.

Blatt, H., 1967. Provenance determinations and recycling of sediments. Journal of Sedimentary Research 37 (4), 1031–1044.

Blatt, H., Middleton, G., Murray, R., 1980. Origin of Sedimentary Rocks. Pearson Prentice Hall, Englewood Cliffs, New Jersey.
Boggs, S.J., 2006. Principles of Sedimentology and Stratigraphy. Pearson Prentice Hall, Upper Saddle River, New Jersey.
Capuzzo, N., Wetzel, A., 2004. Facies and basin architecture of the late carboniferous Salvan-Dorénaz continental basin (Western Alps, Switzerland/France). Sedimentology 51 (4), 675–697.
Catuneanu, O., 2006. Principles of Sequence Stratigraphy. Elsevier, Amsterdam.
Condie, K.C., 1993. Chemical composition and evolution of the upper continental crust: contrasting results from surface samples and shales. Chemical Geology 104 (1–4), 1–37.
Crook, K.A.W., 1974. Lithogenesis and tectonics: the significance of compositional variation in flysch arenites (greywackes). In: Dott, R.H., Shaver, R.H. (Eds.), Modern and Ancient Geosynclinal Sedimentation. Society of Economic Geologists and Paleontologists, pp. 304–310. Special Publication.
Cullers, R.L., 2000. The geochemistry of shales, siltstones and sandstones of Pennsylvanian–Permian age, Colorado, USA: implications for provenance and metamorphic studies. Lithos 51 (3), 181–203.
Cullers, R.L., 2002. Implications of elemental concentrations for provenance, redox conditions, and metamorphic studies of shales and limestones near Pueblo, CO, USA. Chemical Geology 191 (4), 305–327.
Dickinson, W.R., 1985. Interpreting provenance relations from detrital modes of sandstones. In: Zuffa, G.G. (Ed.), Provenance of Arenites. Nato Advanced Study Institute Series, Reidel Publishing Company, Dordrecht, pp. 333–362.
Dickinson, W.R., et al., 1983. Provenance of North American Phanerozoic sandstones in relation to tectonic setting. Geological Society of America Bulletin 94 (2), 222–235.
Dickinson, W.R., Suczek, C.A., 1979. Plate tectonics and sandstone compositions. AAPG Bulletin 63 (12), 2164–2182.
Dupuis, C., et al., 2006. Geochemistry of sedimentary rocks from mélange and flysch units south of the Yarlung Zangbo suture zone, Southern Tibet. Journal of Asian Earth Sciences 26 (5), 489–508.
Falahatgar, M., Mosaddegh, H., Parvanehnejad Shirazi, M., 2012. Foraminiferal Biostratigraphy of the Mobarak Formation (Lower Carboniferous) in Kiyasar Area, SE Sari, Northern Iran. Acta Geologica Sinica - English Edition 86 (6), 1413–1425.
Falcon, N.L., 1969. Problems of the relationship between surface structure and deep displacements illustrated by the Zagros Range. In: Kent, P.E., Satterthwaite, G.E., Spencer, A.M. (Eds.), Time and Place in Orogeny. Geological Society of London, Special Publication, pp. 9–22.
Falcon, N.L., 1974. Southern Iran: Zagros Mountains. In: Spencer, A.M. (Ed.), Mesozoic-Cenozoic Orogenic Belts: Data for Orogenic Studies. Geological Society of London, Special Publication, pp. 199–211.
Fedo, C.M., Eriksson, K.A., Krogstad, E.J., 1996. Geochemistry of shales from the Archean (~3.0 Ga) Buhwa Greenstone Belt, Zimbabwe: Implications for provenance and source-area weathering. Geochimica et Cosmochimica Acta 60 (10), 1751–1763.
Folk, R.L., 1974. Petrology of Sedimentary Rocks. Hemphill Publishing Company, Austin, Texas.
Ghazi, S., Mountney, N.P., 2009. Facies and architectural element analysis of a meandering fluvial succession: the Permian Warchha Sandstone, Salt Range, Pakistan. Sedimentary Geology 221 (1–4), 99–126.
Hayashi, K.-I., Fujisawa, H., Holland, H.D., Ohmoto, H., 1997. Geochemistry of ~1.9 Ga sedimentary rocks from northeastern Labrador, Canada. Geochimica et Cosmochimica Acta 61 (19), 4115–4137.
Herron, M.M., 1988. Geochemical classification of terrigenous sands and shales from core or log data. Journal of Sedimentary Research 58 (5), 820–829.
Hjellbakk, A., 1997. Facies and fluvial architecture of a high-energy braided river: the Upper Proterozoic Seglodden Member, Varanger Peninsula, Northern Norway. Sedimentary Geology 114 (1–4), 131–161.
Huntsman-Mapila, P., Kampunzu, A.B., Vink, B., Ringrose, S., 2005. Cryptic indicators of provenance from the geochemistry of the Okavango Delta sediments, Botswana. Sedimentary Geology 174 (1–2), 123–148.
Ingersoll, R.V., et al., 1984. The effect of grain size on detrital modes; a test of the Gazzi-Dickinson point-counting method. Journal of Sedimentary Research 54 (1), 103–116.

James, G.A., Wynd, J.G., 1965. Stratigraphic nomenclature of Iranian Oil Consortium Agreement area. AAPG Bulletin 49 (12), 2182–2245.

Junhui, L., Qi'an, M., Shuangfang, L., Defeng, Z., Haibo, W., He, L., Dazhi, Z., 2012. Provenance System Analysis of the Nantun Formation of the Huhehu Sag in Hailar Basin. Acta Geologica Sinica - English Edition 86 (1), 189–198.

Kasanzu, C., Maboko, M.A.H., Manya, S., 2008. Geochemistry of fine-grained clastic sedimentary rocks of the Neoproterozoic Ikorongo Group, NE Tanzania: Implications for provenance and source rock weathering. Precambrian Research 164 (3–4), 201–213.

Khanehbad, M., Moussavi-Harami, R., Mahboubi, A., Nadjafi, M., Mahmudy Gharaie, M.H., 2012. Geochemistry of Carboniferous Sandstones (Sardar Formation), East-Central Iran: Implication for Provenance and Tectonic Setting. Acta Geologica Sinica - English Edition 86 (5), 1200–1210.

Kim, S.B., Kim, Y.G., Jo, H.R., Jeong, K.S., Chough, S.K., 2009. Depositional facies, architecture and environments of the Sihwa Formation (Lower Cretaceous), mid-west Korea with special reference to dinosaur eggs. Cretaceous Research 30 (1), 100–126.

Lacombe, O., Mouthereau, F., Kargar, S., Meyer, B., 2006. Late Cenozoic and modern stress Welds in the western Fars (Iran): Implications for the tectonic and kinematic evolution of central Zagros. Tectonics 25, 1–27.

Luzón, A., 2005. Oligocene-Miocene alluvial sedimentation in the northern Ebro Basin, NE Spain: tectonic control and palaeogeographical evolution. Sedimentary Geology 177 (1–2), 19–39.

McLennan, S.M., 1993. Weathering and global denudation. The Journal of Geology 101 (2), 295–303.

McLennan, S.M., Nance, W.B., Taylor, S.R., 1980. Rare earth element-thorium correlations in sedimentary rocks, and the composition of the continental crust. Geochimica et Cosmochimica Acta 44 (11), 1833–1839.

McLennan, S.M., Taylor, S.R., 1991. Sedimentary rocks and crustal evolution: tectonic setting and secular trends. The Journal of Geology 99 (1), 1–21.

McLennan, S.M., Taylor, S.R., Eriksson, K.A., 1983. Geochemistry of Archean shales from the Pilbara Supergroup, Western Australia. Geochimica et Cosmochimica Acta 47 (7), 1211–1222.

Miall, A.D., 1985. Architectural-element analysis: a new method of facies analysis applied to fluvial deposits. Earth-Science Reviews 22 (4), 261–308.

Miall, A.D., 1996. The Geology of Fluvial Deposits, Sedimentary Facies, Basin Analysis and Petroleum Geology. Springer-Verlag, New York.

Mitchell, A.H.G., Reading, H.G., 1986. Sedimentation and tectonics. In: Reading, H.G. (Ed.), Sedimentary Environments and Facies. Blackwell, Oxford, pp. 471–519.

Mobasher, K., Babaie, H.A., 2008. Kinematic significance of fold- and fault-related fracture systems in the Zagros Mountains, Southern Iran. Tectonophysics 451, 156–169.

Motiei, H., 1994. Geology of Iran, Stratigraphy of Zagros. Geological Survey of Iran Publications, Tehran.

Nath, B.N., Kunzendorf, H., Plüger, W.L., 2000. Influence of provenance, weathering, and sedimentary processes on the elemental ratios of the fine-grained fraction of the bedload sediments from the Vembanad Lake and the adjoining Continental Shelf, Southwest Coast of India. Journal of Sedimentary Research 70 (5), 1081–1094.

Nesbitt, H.W., Young, G.M., 1982. Early Proterozoic climates and plate motions inferred from major element chemistry of lutites. Nature 299 (5885), 715–717.

Ning, Z., Tingdong, L., Guoqing, Y., Shuanhou, Z., Jiantang, C., Muwei, C., 2012. Lithofacies Characters and Significance of the Submarine Fan of the Liufengguan Group in Qinling. Acta Geologica Sinica - English Edition 86 (1), 174–188.

Osae, S., Asiedu, D.K., Banoeng-Yakubo, B., Koeberl, C., Dampare, S.B., 2006. Provenance and tectonic setting of Late Proterozoic Buem sandstones of southeastern Ghana: evidence from geochemistry and detrital modes. Journal of African Earth Sciences 44 (1), 85–96.

Paikaray, S., Banerjee, S., Mukherji, S., 2008. Geochemistry of shales from the Paleoproterozoic to Neoproterozoic Vindhyan Supergroup: Implications on provenance, tectonics and paleoweathering. Journal of Asian Earth Sciences 32 (1), 34–48.

Pettijohn, F.J., Potter, P.E., Siever, R., 1987. Sand and Sandstone. Springer-Verlag, New York.
Potter, P.E., Pettijohn, F.J., 1978. Paleocurrent and Basin Analysis. Academic Press, New York.
Roser, B.P., Korsch, R.J., 1986. Determination of tectonic setting of Sandstone-Mudstone Suites using SiO_2 content and K_2O/Na_2O ratio. The Journal of Geology 94 (5), 635–650.
Sahraeyan, M., Bahrami, M., 2012a. Fluvial cyclicity from Aghajari formation (Upper miocene-pliocene) in folded Zagros Zone, Southwestern Iran. International Journal of Advances in Earth Sciences 1 (2), 58–68.
Sahraeyan, M., Bahrami, M., 2012b. Geochemistry of sandstones from the Aghajari formation, folded Zagros Zone, southwestern Iran: implication for paleoweathering condition, provenance, and tectonic setting. International Journal of Basic and Applied Sciences 1 (4), 390–407.
Sahraeyan, M., Bahrami, M., 2012c. Petrography and provenance of sandstones from the Aghajari Formation, Folded Zagros Zone, southwestern Iran. International Journal of Basic and Applied Sciences 1 (3), 283–298.
Sahraeyan, M., Bahrami, M., Hejazi, S.H., 2013. The Aghajari (upper fars) formation in folded Zagros Zone, Iran: insights to identify facies, architectural elements, fluvial systems, petrography and provenance. Acta Geologica Sinica 87 (4), 1019–1031.
Scholle, P.A., 1979. A Color Illustrated Guide to Constituents, Textures, Cements, and Porosities of Sandstones and Associated Rocks. AAPG Memoir, Tulsa, Oklahoma.
Simpson, E.L., Dilliard, K.A., Rowell, B.F., Higgins, D., 2002. The fluvial-to-marine transition within the post-rift Lower Cambrian Hardyston Formation, Eastern Pennsylvania, USA. Sedimentary Geology 147 (1–2), 127–142.
Spalletti, L.A., Queralt, I., Matheos, S.D., Colombo, F., Maggi, J., 2008. Sedimentary petrology and geochemistry of siliciclastic rocks from the upper Jurassic Tordillo formation (Neuquén Basin, Western Argentina): implications for provenance and tectonic setting. Journal of South American Earth Sciences 25 (4), 440–463.
Stöcklin, J., 1974. Possible ancient continental margins in Iran. In: Burk, C.A., Drake, C.L. (Eds.), The Geology of Continental Margins. Springer-Verlag, New York, pp. 873–887.
Stöcklin, J., 1977. Stratigraphic Lexicon of Iran. Report No. 18. Geological Survey Of Iran, Tehran.
Taylor, S.R., McLennan, S.M., 1985. The Continental Crust: Its Composition and Evolution. Blackwell, Oxford.
Tortosa, A., Palomares, M., Arribas, J., 1991. Quartz grain types in Holocene deposits from the Spanish Central System: some problems in provenance analysis. In: Morton, A.C., Todd, S.P., Haughton, P.D.W. (Eds.), Developments in Sedimentary Provenance Studies. Geological Society of London, Special Publication, pp. 47–54.
Tucker, M.E., 2011. Sedimentary Rocks in the Field: A Practical Guide. John Wiley & Sons, Chichester.
Wronkiewicz, D.J., Condie, K.C., 1987. Geochemistry of Archean shales from the Witwatersrand Supergroup, South Africa: source-area weathering and provenance. Geochimica et Cosmochimica Acta 51 (9), 2401–2416.
Zhang, K.J., 2004. Secular geochemical variations of the Lower Cretaceous siliciclastic rocks from central Tibet (China) indicate a tectonic transition from continental collision to back-arc rifting. Earth and Planetary Science Letters 229 (1–2), 73–89.
Ziegler, M.A., 2001. Late Permian to Holocene paleofacies evolution of the Arabian plate and its hydrocarbon occurrences. GeoArabia 6 (3), 445–505.

CHAPTER

6 ENVIRONMENTAL AND CLIMATIC CONDITIONS DURING THE K–T TRANSITION IN THE CAUVERY BASIN, INDIA: CURRENT UNDERSTANDING BASED ON CHEMOSTRATIGRAPHY AND IMPLICATIONS ON THE KTB SCENARIOS

T. Sugantha[1], Mu. Ramkumar[1,*], V. Balaram[2], Jyotsana Rai[3], M. Satyanarayanan[2]

[1]Department of Geology, Periyar University, Salem, Tamilnadu, India; [2]CSIR, National Geophysical Research Institute, Hyderabad, Telangana, India; [3]Birbal Sahni Institute of Paleobotany, Lucknow, Uttar Pradesh, India; [*]South East Asia Carbonate Research Laboratory (Seacarl), Universiti Teknologi Petronas, Tronoh, Malaysia

6.1 INTRODUCTION

Of the known extinction events of the geological history, the Late Cretaceous extinction event is considered to be very significant (Raup, 1986; Iba et al., 2011), owing to the fact that across the Cretaceous–Tertiary boundary (KTB), a wide variety of organisms as big and diverse as fishes and dinosaurs to as small as marine planktons have either became extinct or dwindled in number of species and population. Documentation of iridium anomaly at the KTB by Alvarez et al. (1980), and Alvarez (1986) followed by reports on the occurrences of similar anomalies from more than 138 geographically separated regions (Goderis et al., 2013) by various workers (Schmitz, 1992; Evans et al., 1993, 1995; Sutherland, 1994; Albertao and Martins, 1996; Shukla et al., 2001; Zhao et al., 2002; Shrivastava and Ahmad, 2005, 2008; Chalapathi Rao et al., 2013) led to the interpretation of major bolide collision (Kring, 2007), asteroid or comet shower during the end Cretaceous as the cause of extinction across the KTB. A slew of studies have followed and advocated a variety of causes and patterns of extinctions across the KTB. Despite considerable discussion for over three decades (Tschudy et al., 1984; Jiang and Gartner, 1986; Saito et al., 1986; Sloan et al., 1986; Hallam, 1987; Keller, 1988a,b, 2011; Keller et al., 2013; Frank and Arthur, 1999; Steuber et al., 2002; Alegret and Thomas, 2004, 2013; Ramkumar et al., 2004a, 2005, 2009, 2010a,b; Arinobu et al., 2005; Lamolda et al., 2005; Paul, 2005; Shrivastava and Ahmad, 2005; Stüben et al., 2005; Kawaragi et al., 2009; Le Loeuff, 2012; Tobin et al., 2012; Gertsch et al., 2013; Mizukami et al., 2013; Ohno et al., 2014; Punekar et al., 2014), consensus on the cause(s) of extinction across the KTB remains elusive and the debate still continues (Keller, 2011, 2012). Prauss (2009) summarized that

Chemostratigraphy. http://dx.doi.org/10.1016/B978-0-12-419968-2.00006-6

although a variety of hypotheses for the end Cretaceous mass extinctions are advocated, they can be categorized into two namely, gradual/stepwise extinction due to the environmental stress resulted from a complex feedback of volcanism, sea-level changes and bolide impact and a sudden and catastrophic extinction event associated with terminal Cretaceous bolide impact (Keller and Abramovich, 2009; Gertsch et al., 2011).

Though the Indian subcontinent houses many sedimentary basins that contain Azoic to Anthropocene stratigraphic records, only a few sections exposing KTB are available for study. Notable Maastrichtian–Danian stratigraphic records among them are those of the Um Sohryngkew River section in Meghalaya, Jhilmili intertrappean section in Madhya Pradesh, Rajahmundry Quarry section in Andhra Pradesh, Anjar section in the Kutch region, Gujarat State, and the Cauvery Basin in South India. The Deccan intertrappean sediments reported from Asifabad, Nizambad, Rangapur, Naskal, and Rajamundari based on the paleomagnetic, paleontological and geochronological studies (Rana, 1999) are considered to be Cretaceous–Tertiary transitional sedimentary successions. In addition, the intertrappean equivalents of Talki Formation in Nagpur (Rana, 1999; Bhandari et al., 1993a,b), Mahabaleshwar trap sequence of Bargi Canal section, Padwar, and Jirabad (Bhandari et al., 1993b), Barmer section in Rajasthan, Jabalpur section in Madhya Pradesh (Chatterjee et al., 2006) and extra peninsular Gongma section in Zanskar (Bhandari et al., 1993a,b) are also considered encompassing Cretaceous–Tertiary transitional sedimentary records. Owing to the paucity of extended-complete stratigraphic sections in India, studies on KTB scenario are also scarce until recent years.

The thick population and diverse faunal assemblage of the Barremian–Danian sedimentary rocks of the Cauvery Basin (Fig. 1) attracted the attention of geoscientists since the eighteenth century. Though the stratigraphic, paleontologic, and geochemical databases of the carbonates of the K–T succession of the onland part of the Cauvery Basin are excellently documented (Rama Rao, 1956; Banerji, 1972; Sastry and Rao, 1964; Sastry et al., 1972; Sundaram and Rao, 1986; Gonvindan et al., 1996; Tewari et al., 1996) that espoused the presence of Cretaceous–Tertiary Boundary (Sastry and Rao, 1964; Sastry et al., 1972; Chandrasekaran and Ramkumar 1995), integrated analyses of facies, geochemistry, and paleontologic data on the continuous and scattered exposures are limited. This state-of-affair is due to many factors including, the occurrences of geographically isolated and discontinuous exposures, occurrences of many barren sequences sandwiched between densely fossiliferous strata, scanty, relatively poorly fossiliferous, and weathered nature of the exposures. Nevertheless, the excellently documented lithological, biotic and other data bases, presumption of more or less continuous depositional records of Maastrichtian–Danian and the recent documentation of onland exposure of Maastrichtian–Danian in the Cauvery Basin (Ramkumar et al., 2010a) necessitate the understanding of the paleoenvironmetal and climatic conditions of the K–T transitional period in the Cauvery Basin and to compare the emerging inferences with that of models on KTB, so that improved understanding on the models on the KTB in general, the Cauvery Basin in particular could be achieved. It may also help check the completeness of the K–T section of the Cauvery Basin. If found affirmative, the strata could be correlated with that of K–T sections located elsewhere.

6.2 GEOLOGICAL SETTING

The Cauvery Basin (Fig. 1) is located between the latitudes 08°30′N and the longitudes 78°30′E in the southern part of the Indian Peninsular shield. This is a structurally elongated pericratonic rift basin (Sastri et al., 1981; Chari et al., 1995) created in response to the fragmentation of the Gondwana super continent

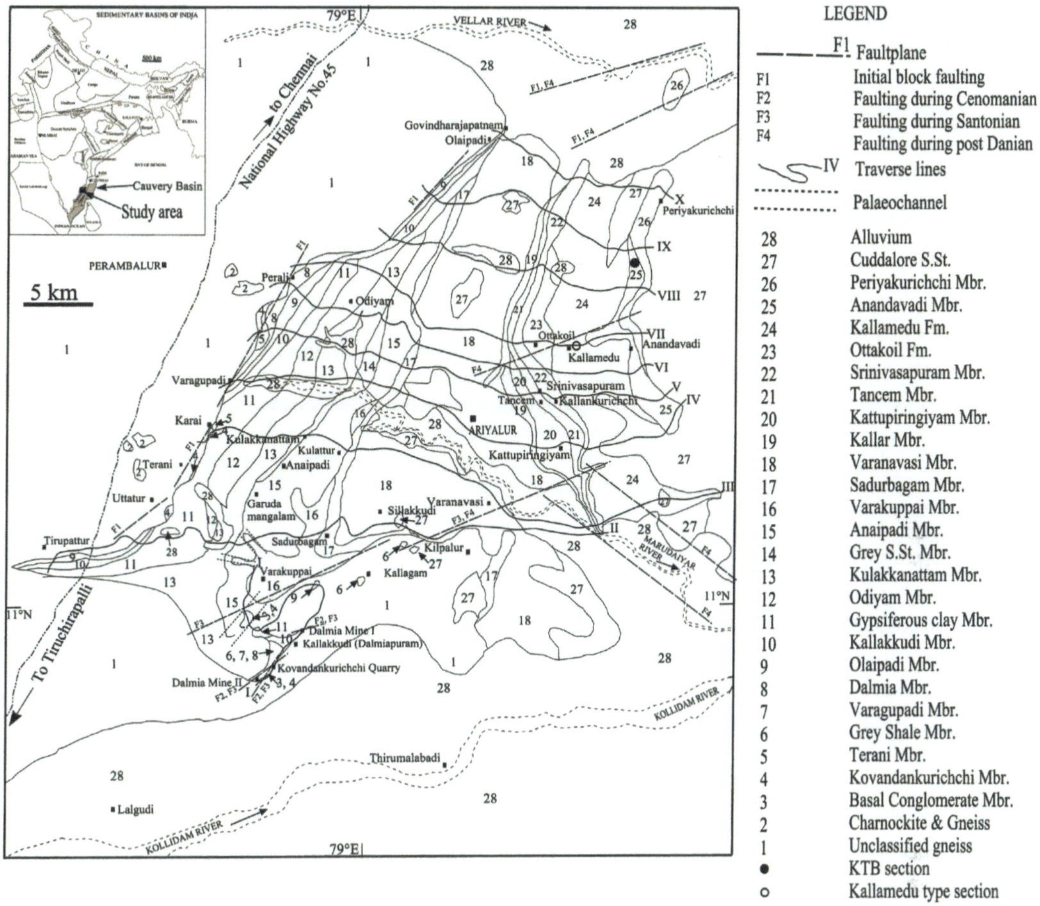

FIGURE 1

Location of the Cauvery Basin and geology of the study area *(after Ramkumar et al. (2004b))*. Serial Nos. 19 to 26 constitute the studied stratigraphic section. The Arabic numerals indicate the traverselines along which samples were collected from exposures.

and rifting between Africa–India–Antarctica (Lal et al., 2009) during the Late Jurassic–Early Cretaceous (Powell et al., 1988; Chari et al., 1995; Jafer, 1996; Chatterjee et al., 2013). Toward the West, it is demarcated by steep, basin margin faults, which separate the crystalline basement rocks from the sedimentary deposits. The basin extends toward the east in the offshore area (Sastri et al., 1981). It covers an exposed area of about 25,000 km^2 onland and 17,500 km^2 in the offshore regions (Sastri et al., 1981) of the Bay of Bengal upto 200 m isobath. The basin forms NE–SW trending half-graben morphology with a regional dip averaging 5–10° to the East and southeast. Typical of the Cretaceous deposits of India, the outcrops of these strata occur discontinuously in the Cauvery Basin, among which, the exposure found in the vicinity of Ariyalur is the largest (covering an area of about 950 km^2) and thickest (~5500 m; Ram Babu and Lakshmi, 2004) and comprises a more or less complete stratigraphic records of Barremian–Danian (Govindan et al., 2000; Ramkumar et al., 2004b). According to Sastry and Rao (1964) and Sastri et al.

Table 1 Lithostratigraphy of the Maastrichitian–Danian Strata, Cauvery Basin

Age	Formation	Member	Thickness (m)
Mio-Pliocene	Cuddalore Sandstone Fm.		4600
	~~Unconformity~~~		
	Niniyur Fm.	Periyakurichchi biostromal Member	26
Danian		Anandavadi arenaceous Member	30
KTB→	~~~Unconformity~~~		
	Kallamedu Fm.		100
	~~Unconformity~~~		
	Ottakoil Fm.		40
Maastrichtian	~~~Unconformity~~~		
		Srinivasapuram gryphean L.St. Member	18
	Kallankurichchi Fm.	Tancem biostromal Member	8
		Kattupiringiyam inoceramus L.St. Member	8
		Kallar arenaceous Member	6
	~~Unconformity~~~		
Campanian	Sillakudi Formation		400

After Ramkumar et al. (2004b).

(1973), the Maastrichtian–Danian succession of the Cauvery Basin is near-complete one while Rama Rao (1956) espoused no significant hiatus between these two ages. The Maastrichtian–Danian boundary in this basin is represented by an unconformity surface and (Nair, 1978; Ramkumar et al., 2004b) across which distinct changes in lithological, faunal and other characteristics are discernible (Sastry and Rao, 1964; Sastry et al., 1972; Chandrasekaran and Ramkumar, 1995).

The present study is confined to the 236-m thick composite section of Maastrichtian–Danian, a continuous section (~17 m thick) reported recently (Ramkumar et al., 2010a) and the type section of the Kallamedu Formation, all of which are located in the Ariyalur–Pondicherry depression of the Cauvery Basin and are accessible for closer examination. The general stratigraphic setup of the Maastrichtian–Danian succession of the Cauvery Basin is presented in Table 1.

6.3 MATERIALS AND METHODS

The present study includes collection of field information on lithology, stratigraphic boundaries, faunal occurrence and association, facies succession and variation, sedimentary and tectonic structures, etc., and rock samples from onland exposure recognized recently in ephemeral stream banks near Niniyur village. It was followed by laboratory study of the collected rock samples for their stratigraphic variation of microscale lithology, textural properties, petrography, mineralogy, and geochemical signatures, etc. (trace, minor and major elemental analyses were performed with XRF (as discussed in Kramar, 1997; Stüben et al., 2002; Ramkumar et al., 2006). The samples of composite section were analyzed in the Institute of Mineralogy and Geochemistry, Karlsruhe Institute of Technology, Germany and the Niniyur section were analyzed in the Geochemistry Division, National Geophysical Research Institute, Hyderabad, India by using the procedure described in Krishna et al., 2007). Based on these data, recognition of depositional processes and controls, the documentation of prevalent sea level changes, and enumeration of relative controls of geological agents over facies distribution and geochemical signatures were made. The inferences were collated with the previous studies on composite section of KTB and type section of the Kallamedu Formation. All these information were analyzed for the occurrences, duration, magnitude and causes of sea level, climatic and environmental changes. Then, the inferences were compared with local, regional and global scale processes and the models on end Cretaceous events with which conclusions are drawn.

6.4 RESULTS

6.4.1 FACIES AND DEPOSITIONAL HISTORY

6.4.1.1 Composite Section

The Maastrichtian–Danian succession of the Cauvery Basin is represented by the Kallankurichchi Formation, the Ottakoil Formation, the Kallamedu Formation and the Niniyur Formation in the stratigraphic order (Table 1). The Kallankurichchi Formation is a carbonate deposit consisting of abundant fauna, typical of intertidal-outer shelf conditions. Sedimentation of the Kallankurichchi Formation commenced by the development of coastal conglomerates as a result of transgression during the Latest Campanian–Early Maastrichtian (Tewari et al., 1996; Ramkumar, 1999, 2004, 2006; Ramkumar et al., 2004a,b; Ayyasamy, 2006). Toward the top, the conglomerate shows reduction in proportion and size of siliciclastics that were increasingly replaced by *gryphea* colonies. In due course concomitant with sea level rise, the gryphean bank had shifted toward former shallower regions and the locations previously occupied by coastal conglomerate have become middle shelf wherein typical *inoceramus* limestone started developing. Break in the sedimentation of *inoceramus* limestone was associated with regression resulting in severe erosion (Ramkumar, 2006) of shell banks and middle shelf deposits and their resedimentation into biostromal deposits. A significant sea level rise following this erosional event is recorded on a basin scale (Raju et al., 1993; Fürsich and Pandey, 1999). Concomitant with this rise, the opportunistic colonizers have created extensive network of *Ophiomorpha irregulaire* all over the substrates of inner shelf (Ramkumar and Sathish, 2009) which was followed by the extensive development of gryphean shell banks (Plate 1(A)). At top, shell fragments and minor amounts of siliciclastics are observed in the Kallankurichchi Formation indicating the onset of regression and higher energy conditions. The occurrence of nondepositional surface at the top of this formation and deposition of shallow marine siliciclastics (Ottakoil Formation) immediately over the carbonates and conformable offlap (Sastry and Rao, 1964; Sastry et al., 1972; Chandrasekaran and Ramkumar 1995; Tewari et al., 1996; Ramkumar et al., 2004b) of much younger fluvial sand deposits (Kallamedu Formation) are all suggestive of gradual regression associated with the establishment of fluvial system during the end Cretaceous.

The Ottakoil Formation is predominantly a siliciclastic deposit, and occurs conformably over the Kallankurichchi Formation. Its well-sorted calcareous sandstones with frequent upward fining sequences, large scale cross-bedding, abundant but patchy occurrences of the stenohaline fauna *Stigmatophygus elatus* (Plate 1(B)) and the extensive ichnofauna suggest the deposition of these rocks in shallow marine environments under the influence of seasonal and or episodic increase in energy conditions. Mitrović-Petrović and Ramamoorthy (1993) inferred shallow, open marine, normal saline characteristics for the depositional environment of this formation. A relatively low energy, open marine environment of deposition for the Ottakoil Formation was envisaged by Tewari et al. (1996). The trace fossil assemblage suggested low rate of deposition in marginal marine regime. The associated ammonites are suggested to be of

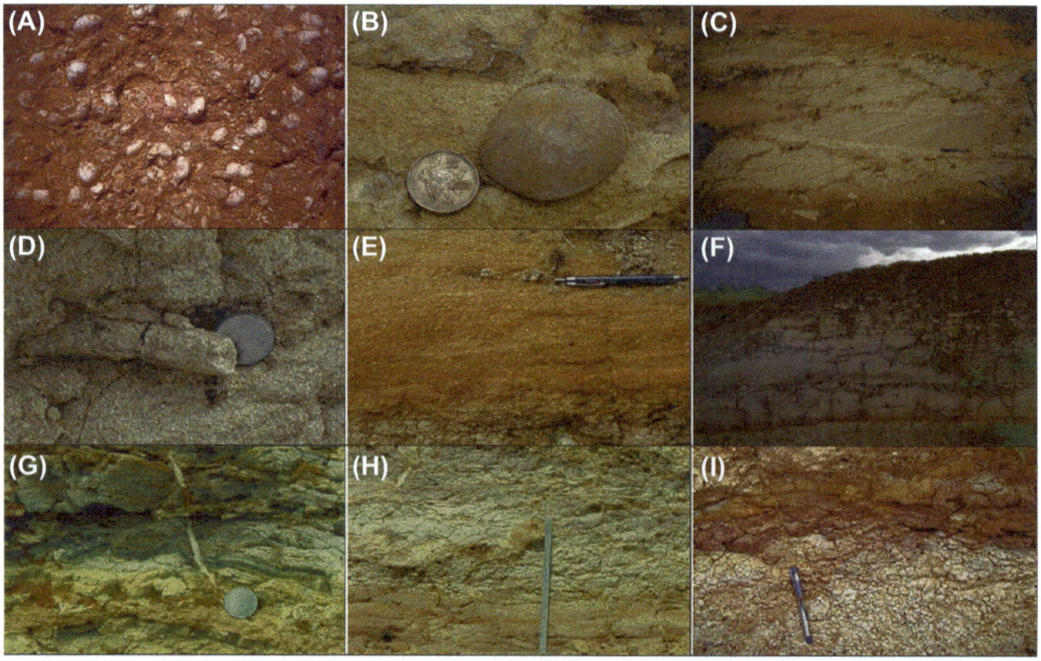

PLATE 1

A. Thick population of insitu gryphean shells that form the bulk of the Srinivasapuram Member of the Kallankurichchi Formation. Location of the photograph: TANCEM mines south of Srinivasapuram village. **B**. Field photograph showing the pale brown colored, thick bedded calcareous sandstone of the Ottakoil Formation that contains the stenohaline echinoderm fauna *Stigmatophygus elatus*. Location of the photograph: Stream section northeast of Ottakoil Village. **C**. Field photograph showing the ross bed sets of climbing ripples found to occur at the type section of the Kallamedu Formation. Similar bedsets forming advancing sand sheets in the active fluvial channels of Modern Krishna and Godavari rivers. Study of modern river processes and geomorphic evolution suggests that ephemeral channels deposit the sediment received from catchment areas in the channel itself as sand sheets to be mobilized and delivered to downstream, coastal and offshore regions only during flood seasons. Prevalence of similar conditions during the deposition of Kallamedu Formation are interpreted based on these criteria. The rocks are unconsolidated, unsorted-poorly sorted sandstones. These have abundant granule to pebble sized siliciclastic grains with typical recycled/resedimented characteristics. **D**. Closer view of a bone fragment found embedded in coarse grained sandstone of the Kallamedu Formation at the type location. The bone and other fossil remains are highly calcretized and weathered, may be due to the intense diagenetic transformation by acidic meteoric waters in meteoric vadose zone of diagenesis. The entire Maastrichtian-Danian succession shows similar intense diagenetic transformation, dissolution and recrystallization phenomena (compare with plate nos. 2.6 and 2.7) suggestive of the prevalence of post-Danian late stage meteoric diagenesis that has uniformly affected the bioclasts, early stage cement and and matrix. Refer Fig.6 wherein the V-index and R-index show uniformly altered nature all through the Maastrichtian and Danian deposits regardless of lithology affirming the post-Danian diagenesis. **E**. Field photograph showing the conformable but erosional surface contact between the silty clay deposits of overbank/flood plain environment and the sand sheet of fluvial environment. Many such cycles are recognizable and are more open marine conditions (locally developed) in an otherwise shallowing basin (Tewari et al., 1996). The moderately diversified nannofossil assemblage, recorded by Rai et al. (2013) supports the interpretation of near shore, open-marine environment of deposition for this formation. These characteristics may indicate simple sea level fall after the deposition of the Kallankurichchi Formation and resultant change in environment of deposition, reduction of depositional bathymetry and influx of siliciclastics as the prime reasons for the deposition of the Ottakoil Formation. Absence of tectonic influence during the change of environmental conditions between the Kallankurichchi and the Ottakoil Formations is also indicated.

The nannofossil assemblage represents low-mid latitude Tethyan province and is globally comparable with nannofossil assemblage recorded from low to mid-latitude. Based on the occurrences of echinoids, Mitrovic-Petrovic and Ramamoorthy (1993) assigned middle Maastrichtian age to this formation. Ayyasamy (1990) and Govindan et al. (1996) recognized this formation to be of *Pachydiscus otacodensis* ammonite Zone and *Globotruncana gansseri* foraminiferal Zone respectively. However, the nannofossil assemblage recorded by Rai et al. (2013) indicated late Late Maastrichtian age to the Ottakoil Formation (see appendix I for a complete checklist and microphotographs of the nannotaxa reported from this formation). The presence of *Ceratolithoides kamptneri* indicates top of UC 20b^{TP} Zone of late Masstrichtian age. The common occurrence of large sized *Arkhangelskiella cymbiformis,* the LAD (Last Appearance Datum) of which marks the base of CC25 nannofossil Zone (Perch-Nielsen, 1985) and the presence of *Ahmuellerella octoradiata* the LAD indicating top of UC 20a^{TP} Zone (Burnett, 1998) and the record of *Microhabdulus murus* the FAD (First Appearance Datum) of which

interpreted as episodic flooding events and deposition in fluvial channel, flood plain and overbank environments. Shallow nature of the fluvial channel, and low depositional topography are also discernible from the facies association and succession. Location of the photograph: Niniyur section. **F**. Field photograph showing the lateral and vertical facies distribution of fining-upward sequences in the cross bedded bedsets. These bed sets have sharp erosional boundary at the bottom and contain coarse granule sized grains that grade to argillaceous siltstone towards top and culminate at ferruginous silty clay. They also grade laterally to buff colored argillaceous siltstone. Together, this facies succession shows the contact relationships between the fluvial channel, flood plain and overbank depositional environments and their facies succession. Location of the Photograph: Type locality of the Kallamedu Formation located northeast of the Kallamedu Village. **G**. Field photograph showing the typical tidalflat/beachface sediments representing alternate lamina of light and dark colored sediment grains (upper part of the photograph) lying over the lagoonal deposits. This facies association together with the alternations of fluvial channel-lagoonal facies association shows the prevalence of episodic stream lengthening and shortening, juxtaposition of these facies belts and low depositional topography. Note that the granulometric analyses of the sediments have also indicated the sediments of riverine origin but were subjected to wave reworking (Fig. 5a and 5b). **H**. Field photograph showing the gradational contact between the fluvial deposits (bottom of the photograph) and the presumably lagoonal deposits (upper part of the photograph). The facies succession shows the development of alternations of fluvial and lagoonal/tidalflat facies suggestive of lengthening and shortening of the fluvial channels in tune with flooding events-short term sea level rise and fall and resultant associated vertical facies succession of fluvial channel-lagoon/tidalflat-fluvial channel. It is to be noted that while the change of lagoon-fluvial channel is marked by erosional contact between the facies types, the change from fluvial channel to lagoon is shown by a gradational contact. **I**. Field photograph showing the occurrences of thin-thick lamina of clay (suspected to be drawn from alteration of volcanic ash) interlayers within the lagoonal facies.

indicates top of UC 20a^{TP} Zone (Rai et al., 2013). All these suggest the zonal placement within CC 25 *A. cymbiformis* Zone (Perch-Nielsen, 1985) equivalent to UC 19 (part) and UC 20b nannofossil zones (Burnett, 1998) of Late Maastrichtian age. The presence of *Chiastozygus litterarius* and *Prediscosphaera cretacea* also supports the zonal placement and age assignment.

The next younger lithostratigraphic unit, the Kallamedu Formation is a 100 m thick siliciclastic deposit with interlayers of argillaceous sandstones, siltstones, clay and silty clay. Thick-bedded, unconsolidated, well-rounded and poorly sorted barren sandstones with varying hues ranging from pale brown to buff characterize this formation. This Formation overlies the Srinivasapuram Member of the Kallankurichchi Formation and the Ottakoil Formation of the Ariyalur Group and underlies the Anandavadi Member of the Niniyur Formation. At the type locality, many bedsets upto 35 cm thick and traceable for over many tens of meters occur (Plate 1(C)) and contain rare-scarce dinosaurian bone fragments (Plate 1(D)). The bedsets grade toward top to medium to thin-bedded deposits consisting of relatively highly argillaceous sandstones and siltstones and finally to a thin layer of ferruginous silty clay. Again, the fluvial channel deposits overlie these rocks (Plate 1(E)) with conformable but erosional surface contact, suggestive of prevalent episodic flooding events, deposition in channel and adjoining floodplain and overbank environmental settings. Laterally continuous, many such cycles are recognizable (Plate 1(F)). Local occurrences of clays and silts with dispersed detrital quartz grains and sandy streaks, nonbedded nature, rare lamination and mud cracks are also common. Toward the top, this formation shows the occurrence of paleosol (Plate 2 (A)) and the establishment of continental conditions. Thus, the upper surface represents a nondepositional and erosional surface that marks the boundary between the Cretaceous and Tertiary periods (Ramkumar et al., 2004b). Earlier works (Mitrovic-Petrovic and Ramamoorthy, 1993; Govindan et al., 1996) have assigned Late Maastrichtian age to the Kallamedu Formation. This formation represents the *Abathomphalus mayoroensis* planktonic foraminiferal Zone (Tripathi and Mamgain, 1987) and is consistent with an Upper Maastrichtian palynomorph assemblage (Venkatachala and Sharma, 1974) that includes *Aquilapollenites bengalensis*, *Cranwellia cauveriensis*, *Araucariacites australis*, *Tricolpites microreticulatus*, and *Triporopollenites minimus* (Masthan, 1978). Samples collected from the lagoonal facies sediments of the Niniyur section have shown the occurrence of typical Late Maastrichtian nannofossil assemblage (*see appendix II for microphotographs and checklist of the nannotaxa reported from this formation by Ramkumar et al., 2010a*). The nannofossil assemblage can be correlated with UC 20dBP Boreal Zone of latest Maastrichtian age as *Corollithion exiguum* is present. The presence of *Pseudomicula quadrata* indicates UC 20a–bTP Tethyan and Tethyan intermediate zones of Upper Maastrichtian age. The rest of the forms recorded belong to Upper Maastrichtian except the record of *Biantholithus cf. sparsus*, which indicates Paleocene or NP1 Zone of Martini, 1971 equated with NNTp1A of Varol, (1998).

The next younger stratigraphic unit, the Anandavadi Member of the Niniyur Formation, rests over the Kallamedu Formation with a distinct disconformity (Nair and Vijayam, 1980; Sastry and Rao, 1964). At the beginning of Danian, transgression took place that covered only the eastern part of the Kallamedu Formation (Ramkumar et al., 2004b). The presence of conformable contact between Anandavadi Member and the Kallamedu Formation and the initiation of carbonate deposition from the beginning of Danian are indicative of absence of fluvial sediment supply and tectonic activity. Fossils in this member are represented by *nautilus*, corals, algae, serpulids, bivalves, and gastropods. Deposition took place in a restricted marine regime under subtidal to intertidal environments (Nair and Vijayam, 1980; Sastry and Rao, 1964). The localized concentrations of

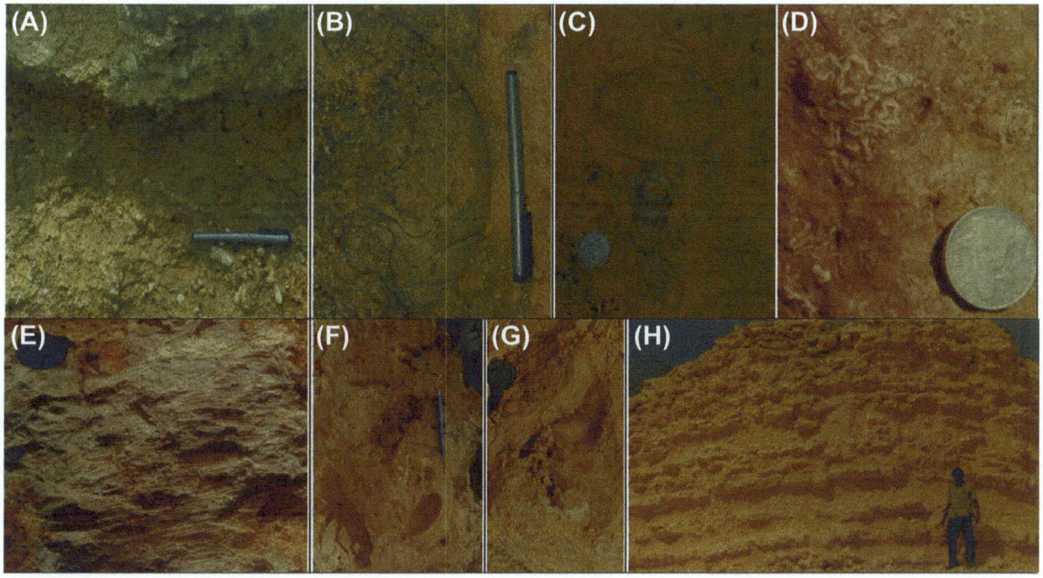

PLATE 2

A. Field photograph showing the palaeosol (central portion of the photograph) sandwiched between the coastal lagoon-tidalflat facies. **B**. Field photograph showing the planar view of the *nautilus* shell embedded in the bioclastic limestone deposit that marks the basal beds of the Danian in the Niniyur section. **C**. Field photograph showing the cross section of the *natilus* shell embedded in the bioclastic limestone deposit. Occurrence of these shells in the bioclastic limestones that rest over the lagoonal facies sediments that contain typical nannofossil taxa typical of Late Maastrichtian are interpreted as affirmative of the occurrence of KTB in the Niniyur section. **D**. Field photograph showing the development of isolated patches of brain coral indicative of shallow, open marine conditions during the deposition of Anandavadi Member of the Niniyur Formation. Location of the photograph: Exposure near Ambapur Village. **E**. Field photograph showing the development of isolated pachy occurrences of tabulate coral mounds in the Anandhavadi Member of the Niniyur Formation. Occurrences of varieties of coral taxa, together with gastropods, bivalves and other microfauna suggests restoration of luxurious environmental conditions for the marine ecosystem after KTB. Occurrences of these soon after the KTB shows restoration of normal ecological conditions early in the Danian. Location of the Photograph: Exposure near Ambapur village. **F**. Field photograph showing the extensive development void spaces as a result of post-depositional dissolution. Due to this, even thick (upto 1cm) bivalve shells are completely dissolved and only void spaces are present. Owing to their aragonitic/high-magnesian mineralogic nature, and also due to rapid burial and absence of syndepositional (early stage marine diagenesis) and marine burial stage (mesodiagenetic) diagenetic processes, these shells might have escaped any alteration and were dissolved during late stage (telodiagenetic alteration) under meteoric waters, leaving only the void spaces. Further, prevalence of warmer climate, high biochemical productivity, and faster rate of deposition are also inferred. Location of the photograph: Basal bioclastic bivalve rich limestone bed in the Niniyur section. **G**. Field photograph showing the large void space of a shell filled by large, clear rhombic crystals of calcite, affirming the prevalence of late stage meteoric phreatic, but oxic conditions of telodiagenesis. Location of the photograph: Ramco Cements Quarry near Periyakurichchi Village. **H**. Field photograph showing the high-frequency marl-limestone cyclic beds of the Periyakurichchi Member of the Niniyur Formation. Location of the photograph: Ramco Cements Quarry near Periyakurichchi Village.

shell fragments, well-lithified coralline limestone and reef-derived talus deposits (Plate 2(D) and (E)) characterize the Anandavadi Member. The increase of sea level and establishment of shallow, wide shelf with open circulation led to the deposition of the Periyakurichchi Member of the Niniyur Formation. The prevalence of high-frequency sea-level fluctuations led to the cyclic marl-limestone deposits that characterize the Periyakurichchi Member. At top, this member has distinct erosional unconformity, which in turn, when interpreted along with the presence of huge thickness of continental sandstone (>4600 m thick Cuddalore sandstone Formation), indicates the restoration of continental conditions in this basin. The global sea level peaks during 73 Ma (±1; Late Campanian), 69.4 Ma (Early to Late Maastrichtian) and 63 Ma (±0.5; Early to Middle Danian) are observed (Raju et al., 1993) to occur in this basin.

6.4.1.2 Niniyur Section

The Niniyur section contains two exposures aligned parallel to each other. Rocks exposed in them are similar except that the second exposure contains the upper part of the Srinivasapuram Member of the Kallankurichchi Formation. The lithofacies characteristics as observed in the Niniyur section are detailed herein.

The bioclastic limestones with abundant *gryphea* and gryphean fragments and noticeable amounts of siliciclastic admixture characterize the Srinivasapuram Member at this location. The boundary between this member and the overlying Kallamedu Formation is a conformable, yet, sharp erosional surface. Immediately over the unconformity surface, there are massive to irregularly bedded, pale yellow colored, poorly sorted fine to coarse sands, represented by overbank deposits (Ramkumar et al., 2010a) followed toward the top by a thin bed of flood deposit with unsorted, recycled sediments showing textural inversion. The overbank deposits also show characteristics of paleosol development, indicative of prevalent hiatus in sedimentation and ephemeral and or episodic nature of the fluvial flow. It shows gradational contact with overlying sediment layers of calcareous, alternate fine silt–fine sand sized thin-thick lamina probably deposited under ephemeral-fluvial fed coastal lagoon that was under the influence of tidal forces seasonally and or episodically (Plate 1(F)–(H)). Within these massive-irregular beds, featureless, reddish, massive, thin-thick lamina of silty clay occur repetitively. Toward top, these beds are followed by irregularly spaced thin-thick lamina of silty clay and clayey silt with various hues ranging from buff, red (Plate 2(I)), purple, etc. Granulometric analyses of these deposits by Ramkumar et al. (2013) suggested the prevalence of shallow ephemeral channels that overflown periodically and inundated coastal marine regions. It is followed toward by resumption of the lagoonal deposits that show gradational contact with beach–tidal flat sediments. The beach–tidalflat sediments show relatively better sorting, alternate lamina of light and heavy minerals, parallel lamination, feebly recognizable ripple marks, etc. At the top, deposition of lagoonal facies with unconsolidated, silty clay and clay layers and recycled sediments of various size characteristics are observed. Occurrences of frequent clayey layers with irregular thicknesses in these deposits are significantly different from other clayey layers in the entire section. These clayey layers also show many hues of colors, perhaps oxidation of either heavy metals such as Fe, Cr, V, Ni, Co etc., that were adhered onto the organic matter. Suspected paleosol layers are also discernible (Plate 2(A)).

Thus, the Kallamedu Formation shows the development of a variety of bedforms and sediment characteristics that can be grouped into four facies associations namely, advancing sand sheet facies of shallow fluvial channel environment, overbank facies deposited adjoining fluvial channels during flooding events, lagoonal facies deposited in the coastal lowlands that were under the influence of fluvial as well as tidal waters and beach–tidalflat facies. The facies succession suggests common sediment starved nature and deposition only during episodic, major flooding events in all these depositional environments. In addition, existence of coastal low lands, shallow nature of fluvial channels, high-frequency sea level fluctuations were also expressed by the facies succession.

These lithofacies and facies succession characteristics suggest the similarity between deltaic sedimentation and progradation characteristics of the Modern deltas of the east coast of India and the Kallamedu Formation as detailed herein.

The east coast of India has vast coastal low lands formed by many rivers through the development of deltaic coastline. These rivers have prograded hundreds of kilometers over the former offshore regions through a process of creation of coastal lagoon–barrier complex and filling the lagoons during flooding seasons and shifting the coastline to newer locales offshore and again creating lagoon–barrier complex (Swami et al., 1990; Singh and Swami, 1996). The lagoons remain sediment starved during lean seasons and experience oxygen poor conditions, leading to the preservation of organic matter and heavy metals adhered onto it. The sediments delivered from catchments in the river channels remain within the channels and move as sand-sheets toward downstream until major flood mobilizes them and delivers them into the coastal lowlands and offshore. During flooding seasons, owing the channel topography and the large sand sheets, banks breach and overbank deposition occurs between channel bank and floodplain (Ramkumar, 2003a; Ramkumar et al., 2000a). The lagoonal sediments show gradational contact with adjoining tidalflat. Owing to the availability of abundant calcareous material and supported by mixing zone environment, beach rocks develop in the beach-tidalflat regions (Ramkumar and Gandhi, 2000; Ramkumar et al., 2000b). All these characteristics reported from the Modern deltas (Swami et al., 1990; Singh and Swami, 1996; Ramkumar, 2003a; Ramkumar et al., 2000a,b) are all observable from the facies associations and facies succession of the Kallamedu Formation.

A typical erosional surface separates the Kallamedu Formation from the overlying Anandavadi Member of the Niniyur Formation. The lowermost bed at this location is a massive, 50 cm thick carbonate unit with variable amounts of siliciclastic admixture, shell fragments and whole shells and articulated *nautilus* (Plate 2(B) and (C)) typical of Danian age. Patchy occurrences of coral mounds (Plate 2(D) and (E)), and thick population of bivalves, preserved as casts and articulated shells (Plate 2(F) and (G)) are common. It is followed toward top by alternate marl/calcareous siltstones and silty bioclastic limestones with variable thicknesses (Plate 2(H)).

Based on the occurrence of nannofossil assemblage in the lagoonal facies of the Kallamedu Formation, Ramkumar et al. (2010a) assigned Late Late Maastrichtian age. Large *nautilus* shells typical of Danian age occur in the massive bioclastic limestone deposits that rest over the Kallamedu Formation at the Niniyur section. These two observations, clearly exhibit the occurrence of K–T section in the onland part of the Cauvery Basin. Considering the thickness of this section (>16 m), it can be an extended Maastrichtian–Danian stratigraphic record.

6.4.2 GEOCHEMICAL ANOMALIES

6.4.2.1 Composite Section

The positive and negative excursions of stable isotopes, strontium isotope and elemental concentrations across the KTB (Fig. 2) as could be observed elsewhere occur in the geochemical profiles of

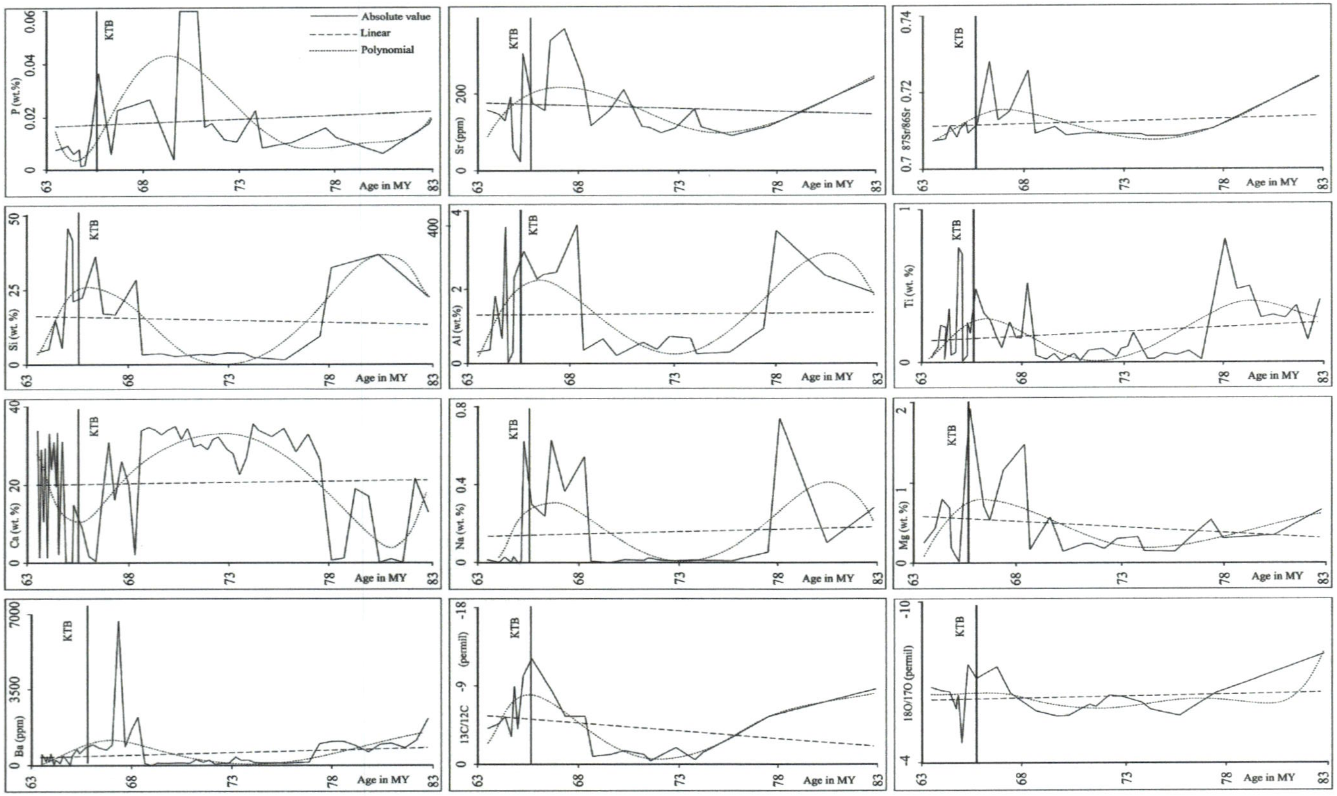

FIGURE 2

Geochemical profiles of the composite section. Note the common anomalies of trace elemental and isotopic anomalies across the KTB.These are influenced by lithological and environmental control as could be observed elsewhere. In addition, the P, Ba, stable isotopes and strontium isotopic anomalies could be observed as they show significantly different nature from the background values of Maastrichtian and Danian and also from that of the entire section signifying immediate causes for their genesis.

Maastrichtian–Danian composite section, affirming the presence of KTB in the onland exposure of the Cauvery Basin. In addition, anomalous and highly different from the background values of P, and Ba elemental concentrations and strontium isotope are recorded. These are detailed herein.

Phosphorus is directly connected with oceanic productivity and has been evidenced as indicative of global changes (Ruttenberg, 1993) on timescales longer than several multiples of its residence time in seawater (Frolich, 1988). As the primary productivity affects the global carbon cycle, changes in fluxes and reservoirs of P may play a critical role in long-term global change. Excursions of P can be directly related to individual eustatic and tectonic events (Hoppie and Garrison, 2001). Several studies have assumed that P weathering and input to the oceans are similar to the weathering and input of strontium (Raymo, 1994). In addition to the distinct negative anomaly across KTB, phosphorus shows distinct positive anomaly during Maastrichtian and has been reported to have been caused by sediment redistribution associated with sea level fall during late Maastrichtian (Ramkumar et al., 2009).

Ba shows double peaked positive excursions preceding the KTB. Occurrences of positive excursions of Ba profiles and absence of coeval excursions of any other element itself suggest that the Ba anomaly is a peculiar phenomenon and is not related to syn- or postdepositional chemical mobilization and redistribution. Except these excursions, Ba does not show any significant shifts either from a linear trend (Fig. 2) or mean value of the entire section (497.67 μg/g) or Maastrichtian average value (672.89 μg/g) or Danian average value (261.78 μg/g). It means that the positive excursions of Ba up to 6593 μg/g are abnormal. Computation of excess Ba content in the sediments according to the formula of Murray and Leinen (1993, 1996) with reference to PAAS (Post-Archaen Average Australian Shale—Taylor and McLennan, 1985) also indicates the enrichment of Ba (Ramkumar et al., 2005). These anomalies represent a 4–10 times higher concentration than average shale (650 μg/g—Taylor and McLennan, 1985).

These positive anomalies of Ba occur in the Ottakoil Formation. This formation is a siliciclastic deposit that experienced shallow marine and regressive environmental conditions. The first peak (Fig. 2) was found in a cross-bedded, fining-upward sequence while the second and major peak occurs in well-sorted sandstones. The rocks contain abundant *S. elatus*, a stenohaline, shallow marine echinoid (Mitrovic-Petrovic and Ramamoorthy, 1993) and an ichnofauna *Ophiomorpha*, which is also typical of shallow marine regions. McManus et al. (1998) suggested that preservation potential in the near-shore regions is lower than the continental margin and further offshore regions. In this context, the occurrence of Ba peaks in shallow marine deposits is itself an intriguing feature. Based on the exclusive occurrences of Ba peaks in monotonously uniform-featured sandstones, it is inferred that the gross lithology could not have influenced them. X-ray diffractograms of rock samples of these sandstones show the presence of Ba-orthoclase. Petrographic observations show stratigraphically uniform distribution of feldspar grains in the Ottakoil Formation, in which case, had the Ba-orthoclase served as the source of Ba anomaly, the positive excursions of Ba should have had a span covering the entire Ottakoil Formation and also that the background values of Ba of the Ottakoil Formation should have been higher than other formations, which in turn is not. Furthermore, Ba has the tendency to adhere onto clay mineral surfaces (Singh, 1978) in which case, significant influx of clay and the adherence of Ba onto clay could be inferred. However, clay admixture in these sandstones seldom exceeds 5%. Considering all the clay present in the sandstones to have adhered Ba on their surfaces signifies availability of free Ba during deposition of the rocks that contain these two peaks. This presumption also calls for weathering of Ba-orthoclase in the continental regions to form clay, transport and deposition of them in the marine basin to cause Ba anomalies. However, occurrence of fresh Ba-orthoclase clasts in the Kallar Member of the Kallankurichchi Formation and feldspathic sandstone in the Ottakoil Formation counters but does not excludes such a possibility. Examination of other properties of the rocks, depositional pattern, sediment influx, sea level fluctuations and facies association etc., revealed a complex interaction of many processes at regional and global scale.

The occurrences of two positive $^{87}Sr/^{86}Sr$ anomalies below the K/T boundary that are different from the general trend of $^{87}Sr/^{86}Sr$ curve (Martin and Macdougall, 1991) are reported Ramkumar et al. (2010b) from the composite section. A pronounced sea level low occurred during Late Maastrichtian on a global scale (Hakansson and Thomsen, 1999) and the same is recorded in the Cauvery Basin too (Raju et al., 1993; Ramkumar et al., 2004b). The sea level curve of the Cauvery Basin shows a long term decreasing trend that culminated at KTB lowstand similar to the observations of Adatte et al. (2002) on K–T section located in Tunisia. Adatte et al. (2002) ascribed middle and high latitude cooling as the reason for the sea level fall. Comparison of the sea level curve and the geochemical profiles of the studied section shows that the positive anomalies of $^{87}Sr/^{86}Sr$ form part of sea level retreats. As deposition in this basin represents a major oscillating carbonate-siliciclastic system coinciding with sea level high-low respectively (Ramkumar, 2004; Ramkumar et al., 2004a,b), the periods of sea level low during glaciation (Wallmann, 2001; Kampschulte et al., 2001) represent advancing of continental river systems that might have brought in higher terrigenous influx due to higher erosion in the continents (Ebneth et al., 2001), the positive anomalies could be interpreted as a direct result of clastic input (Veizer et al., 1999). These views are also ascertained by the observed higher mean value of $^{87}Sr/^{86}Sr$ of the Cauvery Basin strata than the world average. Dessert et al. (2001) have shown that the strontium isotopic values of End-Cretaceous seawater increased synchronous with Deccan volcanism and pre-volcanic values were restored after significant volcanic activity, thus forming a positive anomaly across KTB. They have suggested strong continental weathering as the reason for this spike. Whitford et al. (1996) stated that $^{87}Sr/^{86}Sr$ in rocks that have abundant fluvial source material have a mean of 0.7117 whereas a hydrothermal source would record a mean value of 0.703. According to these estimates also, enhanced continental weathering and fluvial influx that have dominated deposition in the Cretaceous–Tertiary deposits of the Cauvery Basin which in turn influenced by sea level retreats could be inferred. Occurrences of inverse relationships between linear and polynomial trend lines of terrigenous elements Sr, Si, Al, and Ti (Fig. 2), ascertain the influence of continental influx concomitant with sea level reduction. The elemental concentration of Sr shows a broad peak covering the two positive anomalies of $^{87}Sr/^{86}Sr$ curve, coeval with positive excursions of terrigenous elements such as Si and Ti, and thus this observation gains importance. As observed by Martin and Macdougall (1991) in K/T sections that spawn world over, the smoothed trend line (polynomial trend) shows a broad positive anomaly across K/T boundary in the Cauvery Basin. These authors termed the anomaly as a consequence of higher influx of radiogenic Sr to the ocean during deposition. They also opined that, as Deccan volcanism was coeval to K/T boundary event, the weathering of volcanic rocks may have influenced $^{87}Sr/^{86}Sr$ values in the early Cenozoic which is evident in the Fig. 2.

6.4.2.2 Niniyur Section

Though the elemental profiles of the Niniyur section show the occurrences of many positive and negative excursions in absolute concentration profiles, marked change in their patterns across KTB is explicit (Figs 3 and 4) implying the change of environmental and climatic conditions across the boundary and also that the Niniyur section possess the sedimentary record of Late Maastrichtian and Early Danian. It is affirmed by the occurrences of *nautilus* shells (suggestive of Danian) in the massive bioclastic limestone and nannofossil assemblage typical of late Late Maastrichtian in the lagoonal facies at this location. The geochemical signals as observable in the profiles of the Niniyur section are detailed herein.

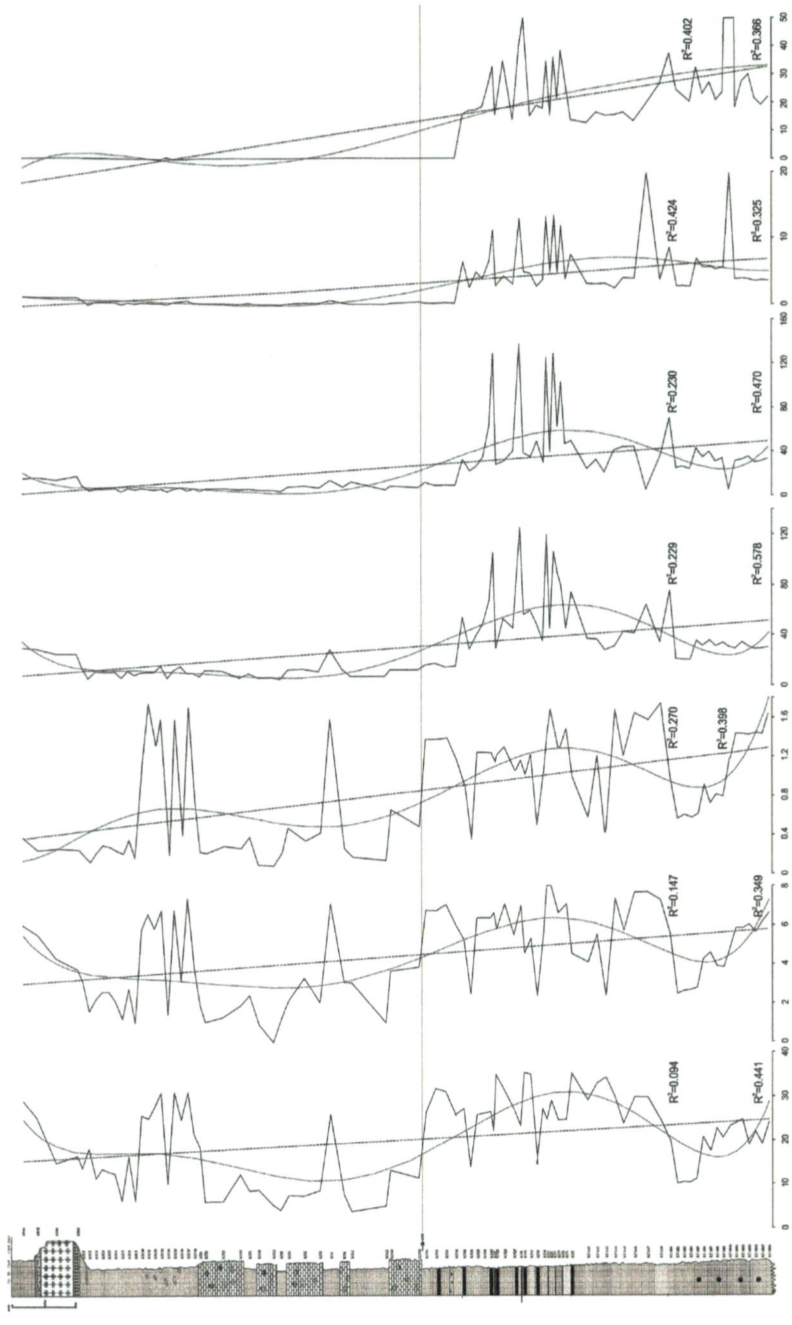

FIGURE 3

Geochemical profiles of major and trace elemental concentrations of the Niniyur section.

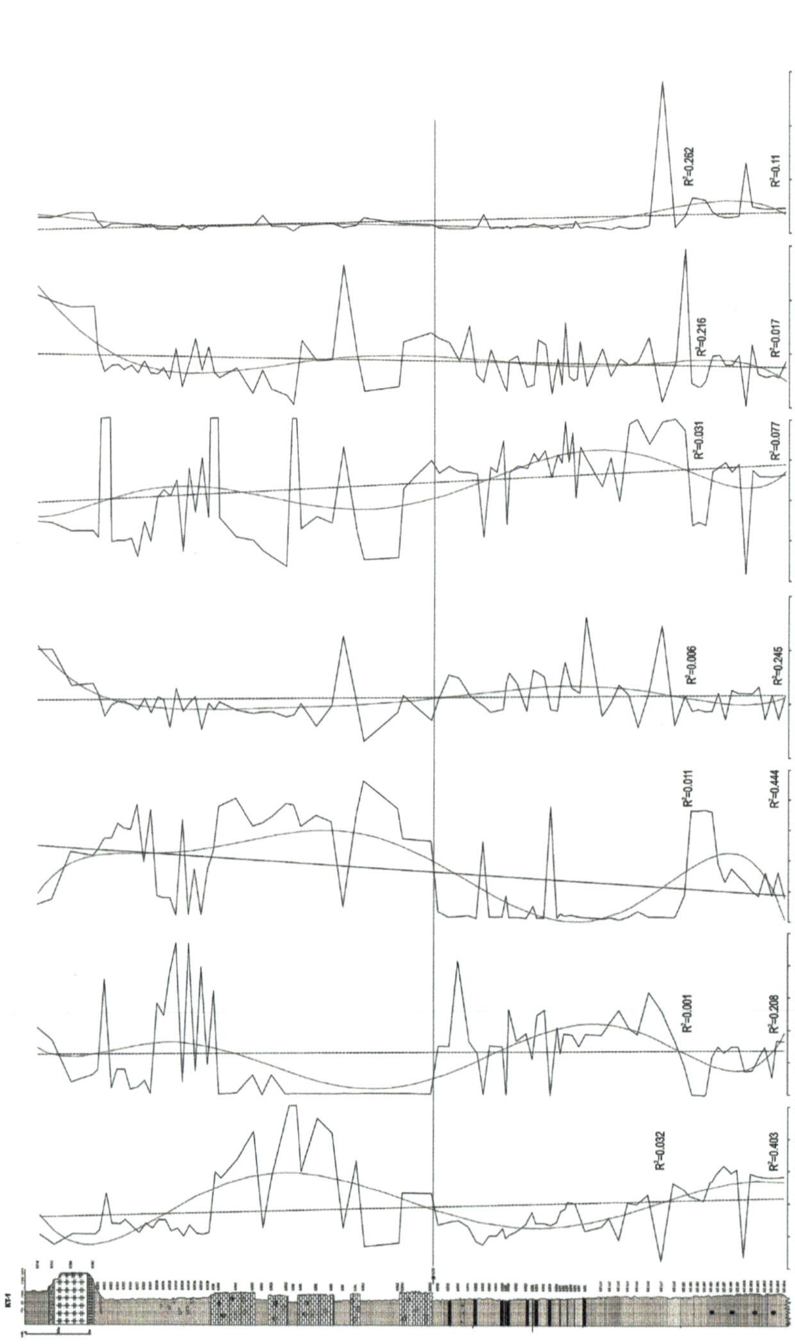

FIGURE 4

Geochemical profiles of major and trace elemental concentrations of the Niniyur section.

Secular trends: The elements Si, Al, and Mg, along with the heavy metals V, Cr, Co, and Ni, show distinct and recognizable decreasing trends while the Sr, and Na, show feeble, but general reducing trends from Maastrichtian to Danian (Figs 3 and 4). On the contrary, the Ca shows a distinct increasing trend (Fig. 4). As such, these linear/secular trends might be the result of lithological control as the Maastrichtian deposits are siliciclastics and the Danian deposits are either bioclastic carbonates and or calcareous in nature. The siliciclastics were deposited under fluvial and associated coastal environments (that were under the influence of fluvial processes episodically) and the carbonates were deposited under the marine conditions. Hence, the control exercised by lithological variation over the stratigraphic variation of elemental concentrations may be inferred. Ramkumar et al. (2011) reported that the all through the Barremian–Danian, the depositional pattern of the Cauvery Basin has been broad scale alternations of siliciclastics and carbonates as a result of glacio-eustatic fluctuations of sea level, which in turn was regulated by the variations in climatic conditions. The nature and pattern of the geochemical profiles of the Niniyur section also affirm the same.

Cyclic trends: The elements Si, Al, Mg, Na, K, V, and Cr show broad scale positive excursions during Maastrichtian and negative excursions during Danian (Figs 3 and 4). Less significant but positive excursions during Maastrichtian and negative excursions during Danian are shown by the Ti, Ba, and Co. The Ca and Sr show inverse relationships with these preferentially detrital elements, i.e., broad scale negative excursions during Maastrichtian and positive excursions during Danian. This behavior may also indicate duel source for Sr. It means that, during Maastrichtian, Sr from continental weathered materials was brought to the depositional sites (detrital source) and significant quantities of Sr was incorporated into the carbonates (biogenic) as well as syndepositional aragonite/calcite precipitation during Danian.

Trends and anomalies in the absolute concentration profiles: Among the geochemical anomalies of the absolute concentrations, the positive excursion of the Ca ranks significant. Though the Maastrichtian deposits are all siliciclastic in nature, the enrichment of the Ca concomitant with proportional depletion of the Si, Al, Mg, Na, and K in silt-sand cyclic laminar deposits that contain the suspected egg casts suggests an interesting phenomenon. The absence of detrital influx during the deposition of these particular sediment layers is also indicated by negative excursions of the Si, Al, Mg, Na, K, and Zr. These rocks may have been deposited adjoining coastal marine regions, perhaps in coastal lagoon conditions. These regions also serve as mixing zone for acidic meteoric water and alkaline sea water in the coastal shallow aquifers that promote faster cementation, either by aragonite and or high-Mg calcite leading to early stage lithification, which in turn might have undergone significant leaching and reprecipitation as more stable low-Mg calcite. Many modern examples with similar mechanisms are considered to be the causative factors for the development of beach rocks. The Krishna (Ramkumar and Gandhi, 2000) and Godavari deltaic coastlines (Ramkumar et al., 2000b) are typical examples of geographically closely located analogs wherein the sediments of the tidal flats and coastal lagoons are fed by the fluvial systems. The sediments from these subenvironments show typical fluvial origin but well-lithified due to their depositional site been located in the mixing zone. It is interesting to note that the discriminant diagrams of granulometric properties of the siliciclastic sediments have shown fluvial character as genetic process (Fig. 5(A)) but show the involvement of wave reworking (Fig. 5(B)) affirming the interpretation of deposition of sediments in coastal lagoon/tidal flat and lithification due to their unique location within mixing zone of the coastal shallow aquifers. The granulometric studies of beach rocks of the modern Krishna and Godavari deltas also show similar trait of fluvial origin despite wave reworking and faster rate of deposition (Ramkumar and Gandhi, 2000; Ramkumar et al., 2000b). Nevertheless, the faster rate of deposition do not corroborate

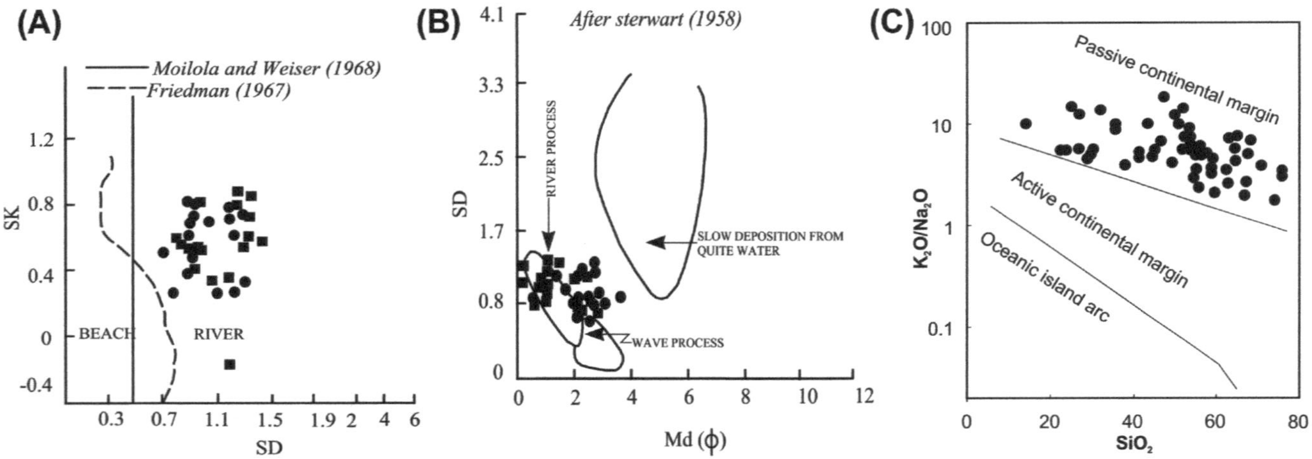

FIGURE 5

Discrimination of the depositional environments, depositional processes and tectonic setting. (A) Discriminant diagram based on discriminant lines of Friedman (1967) and Moiola and Weiser (1968) indicating the exclusively riverine origin for the sediments evidenced by the plot of all the samples in the river fields of both the discriminant lines defined by SK and SD. (B) Discriminant diagram of depositional processes (after Stewart (1958)) indicating the prevalence of riverine environment but reworking by wave process. (C) Discrimination of tectonic setting (after Roser and Korsch (1986), Roser et al. (1996)) of the studied samples. Plot of all the samples in the passive continental margin strongly advocates for the tectonic quiescence during the Late Maastrichtian and the Danian. This inference, along with the observed cyclic nature of lithofacies succession within each depositional environment and the occurrences of short-term, high-frequency excursions in the absolute concentration geochemical profiles of the entire studied section, suggest the overriding influence of sealevel fluctuations and the coupled nature of the depositional pattern of this basin with the global cycles of glacio-eustacy and climate.

with the signals expressed by the negative excursions of detrital elements associated with the positive excursion of Ca (Figs 3 and 4). This phenomenon could be explained by sediment starved nature of the basin, mobilization of sediments and delivery only during the flood seasons, and proximity between source and depositional sites so that the fluvial character is retained by the sediments. All these could be explained by a complex mechanism and sequence of events namely, delivery of the sediments in lagoonal and tidal flat depositional sites, insignificant wave reworking of the delivered sediments during seasonal tides and or short-term sea level rise to impart wave reworking character to the sediments and associated introduction of mixing-zone of diagenesis wherein, the advancing alkaline water mixed with the coastal meteoric water aquifer to promote rapid and intense calcite cementation, due to which, the detrital elements appear to be depleted in the geochemical profiles. These conditions might have been terminated following the lengthening of fluvial channel and deposition of the flood deposit over the facies succession that records positive excursion of Ca (refer lithological profile and Ca—Fig. 4). Hence, the positive excursion of Ca is interpreted more of a diagenetic phenomena than depositional feature. As the cementation process requires specific Pressure-Temperature conditions and might be under the control of the available voids and cementing material and the limited access provided for the circulation of diagenetic fluids by bedding planes (Ramkumar and Gandhi, 2000; Ramkumar et al., 2000b and references cited therein), the diagenetic calcitic dilution of depositional detrital elements appear occurring in a specific bed, producing a positive Ca anomaly.

Very low range of Sr in these rocks (~150 μg/g) affirms the inference of syndepositional lithification in meteoric waters as marine carbonates incorporate significant amounts of Sr in their Ca^{2+} lattice that range upto 700 μg/g for calcite and >2000 μg/g for aragonite (Tucker and Wright, 1990; Ramkumar, 1996a). The general subdued nature of Ca during Maastrichtian and its predominance during Danian are distinctly observable. Interruptions of these general natures of Ca occur as a result of high-frequency fluctuations of positive excursions of Si, Al, Mg, Na, and K, owing to the episodic detrital influx in terms of quartz, feldspar and clay. Influx of these detrital materials episodically into the depositional basin may also be indicative of episodic but short-term, high-frequency climatic cooling, source area weathering, and the ephemeral nature of fluvial activity. The extreme positive excursions of Ca also coincide with positive shifts of Sr, suggestive of biogenic/syndepositional incorporation of Sr into these carbonates and the preservation of primary geochemical signals of the rocks despite observable depletion of mobile elements. This inference also affirms that, despite the diagenetic dilution event, primary signals are not completely obliterated.

The high-frequency fluctuations of the heavy metals namely, V, Cr, Co, and Ni, in association with the cyclic silty-clay lamina require attention. The adherence of heavy metals onto the organic matter which in turn get accumulated in regions that experience low energy, low depositional rate, and oxygen-poor conditions of deposition is a commonly observable feature in many modern deltaic regimes (Ramkumar, 2003b, 2007; Ramkumar and Vivekananda Murty, 2000; Ramkumar et al., 2000c, 2001) and the prevalence of similar phenomena episodically could be inferred from the positive excursions of these heavy metals during the Late Maastrichtian in ephemerally fed, coastal lagoons. These clay layers were buried under cyclic influx of detrital materials that show alternate enrichment of Zr and Si, affirming the inference of ephemeral and periodic flooding and sediment influx into the coastal lagoons. Subdued, but coeval positive excursions of the absolute concentration profile of P support either higher primary productivity (Ruttenberg, 1993; Spalletti et al., 2001) and or redistribution of P during periods of sea level fluctuation (Compton et al., 1993; Ruttenberg, 1993; Filippeli and Delaney, 1994; Rangel et al., 2000; Hoppie and Garrison, 2001) or both.

Ba shows double peaked positive excursions during Danian. Except these two enrichment events, the studied section shows no other significant enrichments. While the first anomaly is associated with bioclastic limestone deposits rich in bivalves, the younger anomaly is associated with marly limestone deposits, suggesting the absence of lithological preference or control over the enrichments of Ba. Ba also shows high fluctuation from a minimum of 33 to a maximum of 1170 μg/g with a mean value of 190 μg/g in the Danian deposiits. From this background value, the double peaks of Ba record six orders of enrichment than the Danian mean value and 4.88 orders higher than the mean value of the entire section. Thus, the positive excursions might be related to significant, short-term events. The pattern of enrichments is also sudden and without any secular or linear trend. Based on the discussions on the Ba anomalies present in the composite section and the reported Ba anomaly in Danian deposits of Guatemala (Ramkumar et al., 2005), and also that the anomalies are younger than the bioclastic deposits (a period that promoted significant organic matter production and accumulation in the depositional basin and thus may have supported the genesis of methane hydrates) venting of methane hydrate due to sea level oscillations could be inferred as the causative process of these Ba anomalies. The inferences of short-term, high-frequency sea level fluctuations and associated changes in the environmental conditions leading to alternations of marl-limestone also support the presumption of methane venting and resultant Ba enrichment events.

6.5 DISCUSSION

6.5.1 COMPLETENESS OF THE K–T STRATIGRAPHIC RECORDS OF THE CAUVERY BASIN

The onland exposure of the Cauvery Basin contains four formations namely, the Kallankurichchi Formation, the Ottakoil Formation, and the Kallamedu Formation representing Maastrichtian and the Niniyur Formation, composed of Anandavadi and Periyakurichchi members, representing Danian. Together, these expose 236 m thick Maastrichtian–Danian stratigraphic record. Numerous publications (Rama Rao, 1956; Sastry and Rao, 1964; Sastry et al., 1972; Nair, 1978; Nair and Vijayam, 1980; Chandrasekaran and Ramkumar, 1993; Chandrasekaran et al., 1993, 1996; Ramkumar and Chandrasekaran, 1996; Ramkumar, 1996a,b, 1999, 2001, 2003c, 2004, 2006; Ramkumar et al., 1996, 2005, 2010a,b, 2013; Rai et al., 2013) have documented litho, bio, nanno, and chronostratigraphic information together with geochemical, facies, petrographic and other criteria of these formations that unequivocally established the occurrence of Maastrichtian and Danian sedimentary records in the Cauvery Basin. However, patchy, scattered, weathered and discontinuous nature of the exposures, and the occurrences of unfossiliferous zones sandwiched between fossiliferous sedimentary rocks, etc., have thwarted any successful study of KTB and related events in this basin. Though, previous studies have documented that the occurrences of eustatic sea level dominated depositional pattern (Ramkumar et al., 2004b, 2010a,c), vitric tuff (Tripathi and Mamgain, 1987), prevalence of shallow ephemeral fluvial channels (Masthan, 1978; Ramkumar et al., 2013), alternations of humid and dry climatic conditions (Madhavaraju et al., 2002), and many geological events and processes as indicated by the stable isotopic, trace elemental, and strontium isotopic anomalies in the basin which in turn were found to be coeval with similar anomalies in geographically widely separated sedimentary basins located at Israel, New Mexico, and Guatemala, etc., (Ramkumar et al., 2004a, 2005, 2009, 2010b), correlating the events recognized from the Cauvery Basin has been a mirage so far due to the paucity of good exposures representing stratigraphic records across KTB. Nevertheless, one of these anomalies (Ba anomaly in the Ottakoil Formation of Maastrichtian age) has been precisely dated to be of Late Maastrichtian, (CC 25 *A. cymbiformis* Zone equivalent to UC 19 nannofossil Zone of Late Maastrichtian age) by the nannofossil study of the authors (Rai et al., 2013). Yet, the correlation and systematic study of KTB as is being conducted elsewhere is awaited.

The recent report of the occurrence of continuous exposures of Maastrichtian and Danian deposits in the onland part of the Cauvery Basin, near Niniyur Village (Ramkumar et al., 2010a) has rekindled the hopes of locating no-boundary section (*sensu* Rama Rao, 1956) of the Cretaceous–Tertiary in the Cauvery Basin. Though the report has initiated considerable debate on the nature and genetic characteristics of the spherical objects (that had more or less uniform diameter (18–22 cm), occurrences of many clusters with uniform number (7–8) of spherical objects, perfectly circularly aligned fashion of the objects), the occurrences of continuous stratigraphic records of the Srinivasapuram Member of the Kallankurichchi Formation, the Kallamedu Formation and the Anandavadi Member of the Niniyur Formation at this location have not been questioned. In addition, the rock/sediment samples collected from the matrix portion of the spherical objects (egg casts) yielded typical late Late Maastrichtian nannofossil assemblage (Ramkumar et al., 2010a) while the Danian limestones contain *nautilus* shells in appreciable numbers, confirming the exposure to be of continuous stratigraphic record of the KTB.

The lithofacies characteristics of the rocks of the Niniyur section, and the geochemical characteristics and the anomalies present in this section have shown that, the depositional pattern of the rocks of this section are consistent with the depositional pattern of the Cauvery Basin, namely, alternation of siliciclastic-carbonate deposits as a result of sea level fluctuations of various orders, stacked nature of higher order cycles within next level of cycles, sediment starved nature of the basin, recycling of the older sediments, occurrences of either positive or negative anomalies of trace elements across the KTB, and occurrences of many geochemical anomalies, far away from KTB and their co-occurrences with similar anomalies with that of other well-constrained KTB sections such as located in Israel, New Mexico and Guatemala. These traits of the Niniyur section support the view that this could be an extended stratigraphic record of KTB and can be studied in detail for correlating the onland strata of the Cauvery Basin with that of KTB sections located elsewhere.

6.5.2 ENVIRONMENTAL AND CLIMATIC CONDITIONS

The Cauvery Basin was located at low-middle latitudes during the Late Cretaceous and was an island continent akin to the present day Australia. Owing to the paleogeographic location and the Island-continent condition, and in the absence of any significant tectonic influence over the depositional system except during Santonian (Ramkumar et al., 2004b, 2006, 2011), the depositional pattern of the Cauvery Basin might have been controlled principally by the sea level fluctuations which were regulated by the variations in climate and resultant glacial advancement and retreat. Further, the nature and quantum of detrital sediment influx were regulated by the glacio-eustatic sea level fluctuations in view of the configuration of the Cauvery Basin and limited nature of the provenance areas. These characteristics have enforced sediment-starved nature during the entire basinal history except during the Santonian. The reactivation of basin margin faults during Santonian brought the positive areas under the influence of marine transgression. Thus, the geochemical anomalies of the composite section were created either by sea level fluctuation directly (P anomaly, Ramkumar et al., 2009; Ba anomaly, Ramkumar et al., 2005), or by significant detrital influx following sea level fall (first Ba anomaly and Sr isotope anomaly, Ramkumar et al., 2005, 2010b) giving credence to the sea level controlled depositional and geochemical patterns in this basin.

The present observations on the Niniyur section have also revealed the overriding influence of sea level fluctuations and resultant fluvial channel lengthening and shortening, recurrence of such high-frequency cycles during the Late Maastrichtian and also during the Danian and deposition of either silty-clay lamina (Maastrichtian) or limestone-marl alternations (Danian). While the off-seasonal floods, overbank deposition, organic matter and heavy metal rich positive excursions were prevalent in the coastal lagoon environmental settings during the Late Maastrichtian, the detrital mineral (quartz, feldspar and clay) rich high-frequency cycles were prevalent during the Danian as a result of these short-term, high-frequency sea level fluctuations. Examination of the tectonic setting of these deposits with the help of established discriminant diagram (Fig. 5(C)) had revealed the absence of tectonic influence even during the Cretaceous–Tertiary transitional boundary, affirming the absence of tectonic influence and the prevalence of the sea level fluctuation controlled depositional pattern. Figure 5 shows that all the samples in the field of passive continental margin tectonic setting (after Roser and Korsch, 1986; Roser et al., 1996), signifying the predomination of sea level fluctuations and tectonic quiescence in this basin during the K–T transition. The recurrence of release of methane gas from gas hydrates during Danian coincidental with similar event in Guatemala (Ramkumar et al., 2005) as a result of sea level fluctuation is also recorded with the present study.

That the environmental and climatic conditions prevalent during the Late Maastrichtian and the Danian are entirely different are evident from the profiles of PIA (Plagioclase Index of Alteration) and (Chemical Index of Alteration) CIA (Fig. 6), that show the predominance of relatively significant continental weathering and detrital influx in association with and or as a result of cooler climate and lower sea level (albeit with short-term, sharp fluctuations to warmer episodes, sea level rise and cessation of detrital influx) until the end of Cretaceous and reversal of such conditions across the KTB and predominance of warmer climate, sea level high and absence of detrital influx during Tertiary. In addition, absence of continental weathering, prevalence of higher sea level are also indicated during the early phase of late Maastrichtian i.e., during the deposition of lagoon facies. The sharp, sudden negative excursions of CIA and coincidental PIA are associated with red-colored and other (purple, buff, pink, etc.) thick lamina of silty clay present interbedded within the deposits of the lagoonal facies. This observation, together with the occurrences of weathered vitric tuff in the Kallamedu Formation and the proximity of these deposits with the KTB and the main phase of Deccan/Godavari volcanism (Keller et al., 2011), suggest the episodic atmospheric ash fall out during these excursions. The Log-redox profile (Fig. 6) shows that the entire succession had experienced oxic conditions of deposition except the high-frequency cyclic deposits of lagoonal facies wherein episodic recurrence of oxygen-poor conditions, either due to the sea level highs, closure of connectivity with open-ocean and or atmospheric block-out by Deccan volcanic ash could be visualized.

That all these sedimentary records underwent significant late stage diagenesis is evident from the uniformly (regardless of Maastrichtian or Danian deposits) altered nature of them in the profiles of R-index and V-index (Fig. 6). However, despite this alteration, the rocks under study preserved primary signals as reflected in the gross lithological control, anomalies in immobile elements, etc.

6.5.3 IMPLICATIONS ON THE CAUSATIVE FACTORS OF EXTINCTION ACROSS K–T

The mass extinction of fauna and flora during the end Cretaceous has been one of the largest geological events of the Earth. Raup (1986) suggested the recurrence of extinction events on a 26 Ma periodicity. Prokoph et al. (2004) computed the periodicity of major biotic diversity events at the scales of

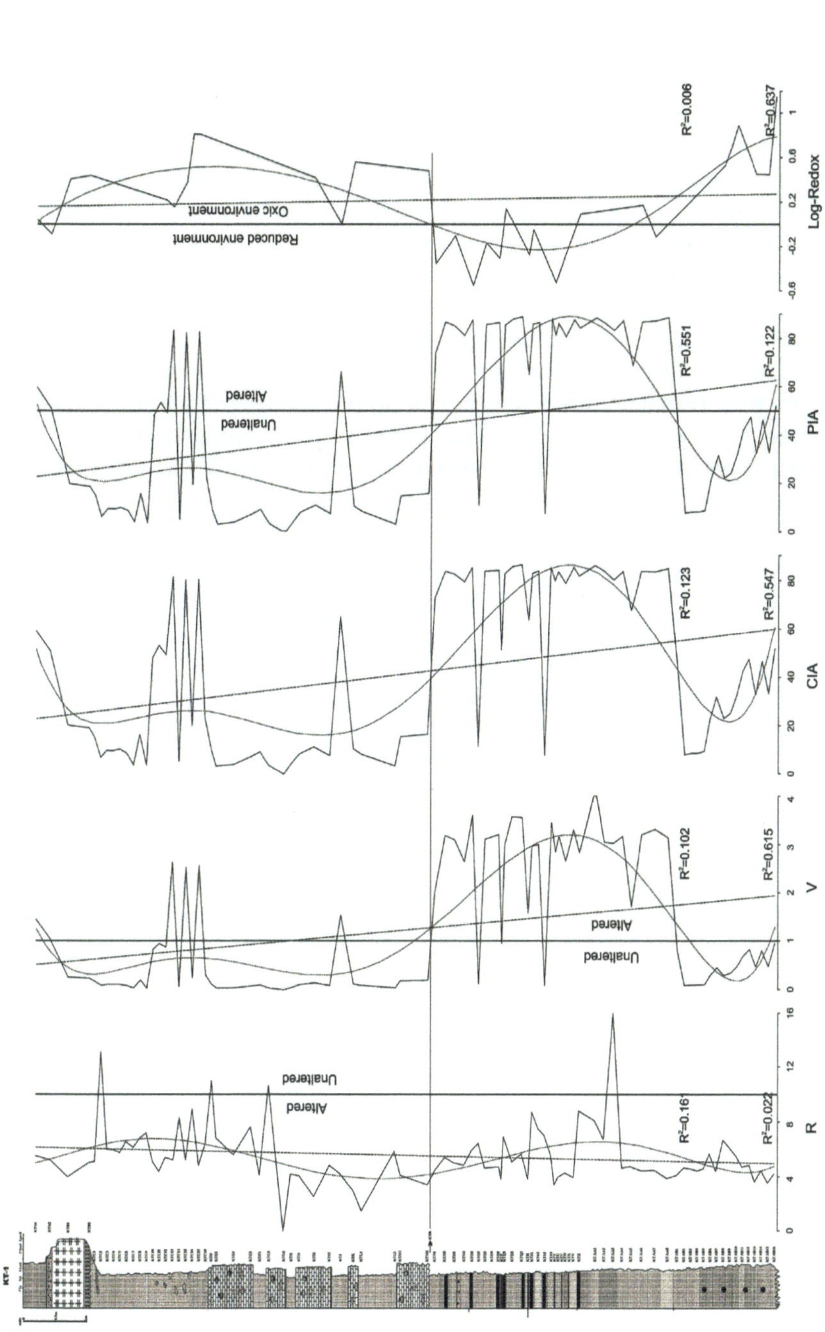

FIGURE 6

Profiles of weathering indices and environmental indicators of Maastrichtian–Danian strata of the Niniyur section.

26–33 Ma based on nannoplankton and planktonic foraminifera and demonstrated coeval stable isotopic and other marine nutrient cycles that countered the premise of 26 Ma extinction periodicity espoused by Raup (1986). The work of Kauffman (1984) suggested that, rather than an instantaneous catastrophe at the close of the Cretaceous, the marine biotic evidences pointed a complex pattern of multiple-extinctions that extended over millions of years. Dickens (2003) stated that the environmental changes in geologic history are considered to be abrupt only if they span <100 kyr.

The environmental and climatic conditions prevalent prior to, during and after the Cretaceous–Tertiary boundary have been the subject matter of research for over three decades. Yet, consensus eludes the geoscientific community. A review on the history and current status of our understanding on the end-Cretaceous environmental and climatic conditions (Sugantha and Ramkumar, 2014) showed that though two major hypotheses were advocated as to the causes of mass extinction namely, bolide impact and Deccan volcanism, there are many other events/factors/mechanisms viz., sea level fluctuation, ocean acidification, fluvial influx, changes in carbon and sulfur budgets of ocean-atmosphere, release of methane from gas hydrates, and many other causes that are either related (directly or indirectly) or independent of the two major events. It is also made plain that, not a single model could convincingly explain all the features recorded in the Late Cretaceous deposits due to the fact that many different processes acted at varying spatiotemporal scales and had their imprints accordingly. The inferences drawn based on the composite and Niniyur sections of the Cauvery Basin data are discussed herein in the light of prevailing causative models on KTB scenarios.

Phosphorus forms essential ingredient for primary production in life cycle and any delay in additional input from terrestrial sources to the marine ecosystems would be devastating, resulting in higher biotic turnover (Tappan, 1967). Phosphorus records negative excursion across the K/T boundary in this basin (Fig. 2). Although many events that have been advocated to occur across K/T transition, a marked biotic turnover owing to nutrient depletion (resulting either from impact scenario or extensive volcanic eruption) has been found to be consistent world over (Tappan, 1967; Peryt et al., 1993). The present observation of nutrient depletion in the Cauvery Basin across K/T boundary affirms the inference of disruption of biotic life cycle by nutrient deprivation. As this climatic change is a global phenomenon, it is not surprising to find major lithological changes associated with various geochemical anomalies across this boundary (Canudo et al., 1991; Peryt et al., 1993). It is also to be noted that the negative anomaly across the boundary turns into positive excursion, signifying early Tertiary recovery of sedimentation and biological system from environmental stress prevalent prior to and across the boundary as could be observed elsewhere (Stüben et al., 2002).

The double peaked positive strontium isotopic excursions were interpreted as the results of combination of various factors such as sea level fall due to global scale climatic change, the exposure of former marine regimes to subaerial conditions and resultant diagenetic release and episodic increase in the influx of siliciclastics (Ramkumar et al., 2010b). Important observation in this inference is that all these changes have occurred prior to Cretaceous–Tertiary boundary. The positive anomalies are short-lived and record a spurt of $^{87}Sr/^{86}Sr$ from background values. These can be linked with major retreat of sea level, which in turn indicates concomitant cooling of global atmosphere. Among global climatic reversals, cooling trends tend to be gradual while warming trends are abrupt (Crowley and North, 1988; Koch et al., 1992) as evident from the profiles of carbon and oxygen isotopes (Fig. 2) of the composite section of the present study. Paleo data imply a significant cooling during the Late Cretaceous followed by a warming during the early Tertiary with a sudden increase in temperature at the K/T boundary

(Wolfe, 1990), which in turn is consistent with the excursions of stable isotopic data of the present study (Fig. 2), nevertheless with less significant shifts.

The record of two positive $^{87}Sr/^{86}Sr$ anomalies preceding K/T boundary, thus supports the views of majority of the authors reviewed earlier that K/T boundary event could not have been caused by a sudden and single event, rather by many events and environmental factors. The marine environmental niches might have been under stress, which in turn were aggravated by the meteoritic impact (in and around Chicxulub in the Yucantan Peninsula) and the Deccan volcanism (on a global scale, *sensu* Glasby and Kunzendorf, 1996; Martin and Macdougall, 1991) and Godavari volcanism (at least in a regional scale). While studying the REE (Rare Earth Elements) of clay minerals of Lameta Formation of Jabalpur and Deccan basalts, Salil et al. (1997) opined that the REE patterns of clays and basalts seem to be identical and interpreted Deccan basalt weathered products to be the provenance of those Maastrichtian clay minerals of Lamata Formation, thus suggesting the prevalence of significant Deccan volcanism well before KTB. As it may involve tens of thousands of years for emplaced basalt to weather and form clay layer to get transported to Late Maastrichtian deposits, the volcanic event might be predating far away (older) from the KTB. In a sense, the occurrences of Sr isotopic anomalies far below KTB could, thus have been influenced either directly by volcanism or by the influx of weathered products of volcanic ash. The occurrence of *Caranosaur* bone fragments followed by clayey sandstones (Fig. 7) presumably derived from weathering of volcanic glass in rocks above the second strontium anomaly signifies the influence of Deccan volcanism in the deposition of Cauvery Basin during terminal Cretaceous Period.

The inference of prevalence of multiple, noncatastrophic events is further substantiated by the presence of Ba anomalies in the Maastrichtian–Danian stratigraphic sections of Israel, NE–Mexico, and Guatemala. During the Maastrichtian, these sections, which contain siliciclastic and carbonate sediments, were parts of shallow to moderately deep-water shelf-slope regions. They represent an expanded, well-preserved lithological sequences of the terminal Maastrichtian and the K/T boundary, including impact-induced spherule layers as well as Ir-anomalies. Stüben et al. (2002, 2005), Keller et al. (2003) and Adatte et al. (2004) documented detailed biostratigraphic, lithological and geochemical characteristics of these sections. The profiles of NE–Mexico record single or juxtaposed double peaked Ba anomalies preceding the K/T boundary, and the expanded section from Israel records a double peaked Ba anomaly similar to the Cauvery Basin section. The Guatemala section also records a Ba anomaly, but during the Danian. The Ba anomalies are not influenced by lithology or any other similar cause. Furthermore, except for the first Ba anomaly in the Israel section, no other coeval trace elemental anomalies could be observed. Together, these observations affirm that, the occurrence of a Ba anomaly is geographically and chronologically widespread, the causes of which may be global or regional. In addition, they also indicate the prevalence of noncatastrophic events, predating the K/T boundary that released Ba into the sedimentary system. Based on these Ba anomalies, Ramkumar et al. (2005) interpreted that the sea level retreat after prolonged organic matter accumulation (deposition of the Kallankurichchi Formation) might have introduced destabilization of gas hydrates, which in turn might have produced Ba anomaly in the Cauvery Basin. The occurrence of first strontium isotopic anomaly coeval with Ba anomaly is interpreted to have resulted due to the sea level retreat and resultant increased detrital influx which, in turn are in tune with the published interpretations. It is interesting to note that the methane venting due to sea level fall has been a recurrent event even after KTB as evident from Ba anomalies in Danian rocks of the Niniyur section (documented by the present study) as well as Guatemala (Ramkumar et al., 2005). Rapid variations of climate, sea level, clay mineral assemblages in sedimentary record of Late Maastrichtian are all observed in the Cauvery Basin similar to many other sedimentary basins and these changes could have contributed toward stepwise extinction of species prior to KTB.

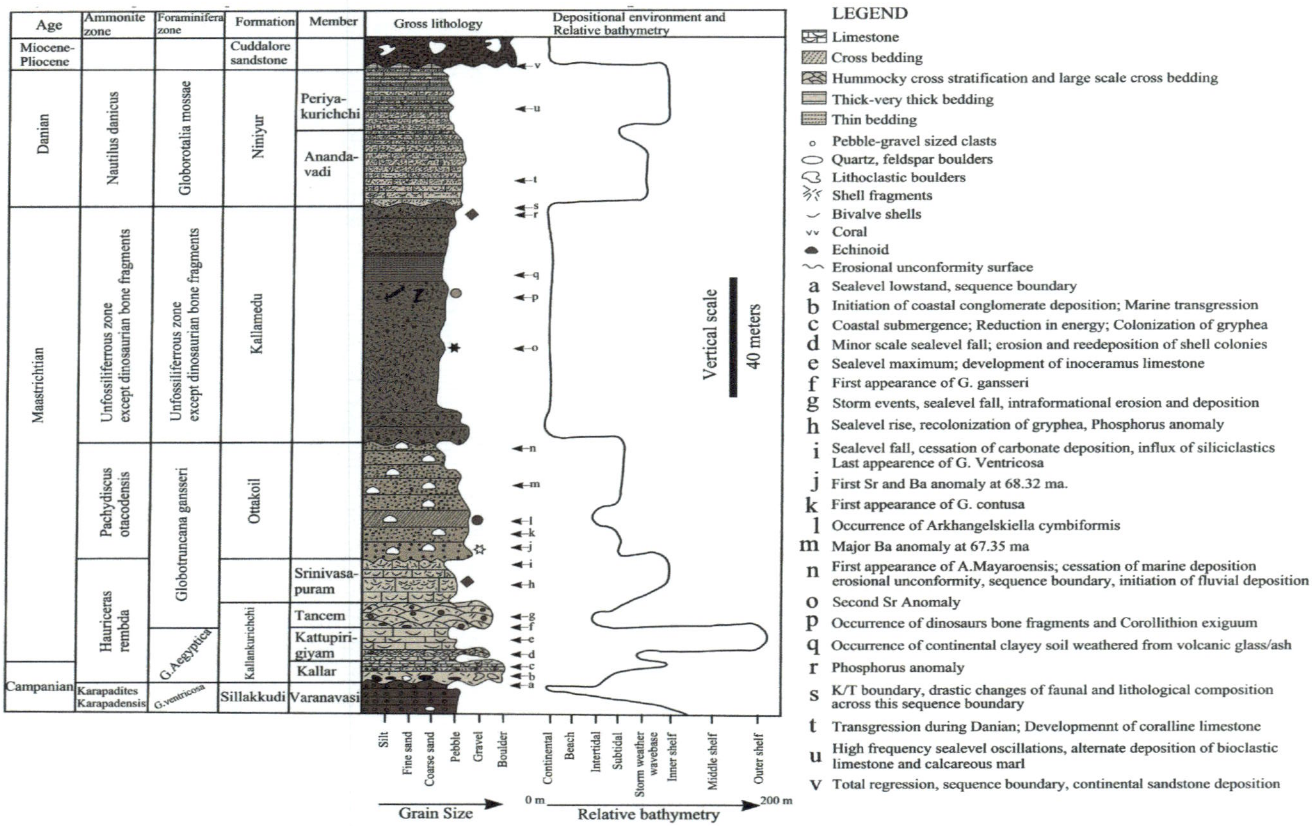

FIGURE 7

Geological and geochemical events inferred from the composite section of the Maastrichtian–Danian strata of the Cauvery Basin.

It seems certain that, although Deccan volcanism commenced during the Early Cretaceous, significant quantities of eruption took place only around 65 Ma while the Ba anomalies preceded this event by about 1.5 Ma. Hence, it is safe to interpret that the Deccan volcanic eruption had not made any direct impact on Ba anomaly. The weathered materials of volcanic ash (presumably spread from Deccan volcanism) form a specific clay layer in the upper portion of the Kallamedu Formation (Fig. 7), which is younger than the Ottakoil Formation. Hence, it is inferred that the Deccan volcanism had made its imprints on sedimentary records of the Cauvery Basin, but the imprints are younger than the Ba anomaly and thus, influence of Deccan volcanism with reference to the Ba anomaly could be ruled out. Recent studies have indicated that oxidation of sapropel materials promote enrichment of Ba in overlying sediments (van Santvoort et al., 1996, 1997; Passier et al., 1998; Dickens, 2001a). Kennedy et al. (2001) and Dickens (2001b,c) suggested that the destabilization of gas hydrates and the release of methane gas into the atmosphere might have been a widespread phenomenon during earth's history and could be traced by negative excursions of $\delta^{13}C$ in sedimentary records. The $\delta^{13}C$ profile of the Cauvery Basin also shows a consistent featureless nature from 75 to 68.7 Ma and a sudden and progressive negative excursion of carbon isotope between 68.7 and 65 Ma, thus adding support to the presumption of emission of methane gas. In conjunction with the recorded sea level change from highstand to lowstand, the occurrence of a Ba anomaly and global climatic change, supported by lithological change and carbon isotopic excursions lead to the presumption of accumulation of volatile hydrocarbons and preserved presumably in the form of gas hydrates (Kvenvolden, 1995; Schmidt et al., 2002) during the deposition of the Kallankurichchi Formation. Sea level fall by global climatic reversal and the subsequent release of methane gas and associated environmental change could have caused or influenced the Ba anomaly. Owing to the cooler climate and falling sea level, no or insignificant continental chemical weathering and persistence of physical weathering might have been introduced, due to which, fresh Ba-orthoclase clasts may have been brought into the ensuing siliciclastic deposits, thereby increasing the background values of Ba concentration, if not influencing the first minor Ba anomaly. Ramkumar et al. (2005) also recorded a chain of events (not necessarily related to one another) namely, sea level highstand and associated carbon burial, climatic cooling, sea level fall, terrigenous sedimentation, Deccan volcanism, climatic deterioration, destabilization of trapped hydrocarbons or gas hydrates and release of methane gas, and resultant Ba anomaly. Although the geographic reach of release of methane gas into the atmosphere and resultant environmental deterioration might have been regional, similar regional events may have occurred as the Ba peaks predating the K/T boundary are recorded in other sections located in Israel and NE–Mexico. These may have aggravated the environmental stress already being experienced by the biota (Glasby and Kunzendorf, 1996) during the end Cretaceous.

The inferences based on textural properties of the Kallamedu Formation, along with the facies succession suggest that there were episodes of high-energy flash floods that recycled the older sediments. While the recycling of older sediments was interpreted as the result of limited catchment area (Ramkumar et al., 2004b, 2010a) and the absence of intensive chemical weathering in the source area (Ramkumar et al., 2011), the occurrences of flood deposits (crevasse splay) and overbank and lagoonal deposits intervened by weathered vitric tuff layers (*sensu* Hansen and Mohabay, 2000) suggest the coeval nature of volcanic activity in the Indian subcontinent and deposition of the fluvial sediments in the Kallamedu Formation. Rao et al. (2002) and Keller et al. (2008) recorded basic volcanism along KTB in the Godavari Basin, which raises an interesting twist to the perceived influence of Deccan volcanism over the depositional records of the Cauvery Basin. As the Godavari Basin is more proximal to the Cauvery Basin than Deccan region, either Deccan volcanism or Godavari volcanism could have influenced depositional episodes of the Cauvery Basin during end-Cretaceous. Further, the presence of unaltered feldspars in these rocks indicates lesser chemical weathering, proximity to source rock and quick burial in the depocenters which in turn, support the inference of change in the depositional pattern during end-Cretaceous in the basin and more particularly, periodic, but irregularly spaced episodes of faster rate of deposition—an interpretation corroborating with the inference of overbank and crevasse splay deposits in the Kallamedu Formation based on textural properties. Ramkumar et al. (2010a) suggested the prevalence of off-seasonal floods, perhaps induced by disturbances in the climatic cycles, related either to volcanism or other causes. This observation, along with the record of low-middle latitude nannofossil assemblage (Rai et al., 2013) and the record of coeval Deccan/Godavari volcanism suggest the role of major flooding and instability of depositional environments vis-à-vis ecological niches as one of the causes prevalent during end-Cretaceous that could have contributed toward environmental stress (*sensu* Abramovich, and Keller, 2002; Keller et al., 2002) in the coastal regions. The absence of well-defined bedding structures, occurrences of massive and very thick beds comprising recycled older sediments, textural inversion and overbank deposition, affirm the prevalence of torrential rain-fed flash floods during the deposition of the Kallamedu Formation and may qualify one of the causes that could have contributed toward dwindling/extinction of taxa across KTB particularly through rapid/sudden change in environmental conditions of depositional sites and destruction of ecological niches, particularly in the Cauvery Basin.

6.6 CONCLUSIONS

Combined analysis of lithofacies characteristics and chemostratigraphy of the composite section and newly identified continuous KTB section located near Niniyur, collation of the previous studies on KTB stratigraphic records of the Cauvery Basin and comparing them with the scenarios suggested as the cause(s) of globally observed Late Cretaceous extinction patterns revealed the following.

- The onland exposures of the Cauvery Basin contain extended stratigraphic records of the Maastrichtian–Danian comparable with counterparts elsewhere.
- The KTB in the Cauvery Basin is recognizable through faunal, lithological and geochemical anomalies comparable with its counterparts located elsewhere. However, the patchy, scattered and weathered nature of the exposures thwarted detailed studies so far. Nevertheless, Niniyur section, as detailed in this paper qualifies to be one of the extended stratigraphic records with recognizable faunal and geochemical markers.
- Prevalence of sea level lowstand, significant chemical weathering, deposition in fluvial channels, adjoining flood plain and overbank regions, coastal lagoon and in tidal flats, predomination of high-frequency fluctuations of oxygenating and oxygen-poor conditions and high-frequency alternations of warm-dry and humid-cooler climatic conditions and out of the turn warm-dry episodes characterize the end Cretaceous of the Cauvery Basin.
- Prevalence of consistently warmer climate, sea level highstand, oxygenating conditions of marine depositional environments during most of the Danian and occurrences of high-frequency alternations of limestone-marl as a result of short-term high-frequency sea level fluctuations characterize the later part of Danian in the Cauvery Basin.
- Different phases of the Deccan volcanism have made their impact on the depositional environments of the Cauvery Basin as evident from the lithofacies characteristics and weathered ash layers in the

Late Cretaceous deposits. The frequency and thickness of the silty clay deposits are suggestive of main and intensive phase of Deccan volcanism to be very close to the KTB.

- Though there are many trace elemental and isotopic anomalies that mark the KTB in the Cauvery Basin, few prominent positive and negative anomalies (Fig. 7), signify multiple events preceding KTB, which are global in nature, intimately coupled with eustatic sea level fluctuations and concomitant with Deccan volcanism. This inference does not support the possibility of sudden extinction due to bolide impact. Instead, the extinction pattern might be due to stepwise incremental environmental stress, caused by multiple events including off-seasonal flooding, destabilization of buried gas hydrates and release of methane gas into the atmosphere, and destruction of ecological niches due to the deteriorating effect of volcanism and associated causes such as acid rain, ash fall out, rapid climatic reversals, acidification of ocean, off-seasonal floods, etc.
- The study has also demonstrated the utility of chemostratigraphy where all other methods have failed and or have limitations. Chemostratigraphy had also proved to be an indispensable tool in understanding prevalent geological processes even in relatively barren, patchy and or weathered exposures and correlate the strata with equivalents located elsewhere and corroborate the depositional and other events reported from geographically widely separated stratigraphic sections.

ACKNOWLEDGMENTS

The study was initiated with the financial grants from Alexander von Humboldt Foundation, Germany, to MR. Prof. Dr. Doris Stüben, and Dr. Zsolt Berner, Head of Laboratories, Institute of Mineralogy, are thanked for scientific support and lively discussions. Partial funding to this work in the form of Research Associateship and Senior Research Associateship was provided to MR by CSIR, New Delhi. UGC, New Delhi is thanked for research project (Maastrichtian-Danian part of the study area) granted to MR. Authors thank the scientific personnel of Institute namely, Dr. Utz Kramar for XRF analytical facilities, Dr. Karotke and Mrs. Oetzel for XRD analyses, Mr. Predrag Zrinjsak for carbon and sulfur analyses and Dr. G. Ott for computing facilities. Shri. T. Sreekumar, Geologist, OFI, Mumbai, is thanked for assistance during the field survey. Permission to collect samples was accorded by the mines managers and geologists of Messers. Alagappa cements, Chettiyar mines Dalmia Cements, Fixit Mines, Pvt. Ltd, Nataraj Ceramics Ltd., Parveen mines and Minerals Ltd., Ramco Cements, Rasi cements, TANCEM Mines, TAMIN mines, Tan-India Mines and Vijay Cements. The Director, Birbal Sahni Institute of Paleobotany (BSIP), Lucknow is thanked for permitting the scientific collaboration between the Department of Geology, Periyar University, Salem and the BSIP, Lucknow. He is also thanked for having provided access to necessary research equipment.

REFERENCES

Abramovich, S., Keller, G., 2002. High stress late Maastrichtian paleoenvironment: inference from planktonic foraminifera in Tunisia. Palaeogeography, Palaeoclimatology, Palaeoecology 178, 145–164.

Adatte, T., Keller, G., Li, L., Stinnesbeck, W., 2002. Late Cretaceous to Early Paleocene climate and sea level fluctuations: the Tunisian record. Palaeogeography, Palaeoclimatology, Palaeoecology 178, 165–196.

Adatte, T., Keller, G., Stüben, D., Harting, M., Kramar, U., Stinnesbeck, W., Abramovitch, S., 2004. Late maastrichtian and K/T paleoenvironment of the Eastern Tethys (Israel): mineralogy, trace element and platinum group elements, biostratigraphy and faunal turnovers. Bulletin de la Societe Geologique de France 176, 35–53.

Albertão, G.A., Martins Jr., P.P., 1996. A possible tsunami deposit at the Cretaceous–Tertiary boundary in Pemambuco, Northeastern Brazil. Sedimentary Geology 104, 189–201.

Algeret, L., Thomas, E., 2004. Benthic foraminifera and environmental turnover across the Cretaceous/Paleogene boundary at Blake Nose (ODP Hole 1049C, Northwestern Atlantic). Palaeogeography, Palaeoclimatology, Palaeoecology 208, 59–83.

Algret, L., Thomas, E., 2013. Benthic foraminifera across the Cretaceous/Paleogene boundary in the Southern Ocean (ODP site 690): diversity, food and carbonate saturation. Marine Micropaleontology 105, 40–51.

Alvarez, L.W., Alvarez, W., Asaro, F., Michel, H.V., 1980. Extraterrestrial cause for Cretaceous–Tertiary extinction. Science 208, 1095–1108.

Alvarez, W., 1986. Toward a theory of impact crises. EOS 67, 649–658.

Arinobu, T., Ishiwatari, R., Kaiho, K., Lamolda, M.A., Seno, H., 2005. Abrupt and massive influx of terrestrial biomarkers into the marine environment at the Cretaceous–Tertiary boundary, Caravaca, Spain. Palaeogeography, Palaeoclimatology, Palaeoecology 224, 1–3.

Ayyasamy, K., 1990. Cretaceous heteromorphy ammonoid biostratigraphy of Southern India. Newsletters on Stratigraphy 22, 111–118.

Ayyasamy, K., 2006. Role of oysters in biostratigraphy: a case study from the Cretaceous of the Ariyalur area, Southern India. Geosciences Journal 10, 237–247.

Banerji, R.K., 1972. Stratigraphy and micropaleontology of the Cauvery Basin. Part I, exposed area. Journal of the Palaeontological Society of India 17, 1–24.

Bhandari, N., Gupta, M., Shukla, P.N., 1993a. Volcanic contribution of Ir and other trace metals near the K/T boundary, India. Chemical Geology 103, 129–139.

Bhandari, N., Shukla, P.N., Castagnoli, G.C., 1993b. Geochemistry of some K/T sections in India. Palaeogeography Palaeoclimatology Palaeoecology 104, 199–211.

Burnett, J., 1998. In: Bown, P.R. (Ed.), Upper Cretaceous in Calcareous Nannofossil Biostratigraphy. Chapman & Hall, Cambridge, pp. 132–199.

Canudo, J.I., Keller, G., Molina, E., 1991. Cretaceous/Tertiary boundary extinction pattern and faunal turnover at Agost and Caravaca, S.E. Spain. Marine Micropaleontology 17, 319–341.

Chalapathi Rao, N.V., Lehmann, B., Balaram, V., 2013. Platinum-group element (PGE) geochemistry of Deccan orangeites, Bastar craton, central India: implication for a non-terrestrial origin for iridium enrichment at the K–Pg boundary. Journal of Asian Earth Sciences 84, 24–33.

Chandrasekaran, V.A., Ramkumar, M., Jacob, M., Saksena, S., 1993. A preliminary note on the occurrence of *Serpula socialis* from the Sillakkudi Formation (Campanian), Ariyalur Group, South India. Journal of Geological Association and Research Center 2, 32–36.

Chandrasekaran, V.A., Ramkumar, M., 1993. On the occurrence of *Planolites* from the Kallankurichchi Formation, Upper Cretaceous (Ariyalur Group), Tiruchy District, South India and its paleoenvironmental significance. Journal of Geological Association and Research Center 2, 31–36.

Chandrasekaran, V.A., Ramkumar, M., 1995. Stratigraphic classification of Ariyalur Group (Upper Cretaceous), Tiruchy District, South India—a review. Journal of Geological Association and Research Centre. Miscellaneous Publication 1, 1–22.

Chandrasekaran, V.A., Ramkumar, M., Jacob, M., 1996. Sediment deformational structures from the Sillakkudi Formation (Campanian) Ariyalur Group, Tiruchy District, Tamil Nadu. Journal of Indian Association of Sedimentologists 15, 43–49.

Chari, M.V.N., Sahu, J.N., Banerjee, B., Zutshi, P.L., Chandra, K., 1995. Evolution of the Cauvery Basin, India from subsidence modeling. Marine and Petroleum Geology 12, 667–675.

Chatterjee, S., Goswami, A., Scotese, C.R., 2013. The longest voyage: tectonic, magmatic, and paleoclimatic evolution of the Indian plate during its northward flight from Gondwana to Asia. Gondwana Research 23, 238–267.

Chatterjee, S., Guven, N., Yoshinobu, A., Donofrio, R., 2006. Shiva structure: a possible KT boundary impact Crater on the western shelf of India. In: Robert, J. (Ed.), Library of Congress Cataloging-in-Publication Data, vol. 50. Special Publications Museum of Texas Tech. University, pp. 1–39.

Compton, J.S., Hodell, D.A., Garrido, J.R., Mallinson, D.J., 1993. Origin and age of phosphorite from the south-central Florida platform: relation of phosphogenesis to sea level fluctuations and $\delta^{13}C$ excursions. Geochimica et Cosmochimica Acta 57, 131–146.

Crowley, T.J., North, G.R., 1988. Abrupt climate change and extinction events in earth history. Science 40, 996–1002.

Dessert, C., Dupre, B., Francois, L.M., Schott, J., Gaillardet, J., Chakrapani, G., Bajpai, S., 2001. Erosion of Deccan Traps determined by river geochemistry: impact on the global climate and the $^{87}Sr/^{86}Sr$ ratio of seawater. Earth and Planetary Science Letters 188, 459–474.

Dickens, G.R., 2001a. Sulfate profiles and barium fronts in sediment on the Blake Ridge: present and past methane fluxes through a large gas hydrate reservoir. Geochimica et Cosmochimica Acta 65, 529–543.

Dickens, G.R., 2001b. On the fate of past gas: what happens to methane released from a bacterially mediated gas hydrate capacitor? Geochemistry, Geophysics, Geosystems 2, 2000GC000131.

Dickens, G.R., 2001c. Carbon addition and removal during the Late Paleocene thermal maximum: basic theory with a preliminary treatment of the isotope record at ODP site 1051, Blake Nose. In: Kroon, D., Norris, R.D., Klaus, A. (Eds.), Western North Atlantic Paleogene and Cretaceous Paleoceanography. Geological Society, London, pp. 293–305.

Dickens, G.R., 2003. Rethinking the global carbon cycle with a large, dynamic and microbially mediated gas hydrate capacitor. Earth and Planetary Science Letters 213, 169–183.

Ebneth, S., Shields, G.A., Veizer, J., Miller, J.F., Shergold, J.H., 2001. High-resolution strontium isotope stratigraphy across the Cambrian-Ordovician transition. Geochimica et Cosmochimica Acta 65, 2273–2292.

Evans, N.J., Ahrens, T.J., Gregoire, D.C., 1995. Fractionation of ruthenium from iridium at the Cretaceous–Tertiary boundary. Earth and Planetary Science Letters 134, 141–153.

Evans, N.J., Gregoire, D.C., Grieve, R.A.F., Goodfellow, W.D., Veizer, J., 1993. Use of platinum-group elements for impactor identification: terrestrial impact craters and Cretaceous–Tertiary boundary. Geochimica et Cosmochimica Acta 57, 3737–3748.

Filippelli, G.M., Delaney, M.L., 1994. The oceanic phosphorus cycle and continental weathering during neogene. Paleocenography 9, 643–652.

Frank, T.D., Arthur, M.A., 1999. Tectonic forcings of Maastrichtian ocean-climate evolution. Paleoceanography 14, 103–117.

Friedman, G.M., 1967. Dynamic processes and statistical parameters compared for size frequency distribution of beach and river sands. Journal of Sedimentary Petrology 37, 327–354.

Froelich, P.N., 1988. Kinetic control of dissolved phosphate in natural rivers and estuaries: a primer on the phosphate buffer mechanism. Limnology and Oceanography 33, 649–668.

Fürsich, F.T., Pandey, D.K., 1999. Genesis and environmental significance of Upper Cretaceous shell Concentrations from the Cauvery Basin, Southern India. Palaeogeography Palaeoclimatology Palaeoecology 145, 119–139.

Gertsch, B., Keller, G., Adatte, T., Berner, Z., 2013. The Cretaceous–Tertiary boundary (KTB) transition in NE Brazil. Journal of the Geological Society 170, 249–262.

Gertsch, B., Keller, G., Adatte, T., Garg, R., Prasad, V., Berner, Z., Fleitmann, D., 2011. Environmental effects of Deccan volcanism across the Cretaceous–Tertiary transition in Meghalaya, India. Earth and Planetary Science Letters 310, 272–285.

Glasby, G.P., Kunzendorf, H., 1996. Multiple factors in the origin of the Cretaceous/Tertiary boundary: the role of environmental stress and Deccan trap volcanism. Geologische Rundschau 85, 191–210.

Goderis, S., Tagle, R., Belza, J., Smit, J., Montanari, A., Vanhaecke, F., Erzinger, J., Claeys, P., 2013. Reevaluation of siderophile element abundances and ratios across the Cretaceous–Paleogene (K–Pg) boundary: implications for the nature of the projectile. Geochimica et Cosmochimica Acta 120, 417–446.

Govindan, A., Ananthanarayanan, S., Vijayalakshmi, K.G., 2000. Cretaceous petroleum system in Cauvery Basin, India. In: Govindan, A. (Ed.), Cretaceous Stratigraphy—An Update. Memoir – Geological Society of India 46, 365–382.

Govindan, A., Ravindran, C.N., Rangaraju, M.K.R., 1996. Cretaceous stratigraphy and planktonic foraminiferal zonation of Cauvery Basin, South India. In: Sahni, A. (Ed.), Cretaceous Stratigraphy and Paleoenvironments. Memoir – Geological Society of India 37, 155–187.

Hakansson, E., Thomsen, E., 1999. Benthic extinction and recovery patterns at the K/T boundary in shallow water carbonates, Denmark. Palaeogeography Palaeoclimatology Palaeoecology 154, 67–85.

Hallam, A., 1987. End-Cretaceous mass extinction event: argument for terrestrial causation. Science 238, 1237–1242.

Hansen, H.J., Mohabay, D.M., 2000. New data on Indian K/T boundaries. In: Govindan, A. (Ed.), Cretaceous Stratigraphy—An Update. Memoir – Geological Society of India 46, 419–420.

Hoppie, B., Garrison, R.E., 2001. Miocene phosphate accumulation in the Cuyama Basin, Southern California. Marine Geology 177, 353–380.

Iba, Y., Mutterlose, J., Tanabe, K., Sano, S., Misaki, A., Kazunobu Terabe, K., 2011. Belemnite extinction and the origin of modern cephalopods 35 m.y. prior to the Cretaceous–Paleogene event. Geology 39, 483–486.

Jafar, S.A., 1996. The evolution of marine Cretaceous basins of India: calibration with nannofossil zones. In: Sahni, A. (Ed.), Cretaceous Stratigraphy and Paleoenvironments. Memoir – Geological Society of India 37, 121–134.

Jiang, M.J., Gartner, S., 1986. Calcareous nannofossil succession across the Cretaceous/Tertiary boundary in east-central Texas. Micropaleontology 32, 232–255.

Kampschulte, A., Bruckschen, P., Strauss, H., 2001. The sulphur isotopic composition of trace sulphates in Carboniferous brachiopods: implications for coeval seawater, correlation with other geochemical cycles and isotope stratigraphy. Chemical Geology 175, 149–173.

Kauffman, E.C., 1984. The fabric of Cretaceous marine extinction. In: Berggren, W.A., Van Couvering, J.A. (Eds.), Catastrophes and Earth History. Princeton University Press, Princeton, pp. 151–246.

Kawaragi, K., Sekine, Y., Kadono, T., Sugita, S., Ohno, S., Ishibashi, K., Kurosawa, K., Matsui, T., Ikeda, S., 2009. Direct measurements of chemical composition of shock-induced gases from calcite: an intense global warming after the Chicxulub impact due to the indirect greenhouse effect of carbon monoxide. Earth and Planetary Science Letters 282, 56–64.

Keller, G., Abramovich, S., 2009. Lilliput effect in Late Maastrichtian planktic foraminifera: response to environmental stress. Palaeogeography Palaeoclimatology Palaeoecology 284, 47–62.

Keller, G., 1988a. Biotic turnover in benthic foraminifera across the Cretaceous/Tertiary boundary at El Kef, Tunisia. Palaeogeography Palaeoclimatology Palaeoecology 66, 153–171.

Keller, G., 1988b. Exinction, survivorship and evolution of planktonic foraminifera across the Cretaceous/Tertiary boundary at El Kef, Tunisia. Marine Micropaleontology 13, 239–263.

Keller, G., 2011. The Cretaceous–Tertiary mass extinction: theories and controversies. Society for Sedimentary Geology Special Publications 100, 7–22.

Keller, G., 2012. The Cretaceous–Tertiary mass extinction, Chicxulub impact, and Deccan volcanism. In: Talent, J.A. (Ed.), Earth and Life, International Year of Planet Earth, pp. 759–793.

Keller, G., Adatte, T., Burns, S.J., Tantawy, A.A., 2002. High-stress paleoenvironment during the late Maastrichtian to early Paleocene in Central Egypt. Palaeogeography Palaeoclimatology Palaeoecology 187, 35–60.

Keller, G., Adatte, T., Gardin, S., Bartolini, A., Bajpai, S., 2008. Main Deccan volcanism phase ends near the K–T boundary: evidence from the Krishna–Godavari Basin, SE India. Earth and Planetary Science Letters 268, 293–311.

Keller, G., Bhowmick, P.K., Upadhya, H., Dave, A., Reddy, A.N., 2011. Deccan volcanism linked to the Cretaceous–Tertiary boundary mass extinction: new evidence from ONGC wells in the Krishna–Godavari Basin. Journal of the Geological Society of India 79, 399–428.

Keller, G., Khozyem, H., Adatte, T., Malarkodi, N., Spangenberg, J.E., Stinnesbeck, W., 2013. Chicxulub impact spherules in the North Atlantic and Caribbean: age constraints and Cretaceous–Tertiary boundary hiatus. Geological Magazine 150, 885–907.

Keller, G., Stinnesbeck, W., Adatte, T., Stüben, D., 2003. Multiple impacts across the Cretaceous–Tertiary boundary. Earth-Science Reviews 62, 327–363.

Kennedy, M., Christie-Blick, N., Sohl, L., 2001. Are Proterozoic cap carbonates and isotopic excursions a record of gas hydrate destabilization following earth's coldest interval? Geology 29, 443–446.

Koch, P.L., Zachos, J.C., Gingerich, P.D., 1992. Correlation between isotope records in marine and continental carbon reservoirs near the Paleocene/Eocene boundary. Nature 358, 319–322.

Kramar, U., 1997. Advances in energy-dispersive X-ray fluorescence. Journal of Geochemical Exploration 58, 73–80.

Kring, D.A., 2007. The Chicxulub impact event and its environmental consequences at the Cretaceous–Tertiary boundary. Palaeogeography Palaeoclimatology Palaeoecology 255, 4–21.

Krishna, A.K., Murthy, N.N., Govil, P.K., 2007. Multi-element analysis of soils by wavelength-dispersive X-ray fluorescence spectrometry. Atomic Spectroscopy 28, 202.

Kvenvolden, K.A., 1995. A review of the geochemistry of methane in natural gas hydrates. Organic Geochemistry 23, 997–1008.

Lal, N.K., Siawal, A., Kaul, A.K., 2009. Evolution of east coast of India–a plate tectonic reconstruction. Journal of the Geological Society of India 73, 249–260.

Lamolda, M.A., Melinte, M.C., Kaiho, K., 2005. Nannofloral extinction and survivorship across the K/T boundary at Caravaca, southeastern Spain. Palaeogeography Palaeoclimatology Palaeoecology 224, 27–52.

Le Loeuff, J., 2012. Paleobiogeography and biodiversity of Late Maastrichtian dinosaurs: how many dinosaurs species went extinct at the Cretaceous–Tertiary boundary? Bulletin De La Societe Geologique De France 183, 547–559.

Madhavaraju, J., Ramasamy, S., Ruffell, A., Mohan, S.P., 2002. Clay mineralogy of the Late Cretaceous and early Tertiary successions of the Cauvery Basin (Southeastern India): implications for sediment source and paleoclimates at the K/T boundary. Cretaceous Research 23, 153–163.

Martin, E.E., Macdougall, J.D., 1991. Seawater Sr isotopes at the Cretaceous/Tertiary boundary. Earth and Planetary Science Letters 104, 166–180.

Martini, E., 1971. Standard Tertiary and Quaternary Calcareous Nannoplankton Zonation. In: Farinacci, A. (Ed.), Proceedings of the II Planktonic Conference, Roma, 1970, Rome, pp. 739–785.

Masthan, S.J., 1978. Depositional environments of Kallamedu sandstone, Masstrichtian, Ariyalur area, South India. Bulletin of the Geological, Mining and Metallurgical Society of India 15, 61–70.

McManus, J., Berelson, W.M., Klinkhammer, G.P., Johnson, K.S., Coale, K.H., Anderson, R.F., Kumar, N., Burdigo, D.J., Hammond, D.E., Brumsack, H.J., McCorkle, D.C., Rushdi, A., 1998. Geochemistry of barium in marine sediments: implications for its use as a paleoproxy. Geochimica et Cosmochimica Acta 62, 3453–3473.

Mitrović-Petrovic, J., Ramamoorthy, K., 1993. Functional morphology of *Stigmatophygus elatus* (Echinoidea: Cassidoiloida) from the Lower Maestrichtian of Southern India. Geološki anali Balkanskoga poluostrva 56, 119–135.

Mizukami, T., Kaiho, K., Oba, M., 2013. Significant changes in land vegetation and oceanic redox across the Cretaceous/Paleogene boundary. Palaeogeography Palaeoclimatology Palaeoecology 369, 41–47.

Moiola, R.J., Weiser, D., 1968. Textural parameters: an evaluation. Journal of Sedimentary Petrology 38, 45–53.

Murray, R.W., Leinen, M., 1993. Chemical transport to the seafloor of the equatorial Pacific ocean across a latitudinal transect at 135°W: tracking sedimentary major, trace and rare earth element fluxes at the equator and the intertropical convergence zone. Geochimica et Cosmochimica Acta 57, 4141–4163.

Murray, R.W., Leinen, M., 1996. Scavenged excess aluminium and its relationship to bulk titanium in biogenic sediment from the central equatorial Pacific ocean. Geochimica et Cosmochimica Acta 60, 3869–3878.

Nair, K.M., 1978. Development of carbonates during Maastrichtian in Cauvery Basin. Bulletin of the Geological, Mining and Metallurgical Society of India 47, 71–80.

Nair, K.M., Vijayam, B.E., 1980. Sedimentology of limestones in Niniyur Formation, Paleocene, Cauvery Basin, South India. Journal of the Geological Society of India 21, 503–510.

Ohno, S., Kadono, T., Kurosawa, K., Hamura, T., Sakaiya, T., Shigemori, K., Hironaka, Y., Sano, T., Watari, T., Otani, K., Matsui, T., Sugita, S., 2014. Production of sulphate-rich vapour during the Chicxulub impact and implications for ocean acidification. Nature Geoscience 7, 279–282.

Passier, H.F., Dekkers, M.J., de Lange, G.J., 1998. Sediment chemistry and magnetic properties in an anomalously reducing core from the eastern Mediterranean sea. Chemical Geology 152, 287–306.

Paul, C.R.C., 2005. Interpreting bioevents: what exactly did happen to planktonic foraminifers across the Cretaceous–Tertiary boundary? Palaeogeography Palaeoclimatology Palaeoecology 224, 291–310.

Perch-Nielsen, K., 1985. In: Bolli, H.M., Saunders, J.B., Perch-Nielsen, K. (Eds.), Mesozoic Calcareous Nannofossils in Plankton Stratigraphy. Cambridge University Press, pp. 329–426.

Peryt, D., Lahodynsky, R., Rocha, R., Boclet, D., 1993. The Cretaceous/Paleogene boundary and planktonic foraminifera in the Flychgosau (Eastern Alps, Austria). Palaeogeography Palaeoclimatology Palaeoecology 104, 239–252.

Powell, C., Mc, A., Roots, S.R., Veevers, J.J., 1988. Pre-break up continental extension in east Gondwanaland and the early opening of the Indian Ocean. Tectonophysics 155, 261–283.

Prauss, L.M., 2009. The K/Pg boundary at Brazos-River, Texas, USA–an approach by marine palynology. Palaeogeography Palaeoclimatology Palaeoecology 283, 195–215.

Prokoph, A.R., Rampino, M.R., El Bilali, H., 2004. Periodic components in the diversity of calcareous plankton and geological events over the past 230 Myr. Palaeogeography Palaeoclimatology Palaeoecology 207, 105–125.

Punekar, J., Keller, G., Khozyem, H., Hamming, C., Adatte, T., Tantawy, A.A., Pangenberg, J.E., 2014. Late Maastrichtianeearly Danian high-stress environments and delayed recovery linked to Deccan volcanism. Cretaceous Research 49, 63–82.

Rai, J., Ramkumar, M., Sugantha, T., 2013. Calcareous nannofossils from the Ottakoil Formation of the Cauvery Basin, South India: implications on age and late Cretaceous environmental conditions. In: Ramkumar, M. (Ed.), On a Sustainable Future of the Earth's Natural Resources. Springer-Verlag, Heidelberg, pp. 109–122.

Raju, D.S.N., Ravindran, C.N., Kalyansundar, R., 1993. Cretaceous cycles of sea level changes in Cauvery Basin, India–a first revision. ONGC Bulletin 30, 101–113.

Ram Babu, H.V., Lakshmi, M.P., 2004. A Reappraisal of the structure and tectonics of the Cauvery Basin (India) from aeromagnetics and gravity. In: Fifth Conference and Exposition on Petroleum Geophysics, Hyderabad, India, pp. 19–23.

Rama Rao, L., 1956. Recent contribution to our knowledge of the Cretaceous rocks of South India. Proceedings of the Indian Academy of Sciences 4, 185–245.

Ramkumar, M., 1996a. Geochemical environments of the Kallankurichchi Formation, Ariyalur Group, Tiruchy District, South India. Indian Journal of Geochemistry 11, 15–26.

Ramkumar, M., 1996b. Occurrence of hardgrounds in the Kallankurichchi Formation (Lower Maestrichtian), Ariyalur Group, Tiruchirapalli Cretaceous sequence, South India and their significance. Indian Journal of Petroleum Geology 5, 83–97.

Ramkumar, M., 1999. Lithostratigraphy, depositional history and constraints on sequence stratigraphy of the Kallankurichchi Formation (Maestrichtian), Ariyalur Group, South India. Annales Géologiques de la Péninsulé Balkanique 63, 19–42.

Ramkumar, M., 2001. Sedimentary structures and depositional conditions of the Kallankurichchi Formation (Lower Maestrichtian), South Indian Cretaceous sequence. Journal of Indian Association of Sedimentologists 20, 85–96.

Ramkumar, M., 2003a. Progradation of the Godavari delta: a fact or empirical artifice? insights from coastal landforms. Journal of the Geological Society of India 62, 290–304.

Ramkumar, M., 2003b. Geochemical and sedimentary processes of mangroves of Godavari delta: implications on estuarine and coastal biological ecosystem. Indian Journal of Geochemistry 18, 95–112.

Ramkumar, M., 2003c. Porosity types in carbonates as indictors of environments of deposition and diagenesis: a case of Kallankurichchi Formation, Ariyalur Group, South India and implication on hydrocarbon prospects. Indian Journal of Petroleum Geology 12, 25–39.

Ramkumar, M., 2004. Lithology, petrography, microfacies, environmental history and hydrocarbon prospects of the Kallankurichchi Formation (Upper Cretaceous, Ariyalur Group, South India). Paleontology, Sedimentology, Facies 12, 77–100.

Ramkumar, M., 2006. A storm event during the Maastrichtian in the Cauvery Basin, South India. Annales Géologiques de la Péninsulé Balkanique 67, 35–40.

Ramkumar, M., 2007. Spatio-temporal variations of sediment texture and their influence on organic carbon distribution in the Krishna estuary. Indian Journal of Geochemistry 22, 143–154.

Ramkumar, M., Chandrasekaran, V.A., 1996. Megafauna and environmental conditions of Kallankurichchi Formation (Lower Maestrichtian), Ariyalur Group, Tiruchy District, South India. Journal of Geological Association and Research Center 4, 38–45.

Ramkumar, M., Chandrasekaran, V.A., Jacob, M., 1996. Late stage meteoric vadose diagenesis in Sillakkudi Formation (Campanian) of Ariyalur Group, Tiruchy District, South India. Journal of Geological Association and Research Center 4, 43–49.

Ramkumar, M., Gandhi, M.S., 2000. Beach rocks in the modern Krishna delta. Journal of Geological Association and Research Center 8, 22–34.

Ramkumar, M., Vivekananda Murty, M., 2000. Distinction of sedimentary environments of the Godavari delta using geochemical and granulometric data through analysis of variance (ANOVA). Indian Journal of Geochemistry 15, 69–84.

Ramkumar, M., Pattabhi Ramayya, M., Gandhi, M.S., 2000b. Beach rock exposures at wave cut terraces of modern Godavari delta: their genesis, diagenesis and indications on coastal submergence and sealevel rise. Indian Journal of Marine Sciences 29, 219–223.

Ramkumar, M., Rajani Kumari, V., Pattabhi Ramayya, M., Gandhi, M.S., Bhagavan, K.V.S., Swamy, A.S.R., 2001. Dynamics of moderately well-mixed tropical estuarine system, Krishna estuary, India: Part I spatio-temporal variations of physico-chemical properties. Indian Journal of Geochemistry 16, 61–74.

Ramkumar, M., Sathish, G., 2009. Paleoenvironmental and sequence stratigraphic significance of the occurrence of *Ophiomorpha irregulaire* in the Kallankurichchi Formation, Ariyalur Group, Cauvery Basin, South India. Paleontology, Sedimentology, Facies 17, 129–137.

Ramkumar, M., Sudha Rani, P., Gandhi, M.S., Pattabhi Ramayya, M., Rajani Kumari, V., Bhagavan, K.V.S., Swamy, 2000a. Textural characteristics and depositional sedimentary environments of the modern Godavari delta. Journal of the Geological Society of India 56, 471–487.

Ramkumar, M., Rajani Kumari, V., Pattabhi Ramyya, M., Vivekananda Murty, M., 2000c. Geochemical characteristics and depositional conditions of coastal sedimentary environments of the modern Godavari delta. Indian Journal of Geochemistry 15, 31–44.

Ramkumar, M., Harting, M., Stüben, D., 2005. Barium anomaly preceding K/T boundary: plausible causes and implications on end Cretaceous events of K/T sections in Cauvery Basin (India), Israel, NE-Mexico and Guatemala. International Journal of Earth Sciences 94, 475–489.

Ramkumar, M., Stüben, D., Berner, Z., Schneider, J., 2004a. Geochemical and isotopic anomalies preceding K/T boundary in the Cauvery Basin, South India: implications for the end Cretaceous events. Current Science 87, 1738–1747.

Ramkumar, M., Stüben, D., Berner, Z., 2004b. Lithostratigraphy, depositional history and sea level changes of the Cauvery Basin, South India. Annales Géologiques de la Péninsulé Balkanique 65, 1–27.

Ramkumar, M., Stüben, D., Berner, Z., 2006. Elemental interrelationships and depositional controls of Barremian-Danian strata of the Cauvery Basin, South India: implications on scales of chemostratigraphic modeling. Indian Journal of Geochemistry 21, 341–367.

Ramkumar, M., Stüben, D., Berner, Z., 2009. Episodes of phosphorus accumulation in the Cauvery Basin, South India: implications on paleoclimate, productivity and weathering. Current Science 97, 251–256.

Ramkumar, M., Anbarasu, K., Sugantha, T., Rai, J., Sathish, G., Suresh, R., 2010a. Occurrence of KTB exposures and dinosaur nesting site near Sendurai, India–an initial report. Ultra Science 22, 573–584.

Ramkumar, M., Stüben, D., Berner, Z., Schneider, J., 2010b. $^{87}Sr/^{86}Sr$ anomalies in Late Cretaceous-Early Tertiary strata of the Cauvery Basin, South India: constraints on nature and rate of environmental changes across K–T boundary. International Journal of Earth Systems Science 119, 1–17.

Ramkumar, M., Stüben, D., Berner, Z., 2010c. Hierarchical delineation and multivariate statistical discrimination of chemozones of the Cauvery Basin, South India: implications on spatiotemporal scales of stratigraphic correlation. Petroleum Science 7, 435–447.

Ramkumar, M., Stüben, D., Berner, Z., 2011. Barremian-Danian chemostratigraphic sequences of the Cauvery Basin, India: implications on scales of stratigraphic correlation. Gondwana Research 19, 291–309.

Ramkumar, M., Sugantha, T., Rai, J., 2013. Facies and textural characteristics of the Kallamedu Formation, Ariyalur Group, Cauvery Basin, South India: implications Cretaceous–Tertiary boundary (KTB) events. In: Ramkumar, M. (Ed.), On the Sustenance of Earth's Resources. Springer-Verlag, Heidelberg, pp. 263–284.

Rana, R.S., 1999. Paleontology and Paleoecology of the intertrappean (Cretaceous–Tertiary transition) beds of the Peninsular India. Journal of the Palaeontological Society of India 35, 105–120.

Rangel, A., Parra, P., Nino, C., 2000. The La Luna Formation: chemostratigraphy and organic facies in the middle Magdelena Basin. Organic Geochemistry 31, 1267–1284.

Rao, A.T., Rao, S., Vijayakumar, V., 2002. Basic volcanism along K–T boundary from Rajahmundry, east coast of India. Journal of the Geological Society of India 60, 583–586.

Raup, D.M., 1986. Periodic extinction of families and genera. Science 231, 833–836.

Raymo, M.E., 1994. The Himalayas, organic carbon burial, and climate in the Miocene. Paleoceanography 9, 399–404.

Raymo, E., Oppo, D.W., Curry, W., 1997. The mid-Pleistocene climate transition: a deepsea carbon isotopic perspective. Paleoceanography 12, 546–559.

Roser, B.P., Cooper, R.A., Nathan, S., Tulloch, A.J., 1996. Reconnaissance sandstone geochemistry, provenance and tectonic setting of the lower Paleozoic terranes of the west coast and Nelson, New Zealand. New Zealand Journal of Geology and Geophysics 39, 1–16.

Roser, B.P., Korsch, R.J., 1986. Determination of tectonic setting of sandstone-mudstone suites using SiO_2 content and K_2O/Na_2O ratio. Journal of Geology 94, 635–650.

Ruttenberg, K.C., 1993. Reassessment of the oceanic residence time of phosphorus. Chemical Geology 107, 405–409.

Saito, T., Yamamoi, T., Kaiho, K., 1986. End-Cretaceous devastation of terrestrial flora in the boreal far East. Nature 323, 253–255.

Salil, M.S., Shrivastava, J.P., Pattanayak, S.K., 1997. Similarities in the mineralogical and geochemical attributes of detrital clays of Maastrichtian Lameta Beds and weathered Deccan basalt, central India. Chemical Geology 136, 25–32.

van Santvoort, P.J.M., de Lange, G.J., Thompson, J., Cussen, H., Wilson, T.R.S., Krom, M.D., Ströhle, K., 1996. Active post-depositional oxidation of the most recent sapropel (S1) in sediments of the eastern Mediterranean. Geochimica et Cosmochimica Acta 60, 4007–4024.

van Santvoort, P.J.M., de Lange, G.J., Langereis, C.G., Dekkers, M.J., 1997. Geochemical and paleomagnetic evidence for the occurrence of missing sapropels in eastern Mediterranean sediments. Paleoceanography 12, 764–777.

Sastri, V.V., Sinha, R.N., Singh, G., Murti, K.V.S., 1973. Stratigraphy and tectonics of sedimentary basins on the east coast of Peninsular India. Bulletin of the American Association of Petroleum Geology 57, 655–678.

Sastri, V.V., Venkatachala, B.S., Narayanan, V., 1981. The evolution of the east coast of India. Palaeogeography Palaeoclimatology Palaeoecology 36, 23–54.

Sastry, M.V.A., Rao, B.R.J., 1964. Cretaceous–Tertiary boundary in South India. In: Proc. Inter. Geol. Cong. XXII on Cretaceous–Tertiary Boundary Including Volcanic Activity. Section 3, Part III, pp. 92–103.

Sastry, M.V.A., Mamgain, V.D., Rao, B.R.J., 1972. Ostracod fauna of the Ariyalur Group (Upper Cretaceous), Trichinopoly District, Tamil Nadu. Journal of the Palaeontological Society of India 40, 1–48.

Schmitz, B., 1992. Chalcophile elements and Ir in continental Cretaceous–Tertiary boundary clays from the western interior of the USA. Geochimica et Cosmochimica Acta 56, 1695–1703.

Schmidt, M., Botz, R., Winn, K., Stoffers, P., Thiessen, O., Herzig, P., 2002. Seeping hydrocarbons and related carbonate mineralisations in sediments south of Lihir Island (New Ireland forearc basin, Papua New Guinea). Chemical Geology 186, 249–264.

Shrivastava, J.P., Ahmad, M., 2005. Compositional studies on organic matter from iridium enriched Anjar intertrappean sediments: Deccan volcanism and paleoenvironmental implications during the Cretaceous/Tertiary boundary. Journal of Iberian Geology 31, 167–177.

Shrivastava, J.P., Ahmad, M., 2008. Trace elemental compositions of iridium entriched illite-smectite assemblages from a K/Pg boundary section in the Anjar area of the Deccan volcanic province; paleoenvironmental implications. Cretaceous Research 31, 592–602.

Shukla, A.D., Bhandari, N., Kushumagar, S., Shukla, P.N., Ghevariya, Z.G., Gopalan, K., Balaram, V., 2001. Geochemistry and Magnetostratigraphy of Deccan flows at Anjar, Kutch. International Journal of Earth Systems Science 110, 111–132.

Singh, I.B., 1978. Microfacies, petrography and mineralogy of the Tertiary rocks of Guvar nala near Narain Sarovar, Kutch, India and their paleoecological significance. Journal of the Palaeontological Society of India 21&22, 78–95.

Sloan, R.W., Rigby Jr., J.K., Van Valen, L.M., Gabriel, D., 1986. Gradual dinosaur extinction and simultaneous ungulate radiation in the Hell Creek Formation. Science 232, 629–633.

Spalletti, L.A., Poire, D.G., Schwarz, E., Veiga, G.D., 2001. Sedimentologic and sequence stratigraphic model of a Neocomian marine carbonate-siliciclastic ramp: Nequen Basin, Argentina. South American Earth Sciences 14, 609–624.

Steuber, T., Mitchell, S.F., Buhl, D., Gunter, G., Kasper, H.U., 2002. Catastrophic extinction of Caribbean rudist bivalves at the Cretaceous–Tertiary boundary. Geology 30, 999–1002.

Stewart Jr., H.B., 1958. Sedimentary reflections on depositional environment in San Migne lagoon, Baja California, Mexico. American Association of Petroleum Geologists Bulletin 42, 2567–2618.

Stüben, D., Kramar, U., Berner, Z., Stinnesbeck, W., Keller, G., Adatte, T., 2002. Trace elements, stable isotopes and clay mineralogy of the Elles II K/T boundary section in Tunisia: indications for sea level fluctuations and primary productivity. Palaeogeography Palaeoclimatology Palaeoecology 178, 321–345.

Stüben, D., Kramar, U., Harting, M., Stinnesbeck, W., Keller, G., 2005. High-resolution geochemical record of Cretaceous–Tertiary boundary sections in Mexico: new constraints on the K/T and Chicxulub events. Geochimica et Cosmochimica Acta 69, 2559–2579.

Sugantha, T., Ramkumar, M. The paradigms of Cretaceous–Tertiary boundary transition: review of current status and its implication on the Maastrichtian-Danian strata of the Cauvery Basin, India. Journal of the Gondwana Geological Society. (In press).

Sundaram, R., Rao, P.S., 1986. Lithostratigraphy of Cretaceous and paleocene rocks of Tiruchirapalli District, Tamil Nadu, South India. Records of the Geological Survey of India 115, 9–23.

Sutherland, F.L., 1994. Volcanism around K/T boundary time-its role in an impact scenario for the K/T extinction events. Earth-Science Reviews 36, 1–26.

Singh, I.B., Swami, A.S.R., 1996. Modern Deltas of East Coast of India. (Technical Report submitted to Oil and Natural Gas Corporation of India, Dehra Dun).

Swami, A.S.R., Rao, M.P., Rao, B.K., 1990. Sediment characteristics of the modern deltas of east coast of India. In: Proc. Inter. Conf. Asian Mar.Geol, pp. 251–265.

Tappan, H., 1967. Primary production, isotopes, extinctions and the atmosphere. Palaeogeography Palaeoclimatology Palaeoecology 4, 187–210.

Taylor, S.R., McLennan, S.M., 1985. The Continental Crust: Its Composition and Evolution. Blackwell, Cambridge. p. 312.

Tewari, A., Hart, M.B., Watkinson, M.P., 1996. A revised lithostratigraphic classification of the Cretaceous rocks of the Trichinopoly District, Cauvery Basin, Southeast India. In: Contributions to the XV Colloquium on Micropaleontology and Stratigraphy, pp. 789–800.

Tobin, T.S., Ward, P.D., Steig, E.J., Olivero, E.B., Hilburn, I.A., Mitchell, R.N., Diamond, M.R., Raub, T.D., Krischvink, J.L., 2012. Extinction pattern, δ^{18} trends, and magnetostratigraphy from a southern high-latitude Cretaceous-Paleogene section links with Deccan volcanism. Palaeogeography Palaeoclimatology Palaeoecology 350–352, 180–188.

Tripathi, C., Mamgain, V.D., 1987. Record of vitric tuff from the Late Cretaceous/Early Paleocene strata of Tiruchirapalli District, Tamil Nadu and its significance. Bulletin of the Indian Geologists Association 20, 9–16.

Tschudy, R.H., Pillmore, C.L., Orth, C.J., Gilmore, J.S., Knight, J.D., 1984. Disruption of the terrestrial plant ecosystem at the Cretaceous/Tertiary boundary, western interior. Science 225, 1030–1032.

Tucker, M.E., Wright, V.P., 1990. Carbonate Sedimentology. Blackwell Scientific Publications, London.

Varol, O., 1998. Paleogene. In: Bown, P.R. (Ed.), Calcareous Nannofossil Biostratigraphy. Kluwer Academic Publishing, Dordrecht, The Netherlands, pp. 200–224.

Veizer, J., Ala, D., Azmy, K., Bruckschen, P., Buhl, D., Bruhn, F., Carden, G.A.F., Diener, A., Ebneth, S., Godderis, Y., Jasper, T., Korte, C., Pawellek, F., Podlaha, O.G., Strauss, H., 1999. $^{87}Sr/^{86}Sr$ and δ^{18} O evolution of Phanerozoic seawater. Chemical Geology 161, 59–88.

Venkatachala, B.S., Sharma, K.D., 1974. Palynology of the Cretaceous sediments from the subsurface of Virdhachalam area, Cauvery Basin. Geophytology 4, 153–183.

Wallmann, K., 2001. Controls on the Cretaceous and Cenozoic evolution of seawater composition, atmospheric CO_2 and climate. Geochimica et Cosmochimica Acta 65, 3005–3025.

Whitford, D.J., Allan, T.L., Andrew, A.S., Craven, S.J., Hamilton, P.J., Korsch, M.J., Trotter, J.A., Valenti, G.L., 1996. Strontium isotope chronostratigraphy and geochemistry of the Darai limestone: Juha 1X well, Papua New Guinea. In: Proc. III PNG Petroleum Convention, Port Moresby, pp. 369–380.

Wolfe, J.A., 1990. Paleobotanical evidence for a marked temperature increase following the Cretaceous/Tertiary boundary. Nature 343, 153–156.

Zhao, Zi-K., Mao, X.-Y., Chai, Z.-F., Gao-Chuang, Y., Kong, P., Ebihara, M., Zhao, Z.-H., 2002. A possible causal relationship between extinction of dinosaurs and K/T iridium enrichment in the Nanxiong Basin, South China: evidence from dinosaur eggshells. Palaeogeography Palaeoclimatology Palaeoecology 178, 1–17.

APPENDIX I NANNOFOSSIL ASSEMBLAGE RECORDED FROM THE OTTAKOIL FORMATION (After Rai et al. (2013))

All the forms are magnified × 2000.

1. *Braarudosphaera bigelowii* (Gran & Braarud) Deflandre, 1947
2A-D, 3A-B. *Arkhangelskiella cymbiformis* Vekshina, 1959
4. *Eiffellithus turriseiffeli* (Deflandre in Deflandre & Fert) Reinhardt, 1965
5A-B. *A. cymbiformis* Vekshina, 1959
6A-B, 7A-B. *E. parallelus* Perch-Nielsen, 1973
8-9. *E. turriseiffeli* (Deflandre in Deflandre & Fert) Reinhardt, 1965
10A-B. *E. gorkae* Reinhardt, 1965
11A-B. *Micula swastika* Stradner and Steinmetz, 1984
12. *M. murus* (Martini) Bukry, 1973
13A-C, 14-16. *M. decussata* Perch- Nielsen, 1973
17A-B, 18A-C, 19A-C. *Stradneria crenulata* (Bramlette & Martini) Nöel, 1970
20-21. *Microrhabdulus undosus* Perch-Nielsen, 1973
22A-B.*Ceratolithoides kamptneri* Bramlette and Martini, 1964
23-24. *Watznuaeria barnesae* (Black) Perch-Nielsen, 1968
25. *Cribrosphaerella ehrenbergii* (Arkhangelsky) Deflandre in Piveteau, 1952
26. *Cribrosphaerella* sp.
27A-B, 28A-B. *Prediscosphaera cretacea* (Arkhangelsky) Deflandre, 1968
29. *P. ponticula* (Bukry) Perch-Nielsen, 1984
30. *P. spinosa* (Bramlette & Martini) Gartner, 1968
31. *Prediscosphaera* sp.
32A-B, 33A-B. *Zygodiscus spiralis* Bramlette & Martini, 1964
34A-B. *Staurolithites crux* (Deflandre & Fert) Caratini, 1963
35A-B. *Z. minimus* Bukry, 1969
36A-B. *Ahmuellerella octoradiata* (Górka) Reinhardt, 1966
37A-B. *Pseudomicula quadrata* Perch-Nielsen, 1985
38A-B. *Cyclagelosphaera deflandrei* (Manvit) Roth, 1973
39. *W. barnesae* (Black) Perch-Nielsen, 1968
40A-*C. petrobrasiella*? *Bownii* Burnett, 1998
41A-B. *Chiastozygus litterarius* (Górka) Manivit, 1971
42A-B. *A. regularis* (Górka) Reinhardt and Górka, 1967

APPENDIX II NANNOFOSSIL ASSEMBLAGE RECORDED FROM THE LAGOONAL FACIES OF THE KALLAMEDU FORMATION IN THE NINIYUR SECTION (After Ramkumar et al., 2010A)

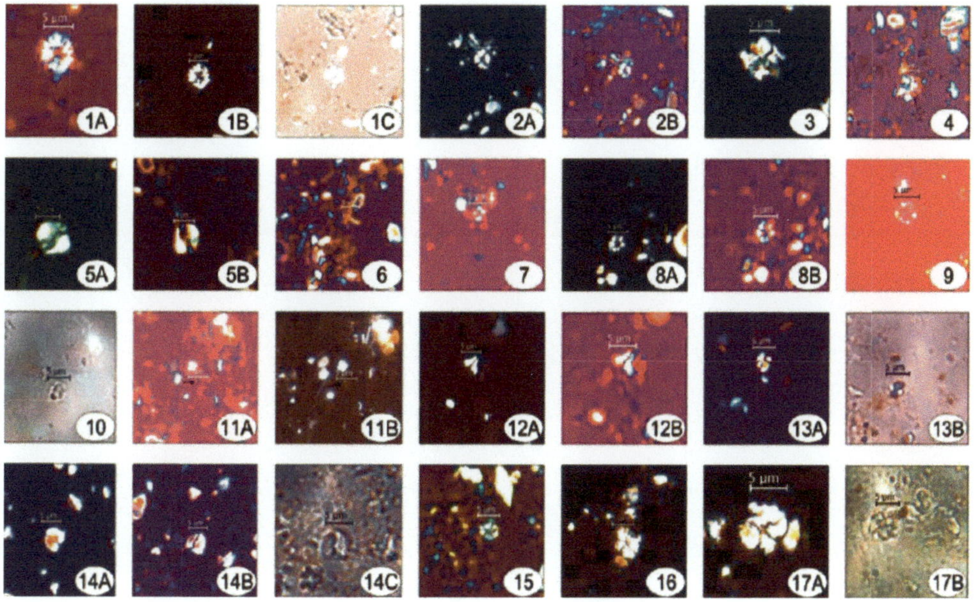

1A-C. *Discorhabdus* ignotus (Gorka, 1957) Perch-Nielsen, 1968.
2A, 2B. *Eprolithus floralis* (Stradner, 1962) Stover, 1966.
3, 4. *Pseudomicula quadrata* Perch-Nielsen, 1985.
5A, 5B. *Holococcolith* sp.
6. *Prediscosphaera* sp.
7. *Uniplanarius* sp.
8A, 8B. *Petrobrasilla*? *bownii* Burnett, 1998b
9. *Eprolithus* sp.
10. *Corollithion exiguum* Stradner, 1961.
11A, 11B. *Ottavianus terrazetus* Risatti, 1973.
12A, 12B. *Lucianorhabdus cf.* L. *maleformis* Reinhardt, 1966.
13A, 13B. *Calculites obscures* (Deflandre, 1959) Prins and Sissingh in Sissing, 1977.
14A-C. *Ceratolithoides pricei* Burnett, 1998a.
15. *Discorhabdus* ignotus (Gorka, 1957) Perch-Nielsen, 1968.
16, 17A, 17B. *Biantholithus cf.* B. *sparsus.*

CHAPTER

7 CRETACEOUS CARBON ISOTOPE STRATIGRAPHY AND CONSTRAINTS ON THE SEDIMENTARY PATTERNS OF THE TURONIAN FOREARC SUCCESSIONS IN HOKKAIDO, NORTHERN JAPAN

Go-Ichiro Uramoto[1], Rui Tahara[2], Hiromichi Hirano[3]

[1]Kochi Institute for Core Sample Research, Japan Agency for Marine-Earth Science and Technology (JAMSTEC), Nankoku, Kochi, Japan; [2]Redpath Museum, McGill University, Montreal, QC, Canada; [3]Department of Earth Sciences, Waseda University, Tokyo, Japan

7.1 INTRODUCTION

Marked stratigraphic fluctuations in the carbon isotopes of carbonates, and marine and terrestrial organic matter, provide a time-control reference for Cretaceous marine strata (e.g., Jenkyns et al., 1994; Stool and Schrag, 2000; Jarvis et al., 2006; Nunn et al., 2010). A number of published studies attempted correlations of Cretaceous (as determined by biostratigraphic calibration) marine and terrestrial $\delta^{13}C$ records and revealed a similarity between the $\delta^{13}C$ profiles of marine and terrestrial materials (e.g., Hasegawa, 1997, 2003; Ando et al., 2002, 2003; Hasegawa et al., 2003; Robinson and Hesselbo, 2004; Gröcke et al., 2005; Ando and Kakegawa, 2007; Uramoto et al., 2007, 2009, 2013; Nunn et al., 2010; Nemoto and Hasegawa, 2011). In particular, investigation of a geological succession that contains both marine carbonates and terrestrial organic matter showed that distinctive $\delta^{13}C$ excursions appear to be synchronous in both the marine and terrestrial data (Nunn et al., 2010). This synchronicity provides not only a strong evidence for a $\delta^{13}C$ linkage between Cretaceous oceanic and atmospheric carbon reservoirs, but also global chemostratigraphic horizons for the Cretaceous successions. Accumulation of Cretaceous marine and terrestrial carbon isotope values and global correlation of these data is essential for the establishment of high-resolution event stratigraphy of the Cretaceous succession.

The forearc siliciclastic successions of the Cretaceous Yezo Group (Fig. 1) are exposed along a 1200-km long belt that extends from southern Hokkaido in Japan, northward to Sakhalin Island in Russia, and along the northwestern Pacific margin (Takashima et al., 2004). Because of the extensive distribution with abundant occurrences of marine macro- and microfossils, the Yezo Group has been employed as a Cretaceous reference unit for the circum-Pacific region. In addition, because the northwestern Pacific margin is located on a critical plate convergent zone and a tectonically active setting (Taira,

Chemostratigraphy. http://dx.doi.org/10.1016/B978-0-12-419968-2.00007-8

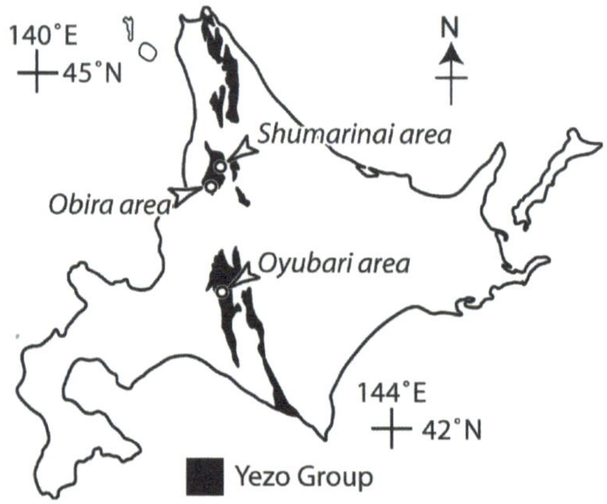

FIGURE 1

Map showing the distribution of the Yezo Group in Hokkaido, Northern Japan (from Takashima et al., 2004) and the localities mentioned in this paper.

2001), the sedimentation rate of the Yezo Group (200–400 m/m.y; Kaiho and Hasegawa, 1994; Nemoto and Hasegawa, 2011) is much higher than that of the Cretaceous type sections in Europe and North America. Detailed analysis of the stratigraphy of the Yezo Group can provide invaluable insight for the interrelationships between tectonics and sedimentary patterns of the Cretaceous successions. In this regard, the Turonian Stage of the Yezo Group is important for identifying the tectonic control of the sedimentary patterns, as the episodic deformation and the formation of a confined basin within the Yezo forearc basin was initiated at about 90 Ma (Tamaki et al., 2009). Unfortunately, the spatiotemporal sedimentary patterns related to the tectonic events of the Yezo Group have not been clearly understood, due to limited age control. Detailed chronostratigraphy for the Turonian Yezo Group is therefore required to clarify spatiotemporal aspects of sedimentary patterns.

Over the past two decades, the overall stratigraphic Turonian $\delta^{13}C$ fluctuations recorded in sedimentary organic matter have been documented from many sections of the Yezo Group (Hasegawa and Saito, 1993; Hasegawa, 1997, 2003; Hasegawa and Hatsugai, 2000; Hasegawa et al., 2003; Tsuchiya et al., 2003; Uramoto et al., 2009, 2013; Hayakawa and Hirano, 2013). Bulk of the organic matter preserved in the Yezo Group is terrestrial in origin (Hasegawa, 2001). Based on the carbon isotope values of terrestrial organic matter, chemostratigraphies of the Yezo Group have been established by assuming an isotopic linkage of marine and terrestrial carbon reservoirs within the Cretaceous Earth surface subsystems (e.g., Hasegawa, 1997, 2003; Uramoto et al., 2009, 2013). Thus, the carbon isotope stratigraphy of terrestrial organic matter can provide high-resolution correlation tools for the Yezo Group.

To constrain the sedimentary patterns of the Turonian Yezo Group (Fig. 1) in northern Japan, we have summarized the recent carbon isotope stratigraphy of the terrestrial organic matter. The Obira, Shumarinai, and Oyubari areas were selected for examination of sedimentary patterns of the Yezo Group because of their stratigraphic completeness and well-established biostratigraphic framework (Hasegawa and Saito, 1993; Hasegawa, 1997, 2003; Tsuchiya et al., 2003; Uramoto et al., 2009, 2013). A compilation of published paleontological and carbon isotope data from the study areas and revised bio- and chemostratigraphic classification of the Yezo Group are presented. We then compared the Turonian $\delta^{13}C$ profile in Japan with the European reference $\delta^{13}C$ profile of marine carbonates (Jarvis et al., 2006). The spatiotemporal characteristics of the sedimentary patterns of the Turonian Yezo Group are discussed, based on regional and global chemostratigraphic correlations.

7.2 GEOLOGICAL SETTING

The studied successions in the Obira and Shumarinai areas consist of the following conformable lithostratigraphic units (in ascending stratigraphic order): the Tenkaritoge, Saku, and Haborogawa Formations (Fig. 2) (Funaki and Hirano, 2004; Oizumi et al., 2005). The Tenkaritoge Formation is dominated by dark-grey bioturbated siltstone, with lesser intercalations of coarser clastic rocks. The Saku Formation consists of alternating beds of turbiditic sandstone and siltstone. The Haborogawa Formation consists of dark-grey bioturbated siltstone. In the Obira area, thickly bedded sandstone (>1 m) and slump deposits characterize the lower part of the Haborogawa Formation (Funaki and Hirano, 2004). The depositional environment is inferred to have been the outer shelf, based on the predominance of offshore mudstone (Takashima et al., 2004).

Based on the occurrences of age-diagnostic ammonoids, inoceramid bivalves, and planktonic foraminifers, the studied successions in the Obira and Shumarinai areas are determined to be Late Cenomanian to Early Coniacian in age. Nishida et al. (1998b) and Funaki and Hirano (2004) reported the occurrence of Late Cenomanian indicator *Inoceramus pictus minus* Matsumoto in the basal part of the studied successions. The Cenomanian–Turonian boundary is recognized based on integration of planktonic foraminifera and macrofossil biostratigraphy in the Tenkaritoge Formation (Shumarinai area: Nishida et al., 1998a,b; Sekiya et al., 2009; Obira area: Hasegawa and Saito, 1993; Funaki and Hirano, 2004; Nemoto and Hasegawa, 2011). Occurrences of global age-diagnostic macrofossils such as *Collignoniceras woollgari* Mantell, *Subprionocyclus* cf. *neptuni* (Geinitz), and *Romaniceras deverianum* (d'Orbigny) characterize the Turonian successions (Tanabe et al., 1977; Nishida et al., 1998b; Funaki and Hirano, 2004). The Turonian–Coniacian boundary is located within the Haborogawa Formation. The boundary is situated between the stratigraphic levels of the occurrence of inoceramids *Mytiloides incertus* Jimbo and *Cremnoceramus waltersdorfensis hannovrensis* Heinz in the Shumarinai area (Nishida et al., 1998b) and is situated between the stratigraphic levels of the occurrence of *Barroisiceras* spp. and *Didymotis costatus* (Fric) in the Obira area (Tanabe et al., 1977; Matsumoto et al., 1981; Funaki and Hirano, 2004).

The studied sections in the Oyubari area consist of two conformable lithostratigraphic units: the Saku Formation and the overlying Kashima Formation (Takashima et al., 2004) (Fig. 2). The Saku Formation is composed mainly of alternating beds of turbiditic sandstone and mudstone. Large-scale sedimentary packages of slump deposits are present in the middle part of the Turonian in the southern Oyubari area ("Slump zone" in Fig. 2). The Kashima Formation is composed of mudstone. The depositional environment is inferred to have been the continental slope, based on the morphologic characters of calcareous benthic foraminifera (Kaiho et al., 1993).

Based on occurrences of age-diagnostic ammonoids, inoceramid bivalves, and planktonic foraminifers the studied sections in the Oyubari area are assigned to Late Cenomanian to latest Turonian age. Kawabe (2000) and Kurihara and Kawabe (2003) reported the occurrence of Cenomanian indicator *Calycoceras* sp. in the basal part of the studied successions. The Cenomanian–Turonian

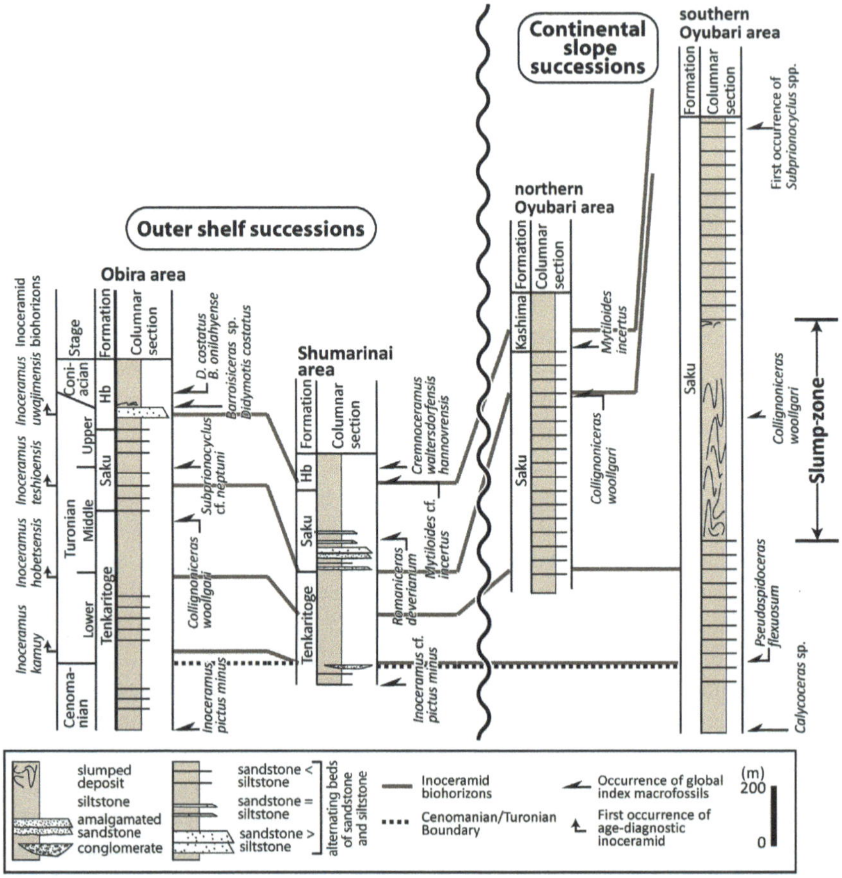

FIGURE 2

Composite columnar sections and stratigraphic distributions of age-diagnostic macrofossils of the Turonian Stage in the Obira, Shumarinai, and Oyubari areas in Hokkaido, northern Japan. Paleontological data are sourced as follows: Shumarinai area (Nishida et al., 1996, 1998b; Sekiya et al., 2009); Obira area (Tanabe et al., 1977; Matsumoto et al., 1981; Funaki and Hirano, 2004; Kaneko and Hirano, 2005; Oizumi et al., 2005; Nishimura et al., 2006; Kurihara et al., 2012); Oyubari area (Matsumoto, 1971; Hirano et al., 1977; Hirano, 1995; Kawabe, 2000; Tsuchiya et al., 2003; Kurihara and Kawabe, 2003).

boundary is recognized based on integration of planktonic foraminifera and macrofossil biostratigraphy in the Saku Formation (Hirano, 1995; Hasegawa, 1997). Occurrences of global age-diagnostic macrofossils such as *Pseudaspidoceras flexuosum* Powell, *C. woollgari* Mantell, and *Subprionocyclus* spp. characterize the Turonian successions (Hirano, 1995; Kawabe, 2000; Tsuchiya et al., 2003). In the upper part of the studied sections, Tsuchiya et al. (2003) reported the occurrence of *M. incertus* in the Kashima Formation, which occurs worldwide in the uppermost part of the Upper Turonian successions (Matsumoto and Noda, 1983; Voigt, 1995; Takahashi, 2005).

7.3 DIAGENESIS AND LOCAL EFFECTS ON $\delta^{13}C$ PROFILES OF TERRESTRIAL ORGANIC CARBON

The terrestrial origin and thermal immaturity of sedimentary organic carbon in the Yezo Group has been well established through visual characterization of kerogen (Hasegawa, 1997, 2001; Hasegawa and Hatsugai, 2000) and Rock-Eval pyrolysis (Hasegawa and Saito, 1993; Uramoto et al., 2009, 2013). The terrestrial organic matter in the Hokkaido samples is composed of plant fragments inferred to have been strongly mixed during transportation from hinterland in East Asia to the Yezo forearc Basin (Hasegawa, 2001). In addition, the bulk samples show homogeneity and no locally high concentrations of marine algal matter. Bulk analysis of sedimentary organic matter is the most suitable method for studying the carbon isotope stratigraphy of terrestrial organic matter from the sections of the Yezo Group.

7.4 CARBON ISOTOPE STRATIGRAPHY

Uramoto et al. (2013) reported that bulk $\delta^{13}C$ records of the terrestrial organic matter for the Turonian succession in the Yezo Group (Hokkaido, Japan) showed not only similar trends with the global oceanic and atmospheric systems, but also varying patterns at a specific stratigraphic horizon. Here, we summarize the regional and global correlations of $\delta^{13}C$ trends and then state the implications of varying $\delta^{13}C$ patterns for sedimentary patterns within the Yezo forearc Basin.

7.4.1 CORRELATION OF TERRESTRIAL $\delta^{13}C$ RECORDS WITH REFERENCE $\delta^{13}C$ RECORDS OF MARINE CARBONATES

Uramoto et al. (2013) conducted chemostratigraphic correlations of Turonian $\delta^{13}C$ records of terrestrial organic matter ($\delta^{13}C_{TOM}$) in northern Japan. A notable feature of Turonian $\delta^{13}C_{TOM}$ fluctuations is the occurrence of three positive $\delta^{13}C$ anomalies with amplitudes of ~0.5–1.0‰ in the Middle–Upper Turonian. The three positive fluctuations are labeled the YT1, YT2, and YT3 events, in ascending stratigraphic order (Fig. 3). The maximum $\delta^{13}C_{TOM}$ value (approximately −23.5‰) of the YT1 positive excursion is recorded in association with first occurrence (FO) of *Inoceramus hobetsensis* (Fig. 3). The presence of the Middle Turonian ammonoid *C. woollgari* in the Yezo Group (Tsuchiya et al., 2003; Funaki and Hirano, 2004) helped fixation of age datum to the YT1 as a Middle Turonian $\delta^{13}C_{TOM}$ event. The YT2 event (Fig. 3) is found below the FO of *Inoceramus teshioensis* and Upper Turonian index ammonoids such as *Subprionocyclus* spp. (Kawabe, 2000; Funaki and Hirano, 2004). The YT3 event is characterized by the FO of *Inoceramus uwajimensis* (Fig. 3) and is marked by the presence of inoceramid *Mytiloides incertus* (Nishida et al., 1998b), which occurs worldwide in the uppermost Turonian successions (Matsumoto and Noda, 1983; Voigt, 1995; Takahashi, 2005). These chemostratigraphic correlations suggest that seven stratigraphic levels (the base, top, and $\delta^{13}C_{TOM}$ maxima of each YT event, as shown by dashed lines in Fig. 3) can be used as chemostratigraphic datum in the Turonian Stage of the Yezo Group.

In addition, the $\delta^{13}C_{TOM}$ anomalies can be correlated to the reference Turonian $\delta^{13}C_{carbonate}$ event in Europe, as revealed by global biostratigraphy (Fig. 4). The Middle Turonian YT1 event in the Yezo

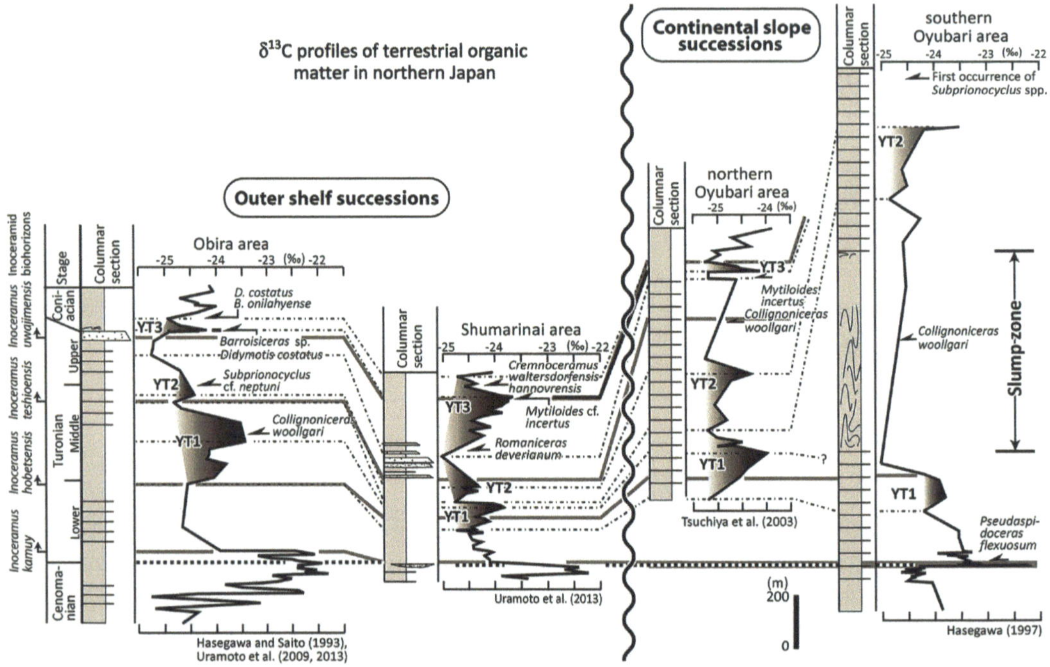

FIGURE 3

Correlation of $\delta^{13}C$ profiles of the terrestrial organic matter ($\delta^{13}C_{TOM}$) of the offshore successions, Turonian Stage of the Yezo Group in Hokkaido, Japan. Paleontological data are sourced as follows: Shumarinai area (Nishida et al., 1996, 1998b; Sekiya et al., 2009); Obira area (Tanabe et al., 1977; Matsumoto et al., 1981; Funaki and Hirano, 2004; Kaneko and Hirano, 2005; Oizumi et al., 2005; Nishimura et al., 2006; Kurihara et al., 2012); Oyubari area (Matsumoto, 1971; Hirano et al., 1977; Hirano, 1995; Kawabe, 2000; Tsuchiya et al., 2003; Kurihara and Kawabe, 2003). See Fig. 2 for legend.

Group can be correlated with the lower Middle Turonian interval from the Lulworth Event to the Round Down Event (Jarvis et al., 2006) based on the presence of the Middle Turonian ammonoid *C. woollgari* (Fig. 4). Positive $\delta^{13}C_{TOM}$ excursions of the YT2 event in the upper Middle Turonian in the Yezo Group are correlated with the Glynde Event to Pewsey Event (Gale, 1996; Jarvis et al., 2006) in Europe, as the YT2 event in the Yezo Group occurs immediately below the FO of Upper Turonian ammonoids such as *Subprionocyclus* spp. (Kawabe, 2000; Funaki and Hirano, 2004; Fig. 4). The YT3 event in the Yezo Group can be correlated with the positive $\delta^{13}C_{carbonate}$ excursion from the Bridgewick Event to Navigation Event (late Turonian events; Gale, 1996; Jarvis et al., 2006) in Europe (Fig. 4). This correlation is supported by the presence of the inoceramid *Mytiloides incertus* in the Yezo Group (Nishida et al., 1998b), which occurs worldwide in the uppermost part of Upper Turonian successions (Matsumoto and Noda, 1983; Voigt, 1995; Takahashi, 2005). These chemostratigraphic correlations suggest that the marked $\delta^{13}C$ variations occurred near-simultaneously in Earth's oceanic and atmospheric systems.

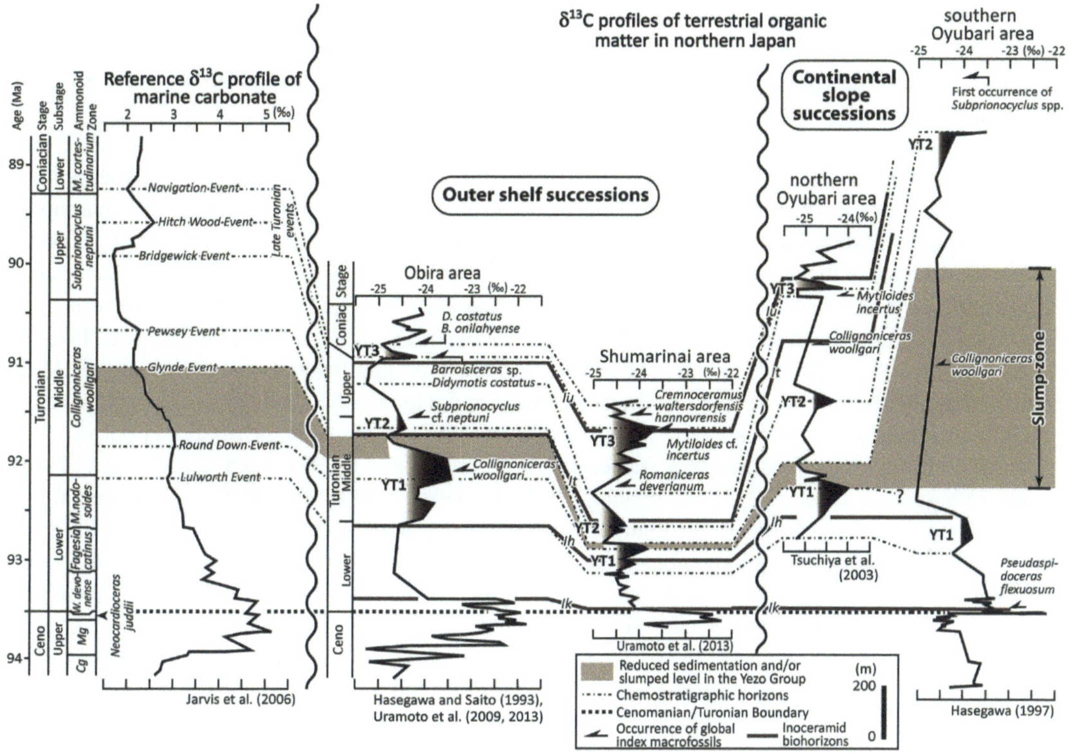

FIGURE 4

Correlation of $\delta^{13}C$ profiles of the terrestrial organic matter ($\delta^{13}C_{TOM}$) of the offshore successions, Turonian Stage of the Yezo Group in Hokkaido, Japan, and the Turonian reference $\delta^{13}C_{carbonate}$ profile in Europe (Jarvis et al., 2006). Dashed lines show the seven chemostratigraphic horizons in the Turonian (Uramoto et al., 2013). Paleontological data are sourced as follows: Shumarinai area (Nishida et al., 1996, 1998b; Sekiya et al., 2009); Obira area (Tanabe et al., 1977; Matsumoto et al., 1981; Funaki and Hirano, 2004; Kaneko and Hirano, 2005; Oizumi et al., 2005; Nishimura et al., 2006; Kurihara et al., 2012); Oyubari area (Matsumoto, 1971; Hirano et al., 1977; Hirano, 1995; Kawabe, 2000; Tsuchiya et al., 2003; Kurihara and Kawabe, 2003). Abbreviations: Cg, Calycoceras guerangeri; Mg, Metoicoceras geslinianum; N, Neocardioceras; W. devonense, Watinoceras devonense; Ik, Inoceramus kamuy; Ih, Inoceramus hobetsensis; It, Inoceramus teshioensis; Iu, Inoceramus uwajimensis; TOM, terrestrial organic matter.

7.4.2 CHEMOSTRATIGRAPHIC CONSTRAINTS ON SEDIMENTARY PATTERNS IN THE YEZO GROUP

The information presented in the previous section demonstrated the utility of the marked $\delta^{13}C$ variations for the global chemostratigraphic correlations. However, varying patterns of $\delta^{13}C_{TOM}$ fluctuations in the Yezo Group are also recognized in the Middle Turonian successions (Fig. 4) and warranted closer examination.

The most notable feature of all the Turonian $\delta^{13}C_{TOM}$ profiles of northern Japan is the absence of a long-term decreasing trend observable in the European reference $\delta^{13}C_{carbonate}$ profile that characterizes the stratigraphic interval between the Round Down Event and the Glynde Event in the Middle Turonian (shown by gray shading in Fig. 4). This stratigraphic interval in the offshore successions of the Yezo Group is characterized by fluctuating $\delta^{13}C_{TOM}$, with sedimentary successions showing reduced sedimentation or large-scale sedimentary packages characteristic of slump deposits (Figs 2–4).

The chemostratigraphic correlations presented in the previous section suggest the $\delta^{13}C_{TOM}$ fluctuations of the upper Middle Turonian Glynde–Pewsey events directly stack on top of the $\delta^{13}C_{TOM}$ fluctuations of the Lulworth–Round Down events in continental shelf successions of the Obira and Shumarinai areas, and in a continental slope succession of the northern Oyubari area (Fig. 4). According to Jarvis et al. (2006), the stratigraphic interval of the gradual decrease in the Middle Turonian $\delta^{13}C_{carbonate}$ values is characterized by a significant time interval of c. 0.7 m.y. Typically, the sedimentation rate of the offshore successions of the Yezo Group is estimated to have been 200–400 m/m.y. (Kaiho and Hasegawa, 1994; Nemoto and Hasegawa, 2011). Therefore, it is highly possible that sedimentation rate was significantly reduced during the Middle Turonian in the shelf-slope successions of the Yezo Group and may also be a reflection of a prolonged sedimentary hiatus related to significant winnowing or bypassing of sediments in the offshore regime.

On the other hand, this sedimentary succession is characterized by increased occurrences of slump deposits (Motoyama et al., 1991; Hasegawa, 1997; Kawabe, 2000), and $\delta^{13}C_{TOM}$ profiles for this interval show no marked change in the southern Oyubari area. Such mass-transport deposits are absent in the Middle Turonian sections of other areas where the sedimentary successions are dominated by siltstone (Figs 2 and 3). The timing of the slumping events in the southern Oyubari area is similar to the confined subsidence in the Yezo forearc Basin that began in the Turonian (Tamaki et al., 2009). The formation of the confined basin might have provided sufficient accommodation space, which is unique to the southern Oyubari area. This localized tectonic activity led to tilting of the sea floor, resulting in intense slumping in the Oyubari area.

The varying pattern of Middle Turonian $\delta^{13}C_{TOM}$ fluctuations in the Yezo Group may reflect spatial variations in sedimentary patterns within the Yezo forearc Basin. It is notable that this time interval coincides with an extensive eustatic sea level fall (Haq et al., 1987). In this regard, the Middle Turonian successions of the Yezo Group are critical for understanding the relative importance of global sea level change, local tectonics, and sedimentary processes in controlling continental margin stratigraphy. A multidisciplinary approach including facies analysis, paleobathymetry, and tectonics is expected to provide more detailed insight into the sedimentary evolution of Cretaceous forearc successions along the northwestern Pacific margin.

7.5 CONCLUSIONS

We have presented the carbon isotope stratigraphy of terrestrial organic matter for the Turonian (Upper Cretaceous) strata of the Yezo Group in Hokkaido, northern Japan. Our regional and global chemostratigraphic correlations suggest that the Turonian $\delta^{13}C_{TOM}$ profiles of the studied sections of the Yezo Group exhibit similarity to the reference $\delta^{13}C$ profiles of marine carbonate in Europe. The similarity in trends indicates a global isotopic linkage of the oceanic and atmospheric systems. Three distinctive positive excursions of $\delta^{13}C_{TOM}$ (the YT1–YT3 events) were identified and these events can be correlated with the reference $\delta^{13}C_{carbonate}$ events from Europe (the Lulworth–Round Down events of the lower Middle Turonian, the Glynde–Pewsey events of the upper Middle Turonian, and the Bridgewick–Navigation events of the Upper Turonian).

However, the gradual $\delta^{13}C_{carbonate}$ records of reference Middle Turonian in Europe over a significant time interval (ca. 0.7 m.y.) is absent in the studied sections due to the reduction of sedimentation rate and prevalent slumping events, hiatus, and synsedimentary tectonics.

Our chemostratigraphic correlations validate the potential of $\delta^{13}C$ records for use in examining not only the global $\delta^{13}C$ trends of oceanic and atmospheric systems, but also sedimentary patterns based on biostratigraphic calibration. This synchroneity demonstrates that chemostratigraphy as a tool transcends the atmosphere-lithosphere-biosphere-hydrosphere boundaries.

REFERENCES

Ando, A., Kakegawa, T., 2007. Carbon isotope record of terrestrial organic matter and occurrence of planktonic foraminifera from the Albian Stage of Hokkaido, Japan: ocean-atmosphere $\delta^{13}C$ trends and chronostratigraphic implications. Palaios 22, 417–432.

Ando, A., Kakegawa, T., Takashima, R., Saito, T., 2002. New perspective on Aptian carbon isotope stratigraphy: data from $\delta^{13}C$ records of terrestrial organic matter. Geology 30, 227–230.

Ando, A., Kakegawa, T., Takashima, R., Saito, T., 2003. Stratigraphic carbon isotope fluctuations of detrital woody materials during the Aptian Stage in Hokkaido, Japan: comprehensive $\delta^{13}C$ data from four sections of the Ashibetsu area. Journal of Asian Earth Sciences 21, 835–847.

Funaki, H., Hirano, H., 2004. Cretaceous stratigraphy in the northeastern part of the Obira area, Hokkaido, Japan. Bulletin of Mikasa City Museum 8, 17–35 (in Japanese with English abstract).

Gale, A.S., 1996. Turonian correlation and sequence stratigraphy of the Chalk in southern England. In: Hesselbo, S.P., Parkinson, D.N. (Eds.), Sequence Stratigraphy in British Geology, vol. 103. Geological Society Special Publication, pp. 177–195.

Gröcke, D.R., Price, G.D., Robinson, S.A., Baraboshkin, E.Y., Mutterlose, J., Ruffell, A.H., 2005. The Upper Valanginian (Early Cretaceous) positive carbon-isotope event recorded in terrestrial plants. Earth and Planetary Science Letters 240, 495–509.

Haq, B.U., Hardenbol, J., Vail, P.R., 1987. Chronology of fluctuating sea levels since the triassic. Science 235, 1156–1167.

Hasegawa, T., 1997. Cenomanian–Turonian carbon isotope events recorded in terrestrial organic matter from northern Japan. Palaeogeography, Palaeoclimatology, Palaeoecology 130, 251–273.

Hasegawa, T., 2001. Predominance of terrigenous organic matter in Cretaceous marine fore-arc sediments, Japan and Far East Russia. International Journal of Coal Geology 47, 207–221.

Hasegawa, T., 2003. A global carbon-isotope event in the Middle Turonian (Cretaceous) sequences in Japan and Russian Far East. Proceedings of Japan Academy Series B 79, 141–144.

Hasegawa, T., Hatsugai, T., 2000. Carbon-isotope stratigraphy and its chronostratigraphic significance for the Cretaceous Yezo Group, Kotanbetsu area, Hokkaido, Japan. Paleontological Research 4, 95–106.

Hasegawa, T., Pratt, L.M., Maeda, H., Shigeta, Y., Okamoto, T., Kase, T., Uemura, K., 2003. Upper Cretaceous stable carbon isotope stratigraphy of terrestrial organic matter from Sakhalin, Russian Far East: a proxy for the isotopic composition of paleoatmospheric CO_2. Palaeogeography, Palaeoclimatology, Palaeoecology 189, 97–115.

Hasegawa, T., Saito, T., 1993. Global synchroneity of a positive carbon isotope excursion at the Cenomanian/Turonian boundary: Validation by calcareous microfossil biostratigraphy of the Yezo Group, Hokkaido, Japan. Island Arc 3, 181–191.

Hayakawa, T., Hirano, H., 2013. A revised inoceramid biozonation for the Upper Cretaceous based on high-resolution carbon isotope stratigraphy in northwestern Hokkaido, Japan. Acta Geologica Polonica 63, 239–263.

Hirano, H., 1995. Correlation of the Cenomanian/Turonian boundary between Japan and Western Interior of the United States in relation with oceanic anoxic events. Journal of the Geological Society of Japan 101, 13–18.

Hirano, H., Matsumoto, T., Tanabe, K., 1977. Mid-Cretaceous stratigraphy of the Oyubari area, central Hokkaido. Palaeontological Society of Japan Special Papers 21, 1–10.

Jarvis, I., Gale, A.S., Jenkyns, H.C., Pearce, M.A., 2006. Secular variation in Late Cretaceous carbon isotopes: a new $\delta^{13}C_{carbonate}$ reference curve for the Cenomanian–Campanian (99.6–70.6 Ma). Geological Magazine 143, 561–608.

Jenkyns, H.C., Gale, A.S., Corfield, R.M., 1994. Carbon- and oxygen-isotope stratigraphy of the English Chalk and Italian Scaglia and its palaeoclimatic significance. Geological Magazine 131, 1–34.

Kaiho, K., Fujiwara, O., Motoyama, I., 1993. Mid-Cretaceous faunal turnover of intermediate-water benthic foraminifera in the northwestern Pacific Ocean margin. Marine Micropaleontology 23, 13–49.

Kaiho, K., Hasegawa, T., 1994. End-Cenomanian benthic foraminiferal extinctions and oceanic dysoxic events in the northwestern Pacific Ocean. Palaeogeography, Palaeoclimatology, Palaeoecology 111, 29–43.

Kaneko, M., Hirano, H., 2005. Dinoflagellate cyst assemblages and reconstruction of primary productivity across the Cenomanian/Turonian boundary in the Obira area, northwestern Hokkaido, Japan. Bulletin of the Mikasa City Museum 9, 27–39.

Kawabe, F., 2000. Cretaceous stratigraphy in the Oyubari area, central Hokkaido, Japan. Bulletin of National Science Museum Tokyo Series C26, 9–56.

Kurihara, K., Kawabe, F., 2003. Molluscan faunal changes across the Cenomanian/Turonian boundary: a comparison between the Oyubari area, Hokkaido, Japan and the Western Interior, USA. Fossils (Palaeontological Society of Japan) 74, 36–47 (in Japanese with English abstract).

Kurihara, K., Toshimitsu, S., Hirano, H., 2012. Ammonoid biodiversity changes across the Cenomanian–Turonian boundary in the Yezo Group, Hokkaido, Japan. Acta Palaeontologica Polonica 57, 749–757.

Matsumoto, T., 1971. Collignoceratidae from Hokkaido Part 5 (Studies of the Cretaceous Ammonites from Hokkaido and Saghalien XXIII). Series D Geology, 21,Memoirs of the Faculty of Science, Kyushu University. 129–162.

Matsumoto, T., Muramoto, K., Hirano, H., Takahashi, T., 1981. Some coniacian ammonites from hokkaido (Studies of the cretaceous ammonites from Hokkaido-XL). Transactions and Proceedings Palaeontological Society Japan, New Series 121, 51–73.

Matsumoto, T., Noda, M., 1983. Restudy of *Inoceramus incertus* Jimbo with special reference to its biostratigraphic implications. Proceedings of Japan Academy Series B 59, 109–112.

Motoyama, I., Fujiwara, O., Kaiho, K., Murota, T., 1991. Lithostratigraphy and calcareous microfossil biochronology of the Cretaceous strata in the Oyubari area, Hokkaido, Japan. Journal of the Geological Society of Japan 97, 507–527 (in Japanese with English abstract).

Nemoto, T., Hasegawa, T., 2011. Submillennial resolution carbon isotope stratigraphy across the Oceanic Anoxic Event 2 horizon in the Tappu section, Hokkaido, Japan. Palaeogeography, Palaeoclimatology, Palaeoecology 309, 271–280.

Nishida, T., Matsumoto, T., Inoue, Y., 1998a. Foraminifera from the Cretaceous system in the Shumarinai valley of Hokkaido. Journal of Faculty of Culture and Education, Saga University 3, 301–331 (in Japanese with English abstract).

Nishida, T., Matsumoto, T., Kawasita, Y., Egashira, N., Aizawa, J., 1998b. Characteristics of the Cretaceous stratigraphy in the Shumarinai valley of Hokkaido. Journal of Faculty of Culture and Education, Saga University 2, 143–181 (in Japanese with English abstract).

Nishida, T., Matsumoto, T., Yokoi, K., Kawashita, Y., Kyuma, Y., Egashira, N., Aizawa, J., Maiya, S., Ikuji, Y., Yao, A., 1996. Biostratigraphy of the Cretaceous Middle Yezo Group in the Soeushinai area of Hokkaido–with special reference to the transitional part from Lower to Upper Cretaceous. Journal of Faculty of Education, Saga University 44, 65–149 (in Japanese with English abstract).

Nishimura, T., Shigeta, Y., Maeda, H., 2006. Ontogenetic shell development of a Cretaceous desmoceratine ammonoids "*Tragodesmoceroides subcostatus*" Matsumoto, 1942 from Hokkaido, Japan. Paleontological Research 10, 11–28.

Nunn, E.V., Price, G.D., Gröcke, D.R., Baraboshkin, E.Y., Leng, M.J., Hart, M.B., 2010. The Valaginian positive carbon isotope event in Arctic Russia: Evidence from terrestrial and marine isotope records and implications for global carbon cycling. Cretaceous Research 31, 577–592.

Oizumi, M., Kurihara, K., Funaki, H., Hirano, H., 2005. Upper cretaceous stratigraphy in the Obira area, Hokkaido, Japan. Bulletin of Mikasa City Museum 9, 11–26 (in Japanese with English abstract).

Robinson, S.A., Hesselbo, S.P., 2004. Fossil-wood carbon-isotope stratigraphy of the non-marine Wealden group (Lower Cretaceous, southern England). Journal of the Geological Society 161, 133–145.

Sekiya, T., Murakami, R., Uramoto, G.-I., Hirano, H., 2009. Upper Cretaceous Macrofossil Assemblages in the Shumarinai Area, Hokkaido, Japan, vol. 57, Academic Studies, Biology and Earth Science, School of Education, Waseda University. pp. 25–31 (in Japanese with English abstract).

Stoll, H.M., Schrag, D.P., 2000. High-resolution stable isotope records from the Upper Cretaceous records of Italy and Spain: glacial episodes in a greenhouse planet? Geological Society of America Bulletin 112, 308–319.

Taira, A., 2001. Tectonic evolution of the Japanese island arc system. Annual Review of Earth and Planetary Sciences 29, 109–134.

Takahashi, A., 2005. Diversity changes in Cretaceous inoceramid bivalves of Japan. Paleontological Research 9, 217–232.

Takashima, R., Kawabe, F., Nishi, H., Moriya, K., Wani, R., Ando, H., 2004. Geology and stratigraphy of forearc basin sediments in Hokkaido, Japan: Cretaceous environmental events on the north-west Pacific margin. Cretaceous Research 25, 365–390.

Tamaki, M., Tsuchida, K., Itoh, Y., 2009. Geochemical modeling of sedimentary rocks in the central Hokkaido, Japan: episodic deformation and subsequent confined basin-formation along the eastern Eurasian margin since the Cretaceous. Journal of Asian Earth Sciences 34, 198–208.

Tanabe, K., Hirano, H., Matsumoto, T., Miyata, Y., 1977. Stratigraphy of the Upper Cretaceous Deposits in the Obira Area, Northwestern Hokkaido, vol. 12, Science Report Department of Geology, Kyushu University. pp. 181–202 (in Japanese with English abstract).

Tsuchiya, K., Hasegawa, T., Pratt, L.M., 2003. Stratigraphic relationship between diagnostic carbon isotope profiles and inoceramid biozones from the Yezo Group, Hokkaido, Japan. Journal of the Geological Society of Japan 109, 30–40 (in Japanese with English abstract).

Uramoto, G.-I., Abe, Y., Hirano, H., 2009. Carbon isotope fluctuations of terrestrial organic matter for the Upper Cretaceous (Cenomanian–Santonian) in the Obira area, Hokkaido, Japan. Geological Magazine 146, 761–774.

Uramoto, G.-I., Fujita, T., Takahashi, A., Hirano, H., 2007. Cenomanian (Upper Cretaceous) carbon isotope stratigraphy of terrestrial organic matter for the Yezo Group, Hokkaido, Japan. Island Arc 16, 465–478.

Uramoto, G.-I., Tahara, R., Sekiya, T., Hirano, H., 2013. Carbon isotope stratigraphy of terrestrial organic matter forthe Turonian (Upper Cretaceous) in northern Japan: implications for ocean–atmosphere $\delta^{13}C$ trends during the mid-Cretaceous climatic optimum. Geosphere 9, 355–366.

Voigt, S., 1995. Palaeobiogeography of early Late Cretaceous inoceramids in the context of a new global palaeogeography. Cretaceous Research 18, 343–356.

CHAPTER

8 GEOCHEMISTRY OF LATE CRETACEOUS SEDIMENTARY ROCKS OF THE CAUVERY BASIN, SOUTH INDIA: CONSTRAINTS ON PALEOWEATHERING, PROVENANCE, AND END CRETACEOUS ENVIRONMENTS

J. Madhavaraju

Estación Regional del Noroeste, Instituto de Geología, Universidad Nacional Autónoma de México, Hermosillo, Sonora, México

8.1 INTRODUCTION

The geochemical composition of sedimentary rocks provides information on provenance, tectonic setting, and paleoclimatic conditions (Bhatia, 1983; Taylor and McLennan, 1985; Cingolani et al., 2003; Nagarajan et al., 2007; Jafarzadeh and Hosseini-Barzi, 2008; Absar et al., 2009; Raza et al., 2012; Armstrong-Altrin et al., 2013). Major and trace element compositions or ratios of sandstones and fine-grained sediments have been used to decipher provenance and tectonic setting (Taylor and McLennan, 1985; Bhatia and Crook, 1986; Roser and Korsch, 1986; Cullers, 1995, 2000; Mongelli et al., 1996). Elemental concentrations of sediments and sedimentary rocks have also been used to study paleoweathering trends of siliciclastic deposits (Fedo et al., 1996; Nath et al., 2000). The elemental ratios of key trace elements, the rare earth element (REE) concentrations, REE patterns, and the magnitude of Eu anomalies in sandstones and shales provide useful information regarding provenance (Bhatia, 1985; Wronkiewicz and Condie, 1989; Feng and Kerrich, 1990; Cullers, 1995; Nath et al., 2000). The effects of weathering on trace element distributions have been reported in several geochemical studies of chemical weathering under variable climatic conditions (e.g., Nesbitt, 1979; Condie et al., 1995). Volcanism may affect contemporary sedimentary environments in many ways (Cox, 1988). The Indian subcontinent experienced basic volcanism that resulted in the formation of the Deccan volcanic province during the latest Cretaceous (Subba Rao, 1999). Mantle plume-related volcanism results in topographic and atmospheric changes that produce significant landscape and environmental changes (Tandon, 2002). Records of the influence of volcanism on landscape, sediment supply, and biotas should be preserved in sedimentary sequences contemporary with volcanism (White, 2000).

Cretaceous sedimentary rocks occur in the Cauvery Basin, which lies adjacent to the Deccan Traps in southern India (Fig.1). The provenance record of a sedimentary succession of Campanian-Maastrichtian age (Ariyalur Group) in the Cauvery Basin, southeastern India, is reported here with the help of major, trace, and

Chemostratigraphy. http://dx.doi.org/10.1016/B978-0-12-419968-2.00008-X

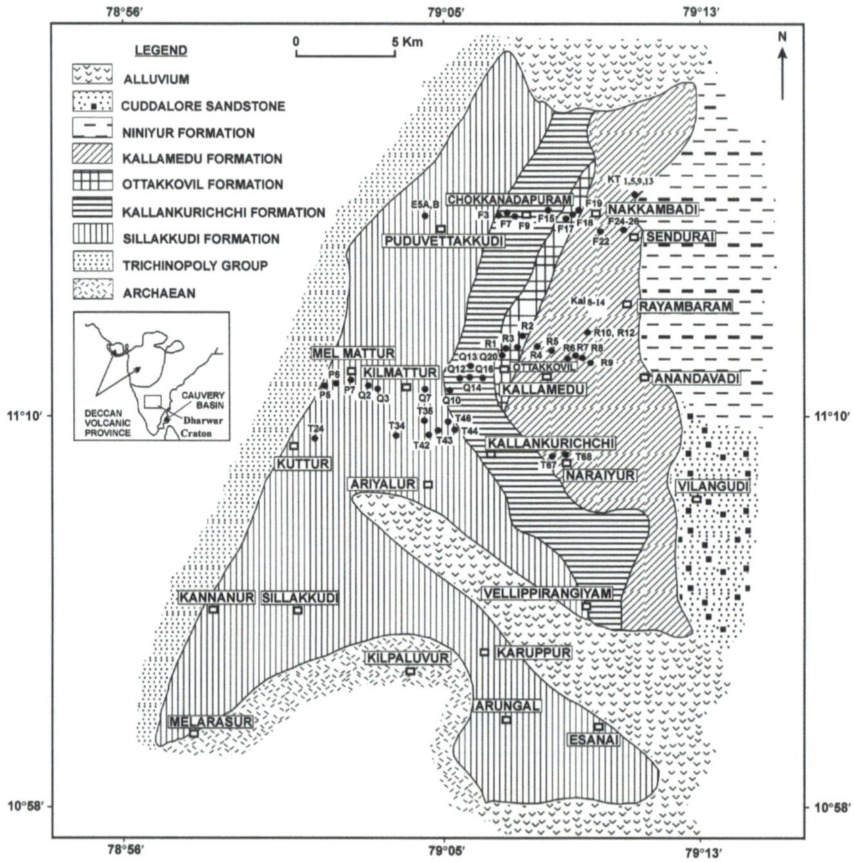

FIGURE 1

Geological map of the Ariyalur area (after Madhavaraju and Ramasamy (1999b)), showing sample locations.

rare earth elemental geochemistry. Many studies have focused on the palaeontology and sedimentology of the Cauvery Basin, but systematic study of geochemistry to constrain on the provenance and tectonic history is scarce. The objectives of this chapter are to examine the extent of weathering that occurred in the source regions of the Cauvery Basin, to identify the probable source rocks of the Ariyalur Group and to determine if Deccan volcanism had any influence on the contemporary sedimentary rocks of the Cauvery Basin. To determine the probable source rock for the Ariyalur Group, a comparison of the geochemical data generated for the present study and the published geochemical data on gneisses, granodiorites, and granites of the Dharwar Craton (Allen, 1985; Jayananda et al., 2000) and Deccan basalts (Melluso et al., 2004) was made.

8.2 GEOLOGY AND STRATIGRAPHY

The Cauvery Basin is the southernmost basin situated along the eastern margin of southern India. It is considered to be a rift basin that developed during the Mesozoic (Prabhakar and Zutshi, 1993). The basin fillings represent a passive margin sedimentary system that developed in response to the separation of the Indian subcontinent from Australia and Antarctica during the late Jurassic to earliest Cretaceous time (Powell et al., 1988). The Cretaceous sedimentary rocks are distributed in five geographic areas in the Cauvery Basin; from north to south, the Pondicherry, Vridhachalam, Ariyalur, Tanjore, and Sivaganga areas. The Cretaceous-Paleocene rocks are well exposed in the Ariyalur area (Fig. 1). Much work has been carried out on these sedimentary rocks, including studies of stratigraphy, palaeontology, and depositional environments (Sastry et al., 1972; Sundaram and Rao, 1986; Ramasamy and Banerji, 1991; Govindan et al., 1996; Madhavaraju and Ramasamy, 1999a,b,c; Madhavaraju et al., 2002, 2004, 2014; Ramkumar et al., 2004, 2005, 2006, 2010), but there has been no systematic study of paleoweathering, provenance, and tectonic setting. Based on lithological and paleontological observations, Blanford (1862) classified the exposed Cretaceous sedimentary deposits into three groups with decreasing age, namely, the Uttatur, Trichinopoly, and Ariyalur groups. The Ariyalur Group (Campanian-Maastrichtian) covers approximately 550 km^2 and disconformably overlies the Trichinopoly Group (Fig. 1). The Ariyalur Group is subdivided into four formations, namely, the Sillakkudi, Kallankurichchi, Ottakkovil, and Kallamedu formations in ascending order (Sastry et al., 1972, Table 1).

The Sillakkudi Formation has the largest areal extent. It is composed predominantly of fine to coarse sandstone alternating with multicolored claystone and sandy limestone deposited in a near-shore marine environment (Sundaram and Rao, 1986; Madhavaraju, 1996; Madhavaraju and Ramasamy, 1999b). The Kallankurichchi Formation is composed of conglomerate, sandstone, and limestone. It has been subdivided into two units on the basis of lithology, the lower sandstone unit, and the upper carbonate unit (Sastry et al., 1972). The sandstone is interbedded with claystone, and these rocks were deposited in a near-shore environment (Sundaram and Rao, 1986; Madhavaraju and Ramasamy, 1999a,b).

Table 1 Lithostratigraphy of the Ariyalur Group of the Cauvery Basin

Group	Formation	Lithology	Thickness (m)	Age
Niniyur		Calcareous sandstone, fossiliferous limestone, and clay	50–100	Danian
Ariyalur	Kallamedu	Unfossiliferous fine- to coarse-grained sandstone interbedded with siltstone, claystone, shale, and marl	60–100	Maastrichtian
	Ottakkovil	Fossiliferous calcareous sandstone interbedded with claystone	0–60	
	Kallankurichchi	Fossiliferous calcareous conglomeratic sandstone, fossiliferous calcareous sandstone interbedded with claystone, sandy fossiliferous limestone, fossiliferous limestone, and marl	10–40	
	Sillakkudi	Unfossiliferous calcareous sandstone, fossiliferous calcareous sandstone interbedded with claystone and thin band of sandy limestone	300–500	Campanian
Trichinopoly				Late Turonian to Santonian

Modified after Sastry et al. (1972).

The carbonate rocks include numerous orbitoids and other larger foraminifera indicating that these carbonates were deposited in a shallow marine environment (Sastry et al., 1972; Hart et al., 2000).

The Ottakkovil Formation is mainly composed of sandstone interbedded with claystone. This formation thins toward the north of the basin due to continued uplift of the basement, indicating that it was deposited during a regressive phase (Tewari et al., 1996). The Kallamedu Formation marks the end of Cretaceous sedimentation in the Cauvery Basin and is considered a contemporary sedimentary deposit of the Deccan Traps. The Kallamedu Formation is mainly composed of white to gray unfossiliferous calcareous sandstone, friable to moderately consolidated fine to medium sandstone, siltstone, and red to green/gray shale. A band of vitric tuff of the Deccan provenance occurs in the clastic rocks of the Kallamedu Formation (Tripathi and Mamgain, 1987). Sandstones yield fragments of dinosaur bones near Kallamedu village (Yadagiri and Ayyasami, 1987). The weathered products of volcanic ash from Deccan volcanism form a discrete layer in the upper part of the Kallamedu Formation (Ramkumar et al., 2005; Madhavaraju and Lee, 2010). Sandstones showing various types of cross-bedding represent rapid and shallow-water deposition in shallow channels, perhaps in fluvial environment (Govindan et al., 1996; Tewari et al., 1996; Madhavaraju and Ramasamy, 1999b).

8.3 MATERIALS AND METHODS

Forty-four samples (sandstones, siltstones, claystones, and shales) were selected for major-element analysis. Major elements were determined using an X-ray fluorescence spectrometer with a rhodium X-ray source. The accuracies for SiO_2, Al_2O_3, K_2O, and MnO are better than ±1%, those for Fe_2O_3, CaO, and MgO are better than ±4%, and Na_2O, P_2O_5, and TiO_2 are better than ±5%. United States Geological Survey standard MAG-1 was used as a standard material. Twenty-six samples were analyzed for trace and REEs using an inductively coupled plasma mass spectrometer. One hundred milligrams of each sample was digested with $HF+HNO_3+HClO_4$ acid mixture and heated at 200 °C. The contents were dissolved with 10 ml of 1:1 HNO_3 and the volume was made to 100 ml for trace and REE analyses. Rhodium and bismuth were used as internal standards to monitor signal drift. To determine the accuracy of the results, MAG-1, GSR4, and GSR6 were used for calibration. The precision of all the trace and REEs was better than ±5%. REEs in samples were normalized by using a chondrite value (Taylor and McLennan, 1985).

8.4 RESULTS

8.4.1 MAJOR ELEMENTS

Based on the geochemical classification diagram of Herron (1988), the sandstones and siltstones of the Ariyalur Group are classified as lith-arenite, wacke, Fe-sand, and arkose (Fig. 2(A)) and the claystone and shale samples are classified as wacke, shale, and Fe-shale (Fig. 2(B)). The major elemental concentrations of individual samples are given in Table 2. The studied rocks exhibit wide variations in almost all major elements. Average SiO_2 content is greater in the Kallamedu sandstones (60.2±16.9, n=9) than the Sillakkudi (47.8±9.5, n=12), Kallankurichchi (38.4±5.5, n=6), and Ottakkovil (42.2±5.0, n=5) formations. Slight enrichment of Al_2O_3 is noted in the Kallamedu than other formations. The sandstones and claystones of the Ariyalur Group contain more K_2O than Na_2O (Table 2). The observed variations in K_2O and Na_2O contents are mainly due to the relative proportions of K-feldspar and plagioclase. The sandstones and claystones among all the studied samples have low TiO_2 contents, except the sandstones, claystones, and shale samples collected from the upper Kallamedu Formation (Table 2).

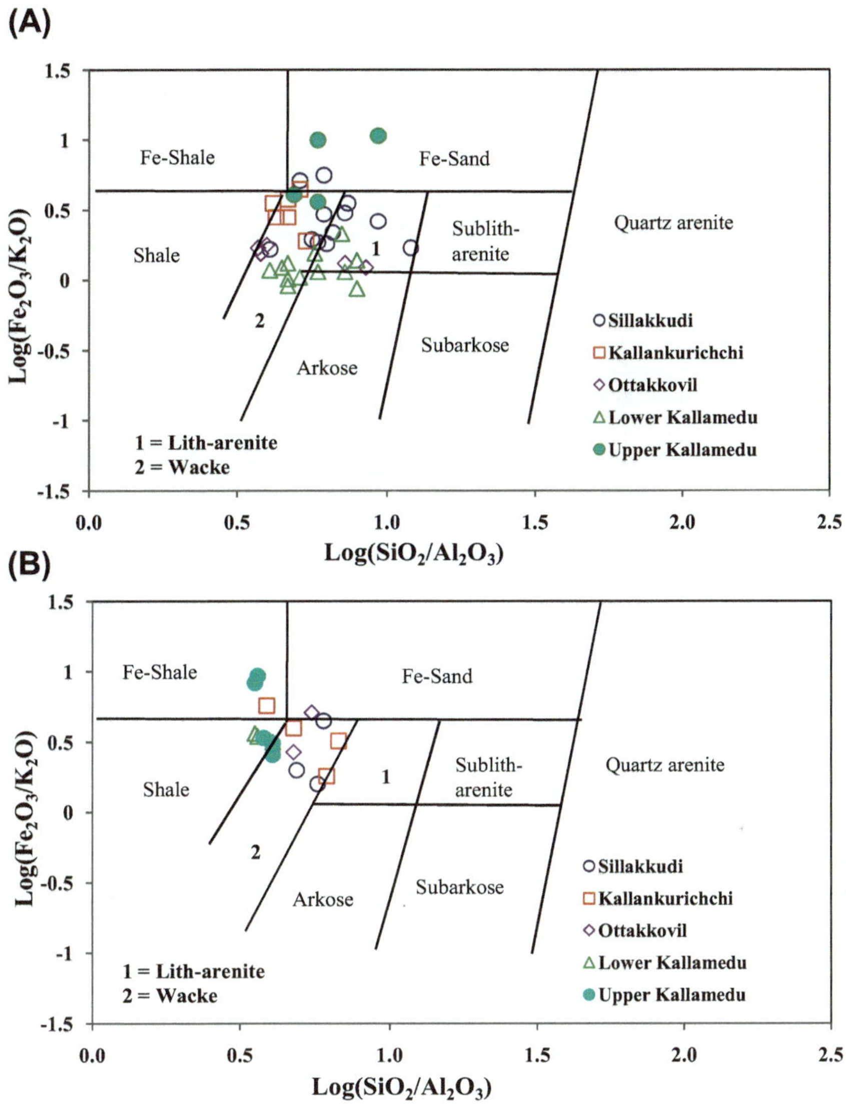

FIGURE 2

(A) Geochemical classification of sandstones and siltstones of the Ariyalur Group using log(SiO_2/Al_2O_3) versus log(Fe_2O_3/K_2O) diagram (after Herron (1988)), (B) Geochemical classification of claystones and shales using Herron's diagram.

Table 2 Major Elements Concentrations (wt%) and Elemental Ratios of Clastic Rocks of Ariyalur Group[a]

	SiO_2	Al_2O_3	Fe_2O_3	CaO	MgO	Na_2O	K_2O	MnO	TiO_2	LOI	Total	CaO*
Upper Kallamedu Sandstone[b]												
KT1	69.9	11.8	2.68	0.68	1.76	0.49	0.74	0.01	0.65	8.88	97.64	0.68
KT5	66.7	13.5	4.11	0.74	1.97	0.35	1.02	0.03	0.82	10.70	99.96	0.74
KT9	74.1	8.0	5.52	0.76	1.09	0.56	0.51	0.06	2.10	7.20	99.85	0.76
KT13	66.7	11.4	5.64	1.09	1.72	0.24	0.57	0.12	2.19	10.03	99.74	1.09
Mean (n=4)	69±4	11±2.3	4.5±1.4	0.8±0.2	1.6±0.4	0.4±0.14	0.7±0.2	0.06±0.05	1.4±0.8	9.2±1.5	99.3±1.1	0.8±0.2
Shale[b]												
Kal8	60.8	16.6	6.15	0.76	3.60	0.94	0.68	0.01	0.77	9.27	99.59	0.76
Kal9	64.1	16.8	3.45	0.71	3.62	0.45	1.03	0.01	0.62	8.85	99.65	0.71
Kal11	60.8	17.2	6.40	0.70	3.75	0.40	0.77	0.01	0.75	8.92	99.63	0.70
Kal12	62.4	15.2	2.32	0.69	3.72	0.64	0.90	0.01	0.68	12.99	99.58	0.69
Kal14	61.9	15.3	3.23	0.88	3.55	0.59	1.04	0.01	0.61	12.54	99.59	0.88
Mean (n=5)	62±1	16±1	4.3±2	0.8±0.1	3.6±0.1	0.6±0.2	0.9±0.2	0.01±0	0.7±0.1	10.5±2	99.6±0.03	0.8±0.1
Lower Kallamedu Sandstone												
F22	52.1	8.7	3.21	14.30	3.27	0.62	2.80	0.03	0.48	13.50	99.01	1.96
F25	76.0	9.6	2.63	1.84	1.00	0.70	3.00	0.06	0.30	3.10	98.23	1.58
R4	74.1	9.3	3.84	2.50	0.80	1.00	2.76	0.04	0.32	4.30	98.96	2.25
R6	72.3	10.1	4.76	2.65	1.21	1.20	2.20	0.02	0.39	4.15	98.68	2.48
R8	40.4	7.1	3.04	23.20	1.74	0.58	1.95	0.04	0.12	21.4	99.57	2.22
R9	72.6	10.0	3.92	2.70	1.00	1.22	3.40	0.03	0.30	4.05	99.22	2.48
T67	41.4	8.1	2.81	22.50	1.80	0.92	2.67	0.04	0.50	19.20	99.94	2.40
T68	37.26	8.3	3.20	24.60	2.01	1.14	2.60	0.05	0.37	20.4	99.93	1.50
Mean (n=8)	58±17	9±1	3.4±0.7	11.8±10.5	1.6±0.8	0.9±0.3	2.7±0.4	0.04±0.01	0.35±0.11	11.3±8.2	99.2±0.6	2.1±0.4
Siltstone												
F24	65.6	14.1	3.80	2.50	3.00	1.05	4.18	0.04	0.27	4.50	99.04	2.03
F26	66.2	14.2	5.30	2.80	1.55	1.00	4.00	0.05	0.39	3.92	99.41	2.25

R5	66.7	16.3	4.75	2.25	1.10	1.90	4.02	0.03	0.37	2.40	98.82	2.00
R7	67.9	14.5	4.40	3.34	0.91	1.59	4.30	0.04	0.44	2.65	100.07	2.71
Mean (n=4)	67±1	15±1	4.6±0.6	2.7±0.5	1.6±0.9	1.4±0.4	4.1±0.14	0.04±0.01	0.37±0.08	3.4±1	99.6±0.5	2.3±0.3
Shale[b]												
Kal1	54.06	15.1	12.21	1.00	3.88	0.68	3.35	0.04	0.85	8.62	99.82	1.00
Kal2	54.74	15.1	11.71	1.00	4.09	0.43	3.37	0.03	0.87	8.49	99.92	1.00
Kal3	55.09	15.3	11.48	1.03	3.96	0.29	3.33	0.04	0.84	8.54	99.92	1.03
Kal5	53.08	12.9	8.14	4.08	3.87	0.56	2.60	0.17	0.61	12.94	98.98	1.08
Mean (n=4)	54±1	15±1	11±2	1.8±1.5	4.0±0.1	0.5±0.2	3.2±0.4	0.07±0.06	0.8±0.1	9.6±2.2	99.7±0.5	1±0.04
Ottakkovil Sandstone												
F17	44.0	6.1	2.80	22.40	1.20	1.00	2.10	0.06	0.42	19.40	99.48	1.14
F18	40.1	10.0	4.50	22.30	1.00	1.05	2.52	0.05	0.30	18.05	99.87	2.90
Q20	48.1	5.7	2.71	21.60	1.25	1.20	2.21	0.05	0.40	16.90	100.12	0.86
R1	42.5	11.3	4.00	19.80	0.50	2.10	2.60	0.03	0.38	16.60	99.81	2.55
R2	37.8	10.2	4.20	23.60	0.45	1.84	2.50	0.04	0.24	19.04	99.91	2.62
Mean (n=5)	42.5±4	9±3	3.6±0.8	21.9±1.4	0.9±0.4	1.4±0.5	2.4±0.2	0.05±0.01	0.35±0.08	18±1.3	99.8±0.2	2±0.9
Claystone												
F19	64.0	13.3	8.92	3.10	0.75	0.88	3.30	0.02	0.55	4.70	99.52	0.85
R3	68.5	12.4	7.60	1.75	1.20	1.15	1.50	0.02	0.50	5.10	99.68	1.70
Mean (n=2)	66±5	13±0.7	8.3±0.9	2.4±0.9	1.0±0.3	1.4±0.2	2.4±1.3	0.02±0	0.53±0.04	4.9±0.3	99.6±0.3	1.3±0.6
Kallankurichchi Sandstone												
F3	38.0	8.3	5.40	24.80	0.50	1.00	1.90	0.27	0.55	18.80	99.52	1.91
F7	34.0	7.9	6.10	25.00	0.92	1.00	2.15	0.10	0.40	21.90	99.47	2.55
F9	35.1	8.5	5.70	22.20	4.65	2.00	1.60	0.12	0.60	19.20	99.67	2.01
Q10	40.8	8.8	7.45	21.35	0.60	0.80	1.95	0.35	0.62	17.60	100.32	2.36
Q12	38.0	7.4	4.90	25.20	0.75	1.02	1.10	0.11	0.37	21.25	100.10	2.31
Q14	48.2	8.9	2.50	21.50	0.30	1.24	1.30	0.16	0.50	15.70	100.30	2.34
Mean (n=6)	39±5	8±0.6	5.3±1.6	23±1.8	1.3±1.7	1.2±0.4	1.7±0.4	0.19±0.1	0.51±0.1	19.1±2.3	99.9±0.4	2.2±0.2

Continued

Table 2 Major Elements Concentrations (wt%) and Elemental Ratios of Clastic Rocks of Ariyalur Group[a]—cont'd

	SiO_2	Al_2O_3	Fe_2O_3	CaO	MgO	Na_2O	K_2O	MnO	TiO_2	LOI	Total	CaO*
Claystone												
F8	64.5	10.5	4.60	8.80	1.20	0.58	2.50	0.06	0.40	6.40	99.54	0.40
F15	65.0	9.6	7.80	5.40	0.60	0.40	2.40	0.03	0.32	7.85	99.40	0.55
Q13	52.3	13.4	13.00	6.70	3.10	0.48	2.27	0.05	0.41	7.60	99.31	1.25
Q16	56.5	11.7	10.40	6.10	0.50	0.56	2.60	0.04	0.30	9.90	98.60	0.72
Mean (n=4)	60±6	11±2	9.0±5.7	6.8±1.5	1.4±1.2	0.5±0.1	2.4±0.14	0.05±0.01	0.36±0.06	7.9±1.5	99.2±0.4	0.7±0.4
Sillakkudi Sandstone												
E5A	42.5	6.9	6.40	22.00	0.25	1.10	2.15	0.05	0.40	18.10	99.85	2.18
P5	35.8	6.1	4.75	26.70	1.10	1.26	2.63	0.08	0.45	21.00	99.87	1.49
P7	53.4	7.4	4.50	18.45	0.50	0.70	1.50	0.03	0.27	12.92	99.67	2.42
Q2	44.3	11.0	6.00	17.50	0.94	0.74	3.60	0.06	0.54	14.00	99.68	2.35
Q7	39.3	5.3	6.20	23.70	2.00	0.60	1.74	0.04	0.51	20.40	99.79	1.71
T24	38.6	6.9	6.80	23.30	1.60	1.30	3.50	0.08	0.44	17.30	99.82	1.57
T34	40.4	6.5	7.90	23.80	0.60	0.62	1.40	0.07	0.36	18.05	99.70	1.95
T35	48.5	8.2	8.30	16.40	2.30	0.90	4.47	0.04	0.20	10.41	99.72	0.80
T42	58.4	11.5	6.00	12.30	0.10	1.17	1.18	0.05	0.23	8.40	99.33	2.92
T43	56.3	8.6	3.20	16.35	0.75	0.80	1.45	0.03	0.27	11.90	99.65	2.09
T44	60.2	6.4	4.75	14.00	1.72	1.40	1.80	0.04	0.42	9.40	100.13	1.30
T46	59.1	4.9	2.50	16.60	1.80	0.58	1.49	0.04	0.20	12.70	99.91	1.38
Mean (n=12)	48±9	7.4±2.2	5.7±1.7	19±4.5	1.1±0.8	1±0.3	2.2±1.2	0.05±0.03	0.4±0.1	15±5	99.9±0.1	1.9±0.6
Claystone												
E5B	58.0	11.8	5.60	11.20	2.10	0.50	2.80	0.03	0.40	7.35	99.78	0.72
P6	72.5	12.6	4.40	2.20	1.00	1.21	2.75	0.03	0.36	2.20	99.25	0.71
Q3	71.2	11.7	6.25	1.90	2.25	0.71	1.40	0.06	0.31	3.60	99.38	1.23
Mean (n=3)	66±7	12±0.4	5.5±1.1	5±6	1.8±0.6	0.8±0.5	2.3±0.6	0.04±0.02	0.4±0.1	5±2	98.9±0.5	1.2±0.8

	CIA	PIA	SiO_2/Al_2O_3	K_2O/Na_2O	K_2O/Al_2O_3	Al_2O_3/TiO_2
Upper Kallamedu Sandstone						
KT1	80.65	84.36	5.91	1.51	0.06	19.06
KT5	81.78	86.58	4.90	2.90	0.08	16.46
KT9	73.68	76.31	9.30	0.91	0.06	3.80
KT13	79.27	81.97	5.84	2.38	0.05	5.21
Mean (n=4)	79±4	82±4	6.5±2	1.9±0.9	0.06±0.01	11±8
Shale						
Kal8	81.93	84.42	3.65	0.72	0.04	21.61
Kal9	84.22	88.50	3.83	2.29	0.06	27.02
Kal11	86.03	89.40	3.54	1.93	0.04	22.89
Kal12	82.31	86.06	4.10	1.41	0.06	22.38
Kal14	80.57	84.60	4.06	1.76	0.07	25.00
Mean (n=5)	83±2	87±2	3.8±0.2	1.6±1.6	0.05±0.01	24±2
Lower Kallamedu Sandstone						
F22	52.24	53.24	5.99	4.52	0.32	18.1
F25	52.89	54.05	7.92	4.29	0.31	33.0
R4	52.36	53.55	7.97	2.76	0.30	29.1
R6	53.78	55.09	7.16	1.83	0.22	25.9
R8	49.89	49.84	5.69	3.36	0.27	59.2
R9	50.84	51.29	7.26	2.79	0.34	33.3
T67	48.36	47.49	5.11	2.90	0.33	16.2
T68	52.68	54.27	4.49	2.28	0.31	22.4
Mean (n=8)	52±2	52±3	6.5±	3.1±	0.3±	30±
Siltstone						
F24	57.30	61.77	4.65	3.98	0.30	52.2
F26	58.45	63.11	4.66	4.00	0.28	36.4
R5	59.54	64.18	4.09	2.12	0.25	44.1
R7	57.30	61.77	4.68	2.70	0.30	33.0
Mean (n=4)	58±1	62±1	4.5±	3.2±	0.28±	41±

Continued

Table 2 Major Elements Concentrations (wt%) and Elemental Ratios of Clastic Rocks of Ariyalur Group[a]—cont'd

	CIA	PIA	SiO_2/Al_2O_3	K_2O/Na_2O	K_2O/Al_2O_3	Al_2O_3/TiO_2
Shale						
Kal1	69.75	79.67	3.58	4.93	0.22	17.75
Kal2	71.12	81.98	3.62	11.48	0.22	17.40
Kal3	72.00	83.31	3.59	11.48	0.22	18.25
Kal5	69.34	84.26	4.12	3.59	0.20	21.13
Mean (n=4)	71±1	82±2	3.7±0.3	8±4	0.22±0	19±2
Ottakkovil Sandstone						
F17	50.56	50.88	7.21	2.10	0.34	14.5
F18	51.41	51.93	4.01	2.40	0.25	33.3
Q20	48.35	47.23	8.44	1.84	0.39	14.3
R1	50.94	51.27	3.76	1.24	0.23	29.7
R2	49.15	48.85	3.71	1.36	0.25	42.5
Mean (n=5)	50±1	50±2	5.4±	1.8±	0.29±	27±10
Claystone						
F19	67.51	76.96	4.81	3.75	0.25	24.2
R3	68.38	72.56	5.54	1.30	0.12	24.7
Mean (n=2)	68±0.6	75±3	5.2±	2.5±	0.19±	24±
Kallankurichchi Sandstone						
F3	52.30	53.19	4.58	1.90	0.23	15.1
F7	48.17	47.48	4.30	2.15	0.27	19.8
F9	51.10	51.36	4.13	0.80	0.19	14.2
Q10	53.22	54.23	4.64	2.44	0.22	14.2
Q12	52.42	52.88	5.14	1.08	0.15	20.0
Q14	53.94	54.69	5.42	1.05	0.15	17.8
Mean (n=6)	52±2	52±3	4.7±	1.6±	0.2±	17±

Claystone						
F8	70.38	82.55	6.14	4.31	0.24	26.3
F15	69.56	81.15	6.77	6.00	0.25	30.0
Q13	71.32	78.55	3.90	4.73	0.17	32.7
Q16	68.35	78.66	4.83	4.64	0.22	39.0
Mean (n=4)	70±1	80±2	5.5±	4.9±	0.22±	32±
Sillakkudi Sandstone						
E5A	46.96	45.73	6.16	1.95	0.31	17.3
P5	44.93	41.74	6.36	2.09	0.43	13.6
P7	50.68	50.85	7.22	2.14	0.20	27.4
Q2	53.34	55.56	4.03	4.86	0.33	20.4
Q7	46.94	45.58	7.42	2.90	0.33	10.4
T24	44.87	40.21	5.59	2.69	0.51	15.7
T34	50.78	50.95	6.22	2.26	0.22	18.1
T35	49.36	48.23	5.91	4.97	0.55	41.0
T42	56.70	57.70	5.08	1.01	0.10	50.0
T43	56.62	58.10	6.55	1.81	0.17	31.9
T44	51.37	51.89	9.41	1.29	0.28	15.2
T46	47.96	47.12	12.06	2.57	0.30	24.5
Mean (n=12)	50±4	50±6	6.8±	2.6±	0.31±	24±
Claystone						
E5B	69.71	80.05	4.92	5.60	0.24	29.5
P6	67.89	73.69	5.75	2.27	0.22	35.0
Q3	70.05	75.26	6.09	1.97	0.12	37.7
Mean (n=3)	69±1	76±3	5.6±	3.3±	0.19±	34±

[a] *Before data presentation, an attempt was made to round the data to the number of significant digits as suggested by Verma (2005).*
[b] *Data from Madhavaraju and Lee (2010).*
CaO = CaO in silicate phase.*

8.4.2 TRACE ELEMENTS

The trace element concentrations of sandstones and fine-grained sediments of the Ariyalur Group are given in Table 3. Most of the sandstone samples are depleted (Fig. 3(A)) in trace elements (except Ni, Cr, and Ba) relative to average upper continental crust (UCC). Most of the trace elements show slight variations between formations, whereas some trace elements such as Co, Cr, Rb, Y, Hf, and Th are more variable. Rb, Pb, Nb, and U are highly depleted with respect to other elements such as Cr, Ba, and Y (Fig. 3(A)). Fine-grained (claystone and shale) samples show marked enrichment in Co, Ni, Cr, and Ba relative to UCC, whereas other elements including Sr, Rb, Zr, and U are highly—slightly depleted (Fig. 3(B)). Rb shows moderate positive correlation with K_2O ($r=0.65$), suggesting that abundances of alkali feldspar and illitic clay control these elements.

8.4.3 RARE EARTH ELEMENTS

REE concentrations are listed in Table 4. Slight variations in ΣREE content are observed in sandstones of the Sillakkudi (121 ppm), Kallankurichchi (141 ppm), Ottakkovil (142 ppm), and Kallamedu (122 ppm) formations. Two samples of the Sillakkudi sandstone (T44 and T46) are depleted in most elements, including REEs, relative to other sandstones in that formation. The fine-grained samples of the Ariyalur Group exhibit high ΣREE content (116–321 ppm), except for two Kallamedu shale samples that show low ΣREE content (45 and 46 ppm).

Chondrite-normalized REE plots (Fig. 4(A)–(I)) of the sandstones from the Sillakkudi Formation show two types of REE patterns: (1) LREE-enriched and flat HREE pattern with relatively little fractionation ($La_N/Yb_N=6.75$–9.20) and (2) LREE-enriched and HREE-depleted pattern with strong fractionation of REE ($La_N/Yb_N=12.20$–22.43). The sandstones of the Kallankurichchi Formation show LREE-enriched and more or less flat HREE patterns with low to moderate fractionation ($La_N/Yb_N=6.94$–14.97). The Ottakkovil Formation exhibits LREE-enriched and HREE-depleted patterns with marked fractionation ($La_N/Yb_N=16.80$–39.12). The Kallamedu Formation shows two types of REE patterns similar to the Sillakkudi Formation: (1) LREE-enriched and flat HREE pattern with lesser fractionation ($La_N/Yb_N=9.41$) and (2) LREE-enriched and HREE-depleted pattern with greater REE fractionation ($La_N/Yb_N=12.67$–19.20). The sandstones of the Ottakkovil Formation exhibit negative Eu anomalies (0.88), whereas the Sillakkudi (1.18), the Kallankurichchi (1.23), and the Kallamedu (1.25) formations show positive Eu anomalies. Fine-grained samples of the Sillakkudi, Ottakkovil, and Kallamedu formations have LREE-enriched and HREE-depleted patterns with marked REE fractionation (La_N/Yb_N: 26.08–54.45, 20.99, 16.33–17.85, respectively), whereas the claystones of the Kallankurichchi Formation show two types of REE patterns similar to the Sillakkudi and Kallamedu formations: (1) LREE-enriched, flat HREE pattern, and mild fractionation ($La_N/Yb_N=7.54$–10.41) and (2) strongly fractionated LREE-enriched and HREE-depleted patterns ($La_N/Yb_N=19.07$–25.16).

8.5 DISCUSSION

8.5.1 GEOCHEMICAL INDICATORS OF WEATHERING

In sedimentary rocks, weathering intensity may be evaluated by examining the relationship among alkali and alkaline earth elements (Nesbitt and Young, 1982, 1996). The dominant process occurring

Table 3 Trace Elements Concentrations (ppm) in the Clastic Rocks of Ariyalur Group

	Sc	V	Cr	Co	Ni	Rb	Sr	Y	Zr	Nb	Ba	Hf	Pb	Th	U	Cr/Th	Cr/Zr	Zr/Sc	Th/Sc	Th/U	Th/Co
Upper Kallamedu Sandstone[a]																				
KT1	2.11	36	622	4	15	12	186	3	95	4.22	259	3.39	11	2.43	2.02	256	6.50	45.00	1.60	1.20	0.61
KT5	3.38	32	41	9	17	10	205	3	133	1.37	318	3.70	11	6.56	1.31	6.25	0.31	39.35	1.94	5.00	0.73
KT9	5.60	85	92	19	21	9	140	6	155	9.18	266	4.13	12	9.99	1.34	9.21	0.59	27.68	1.78	7.50	0.53
KT13	7.54	99	146	43	28	9	161	9	133	5.41	448	3.76	17	10.64	1.26	13.72	1.08	17.64	1.41	8.44	0.25
Shale[a]																					
Kal8	13.80	119	160	10	49	13	319	3	42	5.31	181	1.73	10	5.32	0.90	30.08	3.81	3.04	0.39	5.91	0.53
Kal9	6.58	71	164	5	25	10	329	3	85	4.15	334	2.72	10	3.69	0.78	44.44	1.93	12.92	0.56	4.73	0.74
Kal11	13.26	92	130	8	43	13	257	3	39	1.45	184	1.37	9	4.90	0.98	26.53	3.33	2.94	0.37	5.00	0.61
Kal12	4.11	38	244	4	17	10	257	3	73	1.06	281	1.94	12	3.18	0.80	76.73	3.34	17.76	0.77	3.98	0.80
Kal14	5.76	36	94	6	35	17	251	4	49	1.40	302	1.21	9	3.45	1.37	27.25	1.92	8.51	0.60	2.52	0.58
Lower Kallamedu Sandstone																					
R4	4.66	44	38	5	23	74	205	11	157	4.71	997	1.22	15	9.04	0.82	4.20	0.24	33.69	1.94	11.02	1.81
R6	4.80	–	41	5	26	81	201	12	160	5.10	987	1.31	15	9.32	0.85	4.40	0.26	33.33	1.94	10.96	1.86
T67	5.31	57	35	4	28	77	271	21	120	6.92	528	5.08	8	4.53	0.69	7.73	0.29	22.60	0.85	6.57	1.13
T68	4.47	16	31	4	20	73	269	21	134	6.13	665	4.05	17	7.62	0.88	4.07	0.23	29.98	1.70	8.66	1.91
Siltstone																					
F26	8.01	35	54	8	37	88	302	22	267	12.38	1316	7.65	9	11.45	1.70	4.72	0.20	33.33	1.43	6.74	1.43
R5	6.20	52	64	3	38	91	382	7	142	4.75	1686	1.30	19	5.47	0.97	11.70	0.45	22.90	0.88	5.64	1.82
Shale[a]																					
Kal1	13.72	169	299	42	103	65	122	10	62	8.20	352	2.02	12	8.93	0.54	33.48	4.82	4.52	0.65	16.54	0.21
Kal2	12.87	184	172	34	113	69	121	9	69	8.66	301	2.51	11	9.95	0.59	17.59	2.49	5.36	0.77	16.19	0.29
Kal3	12.68	130	194	50	102	62	114	7	51	7.50	311	1.79	10	8.31	0.55	23.35	3.80	4.02	0.66	15.11	0.17
Kal5	10.40	112	102	41	94	52	149	8	44	16.32	528	1.55	9	7.39	0.57	13.80	2.32	4.20	0.71	12.96	0.18

Continued

Table 3 Trace Elements Concentrations (ppm) in the Clastic Rocks of Ariyalur Group—cont'd

	Sc	V	Cr	Co	Ni	Rb	Sr	Y	Zr	Nb	Ba	Hf	Pb	Th	U	Cr/Th	Cr/Zr	Zr/Sc	Th/Sc	Th/U	Th/Co
Ottakkovil Sandstone																					
F18	4.51	16	27	3	25	64	406	20	167	7.19	1396	4.93	8	17.90	1.18	1.51	0.16	37.03	3.97	15.17	5.96
R2	6.33	17	38	4	20	55	409	21	155	8.08	888	5.23	8	9.03	1.32	4.21	0.25	24.29	1.43	6.84	2.26
Claystone																					
R3	7.96	78	54	12	22	76	269	17	86	9.77	1383	2.44	23	18.38	1.28	2.94	0.63	10.80	2.31	14.36	1.53
Kallankurichchi Sandstone																					
F3	5.80	–	65	5	35	39	261	30	204	6.08	1109	2.11	10	5.10	1.29	12.75	0.32	35.17	0.88	3.95	1.02
F7	6.29	38	71	15	33	36	289	33	228	5.99	1206	2.22	10	5.12	1.34	13.87	0.31	36.25	0.81	3.82	0.34
Q10	6.18	43	99	4	32	47	239	35	225	5.61	1173	1.93	12	5.27	1.42	18.79	0.44	27.51	0.85	3.71	1.32
Q14	2.75	19	40	2	16	46	313	22	178	1.12	815	0.45	9	4.70	0.28	8.51	–	47.47	1.71	9.64	2.35
Claystone																					
F8	5.48	15	44	13	19	32	178	13	96	6.24	1113	2.34	9	6.33	1.08	6.95	0.46	17.52	1.16	5.86	0.49
F15	5.95	21	52	4	21	36	191	8	50	9.50	1223	0.63	31	8.98	0.72	5.79	1.04	8.40	1.51	12.47	2.25
Q13	9.26	18	143	11	70	57	227	8	62	5.94	637	1.71	11	7.92	1.18	18.06	0.69	6.70	0.86	6.71	0.72
Q16	8.15	35	60	11	24	31	165	18	56	9.14	1156	2.47	19	17.35	1.20	3.46	1.07	6.87	2.13	14.46	1.58
Sillakkudi Sandstone																					
E5A	6.82	27	46	19	35	44	207	77	245	9.09	1312	8.21	7	11.32	1.31	4.06	0.19	35.92	1.66	8.64	0.60
P5	8.62	57	33	8	20	59	317	46	170	5.21	2404	1.66	13	10.62	1.85	3.11	0.19	19.72	1.23	5.74	1.33
Q7	4.39	56	198	10	31	64	192	18	98	4.17	1299	0.66	17	7.01	1.38	28.25	1.00	22.32	1.60	2.18	0.30
T24	6.24	23	38	5	19	54	412	22	198	12.15	1394	6.08	8	15.58	1.55	2.44	0.19	31.73	2.50	10.05	3.12
T34	6.73	35	41	8	30	49	129	30	172	17.58	1278	6.05	9	21.77	5.85	1.88	0.24	25.56	3.23	3.72	2.72
T35	6.73	35	46	10	17	117	261	17	175	8.70	1261	4.86	9	8.16	0.99	5.64	0.26	26.00	1.21	8.24	0.82
Claystone																					
E5B	5.67	13	38	4	19	66	137	13	95	10.04	1297	3.96	30	31.23	1.19	1.22	0.40	16.75	5.51	26.24	7.81
Q3	8.97	94	67	9	27	126	261	19	93	14.76	1047	2.69	18	26.40	2.13	2.54	0.72	10.37	2.94	12.39	2.93

[a] *Data from Madhavaraju and Lee (2010).*

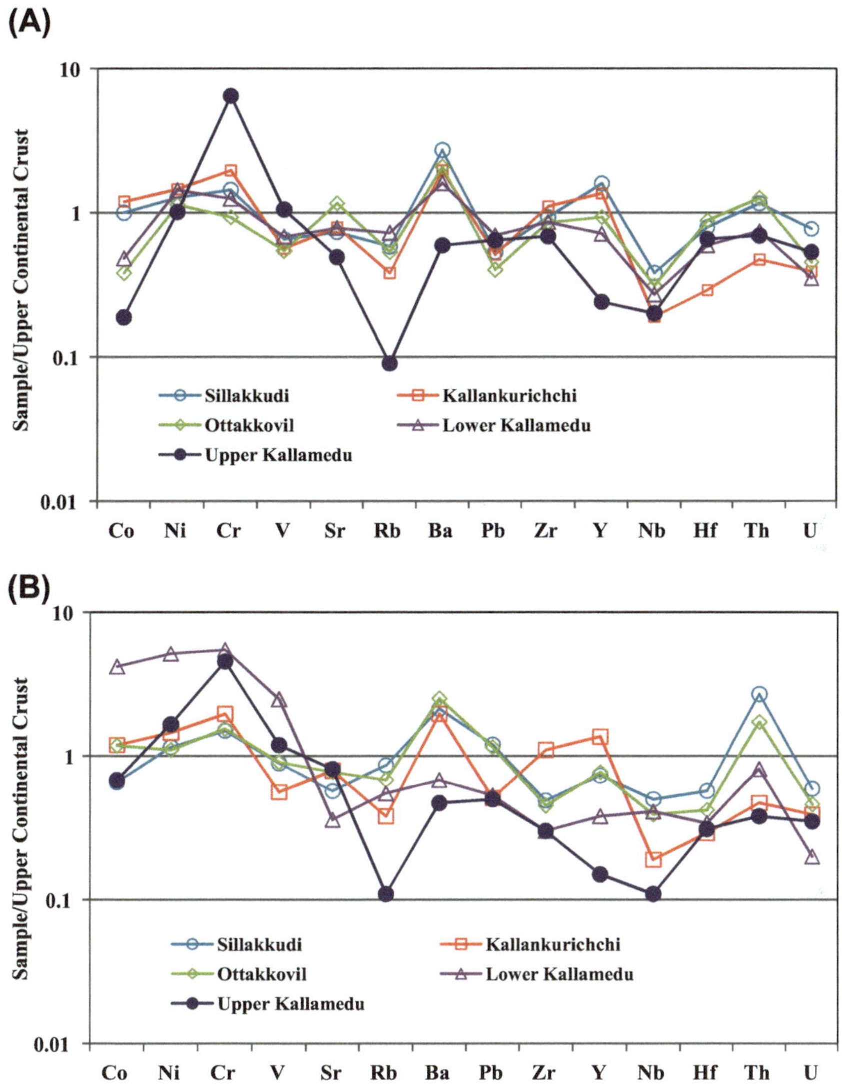

FIGURE 3

(A) The average elemental concentrations of sandstones and siltstones of Ariyalur Group are normalized to UCC (Upper Continental Crust, Taylor and McLennan (1985)), (B) The average elemental concentrations of claystones and shales of the Ariyalur Group are normalized to UCC.

during the chemical weathering of the upper crust is the alteration of feldspar and subsequent formation of clay minerals. The resultant chemical changes may be preserved in sedimentary rocks (Nesbitt and Young, 1982; McLennan et al., 1993; Nath et al., 2000) and can provide useful information regarding paleoweathering of the source rocks. The intensity of chemical alterations may be measured quantitatively using the Chemical Index of Alteration (CIA) proposed by Nesbitt and Young (1982).

Table 4 Rare Earth Elements Concentrations (ppm) and Elemental Ratios for Clastic Rocks of Ariyalur Group

	La	Ce	Pr	Nd	Sm	Eu	Gd	Dy	Er	Yb	Lu	∑REE	∑LREE/∑HREE	La/Sc	La/Co	Eu/Eu[a]	Ce/Ce[a]	$(Gd/Yb)_N$	$(La/Yb)_N$
Upper Kallamedu Sandstone[a]																			
KT1	11.24	19.89	2.37	11.92	1.42	0.55	1.32	0.71	0.40	0.39	0.08	50.29	16.15	5.32	2.60	1.23	0.87	2.75	19.51
KT5	14.32	31.70	3.00	16.20	1.59	0.40	1.40	0.76	0.48	0.53	0.11	70.49	20.37	4.24	1.54	0.82	1.09	2.14	18.23
KT9	22.79	58.57	4.87	28.29	2.53	0.61	2.37	1.36	0.85	0.90	0.16	123.30	20.75	4.07	1.23	0.76	1.26	2.13	17.11
KT13	32.40	189.56	7.49	66.83	4.54	1.29	4.94	2.36	1.23	1.13	0.19	311.96	30.54	4.29	0.76	0.83	2.77	3.54	19.36
Shale[a]																			
Kal8	10.10	18.13	2.18	10.61	1.21	0.32	1.07	0.66	0.43	0.52	0.11	45.34	15.14	0.73	1.03	0.86	0.87	1.67	13.10
Kal9	8.97	17.12	1.95	9.67	1.05	0.31	0.91	0.59	0.43	0.54	0.11	41.62	15.02	1.36	1.72	0.97	0.92		
Kal11	10.30	18.16	2.21	10.71	1.25	0.36	1.15	0.68	0.42	0.50	0.11	45.85	14.91	0.78	1.24	0.90	0.86	1.86	13.90
Kal12	8.41	14.39	1.73	8.31	0.93	0.31	0.92	0.55	0.38	0.46	0.10	36.49	10.59	2.05	2.28	1.02	0.85		
Kal14	10.79	15.52	2.56	11.00	1.56	0.48	1.24	0.80	0.50	0.57	0.12	45.62	12.79	1.87	1.80	1.06	0.67		
Lower Kallamedu Sandstone																			
R4	24.67	44.93	4.71	16.03	2.81	1.63	2.27	1.72	0.90	0.97	0.15	100.79	15.50	5.29	5.33	1.97	1.00	1.90	17.19
R6	28.12	54.10	5.23	18.05	3.92	1.60	2.41	1.82	0.94	0.99	0.18	117.36	17.26	5.86	5.96	1.59	1.04	1.97	19.20
T67	22.88	41.29	4.80	15.80	3.38	1.25	3.26	3.94	1.98	1.22	0.28	100.08	8.25	4.31	5.12	1.15	0.94	2.16	12.67
T68	28.71	53.12	5.88	21.81	7.51	1.12	3.68	3.71	1.95	2.06	0.09	129.64	10.19	6.42	7.25	0.65	0.98	1.45	9.41
Siltstone																			
F26	39.14	73.29	7.88	28.29	11.13	1.20	3.41	2.41	1.36	1.11	0.07	167.08	18.84	4.89	4.99	0.60	1.03	2.49	23.81
R5	26.60	51.30	5.34	26.08	3.29	1.30	2.74	2.67	1.21	0.91	0.15	116.24	13.97	4.29	7.78	1.32	1.03	2.44	19.75
Shale[a]																			
Kal1	37.79	81.07	8.00	42.97	4.78	1.17	4.53	2.55	1.44	1.43	0.24	185.97	17.14	2.75	0.90	0.77	1.05	2.56	17.85
Kal2	33.28	55.70	6.98	33.48	4.27	1.01	3.87	2.42	1.38	1.40	0.24	144.03	14.36	2.59	0.98	0.76	0.86	2.24	16.05
Kal3	29.00	51.65	5.98	29.41	3.39	0.80	3.14	1.82	1.09	1.20	0.21	127.69	16.01	2.29	0.59	0.75	0.88	2.12	16.33
Kal5	24.99	56.70	5.04	28.25	2.96	0.78	2.82	1.68	1.02	1.07	0.19	125.50	17.40	2.40	0.60	0.83	1.13	2.14	15.80

Ottakkovil Sandstone																			
F18	42.00	75.45	8.98	27.24	5.52	1.25	4.67	3.78	1.51	1.69	0.21	172.30	13.42	9.31	13.00	0.75	0.98	2.24	16.80
R2	25.41	48.94	5.66	18.44	3.43	1.06	3.50	3.16	2.08	0.44	0.26	112.38	10.79	4.01	5.83	1.01	0.98	6.46	39.12
Claystone																			
R3	39.13	103.02	8.53	31.53	4.97	1.26	4.11	3.60	1.91	1.26	0.25	199.57	16.82	4.92	3.31	0.85	1.32	2.64	20.99
Kallankurichchi Sandstone																			
F3	42.76	54.28	7.64	30.95	5.43	1.82	6.01	6.89	4.22	4.10	0.42	164.52	6.52	7.37	8.56	0.97		1.19	7.05
F7	41.26	59.43	8.11	31.75	5.21	1.78	6.34	7.09	3.94	4.02	0.38	169.31	6.70	6.56	2.74	0.95	0.78	1.28	6.94
Q10	37.12	60.51	7.14	29.05	5.74	1.95	4.97	5.09	3.15	2.20	0.31	157.23	8.88	4.54	9.28	1.12	0.92	1.83	11.40
Q14	18.62	26.21	3.18	12.31	2.45	1.45	2.40	2.52	1.79	0.84	0.19	71.96	8.11	4.97	7.27	1.87	0.82	2.31	14.97
Claystone																			
F8	38.73	62.50	8.02	30.44	5.26	1.93	4.07	6.17	3.29	3.47	0.37	164.25	8.34	7.07	3.00	1.27	1.26	0.95	7.54
F15	27.75	68.53	6.56	25.00	7.55	0.84	3.24	2.79	1.81	1.80	0.40	146.27	13.49	4.66	1.96	0.52	1.22	1.46	10.42
Q13	26.41	64.03	5.85	21.70	3.91	1.05	2.78	2.07	1.19	0.71	0.15	129.85	17.67	2.85	2.40	0.97	1.24	3.17	25.16
Q16	36.40	91.47	7.94	37.58	6.39	1.22	3.86	2.03	2.12	1.29	0.21	190.51	18.90	4.47	3.24	0.75	1.29	2.43	19.07
Sillakkudi Sandstone																			
E5A	45.17	67.45	9.63	35.91	6.66	1.97	7.29	8.15	4.87	4.17	0.41	191.68	6.62	6.62	2.37	0.86	0.78	1.42	7.32
P5	32.13	62.85	6.91	27.92	5.45	1.74	5.79	5.95	3.74	2.89	0.53	155.90	7.16	3.73	4.21	0.93	1.01	1.62	7.52
Q7	18.46	32.13	3.37	13.40	2.61	1.18	2.56	2.45	1.43	0.97	0.20	78.76	9.19	4.21	1.74	1.39	0.98	2.14	12.86
T24	31.79	63.55	7.01	24.94	4.25	1.05	3.80	3.38	2.58	2.54	0.14	144.83	10.56	5.09	6.69	0.80	1.02	1.22	8.46
T34	30.86	62.97	7.15	25.65	4.24	0.80	3.98	4.89	2.54	3.09	0.38	146.55	8.80	4.59	4.06	0.60	1.02	1.04	6.75
T35	28.34	56.94	6.11	21.99	4.04	1.50	3.30	3.59	1.99	2.08	0.24	130.12	10.48	4.21	2.69	1.26	1.05	3.34	9.20
Claystone																			
E5B	64.54	150.39	15.53	56.45	17.22	1.38	5.74	6.73	1.66	0.80	0.31	320.75	19.96	11.38	15.11	0.42	1.14	5.81	54.45
Q3	57.14	110.02	12.24	45.56	6.32	1.40	5.10	4.00	1.95	1.48	0.28	245.49	18.05	6.37	6.42	0.75	1.05	2.79	26.08

[a]*Data from Madhavaraju and Lee (2010).*

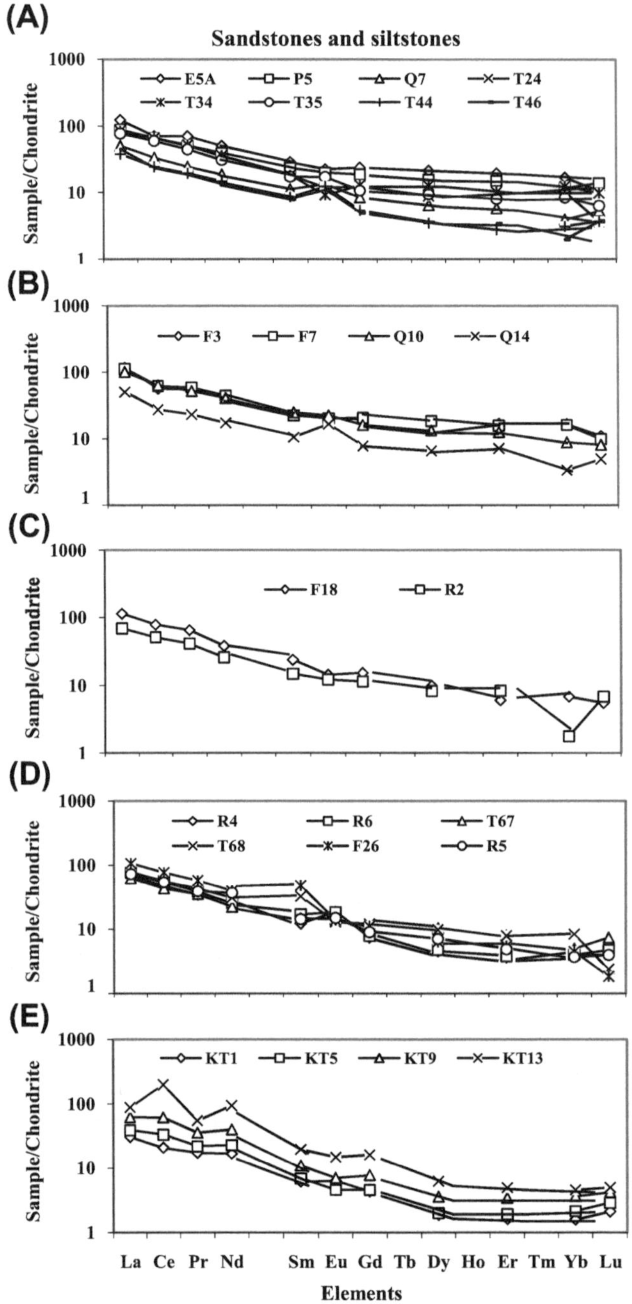

FIGURE 4

(A) REE patterns of sandstones of Sillakkudi Formation, (B) Chondrite-normalized REE patterns of sandstones from the Kallankurichchi Formation, (C) REE patterns of individual sandstones of the Ottakkovil Formation, (D) REE patterns of sandstones and siltstones of Lower Kallamedu Formation, (E) REE patterns of sandstones of Upper Kallamedu Formation, (F) Chondrite-normalized REE patterns of claystones of the Sillakkudi Formation, (G) REE patterns of claystones of the Kallankurichchi Formation, (H) REE patterns of claystones of the Ottakkovil Formation, (I) REE patterns of shales of the Lower Kallamedu Formation, and (J) REE patterns of shales of the Upper Kallamedu Formation.

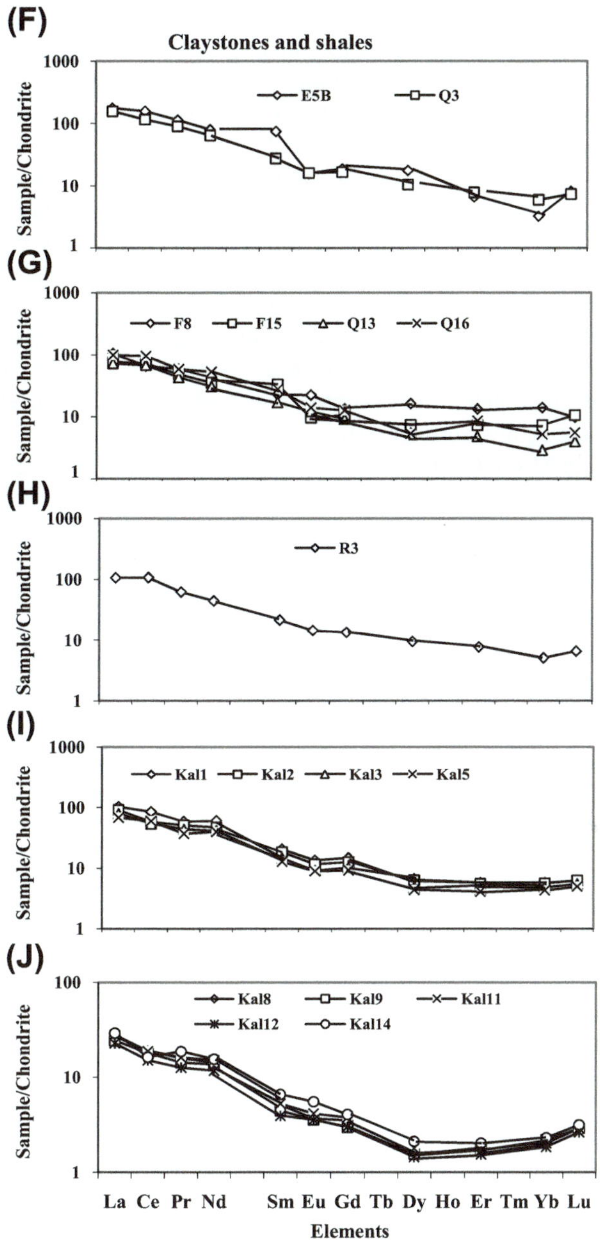

FIGURE 4 CONT'D

Unweathered igneous rocks yield CIA values of 50 or less. CIA values for average shales range from 70 to 75, whereas the residual clays with high kaolinite or gibbsite contents have CIA values of up to 100 (Nesbitt and Young, 1982). The Sillakkudi, Kallankurichchi, Ottakkovil, and lower Kallamedu sandstones (including siltstones) have low to moderate CIA values, whereas the sandstone from the upper Kallamedu Formation shows high CIA values (Table 2). The claystones from the Sillakkudi, Kallankurichchi, and Ottakkovil formations have the least variation in CIA values, whereas the upper Kallamedu shales have greater such variations (Table 2). The average CIA values for sandstones of the Ariyalur Group are as follows: Sillakkudi (58 ± 8, n=12), Kallankurichchi (60 ± 5, n=6), Ottakkovil (54 ± 5, n=5), lower Kallamedu (61 ± 3, n=12, including siltstones), and upper Kallamedu (84 ± 5). The average CIA values for claystones and shale samples are as follows: Sillakkudi (71 ± 5, n=3), Kallankurichchi (72 ± 1, n=4), Ottakkovil (69 ± 2, n=2), lower Kallamedu (75 ± 2, n=4), and upper Kallamedu (84 ± 3, n=5). The average CIA values of the Sillakkudi, Kallankurichchi, Ottakkovil, and lower Kallamedu rocks are similar to the average shale values proposed by Nesbitt and Young (1982), whereas those from upper Kallamedu shales are significantly greater.

The data were plotted in Al_2O_3–CaO^*+Na_2O–K_2O (A–CN–K) space (Fig. 5) to determine the weathering pattern. Average values of tonalitic gneiss (Allen, 1985), granodiorite, and granite (Jayananda et al., 2000) from the Dharwar Craton (source rocks), basalt from the Deccan Traps (Melluso et al., 2004), and PAAS and UCC (Taylor and McLennan, 1985) were also plotted for comparison. The Sillakkudi, Kallankurichchi, Ottakkovil, and lower Kallamedu sandstones plot slightly away from the feldspar join, also do tonalitic gneiss, granodiorite, and granitic rocks, suggesting that the sediments were subjected to a low to moderate degree of chemical weathering. This inference is also supported by the values of the Plagioclase Index of Alteration (PIA; Fedo et al., 1995) of the Sillakkudi, Kallankurichchi, Ottakkovil, and lower Kallamedu sandstones (Table 2). The upper Kallamedu sandstones plot nearer to the A apex, suggesting that the sediments were affected by intense chemical weathering in the source region. In the A–CN–K diagram, the Sillakkudi, Kallankurichchi, and Ottakkovil claystones and shales plot near the average shale (Post Archean Australian Shale (PAAS); CIA: 70), which suggests that the weathering had not proceeded to the stage where alkali and alkaline earth elements were removed from the clay minerals (Taylor and McLennan, 1985).

The distinct contrast in CIA values between the upper Kallamedu and underlying formations suggests variation in source area weathering. The TiO_2 content also increases suddenly in the sandstone sequence of the upper Kallamedu Formation. It is exclusively present in the detrital fraction. The sudden increase might have been influenced by high silicate weathering and detrital input. Vigorous volcanism produced the Deccan Traps during the Late Cretaceous to early Tertiary time (Venkatesan et al., 1993; Baksi, 1994). Volcanism produces enormous amounts of gases and dusts that can be transported globally by strong winds, initiating acid rain worldwide and enhancing weathering (Ebihara and Miura, 1996). The upper Kallamedu Formation was deposited during the later part of Late Maastrichtian. The source area of the Cauvery Basin may thus have been affected by acid rain, which initiated more chemical weathering. Hence, the observed high CIA values and intense chemical weathering signatures in the upper Kallamedu Formation may have been influenced by the Deccan volcanism.

8.5.2 PROVENANCE

The geochemical characteristics of clastic rocks have been used to decipher the provenance of their source regions (Taylor and McLennan, 1985). The SiO_2/Al_2O_3 ratio is sensitive to sediment recycling

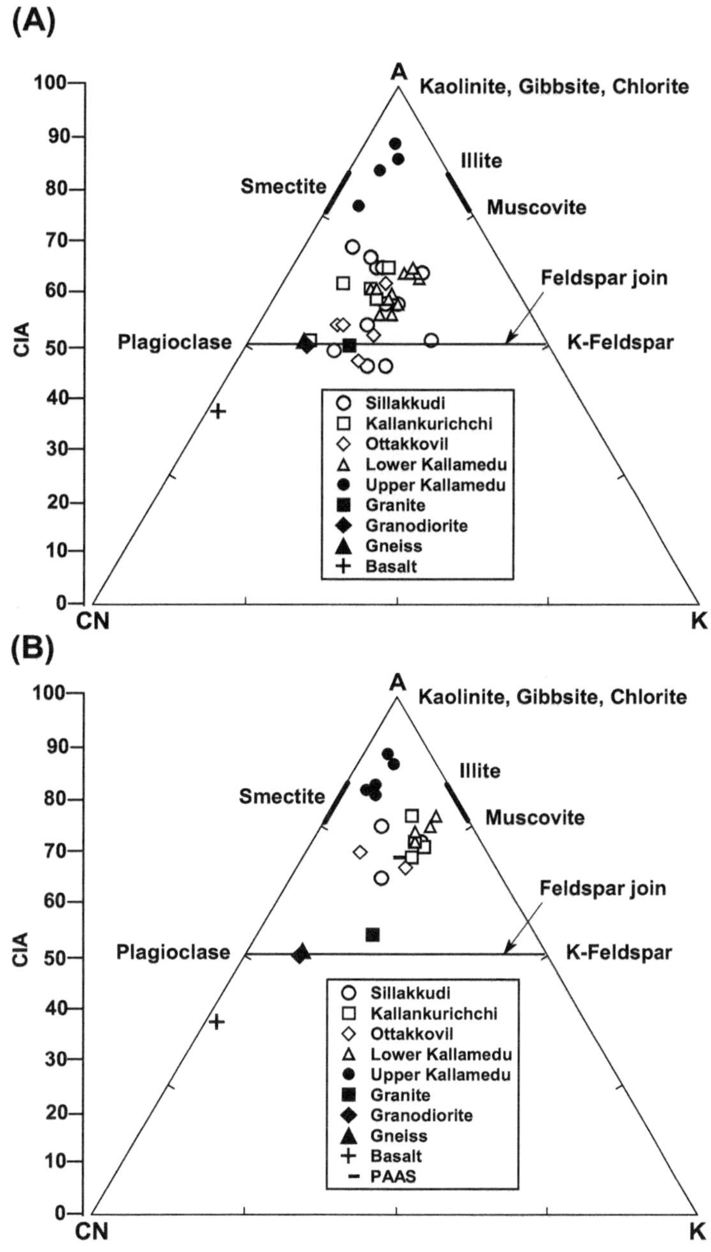

FIGURE 5

(A) Major element composition of sandstones and siltstones of the Ariyalur Group plotted in Al_2O_3–(CaO^*+Na_2O)–K_2O (A–CN–K) diagram (after Nesbitt and Young (1982)), (B) Major element composition of claystone and shales of the Ariyalur Group plotted in A–CN–K diagram.

and the weathering process and can be used as an indicator of sediment maturity (Roser and Korsch, 1986). The average SiO_2/Al_2O_3 ratios of the unaltered igneous rocks range from ~3.0 (basic) to ~5.0 (acidic), while values >5.0 in sediments are an indication of progressive maturity (Roser et al., 1996).

The Sillakkudi, Ottakkovil, lower Kallamedu, and upper Kallamedu sandstones have higher SiO_2/Al_2O_3 ratios (6.8±2.3, 5.6±2.5, 6.4±1.2, 6.5±2, respectively) than the Kallankurichchi equivalents (4.7±0.6). The fine-grained samples from the Sillakkudi, Kallankurichchi, and Ottakkovil formations also have high SiO_2/Al_2O_3 values (5.6±0.6, 5.6±1.4, 6.5±1.4, respectively), but those in the upper Kallamedu Formation have low values (4±0.5). The high average SiO_2/Al_2O_3 ratios in all except the shale samples from the Kallamedu Formation suggest that most of the sediments were derived from acidic source rocks. The upper Kallamedu shales have low average SiO_2/Al_2O_3 ratios, which indicate that significant amounts of basic sediments were presumably contributed by the volcanic source rocks. Attention is drawn to the low TiO_2 values of all the samples except the upper part of the Kallamedu Formation. Significant mobilization of Ti is common during the weathering of ferromagnesian minerals (Gardner, 1980). Deccan volcanism is known to have occurred in the Indian subcontinent during the Late Cretaceous (McLean, 1985; Subba Rao, 1999). Basalts from the Deccan Traps volcanic province contain 2.15% TiO_2 on average (Mahoney et al., 2000). Based on these precepts, the enrichment of TiO_2 in the clastic rocks of the upper Kallamedu Formation could be interpreted as the result of influx from either Deccan volcanism or weathering products of basic volcanic rocks produced by Deccan volcanism.

Clastic rocks of the Ariyalur Group show large variations in Cr and Ni contents and Cr/Th ratios (Tables 3 and 4). The sandstones, siltstones, and claystones from the Sillakkudi, Kallankurichchi, Ottakkovil, and lower Kallamedu formations have low Cr/Th ratios (<19; average 7±5.7, n=27) except one sample (T44). In contrast, sandstones and shales from upper Kallamedu Formation have high Cr/Th ratios (6.25–256, 43±75, n=10). The average of Cr/Th ratios (excluding upper Kallamedu Formation) is 7.0, which is higher than the average value of UCC (Cr/Th=3.5) and similar to PAAS value (Cr/Th=7.53)—both these values are from Taylor and McLennan (1985). The sediments derived from silicic sources have Cr/Th ratios ranging from 4 to 15, while those derived from basic sources have Cr/Th ratios of 25–100. In general, the sandstones, siltstones, and claystones of the Ariyalur Group (except the upper Kallamedu Formation) show low Cr/Th ratios, which indicate the silicic nature of the source rocks. The sandstones and shales from the upper Kallamedu Formation have high Cr (average 196 ppm) and Ni (average 58 ppm) contents and high Cr/Th ratios (average 43), suggesting a source dominated by mafic rocks. Almost all the rock types in the Dharwar Craton of southern India show significant enrichment in Cr and Ni (Radhakrishna and Naqvi, 1986). High concentrations of Cr and Ni have also been measured in the Archean sedimentary rocks (Radhakrishna and Naqvi, 1986) of the Dharwar Craton (the source rocks of the Cauvery Basin) of southern India. High concentrations of Ni (50–130 ppm) and Cr (112–225 ppm) have been reported from the flood plain sediments of the Cauvery River (Singh and Rajamani, 2001). If the source rocks in the Dharwar Craton were contributing Cr and Ni to the latest Late Maastrichtian sedimentary rocks, then those elements should be typically enriched in the Campanian, Maastrichtian, and still older sediments of the Cauvery Basin. However, the Campanian and Maastrichtian sedimentary rocks in the Cauvery Basin show low abundances of both Cr and Ni. Elevated concentrations are observed only in the upper Kallamedu Formation (later part of the Late Maastrichtian).

Although volcanic activity in the Deccan region began during the Early Cretaceous, extensive volcanic eruptions occurred around 65 Ma (McLean, 1985). At the end of the Cretaceous, India experienced an intensive volcanic activity that resulted in the formation of the Deccan Traps volcanic province (Subba Rao, 1999). This intense volcanism had produced topographic and atmospheric modifications that resulted in significant landscape and environmental changes, manifested as drainage adjustments, changes in sediment supply, and changes in distribution of sediment and biotas (Tandon, 2002). The changes induced by the intense volcanism should be preserved in the stratigraphic records of sedimentary sequences contemporary to volcanism (White, 2000). Our observations thus indicate that the higher concentrations of Cr and Ni and high Cr/Th ratios are due to the incorporation of basaltic materials produced by Deccan Traps volcanism. This conclusion is also supported by Th/Sc, Th/Co, and La/Co ratios in the clastic rocks of the upper Kallamedu Formation (Tables 3 and 4).

All the four formations of the Ariyalur Group show similar REE pattern, i.e., LREE-enriched and HREE-depleted patterns with variable Eu anomalies (Fig. 4) except the shales and sandstones of the upper Kallamedu Formation. Post-Archean rocks are characterized by negative europium anomalies (Taylor and McLennan, 1985) due to their derivation from negative anomaly-bearing, upper continental potash-rich granitic rocks. Positive europium anomalies are rare in post-Archean sedimentary rocks, mainly because of the more evolved nature of the crust (Taylor and McLennan, 1985). The sedimentary barite (Eu^{2+} substitutes for Ba^{2+}), some iron formations, and sulfides (via interaction of hydrothermal sulfides with felsic volcanic rocks) generally exhibit positive europium anomalies and are considered to be exceptions (Cullers and Graf, 1984). In the present study, no correlation is observed between Eu/Eu* and Ba content, nor any evidence of hydrothermal activity in the studied clastic rocks. Positive Eu anomalies have been reported from carbonate-rich river sediments, where Eu^{2+} replaces Sr^{2+} in carbonate minerals (Leleyter et al., 1999), but no correlation exists between the CaO and the Eu/Eu* ($r=0.08$). Hence, the positive Eu anomalies were not produced by syn-sedimentary or postsedimentary processes, but reflect the REE geochemistry of the source rock from which these clastic rocks were derived.

Positive Eu anomalies are generally found in the Precambrian rocks (tonalite-trondhjemite-gneiss (TTG), granodiorite, and quartz diorite). TTG are mainly confined to the Archean and their positive anomalies are due mainly to hornblende-melt equilibria rather than to feldspar accumulation (Cullers and Graf, 1984). The geochemical data of the Ariyalur Group were compared with the tonalitic gneiss, granodiorite, and granitic rocks of Dharwar Craton of southern India to determine the source rock characteristics. The Dharwar Craton consists of two parts: the older western Dharwar Craton (3.3–2.7 Ga) and the younger eastern Dharwar Craton (3.0–2.5 Ga) (Chadwick et al., 2000). The western Dharwar Craton comprises TTG peninsular gneisses and volcano-sedimentary greenstone belts, whereas the eastern Dharwar Craton consists of Upper Archean granitic rocks with minor TTG and thin narrow elongated greenstone belts (Jayananda et al., 2000).

The relative contribution of tonalitic gneiss, granodiorite, and granitic rocks for the Ariyalur Group was modeled using the REE data. The REE abundances of the tonalitic gneiss (after Allen, 1985) and REE abundances of granodiorite and granite (after Jayananda et al., 2000) were used for modeling. The mixing parameters and the results are given in Table 5. The modeled REE pattern (Fig. 6) is nearly identical to the average Sillakkudi, Kallankurichchi, Ottakkovil, and lower Kallamedu rocks. The REE modeling revealed that the average Ariyalur Group rocks (except the upper Kallamedu Formation) could be represented by a mixture of 40% tonalitic gneiss, 40% granite, and 20% granodiorite.

The upper Kallamedu rocks show LREE-enriched and HREE-depleted patterns with or without Eu and Ce anomalies. The upper Kallamedu shales (Kal1, Kal2, Kal3, and Kal5) show high Fe and Mg contents. The smectite in the Lameta Formation (Late Maastrichtian) also exhibits high Fe and Mg content (Salil et al., 1997). The upper Kallamedu shales were deposited predominantly in fluvial

Table 5 Results from Mixing Calculations of End Members Compared with the Average Values of Clastic Rocks of the Sillakkudi, Kallankurichchi, Ottakkovil, and Lower Kallamedu Formations

	Average Values								Mixing End Members						Mixing Results 40:40:20	
	Sillakkudi		Kallankurichchi		Ottakkovil		Lower Kallamedu		Tonalite		Granite		Granodiorite			
Elements	ppm	N*	ppm	N*	ppm	N*	ppm	N*	ppm	N*	ppm	N*	ppm	N*	ppm	N*
La	38.55	105.04	33.63	91.63	35.51	96.76	29.52	80.44	30.00	81.74	70.22	191.34	39.12	106.60	47.48	129.36
Ce	75.79	79.20	60.87	63.61	75.80	79.21	56.32	58.85	56.00	58.52	131.7 5	137.67	66.00	68.97	88.32	92.29
Pr	8.49	61.97	6.81	49.71	7.72	56.35	5.98	43.65	–	–	13.51	98.62	6.24	45.55	–	–
Nd	31.48	44.28	27.35	38.47	25.74	36.20	26.02	36.60	22.00	30.94	46.43	65.30	19.29	27.13	31.21	43.90
Sm	6.35	27.49	5.24	22.68	4.64	20.09	4.74	20.52	3.40	14.72	7.99	34.59	2.54	11.00	5.07	21.95
Eu	1.38	15.86	1.51	17.36	1.19	13.68	1.19	13.68	1.00	11.49	1.09	12.39	0.81	9.26	0.99	11.38
Gd	4.70	15.51	4.21	13.76	4.09	13.37	3.21	10.49	2.98	9.74	6.07	19.82	1.98	6.46	4.01	13.11
Tb	–	–	–	–	–	–	–	–	0.45	7.76	0.86	14.79	0.21	3.54	0.56	9.66
Dy	4.89	12.83	4.33	11.36	3.51	9.21	2.47	6.48	–	–	4.49	11.78	1.01	2.64	–	–
Er	2.60	10.44	2.69	10.80	1.83	7.35	1.33	5.34	–	–	2.33	9.34	0.46	1.85	–	–
Yb	2.25	9.07	2.30	9.27	1.13	4.56	1.24	5.00	1.00	4.03	2.27	9.16	0.42	1.68	1.40	5.65
Lu	0.31	8.14	0.30	7.87	0.24	6.30	0.18	4.72	0.17	4.46	0.33	8.73	0.07	1.84	0.21	5.51
Cr/Th	3.55		7.30		2.63		5.54	3.34	2.75		0.91		2.14		1.97	
La/Sc	5.28		5.09		5.66		5.08	15.83	6.00		17.56		7.82		7.90	
La/Co	3.99		2.51		5.49		5.85	1.98	3.00		14.04		7.80		4.79	
Th/Sc	2.12		1.20		2.41		1.42	1.04	1.60		13.60		1.60		2.09	
$(La/Sm)_N$		3.91		4.04		4.82				5.55		5.53		9.69		6.23
$(La/Yb)_N$		11.69		9.88		21.22				20.28		20.89		63.45		24.68
$(Gd/Yb)_N$		1.68		1.48		2.93				2.42		2.16		3.85		2.46
Eu/Eu*		0.85		0.98		0.84				0.96		0.47		1.10		0.84

N = Chondrite normalized value.*

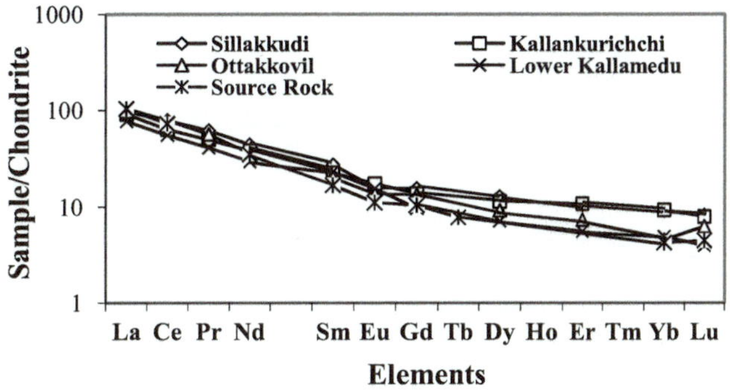

FIGURE 6

Results from mixing calculations for the REEs plotted with average values of the Sillakkudi, Kallankurichchi, Ottakkovil, and Lower Kallamedu formations (see Table 5 for mixing parameters).

environments that were not conducive for the formation of authigenic Mg-rich minerals. The presence of a negative Ce anomaly and high Mg and Fe contents, suggest that those changes were influenced by the Deccan volcanism. The study of ferromagnesian trace elements (Cr, Ni, Co, and Sc), SiO_2/Al_2O_3 ratio, and certain elemental ratios (Th/Co, Cr/Th, Cr/Zr, and La/Co) also indicate that a certain amount of basic materials was contributed by the Deccan volcanism. Thus, basalt has also been included in the mixing calculation to identify the amount of mafic components incorporated in the upper part of the Kallamedu Formation. The REE abundances of basalt taken from Melluso et al. (2004) were used for modeling, and the mixing calculations are given in Table 6. The results show that the average upper Kallamedu rocks represent a mix of sediments derived from a provenance consisting of 60% tonalitic gneiss and 40% basalt (Fig. 7).

8.6 CONCLUSIONS

Documentation of major, trace, and rare earth elemental geochemistry of the Late Cretaceous rocks of the Cauvery Basin, computation of weathering indices, and comparison of the geochemical data with UCC and published data on source rocks have revealed that there exists a difference of patterns of geochemical traits between siliciclastic and carbonate rocks and between fine-grained and other rock types. In addition, all the rocks of the Sillakkudi, Kallankurichchi, Ottakkovil, and lower part of the Kallamedu Formation show relatively uniform geochemical characteristics, however, the upper part of the Kallamedu Formation differ from other formations and show distinctive geochemical patterns. It is perceived by various geochemical compositional traits, elemental and computed ratios, and REE modeling that, the Deccan volcanism, either directly or indirectly by weathered products of volcanic rocks had influenced weathering pattern in the source area and type and quantum of sediment influx into the depositional environments during the deposition of the upper Kallamedu Formation. It is also revealed that the source area experienced less or no significant chemical weathering during Campanian-Maastrichtian, except during the late Maastrichtian, at which time, the rocks of the upper Kallamedu Formation were deposited while the source area experienced significant and intense chemical weathering.

Table 6 Results from Mixing Calculations of End Members Compared with the Average Values of Clastic Rocks of the Upper Kallamedu Formation

	Average Values[a]		Mixing End Members				Mixing Results 60:40	
	Upper Kallamedu		Tonalite		Basalt			
Elements	ppm	N*	ppm	N*	ppm	N*	ppm	ppm
La	12.12	33.02	30.00	81.74	17.24	46.98	24.90	67.85
Ce	24.19	25.28	56.00	58.52	38.98	40.73	59.19	561.85
Pr	2.61	19.05	–	–	5.22	38.10	–	–
Nd	13.34	18.76	22.00	30.94	22.70	31.93	22.28	31.34
Sm	1.44	6.23	3.40	14.72	5.67	24.55	4.31	18.66
Eu	0.42	4.83	1.00	11.49	1.75	20.11	1.30	14.94
Gd	1.30	4.25	2.98	9.74	5.96	19.48	4.17	13.63
Tb	–	–	0.45	7.76	0.97	16.72	0.66	11.38
Dy	0.76	1.99	–	–	5.82	15.28	–	–
Er	0.49	1.97	–	–	2.96	11.89	–	–
Yb	0.55	2.22	1.00	4.03	2.63	10.60	1.65	6.65
Lu	0.11	2.89	0.17	4.46	0.39	10.24	0.26	6.82
Cr/Th	26.34		2.75		80		15.50	
La/Sc	2.35		6.00		0.54		2.29	
La/Co	0.83		3.00		–		–	
Th/Sc	0.81		1.60		0.10		0.54	
$(La/Sm)_N$		5.21		5.55		1.91		4.60
$(La/Yb)_N$		16.52		20.28		4.43		14.38
$(Gd/Yb)_N$		2.22		2.42		1.84		2.18
Eu/Eu[a]		0.82		0.96		0.92		0.92

[a]*n = 8 (except KT13 that shows abnormal values of rare earth elements).*
N = Chondrite normalized value.*

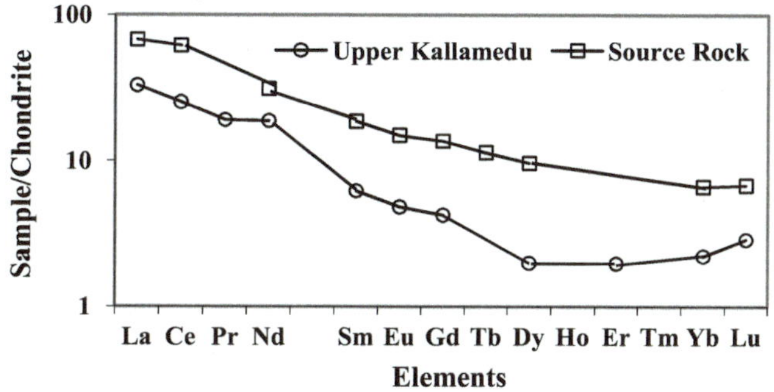

FIGURE 7

Results from mixing calculations for the REEs plotted with average values of Upper Kallamedu Formation (see Table 6 for mixing parameters).

REFERENCES

Absar, N., Raza, M., Roy, M., Naqvi, S.M., Roy, A.K., 2009. Composition and weathering conditions of Paleoproterozoic upper crust of Bundelkhand craton, Central India: records from geochemistry of clastic sediments of 1.9 Ga Gwalior Group. Precambrian Research 168, 313–329.

Allen, P., 1985. The geochemistry of the Amphibolite-Granulite facies transition in Central South India (Unpublished Ph.D. thesis). New Mexico Institute of Mining and Technology, Socorro, New Mexico, USA.

Armstrong-Altrin, J.S., Nagarajan, R., Madhavaraju, J., Rosalez-Hoz, L., Lee, Y.I., Balaram, V., Adriana Cruz-Martínez, A., Avila-Ramírez, G., 2013. Geochemistry of the Jurassic and upper Cretaceous shales from the Molango Region, Hidalgo, eastern Mexico: implications for source-area weathering, provenance, and tectonic setting. Comptes Rendus Geoscience 345, 185–202.

Baksi, A.K., 1994. Geochronological studies on whole-rock basalts, Deccan Traps, India: evaluation of the timing of volcanism relative to the K–T boundary. Earth and Planetary Science Letters 121, 43–56.

Bhatia, M.R., 1983. Plate tectonics and geochemical composition of sandstones. Geology 91, 611–627.

Bhatia, M.R., 1985. Rare earth element geochemistry of Australian Palaeozoic graywackes and mudrocks: provenance and tectonic control. Sedimentary Geology 45, 97–113.

Bhatia, M.R., Crook, K.A.W., 1986. Trace element characteristics of graywackes and tectonic setting discrimination of sedimentary basins. Contributions to Mineralogy and Petrology 92, 181–193.

Blanford, H.F., 1862. On the Cretaceous and other rocks of South Arcot and Trichinopoly Districts. Memoirs of the Geological Survey of India 4, 1–217.

Chadwick, B., Vasudev, V.N., Hegde, G.V., 2000. The Dharwar craton, southern India, interpreted as the result of Late Archaean oblique convergence. Precambrian Research 99, 91–111.

Cingolani, C.A., Manassero, M., Abre, P., 2003. Composition, provenance, and tectonic setting of Ordovician siliciclastic rocks in the San Rafael block: southern extension of the Precordillera crustal fragment, Argentina. Journal of South American Earth Sciences 16, 91–106.

Condie, K.C., Denkate, J., Cullers, R.L., 1995. Behaviour of rare earth elements in a palaeoweathering profile on granodiorite in the front range, Colorado, U.S.A. Geochimica et Cosmochimica Acta 59, 279–294.

Cox, K.G., 1988. Mass extinction: gradual volcanic catastrophes? Nature 333, 802.

Cullers, R.L., 1995. The controls on the major and trace element evolution of shales, siltstones and sandstones of Ordovician to Tertiary age in the Wet Mountain region, Colorado, U.S.A. Chemical Geology 123, 107–131.

Cullers, R.L., 2000. The geochemistry of shales, siltstones and sandstones of Pennsylvanian—Permian age, Colorado, U.S.A. implications for provenance and metamorphic studies. Lithos 51, 181–203.

Cullers, R.L., Graf, J., 1984. Rare earth element in igneous rocks of the continental crust: intermediate and silicic rocks, ore petrogenesis. In: Henderson, P. (Ed.), Rare Earth Geochemistry. Elsevier, pp. 275–316.
Ebihara, M., Miura, T., 1996. Chemical characteristics of the Cretaceous-Tertiary boundary layer at Gubbio, Italy. Geochimica et Cosmochimica Acta 60, 5133–5144.
Fedo, C.M., Eriksson, K., Krogstad, E.J., 1996. Geochemistry of shale from the Archaean (~3.0 Ga) Buhwa Greenstone belt, Zimbabwe: implications for provenance and source area weathering. Geochimica et Cosmochimica Acta 60, 1751–1763.
Fedo, C.M., Nesbitt, H.W., Young, G.M., 1995. Unravelling the effects of potassium metasomatism in sedimentary rocks and paleosols, with implications for paleoweathering conditions and provenance. Geology 23, 921–924.
Feng, R., Kerrich, R., 1990. Geochemistry of fine-grained clastic sediments in the Archaean Abitibi greenstone belt, Canada, implication for provenance and tectonic setting. Geochimica et Cosmochimica Acta 54, 1061–1081.
Gardner, L.R., 1980. Mobilization of Al and Ti during weathering—isovolumetric geochemical evidence. Chemical Geology 30, 151–165.
Govindan, A., Ravindran, C.N., Rangaraju, M.K., 1996. Cretaceous stratigraphy and planktonic foraminiferal zonation of Cauvery Basin, South India. In: Sahni, A., (Ed.), Cretaceous Stratigraphy and Palaeoenvironments. Journal of the Geological Society of India 37, pp. 155–187.
Hart, M.B., Bhaskar, A., Watkinsin, M.P., 2000. Larger foraminifera from the Upper Cretaceous of the Cauvery Basin, S.E. India. Memoir Geological Society of India 46, 159–171.
Herron, M.M., 1988. Geochemical classification of terrigenous sands and shales from core or log data. Journal of Sedimentary Petrology 58, 820–829.
Jafarzadeh, M., Hosseini-Barzi, M., 2008. Petrography and geochemistry of Ahwaz sandstone member of Asmari Formation, Zagros, Iran: implications on provenance and tectonic setting. Revista Mexicana de Ciencias Geológicas 25, 247–260.
Jayananda, M., Moyen, J.F., Martin, H., Peucat, J.J., Auvray, B., Mahabaleswar, B., 2000. Late Archaean (2550–2520 Ma) juvenile magmatism in the Eastern Dharwar craton, southern India: constraints from geochronology, Nd–Sr isotopes and whole rock geochemistry. Precambrian Research 99, 225–254.
Leleyter, L., Probst, J.L., Depetris, 1999. Rare earth element speciation in river sediments. Journal of Conference Abstracts, Symposium on Aqueous Geochemistry, 4, 1.
Madhavaraju, J., 1996. Petrofacies, geochemistry and depositional environments of Ariyalur group of sediments, Tiruchirapalli Cretaceous, Tamil Nadu. (Ph.D thesis (Unpublished), University of Madras). 160 pp.
Madhavaraju, J., Kolosov, I., Buhlak, D., Armstrong-Altrin, J.S., Ramasamy, S., Mohan, S.P., 2004. Carbon and oxygen isotopic signatures in Albian–Danian limestones of Cauvery Basin, Southeastern India. Gondwana Research 7, 519–529.
Madhavaraju, J., Lee, Y.I., 2010. Influence of Deccan volcanism in the sedimentary rocks of Late Maastrichtian–Danian age of Cauvery Basin Southeastern India: constraints from geochemistry. Current Science 98, 528–537.
Madhavaraju, J., Ramasamy, S., 1999a. Rare earth elements in limestones of Kallankurichchi formation of Ariyalur Group, Tiruchirapalli Cretaceous, Tamil Nadu. Journal of the Geological Society of India 54, 291–301.
Madhavaraju, J., Ramasamy, S., 1999b. Microtextures on quartz grains of Campanian—Maastrichtian sediments of Ariyalur Group of Tiruchirapalli Cretaceous, Tamil Nadu—implication on depositional environments. Journal of the Geological Society of India 54, 647–658.
Madhavaraju, J., Ramasamy, S., 1999c. Origin and occurrence of clay minerals in the Ariyalur Group of sedimentary rocks, Cauvery Basin, Tamil Nadu, India. Journal of the Nepal Geological Society 19, 41–48.
Madhavaraju, J., Ramasamy, S., Ruffell, A., Mohan, S.P., 2002. Clay mineralogy of the Late Cretaceous and Early Tertiary Successions of the Cauvery Basin (southeastern India): implications for sediment source and palaeoclimates at the K/T boundary. Caries Research 23, 153–163.
Madhavaraju, J., Sial, A.N., Rakhinath, R., Ramasamy, S., Lee, Y.I., Ramachandran, A., 2014. Carbon, oxygen and strontium isotopic signatures in Maastrichtian-Danian limestones of the Cauvery Basin, South India. Geoscience Journal, http://dx.doi.org/10.1007/s12303-014-0039-1.

Mclean, D.M., 1985. Deccan traps mantle degassing in the terminal Cretaceous marine extinctions. Caries Research 6, 235–259.

Mclennan, S.M., Hemming, S., McDaniel, D.K., Hanson, G.N., 1993. Geochemical approaches to sedimentation, provenance and tectonics. Geological Society of America 284, 21–40 Special Paper.

Melluso, L., Barbieri, M., Beccaluva, L., 2004. Chemical evolution, petrogenesis, and regional chemical correlations of the flood basalt sequence in the central Deccan Traps, India. Proceedings of the Indian Academy of Science 113, 587–603.

Mohaney, J.J., Sheth, H.C., Chandrasekharam, D., Peng, Z.X., 2000. Geochemistry of flood basalts of the Toranmal section, Northern Deccan Traps, India: implications for regional Deccan stratigraphy. Journal of Petrology 41, 1099–1120.

Mongelli, G., Cullers, R.L., Muelheisen, S., 1996. Geochemistry of Late Cretaceous-Oligocene shales form the Varicolori Formation, Southern Apennines Italy: implications for mineralogical, grain-size control and provenance. European Journal of Mineralogy 8, 733–754.

Nagarajan, R., Madhavaraju, J., Nagendra, R., Armstrong-Altrin, J.S., Moutte, J., 2007. Geochemisrty of Neoproterozoic shales of the Rabanpalli Formation Bhima Basin, Northern Karnataka, Southern India: implications for provenance and paleoredox conditions. Revista Mexicana de Ciencias Geológicas 24, 20–30.

Nath, B.N., Kunzendorf, H., Pluger, W.L., 2000. Influence of provenance, weathering, and sedimentary process on the elemental ratios of the fine-grained fraction of the bedload sediments from the Vembanad Lake and the adjoining continental shelf, Southwest Coast of India. Journal of Sedimentary Research 70, 1081–1094.

Nesbitt, H.W., 1979. Mobility and fractionation of rare earth elements during weathering of a granodiorite. Nature 279, 206–210.

Nesbitt, H.W., Young, G.M., 1982. Early Proterozoic climates and plate motions inferred from major element chemistry of lutites. Nature 299, 715–717.

Nesbitt, H.W., Young, G.M., 1996. Petrogenesis of sediments in the absence of chemical weathering: effects of abrasion and sorting on bulk composition and mineralogy. Sedimentology 43, 341–358.

Powell, C.McA., Roots, S.R., Veevers, J.J., 1988. Pre-break up continental extension in East Gondwanaland and the early opening of the eastern Indian Ocean. Tectonophysics 155, 261–283.

Prabhakar, K.N., Zutshi, P.L., 1993. Evolution of southern part of Indian east coast basins. Journal of the Geological Society 41, 215–230.

Radhakrishna, B.P., Naqvi, S.M., 1986. Precambrian continental crust of India and its evolution. Journal of Geology 94, 145–166.

Ramasamy, S., Banerji, R.K., 1991. Geology, petrography and stratigraphy of pre-Ariyalur sequence in Tiruchirapalli District, Tamil Nadu. Journal of the Geological Society 37, 577–594.

Raza, M., Ahmad, A.H.M., Khan, M.S., Khan, F., 2012. Geochemistry and detrital modes of Proterozoic sedimentary rocks, Bayana Basin, north Delhi fold belt: implications for provenance and source-area weathering. International Geology Review 54, 111–129.

Ramkumar, M., Stüben, D., Berner, Z., Schneider, J., 2004. Geochemical and isotopic anomalies preceding K/T boundary in the Cauvery Basin, South India: implications for the end Cretaceous events. Current Science 87, 1738–1747.

Ramkumar, M., Harting, M., Stüben, D., 2005. Barium anomaly preceding K/T boundary: plausible causes and implications on end Cretaceous events of K/T sections in Cauvery Basin (India), Israel, NE-Mexico and Guatemala. International Journal of Earth Sciences 94, 475–489.

Ramkumar, M., Stüben, D., Berner, Z., 2006. Elemental interrelationships and depositional controls of Barremian-Danian strata of the Cauvery Basin, South India: implications on scales of chemostratigraphic modelling. Journal of Geochemistry 21, 341–367.

Ramkumar, M., Stüben, D., Berner, Z., Schneider, J., 2010. $^{87}Sr/^{86}Sr$ anomalies in Late Cretaceous-Early Tertiary strata of the Cauvery Basin, south India: constraints on nature and rate of environmental changes across K-T boundary. Journal of Earth Systems Science 119, 1–17.

Roser, B.P., Cooper, R.A., Nathan, S., Tulloch, A.J., 1996. Reconnaissance sandstone geochemistry, provenance and tectonic setting of the lower Paleozoic terranes of the West Coast and Nelson, New Zealand. New Zealand Journal of Geology and Geophysics 39, 1–16.

Roser, B.P., Korsch, R.J., 1986. Determination of tectonic setting of sandstone-mudstone suites using SiO_2 content and K_2O/Na_2O ratio. Journal of Geology 94, 635–650.

Salil, M.S., Srivastava, J.P., Pattanayak, S.K., 1997. Similarities in the mineralogical and geochemical attributes of detrital clays of Maastrichtian Lameta Beds and weathered Deccan basalt, Central India. Chemical Geology 136, 25–32.

Sastry, M.V.A., Mamgain, V.D., Rao, B.R.J., 1972. Ostracod fauna of the Ariyalur Group (Upper Cretaceous), Tiruchirapalli District, Tamil Nadu. Part I. Lithostratigraphy of the Ariyalur Group. Memoirs of the Geological Survey of India 40, 1–48 Palaeontologica Indica, New Series.

Singh, P., Rajamani, V., 2001. Geochemistry of the floodplain sediments of the Kavery River, Southern India. Journal of Sedementary Research 71, 50–60.

Subba Rao, K.V. (Ed.), 1999. Deccan Volcanic Province, vol. 43. Memoirs of the Geological Survey of India, 947 pp.

Sundaram, R., Rao, P.S., 1986. Lithostratigraphy of Cretaceous and Palaeocene rocks of Tiruchirapalli District, Tamil Nadu, South India. Geological Survey of India 115, 9–23.

Tandon, S.K., 2002. Records of the influence of Deccan volcanism on contemporary sedimentary environments in Central India. Sedimentary Geology 147, 177–192.

Taylor, S.R., McLennan, S.M., 1985. The Continental Crust; Its Composition and Evolution. Blackwell, Oxford. 312 pp.

Tewari, A., Hart, M.B., Watkinson, M.P., 1996. A revised lithostratigraphic classification of the Cretaceous rocks of the Trichinopoly District, Cauvery Basin, South India. In: Pandey, J., Azmi, R.J., Bhandari, A., Dave, A. (Eds.), Contributions to the XVth Indian Colloquium on Micropalaeontology and Stratigraphy, pp. 789–800.

Tripathi, C., Mamgain, V.D., 1987. Record of vitric tuff from the Late Cretaceous/Early Palaeocene strata of Tiruchirapalli District, Tamil Nadu and its significance. Bulletin of the Indian Geologists' Association 20, 9–16.

Venkatesan, T.R., Pande, K., Gopalan, K., 1993. Did Deccan volcanism pre-date the Cretaceous/Tertiary transition. Earth and Planetary Science Letters Supports Open Access Earth and Planetary Science Letters 119, 181–189.

Verma, S.P., 2005. Estadística Básica para el Manejo de Datos Experimentales: Aplicación en la Geoquímica (Geoguimiometría). Universidad Nacional Autónoma de México, Mexico, D.F, 186 pp.

White, N., 2000. Generating Epeirogenic Uplift and Denudation during Continental Break-up. Abstracts, Penrose 2000, Volcanic Rifted Margins, London, 87.

Wronkiewicz, D.J., Condie, K.C., 1989. Geochemistry and provenance of sediments from the Pongola Supergroup, South Africa: evidence for a 3.0-Ga-old continental craton. Geochimica et Cosmochimica Acta 53, 1537–1549.

Yadagiri, P., Ayyasami, K., 1987. A carnosaurian dinosaur from the Kallamedu Formation (Maastrichtian horizon), Tamil Nadu. Geological Survey of India 11, 523–528 Special Publication.

CHAPTER

9 A CHEMOSTRATIGRAPHIC MODEL FOR THE DEVELOPMENT OF PARASEQUENCES AND ITS APPLICATION TO SEQUENCE STRATIGRAPHY AND PALEOCEANOGRAPHY, CRETACEOUS WESTERN INTERIOR BASIN, USA

Philip Morath[1], Leah Calvert[2], Timothy White[3]

[1]EQT Production, Pittsburgh, PA, USA; Department of Geosciences, The Pennsylvania State University, University Park, PA, USA; [2]Barker Lemar Companies, West Des Moines, IA, USA; Department of Earth & Environmental Sciences, The University of Iowa, Iowa City, IA, USA; [3]EMS Earth and Environmental Systems Institute, The Pennsylvania State University, University Park, PA, USA

9.1 INTRODUCTION

The Cretaceous Western Interior Basin (KWIB) has long been the focus of intensive geological research primarily due to the abundant fossil fuel deposits contained within the foreland basin and secondarily due to its stratigraphic record of paleoclimatic and paleoceanographic change and the presence of vast dinosaur-fossil-rich deposits. A Cretaceous Western Interior Seaway (KWIS) once spanned from the Gulf of Alaska through the Northern Great Plains of Canada, most of the Rocky Mountain region, Colorado Plateau, and Great Plains of the United States to the Gulf of Mexico and northern Mexico. This great epicontinental sea was subject to at least nine second-order eustatic sea level cycles (Kauffman and Caldwell, 1993).

The paleosol-bearing Dakota Formation along the eastern margin of the KWIB may provide the best continental stratigraphic record for assessing the paleoclimatic and paleoceanographic effects on terrestrial depositional environments due simply to its existence—no other terrestrial sections exist along the tectonically quiescent eastern margin of the basin. The formation has been the focus of intensive stable carbon and oxygen isotopic analyses of siderite spherules contained therein (e.g., Ludvigson et al., 1998; White et al., 2001, 2005). Those studies and others provided insight into the terrestrial conditions of the Mid-Cretaceous North America as well as information on coeval atmospheric hydrologic conditions and supported subsequent global climate model simulations of that important greenhouse episode (e.g., Poulsen et al., 2007).

Chemostratigraphy. http://dx.doi.org/10.1016/B978-0-12-419968-2.00009-1

The Dakota Formation has been chronostratigraphically resolved at ~0.5–1 Ma accuracy in the early Cenomanian and at ~1–2.5 Ma in the late Albian (Brenner et al., 2000). This resolution is too coarse to facilitate an accurate assessment of the higher frequency climate cycles in the paleosol record. However, a more detailed chronostratigraphy exists for coeval, predominantly marine sections in the axial portion of the Western Interior Basin by combining the results of Obradovich (1993), Kauffman et al. (1993), and Gradstein et al. (2004). Thus, in this study, high-resolution, parasequence-scale correlation of the mid-Cretaceous eastern margin paleosol-bearing sections was made to the well-dated coeval marine strata in the basin center to apply the marine chronostratigraphy to the paleosols and to assess the potential causes of similar stacking patterns on either margin of the basin. To facilitate this goal, various cores and composite outcrop sections were used (Fig. 1). These sites transect an area once dominated by nonmarine deposition in the east to an area of mostly marine deposition in the west. The data collected include field- and core-based lithologic and stratigraphic descriptions, and carbonate content (%$CaCO_3$), total organic carbon content (%TOC), and total sulfur content (%TS). Variations in these parameters are demonstrated and then interpreted to reflect changes in the relative sea level.

A parasequence model for the development of geochemical facies is applied to interpret changes in the development of accommodation space and to develop a relative sea level curve for the KWIS. Stratigraphic evidence is presented for cyclic, 5th order sea level fluctuations. We considered various processes for the development of similar high-frequency stacking patterns in the KWIB and concluded based on the convergence of multiple lines of evidence from an array of facies from disparate locales that one mechanism is viable to describe the observed stacking patterns: sea level fluctuations caused by global variations in paleoclimate.

9.2 GEOLOGICAL SETTING

9.2.1 CRETACEOUS WESTERN INTERIOR BASIN

The KWIB formed during the Sevier Orogeny in response to crustal loading associated with convergence of the North American and Farallon plates that began in the middle Jurassic (DeCelles et al., 1995; DeCelles and Currie, 1996). Initial inundation of the central portion of the North America by the KWIS also began at this time as a southward extension of the northern Boreal Ocean and a northward extension of the circumequatorial Tethyan Sea (proto-Gulf of Mexico). This inundation was due to the presence of the down-warped basin coupled with eustatic sea levels 40–100 m higher than present day (Haq et al., 1987). The elevated eustatic sea levels have been considered to be due to increased rates of oceanic lithosphere production, a lack or paucity of continental-scale glaciation, and thermal expansion of water due to globally warmer temperatures (Hays and Pitman, 1973; Coe, 2003). The two arms of the KWIS were joined during the middle to late Albian (Stelck and Koke, 1987; Stelck, 1991). The extent of the KWIS waxed and waned throughout the remainder of the Cretaceous and finally withdrew from central North America during the Maastrichtian (Kauffman, 1985; Kauffman and Caldwell, 1993). During maximum transgression, the KWIS extended latitudinally from Arctic Canada to the proto-Gulf of Mexico and longitudinally from southwestern Utah-northeastern Nevada to Minnesota-Iowa (Kauffman, 1985). A KWIS connection to the Labrador Sea through central Canada and the Hudson Bay region began as early as the late Albian sea level highstand (White et al., 2000).

On the KWIS relative sea level curve established by Kauffman and Caldwell (1993), one second-order transgressive–regressive cycle and nine third-order transgressive–regressive cycles exist from the

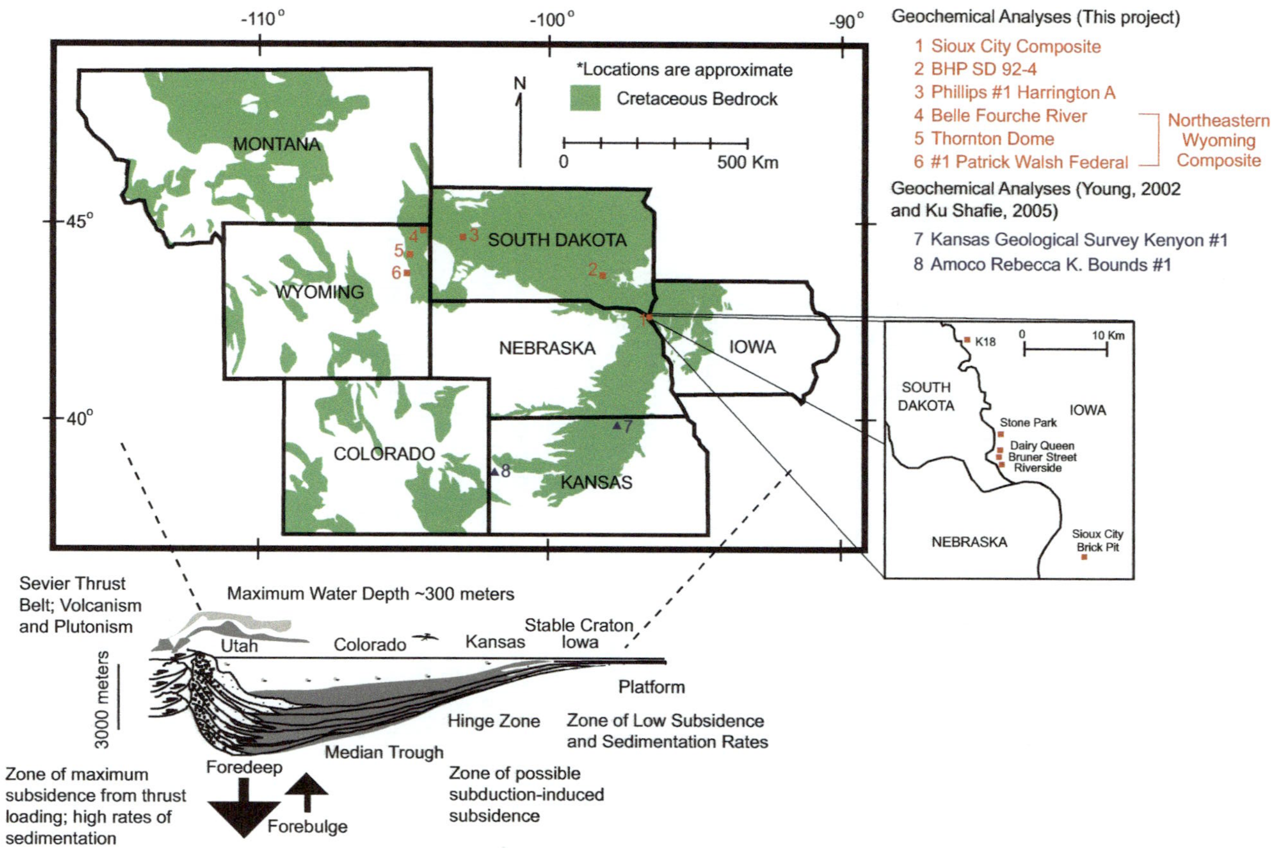

FIGURE 1

Location map and schematic cross-section of the Cretaceous Western Interior Basin. Locations of chemostratigraphic data and lithological descriptions generated here are labeled in red. Data for locations labeled in blue was generated by Young (2002) and Ku Shafie (2005). The distribution of Cretaceous bedrock is shown in light green on the larger map. The outcrops used for the Sioux City Composite section are labeled on the inset to the right.

Modified from White (1999).

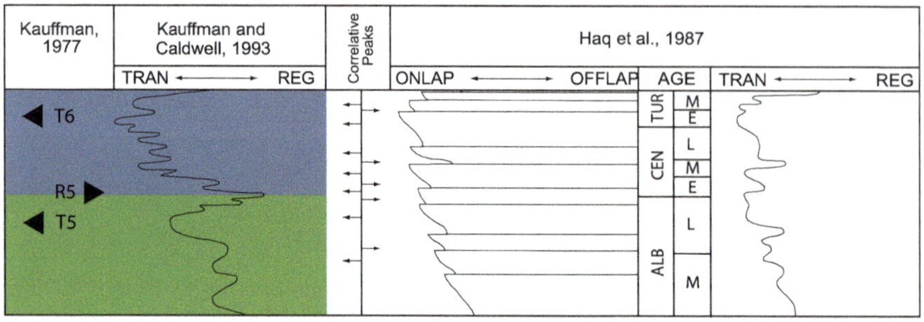

FIGURE 2

Comparison of the Albian–Turonian portion of the relative sea level curve established by Kauffman and Caldwell (1993) for the Cretaceous Western Interior Basin to the onlap–offlap and eustatic sea level curve of Haq et al. (1987) and the generalized system of second-order transgressive–regressive curves of Kauffman (1977). R5 marks the end of the Kiowa–Skull Creek cyclothem (green) and the beginning of the Greenhorn cyclothem (blue). ALB = Albian; CEN = Cenomanian; TUR = Turonian; L = Late; M = Middle; E = Early; TRAN = Transgression; REG = Regression.

Modified from Kauffman and Caldwell (1993).

late Albian to the late Cenomanian (Fig. 2). A major shoreline regression, termed R5 by Kauffman (1977), occurred at the Albian–Cenomanian boundary and marks the end of the Kiowa–Skull Creek second-order sea level cycle (Fig. 2). This lowstand is recognized throughout the KWIB (Kauffman and Caldwell, 1993) as well as worldwide (Immenhauser and Scott, 1999). After this Albian–Cenomanian lowstand, eustatic sea level rose (marking the beginning of the Greenhorn second-order sea level cycle) and reached a maximum highstand during the Turonian, termed T5 by Kauffman (1977) (Fig. 2).

Eustatic sea level changes attributed to increased seafloor spreading rates (Hays and Pitman, 1973) have been suggested as the causes for second-order sea level fluctuations observed within the KWIB (Kauffman and Caldwell, 1993). This hypothesis is supported by the correlation between the relative sea level curve for the KWIS and the eustatic sea level curve of Haq et al. (1987; Kauffman and Caldwell, 1993). However, increased seafloor spreading rates cannot be invoked to explain the higher-frequency relative sea level fluctuations that occurred during the Kiowa–Skull Creek and Greenhorn cyclothems described elsewhere or the parasequence-scale changes in relative sea level documented in this study.

9.2.2 STRATIGRAPHIC UNITS

Mid-Cretaceous strata in the northwestern Iowa, the eastern South Dakota, and the Kansas studied here include (from oldest to youngest) the Woodbury Member of the Dakota Formation, the 'Graneros' Shale, and the Greenhorn Formation. Coeval strata in the western Kansas include (from oldest to youngest): the Purgatoire Formation, the Upper and Lower Dakota Formation, the Graneros Shale, and the Greenhorn Limestone. Coeval strata in the western South Dakota and the northeastern Wyoming include (from oldest to youngest) the Newcastle Sandstone, the Mowry Shale, and the Belle Fourche Shale. For a review and discussion of these stratigraphic units see Cobban and Merewether (1983), Dyman et al. (1994), Hamilton (1994), and Witzke and Ludvigson (1994).

At least three stratigraphic trends are observable in the study area extending from northwestern Iowa to northeastern Wyoming. First, an overall fining-upwards lithological trend exists from the basal upper Albian Dakota and Newcastle sandstones, through the Cenomanian Dakota Formation sandstones/mudstones, Graneros, Mowry, and Belle Fourche shales, and the lower Turonian Greenhorn limestones (Fig. 3). This fining-upward succession formed in response to the expansion and deepening of the seaway during the Greenhorn transgression (Fig. 2). Second, along the transect approximately perpendicular to the eastern shoreline of the KWIB, finer-grained sediments that were deposited in open to shallow marine environments are observed to the west, whereas, mostly coarser-grained sediments that were deposited in fluvial to nearshore environments exist to the east (Fig. 3). Third, a regressive event that occurred at the Albian–Cenomanian boundary led to unconformity development extending to the basin center (Brenner et al., 2000) (Fig. 3).

Ideally, strata equivalent to the Cenomanian–Turonian Bridge Creek Limestone of the Greenhorn Formation, long known to hold a record of orbital variations and Milankovitch cyclicity (e.g., Fischer, 1980; Fischer et al., 1985), would be correlated to more terrestrial units in order to assess the effects of the Milankovitch cyclicity on the terrestrial realm. However, coeval strata have since been eroded (Witzke et al., 1983). Therefore, the Cenomanian Dakota Formation in the northwestern Iowa may provide the best record for determining the effects of the Milankovitch-scale climate cyclicity on terrestrial environments of deposition.

9.2.3 CHRONOSTRATIGRAPHY

A detailed chronostratigraphy has been established for the KWIS by radiometric age dating of bentonites within a marine biostratigraphic framework (Obradovich, 1993; Kauffman et al., 1993). This marine chronostratigraphy is here applied to the paleosol-bearing sections using chemo- and lithostratigraphic techniques outlined in the next section.

Four bentonites from Obradovich (1993) are useful for tying the marine chronostratigraphy to the paleosol-bearing sections: the Arrow Creek Bentonite at the base of the Mowry Shale, the Clay Spur Bentonite at the top of the Mowry Shale, the Thatcher Bentonite found below the Thatcher Limestone, and the Soap Creek Bentonite, biostratigraphically equivalent to the X-bentonite in Kansas because the biostratigraphic zones that these bentonites were deposited within are equivalent in age (see Hattin, 1968; Gradstein et al., 2004). The ages of these bentonites were first published by Obradovich (1993) and later revised by Gradstein et al. (2004) using a more accurate age standard.

In addition, the inoceramid species *I. prefragilis stephensoni* has been observed to occur in the lower "Graneros" Shale and uppermost Dakota Formation in northwestern Iowa (Cobban and Merewether, 1983). The outcrop in which they were found was not used in this study, but Witzke and Ludvigson (1987) used lithostratigraphic techniques to correlate that section to outcrops used in this study. The age of this inoceramid zone, originally defined by Kauffman et al. (1993), was refined from 94.5 to 94.8 Ma using new ages from Gradstein et al. (2004). This is also useful for advancing a chronostratigraphy for the terrestrial eastern margin sections.

9.3 METHODOLOGY

High-resolution parasequence-scale correlation of upper Albian through lower Turonian strata along the transect from the eastern margin to the axial basin of the KWIB were accomplished using geochemical analyses and lithological descriptions of the outcrops and cores listed in Fig. 1. Standard geochemical methods were applied to determine %TS, %TOC, and %$CaCO_3$. All the stratigraphic

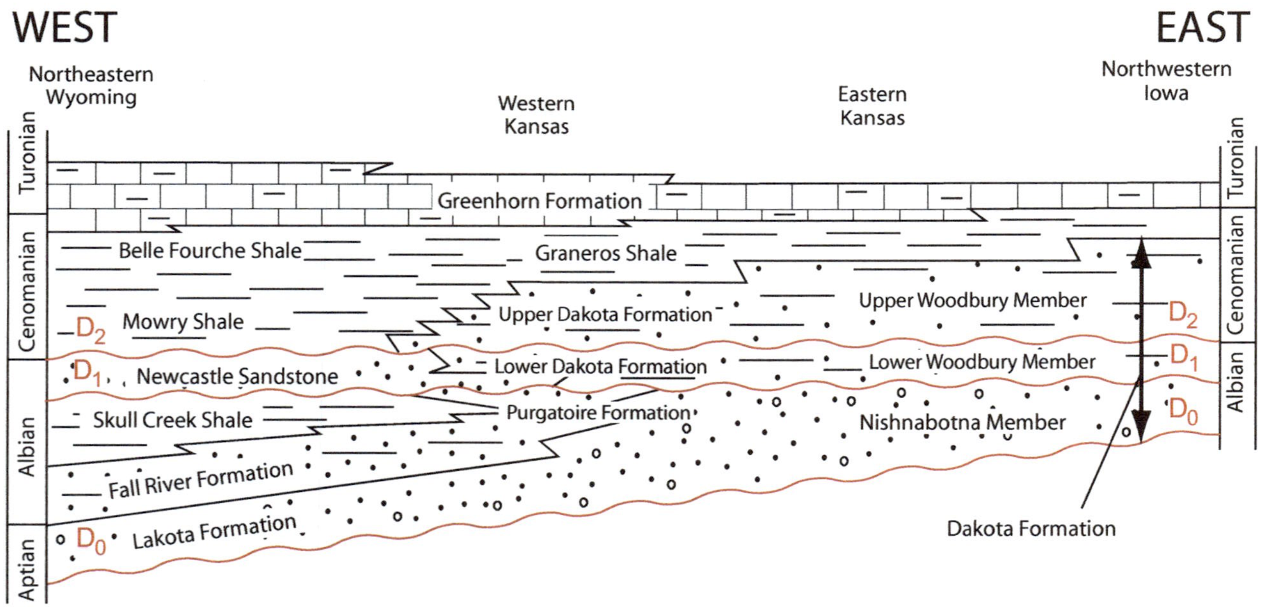

FIGURE 3

Lithostratigraphic relationship of strata deposited in the Western Interior Basin during the mid-Cretaceous along an offshore–onshore transect from northwestern Iowa to northeastern Wyoming. Sequence boundaries are labeled in red; note sequence boundary notation *(From Brenner et al. (2000).)* also in red.

Modified from Witzke et al. (1983).

study sections and cores were sampled at 0.3 m intervals. A detailed description of carbonate coulometry is outlined in Engleman et al. (1985). Total carbon and TS values were obtained for this study using a LECO SC-144DR Carbon and Sulfur Analyzer (LECO Corporation, 2003). A more detailed discussion of the methods can be found in Morath (2007).

9.3.1 APPLICATION OF THE PARASEQUENCE MODEL

White (1999, 2004) developed and applied a parasequence model for geochemical facies based primarily on lithologic and organic petrographic descriptions, $CaCO_3$, TOC and TS content, and Hydrogen Index (HI) values from Rock Eval pyrolysis. The conceptual model is based on the stratigraphic model for prograding parasequences described in Van Wagoner et al. (1990). The methodology developed by White (1999, 2004) presumed that marine flooding surfaces in offshore progradational cycles typically exist as horizons of elevated $CaCO_3$ and TOC and horizons of low TS, while in nearshore progradational cycles marine flooding surfaces exist as horizons of elevated $CaCO_3$, TOC, and TS (described later; Fig. 4). In most instances, these chemostratigraphically defined parasequences represent subtle coarsening-upwards packages that maintain unique geochemical signatures into shoreface facies. Parasequence correlation is built within this model by pattern matching techniques applied to the geochemical profiles. For example, White (2004) used a $CaCO_3$-TOC-TS dataset combined with litho- and biostratigraphic information to refine stratigraphic sequences with important chronostratigraphic implications in Ocean Drilling Program Sites 1168, 1170, 1171, and 1172, even though the geochemical datasets were sampled at a coarse (>1 m) scale. Ku Shafie (2005) used a high-resolution (~0.3 m) $CaCO_3$-TOC-TS-HI dataset to correlate mid-Cretaceous strata across Kansas at a parasequence scale

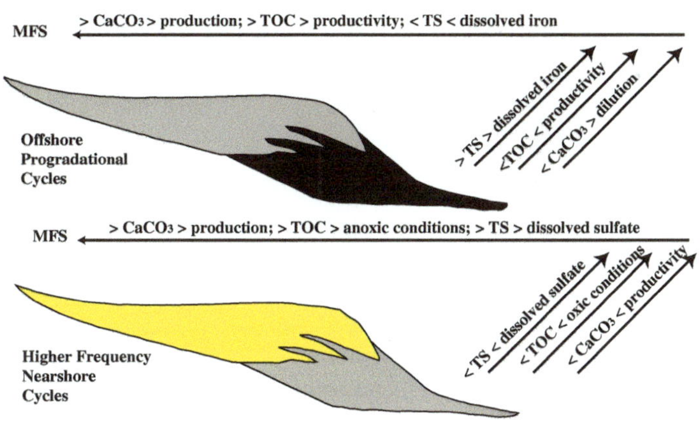

FIGURE 4

Generalized "slug model" for parasequence-scale variations in geochemical facies used in this study *(Modified from White (1999).)*. Geochemical characteristics of marine flooding surfaces are listed above the graphic. Geochemical characteristics of prograding geochemical cycles are listed to the right of the graphic. Note the following: Black = shale; Gray = siltstone; Yellow = sand; $CaCO_3$ = carbonate; TOC = total organic matter; TS = total sulfur; MFS = marine flooding surface.

using the same method for parasequence delineation as those used in this study. In the Amoco Rebecca K. Bounds #1 core (western Kansas), the geochemically defined parasequences of Ku Shafie (2005) correlate to the coarsening and fining-upward cycles delineated by Scott et al. (1998). The correlation of geochemically defined parasequences in Ku Shafie (2005) to lithologically defined cycles in Scott et al. (1998) provides some validity to this parasequence delineation method.

The sedimentary geochemical setting of select portions of the ancient seafloor changed as relative sea level fluctuations occurred within the KWIS. By creating a high-resolution chemostratigraphic profile for upper Albian through lower Turonian sections of the KWIB and applying the "geochemical parasequence model," marine-flooding surfaces can be delineated and correlated from the axial portion to the eastern margin of the KWIB. Some of the minor variation observed in the data can be attributed to calibration error. Prominent peaks in geochemical data were given greater consideration as marine flooding surfaces (especially in the wholly marine sections where parasequence delineation using variations in lithology were more difficult). Coeval patterns in the distribution of marine flooding surfaces were identified between the sections and used to establish the parasequence correlations. Flooding surfaces that were only interpreted in one section or for which no supporting lithologic evidence was observable were eliminated when correlating between sites.

Previous researchers have correlated KWIB strata at varying levels of detail (Dyman et al., 1994; Elder and Kirkland, 1994; Elder et al., 1994). Within the study area, Hamilton (1994) used a lithofacies approach to discern three lithologically defined sequences bounded by unconformities in Albian–Cenomanian strata across the Kansas. Brenner et al. (2000) refined this earlier effort using palynological data and new lithological interpretations to correlate the strata from the northwestern Iowa to the western Kansas at a sequence scale. Young (2002) and Ku Shafie (2005) used litho- and chemostratigraphic methods to establish high-resolution, parasequence-scale correlations along a similar transect. Here, parasequence-scale correlation of Cenomanian strata from the axial basin to the eastern margin of the KWIB, including the use of data from Young (2002) and Ku Shafie (2005), has extended the work of these previous authors.

9.4 DISCUSSION

9.4.1 MARKER HORIZONS

Key litho-, chemo-, and chronostratigraphic surfaces were identified to begin the high-resolution correlation. Key surfaces include (from oldest to youngest) the Albian–Cenomanian (D_2 from Brenner et al., 2000) sequence boundary and overlying Arrow Creek bentonite, the Clay Spur bentonite, the Thatcher Limestone, the X-bentonite, and the Bridge Creek Limestone ravinement surface.

In this study, the Arrow Creek and Clay Spur bentonites were identified by their stratigraphic positions at the base of the Mowry Shale (Condon, 2000) and within the base of the Belle Fourche Shale (Morath, 2007), respectively. When the Arrow Creek bentonite was not present, the Albian–Cenomanian boundary was chosen using palynological data from Scott et al. (1998) (for the Bounds core) and regional trends of depletion in oxygen isotopic signatures from paleosol sphaerosiderites in the Kenyon and Bounds cores (based on the boundary pick described in White et al., 2005). The Thatcher Limestone represents the maximum flooding during the middle Cenomanian (Scott et al., 1998). The X-bentonite was deposited in close proximity to the Graneros–Greenhorn contact in western Kansas (Scott et al., 1998). The Bridge Creek Limestone is a member of the Greenhorn Formation and marks the maximum extent of the Greenhorn marine cycle and the maximum aerial extent of the KWIS (Elder and Kirkland, 1985).

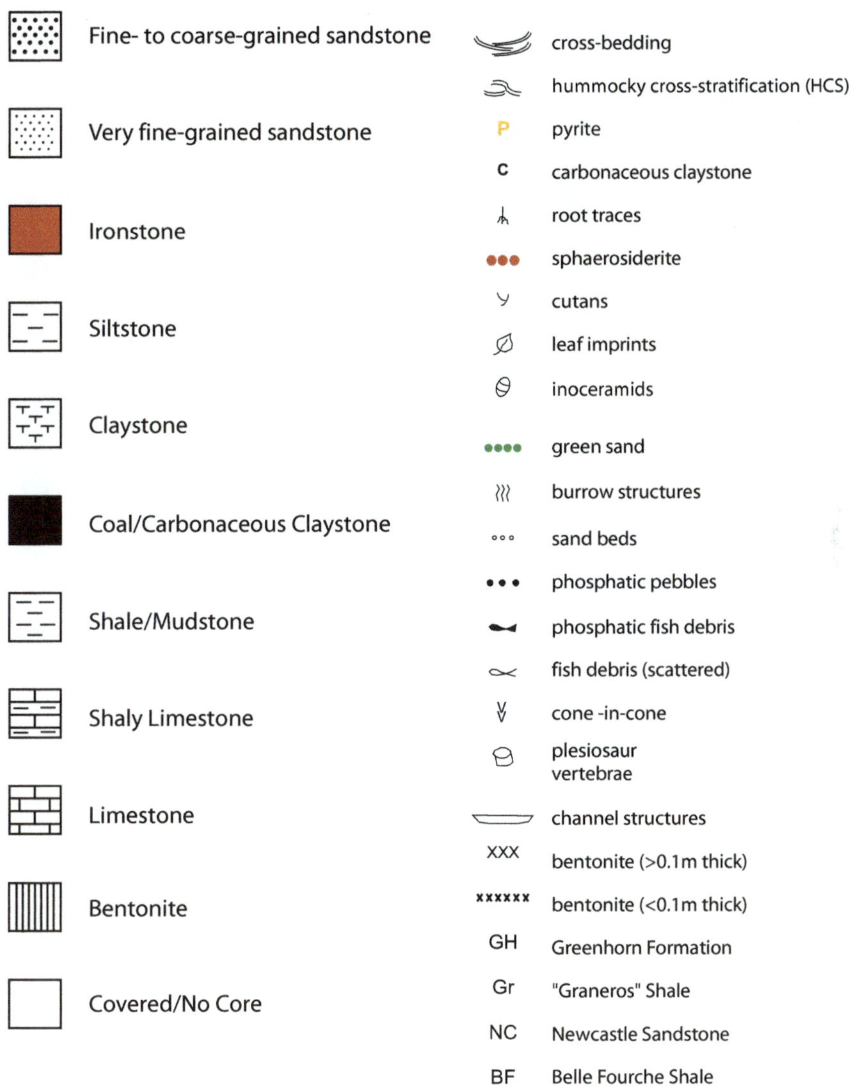

FIGURE 5

Explanation for stratigraphic sections.

These marker horizons were correlated from the axial basin of the KWIS to the eastern margin using geochemical data from previous studies (Young, 2002; Ku Shafie, 2005), coupled with new geochemical, palynological, and lithostratigraphic data presented in Morath (2007). An explanation for the parasequence delineation figures is presented in Fig. 5. The key marker horizons present at each site are shown in Table 1 and in Figs 6–10.

Table 1 Depth/Elevation of Key Marker Horizons Used in This Study. The Parasequence That the Marker Horizon Occurs within is Also Listed. These Key Horizons Were Used to Aid in Correlation of Marine Flooding Surfaces from the Axial Basin to the Eastern Margin of the Cretaceous Western Interior Basin (KWIB)

	A/C Boundary		Clay Spur Bentonite		Thatcher Limestone		X Bentonite		BC Limestone	
Sites	Elev./ Depth (m)	PS#	Elev./ Depth (m)	PS#	Elev./ Depth (m)	PS#	Elev./ Depth (m)	PS#	Elev./ Depth (m)	PS#
Northeastern Wyoming composite	16.9	1	82.6	33	NA		NA		NA	
Amoco Rebecca K Bounds #1 core	388.7	10	254.6	33	331.6	50	319	66	NA	
KGS Kenyon #1 core	108.4	10	61	33	35.7	50	22.9	66	6.7	80
BHP 92-4 core	NA		NA		162.2	50	144.8	66	130.5	80
Sioux City composite	NA		NA		32.1	50	49.2	66	65.4	80

A/C boundary = Albian/Cenomanian boundary; BC limestone = the base of the Bridge Creek Limestone.

9.4.2 PARASEQUENCE CORRELATIONS

The marker horizons discussed above provide a chronostratigraphic framework from which parasequence-scale correlation was built. Initially, geochemical correlation of marine flooding surfaces between the sites was accomplished using obvious and prominent peaks (Figs 6–10). These correlations were then further divided according to the intervals displaying similar geochemical characteristics (i.e., pattern matching). In most instances, the pattern matching was straightforward; however, in some instances, several iterations were necessary. Marine flooding surfaces that could not be correlated between sites, and which did not have a distinct lithological expression, were not considered. At times, a peak in a certain parameter could not be correlated to a peak in the same geochemical parameter at a different site. A good example of this is the Thatcher Limestone flooding surface. In the more distal portions of the KWIB (i.e., Bounds core), the Thatcher Limestone exists as a thin bed of limestone with a large, prominent peak in $CaCO_3$ (Fig. 6). In more proximal sites (i.e., BHP SD 92-4 core and Sioux City Composite section), the Thatcher Limestone flooding event is interpreted to be lithologically and geochemically expressed as green sand containing a large, prominent peak in TS. This interpretation is also based on its relative stratigraphic position within the parasequence-scale framework (Figs 9 and 10).

Using the geochemical model, marine flooding surfaces were delineated first using lithology (i.e., thin beds of carbonate and carbonaceous claystone/coal) and then using prominent peaks in the geochemical parameters. Correlations between sites were first attempted using the marine flooding surfaces delineated in this manner. Peaks in a certain parameter often were correlatable between sites that were proximal to each other (e.g., Sioux Composite to the BHP SD 92-4 core and Kenyon to Bounds core). Patterns observed in the geochemical parameters were also useful when correlating marine

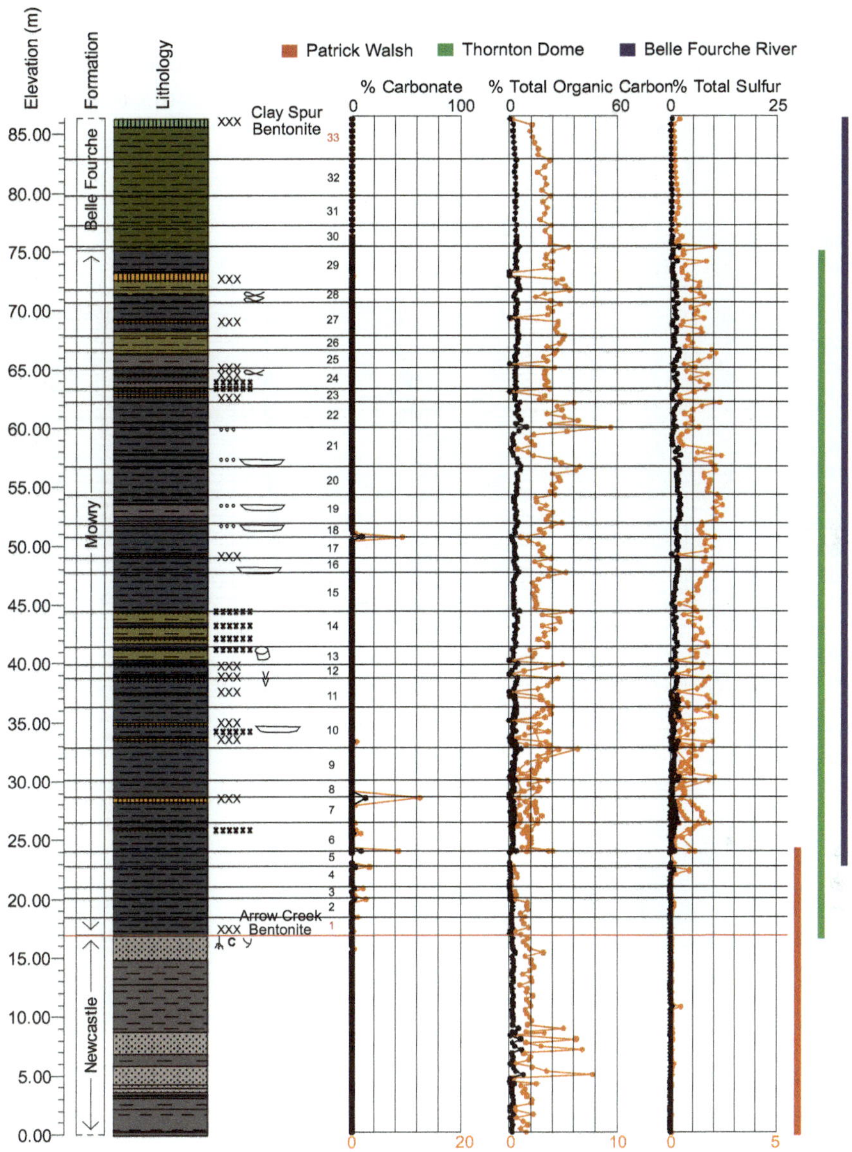

FIGURE 6

Parasequence delineations in the Northeastern Wyoming Composite section. The number listed for each parasequence is equivalent to the number listed in the other parasequence delineation figures. All parasequences for this section are described in Morath (2007). Note that the datasets are plotted in two scales (black = large; orange = small) and that the overlap of the outcrops and core used to construct the composite section is displayed to the right.

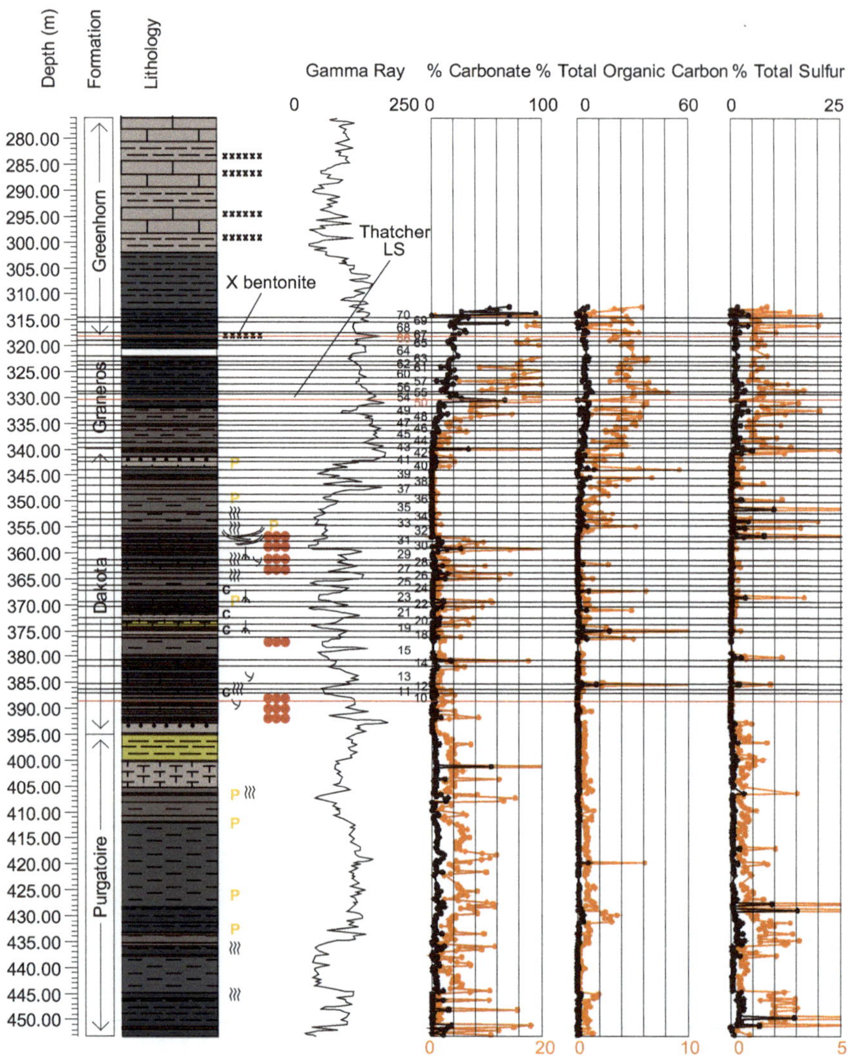

FIGURE 7

Parasequence delineations in the Amoco Rebecca K. Bounds #1 core. The number listed for each parasequence is equivalent to the number listed in the other parasequence delineation figures. All parasequences for this section are described in Morath (2007). Note that the datasets are plotted in two scales (black = large; orange = small).

flooding surfaces, especially between the Kenyon core and Sioux City Composite section. Horizons of equivalent geochemical values (i.e., horizon of low values in every geochemical parameter from parasequence 51 to 57 in the Sioux City Composite and BHP SD 92-4 and Kenyon cores (Figs 8–10)) and elevated sulfur values from parasequence 61 to 66 in the Sioux City Composite section and Kenyon

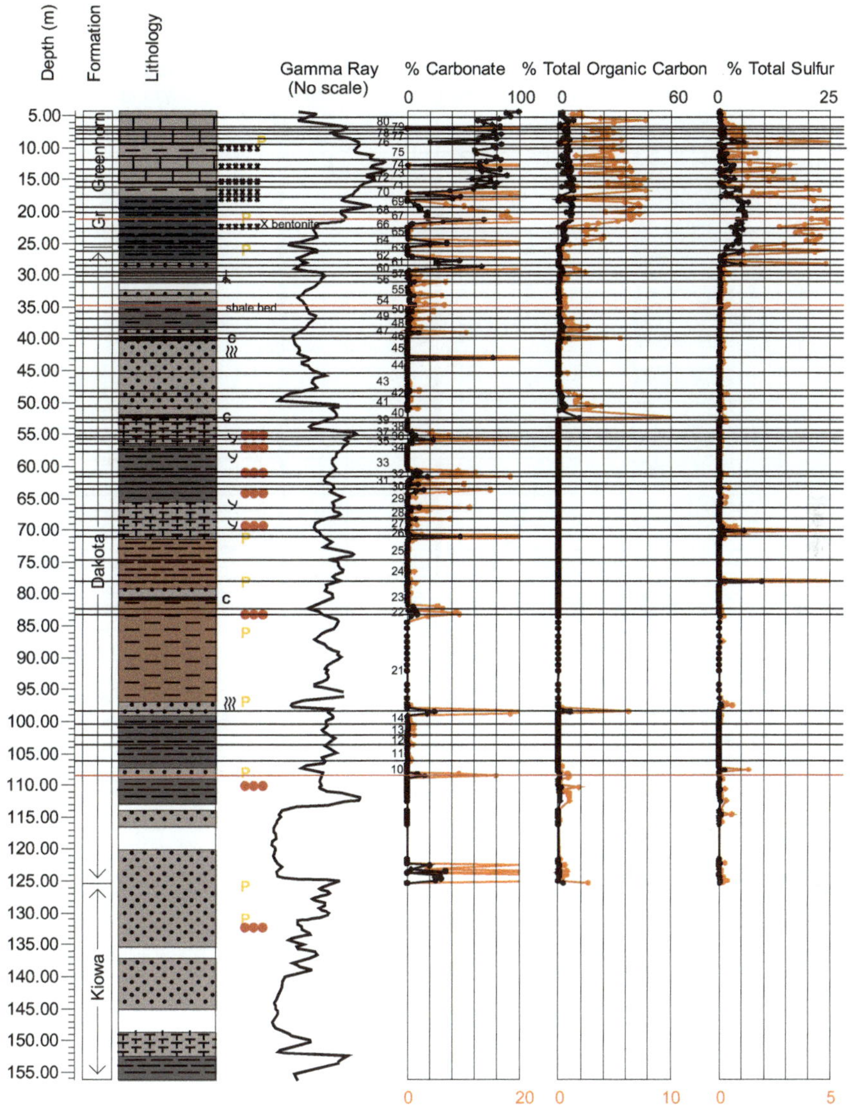

FIGURE 8

Parasequence delineations in the Kansas Geological Survey Kenyon #1 core. The number listed for each parasequence is equivalent to the number listed in the other parasequence delineation figures. All parasequences for this section are described in Morath (2007). Note that the datasets are plotted in two scales (black = large; orange = small).

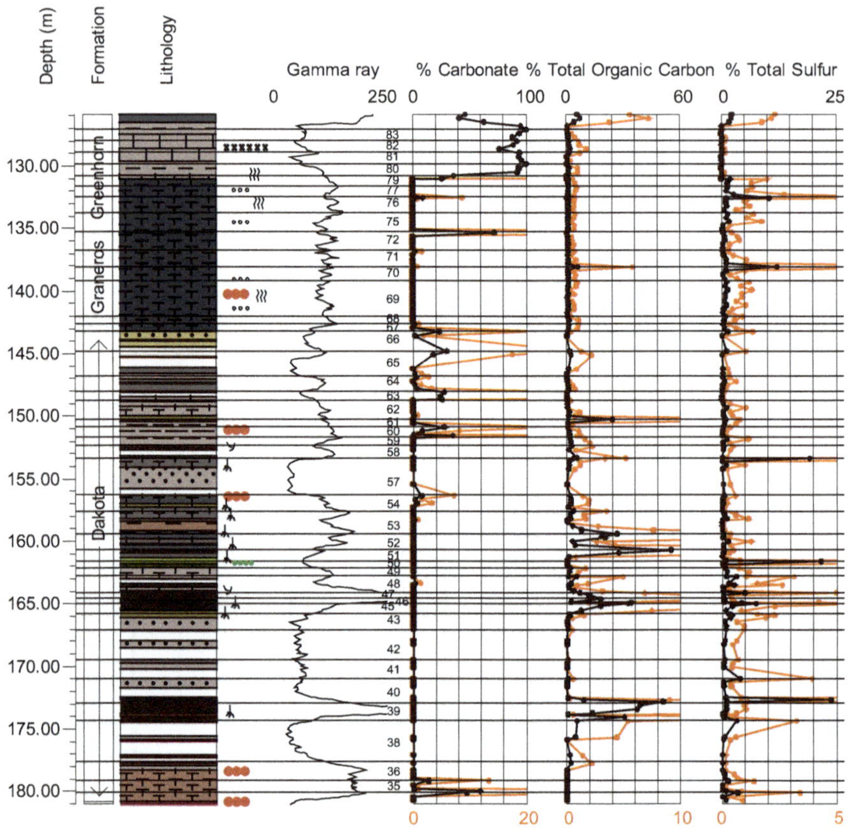

FIGURE 9

Parasequence delineations in the BHP SD 92-4 core. The number listed for each parasequence is equivalent to the number listed in the other parasequence delineation figures. All parasequences for this section are described in Morath (2007). Note that the datasets are plotted in two scales (black = large; orange = small).

core (Figs 8 and 10) were also considered when correlating marine flooding surfaces between the sites. These locations are proximal to each other and are expected to have been deposited by similar sedimentary processes in similar geochemical environments.

The number of peaks between the marker horizons was subsequently counted for each site. If the numbers were equal, then correlations were interpreted to be complete. However, this typically was not the case. Inconsistencies in the number of peaks (marine flooding surfaces) between marker horizons were reconciled by correlating a less-prominent peak at one site to a well-defined peak at another site. If no peaks were encountered, that marine flooding surface was interpreted to have not been formed or to have been eroded. Discontinuous marine flooding surfaces are interpreted to represent unconformities formed due to decreased accommodation space development. More unconformities occur in eastern margin sections due to the increasingly terrestrial environments of deposition, an observation made earlier by Karl (1976); Witzke and Ludvigson (1994), and MacFarlane et al. (1994).

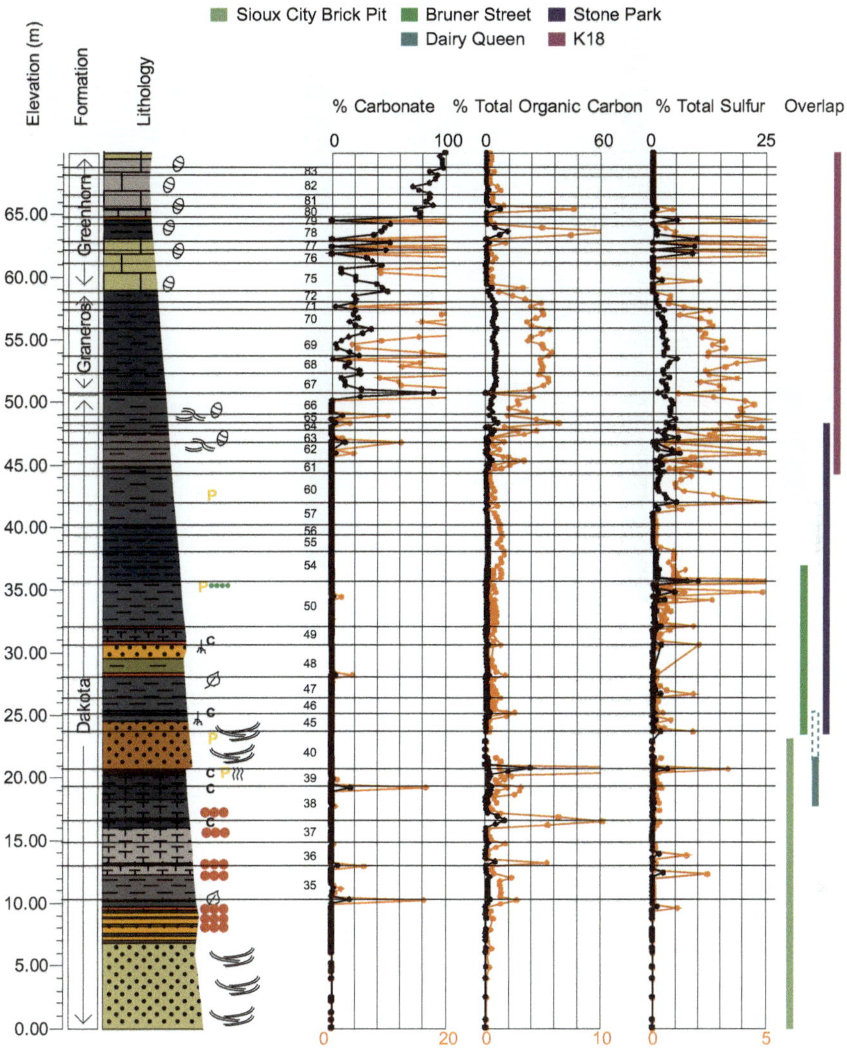

FIGURE 10

Parasequence delineations in the Sioux City Composite section. The number listed for each parasequence is equivalent to the number listed in the other parasequence delineation figures. All parasequences for this section are described in Morath (2007). Note that the datasets are plotted in two scales (black = large; orange = small).

The geochemical character of each parasequence delineated in this study is described in Morath (2007) and is displayed in Figs 6–10. The parasequences are numbered based on the geochemical and lithological correlations. The chemostratigraphic peaks used to delineate marine flooding surfaces along with parasequence thicknesses are also presented in Morath (2007). The study sites are presented

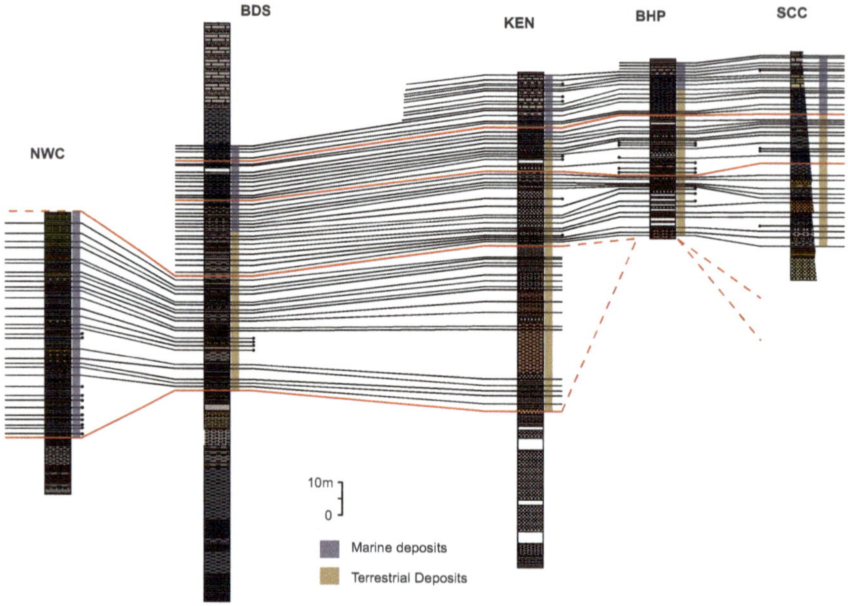

FIGURE 11

Parasequence-scale correlations from the axial portion (northeastern Wyoming) to the eastern margin (northwestern Iowa) of the Cretaceous Western Interior Basin (KWIB). Note the bentonites used in this study are labeled in red and marine flooding surfaces are labeled in black. Discontinuous marine flooding surfaces are marked by a black dot. Note the relative position of marine (blue) versus terrestrial (brown) strata. SCC = Sioux City Composite; BHP = BHP SD 92-4; KEN = Kenyon; BDS = Bounds; NWC = Northeastern Wyoming Composite.

along an offshore–onshore transect from the axial basin (northeastern Wyoming) to the eastern margin (northwestern Iowa) (Fig. 11).

9.4.3 DURATION OF PARASEQUENCES

Updated ages determined by Gradstein et al. (2004) for the Arrow Creek, Clay Spur, Thatcher, and Soap Creek (equivalent to the X-bentonite) bentonites are used to assess the duration of time represented by each geochemically defined parasequence. Here, it is assumed that parasequences were deposited by some cyclic process, so an average time span for deposition was calculated for each parasequence between these key age-dated horizons. This assumption is reasonable since Milankovitch-scale cyclicity has been demonstrated for coeval strata in the KWIB (e.g., Fischer et al., 1985). To obtain an average age for each parasequence, the number of parasequences observed between each bentonite (or equivalent) was divided by the amount of time between the age of the bentonites (with maximum and minimum error). The results are shown in Table 2. The parasequences deposited between the Arrow Creek and Clay Spur bentonites, the Clay Spur and Thatcher bentonites, and the Thatcher and X-bentonites show a mean duration of 43 ± 6 thousand years (ky), 75 ± 6 ky, and 54 ± 5 ky,

Table 2 Parasequence Durations for Cenomanian Strata in the Western Interior Basin

	Age (Ma)	Error (Ma)	Maximum Age (Ma)	Minimum Age (Ma)
Arrow Creek Bentonite (ACB)	98.53	0.51	99.04	98.02
Clay Spur Bentonite (CSB)	97.12	0.69	97.81	96.43
Thatcher Bentonite (TB)	95.84	0.61	96.45	95.23
Soap Creek Bentonite (X)	94.98	0.53	95.51	94.45

	# of Parasequences	Age Difference Using Max. Age (Ma)	Age Difference Using Min. Age (Ma)
ACB to CSB	33	1.23	1.59
CSB to TB	17	1.36	1.20
TB to X	16	0.94	0.78

	Parasequence Duration Using Max. Age (ky)	Parasequence Duration Using Min. Age (ky)
ACB to CSB	37	48
CSB to TB	80	71
TB to X	59	49

	Avg. Parasequence Duration (ky)
ACB to CSB	43±6
CSB to TB	75±6
TB to X	54±5

Ma = million years (age); Ma = million years (duration); ky = thousand years (duration).

respectively. The differences in duration were most likely caused by variations in the quality of the stratigraphic record at each study site. The parasequences between the Arrow Creek and Clay Spur bentonite were deposited in a wholly marine environment that was well preserved and thus probably represent the best record for calculation of parasequence duration. The parasequences that occur above the Clay Spur bentonite were deposited in shallow marine and terrestrial environments where the potential for nondeposition or erosion was greater.

9.4.4 RELATIVE SEA LEVEL FLUCTUATIONS

A generalized relative sea level curve was constructed for each site. Beginning at the base of each section, a determination of the relative "marineness" or "continentality" of each parasequence was made by comparing marine and terrestrial end members. The most terrestrial parasequence (a paleosol within a stack of amalgamated paleosols; e.g., parasequence 15) and the most marine parasequence (e.g., parasequence 80), based on the lithological and geochemical characteristics (Morath, 2007), were used to evaluate the remaining parasequences between these end members. An overall relative rise or fall in the sea level was interpreted based on changes in lithology and/or geochemical parameters above and below a particular parasequence.

The relative sea level curves of the study sites were subsequently overlain and compared (see Morath, 2007 for details) to the KWIS relative sea level curve of Kauffman and Caldwell (1993) and the eustatic sea level curve of Miller et al. (2005). The relative sea level curves generated in this study correlate well to the relative sea level curve for the KWIB, implying correlation with the Haq et al. (1987) curve based on the interpretation of Kauffman and Caldwell (1993). The relative sea level curves also show good correlation with the Miller et al. (2005) eustatic sea level curve. A fall in relative sea level typically coincides with a geochemically defined hiatus.

Inconsistencies were observed between sites in the number of parasequences between marker horizons. These inconsistencies were attributed to the unconformities that formed during falls in the relative sea level, acknowledging that errors in the correlation process may also describe the inconsistencies. The sedimentary response to changes in relative sea level varied between sites and can be attributed to their different paleogeographic positions within the KWIB (see Fig. 1). For example, the parasequences present at the base of the Northeastern Wyoming Composite section (parasequences 1–9) are not observed in the Bounds and Kenyon cores. The early Cenomanian strata in northeastern Wyoming was deposited in a wholly marine environment (Condon, 2000) in the axial portion of the KWIS (Kauffman and Caldwell, 1993), whereas the coeval strata in the Bounds and Kenyon cores were deposited in a fluvial/deltaic setting (MacFarlane et al., 1991) near the eastern paleoshoreline of the KWIS (Hamilton, 1994). Therefore, less accommodation space was available in the Kansas, precluding the deposition of parasequences 1–9 (i.e., nondeposition and/or erosion).

Five horizons of the lowstand deposits were observed in the sections used in this study and are labeled Ce1–Ce5 (Ce—Cenomanian) (Fig. 12). The position of Ce1 (97.70–97.95 Ma) (Fig. 12) corresponds to parasequences 15 to 21 in the Northeastern Wyoming Composite (Fig. 6). The position of the Ce1 lowstand in this site was determined by various factors. For example, the rocks are coarser-grained than adjacent strata and sandstone/bone beds exist. Fewer bentonites are present in this part of the section when compared to the underlying and overlying strata, perhaps due to a higher energy environment of deposition not conducive to bentonite preservation. Also of note are elevated total sulfur values throughout this interval, attributed primarily to the occurrence of pyrite (Morath, 2007, Fig. 6). Three major factors limit primary sedimentary pyrite production: deposition of bacterially decomposable organic matter, sulfate concentration of the water, and the availability of reduced iron minerals that can react with H_2S to form pyrite (Berner and Raiswell, 1984). Since marine systems have abundant sulfate, and the Mowry Shale has elevated TOC contents (Morath, 2007), reduced iron was most likely the limiting factor on pyrite formation in the Mowry Shale. Pyrite formation may have resulted from elevated dissolved iron in the marine system due to the increased inputs from the rivers. Interestingly, based on the geochemical relationships among HI, %TOC, %TS, and %Fe, Davis et al. (1988) stated that the limiting factors on pyrite formation in the Mowry Shale were organic matter type and/or reactive iron content. The Thornton Dome outcrop, which was used to construct the Northeastern Wyoming Composite, was one of the sites used in the Davis et al. (1988) study. These observations, coupled with the presence of unconformities in equivalent strata (discussed in the next section), led to the interpretation of a lowstand in sea level (Ce1).

An early Cenomanian unconformity that can be attributed to Ce1 exists to the east from parasequences 15–20 in the Kenyon core (Fig. 8) and parasequences 16–17 in the Bounds core (Fig. 7), though the missing parasequences in the Bounds core could also have been the result of poor sample recovery. Fewer parasequences were deposited during this lowstand interval in the Kenyon core compared to the Bounds core due to the more landward position of strata in the Kenyon core (i.e., less

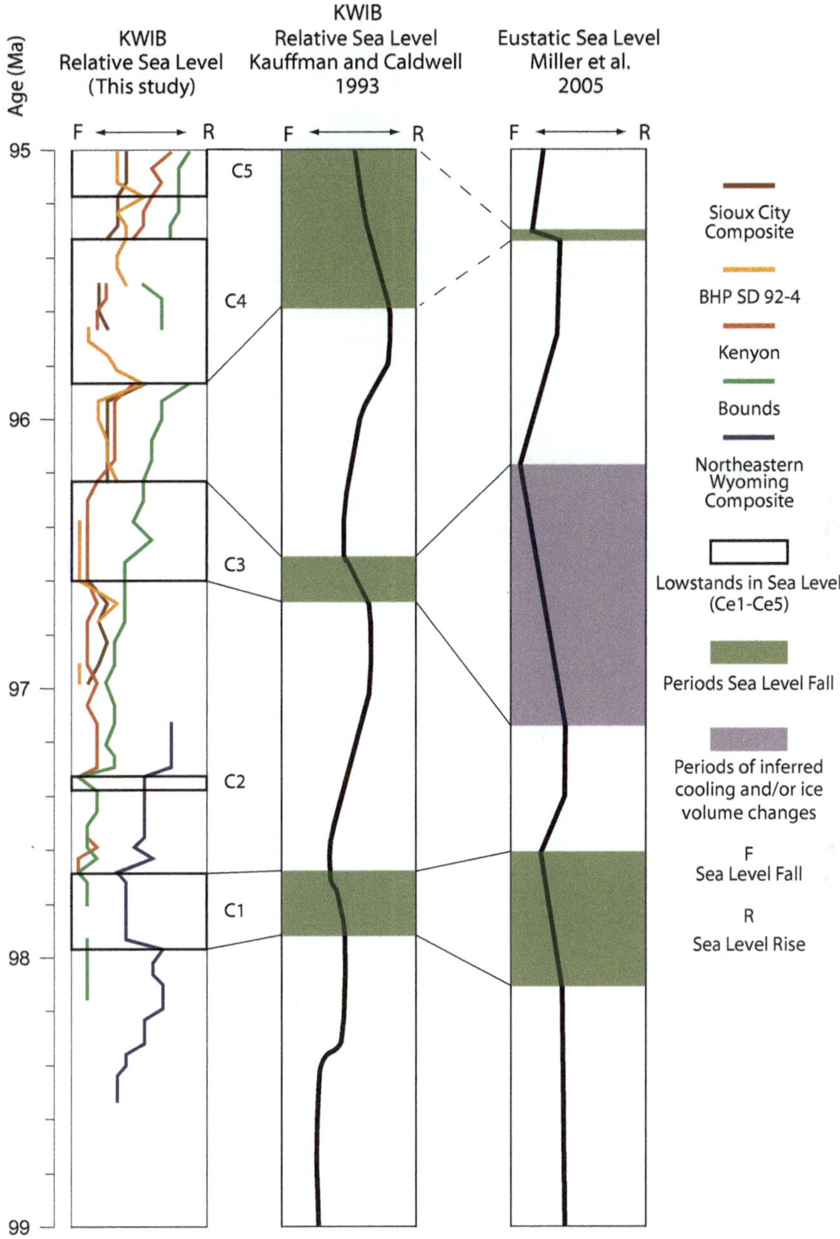

FIGURE 12

Relative sea level curves for the Western Interior Basin (late Albian–late Cenomanian) generated in this study and by Kauffman and Caldwell (1993) compared to the eustatic sea level curve of Miller et al. (2005). Note that lowstands are marked by white and green boxes and labeled Ce1–Ce5. Note the position of the period of inferred cooling and or/ice volume changes from Miller et al. (2005) marked by a blue box.

accommodation space). A horizon of thick amalgamated paleosols exists in both cores at the Ce1 interval, identified by the presence of rooting structures, sphaerosiderites, and mottling. Interestingly, White et al. (2005) determined that a stack of amalgamated early Cenomanian paleosols in the Sioux City Brick Company clay pit formed during a lowstand in relative sea level. This stack of paleosols may have been deposited coeval with Ce1 (Fig. 13).

Parasequences were deposited in the Northeastern Wyoming Composite section and the Bounds and Kenyon cores during Ce2 (97.37–97.39 Ma) (Fig. 12). Parasequences 28–30, which appear to have been deposited during Ce2, exist as a stack of amalgamated paleosols in the Kenyon core 9 (Fig. 8) and were interpreted to be a sequence boundary in the Bounds core (Fig. 7) by Scott et al. (1998). There is no distinct lithological or chemostratigraphic expression of Ce2 in the Northeastern Wyoming Composite section (Fig. 6), which most likely can be attributed to the more distal paleogeographic setting. Interestingly, White et al. (2005) determined a second lowstand in relative sea level occurred during the early Cenomanian in the amalgamated paleosols at Sioux City. This second early Cenomanian lowstand deposit from White et al. (2005) most likely occurred coeval with Ce2 (Fig. 13).

The deposition of parasequences 40–45 is interpreted to have occurred during Ce3 (Fig. 12). Parasequences 40–45 are missing in the Sioux City Composite section (Fig. 10), which is coeval with Ce3 based on the geochemical correlations. A thick unit of sandstone exists in these parasequences within the Kenyon (Fig. 8) and BHP SD 92-4 (Fig. 9) cores, and the Sioux City Composite section (Fig. 10).

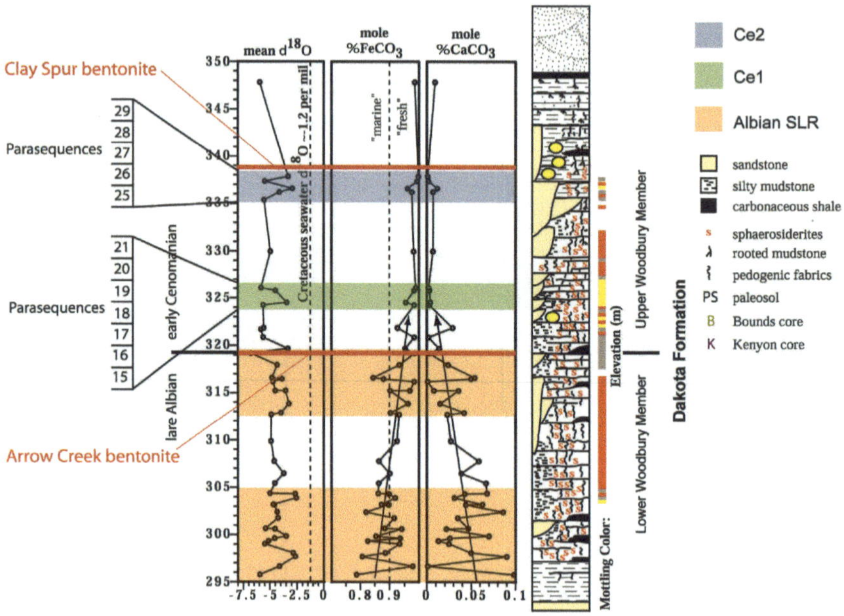

FIGURE 13

Approximate position of Clay Spur and Arrow Creek bentonites based on position of Ce1 and Ce2 in the Sioux City Brick Pit. Note the parasequence numbers that correspond to Ce1 and Ce2, which can be correlated from the axial portion to the eastern margin of the Cretaceous Western Interior Basin (KWIB).

Modified from White et al. (2005).

The thick sandstone unit is bounded by carbonaceous claystone in all of these sections. A sandstone bed also exists in the Bounds core in this interval (Fig. 7). Since parasequences 40–45 are observed in the Bounds, Kenyon, and BHP SD 92-4 cores, these parasequences could be missing in the Sioux City Composite section due to the more landward position and associated higher rates of erosion at this site or omission of strata resulting from recent erosion at the top of the Sioux City Brick Pit highwall.

Three unconformity horizons are present in the Ce4 interval (95.35–95.85 Ma) (Fig. 12). These horizons include parasequences 51–53 and 58–59, respectively, in the Sioux City Composite section and the Kenyon and Bounds cores, and parasequences 55–56 in the BHP SD 92-4 core. Of note, parasequences 51–53 and 58–59 exist within the BHP SD 92-4 core as a stack of amalgamated paleosols (Fig. 9), identified by the occurrence of root traces, cutans, and sphaerosiderites. The presence of numerous paleosol layers in the BHP SD 92-4 may be due to higher avulsion rates associated with deposition during a lowstand in relative sea level. Parasequences 55–56 were interpreted as being lowstand deposits in the Sioux City Composite section, which was deposited on the eastern margin of the KWIB, based on low (<0.5%) total sulfur values (vs higher total sulfur values in the more marine influenced the Northeastern Wyoming Composite section; see Fig. 4) bounded above and below by higher (>1%) total sulfur values and lower total organic carbon values that decrease up-section (Fig. 10). This interval also contains more sandstone than the units above and below in the Sioux City Composite. Rooting structures exist at this horizon in the Kenyon core. Less available accommodation space during the deposition of parasequences 55–56 in the BHP SD 92-4 core may have been due to minor tectonic movements recognized by Weimer (1978) in this area. Also, the depositional environment prior to the deposition of the sandstone in parasequence 57 within the BHP SD 92-4 core may have been erosive and therefore responsible for removing the underlying parasequences. Alternatively, poor core recovery may have led to the misinterpretation of missing marine flooding surfaces.

Parasequences were deposited in all the sections that contain late Cenomanian strata during Ce5 (95–95.17 Ma) (Fig. 12). However, parasequences 64–66, deposited during Ce5 time, exist as fine-grained sandstone at the top of the Dakota Formation in the BHP SD 92-4 core in the eastern South Dakota (Fig. 9). Although thin sand beds exist at the top of the Dakota Formation in the Sioux City Composite section, there is no distinct lithological or chemostratigraphic expression of Ce5 in the other sections. Coarser-grained deposits most likely occurred in the BHP SD 92-4 core because of its closer proximity to the paleoshoreline during Ce5 time.

Three other parasequences are missing in the BHP SD 92-4 core. Parasequences 37 and 44 could not be identified and were most likely missed due to the poor core recovery. Parasequence 78 was most likely eroded (or not deposited). This inference, together with the presence of a coarse-grained, poorly sorted sandstone with rip-up clasts at this interval, interpreted to represent the Bridge Creek Limestone ravinement surface.

The Ce1, Ce3, and Ce4–Ce5 intervals of relative sea level lowstand observed in this study correlate to lowstands in relative sea level on the Kauffman and Caldwell (1993) relative sea level curve for the KWIB, whereas the Ce2 interval does not appear on the Kauffman and Caldwell (1993) curve (Fig. 12). The sections in the Kauffman and Caldwell (1993) study were primarily limited to the central portion of the basin (i.e., Colorado, Kansas, Wyoming), whereas this study included the sections from the eastern margin, which in turn was probably more sensitive to changes in relative sea level due to the more proximal location to the shoreline. This same line of reasoning may apply to the differentiation of two lowstands, Ce4 and Ce5, in this study, where only one lowstand was interpreted in the Kauffman and Caldwell (1993) curve. The disparity in the sea level events between the Kauffman and Caldwell (1993) sea level curve and the sea level curve generated in this study may also be due to the imprecise calculations of parasequence durations due to the incomplete eastern margin sections, especially above the Clay Spur bentonite.

9.4.5 EVOLUTION OF ACCOMMODATION SPACE IN THE KWIB

Fluctuations in the relative sea level played a dominant role in the lithologic and chemostratigraphic evolution of late Albian to Cenomanian strata in the KWIB. However, the cause of these fluctuations is unknown. The relative sea level curve for the KWIS, along with the relative sea level curve generated in this study, correlate well with the eustatic sea level curve of Miller et al. (2005) (Fig. 12). Therefore, variation in eustatic sea level is a plausible mechanism for having caused changes in relative sea level within the study interval (Kauffman and Caldwell, 1993). This observation is supported by more recent reports of synchroneity between disconformities in the upper Turonian–Lower Coniacian strata from Alberta to Colorado that "implies an element of eustatic control on these relative sea-level changes" (Walaszczyk et al., 2014).

Immenhauser and Scott (1999) demonstrated the global correlation of Albian sea level fluctuations (with regional variations on the global trends), while Gale et al. (2002, 2008) determined that the Cenomanian sea level variations were globally synchronous and driven by the long eccentricity cycle (400,000 years). However, the mechanisms proposed in those studies do not explain the higher-order cyclicity observed here. Potential causes for higher-order sea level fluctuations include compaction, basin tectonics, eustatic sea level (e.g., ice volume variation, groundwater storage), and sediment supply. These mechanisms are explored in this section.

According to Milliman and Haq (1996), compaction can lead to subsidence rates of several cm/year, whereas Jelgersma (1996) commented that compaction rates usually do not exceed a few cm/century. Immenhauser (2005) remarked that compaction is nonsystematic and local and therefore cannot be invoked to explain repeated regressive–transgressive events recorded in the Mesozoic strata. While compaction played a role in the creation of accommodation space in the KWIB, its impact would have been spatially and temporally nonuniform, localized, and too small-scale to have caused the regional cyclicity and synchroneity observed in the Cenomanian strata.

Tectonism can also have a profound impact on the accommodation space development within a basin. However, Goldhammer et al. (1987) and Masetti et al. (1991) conjectured that tectonic movements cannot operate on sufficiently rapid time scales to account for high-frequency sea level changes. Interestingly, in the study interval, Bhattacharya and Willis (2001) identified four distinct disconformity-bounded allomembers in the lower portion of the Belle Fourche Shale along the western flank of the Powder River Basin (northeastern Wyoming), and Vakarelov et al. (2006) suggested that minor tectonic pulses locally overshadowed the effects of eustasy and exerted dominant control over the preserved stratal architecture in this region. The results of Vakarelov et al. (2006) indicated that much of the deformation that led to these allomembers occurred at submillion-year frequencies, which appears to contradict the results of Goldhammer et al. (1987) and Masetti et al. (1991). The strata in this region are coeval with the Upper Woodbury Member of the Dakota Formation, but the frequencies of the tectonic effects are a few orders of magnitude higher than those observed in this study.

While tectonic activity created and maintained the structure of the KWIB foredeep (Jordan, 1981), tectonics probably did not extend to the eastern margin of the KWIB (~800 km from the Powder River Basin). Kauffman (1977) and Brenner et al. (2000) noted relatively little subsidence on the eastern margin of the basin. Pang and Nummedal (1995) suggested that rates of subsidence varied throughout the KWIB, with greater subsidence occurring in the present-day Utah and Colorado and more uniform subsidence over larger distances occurring on the cratonic platform. Furthermore, deformation events, like that observed in the KWIB hinterland, occurred randomly and therefore could not have caused cyclic relative sea level changes (Jordan, 1981; Weimer, 1984) observed in the basinal deposits. Therefore, while tectonic movements played a dominant role in accommodation space development in the foredeep, they probably did not drastically and cyclically affect the accommodation space along the eastern margin on the time scales postulated in this study.

The mid-Cretaceous has been suggested as a time of peak warmth (Frakes et al., 1992), with unusually warm polar temperatures (Fischer, 1982). Numerous studies support the idea of an ice-free Cretaceous. For example, atmospheric general circulation model results of Barron and Washington (1985) and Poulsen et al. (1999) simulate the Cretaceous as having been a warm, ice-free world. Other evidence for an ice-free Cretaceous comes from the geologic record. For example, Parrish and Spicer (1988) estimated a mean average temperature range of 10 ± 3 °C for the North Slope of Alaska (~70°–80°N paleolatitude) in the latest Albian to Cenomanian from the presence of subtropical flora. Furthermore, Rich et al. (1988) noted a diverse biota of dinosaurs, invertebrates, vertebrates, and plants in the Aptian–Albian strata from the southeastern coast of present-day Australia (80°S paleolatitude). The presence of such a diverse biota led them to conclude that the early Cretaceous climate of the southern Australia was temperate, humid, and subject to seasonal extremes. Based on these studies, it seems unlikely that large ice sheets existed in the mid-Cretaceous and thus a consideration of other processes that can change eustatic sea level during nonglacial times follows.

An intensified mid-Cretaceous hydrologic cycle (Barron and Washington, 1985; White et al., 2001) could have led to increased water storage within the inland aquifers and lakes. Jacobs and Sahagian (1993) suggested that during nonglacial times, such as the mid-Cretaceous, this mechanism is the dominant Milankovitch-scale eustatic sea level driver. According to Jacobs and Sahagian (1993), lake basins and groundwater aquifers combined could store water equivalent to 4–8 m of sea level. Hay and Leslie (1990) also suggested that Milankovitch-driven changes in groundwater storage could be capable of producing ~10–20 m of eustatic sea level change. However, elevated eustatic sea levels during the mid-Cretaceous would have flooded large portions of the continental interiors, thus eliminating those regions as potential aquifers. Therefore, this mechanism of eustatic sea level change does not seem plausible for this study.

Another method of eustatic sea level change during the ice-free conditions is the expansion and contraction of oceanic waters due to the rise and fall of mean annual temperature. According to Coe (2003), a 1 °C change in the temperature can result in a 1–2 m change in the eustatic sea level. Lehmann et al. (1998) suggested that changes in the storage capacity of the aquifers and lakes, in conjunction with the thermal expansion and contraction of oceanic water, accounted for meter-scale cycles observed in the early Cretaceous carbonates and evaporites of the northeastern Mexico. Therefore, it seems plausible that the combination of these mechanisms could have also caused fluctuations in the relative sea level during the mid-Cretaceous. However, these mechanisms of sea level change during ice-free conditions cannot explain the large (tens of meters) and rapid variations in the relative sea level observed during the mid-Cretaceous (Miller et al., 2005).

Milankovitch-scale fluctuations have been suggested as having affected the relative sea level during the Cretaceous (Barron et al., 1985; Arthur and Dean, 1991; Elder et al., 1994; Gale et al., 2008). While the method used to determine the average duration of parasequences in this study is crude, the duration of parasequence formation between the Arrow Creek and Clay Spur bentonites (43±6ky), which represents the most well-preserved section within this study, is suggestive of a Milankovitch obliquity (41 ky) signal. Cyclic parasequences deposited stratigraphically higher in the study interval between the Clay Spur and Thatcher bentonites, and the Thatcher and X-bentonites, record a mean duration of 75±6 and 54±5 ky, respectively. Various factors could have led to these calculated parasequence durations. For example, if parasequences were eroded in all the sites used in this study during any of the interpreted sea level falls (Ce1–Ce5), then a higher average duration would have been calculated. Erosion would have been more prevalent in the predominantly terrestrial environments along the eastern margin of the KWIB (Scott et al., 1998; Karl, 1976; MacFarlane et al., 1991, 1994; Ravn and Witzke, 1995), therefore the stratigraphic record is less complete. Since the stratigraphic record is less complete, more parasequences may once have existed. Smaller parasequence durations would have been calculated if these missing parasequences were included in the parasequence duration calculations. Also, correlation errors may be responsible for the calculation of longer parasequence duration.

Given the evidence for orbital controls described previously, it may be enlightening to consider other durations. While an incomplete stratigraphic record above the Clay Spur bentonite may have affected the preservation of parasequences, the calculated average parasequence durations from the Clay Spur to the Thatcher bentonites (75±6 ky) and the Thatcher and X-bentonites (54±5 ky) are vaguely suggestive of the Milankovitch eccentricity (100 ky) and obliquity (41 ky) cycles, respectively.

Numerous researchers have suggested the role played by the Milankovitch climate cyclicity in sediment stacking patterns observed in the KWIB. For example, Fischer et al. (1985) suggested obliquity variations as the cause of limestone/marlstone bedding couplets observed in the Bridge Creek Limestone Member of the Cenomanian–Turonian Greenhorn Formation, which is the upper stratigraphic unit studied here. Sageman et al. (1997) also concluded that obliquity was the dominant spectral signal recorded in the Bridge Creek Limestone bedding couplets. The obliquity component of the Milankovitch climate signal has also been suggested as the cause of the relative sea level change in various other Upper Cretaceous deposits in the KWIB (Johnson et al., 2005; Sterzinar and Leckie, 2004).

Interestingly, Brenner et al. (2003) suggested that the long-distance movement of gravel preserved in the late Albian Nishnabotna Member of the Dakota Formation in the central Iowa was directly related to the hydraulic flux and that this flux may have been linked to the climatic cycles. Since obliquity appears to have played a dominant role in sedimentation during the formation of the Cenomanian–Turonian Greenhorn Formation (Sageman et al., 1997), and climatically influenced sedimentation may have occurred during the deposition of the late Albian Nishnabotna Member of the Dakota Formation (Brenner et al., 2003), it seems feasible that the obliquity component of the Milankovitch-scale orbital cyclicity also played a dominant role in controlling the patterns of sedimentation in the Cenomanian nonmarine strata in the KWIB.

According to Miller et al. (2005), only ice volume changes can explain the sea level changes observed during the Cretaceous. Glacio-eustatic cycles occur at higher frequencies considered here and could be invoked to explain eustatic sea level variations during the Cretaceous. In support of this hypothesis, Matthews and Poore (1981) used $\delta^{18}O$ values from tropical marine skeletal remains to infer that there were significant amounts of ice present throughout the Cretaceous, noting that the Cenomanian most likely escaped ice cover. Spicer and Parrish (1990) inferred that permanent ice was likely above 1200 m in Alaska during the Cenomanian. More recently, Bornemann et al. (2008) concluded that $\delta^{18}O$ data from Turonian strata of the tropical Atlantic are consistent with the past presence of ice sheets of about half the size of the modern Antarctic ice cap and that the common assumption that the Cretaceous poles were always ice-free during past episodes of intense global warming should be reconsidered.

At least two studies have concluded that glacioeustasy is the most viable explanation for observations made in the Dakota Formation and coeval strata in North America. First, Plint and Kreitner (2007) attributed multiple upward shoaling sequences in the Early–Late Cenomanian Kaskapau Formation in Canada to ~10 m sea level excursions of durations ranging from ~9 to 74 ky. Second, Koch and Brenner (2009) found that fluvial incision recorded in the Dakota Formation coincided with the sequence boundaries that are correlatable. After consideration and elimination of multiple processes that might explain the stacking patterns and correlations described in this study, we conclude that glacioeustasy is the most viable explanation and concur with Bornemann et al. (2008) in suggesting that the assumption of ice-free poles in the mid-Cretaceous must be reconsidered.

9.4.6 ALLUVIAL-PLAIN STRATAL ARCHITECTURE AND PEDOGENESIS DURING THE MID-CRETACEOUS

White et al. (2005) stated that a mechanism involving climate change seemed plausible to describe the northwestern Iowa paleosol siderite $\delta^{18}O$ record. Variations in paleosol siderite $\delta^{18}O$ within lowstand strata were considered to have recorded climate sensitivity and temporal variation in the intensity of hemispheric-scale moisture transport. The Dakota Formation oxygen isotopic record was formed during the late Albian to early Cenomanian, a portion of which is equivalent to the strata deposited between the Arrow Creek and Clay Spur bentonites (Fig. 13). White et al. (2005) interpreted two lowstands in relative sea level during the early Cenomanian (the shaded regions above the Albian/Cenomanian boundary in Fig. 13). These lowstands in relative sea level are here correlated to Ce1 and Ce2 (Fig. 12) and the parasequences that were deposited during these lowstands in sea level. In this study, the parasequences from this interval were suggestive of an obliquity component of Milankovitch cyclicity. Therefore, it seems plausible that Milankovitch-scale climate cyclicity played a role in causing the variability in the overall third-order lowstand $\delta^{18}O$ values recorded in sphaerosiderites.

According to White et al. (2005), fluctuations in the lowstand isotopic data could be attributed to competing moisture sources during fourth-order waxing and waning of relative sea level through a third-order lowstand. Increased $\delta^{18}O$ values were credited to decreased precipitation rates during a lowstand in sea level, whereas, decreased $\delta^{18}O$ values were ascribed to elevated rates of precipitation. These variable precipitation rates could be attributed to differences in the amount and intensity of insolation caused by the obliquity component of Milankovitch-scale climate cyclicity. For example, during periods of lower and less intense insolation, the intensity and amount of precipitation may have decreased due to the lower rates of evaporation. The opposite would hold true for periods of higher and more intense insolation. Model results from Glancy et al. (1993) suggest that precipitation rates intensified over North America due to the intensification of insolation contrast between the northern and southern hemispheres, supportive of the White et al. (2005) model.

Using the results and interpretations of White et al. (2005) and Glancy et al. (1993), higher relative sea levels and increased amounts of precipitation would have occurred during warmer parts in the obliquity cycle, and lower relative sea levels and decreased amounts of precipitation would have occurred during cooler parts in the obliquity cycle. During warmer periods, eustatic sea level would have been elevated, causing expansion of the KWIS. Alternatively, eustatic sea level would have been lower during the cooler periods, causing the KWIS to withdraw. During these periods of lower eustatic sea level, the tropical Tethys Ocean would have been the dominant source of moisture for precipitation. During periods of elevated eustatic sea level, both the tropical Tethys Ocean and the KWIS would have been the source of moisture. The sphaerosiderites that formed in paleosols within the alluvial-plain during the mid-Cretaceous record variations in the precipitation intensity and source (White et al., 2005). In northwestern Iowa, the $\delta^{18}O$ record obtained from sphaerosiderites within lowstand deposits likely contains a record of Milankovitch-scale climate fluctuations.

9.5 CONCLUSION

Four high-resolution relative sea level curves were constructed using the data and correlations generated for this study. These sea level curves, along with their reference sections, provided the tools for the creation of a new high-resolution relative sea level curve for the KWIB. Cores and outcrop sections from the stable eastern margin of the KWIB were used because tectonics would have played a smaller role in accommodation space development, and a more eustatically influenced sea level signal could be studied. The correlations of these relative sea level curves to the KWIB relative sea level curve produced by Kauffman and Caldwell (1993) and the similarity of the parasequence durations calculated for the lower stratigraphic interval to the obliquity signal (which has been suggested for KWIB strata) provided validation of the 'slug model' of geochemical facies development used in this study. The similarity between the relative sea level curves of the KWIB to the eustatic sea level curve suggests that the variations in the eustatic sea level were the dominant controls on the changes in accommodation space within the KWIB, though the cause of these eustatic sea level fluctuations remains elusive. Considering the results of Miller et al. (2005), it seems likely that an increase in Antarctic ice volume caused at least one of the lowstands in Cenomanian sea level.

The durations of the parasequences calculated in this study are suggestive of Milankovitch-scale climate cyclicity, which appears to be supportive of the presence of ice during the mid-Cretaceous. It also seems likely that Milankovitch-scale changes in insolation contrast played a role in the precipitation rates and in the sedimentation patterns observed in the alluvial-plain sections deposited along the eastern margin of the KWIB. Therefore, changes in eustatic sea level caused by changes in ice volume probably played a role in precipitation and sedimentation patterns in the KWIB. The isotopic record generated from sphaerosiderites that were deposited in a stack of amalgamated paleosols in the alluvial-plain of the KWIB most likely records changes in rainfall patterns that were driven by Milankovitch-scale changes in eustatic sea level.

ACKNOWLEDGMENTS

The authors acknowledge the support of the Petroleum Research Fund of the American Chemical Society (grants 32573-AC8 and 39503-AC8) without which this work would not have been accomplished.

REFERENCES

Arthur, M.A., Dean, W.E., 1991. A holistic approach to cyclomania: examples from Cretaceous pelagic limestone sequences. In: Einsele, et al. (Ed.), Cycles and Events in Stratigraphy. Springer-Verlag, Berlin, pp. 126–166.

Barron, E.J., Arthur, M.A., Kauffman, E.G., 1985. Cretaceous rhythmic bedding sequences: A plausible link between orbital variations and climate. Earth and Planetary Science Letters, vol. 72, pp. 327–340.

Barron, E.J., Washington, W.M., 1985. Warm Cretaceous climates, high atmospheric CO_2 as a plausible mechanism. In: Sundquist, E.T., Broecker, W.S. (Eds.), The Carbon Cycle and Atmospheric CO_2; Natural Variations Archean to Present Geophysical Monograph, vol. 32, pp. 546–553.

Berner, R., Raiswell, R., 1984. C/S method for distinguishing freshwater from marine sedimentary rocks. Geology 12, 365–368.

Bhattacharya, J.P., Willis, B.J., 2001. Lowstand deltas in the Frontier Formation, Powder River Basin, Wyoming: implications for sequence stratigraphic models, U.S.A. AAPG Bulletin 85, 261–294.

Bornemann, A., Norris, R., Friedrich, O., Beckmann, B., Schouten, S., Sinninghe Damste, J., Vogel, J., Hofmann, P., Wagner, T., 2008. Isotopic evidence for glaciation during the Cretaceous Supergreenhouse. Science 319 (5860), 189–192.

Brenner, R.L., Ludvigson, G.A., Witzke, B.J., Zawistoski, A.N., Kvale, E.P., Ravn, R.L., Joecknel, R.M., 2000. Late Albian Kiowa–Skull Creek marine transgression, lower Dakota Formation, Eastern margin of the Western Interior Seaway, U.S.A. Journal of Sedimentary Research 70 (4), 868–878.

Brenner, R.L., Ludvigson, G.A., Witzke, B.J., Phillips, P.L., White, T.S., Ufnar, D.F., González, L.A., Joeckel, R.M., Goettemoeller, A., Shirk, B.J., 2003. Aggradation of gravels in tidally influenced fluvial systems, upper Albian (Lower Cretaceous) on the cratonic margin of the North American Western Interior foreland basin. Cretaceous Research 24 (4), 439–448.

Cobban, W.A., Merewether, E.A., 1983. Stratigraphy and Paleontology of Mid-Cretaceous Rocks in Minnesota and Contiguous Areas. USGS Professional Paper 1253. p. 52.

Coe, A.L., 2003. The Sedimentary Record of Sea-Level Change. New York, NY, p. 287.

Condon, S.M., 2000. Stratigraphic Framework of Lower and Upper Cretaceous Rocks in Central and Eastern Montana. U.S. Geological Survey Digital Data Series DDS-57, p. 12.

Davis, H.R., Byers, C.W., Dean, W.E., 1988. Pyrite formation in the Lower Cretaceous Mowry Shale: effect of organic matter type and reactive iron content. American Journal of Science 288, 873–890.

DeCelles, P.G., Lawton, T.F., Mitra, G., 1995. Thrust timing, growth of structural culminations, and synorogenic sedimentation in type Sevier orogenic belt, western United States. Geology 23 (8), 699–702.

DeCelles, P.G., Currie, B.S., 1996. Long-term sediment accumulation in Middle Jurassic–early Eocene Cordilleran retroarc foreland-basin system. Geology 24 (7), 591–594.

Dyman, T.S., Cobban, W.A., Fox, J.E., Hammond, R.H., Nichols, D.J., Perry Jr., W.J., Porter, K.W., Rice, D.D., Setterholm, D.R., Shurr, G.W., Tysdal, R.G., Haley, J.C., Campen, E.B., 1994. Cretaceous rocks from southwestern Montana to southwestern Minnesota, northern Rocky Mountains, and Great Plains region. In: Shurr, et al. (Ed.), Perspectives on the Eastern Margin of the Cretaceous Western Interior Basin. Geological Society of America Special Paper 287, pp. 5–26.

Elder, W.P., Kirkland, J.I., 1985. Stratigraphy and depositional environments of the Bridge Creek Limestone Member of the Greenhorn Limestone at Rock Canyon Anticline near Pueblo, Colorado. In: Pratt, et al. (Ed.), Fine-Grained Deposits and Biofacies of the Cretaceous Western Interior Seaway: Evidence of Cyclic Sedimentary Processes. SEPM Second Annual Midyear Meeting, Golden, Colorado, Field Trip No. 9, pp. 122–134.

Elder, W.P., Kirkland, J.I., 1994. Cretaceous paleogeography of the southern Western Interior Region. In: Caputo, et al. (Ed.), Mesozoic Systems of the Rocky Mountain Region, USA, pp. 415–440.

Elder, W.P., Gustason, E.R., Sageman, B.B., 1994. Correlation of basinal carbonate cycles to nearshore parasequences in the Late Cretaceous Greenhorn seaway, Western Interior U.S.A. Geological Society of America Bulletin 106, 892–902.

Engleman, E.E., Jackson, L.L., Norton, D.R., 1985. Determination of carbonate carbon in geological materials by coulometric titration. Chemical Geology 53, 125–128.

Fischer, A.G., 1980. Gilbert bedding rhythms and geochronology. In: Yochelson (Ed.), The Scientific Ideas of G. K. Gilbert. Geological Society of America Special Paper 183, pp. 93–104.

Fischer, A.G., 1982. Long-term climate oscillations recorded in stratigraphy. In: Climate in Earth History. National Academy Press, Washington, DC, pp. 97–104.

Fischer, A.G., Herbert, T., Silva, I.P., 1985. Carbonate bedding cycles in Cretaceous pelagic and hemipelagic sequences. In: Pratt, et al. (Ed.), Fine-Grained Deposits and Biofacies of the Cretaceous Western Interior Seaway: Evidence of Cyclic Sedimentary Processes. SEPM Second Annual Midyear Meeting, Golden, Colorado, Field Trip No. 9, pp. 1–10.

Frakes, L.A., Francis, E., Syktus, J.I., 1992. Climate Modes of the Phanerozoic; the History of the Earth's Climate over the Past 600 Million Years. Cambridge University Press. p. 286.

Gale, A.S., Hardenbol, J., Hathway, B., Kennedy, W.J., Young, J.R., Pahnsalkar, V., 2002. Global correlation of Cenomanian (Upper Cretaceous) sequences: evidence for Milankovitch control on sea level. Geology 30 (4), 291–294.

Gale, A., Voigt, S., Sageman, B., Kennedy, W., 2008. Eustatic sea-level record for the cenomanian (Late cretaceous)—extension to the Western Interior Basin, USA. Geology 36 (11), 859–862.

Glancy Jr., T.J., Arthur, M.A., Barron, E.J., Kauffman, E.G., 1993. A paleoclimate model for the North American Cretaceous (Cenomanian–Turonian) epicontinental sea. In: Caldwell, W.G.E., Kauffman, E.G. (Eds.), Evolution of the Western Interior Basin. Geological Association of Canada Special Paper 39, pp. 219–241.

Goldhammer, R.K., Dunn, P.A., Hardie, L.A., 1987. High-frequency glacio-eustatic sea level oscillations with Milankovitch characteristics recorded in Middle Triassic platform carbonates in northern Italy. American Journal of Science 287, 853–892.

Gradstein, F.M., Ogg, J.G., Agterberg, F.P., 2004. A Geological Time Scale 2004. Cambridge Press, Cambridge. p. 500.

Hamilton, V., 1994. Sequence stratigraphy of Cretaceous Albian and Cenomanian strata in Kansas. In: Shurr, et al. (Ed.), Perspectives on the Eastern Margin of the Cretaceous Western Interior Basin. Geological Society of America Special Paper 287, pp. 79–96.

Haq, B.U., Hardenbol, J., Vail, P.R., 1987. Chronology of fluctuating sea levels since the Triassic. Science 235, 1156–1167.

Hattin, D.E., 1968. Plesiacanthoceras wyomingense (Reagan) from Graneros Shale and Greenhorn Limestone (Upper Cretaceous) of Central Kansas. Journal of Paleontology 42 (4), 1084–1090.

Hay, W.W., Leslie, M.A., 1990. Could possible changes in global groundwater reservoir cause eustatic sea level fluctuations? In: Revelle, R. (Ed.), Sea Level Change. National Research Council Studies in Geophysics, National Academy Press, Washingston, pp. 161–170.

Hays, J.D., Pitman, W.C., 1973. Lithospheric plate motion, sea level changes, and climatic and ecological consequences. Nature 194 (4270), 1121–1132.

Immenhauser, A., 2005. High-rate sea-level change during the Mesozoic: new approaches to an old problem. Sedimentary Geology 175, 277–296.

Immenhauser, A., Scott, R.W., 1999. Global correlation of middle Cretaceous sea-level events. Geology 27 (6), 551–554.

Jacobs, D.K., Sahagian, D.L., 1993. Climate-induced fluctuations in sea level during non-glacial times. Nature 361, 710–712.

Jelgersma, S., 1996. Land subsidence in coastal lowlands. In: Milliman, J.D., Haq, B.U. (Eds.), Sea Level Rise and Coastal Subsidence: Causes, Consequences, and Strategies Coastal Systems and Continental Margins, vol. 2, pp. 47–62.

Johnson, K.A., Swift, D.J.P., Grosch, C., 2005. Correlation and cause of fifth order cycles within the Upper Cretaceous Eagle Formation, Bighorn Basin of Wyoming. Abstracts with Programs—Geological Society of America 37 (7), 337.

Jordan, T.E., 1981. Thrust loads and foreland basin evolution, Cretaceous, Western United States. AAPG Bulletin 65 (12), 2506–2520.

Karl, H.A., 1976. Depositional history of Dakota Formation (Cretaceous) Sandstones, Southeastern Nebraska. Journal of Sedimentary Petrology 64, 124–131.

Kauffman, E.G., 1977. Geological and biological overview: Western Interior Cretaceous basin. In: Kauffman, E.G. (Ed.), Cretaceous Facies, Faunas and Paleoenvironments across the Western Interior Basin: Field Guide: North American Paleontological Convention II Mountain Geologist, vol. 14, pp. 75–99, No. 3–4.

Kauffman, E.G., 1985. Cretaceous evolution of the Western Interior Basin of the United States. In: Pratt, et al. (Ed.), Fine-Grained Deposits and Biofacies of the Cretaceous Western Interior Seaway: Evidence of Cyclic Sedimentary Processes. SEPM Field Trip Guidebook, No. 4, p. 10.

Kauffman, E.G., Caldwell, W.G.E., 1993. The Western Interior basin in space and time. In: Caldwell, W.G.E., Kauffman, E.G. (Eds.), Evolution of the Western Interior Basin. Geological Association of Canada Special Paper 39, pp. 1–30.

Kauffman, E.G., Sageman, B.B., Kirkland, J.I., Elder, W.P., Harries, P.J., Villamil, T., 1993. Molluscan biostratigraphy of the Cretaceous Western Interior Basin, North America. In: Caldwell, W.G.E., Kauffman, E.G. (Eds.), Evolution of the Western Interior Basin. Geological Association of Canada Special Paper 39, pp. 397–434.

Koch, J., Brenner, R., 2009. Evidence for glacioeustatic control of large, rapid sea-level fluctuations during the Albian–Cenomanian: Dakota formation, eastern margin of Western Interior Seaway, USA. Cretaceous Research 30 (2), 411–423.

Ku Shafie, K.R., 2005. High Resolution Chemostratigraphic Correlations of Marine and Non-marine Strata of the Eastern Margin, Cretaceous Western Interior Seaway of North America (Unpublished B.S. thesis). The Pennsylvania State University, University Park, PA.

LECO Corporation, 2003. Dual Range Sulfur and Carbon Analyzer Applications Bulletin.

Lehmann, C., Oseleger, D.A., Montanez, I.P., 1998. Controls on cyclostratigraphy of Lower Cretaceous carbonates and evaporites, Cupido and Coahuila platforms, northeastern Mexico. Journal of Sedimentary Research 68, 1109–1130.

Ludvigson, G.A., González, L.A., Metzger, R.A., Witzke, B.J., Brenner, R.L., Murillo, A.P., White, T.S., 1998. Meteoric sphaerosiderite lines and their use for paleohydrology and paleoclimatology. Geology 26 (11), 1039–1042.

MacFarlane, P.A., Wade, A., Doveton, J.H., Hamilton, V.J., 1991. Revised Stratigraphic Interpretation and Implications for Pre-Graneros Paleogeography from Test-Hole Drilling in Central Kansas. Kansas Geological Survey Open-file Report 91–1A, p. 73.

MacFarlane, P.A., Doveton, J.H., Howard, R.F., Butler Jr., J.J., Combes, J.M., Collin, D.R., 1994. Aquifer/aquitard units of the Dakota aquifer system in Kansas: methods and delineations and sedimentary architecture effects on groundwater flow and flow properties. Journal of Sedimentary Research 64 (4), 464–480.

Masetti, D., Neri, C., Bosellini, A., 1991. Deep-water asymmetric cycles and progradation of carbonate platforms governed by high-frequency eustatic oscillations (Triassic of the Dolomites, Italy). Geology 19, 336–339.

Matthews, R.K., Poore, R.Z., 1981. Tertiary $\delta^{18}O$ record and glacio-eustatic sea-level fluctuations. Geology 8, 501–504.

Miller, K.G., Wright, J.D., Browning, J.V., 2005. Visions of ice sheets in a greenhouse world. Marine Geology 217, 215–231.

Milliman, J.D., Haq, B.U., 1996. Sea level rise and coastal subsidence: towards meaningful strategies. In: Milliman, J.D., Haq, B.U. (Eds.), Sea Level Rise and Coastal Subsidence: Causes, Consequences, and Strategies Coastal Systems and Continental Margins, vol. 2, pp. 1–9.

Morath, P., 2007. High-Resolution Chemostratigraphic Correlations of Mid-Cretaceous Strata Deposited from the Eastern Margin to the Axial Basin of the Western Interior Basin, Northwestern Iowa to Northeastern Wyoming (Unpublished M.S. thesis). The Pennsylvania State University, University Park, PA.

Obradovich, J.D., 1993. A Cretaceous time scale. In: Caldwell, W.G.E., Kauffman, E.G. (Eds.), Evolution of the Western Interior Basin. Geological Association of Canada Special Paper 39, pp. 379–396.

Pang, M., Nummedal, D., 1995. Flexural subsidence and basement tectonics of the Cretaceous Western Interior Basin, United States. Geology 23, 173–176.

Parrish, J.T., Spicer, R.A., 1988. Late Cretaceous terrestrial vegetation, a near polar temperature curve. Geology 16 (1), 22–25.

Plint, A., Kreitner, M., 2007. Extensive thin sequences spanning Cretaceous foredeep suggest high-frequency eustatic control: Late Cenomanian, Western Canada foreland basin. Geology 35 (8), 735–738.

Poulsen, C.J., Barron, E.J., Johnson, C.C., Fawcett, P., 1999. Links between major climatic factors and regional oceanic circulation in the Mid-Cretaceous. In: Barrera, E., Johnson, C.C. (Eds.), Evolution of the Cretaceous Ocean-Climate System. Geological Society of America Special Paper 287, pp. 111–128.

Poulsen, C., Pollard, D., White, T., 2007. General circulation model simulation of the $\delta^{18}O$ content of continental precipitation in the middle Cretaceous: a model-proxy comparison. Geology 35 (3), 199–202.

Ravn, R.L., Witzke, B.J., 1995. The palynostratigraphy of the Dakota Formation (?Late Albian–Cenomanian) in its type area, Northwestern Iowa and Northeastern Nebraska, USA. Palaeontographica Abteilung B 234, 93–171.

Rich, P.V., Rich, T.H., Wagstaff, B.E., Mason, J.M., Douthitt, C.B., Gregory, R.T., Felton, E.A., 1988. Evidence for low temperatures and biologic diversity in Cretaceous high latitudes of Australia. Science 242 (4884), 1403–1406.

Sageman, B.B., Rich, J., Arthur, M.A., Birchfield, G.E., Dean, W.E., 1997. Evidence for Milankovitch periodicities in Cenomanian–Turonian lithologic and geochemical cycles, Western Interior U.S.A. Journal of Sedimentary Research 67 (2), 286–302.

Scott, R.W., Franks, P.C., Evetts, M.J., Bergen, J.A., Stein, J.A., 1998. Timing of Mid-Cretaceous relative sea level changes in the Western Interior: Amoco No. 1 Bounds Core. In: Dean, W.E., Arthur, M.A. (Eds.), Stratigraphy and Paleoenvironments of the Cretaceous Western Interior Seaway, USA. SEPM Concepts in Sedimentology and Paleontology No. 6, pp. 11–34.

Spicer, R.A., Parrish, J.T., 1990. Late Cretaceous–early Tertiary palaeoclimates of northern high latitudes: a quantitative view. Journal of the Geological Society of London 147, 329–341.

Stelck, C.R., Koke, K.R., 1987. Foraminiferal zonation of the Viking interval in the Hasler Shale (Albian), northeastern British Columbia. Canadian Journal of Earth Sciences 24, 2254–2278.

Stelck, C.R., 1991. Foraminifera of the middle to upper Albian transition (Lower Cretaceous), northeastern British Columbia. Canadian Journal of Earth Sciences 28, 561–580.

Sterzinar, E., Leckie, R.M., 2004. Foraminiferal assemblages and paleoceanography of the Upper Cretaceous (Coniacian) of western Kansas during ocean anoxic event 3. Abstracts with Programs—Geological Society of America 36 (5), 292.

Vakarelov, B.K., Bhattacharya, J.P., Nebrigic, D.D., 2006. Importance of high-frequency tectonic sequences during greenhouse times of earth history. Geology 34 (9), 797–800.

Van Wagoner, J.C., Mitchum, R.M., Campion, K.M., Rahmanian, V.D., 1990. Siliciclastic Sequence Stratigraphy in Well Logs, Cores, and Outcrops. AAPG Methods in Exploration Series, vol. 7, p. 63.

Walaszczyk, I., Shank, J., Plint, A., Cobban, W., 2014. Interregional correlation of disconformities in Upper Cretaceous strata, Western Interior Seaway: Biostratigraphic and sequence-stratigraphic evidence for eustatic change. Geological Society of America Bulletin 126 (3–4), 307–316.

Weimer, R.J., 1978. Influence of Transcontinental arch in Cretaceous sedimentation and petroleum occurrence in Western Interior. AAPG Bulletin 63 (3), 550.

Weimer, R.J., 1984. Relation of unconformities, tectonics, and sea-level changes, Cretaceous of Western Interior, U.S.A. In: Schlee, J.S. (Ed.), Interregional Unconformities and Hydrocarbon Accumulations AAPG Memoir, vol. 36, pp. 7–35.

White, T., 1999. Organic Petrography, Geochemistry, and Sequence Stratigraphy of Upper Cenomanian through Middle Turonian (Cretaceous) Strata in the Western Interior Basin, Utah and Colorado (Unpublished Ph.D. Dissertation). The Pennsylvania State University, University Park, PA.

White, T.S., Witzke, B.J., Ludvigson, G.A., 2000. Evidence for an Albian Hudson arm connection between the Cretaceous Western Interior Seaway of North America and the Labrador Sea. GSA Bulletin 112 (9), 1342–1355.

White, T., Gonzalez, L., Ludvigson, G., Poulsen, C., 2001. Middle Cretaceous greenhouse hydrologic cycle of North America. Geology 29 (4), 363–366.

White, T., 2004. A chemostratigraphic and geochemical facies analysis of strata deposited in an Eocene Australo-Antarctic Seaway: is cyclicity evidence for glacioeustasy? In: The Cenozoic Ocean: Tectonics, Sedimentation, and Climate Change Between Australia and Antarctica AGU Geophysical Monograph Series, vol. 151, pp. 153–172.

White, T., Witzke, B., Ludvigson, G., Brenner, R., 2005. Distinguishing base-level change and climate signals in a Cretaceous alluvial sequence. Geology 33 (1), 13–16.

Witzke, B.J., Ludvigson, G.A., Poppe, J.R., Ravn, R.L., 1983. Cretaceous paleogeography along the eastern margin of the Western Interior seaway, Iowa, southern Minnesota, and eastern Nebraska and North Dakota. In: Reynolds, M.W., Dolly, E.D. (Eds.), Mesozoic Paleogeography of the West-Central United States. Society of Economic Paleontologists and Mineralogists, Rocky Mountain Paleogeography Symposium 2, pp. 225–252.

Witzke, B.J., Ludvigson, G.A., 1987. Cretaceous Exposures, Big Sioux River Valley North of Sioux City Iowa. Geological Society of America Centennial Field Guide—North-central Section. pp. 97–102.

Witzke, B.J., Ludvigson, G.A., 1994. The Dakota Formation in Iowa and its type area. In: Shurr, et al. (Ed.), Perspectives on the Eastern Margin of the Cretaceous Western Interior Basin. Geological Society of America Special Paper 287, pp. 43–78.

Young, L.D., 2002. High Resolution Chemostratigraphic Correlations and the Development of Accommodation Space, in Mid-Cretaceous Strata, Eastern Margin Western Interior Seaway, Central United States (Unpublished M.S. thesis). University of Iowa, Iowa City, IA.

CHAPTER

10 PALEO-REDOX CONDITIONS OF THE ALBIAN-DANIAN CARBONATE ROCKS OF THE CAUVERY BASIN, SOUTH INDIA: IMPLICATIONS FOR CHEMOSTRATIGRAPHY

J. Madhavaraju[1], S.M. Hussain[2], J. Ugeswari[2], R. Nagarajan[3], S. Ramasamy[2], P. Mahalakshmi[2]

[1]Estación Regional del Noroeste, Instituto de Geología, Universidad Nacional Autónoma de México, Hermosillo, Sonora, México; [2]Department of Geology, University of Madras, Chennai, India; [3]Department of Applied Geology, School of Engineering and Science, Curtin University, Miri, Sarawak, Malaysia

10.1 INTRODUCTION

Chemostratigraphy involves the geochemical classification and correlation of sedimentary strata by using major and trace element geochemistry, which is highly helpful when applied to sequences with poor biostratigraphic control (Pearce et al., 1999). Geochemical data can be used to correlate sequences on the basis of stratigraphic variations and this potential has been exploited for some time (Pettijohn, 1975). It has led to the potential development of down-hole geochemical logging tools (Hertzog et al., 1987). However, the use of such tools is considerably expensive and mainly deals with the major elements (Hertzog et al., 1987; Primmer et al., 1990; Pearce et al., 1999). Recently, more focus has been given to the immobile trace elements (Nb, Hf, Zr and Ta) and to certain rare earth elements (REEs) that provide information regarding the source rocks characteristics (Bhatia and Crook, 1986; Taylor and McLennan, 1985; Floyd et al., 1991).

Chemostratigraphy is emerging as one of the important tools in petroleum exploration and stratigraphic correlation (Ramkumar, 1999). The important consideration is to ensure the preservation of the primary geochemical composition of the strata before undertaking such a study (Ramkumar et al., 2006). During the past several decades, chemostratigraphic concept has emerged to be a potential tool for correlation of widely separated strata where other conventional stratigraphic methods fail or in the absence of biostratigraphic information (e.g., Sawaki et al., 2008; Ishikawa et al., 2008; Le Guerroue and Cozzi, 2010; Ramkumar et al., 2010). This concept has been considered as a critical tool for stratigraphic correlation, fixation of geological boundaries, and petroleum exploration (Saltzman, 2002; Schroeder et al., 2004; Bergstoerm et al., 2006; Kouchinsky et al., 2007; Marquillas et al., 2007; Schroeder and Grotzinger, 2007; Handley et al., 2008; Elrick et al., 2009; Ruhl et al., 2009). Several publications have presented either geochemical and/or isotopic variations across a chronological boundary or

Chemostratigraphy. http://dx.doi.org/10.1016/B978-0-12-419968-2.00010-8

geochemical trends of limited chronological span (Brasier and Shields, 2000; Mutti and Bernoulli, 2003; Saylor et al., 2005; Mutti et al., 2006; Nedelec et al., 2007; Kakizaki and Kano, 2009), thus concentrating on the presence/absence of similarity in geochemical profiles for a particular geologic time (Ramkumar et al., 2011).

Many researchers have carried out detailed studies on stratigraphy, sequence stratigraphy, paleontology, clay mineralogy, depositional environments, and tectonic settings of the Cauvery Basin (Sastry et al., 1972; Sundaram and Rao, 1986; Ramasamy and Banerji, 1991; Govindan et al., 1996; Madhavaraju and Ramasamy, 1999a,b, 2001, 2002; Sundaram et al., 2001; Nagendra et al., 2002; Madhavaraju et al., 2002, 2004, 2006). Madhavaraju and Lee (2009) reported the influence of Deccan volcanism in the sedimentary sequences of the Cauvery Basin using major, trace, and REEs. Ramkumar et al. (2011) have undertaken the geochemical studies to describe the chemostratigraphic variations of the sedimentary rocks of the Cauvery Basin. Isotopic studies on bivalve and belemnite shells have been carried out by Zakharov et al. (2011) to understand the paleotemperatures and Cretaceous climatic conditions in the Cauvery Basin. Several studies were carried out on outcrop samples and they mainly dealt with stratigraphic and depositional environment problems. The present study deals with the carbonate rocks of Albian-Danian age (Dalmiapuram Formation: Albian age; Kallankurichchi Formation: Maastrichtian age; Niniyur Formation: Danian age) to understand the variations in paleoredox conditions among different formations.

10.2 GEOLOGY AND STRATIGRAPHY

The Cauvery Basin is a Mesozoic extensional basin formed along the eastern continental margin of Peninsular India and has been classified as a rift basin that was developed as a result of Mesozoic extension during the break up of Gondwanaland (Prabhakar and Zutshi, 1993; Rangaraju et al., 1993). The Cauvery Basin contains 6-km thick sediments that were deposited on Archean basement (Rangaraju et al., 1993) in an epicontinental sea (Ramkumar et al., 2004). The sedimentary rocks of Cretaceous age are well exposed in five isolated areas (Pondicherry, Vridhachalam, Ariyalur, Tanjore, and Sivaganga) of the Cauvery Basin (Banerji, 1972). Of these areas, clastic and carbonate rocks of Cretaceous-Paleocene age are best developed in the Ariyalur area (Figs 1 and 2). Blanford (1862) divided these Cretaceous sedimentary rocks into three distinct groups on the basis of lithological characteristics, i.e., the Uttattur, Trichinopoly, and Ariyalur Groups.

The Uttattur Group encompasses the transgressive, basal part of the Cretaceous succession that onlap the crystalline basement, progressively stepping to the west. It consists of terrestrial, paralic, and shallow marine strata and is developed to a maximum thickness of some 820 m. Earlier interpretations have separated the basal terrestrial facies, identified by Blanford (1862) as the Uttattur plant beds, from the overlying marine sequence. This interval has been commonly termed as the upper Gondwanas, considered to be bounded by unconformities, and assigned a Late Jurassic or Early Cretaceous age. The Cretaceous Uttatur Group has been subdivided into four formations: Terani Formation, Arogyapuram Formation, Dalmiapuram Formation, and Karai Formation (Sundaram et al., 2001). The first three formations have been dated Early Cretaceous and the latter formation into Late Cretaceous.

The Trichinopoly Group comprises a discrete tectonostratigraphic assemblage, separated from the underlying Uttattur Group by a regional unconformity. To the south, in the vicinity of Dalmiapuram,

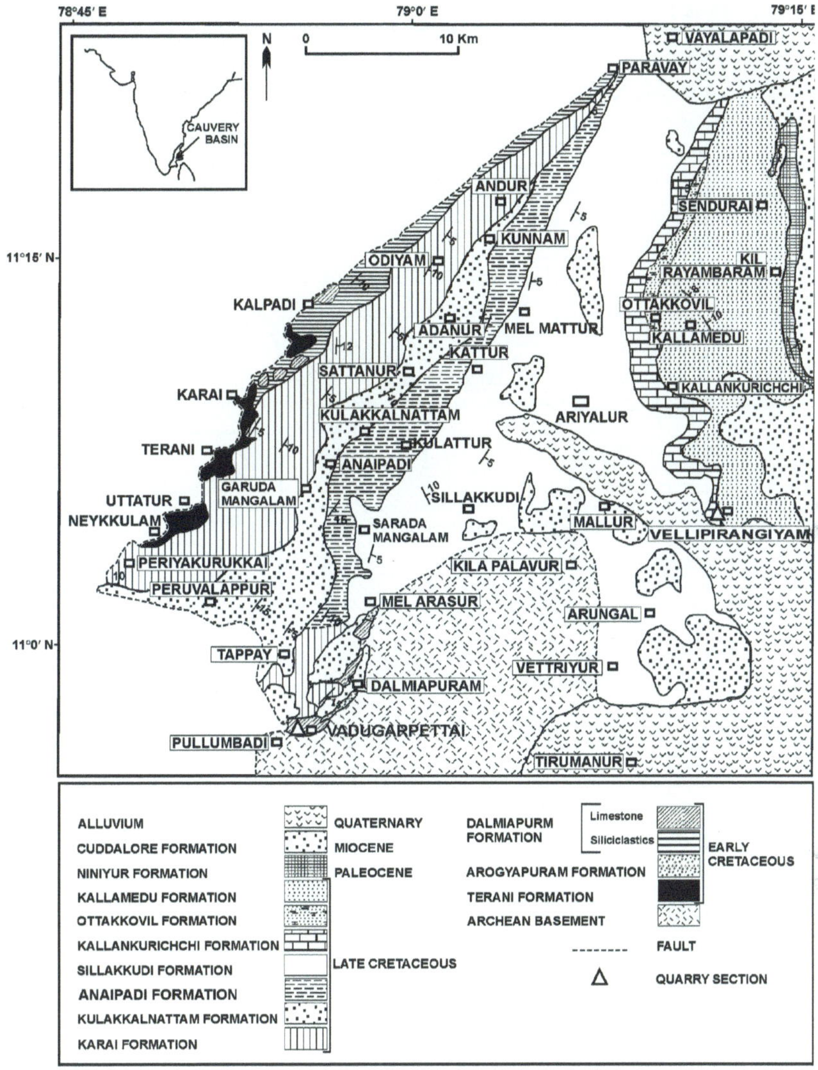

FIGURE 1

Location map of the Vadugarpettai quarry section and Vellipirangiyam quarry section.

Modified after Sundaram et al. (2001).

the separation is marked by an angular unconformity with more steeply dipping Uttattur strata truncated and stepped over by the base of the Trichinopoly Group. The Trichinopoly Group has precedence as do its component Kulakkalnattam and Anaipadi Formations as defined by Sundaram and Rao (1986). The Trichinopoly Group is 490 m thick.

The Ariyalur Group is conformable with the Trichinopoly Group but it oversteps basement along its southern contact, between the Kila Palavur and Vettriyur villages. The Ariyalur Group is more widely

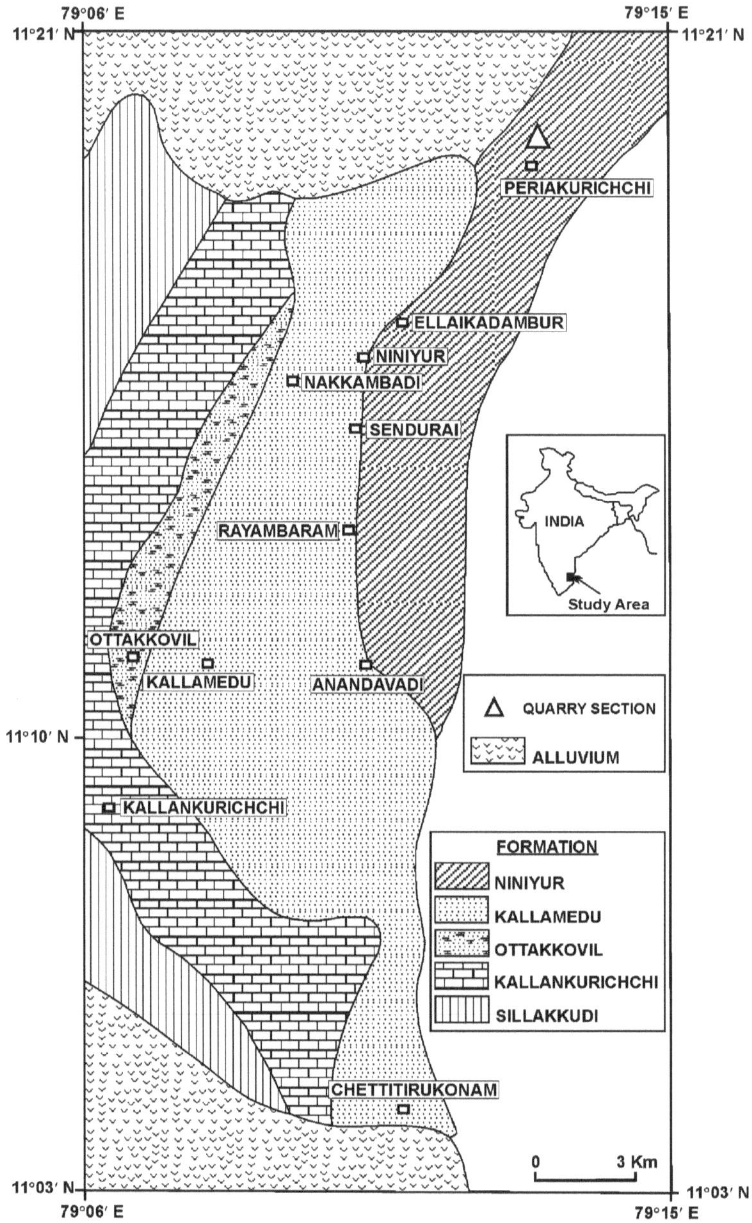

FIGURE 2

Location map of the Periakurichchi quarry section.

After Madhavaraju and Lee (2010).

exposed than the other two. Sastry et al. (1972) subdivided the Ariyalur Group into four formations, namely the Sillakkudi, Kallankurichchi, Ottakkovil, and Kallamedu formations with decreasing age. Later, Sundaram and Rao (1986) studied the Cretaceous-Paleocene rocks and the detailed lithostratigraphic classification. They proposed separate formation status for the Paleocene rocks, the Niniyur Formation. The Niniyur Formation (Early Paleocene) overlies the Kallamedu Formation conformably (Sundaram and Rao, 1986; Sundaram et al., 2001) and comprises calcareous sandstone, sandy claystone, and limestone. Fossils of the Niniyur Formation are indicative of Early Paleocene age.

The carbonate rocks are well exposed in the Dalmiapuram Formation (Uttattur Group; Aptian-Albian age), the Kallankurichchi Formation (Ariyalur Group; Maastrichtian age), and the Niniyur Formation (Danian age). In this study, we have followed the lithostratigraphic classifications proposed by Sastry et al. (1972) and Sundaram et al. (2001). Limestone samples were collected from the Vadugarpettai quarry section (VQS) (Dalmiapuram Formation), the Vellipirangiyam quarry section (VLQS) (Kallankurichchi Formation), and the Periakurichchi quarry section (PQS) (Niniyur Formation).

Detailed lithostratigraphic descriptions of the VQS (Dalmiapuram Formation), the VLQS (Kallankurichchi Formation), and the PQS (Niniyur Formation) are given on the basis of detailed field observations. The observed lithological units are given in the lithostratigraphic columns (Figs 3–5).

The sedimentary rock samples were collected systematically from the VQS of the Dalmiapuram Formation. On the basis of the lithology, it has been divided into two units: the lower gray shale (GS) unit and the upper carbonate/marl unit. The GS unit (9 m thick) is well exposed in the Vadugarpettai quarry. The color of the shale varies from ash gray to dark gray, depending upon the amount of organic matter. The GS is overlain by 8.2 m of coral algal limestone (CAL). The CAL is massive, hard, and pinkish to flesh-red. It contains numerous vugs and cavities. The CAL is overlain by marl bedded limestone (MBL). Thick bands of bedded limestones (BLs) are interbedded with marls, which are soft and off-white to brownish yellow in color. The total thickness of the MBL is about 13 m. The MBL is overlain by 3.5 m of marl. It is soft, powdery, and off-white in color. The bedding contact between the MBL and the marls is well defined in the quarry section. According to earlier studies, the age of the Dalmiapuram Formation is Early to Middle Albian (Ramasamy and Banerji, 1991; Kale and Pansalkar, 1992).

The limestone samples were collected vertically (from base to top) from the VLQS. On the basis of the lithology, it has been divided into five distinct units: lower oyster *Gryphaea* limestone (OGL), *Inoceramus* limestone (IL), upper OGL, fossiliferous limestone (FL), and marl. The basal part of the quarry section exposes the lower OGL unit. It is massive, hard, and yellowish in color. Total thickness of this unit is 8 m. The succeeding bed is IL. It is pale yellow in color and rich in *Inoceramus* fauna. Total thickness of this unit is 3 m. It is overlain by the upper OGL unit, which is hard and yellowish in color and contains numerous oyster and *Gryphaea* shells. Thickness of this unit is 5.2 m. It is succeeded by a thick FL unit (4 m thick), which is massive, hard, and pale yellow in color. The top-most unit of this section is marl, with a total thickness of 3 m. On the basis of the index fossils (foraminifera), an Early Maastrichtian age has been assigned to the Kallankurichchi Formation (Srivastava and Tewari, 1967).

Samples were collected from the PQS on the basis of lithological variations. Five distinct units were identified: FL, sandstone with boulders, limestone interbedded with shale and sandstone, sandstone, and marl. Thickness of this unit is 6 m. The succeeding 7-m thick unit is sandstone, which contains numerous boulders. It is overlain by thick bands of limestone, which are interbedded with shale and sandstone beds. The total thickness of this unit is 12.9 m. These beds are succeeded by a thick

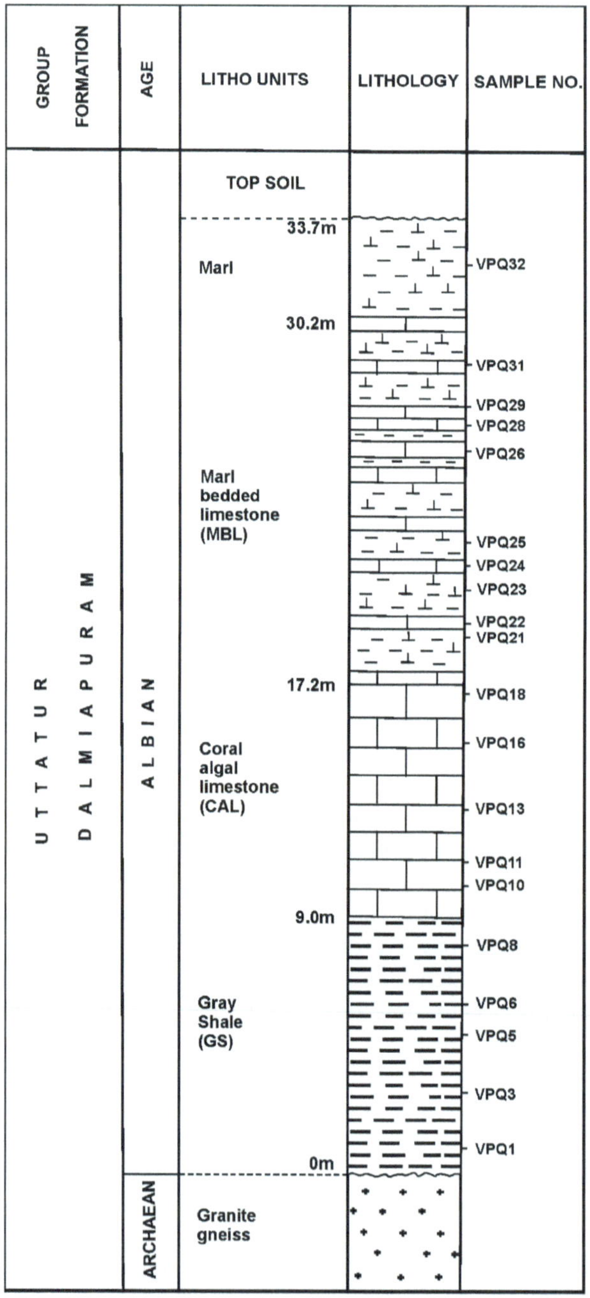

FIGURE 3

Lithostratigraphy of the Vadugarpettai quarry section of the Dalmiapuram Formation.

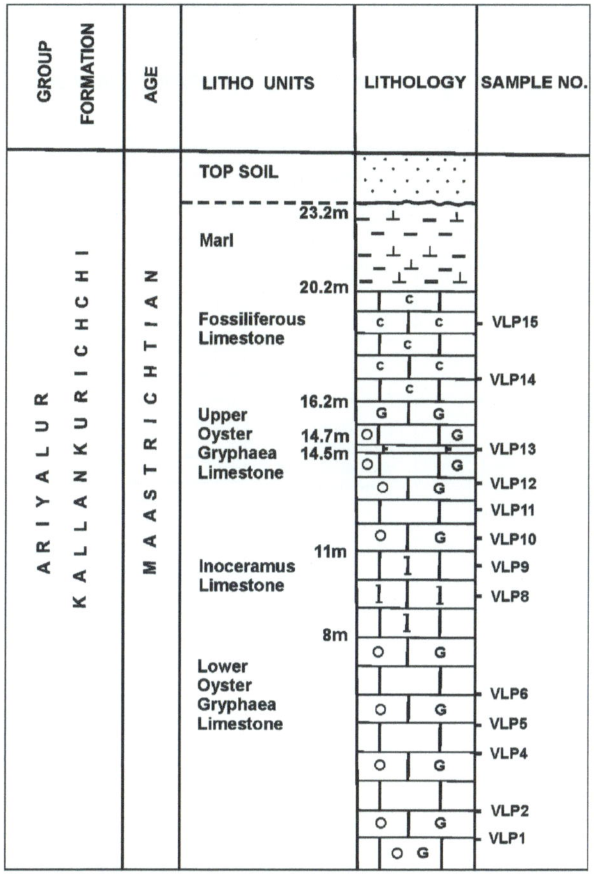

FIGURE 4

Lithostratigraphy of the Vellipirangiyam quarry section of the Kallankurichchi Formation.

sandstone bed having a thickness of 2 m. The top-most unit of this section is marl (1 m thick). The occurrence of ammonites and foraminifera suggests Danian age for the Niniyur Formation (Kossmat, 1897; Mamgain et al., 1968; Sundaram et al., 2001). In addition, strontium isotope studies on the carbonate rocks also support the Danian age (Madhavaraju et al., 2014).

10.3 METHODOLOGY

Samples were collected from three quarry sections: VQS (Early to Middle Albian age); VLQS (Early Maastrichian age); and PQS (Danian age). Forty-five samples (20 samples from VQS, 13 samples from VLQS, and 12 samples from PQS) were selected for geochemical study to understand the variations in paleoredox conditions during Albian, Early Maastrichtian, and Danian age in the Cauvery Basin. The selected samples were washed several times with distilled water, air-dried, and powdered in an agate

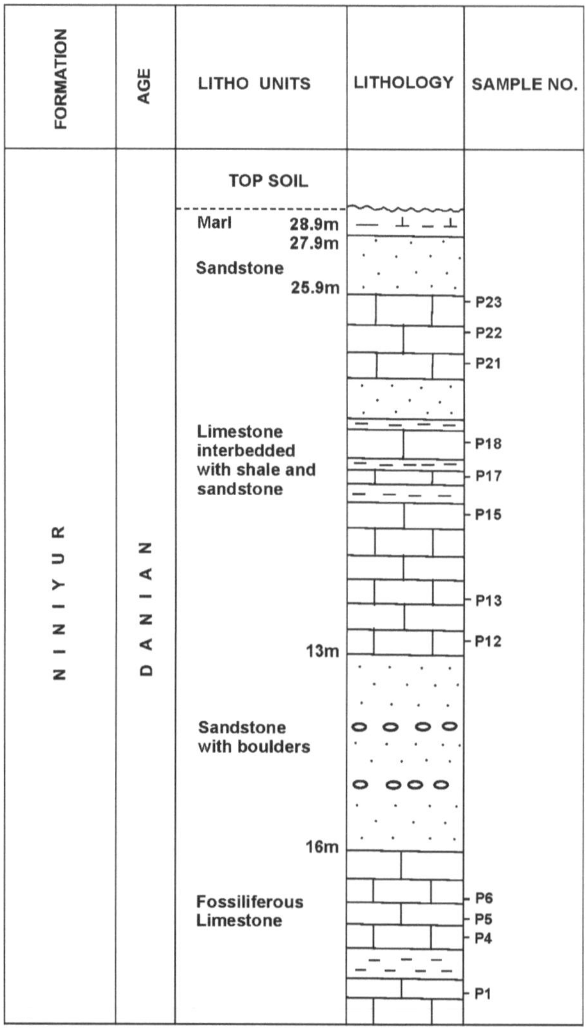

FIGURE 5

Lithostratigraphy of the Periakurichchi quarry section of the Niniyur Formation.

mortar. Then, fused glass beads were prepared for major element analysis using an X-ray fluorescence spectrometer with an Rh-anode X-ray tube as a radiation source. The geochemical standard JGB1 was used to determine data quality. The analytical precision for major elements is better than 1%. Loss on ignition was determined from the weight loss, after heating 1 gm of sample to 1000 °C in a platinum crucible. Trace elements were determined by an Agilent 7500 cc inductively coupled plasma mass spectrometer following the analytical procedures mentioned by Eggins et al. (1997). The geochemical standards IGLa-1 and SDO-1 were used to monitor the analytical reproducibility. The analytical precision errors for V, Cr, Co, Ni, Cu, Zn, Rb, and Sr were better than ±3%, while the analytical accuracy errors of certain trace elements (Ba, Zr, U, and Th) were better than ±6%.

10.4 RESULTS

10.4.1 MAJOR OXIDES

10.4.1.1 Vadugarpettai Quarry Section (Early to Middle Albian)

Major oxide concentrations in the CAL, BL, marl, and GS are listed in Table 1. SiO_2, Al_2O_3, and Fe_2O_3 contents are high in the GS (19.90–34.70%; 2.83–5.95%; 1.00–1.51%, respectively), BL (2.06–5.28%; 0.34–1.01%; 0.10–0.56%, respectively), and marl (11.20–29.10%; 1.66–4.97%; 0.28–1.29%, respectively), whereas low concentrations are found in CAL (0.11–0.72%; 0.06–0.21%; 0.09–0.11%, respectively). Smaller variations are found in CaO content in CAL and BL (54.40–55.50%; 51.30–54.10%, respectively) than marl and GS (31.90–49.70%; 26.90–40.30%, respectively). K_2O and Na_2O contents are low in both CAL (0.03–0.06%; 0.01–0.03%, respectively) and BL (0.12–0.28%; 0.01–0.03%, respectively), whereas higher concentrations are observed in marl (0.41–1.03%; 0.01–0.05%, respectively) and GS (0.071–1.34%; 0.05–0.12%, respectively). Small variations of MnO content are found in CAL, BL, and marl (0.01–0.03%; 0.16–0.21%; 0.27–0.29%, respectively), whereas large variations are found in GS (0.01–0.33%).

The vertical geochemical distributions of major oxides in the VQS are given in Fig. 6. Most of the major oxides are enriched in GS and marl when compared with CAL and BL. SiO_2, Al_2O_3, Fe_2O_3, MgO, K_2O, Na_2O, and TiO_2 show similar distribution patterns, whereas CaO exhibits a different pattern. The geochemical variations suggest that the variations in major oxides in different lithologies may be due to variations in the amount of terrigenous materials in them.

10.4.1.2 Vellipirangiyam Quarry Section (Early Maastrichian)

The concentrations of major oxides in the OGL, IL, and FL are listed in Table 2. Large variations are found in SiO_2, Al_2O_3, Fe_2O_3, and CaO contents in OGL (2.03–13.90%; 0.60–3.81%; 0.54–4.75%; 42.50–53.90%, respectively) whereas the least variations are observed in IL (3.67–4.16%; 0.96–1.00%; 1.05–2.52%; 52.00–52.50%, respectively) and FL (3.25–4.15%; 0.90–0.97%; 1.19–1.73%; 52.70–52.80, respectively). MgO content is higher in OGL (0.39–1.23%) than IL (0.55–0.56%) and FL (0.41–0.43%). Low K_2O contents are observed in OGL, IL, and FL (0.17–0.73%; 0.34–0.46%; 0.25–0.37%, respectively). The contents of Na_2O, MnO, and P_2O_5 are very low in the limestones of Early Maastrichtian age (Table 2). The vertical distributions of major oxides are given in Fig. 7. Most of the major oxides show similar distribution trends except for CaO, which exhibits an opposite pattern.

10.4.1.3 Periakurichchi Quarry Section (Danian)

Table 3 lists the major oxide concentrations of the limestone samples of the PQS. Large variations are found in SiO_2, Al_2O_3, and CaO contents (0.76–14.90%; 0.12–4.09%; 44.10–55.00%, respectively). Small variations are found in Fe_2O_3 and MgO contents (0.09–1.15%; 0.41–1.61%, respectively) in limestones of Danian age. Low concentrations of K_2O, Na_2O, and MnO are observed in the limestones (0.02–0.32%; 0.01–0.12%; 0.16–0.25%, respectively). The vertical geochemical distributions of major

Table 1 Major (%) and Trace (ppm) Elements Concentrations in the Sedimentary Rocks from Vadugarpettai Quarry Section

Rock	Gray Shale					Coral Algal Limestone				
Sample No.	VPQ1	VPQ3	VPQ5	VPQ6	VPQ8	VPQ10	VPQ11	VPQ13	VPQ16	VPQ18
SiO_2	34.70	34.00	19.90	32.00	32.30	0.45	0.42	0.72	0.27	0.11
Al_2O_3	5.53	4.84	2.83	5.23	5.95	0.09	0.06	0.15	0.21	0.09
Fe_2O_3	1.51	1.24	1.00	1.19	1.45	0.10	0.10	0.11	0.09	0.11
CaO	26.90	29.10	40.30	30.40	28.50	54.90	54.40	54.60	55.20	55.50
MgO	2.32	2.15	1.44	2.27	2.56	0.39	0.41	0.21	0.31	0.42
K_2O	1.13	1.07	0.71	1.15	1.34	0.03	0.03	0.06	0.05	0.04
Na_2O	0.12	0.05	0.05	0.08	0.08	0.03	0.02	0.03	0.01	0.01
TiO_2	0.38	0.29	0.15	0.39	0.38	0.01	0.01	0.01	0.01	0.01
P_2O_5	0.08	0.07	0.75	0.07	–	–	0.06	0.19	0.07	0.08
LOI	25.30	26.10	32.70	26.10	25.90	43.50	43.50	43.20	43.80	43.40
Total	97.97	99.22	99.73	99.19	98.46	99.53	99.07	99.32	100.06	99.81
U	1.13	1.17	1.24	1.00	1.25	0.12	0.14	0.76	0.13	0.23
V	95	64	49	66	78	6	5	4	3	2
Mo	0.38	0.28	0.18	0.28	0.26	–	–	–	–	–
Co	6.00	5.00	6.00	9.09	6.00	2.99	2.71	2.77	2.75	2.70
Cu	19.47	14.37	9.66	15.53	18.13	3.14	–	2.06	–	–
Zn	52	43	33	32	41	7	7	8	4	10
Ni	30	25	25	30	26	9	9	9	9	8
Pb	9.66	7.41	5.49	10.76	9.46	0.82	0.56	0.66	0.53	0.58
Cr	84	57	38	60	66	2	2	3	2	3
Ba	827	668	314	975	1206	20	2	2	12	12
Th	9.55	11.69	3.08	5.20	6.49	0.59	0.55	0.59	0.48	0.53
Zr	31	28	24	34	54	2	2	3	2	2
Rb	33.22	26.32	19.00	26.94	29.91	1.18	1.15	1.73	0.89	1.13
U/Th	0.12	0.10	0.40	0.19	0.19	0.21	0.25	1.30	0.27	0.42
Ni/Co	5.00	5.00	4.20	3.30	4.30	3.10	3.20	3.40	3.30	3.00

Rock	Bedded Limestone						Marl			
Sample No.	VPQ22	VPQ24	VPQ26	VPQ28	VPQ29	VPQ31	VPQ21	VPQ23	VPQ25	VPQ32
SiO_2	5.28	5.21	2.73	2.20	2.06	3.29	11.20	29.10	22.90	12.90
Al_2O_3	0.83	0.81	0.52	0.34	1.01	0.44	1.66	4.97	3.06	2.62
Fe_2O_3	0.56	0.22	0.16	0.55	0.13	0.10	0.70	1.29	0.95	0.28
CaO	51.30	51.80	53.50	53.30	54.10	53.10	46.80	31.90	38.50	49.70
MgO	0.97	0.98	0.69	0.85	0.86	0.85	1.08	2.03	1.43	0.74
K_2O	0.28	0.21	0.21	0.19	0.12	0.16	0.41	0.95	0.75	1.03
Na_2O	0.01	0.03	0.01	0.01	0.03	0.01	0.01	0.04	0.03	0.05
TiO_2	0.08	0.05	0.04	0.03	0.01	0.03	0.11	0.33	0.22	0.19
P_2O_5	–	0.06	0.07	0.05	–	–	–	0.07	0.07	–
LOI	40.20	40.30	41.80	42.20	42.10	41.30	37.30	29.10	31.40	32.00
Total	99.51	99.67	99.73	99.72	99.42	99.30	99.27	99.78	99.31	99.51
U	2.15	1.72	0.23	8.36	1.36	0.86	0.68	0.92	0.81	0.78
V	18	14	15	20	13	15	25	65	38	21
Mo	0.09	0.07	–	–	–	–	0.22	0.36	0.28	0.06
Co	3.66	3.65	2.99	3.06	3.50	2.78	4.74	6.49	5.19	4.20
Cu	4.35	3.96	–	3.09	2.00	–	6.90	15.17	9.08	4.00
Zn	16	12	10	6	13	7	17	46	22	12
Ni	11	10	11	11	11	11	16	30	22	13
Pb	2.21	2.14	3.12	1.64	1.43	1.65	4.36	9.77	5.40	9.00
Cr	12	11	11	11	7	9	19	57	36	20
Ba	38	41	15	16	460	456	104	716	223	438
Th	1.38	0.42	0.50	0.71	0.41	0.76	2.55	5.34	3.57	2.46
Zr	11	5	10	7	2	5	22	41	28	53
Rb	7.19	5.54	5.15	5.77	3.91	5.08	10.75	27.10	17.76	23.32
U/Th	1.56	4.13	0.46	11.71	3.36	1.14	0.27	0.17	0.23	0.32
Ni/Co	2.90	2.80	3.60	3.70	3.10	4.00	3.50	4.60	4.20	3.20

–, Below detection limit; LOI, Loss on ignition.

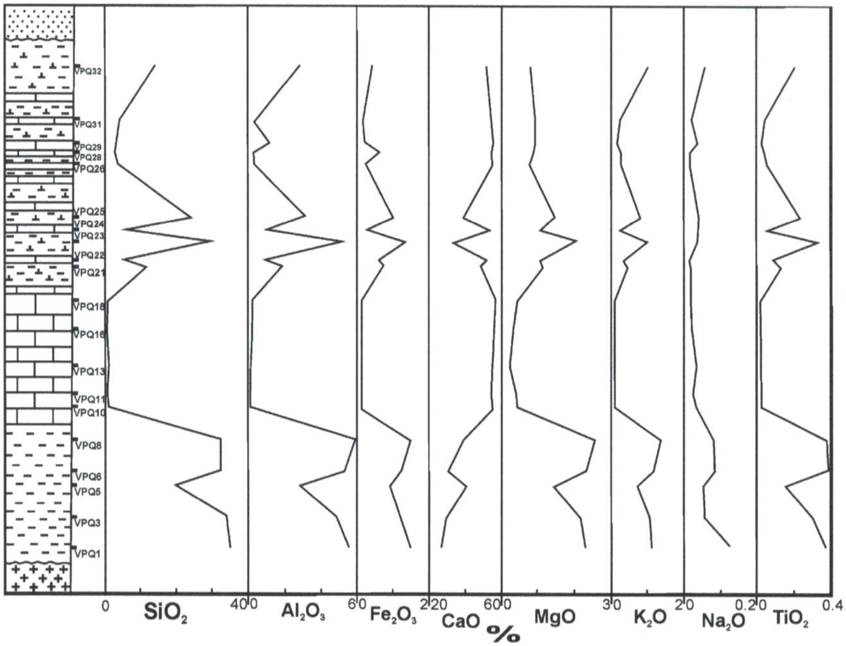

FIGURE 6

Major elements (%) variations in the sedimentary rocks from Vadugarpettai quarry section.

oxides in limestones of PQS are given in Fig. 8. The lower and upper parts of the quarry section are rich in SiO_2 and Al_2O_3 contents, whereas the middle part of the section shows low to moderate contents. Most of the major oxides show similar distribution patterns, except for CaO.

10.4.2 TRACE ELEMENTS

10.4.2.1 Vadugarpettai Quarry Section (Early to Middle Albian)

Trace elemental concentrations of limestone, marl, and GS of the VQS are shown in the Table 1. The concentration of lithophile elements (LILE: Ba and Rb) are higher in GS (314–1206 ppm; 19.00–33.22 ppm, respectively) and marl (104–716 ppm; 10.75–23.32 ppm, respectively) than the CAL (2–20 ppm; 0.89–1.73 ppm, respectively) and BL (15–460 ppm; 3.91–7.19 ppm, respectively). Abnormal values of Ba are observed in the upper part of the GS sequence and lower part of the marl and BL sequences. The ferromagnesian trace elements namely, Cr, V, and Ni are enriched in the GS (38–84 ppm; 49–95 ppm; 25–30 ppm, respectively) and marl (19–57 ppm; 21–65 ppm; 13–30 ppm, respectively), whereas they are depleted in both CAL (2–3 ppm; 2–6 ppm; 8–9 ppm, respectively) and BL (7–12 ppm; 13–20 ppm; 10–11 ppm, respectively). Small variations of Co content are observed in the limestone, GS, and marl (Table 1).

High field strength elements such as Zr, Y, and Th are generally resistant to weathering and alteration processes when compared to other trace elements (Taylor and McLennan, 1985; Bhatia and Crook, 1986; Feng and Kerrich, 1990). Zr content is higher in GS (24–54 ppm) than the CAL, BL, and marl

Table 2 Major (%) and Trace (ppm) Elements Concentrations in the Limestones From Vellipirangiyam Quarry Section

Rock	Oyster *Gryphaea* Limestone									*Inoceramus* Limestone		Fossiliferous Limestone	
Sample No.	**VLP1**	**VLP2**	**VLP4**	**VLP5**	**VLP6**	**VLP10**	**VLP11**	**VLP12**	**VLP13**	**VLP8**	**VLP9**	**VLP14**	**VLP15**
SiO_2	13.90	5.26	9.58	6.70	4.73	2.03	2.55	3.85	6.15	3.67	4.16	3.25	4.15
Al_2O_3	3.81	1.46	1.89	1.59	1.13	0.60	1.86	2.70	1.54	0.96	1.00	0.90	0.97
Fe_2O_3	4.44	1.82	4.75	3.58	2.30	0.60	0.92	0.54	3.35	1.05	2.52	1.73	1.19
CaO	42.50	51.20	47.80	49.10	51.40	53.70	53.90	51.10	49.90	52.50	52.00	52.80	52.70
MgO	1.23	0.67	0.75	0.82	0.58	0.44	0.39	0.59	0.77	0.56	0.55	0.43	0.41
K_2O	0.24	0.26	0.51	0.55	0.42	0.17	0.24	0.38	0.73	0.34	0.46	0.25	0.37
Na_2O	0.02	0.02	0.03	0.04	0.02	0.02	0.03	0.02	0.03	0.02	0.03	0.03	0.04
TiO_2	0.32	0.16	0.02	0.16	0.12	0.05	0.06	0.06	0.13	–	0.12	–	0.10
P_2O_5	–	0.12	0.05	0.05	0.07	0.04	0.04	0.05	–	–	–	0.12	0.10
LOI	32.50	38.60	34.10	37.00	38.60	41.80	40.40	40.60	36.80	40.60	39.00	40.00	39.80
Total	98.96	99.57	99.48	99.59	99.37	99.45	99.99	99.89	99.40	99.70	99.84	99.51	99.83
U	0.71	0.53	0.42	0.58	0.35	0.22	0.33	0.27	0.89	0.53	0.36	0.37	0.27
V	263	133	172	199	108	61	81	41	64	75	54	123	169
Mo	0.31	0.11	0.11	0.34	0.14	0.08	–	–	0.25	0.08	0.12	0.11	0.09
Co	20.48	11.43	29.85	12.05	9.67	6.06	7.91	6.97	23.43	7.16	11.38	13.91	9.12
Cu	4.84	3.14	3.63	5.00	4.45	2.51	2.99	4.00	4.82	12.79	4.02	2.27	3.00
Zn	48	22	30	26	18	8	13	15	23	13	35	17	19
Ni	51	28	41	35	25	15	20	15	24	19	27	35	26
Pb	7.12	4.59	6.19	5.29	3.91	2.08	2.51	2.45	3.90	3.55	3.59	3.65	4.14
Cr	166	60	89	58	61	27	41	27	27	38	27	49	74
Ba	218	49	162	77	102	69	73	91	309	89	139	141	157
Th	7.32	2.82	3.74	3.97	2.93	0.77	2.53	0.74	2.13	0.59	2.77	1.67	3.24
Zr	62	56	22	35	27	9	12	12	26	16	27	18	23
Rb	27.23	10.13	14.65	17.94	15.03	5.65	8.06	12.36	19.15	12.55	18.57	8.35	12.28
U/Th	0.10	0.19	0.11	0.15	0.12	0.28	0.13	0.37	0.42	0.90	0.13	0.22	0.08
Ni/Co	2.50	2.40	1.40	2.90	2.60	2.50	2.50	2.20	1.00	2.60	2.40	2.60	2.90

–, Below detection limit; LOI, Loss on ignition.

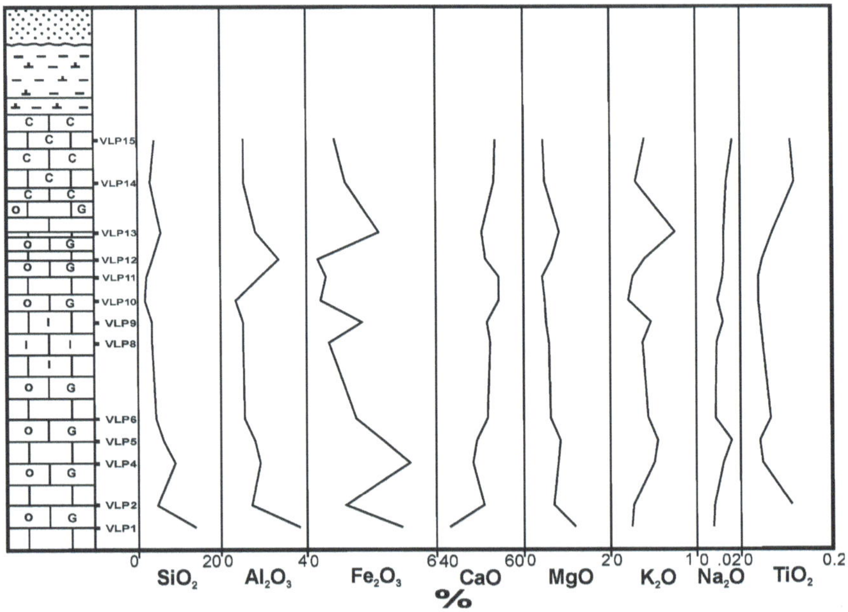

FIGURE 7

Major elements (%) variations in the limestones from Vellipirangiyam quarry section.

(2–3 ppm; 2–11 ppm; 22–53 ppm, respectively). Th content is high in the GS (3.08–11.69 ppm) and marl (2.46–5.34 ppm) than the CAL and BL (0.48–0.59 ppm; 0.41–1.38 ppm, respectively). Al_2O_3 shows positive relationships with immobile trace elements such as Zr and Th, which suggest a terrigenous origin for all these elements. The observed variations in the trace elemental concentrations in GS, CAL, BL, and marl may be due to their lithology.

10.4.2.2 Vellipirangiyam Quarry Section (Early Maastrichtian)

Trace elemental concentrations of the VLQS are listed in Table 2. Large variations in Ba, Co, Cr, and V contents are observed in the OGL (49–309 ppm; 6.06–29.85 ppm; 27–166 ppm; 41–263 ppm, respectively) than in the IL (89–139 ppm; 7.16–11.38 ppm; 27–38 ppm; 54–75 ppm, respectively) and FL (141–157 ppm; 9.12–13.91 ppm; 49–74 ppm; 123–169 ppm, respectively). Zr and Ni contents are higher in the OGL (9–62 ppm; 15–51 ppm, respectively) than the IL (16–27 ppm; 19–27 ppm, respectively) and FL (18–23 ppm; 26–35 ppm, respectively). The observed variations in immobile trace elements such as Zr and Th in OGL, IL, and FL may be due to the inclusion of terrigenous contaminations in the limestones. V, Cr, and Ni contents are higher in the lower and upper part of the quarry section, whereas they are depleted in the middle part of the section. Zr content is higher in the lower part of the section when compared with that of the middle and upper parts of the quarry section.

10.4.2.3 Periakurichchi Quarry Section (Danian)

The concentrations of trace elements in the limestones of the PQS are shown in Table 3. Ba and Zr contents show significant variations in the limestones of the Niniyur Formation (7–1108 ppm;

Table 3 Major (%) and Trace (ppm) Elements Concentrations in the Limestones of the Periakurichchi Quarry Section

Rock	Limestone											
Sample No.	P1	P4	P5	P6	P12	P13	P15	P17	P18	P21	P22	P23
SiO_2	5.64	7.77	2.33	1.71	4.86	0.76	7.56	14.90	2.75	8.58	10.10	3.50
Al_2O_3	1.56	2.20	0.62	1.88	1.04	0.12	2.11	4.09	0.58	1.16	1.28	0.43
Fe_2O_3	0.14	0.18	1.03	0.95	0.49	0.30	1.15	1.02	1.10	0.25	0.22	0.09
CaO	50.80	49.70	52.90	53.30	53.10	54.90	48.80	44.10	53.20	52.70	53.40	55.00
MgO	0.93	0.91	0.49	0.48	0.55	0.41	0.99	1.61	0.63	0.96	0.83	0.51
K_2O	0.12	0.15	0.05	–	0.04	0.02	0.25	0.32	0.09	0.10	0.13	0.04
Na_2O	0.03	0.05	0.03	–	0.02	0.01	0.08	0.12	0.03	0.04	0.06	0.01
TiO_2	0.10	0.14	0.02	0.02	0.12	–	0.17	0.48	0.02	0.10	0.07	0.01
P_2O_5	0.05	0.05	0.05	–	–	0.07	–	–	–	0.06	–	–
LOI	39.70	38.30	42.00	41.30	39.30	43.10	38.20	32.20	41.40	35.40	33.50	40.10
Total	99.07	99.45	99.72	99.64	99.52	99.69	99.31	98.84	99.80	99.35	99.59	99.69
U	1.17	0.96	1.06	0.91	0.94	5.61	0.91	1.41	0.78	1.58	0.96	0.73
V	19	18	14	10	17	13	39	33	35	15	14	10
Mo	2.59	0.19	0.18	<0.05	1.19	0.31	0.60	0.26	0.18	0.38	1.49	0.40
Co	3.69	3.81	3.15	2.83	3.09	3.15	3.38	3.53	3.17	2.98	2.95	3.38
Cu	4.00	2.35	3.01	<2	4.62	2.07	2.71	3.32	3.70	<2	3.63	3.87
Zn	5	4	3	2	5	4	10	13	10	5	5	4
Ni	12	11	11	9	10	10	13	9	13	8	9	10
Pb	2.10	2.22	1.48	1.07	1.84	1.07	3.16	5.04	1.63	1.60	2.42	1.16
Cr	32	23	17	15	31	19	53	66	52	23	20	21
Ba	51	64	13	8	50	7	78	1108	35	930	67	34
Th	0.95	0.97	0.68	0.77	1.32	0.51	1.88	2.11	1.31	0.68	0.74	0.41
Zr	25	15	5	4	20	2	34	90	21	7	13	5
Rb	3.25	3.51	2.08	1.67	3.26	1.29	6.46	7.83	2.79	2.57	3.36	1.80
U/Th	1.23	1.00	1.55	1.18	0.71	11.03	0.48	0.67	0.59	2.32	1.29	1.79
Ni/Co	3.20	2.90	3.40	3.00	3.40	3.10	3.80	2.60	4.00	2.60	3.00	3.00

–, Below detection limit; LOI, Loss on ignition.

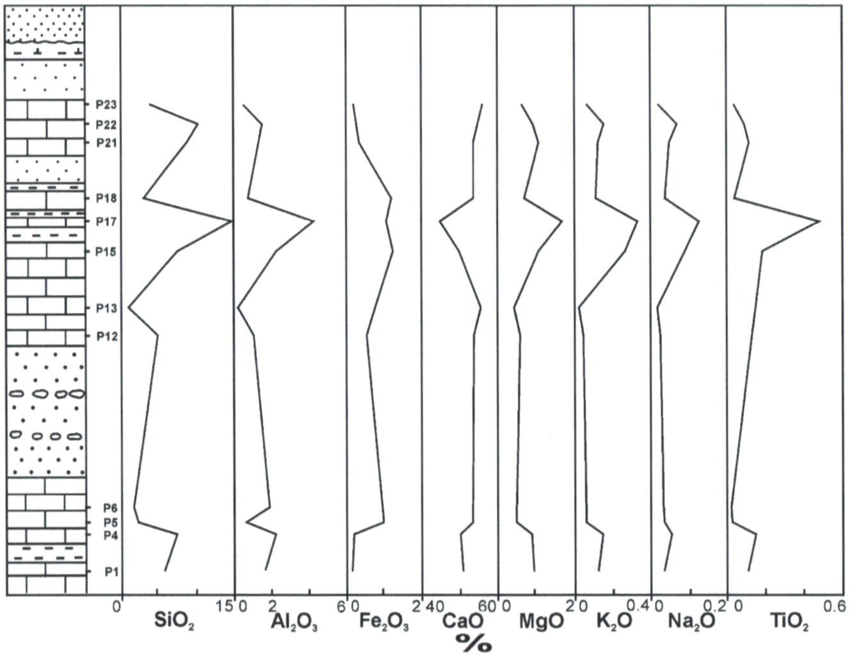

FIGURE 8

Major elements (%) variations in the limestones from Periakurichchi quarry section.

2–90 ppm, respectively). Low concentrations of Co, Ni and Rb are found in the limestones (2.83–3.81 ppm; 8–13 ppm; 1.67–7.83 ppm, respectively). Majority of the limestone samples shows low concentrations of Cr, while a few samples show elevated concentrations (P15: 53 ppm; P17: 66 ppm). Al_2O_3 concentrations show positive correlation with immobile trace elements like Zr and Th. Many elements are depleted in the lower and middle part of the sections. Two abnormal values of Ba coeval with higher concentrations of Cr and Zr are found in the upper part of the sequence.

10.5 REDOX-SENSITIVE TRACE ELEMENTS FOR APPLICATION IN CHEMOSTRATIGRAPHY

Certain elements actively participate in various geochemical processes in the marine environments and they may become authigenically enriched or depleted in sediments depending on the availability of free oxygen during deposition. As a result, the distribution patterns of such elements are extensively used to understand the variations in paleo-redox conditions in both modern marine sediments and ancient rocks (e.g., Calvert and Pedersen, 1993; Canet et al., 2004; Algeo and Maynard, 2004; Tribovillard et al., 2006).

The enrichment of the redox-sensitive trace elements U, V, Mo, Co, and Ni and their ratios U/Th and Ni/Co has significant importance in sedimentological research and is therefore considered to be a tracer of anoxic events through geological time. Enrichment of U, V, and Mo is interpreted as a signature of anoxic and euxinic conditions (Algeo and Maynard, 2004).

Uranium in seawater is present mainly as the uranyl tricarbonate species i.e., U(VI). Reduction of U(VI) to U(IV) occurs under similar pH and alkalinity conditions to those of Fe(III) to Fe(II) reduction in the seawater (Klinkhammer and Palmer, 1991; Crusius et al., 1996; Zheng et al., 2000; Morford et al., 2001; Chaillou et al., 2002; McManus et al., 2005). Enrichment of authigenic U takes place primarily in the sediment and not in the water column, because the reduction of U(VI) to U(IV) is decoupled from the available free H_2S and is not influenced by redox cycling of Fe and Mn in the water column (Algeo and Maynard, 2004; McManus et al., 2005). In the reduced state, removal of uranium from the water column to the sediment may be accelerated by the formation of organometallic ligands in humic acids (Klinkhammer and Palmer, 1991; McManus et al., 2005). Wignall and Maynard (1993) and Algeo and Maynard (2004) also discussed the catalyzing influence of organic substrates on the uranium uptake by the sediments. The accumulation is partly mediated by bacterial sulfate reduction reactions (McManus et al., 2005). In VQS, GSs and marl show moderate content of U (1.00–1.25 ppm; 0.68–0.92 ppm, respectively), whereas CAL shows low to high content of U (0.12–8.36 ppm). In VLQS, the OGL, IL, and FL show low to moderate contents of U (0.22–0.89 ppm; 0.36–0.53 ppm; 0.27–0.37 ppm, respectively; Fig. 9). The limestones of PQS show large variations in U content (0.73–5.61 ppm). Uranium enrichment is found in the lower part of CAL and the lower and upper parts of the MBL. Similarly, enrichment of U is observed in the middle part of the PQS. Low contents of U are found in the sediments deposited in oxygenated conditions in the marine environment (Somayajulu et al., 1994; Madhavaraju and Ramasamy, 1999a), whereas high U contents are found in the sediments from the oxygen minimum zone (Barnes and Cochran, 1990; Klinkhammer and Palmer, 1991; Sarkar et al., 1993; Somayajulu et al., 1994; Nath et al., 1997). The observed enrichment of U in certain intervals of the VQS and middle part of the PQS is due to variations in oxygenation conditions (oxic to suboxic/anoxic).

In the oxic marine environment, vanadium is present as V(V) in the quasiconservative form of vanadate oxyanions. In the hemipelagic and pelagic sediments, vanadium is strongly coupled with the redox cycle of Mn (Hastings et al., 1996). Vanadate adsorbs preferentially onto both Mn- and Fe-oxyhydroxides (Calvert and Piper, 1984; Wehrly and Stumm, 1989) and probably on kaolinite (Breit and Wanty, 1991). Vanadium(V) is reduced to V(IV) and forms vanadyl ions VO^{2-}, related hydroxyl species $VO(OH)_3^-$, and insoluble hydroxides $VO(OH)_2$ under mildly reducing conditions that are favored by the presence of humic and fulvic acids. In the marine environment, the dissolved vanadium is readily bound to the sediment by surface adsorption processes or by formation of organometallic ligands (Emerson and Huested, 1991; Morford and Emerson, 1999). Under highly reduced conditions, the presence of free H_2S released by bacterial sulfate reduction causes V to be further reduced to V(III), which can then be taken up by geoporphyrins or be precipitated as the solid oxide V_2O_3 or hydroxide $V(OH)_3$ phase (Breit and Wanty, 1991; Wanty and Goldhaber, 1992). During the early diagenetic alteration of sediments, V(III) readily substitutes for aluminum in the octahedral sites of authigenic clay minerals or in recrystallizing clay minerals (Breit and Wanty, 1991). Gray shales and marl from the VQS show high content of V (49–95 ppm; 21–65 ppm, respectively), whereas CAL shows low to moderate content of V (2–20 ppm). Slight enrichment of vanadium is found in the lower part of the MBL (Fig. 9). In the VLQS, OGL, IL, and FL show high contents of V (41–263 ppm; 54–75 ppm; 123–169 ppm, respectively). The limestones from PQS show low to moderate contents of V (10–39 ppm). A slight increase in V concentration is found in the middle part of the PQS.

Molybdenum(VI) is found as MnO_4^{2-} in the stable oxidation state of the oxic seawater. Higher concentrations of Mo are observed mainly in MoO_2-rich sediments, where Mo adsorbs on to Mn oxyhydroxides (e.g., Bertine and Turekian, 1973; Crusius et al., 1996). Molybdenum may be considered to be one of the best diagnostic elements for sediment deposition under seawater sulfate-reducing conditions (Dean et al., 1997). Mo concentrations are found enriched in the modern reducing sediments where free H_2S is present, for example, Black Sea reduced sediments containing 2–40 ppm and Saanich Inlet sediments 50–125 ppm (Crusius et al., 1996). Such values are considerably higher than the average Mo concentrations in normal shales (2.6 ppm) (Wedepohl, 1971) and in black shales (10 ppm) (Vine and Tourtelot, 1970). Calvert and Pedersen (1993) suggested that the Mo enrichment can occur both in anoxic surface sediments and in anoxic subsurface sediments. Francois (1988) suggested that the enrichment of Mo may occur in areas of anoxic basins that are more permanently anoxic. Mo may be fixed by sulfide precipitation under anoxic environment (Bertine, 1972), may become associated with the organic matter (Pilipchuk and Volkov, 1974), or may be scavenged by Fe oxides or humic materials (Helz et al., 1996) and with sulfides, especially pyrite (Huerta-Diaz and Morse, 1992). In the VQS, GS, BL, and marl show low concentrations of Mo (0.18–0.38 ppm; 0.07–0.09 ppm; 0.06–0.36 ppm, respectively), whereas the concentration of Mo in CAL is below the detection limit. Similarly, the OGL, IL, and FL of the VLQS also show low contents of Mo (0.08–0.34 ppm; 0.08–0.12 ppm; 0.09–0.11 ppm, respectively). The limestones from the PQS show large variations of Mo contents (0.18–2.59 ppm). Mo enrichment is found in certain intervals of limestones of the PQS (Fig. 9).

U/Th and Ni/Co ratios of sediments can be used to infer the bottom water oxygenation conditions of the depositional environment (Dypvik, 1984; Wignall and Myers, 1988; Hatch and Leventhal, 1992; Jones and Manning, 1994; Rimmer, 2004; Nagarajan et al., 2007; Madhavaraju and Lee, 2009; Madhavaraju and Gonzalez-Leon, 2012). A high U/Th ratio (>1.25) suggests an anoxic environment, whereas low values (<0.75) are associated with oxic environments (Jones and Manning, 1994). In the VQS, both CAL and BL show large variations in U/Th ratio (0.21–1.30; 0.46–11.71, respectively) whereas GS and marl show the least variation (0.10–0.40; 0.17–0.32, respectively). The limestones of the VLQS show smaller variations in U/Th ratio (0.08–0.90), whereas the PQS exhibits large variations (0.48–11.03; Fig. 9.9). The lower part of the CAL and the lower and upper parts of the MBL show higher values of the U/Th ratio (>1.25). Also, high values of the U/Th ratio (>1.25) are found in the lower, middle, and upper parts of the PQS. The observed large variations in U/Th ratio in both VQS and PQS suggest that the fluctuation in U/Th ratio are mainly due to variations in oxygenation levels from oxic to anoxic conditions.

Significant variations in nickel content are observed in the VQS, whereas the limestone exhibits low values (CAL: 8–9 ppm; BL: 10–11 ppm) and slightly higher values in GSs and marl (25–30 ppm; 13–30 ppm, respectively). The limestones from the PQS show low values of Ni (8–13 ppm), whereas the limestones from VLQS show slight increases in Ni contents (15–51 ppm). The Co content fluctuates in a similar manner (Tables 1–3). The Ni/Co ratio has been considered to be a redox indicator (Dypvik, 1984; Dill, 1986). According to Jones and Manning (1994), Ni/Co ratio below five suggests oxic environments, whereas the Ni/Co ratio above five suggests suboxic and anoxic environments. The limestones (CAL and BL) and marls from the VQS (3.00–3.40; 2.80–4.00; 3.20–4.60, respectively), various types of limestones (OGL, IL, and FL) of the VLQS (1.00–2.90; 2.40–2.60; 2.60–2.90, respectively) show low Ni/Co values, whereas the GSs of the VQS (VPQ1, three) show comparatively high Ni/Co content (5.0; Fig. 9), indicating that the GSs were deposited in suboxic environment.

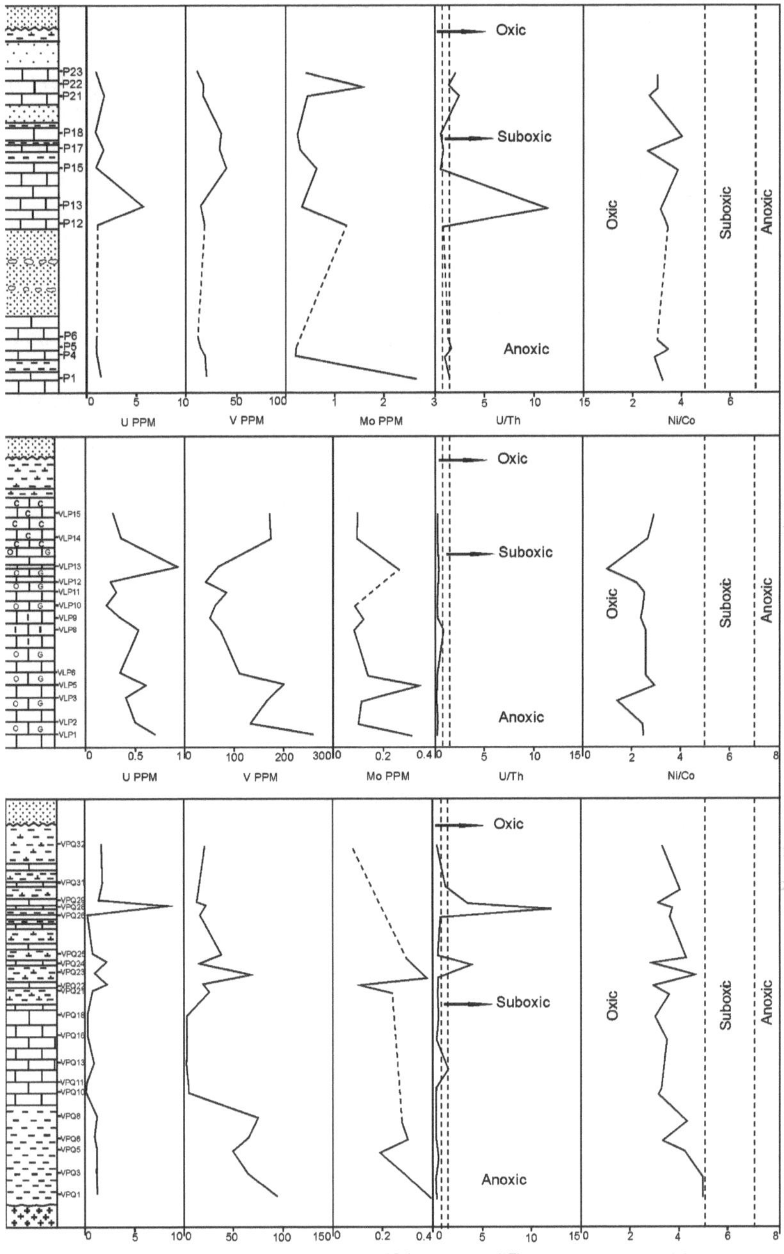

FIGURE 9

Redox sensitive trace elements variations in the limestones from Vadugarpettai quarry, Vellipirangiyam quarry, and Periakurichchi quarry sections.

10.6 CONCLUSIONS

The concentrations of major and trace elements varied considerably between different litho-units of the sedimentary rocks of Albian-Danian age. The enrichment of U and V are observed in the GS and in certain intervals of CAL and MBL of the VQS, lower and upper parts of the OGL of the VLQS and in the lower and middle parts of the limestone sequence of the PQS. Similarly, Mo is enriched in the GS and lower part of MBL (VQS); lower and upper parts of OGL (VLQS); and lower, middle, and upper parts of the PQS. In addition, high values of U/Th are noticed in the CAL and MBL of the Dalmiapuram Formation, OGL of the Kallankurichchi Formation and certain intervals of the PQS. The specific enrichment of U, V, Mo, and U/Th ratio at certain intervals of the studied sections suggests that the Cauvery Basin experienced suboxic-anoxic conditions during the particular intervals of Albian, Early Maastrichtian, and Danian. However, the lack of similar enrichment in Cu, Zn, Ni, Co, and Cr and the Ni/Co ratio may be related to the post-depositional modifications or unknown synsedimentary differences in uptake processes. Hence, the U, V, and Mo elements and the U/Th ratio are considered to be a chemostratigraphic tool to understand variations in oxygenation conditions.

ACKNOWLEDGMENTS

We thank Mr Rufino Lozano-Santa Cruz (XRF analysis), Mrs Elena Lounejeva, and Dr. Juan Pablo Bernal (ICP-MS analysis) for their help in geochemical analyses. We also thank Mr Pablo Peñaflor, ERNO, Instituto de Geología, Universidad Nacional Autónoma de Mexico for powdering of limestone samples for geochemical analyses.

REFERENCES

Algeo, T.J., Maynard, J.B., 2004. Trace-element behavior and redox facies in core shales of Upper Pennsylvanian Kansas-type cyclothems. Chemical Geology 206, 289–318.

Banerji, R.K., 1972. Stratigraphy and micropalaeontology of the Cauvery Basin, Part-I, exposed area. Journal of the Palaeontological Society of India 17, 1–24.

Barnes, U.C., Cochran, J.R., 1990. Uranium removal in oceanic sediments and the oceanic U balance. Earth and Planetary Science Letters 97 (1–2), 94–101.

Bergstoerm, S.M., Saltzman, M.M., Schmitz, B., 2006. First record of the Hirnantian (Upper Ordovician) 13C excursion in the North American Midcontinent and its regional implications. Geological Magazine 143, 657–678.

Bertine, K., 1972. The deposition of molybdenum in anoxic waters. Marine Chemistry 1, 43–53.

Bertine, K.K., Turekian, K.K., 1973. Molybdenum in marine sediments. Geochimica et Cosmochimica Acta 37, 1415–1434.

Bhatia, M.R., Crook, K.A.W., 1986. Trace element characteristics of graywackes and tectonic setting discrimination of sedimentary basins. Contributions to Mineralogy and Petrology 92, 181–193.

Blanford, H.F., 1862. On the Cretaceous and other rocks of South Arcot and Trichinopoly Districts. Memoirs of Geological Survey of India 4 (1), 217.

Brasier, M.D., Shields, G., 2000. Neoproterozoic chemostratigraphy and correlation of the Port Askaig glaciation, Dalradian Supergroup of Scotland. Journal of the Geological Society 157, 909–914.

Breit, G.N., Wanty, R.B., 1991. Vanadium accumulation in carbonaceous rocks: a review of geochemical controls during deposition and diagenesis. Chemical Geology 91, 83–97.

Canet, C., Alfonso, P., Melgarejo, J.C., Belyatsky, B.V., 2004. Geochemical evidences of sedimentary-exhalative origin of the shale-hosted PGE–Ag–Au–Zn–Cu occurrences of the Prades Mountains (Catalonia, Spain): trace-element abundances and Sm–Nd isotopes. Journal of Geochemical Exploration 82, 17–33.

Calvert, S.E., Pedersen, T.F., 1993. Geochemistry of recent oxic and anoxic sediments: implications for the geological record. Marine Geology 113, 67–88.

Calvert, S.E., Piper, D.Z., 1984. Geochemistry of ferromanganese nodules: multiple diagenetic metal sources in the deep sea. Geochimica et Cosmochimica Acta 48, 1913–1928.

Chaillou, G., Anschutz, P., Lavaux, G., Schäfer, J., Blanc, G., 2002. The distribution of Mo, U, and Cd in relation to major redox species in muddy sediments of the Bay of Biscay. Marine Chemistry 80, 41–59.

Crusius, J., Calvert, S., Pedersen, T., Sage, D., 1996. Rhenium and molybdenum enrichments in sediments as indicators of oxic, suboxic, and sulfidic conditions of deposition. Earth and Planetary Science Letters 145, 65–78.

Dean, W.E., Gardner, J.V., Piper, D.Z., 1997. Inorganic geochemical indicators of glacial–interglacial changes in productivity and anoxia on the California continental margin. Geochimica et Cosmochimica Acta 61, 4507–4518.

Dill, H., 1986. Metallogenesis of early Paleozoic graptolite shales from the Graefenthal Horst (northern Bavaria-Federal Republic of Germany). Economic Geology 81, 889–903.

Dypvik, H., 1984. Geochemical compositions and depositional conditions of Upper Jurassic and Lower Cretaceous Yorkshire clays, England. Geological Magazine 121 (5), 489–504.

Eggins, S.M., Woodhead, J.D., Kinsley, L.P.J., Mortimer, G.E., Sylvester, P., McCulloch, M.T., Hergt, J.M., Handler, M.R., 1997. A simple method for the precise determination of ≥40 trace elements in geological samples by ICPMS using enriched isotope internal standardization. Chemical Geology 134, 311–326.

Elrick, M., Molina-Garza, R., Duncan, R., Snow, L., 2009. C-isotope stratigraphy and paleoenvironmental changes across OAE2 (mid-Cretaceous) from shallow-water platform carbonates of southern Mexico. Earth and Planetary Science Letters 277, 295–306.

Emerson, S.R., Huested, S.S., 1991. Ocean anoxia and the concentration of molybdenum and vanadium in seawater. Marine Chemistry 34, 177–196.

Feng, R., Kerrich, R., 1990. Geochemistry of fine grained clastic sediments in the Archean Abitibi greenstone belt, Canada: implications for provenance and tectonic setting. Geochimica et Cosmochimica Acta 54, 1061–1081.

Floyd, P.A., Shail, R., Leveridge, B.L., Franke, W., 1991. Geochemistry and provenance of Rhenohercynian synorogenic sandstones: implications for tectonic environment discrimination. In: Morton, A.C., Todd, S.P., Haughton, P.D.W. (Eds.), Developments in Sedimentary Provenance StudiesSpecial Publication Geological Society of London 57, 173–189.

Francois, R., 1988. A study on the regulation of the concentrations of some trace metals (Rb, Sr, Zn, Pb, Cu, V, Cr, Ni, Mn and Mo) in Saanich Inlet sediments, British Columbia, Canada. Marine Geology 83, 285–308.

Govindan, A., Ravindran, C.N., Rangaraju, M.K., 1996. Cretaceous stratigraphy and planktonic foraminiferal zonation of Cauvery Basin, South India. In: Sahni, A. (Ed.), Cretaceous Stratigraphy and Palaeoenvironments-Memoirs of Geological Society of India 37, 155–187.

Le Guerroue, E., Cozzi, A., 2010. Veracity of Neoproterozoic negative C-isotope values: the termination of the Shuram negative excursion. Gondwana Research 17, 653–661.

Handley, L., Pearson, P.N., McMillan, I.K., Pancost, R.D., 2008. Large terrestrial and marine carbon and hydrogen isotope excursions in a new Paleocene/Eocene boundary section from Tanzania. Earth and Planetary Science Letters. 275, 17–25.

Hastings, D.W., Emerson, S.R., Mix, A.C., 1996. Vanadium in foraminiferal calcite as a tracer for changes in the areal extent of reducing sediments. Paleoceanography 11, 665–678.

Hatch, J.R., Leventhal, J.S., 1992. Relationship between inferred redox potential of the depositional environment and geochemistry of the Upper Pennsylvanian (Missourian) Stark shale member of the Dennis Limestone, Wabaunsee County, Kansan, USA. Chemical Geology 99, 65–82.

Helz, G.R., Miller, C.V., Charnock, J.M., Mosselmans, J.L.W., Pattrick, R.A.D., Garner, C.D., Vaughan, D.J., 1996. Mechanisms of molybdenum removal from the sea and its concentration in black shales: EXAFS evidences. Geochimica et Cosmochimica Acta 60, 3631–3642.

Hertzog, R., Colson, L., Seeman, B., O'Brien, M., Scott, H., McKeon, D., Wraight, P., Garu, J., Ellis, D., Schweitzer, J., Herron, M., 1987. Geochemical Logging with Spectrometry Tools. (SPE Paper 16792).

Huerta-Diaz, M.A., Morse, J.W., 1992. Pyritization of trace metals in anoxic marine sediments. Geochimica et Cosmochimica Acta 56, 2681–2702.

Ishikawa, T., Ueno, Y., Komiya, T., Sawaki, Y., Han, J., Shu, D., Li, Y., Maruyama, S., Yoshida, N., 2008. Carbon isotope chemostratigraphy of a Precambrian/Cambrian boundary section in the Three Gorge area, South China: prominent global-scale isotope excursions just before the Cambrian Explosion. Gondwana Research 14, 193–208.

Jones, B., Manning, D.A.C., 1994. Comparison of geochemical indices used for the interpretation of palaeoredox conditions in ancient mudstones. Chemical Geology 111, 111–129.

Kakizaki, Y., Kano, A., 2009. Architecture and chemostratigraphy of Late Jurassic shallow marine carbonates in NE Japan, western Paleo-Pacific. Sedimentary Geology 214, 49–61.

Kale, A.S., Phansalkar, V.G., 1992. Calcareous nanofossils from the Utatur group, Trichinopoly District, Tamil Nadu, India. Journal of the Paleontological Society of India 37, 85–102.

Klinkhammer, G.P., Palmer, M.R., 1991. Uranium in the oceans: where it goes and why. Geochimica et Cosmochimica Acta 55, 1799–1806.

Kossmat, F., 1897. The Cretaceous deposits of Pondicherri. Records of the Geological Survey of India 30, 51–110.

Kouchinsky, A., Bengtson, S., Pavlov, V., Runnegar, B., Torssander, P., Young, E., Ziegler, K., 2007. Carbon isotope stratigraphy of the Precambrian–Cambrian Sukharikha River section, northwestern Siberian platform. Geological Magazine 144, 609–618.

Madhavaraju, J., González-León, C.M., 2012. Depositional conditions and source of rare earth elements in carbonate strata of the Aptian-Albian Mural Formation, Pitaycachi section, northeastern Sonora, Mexico. Revista Mexicana De Ciencias Geológicas 29, 478–491.

Madhavaraju, J., Hussain, S.M., Guruvappan, M., Ramasamy, S., Mohan, S.P., 2006. Sequence stratigraphy of lower Niniyur Formation of Cauvery Basin, Southern India. Journal of the Geological Society of India 68, 685–694.

Madhavaraju, J., Kolosov, I., Buhlak, D., Armstrong-Altrin, J.S., Ramasamy, S., Mohan, S.P., 2004. Carbon and oxygen isotopic signatures in Albian-Danian limestones of Cauvery Basin, Southeastern India. Gondwana Research 7, 519–529.

Madhavaraju, J., Lee, Y.I., 2009. Geochemistry of the Dalmiapuram Formation of the Uttatur group (Early Cretaceous), Cauvery Basin, Southeastern India: implications on provenance and paleo-redox conditions. Revista Mexicana De Ciencias Geológicas 26, 380–394.

Madhavaraju, J., Lee, Y.I., 2010. Influence of Deccan volcanism in the sedimentary rocks of Late Maastrichtian-Danian age of Cauvery basin, Southeastern India: constraints from geochemistry. Current Sciences 98, 528–537.

Madhavaraju, J., Ramasamy, S., 1999a. Rare earth elements in limestones of Kallankurichchi Formation of Ariyalur Group, Tiruchirapalli Cretaceous, Tamil Nadu. Journal of the Geological Society of India 54, 291–301.

Madhavaraju, J., Ramasamy, S., 1999b. Microtextures on quartz grains of Campanian – Maastrichtian sediments of Ariyalur Group of Tiruchirapalli Cretaceous, Tamil Nadu – implication on depositional environments. Journal of the Geological Society of India 54, 647–658.

Madhavaraju, J., Ramasamy, S., 2001. Clay mineral assemblages and rare earth element distribution in the sediments of Ariyalur Group, Tiruchirapalli District, Tamil Nadu – implication for paleoclimate. Journal of the Geological Society of India 58, 69–77.

Madhavaraju, J., Ramasamy, S., 2002. Petrography and geochemistry of late maastrichtian – early paleocene sediments of Tiruchirapalli Cretaceous, Tamil Nadu – Paleoweathering and provenance implications. Journal of Geological Society of India 59, 133–142.

Madhavaraju, J., Ramasamy, S., Ruffell, A., Mohan, S.P., 2002. Clay mineralogy of the Late Cretaceous and Early Tertiary Successions of the Cauvery Basin (Southeastern India): implications for sediment source and palaeoclimates at the K/T boundary. Cretaceous Research 23, 153–163.

Madhavaraju, J., Sial, A.N., Rakhinath, R., Ramasamy, S., Lee, Y.I., Ramachandran, A., 2014. Carbon, oxygen and strontium isotopic signatures in Maastrichtian-Danian limestones of the Cauvery Basin, South India. Geoscience Journal. http://dx.doi.org/10.1007/s12303-014-0039-1.

Mamgain, V.D., Rao, B.R.J., Sastry, M.V.A., 1968. The Niniyur Group of Trichinopoly, South India (Cretaceous–Tertiary formations of South India). Memoirs of Geological Society of India 2, 85–91.

Marquillas, R., Sabino, I., Sial, A.N., Papa, C., Ferreira, V., Matthews, S., 2007. Carbon and oxygen isotopes of Maastrichtian–Danian shallow marine carbonates: Yacoraite Formation, northwestern Argentina. Journal of South American Earth Science 23, 304–320.

McManus, J., Berelson, W.M., Klinkhammer, G.P., Hammond, D.E., Holm, C., 2005. Authigenic uranium: relationship to oxygen penetration depth and organic carbon rain. Geochimica et Cosmochimica Acta 69, 95–108.

Morford, J.L., Emerson, S., 1999. The geochemistry of redox sensitive trace metals in sediments. Geochimica et Cosmochimica Acta 63, 1735–1750.

Morford, J.L., Russell, A.D., Emerson, S., 2001. Trace metal evidence for changes in the redox environment assocoated with the transition from terrigenous clay to diatomaceous sediments, Saanich Inlet, BC. Marine Geology 174, 355–369.

Mutti, M., Bernoulli, D., 2003. Early marine lithification and hardground development on a Miocene ramp (Maiella, Italy): key surfaces to track changes in trophic resources in non-tropical carbonate settings. Journal of Sedimentary Research 73, 296–308.

Mutti, M., John, C.M., Knoerich, A.C., 2006. Chemostratigraphy in Miocene heterozoan carbonate settings: applications, limitations and perspectives. Geological Society of London Special Publications 255, 307–322.

Nagarajan, R., Madhavaraju, J., Nagendra, R., Armstrong-Altrin, J.S., Moutte, J., 2007. Geochemistry of Neoproterozoic shales of the Rabanpalli Formation, Bhima Basin, Northern Karnataka, Southern India: implications for provenance and paleoredox conditions. Revista Mexicana De Ciencias Geológicas 24 (2), 150–160.

Nagendra, R., Nagendran, G., Narasimha, K., Jaiprakash, B.C., Nallapa Reddy, A., 2002. Sequence stratigraphy of Dalmiapuram Formation, Kallakkudi quarry – II, South India. Journal of the Geological Society of India 59, 249–258.

Nath, B.N., Bau, M., Ramlingeswara Rao, B., Rao, ChM., 1997. Trace and rare earth elemental variation in Arabian Sea sediments through a transect across the oxygen minimum zone. Geochimica et Cosmochimica Acta 61, 2375–2388.

Nedelec, A., Affaton, P., France-Lanord, C., Charriere, A., Alvaro, J., 2007. Sedimentology and chemostratigraphy of the Bwipe Neoproterozoic cap dolostones (Ghana, Volta Basin): a record of microbial activity in a peritidal environment. Comptes Rendus Geoscience 339, 223–239.

Pearce, T.J., Besly, B.M., Wray, D.S., Wright, D.K., 1999. Chemostratigraphy: a method to improve interwell correlation in barren sequences – a case study using onshore Duckmantian/Stephanian sequences (West Midlands, U.K.). Sedimentary Geology 124, 197–220.

Pettijohn, F.J., 1975. Sedimentary Rocks, third ed. Harper Row, New York.

Pilipchuk, M.F., Volkov, I.I., 1974. Behaviour of molybdenum in processes of sediment formation and diagenesis in the Black Sea. In: Degens, E.T., Ross, D.A. (Eds.), The Black Sea—Geology, Chemistry and BiologyMemoir of American Association of Petroleum Geologists 542–553.

Prabhakar, K.N., Zutshi, P.L., 1993. Evolution of southern part of Indian east coast basins. Journal of the Geological Society of India 41, 215–230.

Primmer, T.J., Kerr, S.S., Myers, K.J., 1990. Feasibility of insitu elemental analysis in mudrock evaluation. In: Hurst, A., Lovell, M.A., Morton, A.C. (Eds.), Geological Applications of Wireline LogsSpecial Publication Geological Society of London 48, 203–210.

Ramasamy, S., Banerji, R.K., 1991. Geology, petrography and stratigraphy of pre-Ariyalur sequence in Tiruchirapalli District, Tamil Nadu. Journal of the Geological Society of India 37, 577–594.

Ramkumar, M., 1999. Role of chemostratigraphic technique in reservoir characterization and global stratigraphic correlation. Indian Journal of Geochemistry 14, 33–45.

Ramkumar, M., Stüben, D., Berner, Z., 2004. Lithostratigraphy, depositional history and sea level changes of the Cauvery Basin, South India. Annales Géologiques De La Péninsule Balkanique 65, 1–27.

Ramkumar, M., Stüben, D., Berner, Z., 2006. Elemental interrelationships and depositional controls of Barremian–Danian strata of the Cauvery Basin, South India: implications on scales of chemostratigraphic modelling. Indian Journal of Geochemistry 21, 341–367.

Ramkumar, M., Berner, Z., Stüben, D., 2010. Hierarchical delineation and multivariate statistical discrimination of chemozones of the Cauvery Basin, South India: implications on spatio-temporal scales of stratigraphic correlation. Petroleum Science 7, 435–447.

Ramkumar, M., Stuben, D., Berner, Z., 2011. Barremian–Danian chemostratigraphic sequences of the Cauvery Basin, India: implications on scales of stratigraphic correlation. Gondwana Research 19, 291–309.

Rangaraju, M.K., Agarwal, A., Prabhakar, K.N., 1993. Tectono-stratigraphy, structural styles, evolutionary model and hydrocarbon habitat, Cauvery and Palar Basins. In: Biswas, S.K., et al. (Ed.), Proc. second seminar on petroliferous basins of India, vol. 1. KDMIPE, ONGC, Dehra Dun, pp. 371–388.

Rimmer, S.M., 2004. Geochemical paleoredox indicators in Devonian–Mississippian black shales, Central Appalachian Basin (USA). Chemical Geology 206, 373–391.

Ruhl, M., Kueschner, W.M., Krystyn, L., 2009. Triassic–Jurassic organic carbon isotope stratigraphy of key sections in the western Tethys realm (Austria). Earth and Planetary Science Letters 281, 169–187.

Saltzman, M.R., 2002. Carbon and oxygen isotope stratigraphy of the Lower Mississippian (Kinnderhookian–Lower Osagean), western United States: implications for seawater chemistry and glaciation. Geological Society of America Bulletin 114, 96–108.

Sarkar, A., Battacharya, S.K., Sarin, M.M., 1993. Geochemical evidence for anoxic deep water in the Arabian Sea during the last glaciations. Geochimica et Cosmochimica Acta 57 (5), 1009–1016.

Sastry, M.V.A., Mamgain, V.D., Rao, B.R.J., 1972. Ostracod Fauna of the Ariyalur Group (Upper Cretaceous), Tiruchirapalli District, Tamil Nadu. Part I. Lithostratigraphy of the Ariyalur Group. Mem. Geol. Surv. India, Palaeon. Indica, New Ser., vol. 40, 1–48.

Sawaki, Y., Ohno, T., Fukushi, Y., Komiya, T., Ishikawa, T., Hirata, T., Maruyama, S., 2008. Sr isotope excursion across the Precambrian–Cambrian boundary in the Three Gorges area, South China. Gondwana Research 14, 134–147.

Saylor, B.Z., Poling, J.M., Huff, W.D., 2005. Stratigraphic and chemical correlation of volcanic ash beds in the terminal Proterozoic Nama Group, Namibia. Geological Magazine 142, 519–538.

Schroeder, S., Grotzinger, J.P., 2007. Evidence for anoxia at the Ediacaran–Cambrian boundary: the record of redox-sensitive trace elements and rare earth elements in Oman. Journal of the Geological Society 164, 175–187.

Schroeder, S., Schreiber, B.C., Amthor, J.A., Matter, A., 2004. Stratigraphy and environmental conditions of the terminal Neoproterozoic–Cambrian Period in Oman: evidence from sulfur isotopes. Journal of the Geological Society 161, 489–499.

Somayajulu, B.L.K., Yadav, D.N., Sarin, M.M., 1994. Recent sedimentary records from the Arabian Sea. Proceedings of the Indian Academy of Science 103 (2), 315–327.

Srivastava, R.P., Tewari, B.S., 1967. Biostratigraphy of the Ariyalur Stage, Cretaceous of Trichinopoly. Journal of the Palaeontological Society of India 12, 48–54.

Sundaram, R., Rao, P.S., 1986. Lithostratigraphy of Cretaceous and Palaeocene rocks of Tiruchirapalli District, Tamil Nadu, South India. Records of the Geological Survey of India 115, 9–23.

Sundaram, R., Henderson, R.A., Ayyasami, K., Stilwell, J.D., 2001. A lithostratigraphic revision and palaeoenvironmental assessment of the Cretaceous system exposed in the onshore Cauvery Basin, Southern India. Cretaceous Research 22, 743–762.

Taylor, S.R., McLennan, S.M., 1985. The Continental Crust, Its Composition and Evaluation. Blackwell, Oxford.

Tribovillard, N., Algeo, T.J., Lyons, T., Riboulleau, A., 2006. Trace metals as paleoredox and paleoproductivity proxies: an update. Chemical Geology 232, 12–32.

Vine, J.D., Tourtelot, E.B., 1970. Geochemistry of black shale deposits—a summary report. Economic Geology 65, 253–272.

Wanty, R.B., Goldhaber, R., 1992. Thermodynamics and kinetics of reactions involving vanadium in natural systems: accumulation of vanadium in sedimentary rock. Geochimica et Cosmochimica Acta 56, 171–183.

Wedepohl, K.H., 1971. Environmental influences on the chemical composition of shales and clays. In: Ahrens, L.H., Press, F., Runcorn, S.K., Urey, H.C. (Eds.), Physics and Chemistry of the Earth. Pergamon, Oxford, pp. 305–333.

Wehrly, B., Stumm, W., 1989. Vanadyl in natural waters: adsorption, and hydrolysis promote oxygenation. Geochimica et Cosmochimica Acta 53, 69–77.

Wignall, P.B., Maynard, J.R., 1993. The sequence stratigraphy of transgressive black shales. In: Katz, B.J., Pratt, L.M. (Eds.), Source Rocks in a Sequence Stratigraphic FrameworkAAPG Studies in Geology, vol. 37, pp. 35–47.

Wignall, P.B., Myers, K.J., 1988. Interpreting the benthic oxygen levels in mudrocks, a new approach. Geology 16, 452–455.

Zakharov, Y.D., Shigeta, Y., Nagendra, R., Safronov, P.P., Smyshlyaeya, O.P., Popov, A.M., Velivetskaya, T.A., Afanasyeva, T.B., 2011. Cretaceous climate oscillations in the southern palaeolatitudes: new stable isotope evidence from India and Madagascar. Cretaceous Research 32, 623–645.

Zheng, Y., Anderson, R.F., van Geen, A., Kuwabara, J., 2000. Authigenic molybdenum formation in marine sediments: a link to pore water sulfide in the Santa Barbara Basin. Geochimica et Cosmochimica Acta 64, 4165–4178.

CHAPTER

11 TEMPORAL TRENDS OF GEOCHEMISTRY, RELATIVE SEA LEVEL, AND SOURCE AREA WEATHERING IN THE CAUVERY BASIN, SOUTH INDIA

Mu. Ramkumar[1,*], Z. Berner[2]

*[1]Department of Geology, Periyar University, Salem, Tamilnadu, India; *South East Asia Carbonate Research Laboratory (Seacarl), Universiti Teknologi Petronas, Tronoh, Malaysia; [2]Institute of Mineralogy and Geochemistry, Karlsruhe Institute of Technology, Karlsruhe, Germany*

11.1 INTRODUCTION

Enhanced silicate weathering in the continental regions, significant influx of detrital materials to the ocean basins, and the resultant reduction in carbonate accumulation during sea level lowstands and conversely, reduction of detrital influx owing to the limited availability of terrestrial areas, shortening and/or drowning of fluvial channels, and significant carbonate accumulation during sea level highstands, are considered to be the basic tenets of sequence stratigraphy. The sequence stratigraphic concepts were introduced based on the stratal patterns as inferred from marginal siliciclastic deposits and were linked to the interplay between sediment supply, relative sea level oscillations, climate, and tectonics. The responses of siliciclastic and carbonate systems to sediment supply, relative sea level, climate, and tectonics differ due to the fact that while the siliciclastic system depends on sediment influx from landmass (Sandulli and Raspini, 2004), source rock composition, source area weathering, sorting, tectonic setting, diagenesis, and recycling (Alvarez and Roser, 2007), the carbonate system predominantly relies on the sediment production conditions prevailing within the depositional basin. The periods of significant carbonate accumulation accompany enhanced burial of organic matter. The mixed siliciclastic-carbonate system, the result of the interplay between the hydrodynamic factors affecting a particular environment and the sediment-forming minerals (Garcia et al., 2005), poses yet more complexity.

These changes are coupled with climate and atmospheric CO_2 (Kampschulte et al., 2001; Pellenard et al., 2014) and trigger silicate weathering (Wallmann, 2001; Torres et al., 2014) in the continental regions, and promote a cycle of alternate deposition of siliciclastics and carbonates. Atmospheric circulation models indicate that global climate is intimately linked to the atmospheric CO_2 content; increased CO_2 leads to warmer temperatures, changing precipitation patterns, and an overall increase in riverine runoff. Increased CO_2 content in the atmosphere, however, is regulated over geologic time by weathering of silicate rocks. In an innovative study, Van der Meer et al. (2014) demonstrated the tectonic control over the atmospheric CO_2. Long-term climate moderation is commonly attributed to chemical weathering; the higher the temperature and precipitation, the higher the weathering rate. Weathering releases divalent cations to the ocean via riverine transport where they promote the drawdown of CO_2 from the atmosphere by the precipitation and subsequent burial of carbonate minerals (Gislason et al., 2009). Chemical weathering is the major geological and geochemical process leading to mass transfer of upper continental crust to the ocean (Bouchez and Gaillardet, 2014). Depending on the intensity and sprawl of these processes, the cyclic changes may get preserved in the ensuing sedimentary strata at local, regional, and global scales. The temporal scale of which may vary from few thousand years to few or few tens of Ma—a postulate widely utilized in chemostratigraphy. Though literature abounds in documenting the climate and related impact on continental weathering and resultant siliciclastic and carbonate deposition (for example: Price et al., 1998; Price and Velbel, 2003; Hofmann et al., 2001; Pearce et al., 2003; Wortmann et al., 2004; Alvarez and Roser, 2007; Gislason et al., 2009; Boulila et al., 2010), documentary evidences for this phenomenon, particularly on the long-duration, regional and global scale analogues are scarce.

The Cauvery Basin (Fig. 1) is located in the southern part of Indian peninsula and contains a near-complete stratigraphic record of Barremian-Danian. It is one of the most studied basins

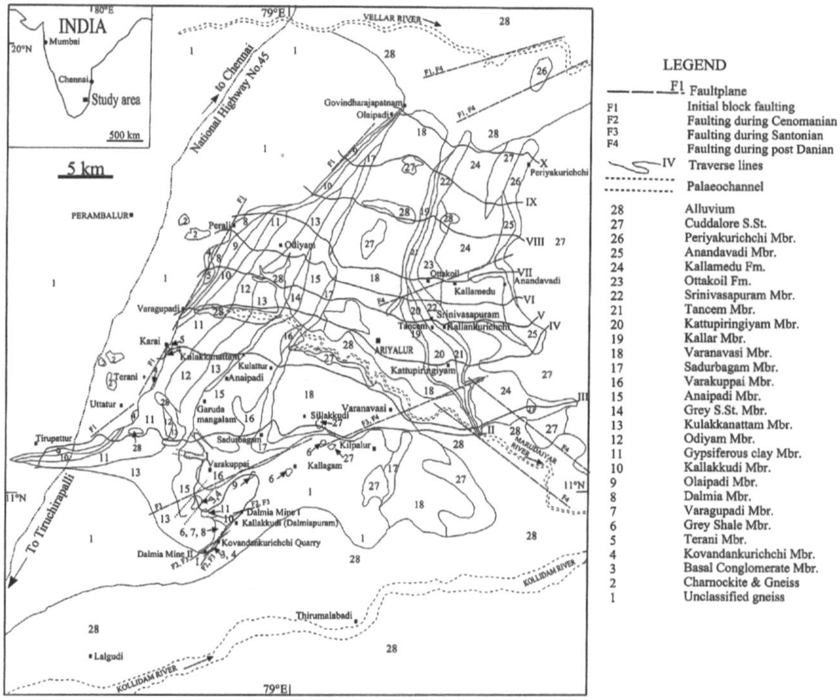

FIGURE 1 Location and geological map of the onland part of the Cauvery Basin.

After Ramkumar et al. (2004a).

(Acharyya and Lahiri, 1991). In a first ever basin-scale temporally long-range chemostratigraphic study on this basin, Ramkumar et al. (2011) recognized six major chemozones, separated by type 1 sequence boundaries and other correlative surfaces coeval with third-order cycles of sea level, which in turn contained high-frequency cycles, probably in the order of 10^4–10^6 years and found to be consistent with the timescale–sea level curve of Gradstein et al. (2004). Though there are disagreements on the connectedness of the Indian subcontinent with other continental plates during Barremian to Danian (Ali and Aitchison, 2008), the enclosed nature of the Indian subcontinent by sea and its behavior as an Island akin to the present-day Australia were not questioned. As the climatic conditions of Island continents are controlled by the temperature of surrounding seawater and as the Cretaceous Period is suggested to have experienced extended greenhouse effect (Veizer et al., 2000; Bice and Norris, 2002; Coccioni and Galeotti, 2003; Herrle et al., 2003; Najarro et al., 2011), it is hypothesized that the changes in the seawater temperature might have affected the glaciers to retreat or advance, causing high-frequency sea level oscillations that in turn influenced the depositional system of the Cauvery Basin. As the provenance area of the Cauvery Basin sediments was confined to adjacently located horsts and hinterland (Ramkumar et al., 2006), the sea level changes might have been exacerbated and reflected in the ensuing sediment records. Thus, the Cauvery Basin offers a unique opportunity to test the relationships between climate, sea level fluctuations, continental weathering, and their implications on the chemostratigraphic study. As the stratigraphic record is the outcome of an exogenic system consisting of geologic setting, changes in sea level, changes in geochemical reactions between the sea, earth, and climate (Srinivasan, 1989), we interpret the strata as a repository of paleoclimatological information. In this chapter, we examined the temporal variation of the geochemical composition for documenting prevalent climatic conditions and the responses of depositional system.

11.2 GEOLOGICAL SETTING

Among the NE-SW trending late Jurassic-early Cretaceous rift basins of the east coast of India (Powell et al., 1988), the Cauvery Basin (Fig. 1) is located at the southern part of the Indian peninsular shield. The evolutionary (Sastri et al., 1981; Prabhakar and Zutchi, 1993; Chari et al., 1995; Lal et al., 2009), stratigraphic (Ramanathan, 1968; Banerji, 1972; Sundaram and Rao, 1986; Sundaram et al., 2001; Ramasamy and Banerji, 1991; Tewari et al., 1996; Ramkumar et al., 2004a, 2005a), paleontologic (Chiplonkar, 1987; Govindhan et al., 1996; Bhatia, 1984; Jafar and Rai, 1989; Kale and Phansalkar, 1992; Kale et al., 2000; Guha, 1987; Guha and Senthilnathan, 1990, 1996; Ramkumar and Chandrasekaran, 1996; Ramkumar et al., 2010a; Rai et al., 2012), and geochemical (Ramkumar et al., 2004b, 2005b, 2006, 2010b, 2010c, 2011; Ramkumar, 2007) characteristics of this basin have been excellently documented. It is one among the Mesozoic extensional basins formed all along the eastern continental margin of the peninsular India and has been classified as a pericratonic rift basin (Sastri et al., 1981; Chari et al., 1995) and is located between latitudes 08°30′N and longitudes 78°30′E. It covers an exposed area of about 25,000 km^2 onland and 17,500 km^2 in the offshore regions (Sastri et al., 1981) of the Bay of Bengal up to 200 m isobath. This basin was created in response to the fragmentation of Gondwana supercontinent and rifting between Africa, India, and Antarctica (Lal et al., 2009) during late Jurassic-early Cretaceous (Powell et al., 1988; Chari et al., 1995; Jafer, 1996; Chatterjee et al., 2013). It continued evolving till the end of Tertiary through rift, pull-apart, shelf sag, and tilt phases (Prabhakar and Zutshi, 1993). It is a structurally elongated basin with NE-SW trending half-graben morphology and a regional dip averaging 5°–10° to the east and southeast. It consists of various subbasins and horsts (Acharya and Lahiri, 1991; Chandra, 1991; Prabhakar and Zulchi, 1993; Chari et al., 1995) namely, Ariyalur–Pondicherry depression, Kumbakonam–Mandanam ridge, Tanjore–Tranquebar–Nagapattinam depression, Pattukottai–Mannargudi–Karaikal ridge, Ramnad–Palk Bay depression, and Mandapam–Delft ridge. Toward west, it is demarcated by steep, basin margin faults, which separate the Archaean shield from the sedimentary deposits.

The sedimentary succession of this basin exceeds 5500 m in thickness (Govindan et al., 2000). Lithofacies associations and fossil data indicate periodic sediment-starved nature of the basin and basin-filling process of depositional pattern (Ayyasamy, 1990). A recent review of previous publications on evolutionary stages and sequence development (Kale, 2011) suggested the availability of a larger accommodation space than the sediment influx all through its evolutionary history. Enumeration of the tectonic structures and depositional history of the Cauvery Basin indicated that after initial block faulting and inception of sedimentation during late Jurassic-early Cretaceous, intensity of the tectonic control over sedimentation was diminutive (Prabakar and Zutchi, 1993; Ramkumar, 1996). Watkinson et al. (2007) recognized three major tectonic stages and resultant stratigraphic groups for this basin: syn-rift Gondwana Group (early Cretaceous), syn-rift Uttatur Group (Albian-Coniacian), and postrift Ariyalur Group (Santonian-Maastrichtian).

Based on the facies characteristics, comprehensive lithostratigraphy of the onland part of this basin was presented by Tewari et al. (1996). It was modified by Sundaram et al. (2001) and finally a systematic revision was made by Ramkumar et al. (2004a) following standard stratigraphic procedures and terminologies. The lithostratigraphic subdivisions (Table 1; Fig. 1) are separated by sequence boundaries and other correlative surfaces (Ramkumar et al., 2011) and are geochemically distinct to the tune of 100% from each other (Ramkumar et al., 2010b). A brief description on facies characteristics of the rocks is presented in Table 2.

Sedimentation in this basin took place in an epicontinental sea and the bathymetry was at shallow—modest levels (<50 m—as indicated by the linear curve in Fig. 2) although variations from supratidal or basinal levels were inferred. Based on foraminifer data, Raju and Ravindran (1990) and Raju et al. (1993) documented six third-order cycles of glacio-eustatic origin. Ramkumar et al. (2004a) constructed a sea level curve for this basin based on lithofacies data, which is similar to the curves presented by Raju et al. (1993) except that it additionally recorded fourth- and higher order sea level cycles (Fig. 2). The global sea level peaks during 104 Ma (early to late Albian), 93.7 Ma (±0.9; middle to late Cenomanian), 92.5 Ma (±1; early to middle Turonian), 86.9 Ma (±0.5; early to late Coniacian), 85.5 Ma (±1; early to late Santonian), 73 Ma (±1; late Campanian), 69.4 Ma (early to late Maastrichtian), and 63 Ma (±0.5; early to middle Danian) were inferred to occur in this basin (Raju and Ravindran, 1990; Raju et al., 1993; Ramkumar et al., 2004a). The third-order cycles are separated by type 1 sequence boundaries (recognized through shift of shoreline crossing shelf break as explicit in lithologic information, contact relationship between strata, evidences of subaerial exposure and erosion, advancement of fluvial channels over former offshore regions, etc.). While the older deposits are characterized by frequent environmental changes and texturally immature mixed siliciclastics and argillites, the younger deposits are characterized by relatively stable depocenters, carbonate deposition and textural inversion, and maturity. At each major unconformity, basement rocks and older sedimentary rocks were eroded and transported to newer depocenters to deposit lithoclastic conglomerates. A major tectonic movement during Santonian that caused transgression in areas that have remained topographically highlands since inception of the basin (Barremian) had changed the depositional pattern from dynamic depocenters to more stable shelf conditions (Ramkumar et al. 2005a). Except this tectonic perturbation, the

Table 1 Lithostratigraphy of the Exposed Part of the Cauvery Basin

Age	Formation	Member	Thickness (m)
Mio- Pliocene	Cuddalore S.St. Fm.		>150
	---Unconformity---		
Danian	Niniyur Fm.	Periyakurichchi biostromal Mbr.	26
		Anandavadi arenaceous Mbr.	30
	---Unconformity---		
Maastrichtian	Kallamedu Fm.		100
	---Unconformity---		
	Ottakoil Fm.		40
	---Unconformity---		
	Kallankurichchi Fm.	Srinivasapuram gryphean L.St. Mbr.	18
		Tancem biostromal Mbr.	8
		Kattupiringiyam inoceramus L.St. Mbr.	8
		Kallar arenaceous Mbr.	6
	---Unconformity---		
Campanian	Sillakkudi Fm.	Varanavasi S.St. Mbr.	270
		Sadurbagam pebbly S.St. Mbr.	80
Santonian		Varakuppai lithoclastic conglomerate Mbr.	45
	---Unconformity---		
Coniacian	Garudamangalam Fm	Anaipadi S.St. Mbr.	215
		Grey S.St. Mbr.	35
		Kulakkanattam S.St. Mbr.	123
	---Unconformity---		
Turonian	Karai Fm.	Odiyam Sandy clay Mbr.	175
		Gypsiferous clay Mbr.	275
	---Unconformity---		
Cenomanian	Dalmiapuram Fm.	Kallakkudi Calcareous S.St. Mbr.	60
		Olaipadi conglomerate Mbr.	65
		Dalmiya biohermal L.St. Mbr.	15
Albian		Varagupadi biostromal L.St. Mbr.	23
		Grey shale Mbr.	7
	---Unconformity---		
Aptian	Sivaganga Fm.	Terani clay Mbr.	30
		Kovandankurichchi S.St. Mbr	24
Barremian		Basal Conglomerate Mbr.	18
	---Unconformity---		
	Basement Rocks (Granitic gneiss, charnockite, pegmatite, etc.)		

After Ramkumar et al. (2004a).

Table 2 Lithofacies Characteristics of the Barremian-Danian Strata of the Cauvery Basin

Formation	Member	Facies Characteristics
Niniyur	Periyakurichchi	Thick arenaceous bioclastic limestone bed followed by medium to thick, parallel, even-bedded recurrent biostromal limestone and marl typify this member. Regionally varying concentrations of shell fragments and whole shells of bivalve, gastropod and remains of amphibia, pisces, algae, foraminifera, and ostracoda are also observed.
	Anandavadi	Isolated coral mounds, impure arenaceous limestone, and lenses of sandstone and clay deposited in a restricted marine regime under subtidal to intertidal regions. Occurrence of localized concentrations of shell fragments, coralline limestone, and reef derived talus deposits are characteristics of this member. This member rests over the Kallamedu Formation with distinct disconformity. At the top, an erosional surface is recognizable.
Kallamedu		Unconsolidated, well-rounded, and poorly sorted barren sands with rare-scarce dinosaurian bone fragments. Toward the top, these grade to medium to thin bedded, relatively highly argillaceous sandstones. Local occurrence of clays and silts with dispersed detrital quartz grains and sandy streaks, nonbedded nature, rare lamination, and mud cracks in them indicate sedimentation as over-bank deposits. Toward the top, this formation shows development of soil and return of continental conditions.
Ottakoil		The rocks are coarse- to medium-sized, well-sorted, fossiliferous, low-angle cross-bedded, and planar to massive bedded sandstones with regionally varying sparse calcareous cement. They also show recurrent fining upward sequences. Abundant *Stigmatophygus elatus* and few trace fossils are indicative of shallow marine environment of deposition. This formation rests over the Kallankurichchi Formation with disconformity and overlapped by the Kallamedu Formation.
Kallanku-richchi	Srinivasapuram	Uniform, parallel, thick- and very thick-bedded gryphean shell banks with *Terebratula*, *Exogyra*, bryozoa, and sponge. Extensive boring in gryphean shells, synsedimentary cementation, colonies of encrusting bryozoa over gryphean shells, and micritization of bioclasts suggest deposition of this member in inner shelf.
	Tancem	Biostromal limestone beds with thin to thick, parallel, even-bedded nature, cross-bedding, normal grading, hummocky cross-stratification, feeding traces, escape structures, and tidal channel structures. Local concentrations of various fossils, sporadic admixture of siliciclastics, and intraclasts are observed.
	Kattupiringiyam	Dusty brown friable carbonate sands with parallel, even, and thick to very thick bedding that contain only *Inoceramus* and bryozoa. This member has diagenetic bedding and abundant geopetal structures filled with millimeter- to centimeter-sized dog tooth spars of low magnesian non-ferroan calcite. This member has nondepositional surface at the bottom and has erosional surface at the top.
	Kallar	Normal-graded conglomerates in which well-rounded clasts of basement rocks, fresh feldspar, resedimented colonies of serpulids, and other older sedimentary rocks that range in size from coarse sand to boulder are observed. Lower contact of this member is an erosional surface. The upper contact is nondepositional surface resulted from marine flooding.

Table 2 Lithofacies Characteristics of the Barremian-Danian Strata of the Cauvery Basin—cont'd

Formation	Member	Facies Characteristics
Sillakkudi	Varanavasi	Featureless, massive, thick- to very thick-bedded, and coarse- to medium-grained sandstones. Occasional very coarse sandstone lenses, pockets of shell hash, intraformational lithoclastic bounders in association with vertical cylindrical burrows, and resedimented petrified wood logs are observed. The rocks rest over the pebbly sandstone member with nondepositional surface Upper surface of this member represents erosional surface associated with regression.
	Sadurbagam	Coarse siliciclastics with abundant marine fauna, shell fragments, and varying proportions of calcareous matrix. At the base, an erosional surface followed by distinct cobble-pebble quartzite conglomerate is observed. The rocks show normal grading; low-angle cross-bedding; massive, thick to medium, even, and parallel bedding. At places, pockets of shell-rich carbonate lenses with abundant siliciclastic admixture are found to occur. Load casts, slump folds, pillow structures and synerasis cracks, and occasional development of algal mounds are also found.
	Varakuppai	It rests over older sedimentary rocks with typical erosional surface. The erosional intensity was such high that, the beds have direct contact with much older Karai Formation. Fluviatile sandstones with well-rounded basement rocks, quartzite and older sedimentary rock boulders in addition to unsorted coarse sand-pebble sized siliciclastics constitute this member. These are typically reverse graded, show cyclic bedding and large-scale cross-bedding, and lack any body fossils. Large-scale cross-bedding, mud drapes, fresh feldspar, and sandstone clasts are also recorded. Toward the top, *thalassinoid* burrows are reported, which indicates gradual submergence of the depocentre by rising sea level.
Garuda mangalam	Anaipadi	Massive and thin-bedded claystones, silty claystones, and clayey sandstones in south that gradually grade to silty clay in south center and thin down. Again, from there, thickness of these beds and sediment grain size increase and contain abundant large ammonites. Further north, these were observed to be clayey siltstones with abundant shell fragments and ammonites.
	Grey sandstone	Highly well-cemented, sorted, and rounded grains giving massive appearance. The beds are cyclic, parallel, even-bedded alternative layers of barren and highly fossiliferous and sandy layers with regionally varying thicknesses. This member rests conformably over the Kulakkanattam Member and has distinct erosional and nondepositional surface. Upper contact is nondepositional surface associated with marine flooding.
	Kulakkanattam	Massive, yellowish brown ferruginous sandstones with abundant admixture of silt and clay. Localized concentrations of shell fragments, bivalves and gastropods, and ammonites are common. It also contains abundant wood fragments with extensive oyster boring. Cross-bedding, channel courses, planar bedding, and feeding traces are also common. The depositional surface was strongly bioturbated and riddled by roots. An angular erosional unconformity separates this member from the underlying Karai Formation.

Continued

Table 2 Lithofacies Characteristics of the Barremian-Danian Strata of the Cauvery Basin—cont'd

Formation	Member	Facies Characteristics
Karai	Odiyam sandy clay	Silty and sandy clays with abundant ammonites. Load structures and syndepositional slump folds are frequently observed. Upper portion of this member has localized pockets of fine sandstone along with ammonites. While the lower contact is conformable with underlying member, upper contact is erosional.
	Gypsiferous clay	Unconsolidated deep marine clays and silty clays. These beds contain thick population of belemnite rostrum and phosphate nodules. While a nondepositional unconformity surface separates this member from the underlying member, upper contact is nondepositional and erosional. From south to north, gradual reduction of thickness, population of belemnite and phosphate nodules, and frequency of gypsum layers are observed. Repetitive occurrences of 1- to 3-cm-thick red and green clay layers (perhaps CORBs) that can be traced for many kilometers are ubiquitous.
Dalmiapuram	Kallakkudi	Fine-coarse sandstones with alternate medium to thick beds of silty clay, calcareous siltstones, bioclastic arenaceous limestone, and gypsiferous clay. In the southern region, these beds show recurrent bands of fining upward sequences of siliciclastics with calcareous cements. Toward northern regions, this member grades to more silty and clayey, but gradation and gypsiferous bands are persistent with an addition of ferruginous silty clay bands.
	Olaipadi	Basinal silty clays and clays in which chaotic blocks are embedded. The beds contain large blocks of angular and subrounded basement rocks, coralline limestone blocks, claystones (lithoclasts of CORBs and greenish claystones typical of Karai Formation.), lithoclasts of older conglomerates, etc. The intercalations are recurrent and show typical Bouma sequences, normal grading, load casts, and channel and scour structures. Toward the top, deep marine clays grade into calcareous siltstone and include granitic cobbles and minor amounts of siliciclastic sands.
	Dalmia	Pure algal and coral facies limestone beds that form reef core. Upper contact of this member is a forced regression surface.
	Varagupadi	Limestone beds typical of reef flank biostromal beds deposited under high-energy conditions. Thin- to thick-bedded, even to parallel bioclastic limestone beds that have drawn their detritus from reefs predominate. These beds are found to be directly overlying the Grey shale Member. The rocks show wackstone to rudstone fabric and have clasts of redeposited boundstones.
	Grey shale	Grey shale beds with frequent thickening upward interbeds of fossiliferous grey limestone and minor to significant admixture of silt-sized siliciclastics. Lower contact of this member is an unconformity surface associated with marine flooding and the upper contact is nondepositional and erosional.
Sivaganga	Terani clay	White to brownish clay and argillaceous siltstone that show transition from the Kovandankurichchi Member. Beds are massive to very thick in nature. Lower contact of this member is nondepositional surface.
	Kovandankurichchi	Grain supported coarsening upward cyclic beds (20–100 cm thick each) of very coarse sandstones that show parallel, even, and thin to thick bedding. Grains are well sorted within each lamina and show rounded to well-rounded shape. These represent recurrent sheet flow deposits probably in a subaqueous fan deltaic environment.
	Basal conglomerate	Recurrent fining upward sequences of lithoclastic conglomerates of fluviatile and coastal marine environments. Lithoclasts are of gneissic basement rocks. Rests over basement rocks with distinct erosional surface. Upper contact is a nondepositional surface.

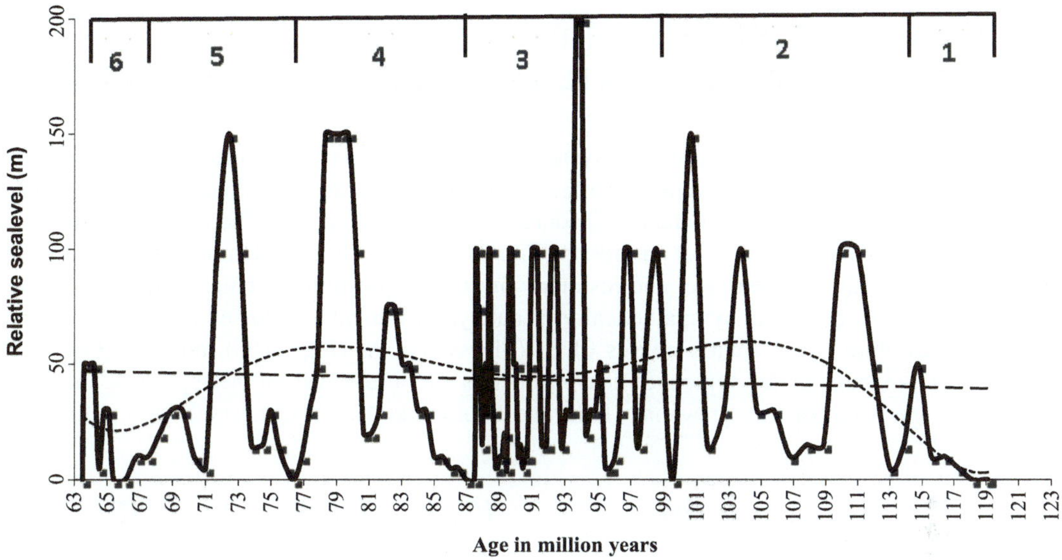

FIGURE 2 Relative sea level (RSL) fluctuations during Barremian-Danian in the Cauvery Basin.

After **Ramkumar et al. (2004a)**

Solid line curve is indicative of absolute values of RSL; --- linear trend; .... polynomial trend. Recognizable sea level cycles are indicated by Arabic numbers.

depositional history of this basin remained primarily under the influence of sea level variations albeit at varying temporal scales.

11.3 MATERIAL AND METHODS

Systematic field mapping in the scale of 1:50,000 was conducted through 10 traverses (Fig. 1), in which 308 locations were logged and sampled and the information on lithofacies, contact relationships, sedimentary and tectonic structures, and occurrences of mega and ichnofossil assemblages were also recorded. A composite stratigraphic profile of Barremian-Danian strata was constructed that allowed selection of 157 rock samples for analyzing trace elemental composition. From these 157 samples, 70 samples were further selected and analyzed by X-ray fluorescence (XRF) for major elemental composition following the procedures discussed in Kramar (1997) and Stüben et al. (2002). Stable isotopic analyses were performed as per the procedures presented in Keller et al. (2004). Analyses of 157 samples for petrography and whole-rock mineralogy and analyses of 70 samples for clay mineralogy were also performed. The geochemical data were interpreted based on the relative abundances with reference to the upper continental crust values and stratigraphic variation of absolute concentrations and weathering indices. The weathering and other indices were computed based on the standard procedures available in the literature. Collation of all these information along with the published data allowed to interpret the prevalent paleoclimatological and environmental conditions, which in turn were corroborated and contrasted in relation with regional and global climatic models.

11.4 RESULTS AND INTERPRETATIONS

11.4.1 INFLUENCE OF RELATIVE SEA LEVEL CHANGES OVER THE DEPOSITIONAL ENVIRONMENTS

The general knowledge about a geochemical system allows establishing a definite number of processes governing the sedimentary system namely, the redox conditions, detrital input, changes in provenance, quantum of sediment influx and climate, etc. (Montero-Serrano et al., 2010). The lithofacies characteristics along with the relative sea level variations (Fig. 2) of the study area demonstrate the prevalent high-frequency variations of sea level and resultant alternate depositions of siliciclastics and carbonates. The alternations are either sympathetically or antisympathetically followed by the stratigraphic variations of absolute concentration curves of Si (Fig. 3(A)), Al (Fig. 3(B)), and Ca (Fig. 3(C)). Decrease in Si and many metals typical of heavy minerals are observed from these profiles and can be interpreted as the result of transgressions (Hild and Brumsack, 1998). Conversely, the reduction in

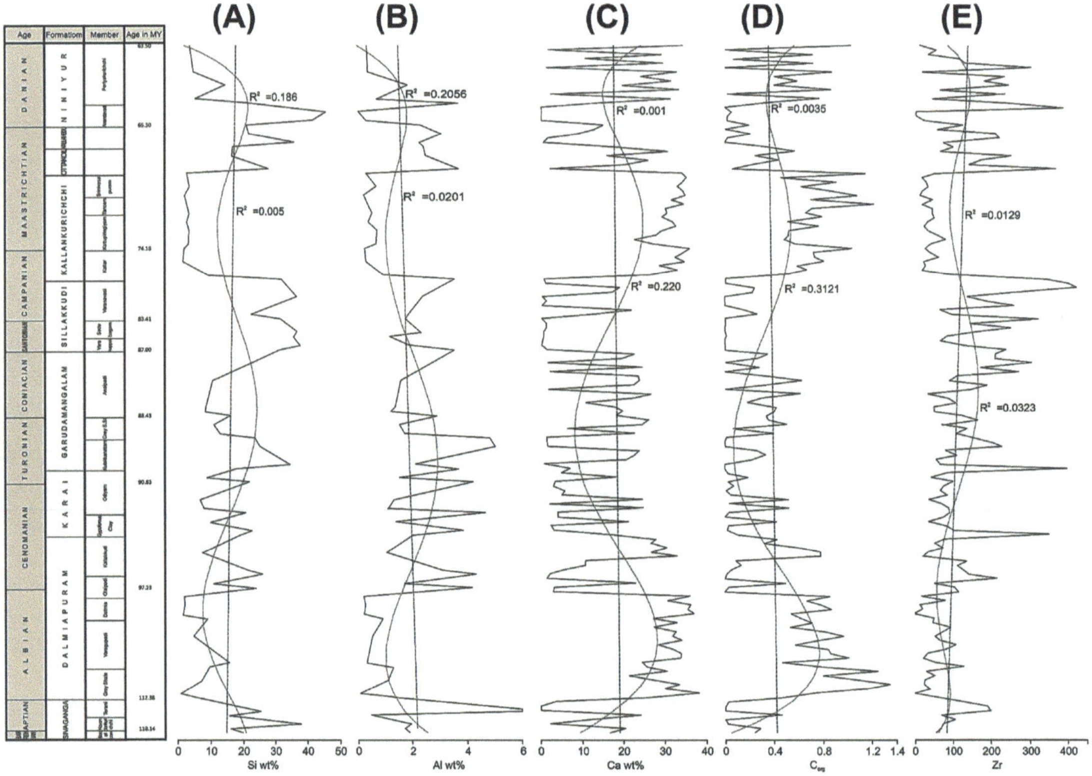

FIGURE 3 Stratigraphic variations of elemental concentrations and organic carbon.

Si, Al, Ca, and C_{org} are given in weight percentage. Zr is given in ppm. Solid line curve is indicative of absolute values; --- linear trend; · ···· polynomial trend. Variations are shown against the chronostratigraphic and lithostratigraphic subdivisions and absolute age.

carbonate and C_{Org} content is found to be associated with regressions. As could be observed elsewhere (Rachold and Brumsack, 2001; Hofmann et al., 2001; Boulila et al., 2010), whenever siliciclastic deposition ceased, carbonate deposition was initiated (Warzeski et al., 1996) in this basin. These lithofacies alternations and inverse relationships between Si-Al and Ca are considered to be influenced by the bathymetric variations, triggered by climatic reversals (Srinivasan, 1989; Fenner, 2001a; Pearce et al., 2003). Keller et al. (2004) reported higher freshwater and siliciclastic influx during wet-humid periods. Sarg (1988) reported starvation of sedimentary basins for detrital sediments during highstands that led to the proliferation of carbonates. Ruffel and Rawson (1994) and Soreghan (1997) opined that dry periods might cause a deficit in detrital supply and favor carbonate deposition. It supports the interpretation of occurrences of sea level highstands during interglacial periods (warmer than glacial periods) and resultant general aridity and deprivation of clastic sediment supply. The glacial periods bring in enhanced terrigenous supply to the depocenters in view of shelf erosion (Kampschulte et al., 2001) and fluvial system advancement (Sarg, 1988; Carter et al., 1991). The Cretaceous Period had experienced extended greenhouse effect (Thiry, 2000) and thus, sea level lowering may have influenced the weathering and erosion (Hay, 1994). Erosion during the periods of lower sea level, influenced also by the proximity to source rocks and adequate slope, could be inferred from the configuration of the Cauvery Basin, occurrences of paleo-channel courses (Fig. 1), and their association with siliciclastic deposits. Occurrences of unaltered lithoclasts and feldspar clasts in rocks that immediately follow regressive surfaces also suggest the prevalent mechanical erosion, rapid and short duration of transport, and quick burial. Such rapid physical erosion and textural immaturity of ensuing sediments could have produced the covariation of Si and Al besides other elements associated with quartz, feldspar, and other silicates, which explains the sympathetic nature of Si-Al during sea level lowstands.

Generally, organic matter in sediments is derived from the remains of aquatic and terrigenous plants, which are microbially, chemically, and physically degraded during their transport to the site of deposition and during early diagenesis. Hence, organic matter preservation depends on environmental conditions such as the presence of oxygen, sulfate, and other electron acceptors in bottom and pore waters. For ancient sedimentary rocks, the close link between organic matter preservation and environment can therefore be used as a key to deduce paleoenvironmental conditions (Littke et al., 1998). Jendrzejewski et al. (2001) commented that the production, accumulation, and preservation of organic matter are very sensitive to oceanographic and climatic changes. Tu et al. (1999) and Pellenard et al. (2014) have shown the existences of intimate relationships between marine organic matter, atmospheric CO_2 content, climate, and sea level. The rise in sea level and resulting increase in shelf area are believed to be mechanisms for increased productivity and burial of organic matter (Hofmann et al., 2001; Pearce et al., 2003). Oxic nature of the depositional waters during the sea level lowstands and associated deposition of siliciclastics thwart accumulation and preservation of organic carbon in the sediments (Friedman and Chakraborty, 1997), thus restricting its higher concentration only with carbonates and relatively lesser oxygenated conditions (Fenner, 2001a) and thereby with higher sea level conditions (Fenner, 2001b). These are evidenced through the C_{Org} profile and lithofacies succession of this basin as could be observed elsewhere. Occurrences of alternations of siliciclastics and carbonates, among which, higher C_{Org}, in carbonates coupled with the records of global signals of sea level highs in this basin permit interpretation of falling pCO_2 due to the organic matter burial, resulting in climatic cooling (Jeans et al., 1991; Ulićny, 1992; Raymo et al., 1997) and sea level decrease (Tu et al., 1999). In due course, this decrease might have promoted shelf and continental erosion, enhanced siliciclastic influx and oxic waters, and reversed the cooling trend, which explain the lithofacies alternations and association of C_{Org} with carbonates.

Jendrzejewski et al. (2001) and Fenner (2001a) reported maximum flooding surfaces with enrichments of total organic content in Albian sediments. During the Cretaceous Period, the Indian plate was surrounded by seawater akin to the present-day Australia. Owing to the prevalent greenhouse effect, the temperature gradient between the tropic and polar regions might have been lower. Due to this lesser temperature gradient, climatic cooling and resultant sea level fall enforced lengthening of fluvial systems to transport detrital material that in turn had balanced the C_{Org} burial through cessation of carbonate deposition. Influx of continental waters coupled with reduced bathymetry significantly increases available oxygen in the waters, increases the mixing (Kampschulte et al., 2001; Jendrzejewski et al., 2001), and promotes the destruction of C_{Org} (McKirdy et al., 2001; Fenner, 2001a). This mechanism may explain the alternate occurrence of siliciclastic and carbonate deposits as a consequence of sea level fluctuations in this basin. Based on the occurrences of all the six global sea level peaks, recognized independently through foraminifer (Raju and Ravindran, 1990; Raju et al., 1993), lithofacies (Ramkumar et al., 2004a) and geochemical data (Ramkumar et al., 2005b, 2010a,b, 2011) influence of eustatic sea level changes and thus global scale climatic fluctuations are inferred.

The C_{Org} (Fig. 3(D)) apparently mimics Ca, explaining its dependence on carbonate deposition which in turn was under the control of climatic and sea level changes (Boulila et al., 2010). Preservation of low C_{Org} except during the Albian, latest Campanian-middle-upper Maastrichtian, and latest Danian is consistent with bioturbation, benthic organisms, and oxygenated waters of deposition. Dependence of C_{Org} availability with carbonate deposition indicates its marine biological origin (Peryt and Wyrwicka, 1993) and low bottom-water oxygenation (Pratt, 1984). Predomination of marine organic matter and the inferred absence of detrital organic matter may indicate the absence of significant continental vegetation. However, occurrence of less than 0.4% in most of the profile could be the result of poor preservation conditions (Jendrzejewski et al., 2001) and also due to lower influx of terrestrial organic matter. This inference is corroborated by the presence of antisympathetic nature of excursions of Al and C_{Org} (compare Fig. 3(B) and (D)).

Significant enrichment of C_{Org} in the Albian Grey shale Member is observed. The 7-m-thick grey shale with frequent interbeds of fossiliferous grey limestone (whose thickness increases upward) indicates definitive evidences of climatic and sea level fluctuations. Based on the study of alternations of organic matter-rich black shale and limestone beds of upper Albian deposits of central Italy, Galeotti et al. (2003) inferred those deposits as the result of orbital forcing-induced changes in hydrological and atmospheric cycles. However, Böhm et al. (2003) cautioned against interpreting limestone-shale alternations as primary differences in sedimentary conditions. Coccioni and Galeotti (2003) reported from the type areas of OAE 1a and OAE 2 that these events were preceded by many organic-rich beds. Initially, two OAEs were introduced for the Cretaceous: the Aptian/Albian OAE 1 and the Cenomanian/Turonian OAE 2. Subsequent investigations led to the recognition of a third OAE in the Coniacian/Santonian and to a further division of the extensive period covered by the OAE 1 into four distinct subevents of ocean-wide dysoxia/anoxia separated by dominantly oxic oceanic conditions: early Aptian OAE 1a, early Albian OAE 1b, and late Albian OAE 1c and 1d (Hofmann et al., 2001). It is observed that OAEs occurred either during transgressive periods at times of maximum flooding and, hence, sea level highstands (e.g., the OAEs 1a, 1b, 1d and OAE 2) or during regressive periods at times of maximum sea level fall (OAE 1c). Keller et al. (2004) are of the opinion that the OAEs result by enhanced organic matter burial, the causes may either be higher primary production or influx of terrestrial organic matter. Since polar ice caps were not present during the mid-Cretaceous (Takashima et al., 2009), these variations probably resulted from feedback mechanisms of the mid-Cretaceous low latitudes involving warm/humid and cool/dry cycles (Herrle et al., 2003). The "greenhouse" model encompasses the sea surface temperature in tropics at about 28 °C and polar region at about 0 °C while the "icehouse model" encompasses 24 °C in the tropics and –6 °C at the poles (Price et al., 1998).

Various authors suggested that the burial of organic carbon was primarily a function of sedimentation rates and organic productivity. They attributed the formation of black shales to increased surface water productivity during warmer and more humid climatic conditions. Increasing runoff accelerates the transfer of nutrients from the continents into the oceans, inducing higher primary productivity in the surface waters (Herrle et al., 2003). In addition, organic-rich horizons have been interpreted to be the result of local oceanographic and topographic factors, resedimentation of organic matter from shallow to deepwater environments, increased supply of terrestrial organic carbon, or sea level falls. In addition to OAE black shales, regionally distributed black shales are common in the Aptian to Albian sediments of the western Tethys. These black shale horizons are partly characterized by a cyclic occurrence and have been interpreted as a result of changes in seasonality. Regardless of their origin, these events indicate large-scale disequilibria of the carbon cycle, which occurred during well-developed greenhouse conditions (Galeotti et al., 2003). Jarvis et al. (2002) are of the opinion that high global temperatures driven by increased levels of volcanogenic CO_2 in the atmosphere precede or are coincident with OAEs because they affect weathering patterns on the continents, increase nutrient fluxes to the oceans and the intensity of wind-driven upwelling, and thereby increase phytoplankton productivity and organic carbon burial. Increased organic carbon burial and reduced carbonate carbon fluxes cause excess CO_2 drawdown and nutrient depletion due to the enhanced organic carbon burial, which reduce global temperatures and terminate the OAE. Bralower et al. (1994) opined that normal marine depositions were interrupted episodically, resulting in depositions of organic-rich sediments during Aptian-Albian, Cenomanian-Turonian, and Coniacian-Santonian. According to these authors, though these events, often called OAE, are recorded from many locations, they have differing spans from less than 1 Ma (Cenomanian-Turonian) to c. 20 Ma (Aptian-Albian) and are not exactly synchronous worldwide, for example, the OAE 2. Though it is often referred to be straddling across the Cenomanian-Turonian boundary, it spans from late Cenomanian to early Turonian (Linnert et al., 2010). It holds true to the Cauvery Basin. Analysis of transcontinental variations of Cenomanian-Turonian OAE by Du Vivier et al. (2014) has also arrived at the similar regional variability. Jati et al. (2010) reported the occurrence of Cenomanian-Turonian boundary from oyster beds that occur below the black shale deposits of Agadir Basin, North Atlantic Margin of Morocco. And there are evidences of mismatch between enhanced C_{Org} and positive excursions of carbon isotope (Takashima et al., 2009). Based on biomarker and geochemical studies of Posidonia black shales, Frimmel et al. (2004) suggested that the intimate coupling of biomarker association with sea level and the onset of black shale deposition coinciding with the first transgressive pulse following the lowstand systems tract emphasize the dependency of organic matter accumulation on basin configuration and morphology for the heavily structured and shallow shelf region of the Tethys margin. Similar climate-controlled nature, basin morphology, and other factors in this epicontinental sea could offer explanation to the absence of synchrony of organic-rich facies in the Cauvery Basin. Though many black/grey shale beds are recognizable, they occur at different levels in this basin, meaning attenuation of local and regional causes.

The variegated clays of the Karai Shale, that also contain the OAE 2, have many centimeter-scale intercalations of red and green shale layers, that are traceable for many kilometers with similar thickness and other lithofacies characteristics. Similar to the observations of Vârban (2003) and Melinte-Dobrinescu and Roban (2011) on CORBs (Cretaceous Oceanic Red Beds) of Vrancea Nappe region, the Karai Shale shows the absence of any black shale and pyrite and the presence of bioturbations both in the red and green shale couplets. Taking cue from the mechanisms suggested for the formation of CORBs, we infer the following: the iron oxide source is represented by continental lateritic soils. During diagenesis, reducing interstitial fluids migrated preferentially in the more permeable carbonate levels overprinting the green color. During redox processes, the oxygenation should be produced by oxygen-rich currents, which induced to the clastic material the red appearance by hematite formation. It is also possible that an arid climate of those times led to the accumulation of red soils on emerged coastal plains; transgression could have led to redeposition of sediments rich in Fe-hydroxides into the marine environment, generating the CORBs that are now preserved. We assume that the chlorites were oxidized on the land at different intervals, related to short regional climate fluctuations, resulting in the alternating red and green shales. A similar lithology consisting of variegated red and green couplets may be produced by the intermittent occurrence of bottom-oxygenated currents. The illite, kaolinite, and the terrigenous clasts were transported from the continent. All these inferences point to a situation wherein the Karai Shale deposition was under the influence of high-frequency climatic cycles.

The high-frequency fluctuations of Ca, and C_{Org}, during upper part of Danian are the results of limestone-marl alternations as could be observed from the lithofacies succession. Westphal et al. (2000) and Munnecke and Westphal (2004) suggested that the limestone-marl alternations are an artifact of a process of redistribution of calcium carbonate from marl layers to limestone beds by dissolution, migration of ions, and reprecipitation. Böhm et al. (2003) cautioned the use of one-to-one interpretation of climatic cyclicity with limestone-marl rhythmic alternations. As observed by Boulila et al. (2010) from the lower Kimmeridgian limestone-marl alternations of SE France, the rocks under study show the laterally persistent nature of these alternations for over many tens of kilometers, which could not be explained by homogeneous initial sediment composition, predominance of aragonite as original mineralogy, and regionally extensive, but uniform diagenetic transformation. Keller et al. (2004) are of the opinion that the limestone-marl couplets are the results of fluctuations in clastic dilution, water mass stratification, and productivity due to Milankovitch forcing of precipitation and runoff. In tune with the inferences of Boulila et al. (2010), and Keller et al. (2004), we interpret these cyclic deposits as the result of primary environmental cyclicity reflecting Earth's orbital parameters. In addition to these alternations, there are subtler, superimposed long-term cyclicity in the studied section which are characterized by variations of thicknesses of limestone-marl beds. Tucker and Wright (1990) described the factors involved in carbonate production and dilution. According to them, there are many factors that can have an influence on carbonate deposition: sea level change, water depth, nutrient supply, oxygenation, current strength, turbidity and clastic influx, freshwater-riverine influx, rainfall, aridity–humidity (dry–wet) climate cycles, temperature, etc., and many of these are interrelated. Tucker and Wright (1990) postulated three possibilities and in all the possibilities, the role of climatic fluctuation is indicated and is inferred to be the cause of limestone-marl alternations and the thickness variations of limestone-marl beds in the study area.

These variations where more limestone-dominated intervals alternate with less limestone-dominated intervals, also called "bundles" (*sensu* Munnecke and Westphal, 2004), are interpreted to be the result of orbital cycles. In contrast to the high-frequency limestone-marl alternations, these bundles are the result of fluctuations in environmental conditions. Munnecke and Westphal (2004) interpreted the more limestone-dominated parts of bundles as the result of enhanced pelagic productivity, increased import of aragonite-dominated, platform-derived carbonate, and/or a decrease in terrigenous influx. While each of these factors presents different environmental conditions, the high-frequency alternation of Zr (Fig. 3(E)) in resonance with the alternations of Ca and C_{Org} affirms the influence of terrigenous influx in the Cauvery Basin over the formation of bundles. The superimposed nature of the limestone-marl alternations into bundles and superbundles, but not with classical 5:1 or 4:1 ratio may suggest the omission of bundles or the portions thereof.

The abundance of Zr in clastic rocks was interpreted to be the result of detrital influx as well as sediment recycling (Spalletti et al., 2008). Occurrences of generally higher levels of Zr all through the Barremian-Danian with the exception of latest Campanian to middle Maastrichtian (Kallankurichchi Formation) and many episodes of positive excursions over this background value suggest sediment-starved nature of the basin and significant recycling of older sedimentary rocks. While analyzing the stratigraphic variations of clay mineral assemblage of end-Cretaceous-early Tertiary rocks of this basin, Madhavaraju et al. (2002) inferred prevalence of warm-wet and warm-dry seasonal conditions that controlled chemical and physical weathering, respectively. High-frequency alternations of mid-Carboniferous strata of England were interpreted by Tucker et al. (2009) to be the result of millennial-scale arid-humid climatic fluctuations and the bed sets as the result of longer term arid-humid climate and sea level fluctuations due to the precession rhythm. Ti and Zr are generally assumed to represent detrital inputs into a sedimentary basin and their variations should be related with changes in weathering conditions in the hinterland or changing source regions (Bellanca et al., 2002). Zr is mostly concentrated in zircons, which accumulate during sedimentation while less-resistant phases are preferentially destroyed (Alvarez and Roser, 2007). Peak enrichments of Zr during middle-latest Cenomanian (Gypsiferous clay and Odium members), middle Campanian (Varakuppai Member), middle-late Maastrichtian (Ottakoil Formation), and early Danian (Anandavadi Member) may indicate the influx and cessation of terrigenous influx. The influx-cessation might have been controlled by variations in source area weathering and/or a change from more humid to more arid conditions or tectonic movements (Munnecke and Westphal, 2004). Sandulli and Raspini (2004) interpreted the elemental cycles as the precession and obliquity periodicities, and the bundles and superbundles into short and long eccentricity cycles and similar inference could be made to the rocks under study.

Al (Fig. 3(B)) and Sr (Fig. 4(A)) show noticeable reductions of concentrations from older to younger deposits. Reduction of Al (Fig. 3(B)) may be reflective of either reduction of weathering in the continental regions and deprivation of availability of feldspars and clays for transport to the depocenters and/or due to the gross lithological variations of the sedimentary record (Andrew et al., 1996). While gradual deprivation of feldspars and clays is supported by the linear trend line of K (Fig. 4(B)) and X-ray diffraction (XRD) patterns, lithofacies variation on a basin-scale (wherein dominant proportion of siliciclastics in older deposits and carbonates in younger deposits could be observed) suggests the roles of both of these causes. Sr is spatially conservative in the sea and enrichment/depletion of Sr could be linked to prevalent contemporaneous processes (Andersen and Delaney, 2000). Reduction of strontium from older to younger strata could be interpreted as decreasing terrestrial input or sea level or both (Martin and Macdougall, 1991). Existences of broad, double-peaked positive excursions during Albian and sharp, double-peaked positive excursions of Sr during Cenomanian, and associations of these excursions with shale deposits, suggest significant weathering (Bralower et al., 1997) in the provenance and significant influx into the depositional loci. Nevertheless, existence of a general pattern of increase of relative sea level (RSL) from Barremian to Santonian and decrease thenceforth up to Danian (Fig. 2) despite the presence of high-frequency cycles may indicate sea level-controlled nature for the responses of Sr and by implication, the depositional system.

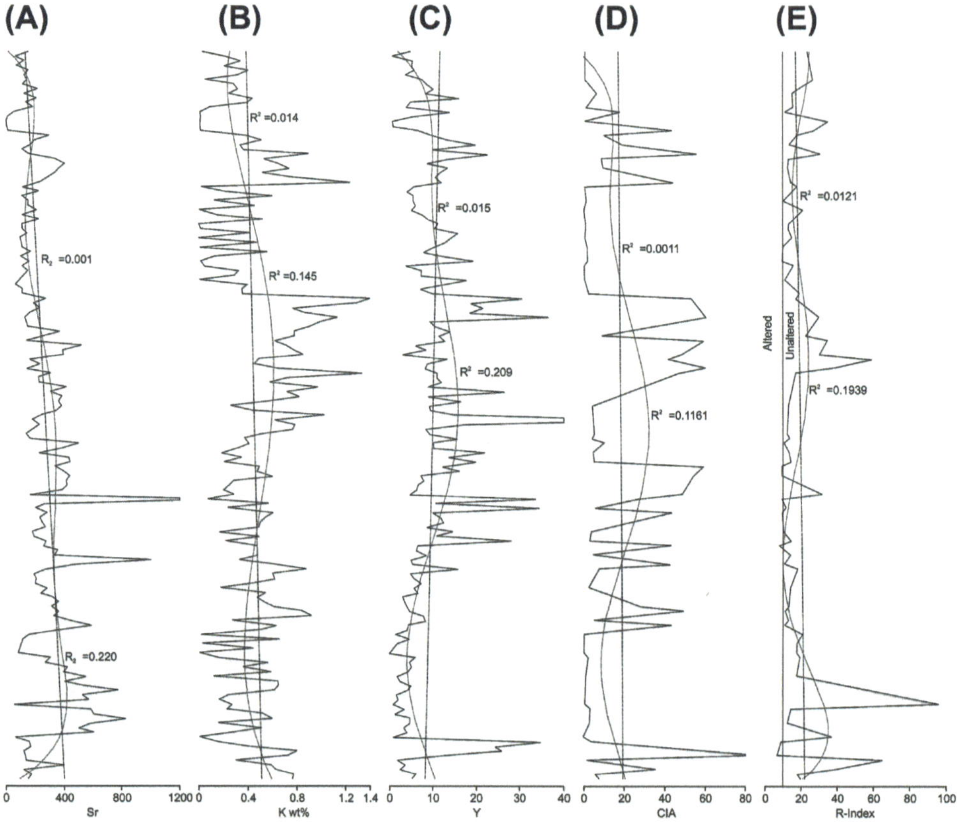

FIGURE 4 Statigraphic variations of elemental concentrations of elemental concentrations, CIA and R-index.

Sr and Y are given in ppm. K is given in weight percentage.

11.4.2 INFLUENCE OF TECTONICS OVER THE DEPOSITIONAL ENVIRONMENTS

The element Y shows a peculiar polynomial peak across Aptian-Albian boundary (Fig. 4(C)), subdued nature during most of the successive period till latter part of middle Albian and gradual increase afterward reaching a significant peak during Coniacian followed by a gradual decrease. In magnitude and scale, it mimics Zr (Fig. 3(E)). As the element Y undergoes little or no diagenetic alterations (Andrew et al., 1996; Das, 1997), presence of its short and significantly prominent peaks exactly coinciding faulting events (as inferred independently in the field structures and contact relationships of beds—Ramkumar et al., 2004a) and the associated change in sedimentation pattern as reflected in lithofacies characteristics indicate influx of Y immediately after major tectonic movements and resultant change in nature, quantum, and composition of detrital influx into the basin. Sedimentation pattern and nature of sediments of the periods between Barremian-Coniacian and Santonian-Danian were different and are reflected in the patterns of Y and Zr during these two time spans. The differences between these elements in terms of temporal resolution may be a consequence of their differential response (Whitford et al., 1996) to prevalent depositional environmental conditions as enforced/introduced by the major tectonic event. Dubicka et al. (2014) also observed significant changes in environmental conditions due to Subhercynian tectonic movements in Ukrine during Coniacian-Santonian. Enrichment of these elements up to Coniacian and their subdued nature after Coniacian could be attributed to the changes brought in by major tectonic activity occurred during Santonian, across which significant changes in proximity of sediment source and nature and quantum of detrital influx were witnessed (Sundaram and Rao, 1986).

11.4.3 TEMPORAL VARIATIONS OF GEOCHEMICAL COMPOSITION VIS-À-VIS DEPOSITIONAL ENVIRONMENTS

Having affirmed the influence of tectonics over depositional system of the Cauvery Basin only across the Coniacian-Santonian boundary, as reflected by the change of lithofacies and geochemical characteristics and the predomination of sea level variations over depositional environments, major and trace elemental abundances of the rocks under study over time with reference to the Upper Continental Crust (UCC) were examined (similar to Alvarez and Roser, 2007; Spalletti et al., 2008) to understand the temporal variations of geochemical composition vis-à-vis depositional environments. Geochemical signatures of the clastic sedimentary rocks provide important sources of information that record different aspects of provenance, tectonic, environmental, and ecological evolution. Although hydraulic sorting, weathering, and diagenesis can alter the geochemical composition of basin sediments, there often remains a strong signature to the original source terrain, reflecting the nature of the exposed continental crust (Yan et al., 2007) and hence comparison of the data of studied samples with those of published values of UCC was attempted. Table 3 presents the mean values of major and trace elements based on the analyses of 70 and 157 samples, respectively.

It follows from Table 3 that the mean values of all the elemental composition (except Ca) of all the studied samples are far below the values of UCC, suggesting sediment-starved nature of the Cauvery Basin. Though previous studies have reported the sediment-starved nature (for example, Ayyasamy, 1990; Ramkumar et al., 2004a, 2005a; Kale, 2011), these data provide affirmative and documentary evidence for this characteristic. The enrichment of Ca with reference to UCC is related with the biohermal and biostromal sediment production. While all the elements are one or more orders lower than the UCC, the elements Ba and Mn are close to the UCC values. Their enrichments are due to the occurrences of unaltered Ba-Mn-rich orthoclase cobble-boulder clasts in the conglomeratic deposits of the basin (Ramkumar, 2007; Ramkumar et al., 2004a, 2005a,b, 2010b). The Ba-Mn-rich orthoclase is a major mineral in granitic gneissic rocks of the provenance area and its occurrence without any alteration in the receiving basin, together with the sediment-starved nature of the basin, suggests proximity between source and depositional basin, predomination of physical erosion, and subdued nature of chemical weathering in the source area. At each of the unconformity surfaces, conglomerate bed with frequent basement rock boulders along with the lithoclasts of older sediments is observed to occur. It indicates the initiation of deposition of conglomeratic sediments after every major break in sedimentation, prevalence of sediment recycling, and sediment-starved nature. Cooccurrences of boulders of basement rocks and lithoclasts of older sedimentary rocks may indicate sourcing of sediments from basement region as well as former depositional sites. This feature also indicates the prevalent shift of principal depositional loci from west to east over an interval since Barremian and the persistence of provenance area as a positive region all through the depositional history. All these observations may also indicate the prevalence of glacial periods and their associated lengthening of fluvial channels,

Table 3 Abundances of Major and Trace Elements in the Rocks under Study with Reference to the Upper Continental Crustal Values Presented in Yan et al. (2007)

	UCC	Basin	1	2	3	4	5	6	7	8	9	10
Si	30.70	16.34	18.03	21.40	7.90	16.14	19.00	15.05	35.64	17.47	12.00	19.67
Al	8.00	1.81	1.82	2.60	0.91	2.61	2.92	1.93	1.78	1.55	1.44	1.01
Ca	2.99	18.37	18.63	9.78	29.07	13.36	13.39	16.19	0.85	18.84	24.46	16.42
Fe	3.48	1.65	0.46	1.68	0.61	2.32	2.25	3.17	2.04	1.38	1.73	0.50
Na	2.88	0.24	0.53	0.09	0.06	0.26	0.62	0.39	0.28	0.20	0.20	0.02
Mg	1.32	0.59	0.82	0.24	0.52	0.90	0.77	0.82	0.19	0.36	0.60	0.35
K	2.81	0.45	0.77	0.51	0.36	0.51	0.33	0.70	0.65	0.67	0.38	0.19
P	90.00	21.59	10.22	48.81	7.29	41.85	12.41	22.79	12.86	11.70	29.02	6.29
Ti	300.00	227.00	49.35	355.24	72.92	261.71	293.51	398.71	449.40	265.02	142.12	177.79
Mn	60.00	58.18	51.57	98.70	17.51	39.58	96.99	104.29	12.33	66.68	91.16	12.67
Ni	20.00	20.64	58.71	34.62	18.12	27.16	27.71	21.81	10.11	12.00	18.99	13.28
Cu	25.00	15.07	10.97	22.41	14.19	23.20	23.60	17.86	10.17	7.98	9.43	10.61
Zn	71.00	33.58	19.16	52.29	22.25	56.54	50.27	51.94	26.41	26.89	21.88	10.74
Sr	350.00	269.09	161.86	182.45	403.67	319.98	439.41	275.08	286.76	169.11	193.97	111.82
Zr	190.00	110.72	60.47	121.01	53.61	96.36	135.23	152.26	165.92	153.80	88.03	131.22
Ba	550.00	535.98	739.03	645.69	219.59	375.84	658.25	851.44	817.37	721.28	694.63	237.95
S		0.03	0.00	0.00	0.08	0.00	0.03	0.00	0.17	0.01	0.01	0.00
C_{Org}		0.39	0.26	0.14	0.76	0.23	0.18	0.28	0.00	0.37	0.53	0.38

UCC: Upper Continental Crust; Basin: Mean of all the samples of Barremian-Danian strata of the Cauvery Basin. (1) Barremian; (2) Aptian; (3) Albian; (4) Cenomanian; (5) Turonian; (6) Coniacian; (7) Santonian; (8) Campanian; (9) Maastrichtian; (10) Danian. Si, Al, Ca, Fe, Na, Mg, K, S, and C_{Org} are expressed in weight percentage. P, Ti, Mn, Ni, Cu, Zn, Sr, Zr, and Ba are expressed in ppm. Si, Al, Ca, Fe, Na, Mg, K, P: n = 70; C_{Org}, S, Ti, Mn, Ni, Mn, Cu, Zn, Sr, Zr, Ba: n = 157.

erosion of provenance, as well as shelf area quite similar to the discussions of Sarg (1988), Carter et al. (1991), and Kampschulte et al. (2001).

Si showed a decreasing trend of abundance from Santonian > Aptian > Danian > Turonian > Barremian > Campanian > Cenomanian > Coniacian > Maastrichtian > Albian. The trends of other detrital elements are as follows: Zr: Santonian > Campanian > Coniacian > Turonian > Danian > Aptian > Cenomanian > Maastrichtian > Barremian > Albian; Ti: Santonian > Coniacian > Aptian > Turonian > Campanian > Cenomanian > Danian > Maastrichtian > Albian > Barremian; Ba: Santonian > Coniacian > Barremian > Campanian > Maastrichtian > Turonian > Aptian > Cenomanian > Danian > Albian. All the definitive detrital elements are highly enriched in Santonian and highly depleted in Albian. The enriched nature of Si, Ti, Ba and very close value of Zr to that of UCC values during Santonian indicate enhanced influx of detrital sediments immediately following the major tectonic event. Despite the enrichment of Si, Santonian samples contain more than four orders of lower Al, indicating the absence of chemical weathering and simple physical erosion and influx of siliciclastic sediments in the form of unaltered clasts. The Albian in the Cauvery Basin witnessed sea level maximum and coincided with oxygen-poor conditions of deposition and had insignificant terrestrial influx (Ramkumar et al., 2004a) and enhanced carbonate accumulation which in turn are reflected in the geochemical signatures too.

During Maastrichtian, the bioproduced carbonate sedimentation dominated within the basin, which is reflected in the 8.18 orders of enrichment of Ca with reference to UCC. It is also inferred that the alternation of siliciclastics and carbonates as a response to climatic and thus sea level fluctuations is affirmed by the highly enriched nature of Ca in the Albian, followed by Maastrichtian and Danian (which contain bioproduced carbonate deposits higher than others) and highly depleted nature of Ca- and carbonate-hosted elements in Santonian. This inference is further substantiated by the highest values of C_{Org} in Albian followed by Maastrichtian and its absence in the Santonian samples.

11.4.4 STRATIGRAPHIC VARIATIONS OF WEATHERING INDICES

Rock and soil weathering involves physical and chemical processes that may be biotic and/or abiotic. Chemical weathering rates of soil and continental bedrock are complex functions of temperature, humidity, organic activity, exposure, relief and elevation, and sometimes age (Boucot and Gray, 2001). In chemical weathering, both relatively stable secondary minerals and relatively soluble substances are produced. Chemical weathering is an important mechanism driving elemental fractionation away from parental bedrock signatures. The extent of fractionation depends on bedrock and local weathering conditions tied to climate. Stronger chemical weathering is generally associated with warm and humid climates, while more arid climate is generally associated with relatively weak chemical weathering (Yan et al., 2007). The Chemical Index of Alteration (CIA) values help quantify the degree of conversion of primary minerals of the source rocks to clay minerals (Spalletti et al., 2008). In addition to reflecting weathering, the CIA, the Plagioclase Index of Alteration (PIA), and the Vogt's Residual Index (V) may be sensitive to subtle geochemical changes and hence are widely being utilized for demonstrating the impact of climate on bedrock, neotectonic characteristics, and Earth's surface environment over geologic time (Price and Velbel, 2003). Following the standard procedures, we have computed various indices and discuss herein the results in relation with the objectives of the present study.

The source area consists of granitic gneisses in low-lying plains and massive hills of charnockite. These rocks consist of very coarse-grained plagioclase, smaller grains of quartz, hypersthene, and amphibole as major minerals; magnetite, garnet, and biotite as minor minerals; and zircon, rutile, and apatite as accessory phases (Sharma and Rajamani, 2001). Cutting across the gneisses and charnockites, the pegmatitic veins composed of large to very large feldspar crystals (at places ranging up to many tens of centimeters) occur frequently. All the stratigraphic profiles of weathering indices show a common feature that, all through the depositional history, the sediments received by the Cauvery Basin were not well differentiated and chemically weathered. This observation, together with the statement of Sharma and Rajamani (2001), that *The nature and extent of source rock weathering, physical sorting during transport and environmental conditions during deposition at the depocenters exert significant control over sediment geochemistry* affirms the interpretation of proximity to source, and predomination of physical erosion and weathering. Models of climate–chemical weathering feedback, taking account of such experimental data, suggest a 2% to 10% increase in chemical weathering rates for each degree of temperature increase and it is also observed by many studies that runoff and slope control significantly the mechanical erosion (Gislason et al., 2009).

The CIA shows nonsteady nature which may indicate active tectonism in the source area that led to erosion of all the zones of weathering profiles (Lee, 2009). While examining the weathering pattern of basement rocks, Sharma and Rajamani (2001) observed them to be chemically immature, though exposed to the tropic climate and reasoned this nature due to the poor differentiation by streams and multiple stages of tectonic uplift. These authors have also recorded first signs of weathering on plagioclase and hypersthene, followed by formation of kaolinite and smectite as the alteration products; depletion of Ca, Na, and Mg; and enrichment of Al, Fe, and Ti with progressive weathering. Similar observations were made by Sharma and Rajamani (2000) based on the weathering pattern of gneissic rocks. A striking similarity between the source rock (charnockites and gneisses) composition and the recent flood plain sediments of the Cauvery River was reported by Singh and Rajamani (2001).

The idealistic trends in individual profiles of the CIA (Fig. 4(D)) showed the highest alteration during Late Aptian (Terani Clay Member) corresponding with extreme positive excursions of R-index (Fig. 4(E)), V-index (Fig. 5(A)), Ti (Fig. 5(B)), and Al (Fig. 3(B)). In general, the profile of CIA shows a waning and waxing nature, sympathetic to the detrital elemental profiles (Al, Zr, Ti, Y, etc., see Figs 3(B,E), 4(C), 5(B)) and antisympathetic with the Ca (Fig. 3(C)) and C_{Org} (Fig. 3(D)). It is interesting to note that except during the deposition of Terani Clay Member (Aptian), all through the depositional history of the basin, the sediments were nowhere near the CIA value of Post-Archean Australian Shale (70), suggestive of relatively subdued nature of chemical weathering, sediment fractionation, and ultimately, predomination of warm, dry climatic conditions. As chemical weathering on continents is largely controlled by moisture and temperature, a wet and warm climate may enhance the chemical weathering (Yan et al., 2007). Goldberg and Humayun (2010) demonstrated that the higher CIA values are also indicators of humid conditions. Based on this presumption, the positive excursions of CIA in the Cauvery Basin are observed to be associated with siliciclastic and clay deposits and also that the peaks of CIA corroborated exclusively with clay-dominated deposits. The role of source rock weathering and elemental redistribution during diagenesis also can be assessed using the PIA, which quantifies the degree of destruction of plagioclase and identifies albitization (Alvarez and Roser, 2007). The samples under study show that the period from Cenomanian to Coniacian with very high PIA values and a peak value during middle Turonian meaning that the plagioclase was almost totally destroyed by source area weathering during Cenomanian-Coniacian and during other periods, there was no such wholesome alteration. This observation, when compared with the conditions of chemical weathering at lower latitudes listed by Boucot and Gray (2001), and with paleogeographic location of the Cauvery Basin in the lower latitudes during Barremian-Maastrichtian, limited extent of the provenance and the

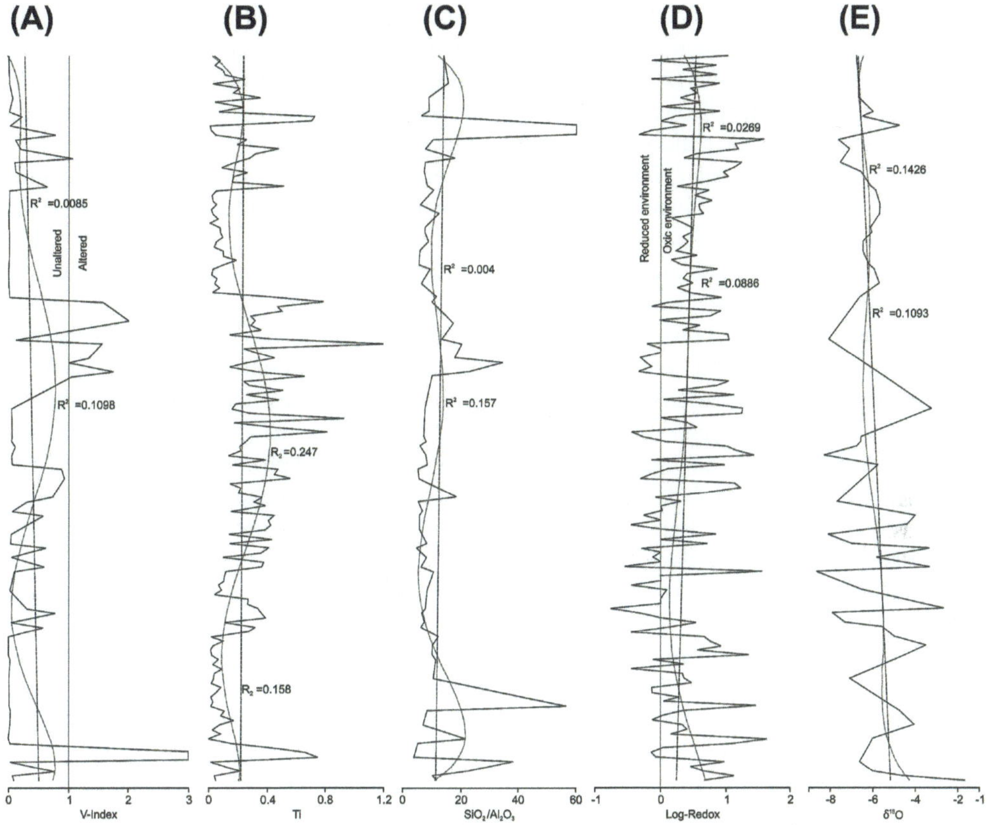

FIGURE 5 Stratigraphic variations of elemental concentration, oxygen isotope and computed elemental ratios.
Ti is shown in weight percentage and oxygen isotope in per mil.

configurations of the depositional basin and provenance, support the inference of prevalent high humidity and low relief that have promoted weaker chemical weathering. This inference is further substantiated by the profiles of Si, Al, K, and SiO_2/Al_2O_3 (Figs 3(A,B), 4(B), 5(C)).

On a temporal scale, the Albian, Coniacian, middle-late Campanian to middle Maastrichtian, and middle-late Danian show least or no chemical weathering. On the contrary, the Aptian, early Cenomanian-early Turonian, Santonian-middle Campanian, and middle Maastrichtian-early Danian show significant weathering. However, these periods of chemical weathering show very high orders of fluctuations while nonweathering periods show a consistently unaltered characteristic. Similar nature could be observed in the other weathering indices, particularly in the V-index (Fig. 5(A)).

The profile of Log–Redox (Fig. 5(D)) shows the predomination of oxic waters of depositional environments all through the depositional history, and very high orders of fluctuation from Barremian-Santonian. These higher orders of fluctuations could have been the result of variability of depositional conditions (Jendrzejewski et al., 2001). Very often, the fluctuations led to reducing conditions; the period from Santonian to Danian showed less-frequent fluctuations, that too within the oxygenating waters. The anoxic–oxic changes may reflect various Earth processes, such as paleoceanographic shift, tectonic movements, and/or climatic fluctuations that changed the balance between carbon sources and sinks in the world ocean (Melinte-Dobrinescu and Roban, 2011). A dent in this trend is observed across the KTB (Cretaceous - Tertiary Boundary). Though fluctuating within oxygenating waters, the samples of Danian show the fluctuation pattern different from that of all other samples, particularly from that of Maastrichtian. Madhavaraju et al. (2002), based on stratigraphic variations of clay mineral assemblages of the late Cretaceous—early Palaeocene of the Cauvery Basin, inferred the prevalence of high-frequency alternations of warm and wet to warm and dry seasons. All these trends and indications on oxygenating and reducing waters are corroborated with the stable isotopic trends and excursions also (Figs 5(E) and 6(A)). It is interesting to note that similar to the pattern of Log–Redox profile, the Barremian-Santonian duration shows high variability, significant change in the pattern across Santonian (refer to the influence of tectonic movement discussed earlier), and a dent in the pattern across KTB. While generally oxygenating waters of deposition are decipherable by the $\delta^{13}C$ curve, fluctuations toward mildly reducing or oxygen-poor conditions are indicated by the frequent positive excursions (Nederbragt et al., 2001) of it and antisympathetic relationship of $\delta^{18}O$ profile. Increased $\delta^{13}C$ values in the sedimentary record are attributed to enrichment in $\delta^{13}C$ of the oceanic carbon reservoir due to enhanced burial of isotopically light organic matter (Arthur and Sageman, 1994). Nederbragt et al. (2001) reported positive excursions of $\delta^{13}C$ during Aptian and Cenomanian. Similar excursions are observable in the Cauvery Basin too. Bergstörm et al. (2010) recorded $\delta^{13}C$ excursions due to eustatic as well as local or regional causes.

As could be observed elsewhere (Bak, 2007; Takashima et al., 2009; Linnert et al., 2010), the $\delta^{13}C$ shows a positive excursion across Cenomanian-Turonian boundary in the Cauvery Basin (Fig. 6(A)). The prebuildup stage (*sensu* Takashima et al., 2009) of this positive excursion records coeval triple C_{Org} positive excursions (Fig. 3(D)). Presence of such multiple positive excursions in many Cenomanian-Turonian sections across the World was interpreted by Takashima et al. (2009) and Jati et al. (2010) to be the result of a global scale phenomenon rather than local factors. The rocks of Turonian-Coniacian document intensive feldspar weathering with a peak weathering during middle Turonian than all other times (Figs 4(D) and 6(B)).

The profile of SiO_2/Al_2O_3 ratio (Fig. 5(C)), an indicator of fluctuations in the detrital composition (Rachold and Brumsack, 2001), remains within a short range and consistent except during the Barremian-middle Albian, lower Turonian, across Coniacian-Santonian boundary, and during lower Danian. While the spike during Coniacian-Santonian boundary is inferred to be due to creation of positive area due to tectonic uplift of provenance, other spikes are inferred to be due to variations in depositional conditions. The profile of K_2O/Na_2O ratio (Fig. 6(C)) traverses all through the depositional history with a value of more than 1 and indicates the prevalence of single provenance area rich in K-feldspar. Synonymy of trends of SiO_2/Al_2O_3 profile and K_2O/Al_2O_3 profile suggests common sources of carrier phases of these elements (Hofmann et al., 2001), i.e., quartz, K-feldspar, and illite. Within this trend, there are frequent bouts of intensive positive excursions, mostly across the lithostratigraphic boundaries. As inferred earlier, it signifies the influx of unaltered feldspar clasts during shift of principal depositional loci and at the initiation of newer lithostratigraphic units. This inference is substantiated by the occurrences of coeval excursions of either of or any two of Zr, Y, and Ti (Figs 3(E), 4(C), 5(B)), respectively. Sympathetic nature between SiO_2, Al_2O_3, K_2O, TiO_2, and Zr was interpreted to be detrital origin by Hild and Brumsack (1998). The K_2O/Al_2O_3 ratio values seldom fall within the 0.3–0.9 range, typical of K-feldspars (Alvarez and Roser, 2007) and fall within the range characteristic of

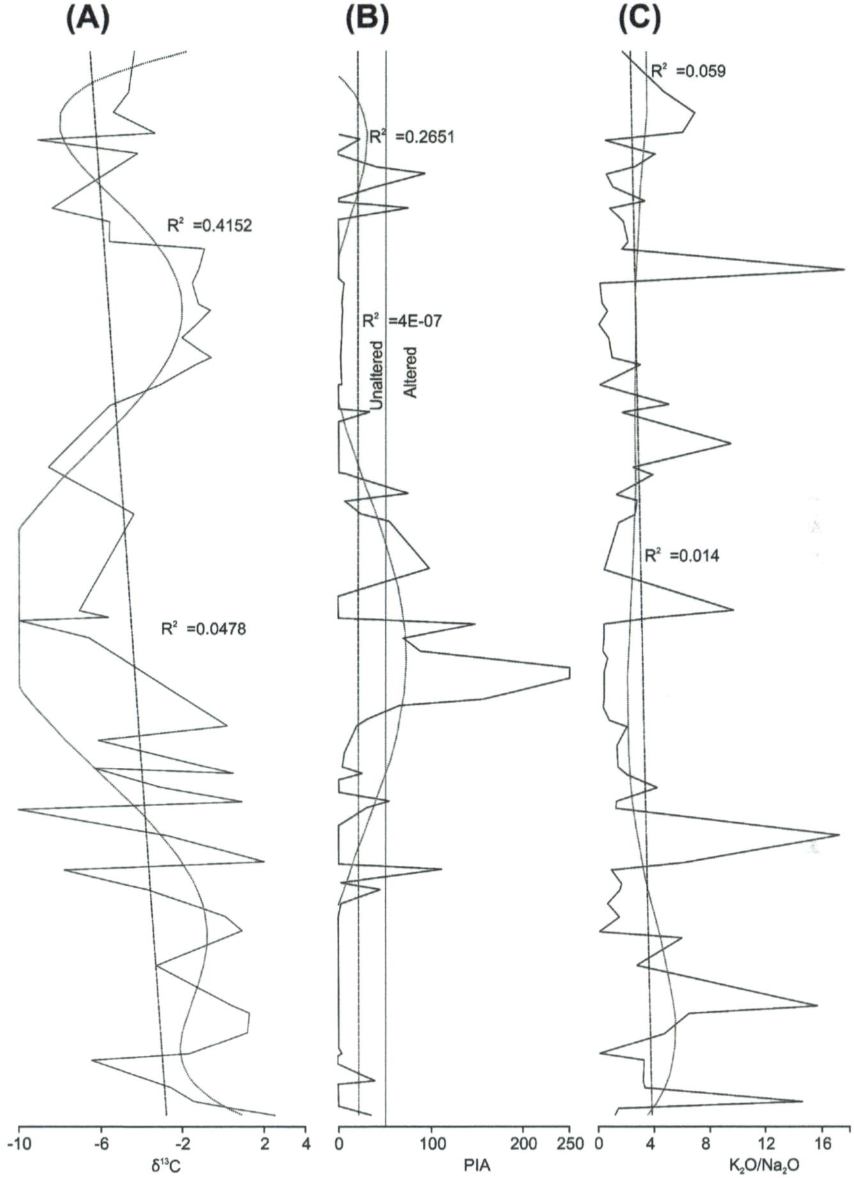

FIGURE 6 Stratigraphic variations of carbon isotope (given in per mil), PIA, and K_2O/Na_2O ratio.

clay (<0.3). The K_2O/Na_2O ratios of almost all the samples are more than 1. Many a times, this ratio exceeds a value of 3 and at times reaches >18, affirming passive margin tectonic setting (Roser and Korsch, 1986; Alvarz and Roser, 2007). These results indicate that the sediments were deposited in a passive margin setting, in either a rift or backarc environment, but the depocenter received detritus primarily from continental sources and thus is likely to have been near the cratonic margin.

11.5 DISCUSSION ON PALEOCLIMATIC TRENDS AND CYCLES

The sedimentary stratigraphic record is the outcome of an exogenic system consisting of geologic setting, changes in sea level, changes in geochemical reactions between the sea and the earth, climate, and processes of sediment formation (Srinivasan, 1989) and diagenesis (Ramkumar, 1999). The ensuing sedimentary records show differences in bulk chemistry as a result of changes in different combinations of minerals, either bioproduced and/or brought into the depocenters, and neoformed during postdepositional times, all of which were influenced by the climatic conditions. According to Veizer et al. (1997), the sedimentation system is dominated by cyclic processes that operate on a hierarchy of temporal and spatial scales on which short-lived events are superimposed. As a consequence, only a net result of cycles and events could be recognized in rock records. With these precepts, the climatic conditions of the Cauvery Basin can be summarized as detailed herein.

The pre-Albian deposits are predominantly continental in nature and may indicate humid climatic conditions (Wortmann et al., 2004), accompanied by chemical weathering. The warm-arid conditions and physical weathering were the norms during the deposition of Olaipadi Member (latest Albian-earliest Cenomanian). As strong monsoonal precipitation often results in intense chemical weathering (Fricke et al., 2010), general subdued nature of chemical weathering in the study area, along with the observations of Sharma and Rajamani (2001) that tectonically active nature of present-day Cauvery catchment areas resulted in poorly chemically weathered nature, could be construed as tectonically active nature of the provenance of the Cauvery Basin during Cretaceous and predomination of warm-arid climatic conditions, punctuated with warm-humid conditions. During the Cretaceous, large variations in environmental conditions are well documented for many oceans and marginal seas as a consequence of large-scale, often global changes. Such worldwide changes should have had an influence on sediment accumulation patterns in most marine environments at different paleolatitudes. However, regional factors such as, for example, tectonic activity in shallow and marginal seas, changing seaways and current systems, and the variability in sediment supply are also important and may sometimes have overridden the global signal (Littke et al., 1998). Based on the configuration of the Cauvery Basin, prevalent close proximity between the provenance and depositional sites, it is perceivable that the modifying factors of lithological composition might have had little or no impact (Spalletti et al., 2008). In addition, if the sediments were deposited during dry periods, the climate does not exercise control over chemical weathering. Thus, the chemical immaturity of the sediments of the Cauvery Basin could have been the result of combination of factors, including, but not limited to, tectonics, basin configuration, proximity between source and depositional site, and general warm-arid climatic conditions.

Changes in clay mineralogy are appreciable in marine sediments and have successfully been used to infer climate change. For example, declines in the relative amounts of kaolinite and increases in smectite are considered as a result of increases in aridity and a "drying out" of the continental source area. Such changes in the early Cretaceous have been tentatively correlated with a cooler climate. Cool climate episodes do not necessarily result in a global increase in continental aridity. At low latitude in the mid-Cretaceous, in fact the opposite occurred (Price et al., 1998). Aluminous clay minerals (such as kaolinite) are formed during intensive chemical weathering in tropical conditions and the same is reflected in the high CIA values (80–100), whereas, in glacial environments, where abrasion is dominant over chemical weathering, common CIA values range from 50 to 70 (Nesbitt and Young, 1982) and the CIA values for average shales range from 70 to 75. When viewed with these estimates, the data of the Cauvery Basin show that except during the Terani clay Member, the CIA values of the studied rocks are far below 80.

This observation, together with the greenhouse conditions of Cretaceous, absence of major glaciation and lack of permanent ice sheets (Takashima et al., 2009) during Cretaceous, and the documentation of low latitude nannofossil assemblage from end-Cretaceous deposits of the basin (Rai et al., 2012), clearly suggests the prevalence of monsoonal-climate and arid-humid alternations.

Based on stable isotopic analyses, Zakharov et al. (2006) opined that the Albian and Turonian times in the Cauvery Basin experienced a range of paleotemperatures from 16.3 °C to 18.5 °C. They also stated that the paleotemperatures were about 19.8–21.2 °C in the early Maastrichtian, during which time India shifted from Warm-Temperature climate zone into the nearby Subtropical zone with the warm and perhaps humid climate. A global sea level low at the Aptian-Albian boundary is recognizable at about 112 Ma (Hancock, 2001). Similar to the platform crisis 1 observed by Schulze et al. (2004) during middle Cenomanian in the west-central Jordan, reduction of carbonate and increase of clay and organic matter coincident with rise in sea level are observed in the Cauvery Basin. However, typical grey shale with highest C_{Org} and grey sandstone with higher C_{Org} occur during early Albian and latest Turonian in this basin. The early Albian event is represented by alternate beds of grey shale and grey bioclastic limestone among which the thickness of limestone progressively thicken toward the top with concomitant reduction of thickness of shale beds (Ramkumar et al., 2004a) and shows similarity with the Middle Cretaceous Event beds described by Coccioni and Galeotti (2003) from Scaglia Bianca Formation of Umbria-Marche Basin. While the increase of carbonate content and resultant limestone beds may be associated with productivity cycles, the clay-rich beds may represent dilution cycles (Böhm et al., 2003; Boulila et al., 2010) and thus, a general increase in productivity and concomitant reduction of dilution cycles could be inferred from this phenomenon. Keller et al. (2004) observed cyclic limestone/shale deposition of the lower Bridge Creek Limestone Member of the Pueblo section and interpreted as a product of orbital cycles with the obliquity index influencing shale deposition through high-latitude precipitation and clastic influx, whereas evaporation and nutrient upwelling of the precessional index control carbonate productivity.

Foraminiferal $\delta^{18}O$ and the paleobotanical record provide convincing evidence that the Cretaceous Period was substantially warmer than today. Globally averaged temperatures were 6–12 °C higher than today with pronounced warming at poles (Zeebe, 2001). Cretaceous sea surface temperature estimates as derived from foraminiferal $\delta^{18}O$ vary from 18 °C to 26 °C for the early Cretaceous (130 Ma), 24.4–25.9 °C for the Cenomanian (97.5–91 Ma), 20–23 °C for the late Campanian-Maastrichtian (74–68 Ma), and 20–21 °C for the late Cretaceous (~66 Ma) (Zeebe, 2001). The pCO_2 of Cretaceous atmosphere has been estimated to be 2–10 times higher than today (Zeebe, 2001). Cretaceous isotope paleothermometry data from low-latitude oceans (Price et al., 1998) indicate that the early Cretaceous was only a few degrees (approximately 2–4 °C) cooler than the middle Cretaceous (Cenomanian).

The possible duration of icehouse interludes within the Cretaceous was such that the entire Barremian to middle of Aptian experienced cold climate; followed by warm climate during middle to upper Aptian and then a period of cold climate during late Aptian. The Aptian to Albian boundary marked a notable change of climate from cold to warm conditions. However, the warm climate was short lived. The followed up cold climate persisted until the upper part of Albian. The latest Albian showed warm climate that persisted through Cenomanian and until the lower Turonian (Price et al., 1998). A few thousand to 2 Ma of long cold periods separated these warm equable climatic periods. Hay (2011) commented that the modern pole-to-equator sea level temperature difference is about 50 °C; that of the mid-Cretaceous ranged from 30 °C to as low as 24 °C, implying a much more equable climate. This may have been caused by reduction of the ice-forced albedo of the polar regions, more efficient meridional energy transport by the atmosphere and ocean, and increased atmospheric greenhouse gas concentrations. Coccioni and Galeotti (2003) reported that the middle Cretaceous (Barremian-Turonian) "greenhouse" world experienced several breaks in the carbon cycle. These events were associated with significant changes in the ocean-climate system (Najarro et al., 2011) since burial of excessive carbon drew down CO_2, apparently might have initiated a global temperature decline from relative maxima. Bellanca et al. (2002) recognized the occurrences of major hiatuses during Barremian, most of the upper Aptian, the entire lower Albian, and part of the middle Albian in the Tethyan region. Similar to the lightening trend of $\delta^{18}O$ observed through Barremian-lower Aptian in the Tethyan region based on whole-rock analyses, attempted by Bellanca et al. (2002), the Cauvery Basin data also show a clear lightening trend followed by opposite trend up to upper Aptian and fluctuations up to the late Albian.

On a basin scale, a gradual reduction of sea level as a result of climatic cooling from Albian to Santonian and then onward a gradual rise of sea level—climatic warming up to the end of Cretaceous followed by a sea level fall—cooling until middle Danian, and finally rise of sea level—warming could be observed. These trends are consistent with the sea level curve of Gradstein et al. (2004) and are synchronously reflected in the profiles of C_{Org} and lithofacies. It is generally followed that the cool, wet conditions favor enhanced delivery of clastic detritus while the warm, dry phases favor reduced clastic sediment delivery and biogenic carbonate generation in the basin (Miall, 2009). Within these major cyclic changes, there are many minor cycles. The mid-Cretaceous (120–90 Ma) was characterized by warm, equable climatic conditions that differed qualitatively from modern icehouse climate. High sea-floor spreading rates (Larson, 1991), high atmospheric CO_2 concentrations (Berner, 1983; Caldeira and Rampino, 1991), and high sea level (Pitman, 1978; Haq et al., 1987) have had a major influence on the prevalent climatic conditions. The middle Cretaceous is one of the warmest climate intervals of the Phanerozoic (Veizer et al., 2000; Bice and Norris, 2002; Coccioni and Galeotti, 2003). Coccioni and Galeotti (2003), based on their extensive biotic, lithological, geochemical, and stable isotopic study, concluded that at about 92 Ma, significant climatic change occurred, giving rise to middle Cretaceous Event and the same holds good for the Cauvery Basin. Coccioni and Galeotti (2003) also stated that both the biotic and sedimentological records from the classical Tethyan setting of Gubbio strongly suggest that the middle Cretaceous event represented a turning point in the climatic and oceanographic evolution of the last part of the mid-Cretaceous. Many models assume a positive correlation between atmospheric CO_2 and surface temperature, thus viewing changes in atmospheric CO_2 as playing the critical role in regulating climate temperature (Boucot and Gray, 2001). Patterns of stratigraphic variation in carbon isotope composition correlate over diverse oceanographic settings, not only during major 2‰ events like OAE 1a and OAE 1b (Aptian) and OAE 2 (latest Cenomanian) but also for small excursions like the one spanning the Albian-Cenomanian boundary. This strongly argues for a global control on carbon isotope stratigraphy, through $\delta^{13}C$ enrichment of the global CO_2 reservoir, rather than local variation in surface water productivity.

The global climatic cooling during middle Turonian (Pearce et al., 2003) is exhibited in the Cauvery Basin with a broad positive excursion of CIA, $\delta^{18}O$, pulses of positive excursions of C_{Org} and Ti, perhaps indicative of significant terrestrial weathering, influx of terrestrial organic matter (as also indicated by the petrified wood fragments of sizes up to few tens of centimeters) and heavy minerals. The Campanian (83.5–71.3 Ma) is the longest stage of the late Cretaceous, and represents a major turning point in Earth history. High and declining sea levels were accompanied by falling global surface temperatures leading to a transition from typical Mesozoic to "modern" oceanic circulation (Jarvis et al., 2002). The climate of the late Cretaceous was probably one of the warmest of the last 200 Ma. However, climate cooling was initiated at least by the Campanian, and possibly earlier at high latitudes. The well-documented long-term cooling trend that started in the late Campanian and continued into the early Maastrichtian was interrupted by a short warming event between ~65.6 and 66 Ma. The Maastrichtian was characterized by a cooler but variable climate with two periods of warming in the late Maastrichtian inferred from microfossil and geochemical studies (Sheldon et al., 2010). Though several authors suggested that flood basalt eruptions were likely to have caused greenhouse warming via CO_2 degassing (Thibault and Gardin, 2010), estimations of CO_2 and SO_2 degassing and subsequent paleotemperature models are somewhat conflicting with a cause-and-effect scenario between Deccan volcanism and the end-Cretaceous warming. According to Zakharov et al. (2011), the paleoclimatic data of Bering area based on $\delta^{18}O$ paleotemperature calculations and paleofloral assemblages, existences of warm climatic conditions in the late Albian, latest Cenomanian, Coniacian, Santonian to early Campanian, latest Campanian, and also late early Maastrichtian could be observed. In his argument on the return of "Cretaceous greenhouse conditions," Hay (2011) commented that, the Cretaceous was long regarded as essentially ice free, but sequence stratigraphic studies suggested large, rapid falls of sea level of the order of 50 m or more during the Cretaceous that could only be due to the buildup of ice on land. Hay (2011) has recently reviewed the arguments for and against the presence of ice in the Cretaceous. Dropstones, lonestones, and glendonites suggest that there was sea ice in high-latitude regions during the early Cretaceous and Maastrichtian. Periods of glaciation in the Turonian, Coniacian, Campanian, and Maastrichtian have been recently hypothesized and there are opinions suggesting that late Cretaceous paleotemperatures were too warm for significant ice sheet formation (Takashima et al., 2009; Sheldon et al., 2010). Jenkyns et al. (2004) estimated early–mid Maastrichtian Arctic Ocean (80°N) and suggested an equator-to-pole temperature gradient of 15 °C, inferring sea surface temperatures of 30 °C for low latitudes and 17 °C for the English Chalk. In addition to progressively cooler conditions, there is also increasing evidence that Maastrichtian climate was variable. The C_{Org} data of the Cauvery Basin aptly show significant to prominent positive excursions during these warmer periods (Fig. 3(D)). As could be observed elsewhere (Schlüter et al., 2008), a prominent negative excursion of $\delta^{13}C$ across KTB and corroborating sea level fall are observed in the Cauvery Basin. The negative excursion of almost 3‰ $\delta^{18}O$ may reflect a decrease of the marine primary productivity due to biological crisis at the KTB (Ali and Aitchison, 2008).

11.6 CONCLUSIONS

- The Barremian-Danian strata of the Cauvery Basin record all the six third-order sea level cycles within which many high-frequency cycles could be recognized and reflected in the lithofacies alternations and enrichment–depletion patterns of sensitive geochemical proxies.
- The depositional pattern of the Cauvery Basin was under the predominant influence of eustatic sea level fluctuations and climatic cycles.
- Given cognizance to the lithofacies characteristics, areal extent of deposits, and the geochemical characteristics, the Campanian is found to have experienced prolonged and the coldest climate while the Albian and Maastrichtian experienced the warmest climate.
- The lithofacies and stratigraphic variations of geochemical characteristics stand testimony to the varying temporal scales of climatic fluctuations ranging from third order to Earth's orbital cycles.
- The chemical weathering of the source rocks was at subdued level while the physical weathering dominated all through the depositional history of the Cauvery Basin, and these were under the

influence of a combination of sea level, climate and tectonics from Barremian to Santonian and thenceforth until Danian under the influence of prevalent climatic conditions.

- On a basinal scale, a gradual warming from Barremian to Santonian within which high-frequency cycles and thenceforth up to Danian a gradual cooling are observed. In this general backdrop, the very high fluctuations observed across the KTB are truly a perturbation.
- The depositional as well as climatic conditions were highly fluctuating prior to the Santonian and persistent and secular trends were prevalent since Santonian.
- After a break in sedimentation, as represented by lithostratigraphic boundaries, resumption of sedimentation was represented by lithoclastic conglomerates, the sediments for which were sourced from older depositional sites as well as basement provenance. It indicates four inferences:
  - The basement provenance remained a positive area.
  - Immediately after every lithostratigraphic boundary, the older depositional site became positive area.
  - There was a time lag between resumption of sedimentation and actual sediment production (in the case of carbonates)/accumulation (in the case of siliciclastics).
  - The initial inundation of the continental regions was farthest into the continent, which gradually decreased over time.
- A general cooling trend from Albian (warmest and thus had deeper bathymetry than all other times under study and had also experienced significant stratified nature of seawater), followed by Maastrichtian (warmer and the conditions persisted for a longer duration than other times), and finally Danian (warm, but high-frequency variations were witnessed).
- These observations and inferences have implications on chemostratigraphic and paleoclimatic studies. As the recycling of older sediments and the time lag between initiation of sedimentation and actual production-accumulation of sediments may not be in sync with the newer set of environmental-climatic conditions, interpretation of chemozones and climatic cycles based on the rock record deposited at the initial stages of resumed sedimentation may not be actual representation of the climatic conditions prevalent. Care has to be taken to avoid this potential error.

ACKNOWLEDGMENTS

The study was initiated with the financial grants from Alexander von Humboldt Foundation, Germany, to MR. Prof. Doris Stüben, Institute of Mineralogy and Geochemistry, University of Karlsruhe, Germany, is thanked for her invitation to MR to conduct collaborative research. Partial funding to this work in the form of Research Associateship and Senior Research Associateship was provided to MR by Council of Scientific and Industrial Research, New Delhi, and University Grants Commission, New Delhi, in the form of research project (Maastrichtian-Danian part of the study area). Authors thank the scientific personnel of the Institute namely, Dr. Utz Kramar for XRF analytical facilities, Dr. Karotke and Mrs. Oetzel for XRD analyses, Mr. Predrag Zrinjsak for carbon and sulfur analyses, and Dr. G. Ott for computing facilities. Shri. T. Sreekumar, Geologist, OFI, Mumbai, is thanked for assistance during the field survey. Permission to collect samples was accorded by the mines managers and geologists of Messrs, Alagappa Cements, Chettiyar Mines, Dalmia Cements, Fixit Mines Pvt. Ltd, Nataraj Ceramics Ltd, Parveen Mines and Minerals Ltd, Ramco Cements, Rasi Cements, TANCEM Mines, TAMIN Mines, Tan-India Mines, and Vijay Cements. Shri T. Sugantha, Research Scholar, Department of Geology, Periyar University, helped in final drawing of the profiles.

REFERENCES

Acharyya, S.K., Lahiri, T.C., 1991. Cretaceous palaeogeography of the Indian subcontinent: a review. Cretaceous Research 12, 3–26.

Ali, J.R., Aitchison, J.C., 2008. Gondwana to Asia: plate tectonics, paleogeography and the biological connectivity of the Indian sub-continent from the Middle Jurassic through latest Eocene (166–35 Ma). Earth-Science Reviews 88, 145–166.

Alvarez, N.O.C., Roser, B.P., 2007. Geochemistry of black shales from the Lower Cretaceous Paja Formation, Eastern Cordillera, Colombia: source weathering, provenance, and tectonic setting. Journal of South American Earth Sciences 23, 271–289.

Andersen, G.H., Delaney, M.L., 2000. Lithologic controls on calcite recrystallisation in Cenozoic deep sea sediments. Marine Geology 163, 109–124.

Andrew, A.S., Whitford, D.J., Hamilton, P.J., Scarano, S., Buckley, M., 1996. Application of chemostratigraphy to petroleum exploration and field appraisal. An example from the Surat basin. In: Proceedings of Seminar on Asia Pacific Oil and Gas, pp. 421–429.

Arthur, M.A., Sageman, B.B., 1994. Marine black shales: depositional mechanisms and environments of ancient deposits. Annual Review of Earth and Planetary Sciences 22, 499–551.

Ayyasamy, K., 1990. Cretaceous heteromorph ammonoid biostratigraphy of southern India. Newsletters on Stratigraphy 22, 111–118.

Bak, K., 2007. Organic-rich and manganese sedimentation during the Cenomanian–Turonian boundary event in the Outer Carpathian basins; a new record from the Skole Nappe, Poland. Palaeogeography, Palaeoclimatology, Palaeoecology 256, 21–46.

Banerji, R.K., 1972. Stratigraphy and micropalaeontology of the Cauvery Basin. Part I, exposed area. Journal of the Palaeontological Society of India 17, 1–24.

Bellanca, A., Erba, E., Neri, R., Silva, I.P., Sprovieri, M., Tremolada, F., Verga, D., 2002. Palaeoceanographic signifiance of the Tethyan "Livello Selli" (Early Aptian) from the Hybla Formation, northwestern Sicily: biostratigraphy and high-resolution chemostratigraphic records. Palaeogeography, Palaeoclimatology, Palaeoecology 185, 175–196.

Bergstörm, S.M., Young, S., Schmitz, B., 2010. Katian (Upper Ordovician) $\delta^{13}C$ chemostratigraphy and sequence stratigraphy in the United States and Baltoscandia: a regional comparison. Palaeogeography, Palaeoclimatology, Palaeoecology 296, 217–234.

Berner, R.A., 1983. Atmospheric carbon dioxide over Phanerozoic time. Science 249, 1382–1386.

Bhatia, S.B., 1984. Ostracode faunas of the Indian subcontinent-their palaeozoogeographic and palaeoecologic implications. Journal of the Palaeontological Society of India 20, 1–8.

Bice, K.L., Norris, R.D., 2002. Possible atmospheric CO_2 extremes of the Middle Cretaceous (late Albian-Turonian). Palaeoceanography 17. http://dx.doi.org/10.1029/2002PA000778.

Böhm, F., Westphal, H., Bornholdt, S., 2003. Required but disguised: environmental signals in limestone-marl alternations. Palaeogeography, Palaeoclimatology, Palaeoecology 189, 161–178.

Bouchez, J., Gaillardet, J., 2014. How accurate are rivers as gauges of chemical denudation of the earth surface? Geology 42, 171–174.

Boucot, A.J., Gray, J., 2001. A critique of Phanerozoic climatic models involving changes in the CO_2 content of the atmosphere. Earth-Science Reviews 56, 1–159.

Boulila, S., de Rafélis, M., Hinnov, L.A., Gardin, S., Galbrun, B., Collin, P., 2010. Orbitally forced climate and sea-level changes in the Paleoceanic Tethyan domain (marl–limestone alternations, Lower Kimmeridgian, SE France). Palaeogeography, Palaeoclimatology, Palaeoecology 292, 57–70.

Bralower, T.J., Arthur, M.A., Mark Leckie, R., Sliter, W.V., Allard, D.J., Schlanger, S.O., 1994. Timing and paleoceanography of oceanic dysoxia/anoxia in the Late Barremian to Early Aptian (Early Cretaceous). Palaios 9, 335–369.

Bralower, T.J., Fullagar, P.D., Paull, C.K., Dwyer, G.S., Leckie, R.M., 1997. Mid-Cretaceous strontium-isotope stratigraphy of deep-sea sections. Geological Society of America Bulletin 109, 1421–1442.

Caldeira, K., Rampino, M.R., 1991. The mid-Cretaceous super plume, carbon dioxide, and global warming. Geophysical Research Letters 18, 987–990.

Carter, R.M., Abbott, S.T., Fulthorpe, C.S., Haywick, D.W., Henderson, R.A., 1991. Application of global sea level and sequence stratigraphic models on Southern Hemisphere Neogene strata from New Zealand. International Association of Sedimentologists 12 (Spl. Pub.), 41–65.

Chari, M.V.N., Sahu, J.N., Banerjee, B., Zutshi, P.L., Chandra, K., 1995. Evolution of the Cauvery Basin, India from subsidence modeling. Marine and Petroleum Geology 12, 667–675.

Chandra, P.K., 1991. Sedimentary basins of India. In: Presidential Address of VIII Convention of Ind. Geol. Cong., pp. 1–31.

Chatterjee, S., Goswami, A., Scotese, C.R., 2013. The longest voyage: tectonic, magmatic, and paleoclimatic evolution of the Indian plate during its northward flight from Gondwana to Asia. Gondwana Research 23, 238–267.

Chiplonkar, G.W., 1987. Three decades of invertebrate palaeontology and biostratigraphy of marine Cretaceous rocks of India. Geological Survey of India 11 (Spl. Pub.), 305–339.

Coccioni, R., Galeotti, S., 2003. The mid-Cenomanian event: prelude to OAE 2. Palaeogeography, Palaeoclimatology, Palaeoecology 190, 427–440.

Das, N., 1997. Chemostratigraphy of sedimentary sequences: a review of the state of the art. Journal of the Geological Society of India 49, 621–628.

Du Vivier, A.D.C., Selby, D., Sageman, B.B., Jarvis, I., Grocke, D.R., Voigt, S., 2014. Marine $^{187}Os/^{188}Os$ isotope stratigraphy reveals the interaction of volcanism and ocean circulation during Oceanic Anoxic Event 2. Earth and Planetary Science Letters 389, 23–33.

Dubicka, Z., Peryt, D., Szuszkiewicz, M., 2014. Foraminiferal evidence for paleogeographic and paleoenvironmental changes across the Coniacian-Santonian boundary in western Ukrine. Palaeogeography, Palaeoclimatology, Palaeoecology 401, 43–56.

Fenner, J., 2001a. Palaeoceanographic and climatic changes during the Albian: summary of the results from the Kirchrode boreholes. Palaeogeography, Palaeoclimatology, Palaeoecology 174, 287–304.

Fenner, J., 2001b. The Kirchrode I and II boreholes: technical details and evidence on tectonics and on palaeoceanographic development during Albian. Palaeogeography, Palaeoclimatology, Palaeoecology 174, 33–65.

Fricke, H.C., Foreman, B.Z., Sewall, J.O., 2010. Integrated climate model-oxygen isotope evidence for a North American monsoon during the late Cretaceous. Earth and Planetary Science Letters 289, 11–21.

Friedman, G.M., Chakraborty, C., 1997. Stable isotopes in marine carbonates: their implications for the paleoenvironment with special reference to the Proterozoic Vindhyan carbonates (Central India). Journal of the Geological Society of India 50, 131–159.

Frimmel, A., Oschmann, W., Schwark, L., 2004. Chemostratigraphy of the Posidonia black shale, SW Germany influence of sea-level variation on organic facies evolution. Chemical Geology 206, 199–230.

Galeotti, S., Sprovieri, M., Coccioni, R., Bellanca, A., Neri, R., 2003. Orbitally modulated black shale deposition in the upper Albian Amadeus Segment (central Italy): a multi-proxy reconstruction. Palaeogeography, Palaeoclimatology, Palaeoecology 190, 441–458.

Garcia, T., Velo, A., Fernandez-Bastero, S., Gago-Duport, L., Santos, A., Alejo, I., Vilas, F., 2005. Coupled transport-reaction pathways and distribution patterns between siliciclastic-carbonate sediments at the Ria de Vigo. Journal of Marine Systems 54, 227–244.

Gislason, S.R., Oelkers, E.H., Eiriksdottir, E.S., Kardjilov, M.I., Gisladottir, G., Sigfusson, B., Snorrason, A., Elefsen, S., Hardardottir, J., Torssander, P., Oskarsson, N., 2009. Direct evidence of the feedback between climate and weathering. Earth and Planetary Science Letters 277, 213–222.

Goldberg, K., Humayun, M., 2010. The applicability of the chemical index of alteration as a paleoclimatic indicator: an example from the Permian of the Paraná Basin, Brazil. Palaeogeography, Palaeoclimatology, Palaeoecology 293, 175–183.

Govindhan, A., Ravindran, C.N., Rangaraju, M.K.R., 1996. Cretaceous stratigraphy and planktonic foraminiferal zonation of Cauvery Basin, South India. In: Sahni, A. (Ed.), Cretaceous Stratigraphy and Palaeoenvironments, vol. 37. Memoir– Geological Society of India, pp. 155–187.

Govindan, A., Ananthanarayanan, S., Vijayalakshmi, K.G., 2000. Cretaceous petroleum system in Cauvery Basin, India. In: Govindhan, A. (Ed.), Cretaceous Stratigraphy—An Update, vol. 46. Memoir– Geological Society of India, pp. 365–382.

Guha, A.K., 1987. Palaeoecology of some upper Creaceous sediments of India—an approach based on bryozoan. Geological Survey of India 11 (Spl. Pub.), 419–429.

Guha, A.K., Senthilnathan, D., 1990. Onychocellids (Bryozoa: Cheilostomata) from the Ariyalur carbonate sediments of South India. Journal of the Palaeontological Society of India 35, 41–51.

Guha, A.K., Senthilnathan, D., 1996. Bryozoan fauna of the Ariyalur Group (Late Cretaceous) Tamil Nadu and Pondicherry, India. Palaeontologia India 49, 2–17.

Gradstein, F.M., Ogg, J.G., Smith, A.G. (Eds.), 2004. A New Geologic Time Scale 2004. Cambridge University Press, Cambridge. p. 464.

Hancock, J., 2001. A proposal for a new position for the Aptian/Albian boundary. Cretaceous Research 22, 677–683.

Haq, B.U., Hardenbol, J., Vail, P.R., 1987. Chronology of fluctuating sea levels since the Triassic. Science 235, 1156–1167.

Hay, W.W., 1994. Pleistocene-Holocene fluxes are not the Earth's norm. In: Hay, W.W., Usselman, T. (Eds.), Material Fluxes on the Surface of the Earth. National Academy Press, pp. 15–27.

Hay, W.W., 2011. Can humans force a return to a "Cretaceous" climate? Sedimentary Geology 235, 5–26.

Herrle, J.O., Pross, J., Friedrich, O., Kossler, P., Hemleben, C., 2003. Forcing mechanisms for mid-Cretaceous black shale formation: evidence from the Upper Aptian and Lower Albian of the Vocontian Basin (SE France). Palaeogeography, Palaeoclimatology, Palaeoecology 190, 399–426.

Hild, E., Brumsack, H.J., 1998. Major and minor element geochemistry of Lower Aptian sediments from the NW German basin (core Hoheneggelsen KB 40). Cretaceous Research 19, 615–633.

Hofmann, P., Ricken, W., Schwark, L., Leythaeuser, D., 2001. Geochemical signature and related climatic-oceanographic processes for early Albian black shales: site 417D, North Atlantic Ocean. Cretaceous Research 22, 243–257.

Jati, M., Grosheny, D., Ferry, S., Masrour, M., Aoutem, M., Icame, N., Gauthier-Lafaye, F., Desmares, D., 2010. The Cenomanian–Turonian boundary event on the Moroccan Atlantic margin (Agadir Basin): stable isotope and sequence stratigraphy. Palaeogeography, Palaeoclimatology, Palaeoecology 296, 151–164.

Jafer, S.A., 1996. The evolution of marine Cretaceous basins of India: calibration with nannofossil zones. In: Sahni, A. (Ed.), Cretaceous Stratigraphy and Palaeoenvironments, vol. 37. Memoir– Geological Society of India, pp. 121–134.

Jafar, S.A., Rai, J., 1989. Discovery of Albian nannoflora from type Dalmiapuram Formation, Cauvery Basin, India-Palaeooceanographic remarks. Current Science 58, 358–363.

Jarvis, I., Mabrouk, A., Moody, R.T.J., de Cabrera, S., 2002. Late Cretaceous (Campanian) carbon isotope events, sea-level change and correlation of the Tethyan and Boreal realms. Palaeogeography, Palaeoclimatology, Palaeoecology 188, 215–248.

Jeans, C.V., Long, D., Hall, M.A., Bland, D.J., Conford, C., 1991. The geochemistry of the Plenus marls at Dover, England: evidence of fluctuating oceanographic conditions and of glacial control during the development of the Cenomanian-Turonian $\delta^{13}C$ anomaly. Geological Magazine 128, 603–632.

Jendrzejewski, L., Littke, R., Rullkoetter, K., 2001. Organic geochemistry and depositional history of upper Albian sediments from the Kirchrode I borehole, northern Germany. Palaeogeography, Palaeoclimatology, Palaeoecology 174, 107–120.

Jenkyns, H.C., Forster, A., Schouten, S., Sinninghe Damsté, J.S., 2004. High temperatures in the late Cretaceous Arctic Ocean. Nature 432, 888–892.

Kale, A.S., 2011. Comments on "Sequence surfaces and paleobathymetric trends in Albian to Maastrichtian sediments of Ariyalur area, Cauvery Basin, India" from Nagendra, Kannan, Sen, Gilbert, Bakkiaraj, Reddy, and Jaiprakash. Marine and Petroleum Geology 28, 1252–1259.

Kale, A.S., Phansalkar, V.G., 1992. Calcareous nannofossils from the Uttatur group, Trichinopoly District, Tamil Nadu, India. Journal of the Palaeontological Society of India 37, 85–102.

Kale, A.S., Lotfalikani, A., Phansalkar, V.G., 2000. Calcareous nanofossils from the Uttatur Group of Trichinopoly Cretaceous, South India. In: Govindan, A. (Ed.), Cretaceous Stratigraphy—An Update, vol. 46. Memoir– Geological Society of India, pp. 213–227.

Kampschulte, A., Bruckschen, P., Strauss, H., 2001. The sulphur isotope composition of trace sulphates in Carboniferous brachiopods: implications for coeval seawater correlation with other geochemical cycles and isotope stratigraphy. Chemical Geology 175, 149–173.

Keller, G., Berner, Z., Adatte, T., Stueben, D., 2004. Cenomanian–Turonian and $\delta^{13}C$, and $\delta^{18}O$, sea level and salinity variations at Pueblo, Colorado. Palaeogeography, Palaeoclimatology, Palaeoecology 211, 19–43.

Kramar, U., 1997. Advances in energy-dispersive X-ray fluorescence. Journal of Geochemical Exploration 58, 73–80.

Lal, N.K., Siawal, A., Kaul, A.K., 2009. Evolution of east coast of India—a plate tectonic reconstruction. Journal of the Geological Society of India 73, 249–260.

Larson, R.L., 1991. Geological consequences of superplumes. Geology 19, 963–966.

Lee, Y.I., 2009. Geochemistry of shales of the upper Cretaceous Hayang Group, SE Korea: implications for provenance and source weathering at an active continental margin. Sedimentary Geology 215, 1–12.

Linnert, C., Mutterlose, J., Erbacher, J., 2010. Calcareous nannofossils of the Cenomanian/Turonian boundary interval from the Boreal Realm (Wunstorf, northwest Germany). Marine Micropalaeontology 74, 38–58.

Littke, R., Jendrzejewski, L., Lokay, P., Shuangqing, W., Rullkoetter, J., 1998. Organic geochemistry and depositional history of the Barremian-Aptian boundary interval in the Lower Saxony Basin, northern Germany. Cretaceous Research 19, 581–614.

Madhavaraju, J., Ramasamy, S., Ruffell, A., Mohan, S.P., 2002. Clay mineralogy of the late Cretaceous and early Tertiary successions of the Cauvery Basin (southeastern India): implications for sediment source and palaeoclimates at the K/T boundary. Cretaceous Research 23, 153–163.

Martin, E.E., Macdougall, J.D., 1991. Seawater Sr isotopes at the Cretaceous/Tertiary boundary. Earth and Planetary Science Letters 104, 166–180.

McKirdy, D.M., Burgess, J.M., Lemon, N.M., Yu, X., Cooper, A.M., Gostin, V.A., Jenkins, R.J.F., Both, R.A., 2001. A chemostratigraphic overview of the late Cryogenian interglacial sequence in the Adelaide fold thrust belt, south Australia. Precambrian Research 106, 149–186.

Melinte-Dobrinescu, M.C., Roban, R., 2011. Cretaceous anoxic–oxic changes in the Moldavids (Carpathians, Romania). Sedimentary Geology 235, 79–90.

Miall, A.D., 2009. Correlation of sequences and the global eustasy paradigm: a review of current data. In: CSPG CSEG CWLS Convention, Frontiers + Innovation. Alberta, Canada, pp. 123–126.

Montero-Serrano, J.C., Palarea-Albaladejo, J., Martín-Fernández, J.A., Martínez-Santana, M., Gutiérrez-Martín, J.V., 2010. Sedimentary chemofacies characterization by means of multivariate analysis. Sedimentary Geology 228, 218–228.

Munnecke, A., Westphal, H., 2004. Shallow water aragonite recorded in bundles of limestone-marl alternations—the upper Jurassic of SW Germany. Sedimentary Geology 164, 191–202.

Najarro, M., Rosales, I., Martin-Chivelet, J., 2011. Major palaeoenvironmental perturbation in an early Aptian carbonate platform: prelude of the Oceanic Anoxic Event 1a? Sedimentary Geology 235, 50–71.

Nederbragt, A.J., Fiorentino, A., Klosowska, B., 2001. Quantitative analysis of calcareous microfossils across the Albian-Cenomanian boundary oceanic anoxic event at DSDP site 547 (North Atlantic). Palaeogeography, Palaeoclimatology, Palaeoecology 166, 401–421.

Nesbitt, H.W., Young, G.M., 1982. Early Proterozoic climates and plate motions inferred from major element chemistry of lutites. Nature 299, 715–717.

Pearce, M.A., Jarvis, I., Swan, A.R.H., Murphy, A.M., Tocher, B.A., Edmunds, W.M., 2003. Integrating palynological and geochemical data in a new approach to palaeoecological studies: upper Cretaceous of the Banterwick Barn Chalk borehole, Berkshire, UK. Marine Micropaleontology 47, 271–306.

Pellenard, P., Tramoy, R., Puceat, E., Huret, E., Martinez, M., Bruneau, L., Thierry, J., 2014. Carbon cycle and seawater palaeotemperature evolution at the Middle-Late Jurassic transition, eastern Paris Basin (France). Marine and Petroleum Geology 53, 30–43.

Peryt, D., Wyrwicka, K., 1993. The Cenomanian/Turonian boundary event in Central Poland. Palaeogeography, Palaeoclimatology, Palaeoecology 104, 185–197.

Pitman III, W.C., 1978. Relationship between eustacy and stratigraphic sequences of passive margins. Geological Society of America Bulletin 89, 1389–1403.

Powell, C.Mc.A., Roots, S.R., Veevers, J.J., 1988. Pre-break up continental extension in east Gondwanaland and the early opening of the Indian Ocean. Tectonophysics 155, 261–283.

Prabhakar, K.N., Zutshi, P.L., 1993. Evolution of southern part of Indian east coast basins. Journal of the Geological Society of India 41, 215–230.

Pratt, L.M., 1984. Influence of paleoenvironmental factors on preservation of organic matter in middle Cretaceous Greenhorn Formation, Puebio, Colorado. American Association of Petroleum Geologists Bulletin 68, 1146–1159.

Price, G.D., Valdes, P.J., Sellwood, B.W., 1998. A comparison of GCM simulated Cretaceous "greenhouse" and "icehouse" climates: implications for the sedimentary record. Palaeogeography, Palaeoclimatology, Palaeoecology 142, 123–138.

Price, J.R., Velbel, M.A., 2003. Chemical weathering indices applied to weathering profiles developed on heterogeneous felsic metamorphic parent rocks. Chemical Geology 202, 397–416.

Rachold, V., Brumsack, Hans-Jurgen, 2001. Inorganic geochemistry of Albian sediments from the Lower Saxony Basin, NW Germany: palaeoenvironmental constraints and orbital cycles. Palaeogeography, Palaeoclimatology, Palaeoecology 174, 121–143.

Rai, J., Ramkumar, M., Sugantha, T., 2012. Calcareous nannofossils from the Ottakoil Formation, Cauvery Basin, South India: implications on age, biostratigraphic correlation and palaeobiogeography. In: Ramkumar, M. (Ed.), On the Sustenance of Earth's Resources. Springer-Verlag, Heidelberg, pp. 109–122.

Raju, D.S.N., Ravindran, C.N., 1990. Cretaceous sea level changes and transgressive/regressive phases in India—a review. In: Proceedings on Cretaceous Event Stratigraphy and the Correlation of Indian Non-marine Strata, Contributions to Seminar Cum Workshop on IGCP-216, Chandigarh, pp. 38–46.

Raju, D.S.N., Ravindran, C.N., Kalyansundar, R., 1993. Cretaceous cycles of sea level changes in Cauvery Basin, India—a first revision. Oil and Natural Gas Corporation Bulletin 30, 101–113.

Ramanathan, S., 1968. Stratigraphy of the Cauvery Basin with reference to its oil prospects. Cretaceous-Tertiary of South India, vol. 2, Memoir– Geological Society of India, pp. 153–167.

Ramasamy, S., Banerji, R.K., 1991. Geology, petrography and stratigraphy of pre Ariyalur sequence in Tiruchirapalli District, Tamil Nadu. Journal of the Geological Society of India 37, 577–594.

Ramkumar, M., 1996. Evolution of Cauvery Basin and tectonic stabilization of parts of South Indian shield—insights from structural and sedimentologic data. Journal of the Geological Association and Research Centre 4 (Misc. Pub.), 1–15.

Ramkumar, M., 1999. Role of chemostratigraphic technique in reservoir characterization and global stratigraphic correlation. Indian Journal of Geochemistry 14, 33–45.

Ramkumar, M., 2007. Dolomitic limestone in the Kallankurichchi Formation (Lower Maastrichtian) Ariyalur Group, South India. Journal of Earth Science 1, 7–21.

Ramkumar, M., Chandrasekaran, V.A., 1996. Megafauna and environmental conditions of Kallankurichchi Formation (Lower Maestrichtian), Ariyalur Group, Tiruchy district, South India. Journal of the Geological Association and Research Centre 4, 38–45.

Ramkumar, M., Stüben, D., Berner, Z., 2004a. Lithostratigraphy, depositional history and sea level changes of the Cauvery Basin, South India. Annales Géologiques de la Péninsulé Balkanique 65, 1–27.

Ramkumar, M., Stüben, D., Berner, Z., Schneider, J., 2004b. Geochemical and isotopic anomalies preceding K/T boundary in the Cauvery Basin, South India: implications for the end Cretaceous events. Current Science 87, 1738–1747.

Ramkumar, M., Subramanian, V., Stüben, D., 2005a. Deltaic sedimentation during Cretaceous Period in the Northern Cauvery Basin, South India: facies architecture, depositional history and sequence stratigraphy. Journal of the Geological Society of India 66, 81–94.

Ramkumar, M., Harting, M., Stüben, D., 2005b. Barium anomaly preceding K/T boundary: possible causes and implications on end Cretaceous events of K/T sections in Cauvery Basin (India), Israel, NE-Mexico and Guatemala. International Journal of Earth Sciences 94, 475–489.

Ramkumar, M., Stüben, D., Berner, Z., 2006. Elemental interrelationships and depositional controls of Barremian-Danian strata of the Cauvery Basin, South India: implications on scales of chemostratigraphic modelling. Indian Journal of Geochemistry 21, 341–367.

Ramkumar, M., Anbarasu, K., Sugantha, T., Rai, J., Sathish, G., Suresh, R., 2010a. Occurrences of KTB exposures and dinosaur nesting site near Sendurai, India: an initial report. International Journal of Physical Sciences 22, 573–584.

Ramkumar, M., Berner, Z., Stüben, D., 2010b. Hierarchical delineation and multivariate statistical discrimination of chemozones of the Cauvery Basin, South India: implications on spatio-temporal scales of stratigraphic correlation. Petroleum Science 7, 435–447.

Ramkumar, M., Stüben, D., Berner, Z., Schneider, J., 2010c. $^{87}Sr/^{86}Sr$ anomalies in late Cretaceous-early Tertiary strata of the Cauvery Basin, South India: constraints on nature and rate of environmental changes across K–T boundary. Journal of Earth System Science 119, 1–17.

Ramkumar, M., Stueben, D., Berner, Z., 2011. Barremian–Danian chemostratigraphic sequences of the Cauvery Basin, India: implications on scales of stratigraphic correlation. Gondwana Research 19, 291–309.

Raymo, E., Oppo, D.W., Curry, W., 1997. The mid-Pleistocene climate transition: a deepsea carbon isotopic perspective. Paleoceanography 12, 546–559.

Roser, B.P., Korsch, R.J., 1986. Determination of tectonic setting of sandstone–mudstone suites using SiO_2 content and K_2O/Na_2O ratio. Journal of Geology 94, 635–650.

Ruffel, A.H., Rawson, P.F., 1994. Paleoclimate control on sequence stratigraphic patterns in the late Jurassic to mid-Cretaceous, with a case study from eastern England. Palaeogeography, Palaeoclimatology, Palaeoecology 110, 43–54.

Sandulli, R., Raspini, A., 2004. Regional to global correlation of lower Cretaceous (Hauterivian-Barremian) shallow-water carbonates of the southern Appennines (Italy) and Dinarides (Montenegro), southern Tethyan margin. Sedimentary Geology 165, 117–153.

Sarg, J.F., 1988. Carbonate sequence stratigraphy. Sea Level Changes—An Integrated Approach, SEPM Special Publication vol. 42, 155–181.

Sastri, V.V., Venkatachala, B.S., Narayanan, V., 1981. The evolution of the east coast of India. Palaeogeography, Palaeoclimatology, Palaeoecology 36, 23–54.

Schulze, F., Marzouk, A.M., Bassiouni, M.A.A., Kuss, J., 2004. The late Albian-Turonian platform succession of west-central Jordan: stratigraphy and crisis. Cretaceous Research 25, 709–737.

Schlüter, M., Steuber, T., Parente, M., Mutterlose, J., 2008. Evolution of a Maastrichtian–Paleocene tropical shallow-water carbonate platform (Qalhat, NE Oman). Facies 54, 513–527.

Sharma, A., Rajamani, V., 2000. Weathering of gneissic rocks in the upper reaches of Cauvery River, South India: implications to neotectonics of the region. Chemical Geology 166, 203–223.

Sharma, A., Rajamani, V., 2001. Weathering of charnockites and sediment production in the catchment area of the Cauvery River, southern India. Sedimentary Geology 143, 169–184.

Sheldon, E., Ineson, J., Bown, P., 2010. Late Maastrichtian warming in the Boreal Realm: calcareous nannofossil evidence from Denmark. Palaeogeography, Palaeoclimatology, Palaeoecology 295, 55–75.

Singh, P., Rajamani, V., 2001. Geochemistry of Kaveri flood plain sediments, southern India. Journal of Sedimentary Research 71, 50–60.

Soreghan, G.S., 1997. Walther's law, climate changes and upper Palaeozoic cyclostratigraphy in the ancestral Rocky mountains. Journal of Sedimentary Research 67, 1002–1004.

Spalletti, L.A., Queralt, I., Matheos, S.D., Colombo, F., Maggi, J., 2008. Sedimentary petrology and geochemistry of siliciclastic rocks from the upper Jurassic Tordillo Formation (Neuquén Basin, western Argentina): implications for provenance and tectonic setting. Journal of South American Earth Sciences 25, 440–463.

Srinivasan, M.S., 1989. Recent advances in Neogene planktonic foraminiferal biostratigraphy, chemostratigraphy and paleoceanography, Northern Indian Ocean. Journal of the Palaeontological Society of India 34, 1–18.

Sundaram, R., Rao, P.S., 1986. Lithostratigraphy of Cretaceous and Palaeocene rocks of Tiruchirapalli District, Tamil Nadu, South India. Records of the Geological Survey of India 115, 9–23.

Sundaram, R., Henderson, R.A., Ayyasami, K., Stilwell, J.D., 2001. A lithostratigraphic revision and palaeoenvironmental assessment of the Cretaceous system exposed in the onshore Cauvery Basin, southern India. Cretaceous Research 22, 743–762.

Stüben, D., Kramar, U., Berner, Z., Stinnesbeck, W., Keller, G., Adatte, T., 2002. Trace elements, stable isotopes and clay mineralogy of the Elles II K/T boundary section in Tunisia: indications for sea level fluctuations and primary productivity. Palaeogeography, Palaeoclimatology, Palaeoecology 178, 321–345.

Takashima, R., Nishi, H., Hayashi, K., Okada, H., Kawahata, H., Yamanaka, T., Fernando, A.G., Mampuku, M., 2009. Litho-, bio- and chemostratigraphy across the Cenomanian/Turonian boundary (OAE 2) in the Vocontian Basin of southeastern France. Palaeogeography, Palaeoclimatology, Palaeoecology 273, 61–74.

Tewari, A., Hart, M.B., Watkinson, M.P., 1996. A revised lithostratigraphic classification of the Cretaceous rocks of the Trichinopoly district, Cauvery Basin, Southeast India. In: Pandey, J., Azmi, R.J., Bhandari, A., Dave, A. (Eds.), Contributions to the XV Indian Colloquium on Micropalaeontology and Stratigraphy, pp. 789–800.

Thibault, N., Gardin, S., 2010. The calcareous nannofossil response to the end-Cretaceous warm event in the Tropical Pacific. Palaeogeography, Palaeoclimatology, Palaeoecology 291, 239–252.

Thiry, M., 2000. Palaeoclimatic interpretation of clay minerals in marine deposits: an outlook from the continental origin. Earth-Science Reviews 49, 201–221.

Torres, M.A., West, A.J., Li, G., 2014. Sulphide oxidation and carbonate dissolution as a source of CO_2 over geological timescales. Nature 507, 346–349.

Tu, T.T.N., Bocherens, H., Mariotti, A., Baudin, F., Pons, D., Broutin, J., Derenne, S., Largeau, C., 1999. Ecological distribution of Cenomanian terrestrial plants based on $\delta^{13}C/\delta^{12}C$ ratios. Palaeogeography, Palaeoclimatology, Palaeoecology 145, 79–93.

Tucker, M.E., Gallagher, J., Leng, M.J., 2009. Are beds in shelf carbonates millennial-scale cycles? An example from the mid-Carboniferous of northern England. Sedimentary Geology 214, 19–34.

Tucker, M.E., Wright, V.P., 1990. Carbonate Sedimentology. Blackwell Scientific Publications, London.

Ulicny, D., 1992. Discussion on the fluctuating oceanographic conditions and glacial control across the Cenomanian-Turonian boundary. Geological Magazine 129, 637–640.

Van der Meer, D.G., Zeebe, R.E., van Hinsbergen, D.J.J., Sluijs, A., Spakman, W., Torswik, T.H., 2014. Plate Tectonic Controls on Atmospheric CO_2 Levels since Triassic. PNAS 111, 4380–4385.

Vârban, B.L., 2003 Analiza sedimentologică a secvenţelor ciclice de vârstă Cretacic superior din Moldavide: reconstituiri paleoambientale (Ph.D thesis). University of Bucharest, Romania.

Veizer, J., Bruckschen, P., Pawellek, F., Diener, A., Podlaha, O.G., Carden, G.A.F., Jasper, T., Korte, C., Strauss, H., Azmy, K., Ala, D., 1997. Oxygen isotope evolution of Phanerozoic seawater. Palaeogeography, Palaeoclimatology, Palaeoecology 132, 59–172.

Veizer, J., Godderis, Y., Francois, L.M., 2000. Evidence for decoupling of atmospheric CO_2 and global climate during the Phanerozoic eon. Nature 408, 698–701.

Wallmann, K., 2001. Controls on the Cretaceous and Cenozoic evolution of seawater composition, atmospheric CO_2 and climate. Geochimica et Cosmochimica Acta 65, 3005–3025.

Warzeski, E.R., Cunningham, K.J., Ginsburg, R.N., Anderson, J.B., Ding, Z., 1996. A Neogene mixed siliciclastic and carbonate foundation for the Quaternary carbonate shelf, Florida Keys. Journal of Sedimentary Research 66, 788–800.

Watkinson, M.P., Hart, M.B., Joshi, A., 2007. Cretaceous tectonostratigraphy and the development of the Cauvery Basin, Southeast India. Petroleum Geoscience 13, 181–191.

Westphal, H., Head, M.J., Munnecke, A., 2000. Differential diagenesis of rhythmic limestone alternations supported by palynological evidence. Journal of Sedimentary Research 70, 715–725.

Whitford, D.J., Allan, T.L., Andrew, A.S., Craven, S.J., Hamilton, P.J., Korsch, M.J., Trotter, J.A., Valenti, G.L., 1996. Strontium isotope chronostratigraphy and geochemistry of the Darai limestone: Juha 1 X well, Papua New Guinea. In: Proceedings of III PNG Convention, pp. 369–380.

Wortmann, U.G., Herrle, J.O., Weissert, H., 2004. Altered carbon cycling and coupled changes in early Cretaceous weathering patterns: evidence from integrated carbon isotope and sandstone records of the western Tethys. Earth and Planetary Science Letters 220, 69–82.

Yan, Y., Xia, B., Lin, G., Cui, X., Hu, X., Yan, P., Zhang, F., 2007. Geochemistry of the sedimentary rocks from the Nanxiong Basin, South China and implications for provenance, paleoenvironment and paleoclimate at the K/T boundary. Sedimentary Geology 197, 127–140.

Zakharov, Y.D., Popov, A.M., Shigeta, Y., Smyshlyaeva, O.P., Sokolova, E.A., Nagendra, R., Velivetskaya, T.A., Afanasyeva, T.B., 2006. New Maastrichtian oxygen and carbon isotope record: additional evidence for warm low latitudes. Geosciences Journal 10, 347–367.

Zakharov, Y.D., Shigeta, Y., Popov, A.M., Velivetskaya, T.A., Afanasyeva, T.B., 2011. Cretaceous climatic oscillations in the Bering area (Alaska and Koryak Upland): isotopic and palaeontological evidence. Sedimentary Geology 235, 122–131.

Zeebe, R.E., 2001. Seawater pH and isotopic paleotemperatures of Cretaceous oceans. Palaeogeography, Palaeoclimatology, Palaeoecology 170, 49–57.

CHAPTER

CHEMOSTRATIGRAPHY OF THE DHOSA OOLITE MEMBER (OXFORDIAN), KACHCHH BASIN, WESTERN INDIA: IMPLICATIONS FOR COMPLETENESS OF THE STRATIGRAPHIC RECORD AND CORRELATION WITH GLOBAL OOLITE PEAK

12

Mu. Ramkumar[1,*], Matthias Alberti[2], Franz T. Fürsich[2]

*[1]Department of Geology, Periyar University, Salem, Tamilnadu, India; [2]GeoZentrum Nordbayern, Fachgruppe Paläoumwelt, Universität Erlangen-Nürnberg, Erlangen, Germany; *South East Asia Carbonate Research Laboratory (Seacarl), Universiti Teknologi Petronas, Tronoh, Malaysia*

12.1 INTRODUCTION

Ooids and oolitic rocks are of special interest due to their specific conditions of formation and reservoir properties (Tucker and Wright, 1990). In a classic treatise on carbonates, Flügel (1982) stated that ooids are among the limestone particles most important for microfacies interpretations, but are also the most controversial in regard to origin as they form in a wide variety of environmental conditions. Inherent upon their size and shape and environment of origin, these are highly mobile. In addition, due to their less stable mineralogy, myriad varieties of neomorphic and late-stage alterations are imparted. It appears that despite decades of research and excellent reviews on ooids, no single model or a combination of mechanisms could explain the origin of ooids in geological record (Land et al., 1979; Peryt, 1983). Nevertheless, interpretation of ancient ooids unravels vital information pertaining to energy and hydrodynamic conditions prevalent, ambient seawater chemistry, and climate that are crucial in understanding ancient environments, which in turn promotes academic and commercial interest on the study of ooids.

The Dhosa Oolite Member (DOM) of the Chari Formation, Western Kachchh Basin (Fig. 1) of India is known for its ferruginous ooid-bearing carbonates deposited during Oxfordian (Table 1) under transgressive systems tract (Kulkarni and Borkar, 2000). Occurrences of ooids coinciding with globally recorded oolite peak, condensed sections, and hard grounds of Kachchh Basin were reported earlier (Singh, 1989; Fürsich and Oschmann, 1993; Fürsich et al., 1991, 1992, 2001, 2004, 2005). Yet, there exists a gap in understanding its conditions of origin and implications on paleoenvironmental

Chemostratigraphy. http://dx.doi.org/10.1016/B978-0-12-419968-2.00012-1

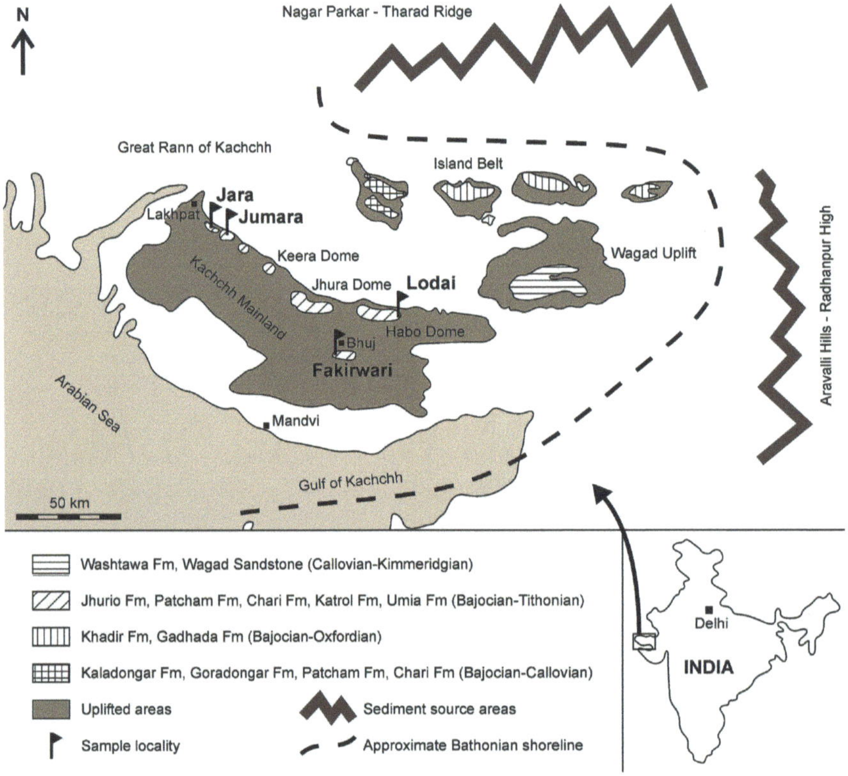

FIGURE 1

Geological map of the study area with position of the studied sections.

After Fürsich et al. (2004, 2005).

and climatological conditions, besides its relationship with global occurrences of oolite during Oxfordian. As the Jurassic deposits of the Kachchh Basin have been adjudged to be the result of eustatic sea-level changes and have unique depositional features different from that of adjacently located Jaisalmer Basin (Pandey et al., 2006, 2009), understanding the conditions of origin of these ooid-bearing rocks gains further importance. Lack of such information poses constraints on precise correlation of these deposits with counterparts elsewhere. In addition, owing to its condensed nature, occurrences of omission and hardground surfaces, mixed fauna typical of many biozones and resedimented nature of sediments, and correlation of this member with regional and global equivalents are often found to be complex and tenuous.

Chemostratigraphy, relatively a younger branch of geoscience is founded on the principle that chemical composition of sediments is a faithful recorder of environment in which the sediments were deposited and reflect the change in climate over time (Ramkumar, 1999). It is widely used to complement sequence stratigraphy and found to be useful in subdividing relatively unfossiliferous and monotonous thick successions (Pearce et al., 1999) distributed over large areas (e.g., Pearce and Jarvis, 1991).

Table 1 Lithostratigraphy of the Jurassic Rocks of Western Kachchh

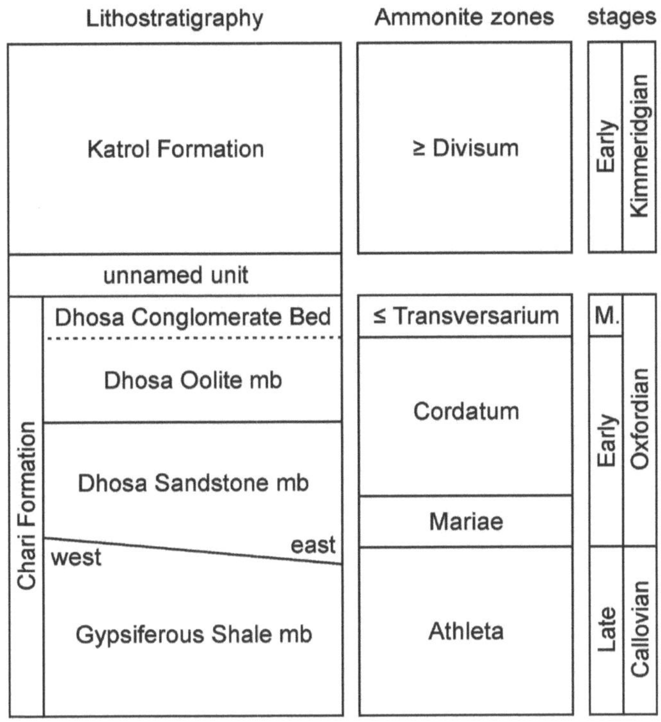

The Present Study is Concerned with the Part of the Chari Formation (Callovian-Oxfordian).
After Waagen (1873–1875), Fürsich et al. (2001), and Pandey et al. (2006, 2009).

Chemostratigraphy involves the examination of stratigraphic variations of elemental and isotopic geochemistry for the characterization of sedimentary sequences (Das, 1997). This tool is based on the sedimentary record of changes in certain elements with time. Many studies (e.g., Saraswati and Ramesh, 1992; Pearce et al., 1999; Bekker et al., 2001; Pujol et al., 2006; Kakizaki and Kano, 2009; Ramkumar et al., 2005, 2009, 2010a,b, 2011) utilized this tool for understanding sedimentation events, stratigraphic correlation, and environmental changes across specific time slices of Earth's history, etc. With the exception of purely bioproduced sediments, marine sediments are largely physical mixtures of a few lithological end-members and these end-members lie within a fairly restricted range of chemical composition (Plank and Langmuir, 1998) that forms the basis of assigning "geochemical signatures" to unique lithological types, sedimentation events, etc. (Ramkumar, 1999). Examination of patterns of stratigraphic variation of sediment geochemistry helps understand interactions between earth's processes of varying temporal scales. This paper documents chemostratigraphic characteristics of upper Callovian-Oxfordian strata of the Kachchh Basin, western India to identify synchronicity and to assess spatial and temporal scales of chemical signatures to constrain on their relationship with Oxfordian global oolite peak.

12.2 GEOLOGICAL SETTING

The DOM is exposed for about 100 km along its strike direction in the Kachchh mainland. It is easily recognizable in the field by its red-colored, well-cemented cap rock and massive amorphous iron crusts at top. Good exposures are found in Jara, Jumara, Keera, Jhura, and Habo domes/areas, located between southeast of Lakhpat and east of Bhuj (Fig. 1). The DOM rests over the Dhosa Sandstone Member with a distinct erosional/nondepositional surface. This surface is a chronological boundary between Callovian-Oxfordian (Fürsich et al., 2001) and also a transgressive systems tract surface (Pandey et al., 2009). Among various exposures examined, three sections located in Lodai, Fakirwara, and Jumara that contain condensed sections of the DOM are discussed in this paper. The Lodai section (LDS) is a composite section located at northeast of Bhuj in the Habo dome (Fig. 1). The rocks are exposed close to the road 3 km south of Lodai, over a lateral distance of 320 m, showing the occurrences of the Gypsiferous Shale, Dhosa Sandstone, and DOMs of the Chari Formation and basal part of the Katrol Formation (Fig. 2). The Fakirwari section (FWS) is located at about 6 km south of Bhuj, along the road to Mundra, exposing the DOM and other members of the Chari Formation (Fig. 2). The Jumara section (JMS) is located at about 20 km northeast of Matanomadh close to the Jumara village and exposes the DOM and basal part of the Katrol Formation (Fig. 2).

The Gypsiferous Shale Member is composed of bioclastic calcareous very fine silt interspersed with bioclastic arenaceous packstone and occasional arenaceous peloidal bioclastic packstone. These interlayers are found without any break in sedimentation and resulted by increased influx of bioclasts of bivalves and echinoderms. The rock shows discontinuous, but bedding parallel quartz rich (<40%) and mud-rich layers. Bedding parallel dark insoluble residue/organic matter, lenticular quartz grains, and angular bioclasts are observed frequently. In the LDS, this member is represented by very finely laminated, marly/argillaceous very fine silt. The bioclastic arenaceous packstones are composed of 50% grains and 50% argillaceous and micritic matrix. The monocrystalline quartz silt is randomly oriented, lenticular and equant in shape, angular, sorted, and confined within fine to very fine silt size classes. The bioclasts, particularly finer ones, show completely micritized character. Other bioclasts are lath shaped and angular. They show mouldic nature and are filled with fine to medium sized spar. The peloids are subangular to subrounded and are poorly to moderately sorted. Very few micritized echinoderm fragments are also observed. The matrix is micritic, organic rich, and shows calcite sparitization.

The rocks of the Dhosa Sandstone Member are marly/argillaceous coarse silt—fine sandstones and contain varying abundance of bioclasts. The micritic and argillaceous matrix is present only in intergranular porosity. Only a dark-colored insoluble residue could be observed as matrix in most of the samples owing to the compaction and dissolution. However, the organic matter rich micritic matrix forms about 25% of the rock.

In the LDS, the DOM comprises bioclastic wacke-pack-floatstones, bioclastic oolitic wacke-pack-floatstones, calcareous siltstones, bioclastic-oolitic siltstones, and stromatolitic oolitic siltstones, almost distributed evenly. Peloidal and arenaceous varieties are comparatively abundant than the FWS. Towards top, this member shows a gradual shift from carbonates to arenaceous varieties and finally into oolitic siltstone.

In the FWS, the DOM is predominantly a bioclastic and oolitic wacke-pack-floatstone with varying amounts of quartz silt. The variability of siliciclastic admixture results in arenaceous varieties of these

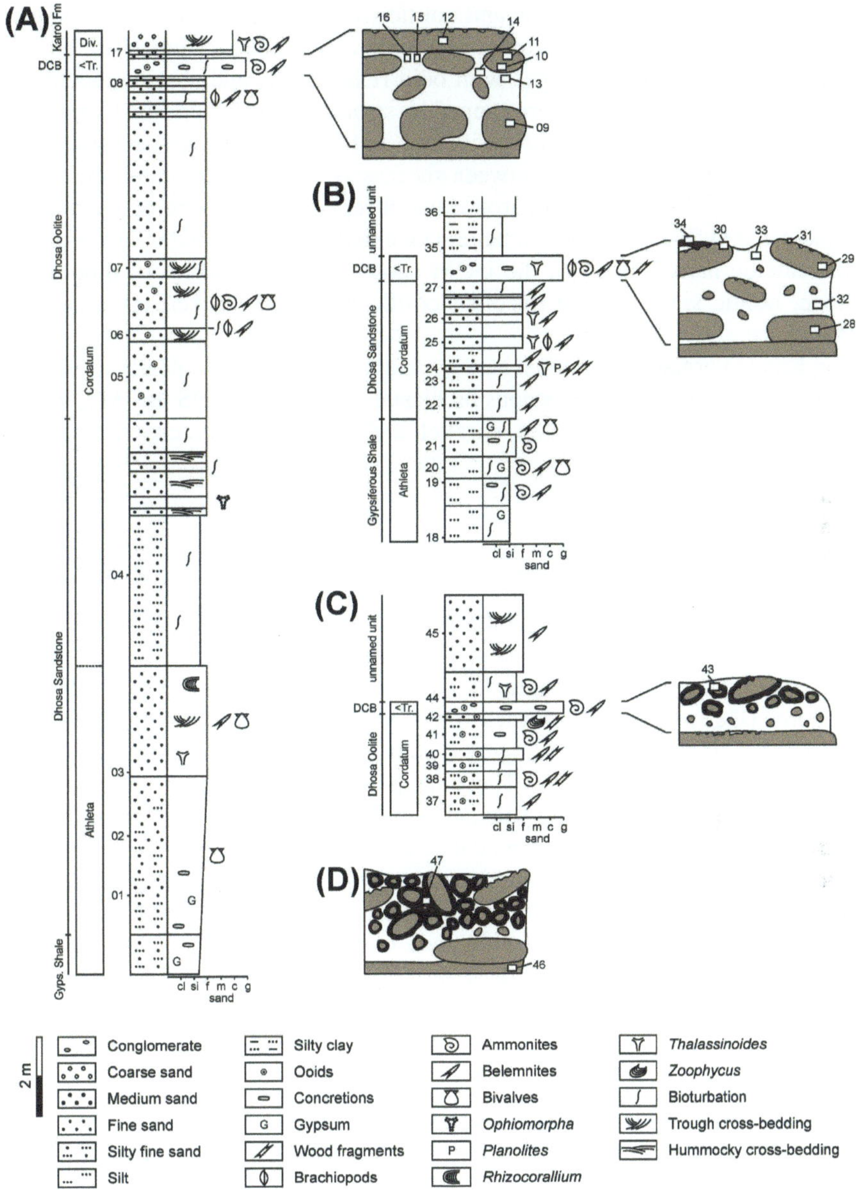

FIGURE 2

Lithologs of the three studied sections and position of geochemical samples. (A) Lodai section; (B) Fakirwari section; (C) Jumara section; (D) Jara dome; Tr. Transversarium zone; Bif. Bifurcatus zone; Hyp. Hypselocyclum zone; Div. Divisum zone; DCB. Dhosa Conglomerate Bed.

carbonate petrographic types and bioclastic and oolitic siltstones. Distribution of ooids in these rocks is highly varying that resulted in the occurrence of alternate cyclic beds of bioclastic limestones, arenaceous limestones, and calcareous siltstones with ooid rich, scarce, and absent characters. At top, the bioclastic and oolitic carbonate varieties are capped by stromatolites. In addition to mud, ooid, bioclast, and quartz silt, many intraclasts of calcareous siltstone and bioclastic wacke-packstone, rich in organic matter and iron oxide, are found trapped between successive layers of stromatolites. The bioclasts and ooids in these intraclasts show features of iron-mineral replacement, probably by siderite, marcasite, and ankerite (indicated by X-ray diffraction pattern) while the bioclasts in the host rock typically show low-magnesian calcitic nature. In JMS, the DOM shows the development of predominantly oolitic wacke/packstone, oolitic siltstone, and oncoidal oolitic bioclastic wacke/packstone in the order of decreasing abundance.

In FWS, the Katrol Formation is represented by friable, argillaceous/calcareous very fine siltstone. It grades to bioclastic siltstone and arenaceous bioclastic floatstone in LDS. In JMS, it shows the presence of dusty brown-colored, alternate thin, parallel, and even bedded bioclastic siltstone and argillaceous siltstone.

12.3 MATERIALS AND METHODS

Systematic field survey was conducted in the Kachchh Basin to log available exposures of the DOM for lithofacies variation, contact relationships, faunal occurrence and association, and sedimentary structures. From three well-preserved and geographically separated sections namely, LDS, FWS, and JMS and from Jara dome (Fig. 1) a total of 48 rock samples were collected (20 from FWS, 17 from LDS, and 9 from JMS and 2 from Jara dome). Figure 2 shows the lithologs of these sections and litho and biostratigraphic positions of these samples. These samples were subjected to major and trace elemental analyses through XRF method following standard laboratory procedures (Kramar, 1997). Data were examined in terms of comparison of geochemical compositions of stratigraphic units with Upper Continental Crust Composition (UCC); stratigraphic and regional variation of elemental composition; computation of various ratios of elements; corroboration of observed anomalies; shift in stratigraphic trends with field, petrographic, mineralogic, and published data and through multivariate statistical factor analysis. The interpretations made through these analyses were then discussed the context of voracity of correlating the DOM with Oxfordian global oolite peak.

12.4 RESULTS AND INTERPRETATIONS

12.4.1 GEOCHEMICAL CHARACTERISTICS WITH REFERENCE TO UPPER CONTINENTAL CRUST COMPOSITION

Mean values of the geochemical data, tabulated according to individual stratigraphic members, stratigraphic sections, and all the rocks studied are listed in the Table 2. Comparison of these data with the mean values of UCC of Taylor and McLennan (1985) shows significant enrichments of Fe and Ca and depletion of Si, Al, Mg, Na, K, with reference to UCC. It is an indication of sediment starved nature of the basin and less significant chemical weathering in source area. It is ascertained by the enrichments of Co, Cr, Ni, V, Zn, and Zr together with depletion of Ba and Sr. While the

Table 2 Geochemical Characteristics of the Studied Sections and Stratigraphic Members with Reference to UCC

	1	2	3	4	5	6	7	8	9	10	11	12	13	14	15	16	17	18	19
Si	17.28	17.32	20.10	13.62	25.41	12.46	8.10	13.66	27.49	18.30	13.03	10.54	15.76	29.15	26.39	21.60	28.79	6.15	30.70
Ti	0.30	0.264	0.378	0.232	0.61	0.21	0.19	0.21	0.42	0.38	0.19	0.18	0.26	0.50	0.37	0.17	0.74	0.13	0.30
Al	3.34	2.57	4.29	2.91	6.70	2.32	1.91	2.46	4.03	5.16	2.17	2.04	2.92	4.08	3.99	1.34	9.35	2.87	8.00
Fe	7.07	5.84	5.17	11.79	4.03	7.84	8.24	8.03	3.43	12.31	5.63	7.38	10.04	3.44	3.42	17.94	3.86	17.94	3.48
Mn	0.06	0.06	0.06	0.09	0.06	0.07	0.06	0.08	0.04	0.06	0.06	0.07	0.10	0.03	0.04	0.14	0.02	0.06	0.06
Mg	0.58	0.48	0.57	0.72	0.67	0.56	0.60	0.54	0.48	0.82	0.49	0.51	0.63	0.46	0.50	0.47	0.77	1.04	1.32
Ca	14.37	17.29	13.55	8.93	6.68	19.52	24.96	17.75	8.54	2.30	21.80	23.35	10.48	7.14	9.48	3.79	0.33	3.52	2.99
Na	0.19	0.113	0.296	0.122	0.38	0.13	0.08	0.15	0.29	0.16	0.12	0.17	0.14	0.12	0.40	0.05	0.32	0.06	2.88
K	1.00	0.92	1.27	0.754	1.74	0.64	0.32	0.74	1.55	1.37	0.73	0.54	0.77	1.57	1.54	0.56	2.44	0.71	2.81
P	0.18	0.10	0.19	0.24	0.24	0.21	0.15	0.23	0.10	0.10	0.13	0.23	0.26	0.02	0.15	0.10	0.03	0.16	0.09
Ba	246	272	275	148	286	194	127	211	371	263	231	160	176	366	375	385	417	48	550
Ce	96	85	105	91	138	87	57	95	125	61	66	84	112	160	101	12	129	17	64
Co	28	20	26	35	17	31	35	30	16	44	20	38	27	15	17	47	24	62	10
Cr	126	95	107	209	107	127	144	128	92	208	101	115	161	93	91	40	126	375	35
Cu	29	30	21	38	27	27	20	29	26	46	28	21	25	36	19	28	17	84	25
La	33	23	34	47	60	25	16	29	34	55	17	19	44	42	28	25	64	61	30
Ni	61	44	52	73	43	63	63	65	48	89	31	67	64	77	29	67	84	107	20
Pb	15	12	18	18	22	11	5	13	19	28	11	9	16	19	20	4	42	26	17
Sr	166	155	195	128	192	155	172	149	205	127	142	199	132	210	202	84	162	115	350
V	182	155	169	245	137	205	229	199	121	209	184	191	241	86	145	81	221	262	60
Y	35	27	37	43	58	34	28	36	33	30	23	34	46	39	28	14	36	33	22
Zn	188	202	114	319	93	185	216	177	76	505	157	143	273	62	86	1300	129	483	71
Zr	239	258	269	174	320	154	129	160	436	273	164	122	197	574	345	119	530	93	190
Th	18	19	18	18	21	12	11	12	30	23	12	8	20	44	21	7	42	12	10

1, Mean of all the 48 samples; 2, Mean of all the samples of Lodai Section; 3, Mean of all the samples of Fakirwari Section; 4, Mean of all the samples of Jhumara Section; 5, Mean of all the samples of Gypsiferous shale Member from all the sections; 6, Mean of all the samples of Dhosa Oolite Member from all the sections; 7, Mean of all the samples of "matrix" portion of the DOM from all the samples from all the sections; 8, Mean of all the samples of DOM other than matrix portion of all the sections; 9, Mean of all the samples of Dhosa sandstone Member from all the sections; 10, Mean of all the samples of Katrol Formation from all the sections; 11, Mean of all the samples of DOM at Lodai Section; 12, Mean of all the samples of DOM at Fakirwari Section; 13, Mean of all the samples of DOM at Jhumara Section; 14, Mean of all the samples of Dhosa sandstone Member at Lodai Section; 15, Mean of all the samples of Dhosa sandstone Member at Fakirwari section; 16, Mean of all the samples of Katrol Formation at Lodai Section; 17, Mean of all the samples of Katrol Formation at Fakirwari Section; 18, Mean of all the samples of Katrol Formation at Jhumara Section; 19, Mean of Upper Continental Crust (Taylor and McLennan 1985). Si, Ti, Al, Fe, Mn, Mg, Ca, Na, K and P are given in weight percentages. Ba, Ce, Co, Cr, Cu, La, Ni, Pb, Sr, V, Y, Zn, Zr, Th are given in ppm.

enrichment of Zr could indicate detrital influx, enrichments of all other elements that occur along with depletion of Sr indicate diagenetic leaching of Sr and simultaneous enrichment of heavy metals into carbonates under reducing environmental conditions. Sectional mean values indicate that while all the sections show depletion of Si, Al, Mg, Na, K, with reference to UCC, the FWS shows Si and Al above mean levels of other sections. It is interesting to note that while all the sections show enrichments of Ca and Fe, the relative enrichment of Ca versus Fe is inversely proportional, meaning diagenetic incorporation of Fe into carbonate phases. It is also an oblique indication to diagenetic leaching of Na and concomitant incorporation of Fe into carbonates. The JMS has least Ca and highest Fe, suggestive of diagenetic enrichment of Fe with simultaneous leaching of Ca. This inference is in tune with the enrichments of Co, Cr, V, Ni, Zn, and highly depleted values of Ba, Sr, Zr, and Th in JMS than other sections, signifying incorporation of heavy metals under anoxic conditions.

Though the stratigraphic members follow the all-data mean data with reference to UCC, there are significant differences, enforced by lithological traits. The Dhosa Sandstone Member followed by the Gypsiferous Shale Member are close to the mean value of Si in UCC, while the matrix portion of DOM shows highest depletion signifying highest detrital influx and cessation of detrital influx, respectively, during deposition of these members. Similarly, predominantly siliciclastic members show enrichment of detrital elements such as Al, Ti, Zr, Ba, etc., together with depleted nature of Ca, Fe, Mn, Co, Cr, etc., than the DOM. Within the DOM, the matrix samples differ from other samples in terms of heightened Ca and related elements and depleted detrital elements. If the values of detrital elements such as Si, Ti, Zr, etc., are taken as a measure, the JMS could be termed as proximal and the FWS as distal from the source during the deposition of DOM, while LDS was located in between these two sections. Cr, V, and Zn are more enriched along with least amounts of Sr in DOM of JMS than other sections, signifying intense diagenetic enrichments of former at the expense of latter.

Geochemical composition of the Katrol Formation follows closely that of UCC albeit with significant difference in terms of magnitudes of Fe, Na, Cr, V, Zn, and Sr than other stratigraphic members. Very high enrichment of Fe and depletion of Ca in the Katrol Formation are significant. The element Zr, an indicator of terrigeneous influx, shows highly depleted nature in DOM in the FWS section, while the Dhosa Sandstone Member, followed by the Katrol Formation in FWS show highest and higher enrichments, respectively. There is a decreasing trend of Zr and Si from FWS > LDS > JMS in the Katrol Formation that may indicate the proximity of the FWS and distal nature of the JMS from source during the deposition of the Katrol Formation. This is different from the conditions prevalent during deposition of the DOM. It could indicate either tectonic activity during the intervening period and/or change in flow conditions and course of channels in which sediment influx took place.

12.4.2 STRATIGRAPHIC VARIATIONS OF ABSOLUTE CONCENTRATIONS AND TRENDS IN GEOCHEMICAL PROFILES

Geochemical profiles of absolute elemental concentrations were drawn and compared with each other to understand their stratigraphic variation. Linear and polynomial trendlines were also superimposed over these in order to elicit information on secular and or cyclic variation of geochemical composition. The inferences were supplemented with stratigraphic variation of quantitative data on modal composition of petrographic components.

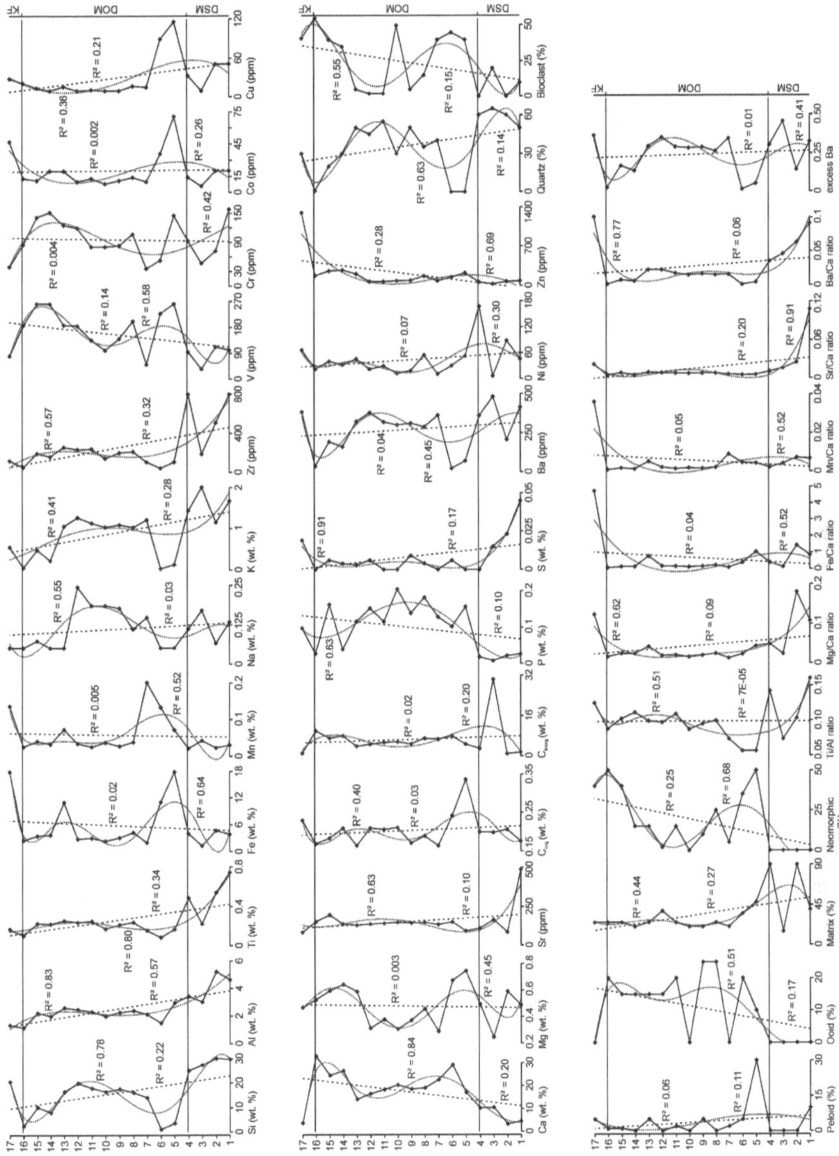

FIGURE 3

Profiles of selected elemental compositions, petrographic variables, and elemental ratios of the Lodal Section.

The geochemical profiles of all the studied sections (LDS—Fig. 3, FWS—Fig. 4, and JMS—Fig. 5) show a general lithological-lithostratigraphic control, in terms of change in their pattern across the lithological-lithostratigraphic boundaries. Uniformities among select group of elements namely Si–Al–Ti–K–Zr and Ca–Fe–Mn–V–P–Zn are observed. Profile of the petrographic variable quartz shows similarity of its trend with Si group of elements while the profiles of bioclast-ooid-neomorphic spar

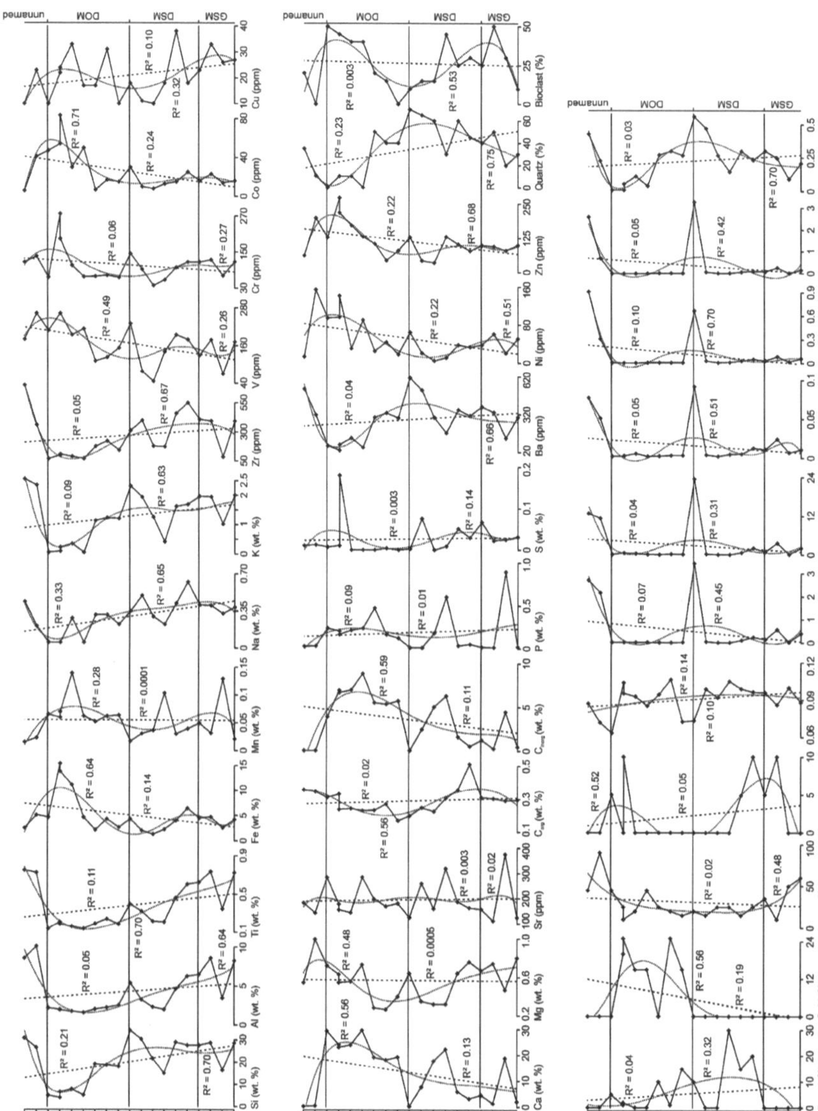

FIGURE 4

Profiles of selected elemental compositions, petrographic variables, and elemental ratios of the Fakirwari Section.

show similarity with Ca group of elements. The uniformities of trends within Si and Ca groups of elements and petrographic variables confirm the inference of lithological and lithostratigraphic control over whole-rock geochemistry, as could be found elsewhere. However, the featureless nature of Mg/Ca, Fe/Ca, Sr/Ca, Mn/Ca, and Ba/Ca ratios might have resulted from diagenesis, probably during late stage that had imparted certain uniformity regardless of lithostratigraphic units. As per the linear trendlines,

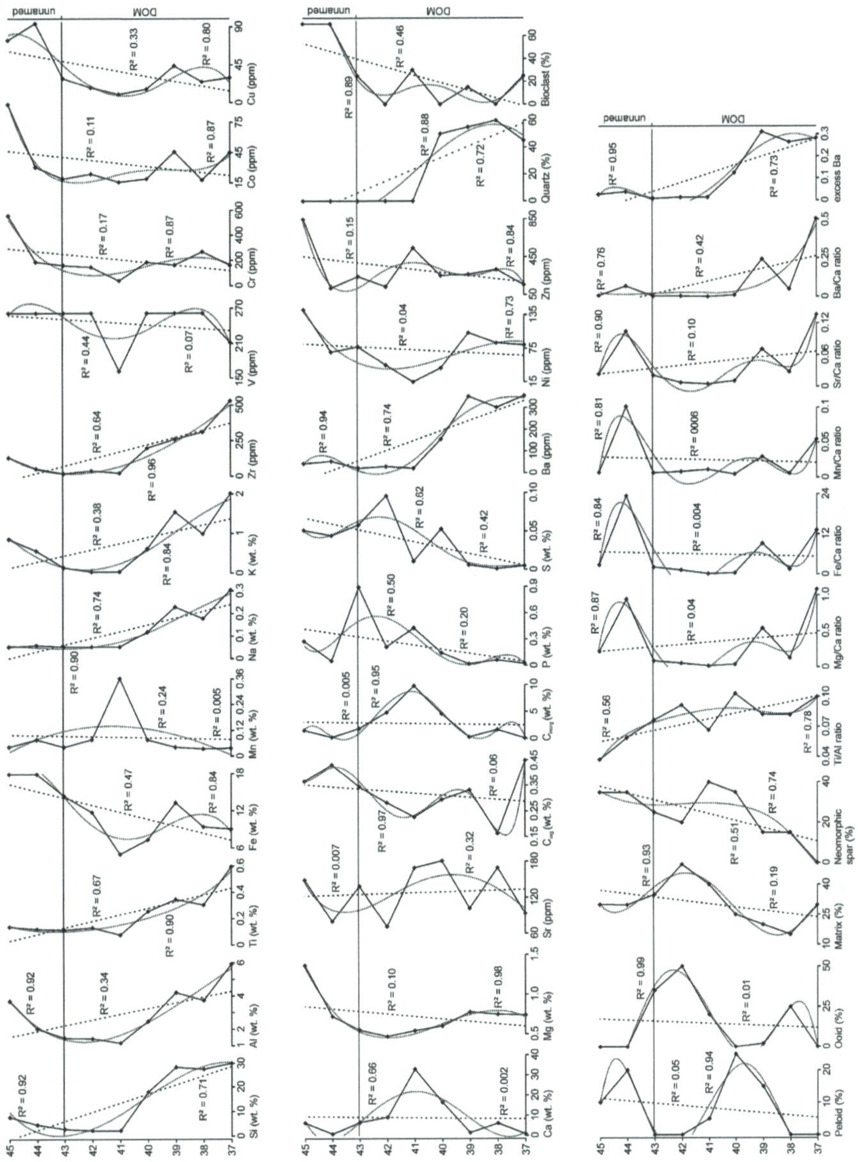

FIGURE 5

Profiles of selected elemental compositions, petrographic variables, and elemental ratios of the Jumara Section.

the Si and detrital elements such as Al, Ti, Zr, K, Na (partly), Mg (partly), Sr (partly), and Ba (partly) show a decreasing trend from older to younger strata in all the studied sections albeit with difference in magnitude (Figs 3–5). Reverse is the case observed with reference to Ca and associated group of elements.

Among the absolute concentration profiles of LWS (Fig. 3), positive excursions of Ca, Mg, C_{Org}, Cu, Co, Cr, V, Mn, and Fe along with bioclast, peloid, ooid, and neomorphic spar and concomitant negative excursions of quartz, Si, Na, K, and Ba in the DOM stand prominent than others. These excursions indicate depositional control over elemental associations, lithological control over enrichment-depletion of elements in host rocks during diagenesis and late-stage diagenetic incorporation of heavy metals within carbonate phases. The samples collected from "matrix" of the Dhosa cap bed that were interpreted to be recycled sediments during sea-level retreat show a slight to distinct difference than the other samples of the DOM, as explicit in almost all the profiles (sample Nos. 13, 14, 15, and 16 in Fig. 3). The samples from "matrix" (Sample Nos. 31, 32, and 33 in Fig. 4) of DOM in the FWS also show difference, especially the positive anomalies of S, Ni, Cu, Co, Fe, ooid, and neomorphic spar and negative anomalies of Si, K, and quartz. Phosphorus shows three positive anomalies with gradually reducing magnitudes from bottom to top in the FWS that are coincidental with reduced siliciclastic influx as indicated by corresponding decrease of Na, K, Zr, Ti, Al, and Si (this is prominent in the Gypsiferous Shale Member than other members). The prominent positive excursions of bioclast, C_{Inorg}, neomorphic spar along with Ca, and Mn corresponding to negative excursions of C_{Org}, Fe, V, S, and Ni in the JMS (Fig. 5), which correspond with the occurrences of Mn-rich carbonate (as revealed from X-ray diffraction patterns), may have been associated with organic matter decomposition. At the upper part of the Dhosa Sandstone Member, a positive Ba anomaly (explicit in FWS—Fig. 4 and subdued in LDS—Fig. 3), coinciding with positive anomalies of Al, Ti, K, Mg, and Ba_{Excess} is observed. In addition, a prominent negative anomaly of Ba in the lower part of the DOM and a smaller positive anomaly at the boundary between the DOM and the Katrol Formation at FWS (Fig. 4) are observed to occur.

12.4.3 WEATHERING INDICES AND ENVIRONMENTAL INDICATORS

Various weathering indexes and other ratios were computed and plotted in stratigraphic order (Figs 6 and 7). The Ruxton Weathering Index (RWI) that tracks the relative proportion of Silica and Alumina oxides and the V-Index that tracks the clay mineral alteration at source area show varying natures of weathering between three sections studied. Change in trends of these indices across stratigraphic boundaries in all the studied sections is also observed. The samples from "matrix" of the DOM show altered nature than other samples, as explicit in RWI of the LDS and the FWS. The Chemical Index of Alteration (CIA) and the A-CNK-FM ratio (also known as Chemical Index of Weathering or CIW) measure the conversion of feldspars into clays and the Plagioclase Index of Alteration (PIA) measures alteration of plagioclase feldspars into clays. Application of these indices to the samples under study had indicated that all these three indices show subdued weathering (Fig. 6) during the deposition of the DOM and heightened weathering during the deposition of the Katrol Formation, the Dhosa Sandstone Member, and the Gypsiferous Shale Member. Predomination of K-feldspars and illite in the rocks, absence of alkali feldspars, and spatial variation of detrital influx to different parts of the depositional basin are indicated by the profiles of K_2O/Al_2O_3 ratio (alkali feldspars: 0.5–1; illite: 0.19–0.3; ~0.0 other clay minerals and other feldspars) and K_2O/Na_2O ratio (Fig. 7) that are in conformity with the X-ray diffraction data. These inferences are also supported by the Cr/Th ratio profiles that indicated predomination of felsic source rocks and absence of mafic rocks (Fig. 7). Ratio ranging from 1.3 to 1.5,

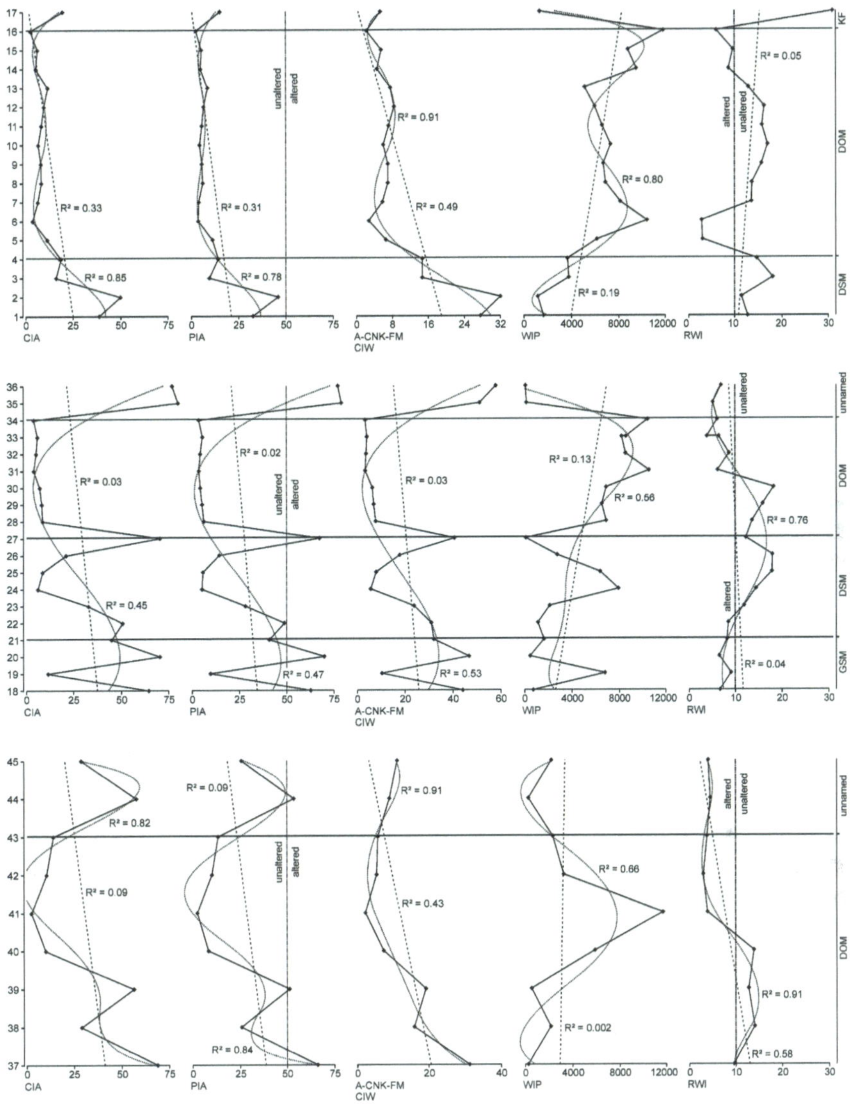

FIGURE 6

Profiles of weathering indices of the studied sections.

if associated with high Cr and Ni content, is indicative of mafic rock source, while more than 10 may indicate ultramafic source (Madhavaraju and Il Lee, 2009). The rocks under study show a variation of Cr/Ni ratio from 2 to 8 and most of the samples fall within a range of 2–3 (Fig. 8). Based on these, diagenetic enrichment of heavy metals in these rocks (particularly positive excursions of Cr/Th ratio

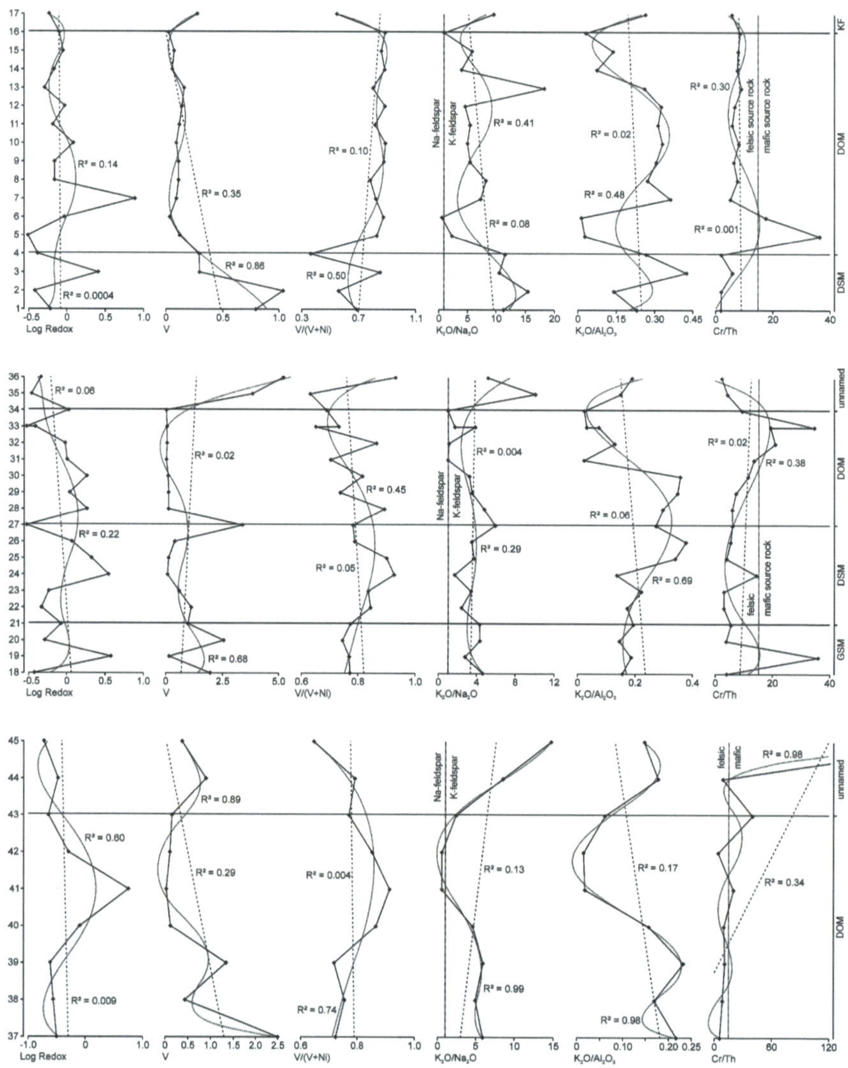

FIGURE 7

Profiles showing the stratigraphic variations of redox potential, elemental ratios, and source rock characteristics of the studied sections.

within the DOM, the Katrol Formation, and the Gypsiferous shale—Fig. 7) and absence of mafic source rock are ascertained. The Mn/Al ratios of the rocks are low when compared with PAAS (8.5) and very low when compared with anoxic basin (31.2). Hence, deposition under anoxic conditions could be ruled out. These profiles of Mn/Al also show regional variability as evident from the profiles of various sections (Fig. 8). On the other hand, according to the V/V + Ni ratio, all the samples (Fig. 7) except few fall in the field of reducing environment. However, majority of the samples fall within a narrow range of 0.6–0.8 resulted by reducing conditions of late-stage diagenesis.

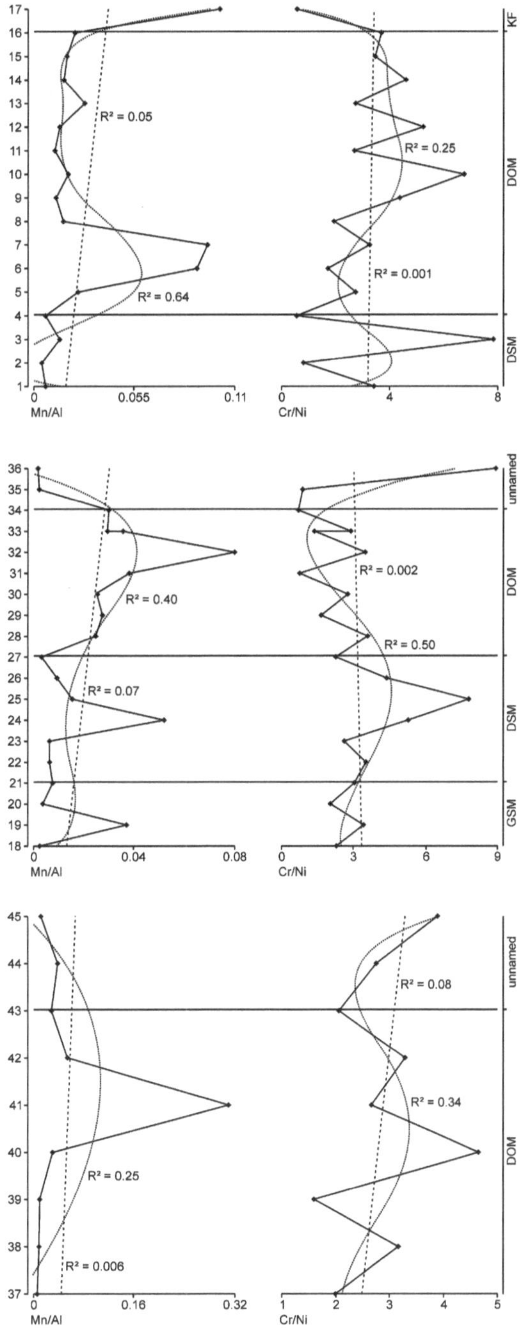

FIGURE 8

Profiles showing the stratigraphic variations of Mn/Al ratio and Cr/Ni ratio in the studied sections.

12.5 DISCUSSION

12.5.1 TEMPORAL VARIATIONS AND COUPLED NATURE OF DEPOSITIONAL ENVIRONMENTS, CLIMATE, SOURCE AREA WEATHERING, AND RELATIVE SEA LEVEL

Stratigraphic record of Phanerozoic shows peak abundances of oolites at specific time spans (Tucker and Wright, 1990) namely, during Late Cambrian, Late Mississippian, Late Jurassic, and Holocene (Wilkinson et al., 1985). These authors and other independent studies (e.g., Strasser and Samankassou, 2003) espoused coincidences of these peaks with first-order cycles of global transgression and regression. Zavar (2006) recorded ooids in the Oxfordian deposits of Poland, found their similarity with strata from France and England and interpreted them to have been formed during sea-level rise. Few authors have opined occurrences of iron-oolites at the base of transgression as a worldwide characteristic as evident from stratigraphic sections of Europe (Jacquin et al., 1998), Tibet Himalayas (Li and Grant-Mackie, 1993), Pakistan (Fatmi, 1972), Poland (Zavar, 2006), India (Gaetani and Garzanti, 1991), and other sections (Sahagian et al., 1996; Heydari, 2000; Wright, 2007). However, ooid accumulation during sea-level lowering and agitated water conditions are considered to be prerequisites by Geiger and Schweigert (2005). Accumulation of Fe-oolite during regression along margins of Paris basin was documented by Curial and Dromart (1998). Leinfelder (1993) recorded occurrence of oolites in shallow-water carbonates deposited near the top of shallowing upward sequences of upper Jurassic Lusitanian Basin, Portugal. From Swiss Jura Mountain sections, Allenbach (2002) reported oolite deposition as a result of accommodation space created by local tectonism. The study of Dromart et al. (2003) documenting stratigraphic, paleontological, and geochemical variations across middle to late Jurassic has also suggested that during this time span, atmospheric $_{P}CO_2$ controlled the Mesozoic climate and had influenced sea level, continental ice formation, carbon burial, etc., which in turn, along with the coupled nature of sea-level oscillations and $_{P}CO_2$ (Moore, 2001) affirms the relationship between sea level and oolite deposition, but its association with either transgressive or regressive part of an eustatic curve remains enigmatic.

The lithofacies and petrographic information of the DOM suggest deposition in a shallow, tectonically stable, sediment starved basin under the influence of sea-level variation. From the characteristics of mean values of all the samples, sectional mean values and mean values of stratigraphic members; lithological control over bulk-chemistry of the rocks; a sediment starved nature of the basin; less significant chemical weathering of source area; episodic influx of detrital sediments; and detrital influx dominated by Si, Al, and K are inferred. Presence of Si–Al–K–Na–Ba detrital signature and absence of Cr and Fe in detrital signature suggest predominantly felsic source rocks. Absence of sediment influx from mafic source rocks is indicated by the Cr/Th ratio (Fig. 7) and Cr/Ni ratio (Fig. 8). During weathering cycle, on exposure to the atmosphere, the mafic rocks rapidly break down and contribute to the depocenters (Sensarma et al., 2008). Absence of such mafic signatures may indicate in an oblique way, absence of tectonic movements during the deposition of the rocks under study and prevalence of only sea-level controlled nature of deposition. It is to be noted at this juncture that, though the quantum of siliciclastics (represented by quartz silt of uniform size and shape) reduced generally at the dawn of DOM, its influx was persistent, signaling positive nature of the source area. Prevalence of positive nature of the source area all through the depositional history of the Chari Formation suggests inactive nature of tectonic movement and domination of relative sea-level change as the cause of facies and gross lithological variations.

From the enrichment-depletion characteristics, lithology was found to be the dominant controlling factor. In such a case, influence of climate and thus sea-level could be construed as the controlling factors of chemical signatures of these rocks. This climate-controlled nature might have also enforced significant reduction of chemical weathering in the source area, as a result of which, reduction of siliciclastic influx to the depocenters accompanied by proliferation of carbonate deposition might have taken place across the transition from the Dhosa Sandstone Member to the DOM.

Judging from the polynomial trends of Si and Ca as indications of relative sea-level change, it could be inferred that the sea-level rise had commenced from the Dhosa Sandstone Member that reached its zenith at near top of the Dhosa limestone Member from where there might have been a rapid fall. In a sense, the sea-level rise was gradual while the rate of fall was high. This indicates sea-level controlled nature of the depositional pattern, gradual increase in sea-level, reduction in detrital influx, and reduction in chemical weathering in the source region. The changes across lithostratigraphic boundaries, as indicated by the profiles of RWI (Fig. 6) and V-index (Fig. 7), together with change in lithology across stratigraphic boundaries, support interpretation of varying climatic conditions and thus varying chemical weathering in source area and resultant variation in detrital influx. Occurrences of DOM samples (except the matrix samples) in unaltered regions of the weathering indices indicate absence of either detrital influx or chemical weathering in the source area during deposition of DOM, or both, which in turn may indicate sea-level highstand and warm dry climate. The noncarbonate deposits (i.e., siliciclastics) depict significant chemical alteration at source area in the profiles of all the studied sections (according to CIA, CIW, PIA—Fig. 6). This fits well with the interpretation of sea-level lows during siliciclastic deposition, significant weathering and detrital influx and humid+cold weather while warm climate, sea-level high, insignificant chemical weathering (as indicated by deep lows in polynomial curves drawn over the profiles of weathering indices—Fig. 6) and scanty detrital influx-promoted carbonate deposition. It also reinforces the interpretation of deposition in this part of the basin as a result of climate and thus under the control of sea-level variations than other processes. As indicated by the CIA, CIW, and PIA, occurrences of heightened weathering during predominantly siliciclastic deposits and lowered weathering during predominantly carbonate deposition are in tune with the inference of warmer climate, sea-level rise and reduction of siliciclastic influx during DOM and comparatively colder climate, and sea-level lowering and significant siliciclastic influx during the deposition of other stratigraphic members. This inference also calls for the coupled nature of climate-sea-level-source area weathering-siliciclastic sediment influx. In addition to remaining within a narrow range of weathering index values, there is no perceptible change between the samples of DOM including the samples collected from matrix, signifying, no or insignificant chemical weathering during the deposition of DOM and sourcing of sediments within the sedimentary basin for recycling that produced the matrix part of the DOM. The profiles of K_2O/Al_2O_3 ratio (Fig. 7) of the LDS and the FWS also indicate a positive excursion of K-feldspars/illite in the midst of the DOM followed by steep decrease during the deposition of recycled sediments (matrix). It is to be noted that while the positive anomaly shows a gradual increase, the negative shift is comparatively steep. As these resedimented rocks resulted during sea-level retreat, attention is drawn to the general nature of sea-level cycles that while the rise tends to be gradual, the fall is always rapid. By implication, the sea-level retreat reported to have caused re-sedimentation of sediments might be linked with regional, if not global cycle and certainly not of local in origin.

12.5.2 PHOSPHORUS ANOMALIES AND SHORT-TERM SEA-LEVEL VIS-À-VIS CLIMATIC FLUCTUATIONS

Phosphorus shows three distinct positive excursions (Fig. 4) in the FWS. Magnitudes of these excursions diminish from oldest (Gypsiferous Shale Member—first and largest anomaly) to the younger lithostratigraphic members (Dhosa Sandstone Member—second and moderate anomaly; DOM—third and smallest anomaly). While the LDS that exposes the Dhosa Sandstone Member, the DOM, and basal beds of the Katrol Formation shows the occurrences of second and third anomalies (Fig. 3), the JMS that exposes the DOM and the Katrol Formation records the occurrence of the third anomaly (Fig. 5). However, P anomalies in these two sections show subdued nature. Given cognizance to the spatial distribution of the studied stratigraphic sections, the phosphorus anomalies could be inferred to have been caused by processes that acted on a regional/basinal scale.

Phosphorus forms an essential ingredient for primary production in life cycle (Tappan, 1967; Munnecke et al., 2010) and is directly connected with oceanic productivity at global scale (Ruttenberg, 1993) on timescales longer than several multiples of its residence time in seawater (Froelich, 1988). As primary productivity affects the global carbon cycle, changes in global climate and relative sea-level may play a critical role in fluxes and reservoirs of P. In this context, and on the basis of earlier studies (Filippelli and Delaney, 1994; Dill et al., 1997; Hoppie and Garrison, 2001; Sageman et al., 2003), the positive excursions of P are identified as a consequence of sea-level fluctuations in the basin. Correspondences between the first and second positive P anomalies and significant reduction of Si, Al, Ti, Zr, and Ba (Fig. 4) support this surmise. The third anomaly also shows similar relationship with detrital elements, but at subdued level. As discussed in previous section, detrital influx into the basin was controlled by the sea-level fluctuations and hence the reduction in detrital elements and enrichment of Ca (Figs 3–5) coinciding with positive anomalies of phosphorus may be linked with the periods of sea-level rise. Comparison of the trends of P and Sr showed that while there are corresponding positive excursions of Sr during the episodes of enhanced accumulation of P, occurrences of unrelated positive excursions of Sr and marked shift in detrital influx during periods of phosphorus accumulation as expressed by negative excursions of typically detrital elements such as Si, Al, Ti, Zr, and Ba negate the possibility of exclusively enhanced terrestrial influx (Raymo, 1994; Raymo et al., 1997) as the sole cause of positive anomalies of phosphorus. As the excursions of P and Sr (Fig. 4) do not correspond with the Mn/Al profile (Fig. 8), diagenetic enrichments of P are not considered. Acidic anoxic conditions remobilize P, under which conditions, enrichment of Fe also occurs (Jaffe et al., 2002). Such a possibility is negated by the absence of sympathetic nature of Fe (Fig. 4).

Occurrences of P anomalies in rocks that have significant shell concentrations and either concretion (first anomaly) or bioturbation (second anomaly) or recycled sediment (third anomaly) may relate them with prevalent sea-level changes. The third anomaly that occurs in the DOM precedes the perceived fall in sea level (Singh, 1989; Alberti et al., 2012) that produced significant reworking of fauna, extensive syndepositional erosion, and re-sedimentation and deposition of chaotic "matrix" of Dhosa cap bed, etc. and supports the interpretation of enhanced phosphorus accumulation as a result of sea-level fluctuation. It is interesting to note that the periods of enhanced P accumulation are coincidental with periods of significant reduction of chemical weathering in the source area as indicated by the corresponding negative anomalies of CIA (Fig. 6), which normally result during periods of sea-level rise/high. Rangel et al. (2000) and Ramkumar et al. (2009) documented the enhanced accumulation of phosphorus in rocks deposited immediately succeeding sea-level changes. Hoppie and Garrison (2001) opined that excursions of P can be directly related to individual eustatic and tectonic events. According to Filippelli and Delaney (1994), fluctuations in sea-level cause changes in shelf area which in turn alters the oceanic P mass balance, resulting in redistribution of P within basin. The periods of sea-level rise/high promote enhanced accumulations of P (Hoppie and Garrison, 2001) due to strong coastal upwelling, which in turn, normally result in higher primary productivity (Derry et al., 1994).

Prevalence of these events could be verified either through strong positive correlation of P with C_{Org} or through similar proxies. Absence of any significant correlation with C_{Org} (Fig. 4), occurrences of very low Ba_{Excess} (sectional mean: 0.230; first anomaly: 0.090; second anomaly:0.147; third anomaly:0.272) and very low Mn/Al ratio of the rocks (sectional mean: 0.022; first anomaly: 0.037; second anomaly:0.052; third anomaly:0.026) when compared with Mn/Al ratio of anoxic basin (31.2) or PAAS (8.5) negate the possibility of relating these positive anomalies with sea-level maximum and enhanced primary productivity.

Hoppie and Garrison (2001) and Spalletti et al. (2001) suggested that phosphorus could get accumulated in oxygen minimum zones in areas of reduced detrital influx. The very low Mn values (129, 103, 51 ppm with reference to first, second and third P anomalies, respectively) of the rocks in which P anomalies are recorded may indicate oxygen poor conditions (Sageman et al., 2003). However, the redox-sensitive elements such as Cr and V do not show corresponding change in oxygen levels during the positive excursions of P. Ti/Al ratio, an indicator of clastic influx into the depositional sites (Sageman et al., 2003) do show slight increase at the first and second anomalies of P, while there is a decrease coinciding with third positive excursion (Fig. 4). However, all through the observed section, the change is within an acute range, signifying smaller scale sea-level fluctuations and resultant subdued detrital influx. Profile of Ti/Al ratio (Fig. 4) shows significant negative shifts "immediately after" positive excursions of phosphorus, implying occurrences of phosphorous anomalies before sea-level maximum/highstand. Sageman et al. (2003) also suggested that increase in Ti/Al ratio can be interpreted to be the result of basinward movement of shoreline. In this context, slight changes in sea-level and consequent shoreline shifts towards continental regions "after" the positive anomalies of phosphorus can be inferred. Considering the broad, but comparatively shallow nature of depositional basin envisaged for these deposits (e.g., Singh, 1989; Ramkumar et al., 2012, 2013), it is possible to envisage impact on sedimentation pattern, sediment type, and geochemical composition of the sediments by changes in sea-level. Based on the review of models on P cycles, Dill et al. (1997) regarded that smaller scale sea-level fluctuations are sufficient to cause significant enrichment-depletion. These observations, together with the antisympathetic nature of phosphorus anomalies with maximum flooding, anoxic conditions, and detrital elements suggest enrichment of P by sediment redistribution during periods of sea-level fluctuations, possibly during sea-level rise, but not necessarily associated with sea-level highstand and/or maximum flooding.

General accompaniment of sea-level rise and corresponding reduction/absence of chemical weathering in the source area may also indicate higher $_{P}CO_2$ in the atmosphere, warm climate, and consequent sea-level rise. Repetitive occurrences of the P anomalies, (i.e., in the Gypsiferous Shale Member, the Dhosa Sandstone Member and the DOM) may support interpreting these in relation with cyclic phenomenon. By implication, these anomalies could be linked with short-term climate-sea-level cycles (Elrick and Scott, 2010).

12.5.3 BARIUM ANOMALIES

A positive anomaly of barium near the top of the Dhosa Sandstone Member (first), a prominent negative anomaly at the lower part of DOM (second), and a smaller negative anomaly (third) at the boundary between the DOM and the Katrol Formation in the profiles of FWS (Fig. 4) are recognized. The first anomaly is coincidental with positive excursions of Na, K, C_{Inorg}, quartz, bioclast, and Ba_{Excess} and prominent negative excursions of Ti, Zr, V, Cr, Cu, Mg, Ni, Matrix, and Ti/Al ratio. The second anomaly is coincidental with negative excursions of Si, K, Na, Zr, quartz, Ti/Al ratio, and Ba_{Excess} and positive excursions of Fe, Mn, V, Cu, Ca, Mg, and Bioclast. The third anomaly is coinciding with negative excursions of Si, V, Cr, Mg, P, quartz, and Ba_{Excess} and positive excursions of bioclast and neomorphic spar. At LDS, the Ba records a positive anomaly at the boundary between the Dhosa Sandstone Member and the DOM and negative shift at the boundary between the DOM and the Katrol Formation (Fig. 3). This negative shift is also recorded in JMS (Fig. 5).

The Ba in all the samples is far below (Table 2) the value of UCC (550 ppm—Taylor and McLennan, 1985) and shows an extended range from as low as 22 ppm to as high as 608 ppm with a mean value of 246 ppm. Occurrences of negative anomalies within this reduced levels and the occurrence of positive anomaly, which is close (608 ppm) to the average UCC value, require examination. As the oceanic residence time of Ba is very short (in the order of 10,000 years—Chan et al., 1976), any change in its influx and sink would be immediately recorded in the sedimentary record and thus, the excursions of Ba might be related with immediately preceding causes. Barium in sediments is being utilized as a proxy for interpretation of modern (Bishop, 1988; Dymond et al., 1992) and paleoceanographic conditions (Francois et al., 1995) and also as a paleoproductivity tracer (Francois et al., 1995; McManus et al., 1998). Despite disagreements for such usage (e.g., Murray and Leinen, 1993), many studies have established a link between relative abundances of Ba and productivity levels (e.g., Lyle et al., 1992; Dymond et al., 1992; Stüben et al., 2002). However, precise mechanisms of Ba precipitation and its resultant incorporation into marine sediments still remain unclear (Schroeder et al., 1997) as it reacts to primary productivity, terrigenous influx, and post-depositional preservation processes (Keller et al., 1998; Stüben et al., 2002). Kasten et al. (1998) have demonstrated that enrichments of Ba in the equatorial Atlantic Ocean sediments coincide with glacial-interglacial transitions. Ramkumar et al. (2005) identified sea-level fluctuations as a result of climatic change, physical weathering in the source area, and resultant positive excursions of Ba in the Cauvery Basin, South India. The oxidation-accumulation cycles of carbon are reflected in the global climatic cycles and this carbon budget is of primary importance for understanding climatic change on all timescales. On a glacial-interglacial timescale, terrestrial vegetation, a very reactive carbon reservoir, is destroyed when ice sheets override it. Growth and destruction of terrestrial vegetation also occurs on the continental shelf area as sea-level falls and rises (Oppo and Fairbanks, 1989). As the enhanced productivity during sea-level high may promote enrichment of Ba in the ensuing sediments and the sea-level lows may promote higher detrital influx and in turn, under favorable conditions, bring in higher Ba into the depositional basin, interpretation of sources of Ba remains complex (Schroeder et al., 1997). Dymond et al. (1992) and Schroeder et al. (1997) have established that much of the Ba in marine sediments remains in the form of calcite bound components. If it holds true for the rocks under study, then Ba should show a coinciding peak with Ca. On the contrary, the profiles of Ca and Ba show sharp negative correlations (Figs 3–5) and thus its relationship with sea-level vis-à-vis climate remain sympathetic. Examination of the excursions of Ba given cognizance to these information reveals that the positive anomaly of Ba was coinciding with significant detrital influx during the short-term sea-level retreat (as indicated by covariation of most of the detrital elements and also the Ti/Al ratio in Figs 3–5, indicating sea-level lowering and consequent shoreline shift), while the negative anomalies are coinciding with significant carbonate accumulation in the basin. Covariation of Ba with K and Na may as explicit in Figs 3–5 signify its influx either in the form of feldspars or clay material weathered from feldspars. Ba has also the tendency to adhere onto clay mineral surfaces (Singh, 1978; Kumar et al., 1995) in which case, significant influx of clay and the adherence of Ba onto clay could be inferred. This presumption also calls for significant weathering in the continental regions to form clay, transport, and deposition in the marine basin to cause Ba anomalies. All these events necessitate prevalent higher rate of sedimentation during the positive anomaly and very low rate of sedimentation during negative anomalies. While the negative Ba anomalies do reflect the concomitant reduction in detrital influx and thereby signify detrital source of Ba, the positive anomaly shows a mixed character as it covaries with bioclast, C_{Inorg}, despite its sympathetic nature with K, Na, etc. Recurrent re-sedimentation associated with detrital influx was interpreted to have prevalent during the deposition of these rocks, which may explain covariation of bioclasts with positive excursion of Ba. Torres et al. (1996) recorded high concentrations of microcrystalline barite in sediments of world oceans and concluded that it resulted from the precipitation of Ba sulfate within microenvironments of decaying biological debris in the water column. Absence of barite (as evidenced from X-ray diffraction pattern) and observation of very low sulfur content in these rocks defy such a possibility.

Thus, it is prudent to interpret the excursions of Ba in the studied rocks as a result of detrital influx and sea-level changes than primary productivity in the marine realm itself. In this context, lower than UCC mean value may signify absence of significant chemical weathering in the source area and a generally sediment starved nature of the sedimentary basin may be as a result of low positive relief of the source area. Such sediment starvation was aggravated during the deep negative Ba anomaly occurred at the lower part of the DOM. Given cognizance to the climate-sea-level-related nature of depletion-enrichment behavior of Ba, its low residence time in marine waters and the occurrence of larger than other negative anomaly, a short-term sea-level high that deprived detrital sediment influx to the depocenters during the deposition of lower part of the DOM could be envisaged. It is important to note that this negative excursion is also coinciding with K_2O/Na_2O and K_2O/Al_2O_3 and is antisympathetic with Cr/Th, signifying dependence of Ba on alkali feldspars and clay. It implies that enrichment-depletion of Ba was under the control of source area weathering, climate, and sea-level and thus, interpretation of sea-level rise/high during deep negative excursion of Ba in tune with negative excursions of detrital elements and positive excursions of Ca and related elements is tenable.

12.5.4 RELATIVE CONTRIBUTION OF DEPOSITIONAL AND DIAGENETIC PROCESSES OVER GEOCHEMICAL SIGNALS

Discussions presented in preceding sections demonstrated that though there are many geochemical signals that are the creations of depositional processes/events, signs of post depositional remobilization of elements are also present within these signals. Such multicausal signals may dent the validity of interpretations on depositional events. For example, the amount of the element Sr in these rocks is nowhere near marine values (in the range of approximately 800–1200 ppm) even in the carbonates (DOM—Table 2), while the elements Fe, V, Cr, Zn, etc., show enrichment. Enrichment of these elements at the expense of Sr occured when the rocks experienced diagenesis (Brand et al., 2003, 2010) under reducing conditions of meteoric regime (Ramkumar et al., 2013). The monotony in the profiles of C_{Inorg}, Mg/Ca, Fe/Ca, Mn/Ca, and Sr/Ca, (Figs 3–5) might have resulted by post-depositional leaching, possibly after the deposition of Katrol Formation as all the stratigraphic units including Katrol Formation show such monotony. Similarly, occurrence of lower Mn/Al ratio far below PAAS and anoxic basin in carbonates and siliciclastics alike (Figs 2–5) may indicate prevalent late-stage diagenesis. This information, together with identification of all the samples to be that of reducing environment by V/V + Ni ratio (Fig. 7), supports meteoric phreatic conditions of diagenesis after the deposition of Katrol Formation that affected all the rocks under study regardless of lithological difference (though not exactly as there are subtle differences between members and sections, etc.). Diagenetic transformation through replacement mode by medium-slightly weak diagenetic waters is indicated by the absence of features characteristic of large-scale dissolution-precipitation and the presence of recognizable diagenetic changes only in susceptible grains. The information also indicates stable mineralogical nature of the host rock and the existence of only a slight difference between diagenetic components (diagenetic fluid and host rock) that affected only the susceptible grains (ooids, micritic matrix, etc.), absence of large-scale dissolution-precipitation mode of diagenesis, and prevalence of only replacement mode of diagenesis. At this juncture, it is imperative to ensure pristine nature of the geochemical signals. If the geochemical signatures are not remobilized in entirety as indicated by weak diagenetic process and the absence of dissolution-precipitation mode of diagenesis, it is necessary to estimate relative contribution of depositional and diagenetic processes over the geochemistry of the rocks.

In order to address this issue, factor analysis was performed following standard procedures (Johnson and Wichern, 1992). The database for this analysis consisted of quantitative measurements of 28 geochemical elements, computed ratio of Ba_{Excess}, and modal composition of the rocks, accounting for 10 parameters, totaling 39 variables. These quantitative data were found to be portraying nongaussian nature, which would eventually thwart any fruitful statistical analyses (Ramkumar, 2001). The measurement units of these variables are also different, in which case, transformation of the data prior to statistical analyses is necessary (Sahu, 1995; Nayak et al., 1997; Ramkumar, 2001; Ramkumar and Guha, 2000; Ramkumar et al., 2010b). The entire dataset was standardized with a transform function $z=(x-m)/s$ (where x is the value of the random variable; m is mean of the random variable; s is standard deviation; z is standardized variable) following the instructions of Davis (1973) and Clark and Hosking (1986). Contrary to the widely scattered raw data, the standardized values have an acute range with a mean value of zero and standard deviation of 1. For statistical analyses of this study, only the standardized dataset is utilized. Statistical analyses were performed with STATISTICA software. Factor analysis segregates the variables into different "sets/groups" based on their contribution to the variance of the data. It is conceived that each group of variables as segregated by factor analysis represents a single underlying construct (factor) that is responsible for the observed correlations (Johnson and Wichern, 1992). The number of factors to be retained for interpretation is chosen by proportion of the sample variances explained and geological reasonableness of the results. Since the original factor loadings may not be readily suitable for interpretation, it is usual to rotate them until a simple structure is achieved. Ideally, it is to bring out a pattern of loadings such that each variable loads highly on a single factor and has small to moderate loadings on the remaining factors (Malmgren and Haq, 1982). Taking cognizance to all these, factor analysis was performed with varimax rotation—a common type of orthogonal rotation used in geological problems (Kumar et al., 1995; Behera et al., 2003; Ramkumar et al., 2006) that allows all the axes of matrices to rotate simultaneously and fix at their optimal positions. Only the factors that have eigen values more than 1 and explaining higher than 5% of variance of data are retained for interpretation. The rotated factor loading shows five factors (Fig. 9) accounting for eigen value of 26.60 and explaining 68.10% of the total variance of the data. Many workers who studied geological problems through statistical analyses, felt content by explaining 70–80% of data variance and attributed rest of the variance to sampling error, analytical error, or random causes (e.g., Sahu, 1995; Kumar et al., 1995; Nayak et al., 1997; Behera et al., 2003; Ramkumar et al., 2006) and thus explanation of 68.10% variance with five factors is deemed to be appropriate.

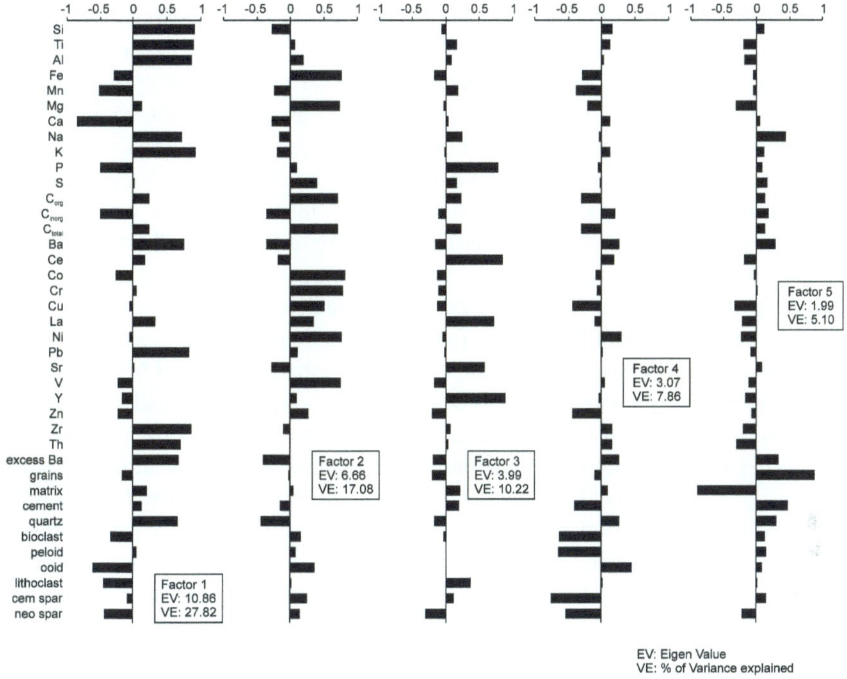

FIGURE 9

Figure showing the factor scores obtained by performing factor analysis with all the samples. Contribution by individual variables to various factors are also shown.

The first factor accounts for an eigen value of 10.86, which explains 27.82% of the data. This factor is positively loaded with K>Si>Ti>Al>Zr>Pb>Na>Ba>Th>Ba_{Excess}>Quartz (listed in the descending order of factor scores) and negatively loaded with Ca>Ooid>Mn>C_{Inorg}>P>Lithoclast. The grouping and scores (negative or positive) strongly indicate the preservation of depositional characteristics in the rocks. This is a perfect combination of petrographic and geochemical variables that espouse (by positive loadings of detrital elements and petrographic components and negative loadings of Ca and associated elements and petrographic components) prevalence of detrital versus authigenic sediment generation system, enforced by sea-level lows and rises, respectively. As this factor alone accounts for nearly half of total variance explained by five factors, domination of depositional signatures than later stage remobilization of elements is interpreted. Predomination of K over Si and the presence of Ba and Ba_{Excess} in the positive loading together with the observations on Ba and CIA (presented in previous sections) support the interpretation of generally sediment starved nature of the basin, coupled nature of climate-sea-level fluctuation-source area weathering and corresponding imprints in the sedimentary record. Interpretation of the sea-level rise/high-supported enhanced accumulation of carbonates that occurred only at the expense of deprivation of detrital influx is affirmed by the negative loadings of Ca and associated elements and petrographic components. Occurrences of lithoclasts and P in this factor could indicate the sediment redistribution events that episodically enriched P in the sediments. It also indicates the preservation and predomination of depositional signatures in the studied rock samples. The fifth factor, which is loaded positively with grains and negatively with matrix (almost in equal values of factor scores) indicates that prevalent energy conditions of depositional regime could also be differentiated in terms of either grain-supported texture or matrix-supported texture. Preservation of such information engrained in the rocks is an evidence for preservation of depositional characteristics without much diagenetic obliteration.

The second factor is positively loaded with Co>Cr>Fe>Ni>V>Mg>C_{Total}>C_{Org}. This factor may indicate the diagenetic involvement of the heavy metals into carbonates, probably under reducing conditions. It is interesting to note that the ooids show positive loadings albeit with low factor score. Their association with this factor may indicate a specific diagenetic event associated with oxidation/reduction of organic matter, during which, the carbonate particles including ooids were enriched with these heavy metals. Involvement of Mg in this factor may indicate prevalence of this diagenesis either with simultaneous expulsion of Mg from the carbonates or in the presence of Mg-rich clay or both. The third factor is positively loaded with Y>Ce>P>La>Sr and very weak positive loading of Lithoclast and may indicate sediment redistribution events and/or diagenetic remobilization of these elements. Presence of weak negative loading of neomorphic spar in this factor may indicate remobilization of these elements during late stage meteoric conditions of neomorphic transformation (as inferred through petrographic observations). Included nature of Sr in this factor and generally leached nature of it in all the samples supports this inference. The fourth factor is negatively loaded with Cement Spar>Peloid>Bioclast>Neomorphic spar indicative of diagenetic characteristics of the rocks, probably dissolution-precipitation event that occurred specifically affecting less stable mineralogic grains and matrix at the marine-phreatic zone.

12.5.5 SCALES OF SEA-LEVEL FLUCTUATIONS, OOIDS IN DOM, AND THEIR RELATIONSHIP WITH GLOBAL OOLITE PEAK

Given cognizance to the influence of sea-level fluctuations and consequent mutually exclusive excursions of Si and Ca groups of elements, it is possible to envisage a gradual increase of sea-level from upper part of the Dhosa Sandstone Member that reached its zenith near top of the DOM. In addition to this larger cycle, there were fluctuations (as indicated by phosphorus anomalies) and sea-level maximum (as indicated by a deep negative anomaly of Ba). The profiles of factor scores of first and fifth factors (Fig. 10) also confirm such trends in terms of fluctuations in sea-level and energy. From these profiles, it is inferred that though there are subtle differences between studied sections as indicated by variations in the patterns of fluctuations in the profiles, a commonality between them exists in terms of

1. distinct changes of sea-level and energy conditions across the boundaries of stratigraphic members and formation in all the studied sections;
2. a deepening trend from the Dhosa Sandstone Member towards the DOM;
3. generally deeper nature and associated low energy conditions of DOM than other stratigraphic members;
4. occurrences of distinct fluctuations of sea-level and energy coinciding with positive excursions of P and the larger negative excursion of Ba;
5. a distinct fall in sea-level and concomitant increase in energy conditions just prior to the "matrix" portion of DOM. It is interesting to note that the matrix portion shows a trend of deepening of sea-level and lowering of energy during its deposition, as explicit in LDS typically. In FWS, a

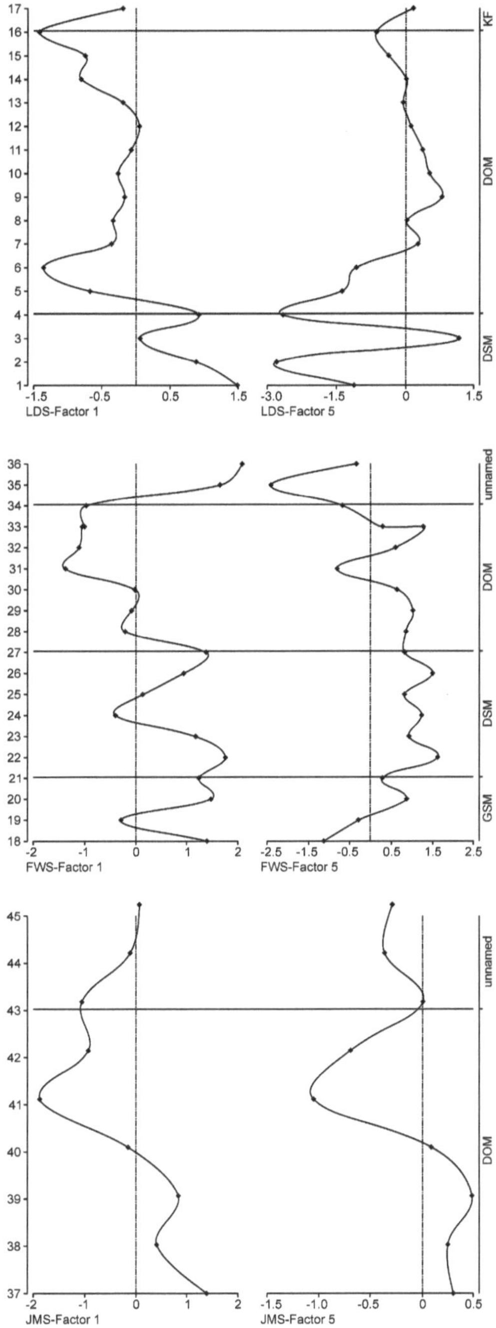

FIGURE 10

Profiles showing the stratigraphic variation of factor scores.

consistent, but relatively deeper sea-level, but inconsistent energy conditions are exhibited, may be as a result of mixing of texturally inconsistent sediments during sediment redistribution and recycling. While JMS also shows a deepening nature and concomitant reduction in energy, paucity of "matrix" portion of DOM do not allow any significant inference; and

6. the shift from DOM to overlying unnamed stratigraphic member and Katrol Formation is sharper than other stratigraphic boundaries. It may either indicate rapid climatic reversals (cooling event?—as the succeeding deposits are siliciclastics) or significant loss of upper beds of DOM through dissolution and erosion, as these beds were left exposed for a longer duration between cessation of DOM and initiation of younger deposits.

All these observations suggest unequivocal presence of depositional signatures in geochemical composition of the rocks, coupled nature of depositional environments-climate-sea-level in the basin and chemical weathering in the source area, which in turn are exhibited by the profiles of factor scores as a collective signature of groups of elements and petrographic components. Establishment of such statistically distinctive geochemical signatures and occurrences of short-term fluctuations of sea-level and elemental composition in rocks that are coupled with climate offer an opportunity to correlate the strata under study, more particularly, the Fe-ooids of the DOM, as they are regionally extensive and have unique lithological and faunal characteristics.

As could be observed elsewhere, broad-scale siliciclastic and carbonate deposition (as indicated by Ca and Si groups of elemental profiles) are promoted by sea-level high and low respectively in this basin too, the role of sea-level that favored ooid formation and deposition in DOM requires consideration. While the sandwiched nature of DOM between siliciclastic rocks (the Dhosa Sandstone Member at bottom and unnamed siliciclastic member and the Katrol Formation at top) indicate a perturbation in sea-level from general trend (also indicated by the geochemical profiles as well as the trends in the factor score profiles) that might have caused switching from predomination of siliciclastics to carbonates, presence of episodic siliciclastic rich interbeds within DOM suggest short-term oscillations. While the long-term perturbation was recorded by earlier studies (Kulkarni and Borkar, 2000; Pandey et al., 2009) which in turn was considered to be eustatic in nature (Fürsich et al., 1991, 2001), the short-term fluctuations within this larger cycle are recorded by the present study. The top of the DOM is interpreted as a maximum flooding surface of a relative sea-level highstand that is in tune with the trends observed in factor score profiles. Data from noncondensed sections of Callovian-Oxfordian suggest global sea-level rise much before the dawn of Oxfordian (Wierzbowski et al., 2009), a fact that could not be verified with other correlative equivalents located elsewhere, owing to their condensed nature and presence of frequent omission surfaces and hardground surfaces in them, but indicated by the trends of Ca and Si groups of elemental profiles and factor score profiles in this study. Absence of shallow-water benthic fauna and sedimentary structures typical of supratidal-intertidal and subaerial exposure in the Spanish Iberian sections were interpreted as a phase of sediment starvation and maximal flooding in the uppermost Callovian and lowermost Oxfordian (Norris and Hallam, 1995). The geochemical and factor score profiles of the study area have indicated initiation of sea-level rise much before the dawn of DOM. Regardless of merit of this interpretation, association of enhanced ooid production with either sea-level high or transgression in this basin is ascertained.

While reviewing the carbonate factories of Phanerozoic, Pomar and Hallock (2008), based on late Jurassic sections distributed Worldover, stated that this time slice was dominated by carbonate production in shallow seas and conspicuous for resedimentation process and basinward transport of lime mud and other sediments. These authors also relied on the forward computer modeling and other facies studies (Pomar and Hallock, 2008 and references cited therein) that this basinward transport during Oxfordian followed a cyclic pattern of 20 Ka and 100 Ka. Recognition of short-term sea-level fluctuations within the Callovian-Oxfordian strata of Kachchh Basin and corresponding geochemical excursions, that too by the elements that have very low residence times (in the order of 10,000 years) and have link with carbon budget and immediately preceding geological causes enhances the potential of correlation of these strata more precisely. This potential is further substantiated by the recognition of larger sea-level cycle and few other short-term cycles.

Despite this potential and stacked nature of short-term cycles, owing to the discontinuous nature of these excursions, they could not be dovetailed to form a continuous chain. Compounded with this is the problem of occurrences of omission surfaces, erosional surfaces, hardground surfaces, sediment recycling events, and varied depositional rates between different geographic locations. Nevertheless, these do not preclude correlation of the individual events to be recognized from regional counterparts, provided sufficient caution is exercised. It also offers an opportunity to reexamine the studied sections through very close intervals of sampling to establish a more complete geochemical profile and high-resolution chemostratigraphy.

12.6 CONCLUSIONS

Sedimentation in the Kachchh Basin during Callovian-Oxfordian times was coupled with climate and sea-level and resultant chemical weathering in the source area and detrital influx. Sediment starved nature of the basin was the norm during the depositional history of studied sections. Owing to the prevalent coupling, broad, shallow, and monotonous topography of the receiving basin and limited altitudinal difference between the source area and receiving basin, slight changes in the sea-level could make significant impact on the depositional pattern. As a result, there were many erosional and nondepositional events that in turn thwart reconstruction of depositional history precisely.

A broad scale alternation of siliciclastic and carbonate-dominated depositional pattern, as explicit in the Ca and Si groups of elemental profiles that recorded significant climatic changes across the boundaries of perceived lithostratigraphic members, in addition to perturbations in terms of positive and negative excursions of elements is recognized with this study. Geochemical profiles and their characteristics unequivocally indicate that DOM was formed during the global eustatic cycle. The maximum flooding occurred globally during Oxfordian is also recorded by this member as reported in earlier studies. While this may offer an excellent opportunity to correlate the strata under study, particularly the DOM precisely with strata elsewhere, owing to the occurrences of omission surfaces, hard ground surfaces, erosional and re-sedimentation events, present study opines that correlation of DOM merely by its similarity of facies characters and occurrences of ooids with Fe-ooid deposits of Oxfordian is wrought with potential errors.

ACKNOWLEDGMENTS

This work has been supported by the German Research Foundation (DFG) Research Grant No. Fü/131. Research stays of MR at the Erlangen University was financially supported by Alexander von Humboldt Foundation, which is gratefully acknowledged.

REFERENCES

Alberti, M., Fürsich, F.T., Pandey, D.K., Ramkumar, M., 2012. Stable isotope analyses of belemnites from the Kachchh Basin, western India: Paleoclimatic implications for the Middle to Late Jurassic transition. Facies 58, 261–278.

Allenbach, R.P., 2002. The ups and downs of 'Tectonic Quiescence'—recognizing differential subsidence in the epicontinental sea of the Oxfordian in the Swiss Jura Mountains. Sedimentary Geology 150, 323–342.

Behera, B.K., Sahu, B.K., Bagchi, J., 2003. Factor analysis of sphericity and roundness statistics and correspondence analysis of size statistics of the sediments of Gujarat coast. Journal of the Geological Society of India 62, 51–62.

Bekker, A., Kaufman, A.J., Karhu, J.A., Beukes, N.J., Swart, Q.D., Coetzee, L.L., Eriksson, K.A., 2001. Chemostratigraphy of the Paleoproterozoic Duitschland Formation, South Africa: implications for coupled climate change and carbon cycling. American Journal of Science 301, 261–285.

Bishop, J.K.B., 1988. The barite-opal-organic carbon association in oceanic particulate matter. Nature 331, 341–343.

Brand, U., Logan, A., Hiller, N., Richardson, J., 2003. Geochemistry of modern brachiopods: applications and implications for oceanography and paleoceanography. Chemical Geology 198, 305–334.

Brand, U., Azmy, K., Tazawa, J., Sano, H., Buhl, D., 2010. Hydrothermal diagenesis of Paleozoic seamount carbonate components. Chemical Geology 278, 173–185.

Chan, L.H., Edmond, J.M., Stallard, R.F., Broecker, W.S., Chung, Y.C., Weiss, R.F., Ku, T.L., 1976. Radium and barium at GEOSECS stations in the Atlantic and Pacific. Earth and Planetary Science Letters 32, 1258–267.

Clark, W.A.V., Hosking, P.C., 1986. Statistical Methods for Geographers. John Wiley & Sons. Inc., New York. 90 pp.

Curial, A., Dromart, G., 1998. Apport de la sismique-reflexion et des forages a' l'analyse séquentielle et géométrique des corps sédimentaires et de ses encaissants: Journées scientifiques Agence Nationale de la gestion des Déchets Radioactifs, Bar-le-Duc, Atlas des Posters. pp. 25–27.

Davis, J.C., 1973. Statistics and Data Analysis in Geology. John Wiley & Sons, New York. 148 pp.

Das, N., 1997. Chemostratigraphy of sedimentary sequences: a review of the state of the art. Journal of the Geological Society of India 49, 621–628.

Derry, L.A., Brasier, M.D., Corfield, R.M., Rozanov, A.Yu, Zhuravlev, A.Y., 1994. Sr and C isotopes in lower Cambrian carbonates from the Siberian craton: a paleoenvironmental record during the 'Cambrian explosion'. Earth and Planetary Science Letters 128, 671–681.

Dill, G., Scheel, M., Köthe, A., Botz, R., Henjes-Kunst, F., 1997. An integrated environment analysis – lithofacies, chemofacies, biofacies of the Oligocene calcareous-siliciclastic shelf deposits in northern Germany. Palaeogeography, Palaeoclimatology, Palaeoecology 131, 145–174.

Dromart, G., Garcia, J., Gaumet, F., Picard, S., Rousseau, M., Atrops, F., Lecuyer, C., Sheppard, S.M.F., 2003. Perturbation of the carbon cycle at the middle/late Jurassic transition: geological and geochemical evidence. American Journal of Science 303, 667–707.

Dymond, J., Suess, E., Lyle, M., 1992. Barium in deep sea sediments: a geochemical indicator of paleoproductivity. Paleoceanography 7, 163–181.

Elrick, M., Scott, L.A., 2010. Carbon and oxygen isotope evidence for high-frequency (10^4–10^5 year) and MY-scale glacio-eustasy in Middle Pennsylvanian cyclic carbonates (Gray Mesa Formation), central New Mexico. Palaeogeography, Palaeoclimatology, Palaeoecology 285, 307–320.

Fatmi, A.N., 1972. Stratigraphy of the Jurassic and lower Cretaceous rocks and Jurassic ammonites from northern areas of Western Pakistan. Bulletin of the British Museum of Geology 20, 299–380.

Filippelli, G.M., Delaney, M.L., 1994. The oceanic phosphorus cycle and continental weathering during Neogene. Paleocenography 9, 643–652.

Flügel, E., 1982. Microfacies Analysis of Limestone. Springer-Verlag, Heidelberg. 633 pp.
Francois, R., Honjo, S., Manganini, S.J., Ravizza, G.E., 1995. Biogenic barium fluxes to the deep sea: implications for paleoproductivity reconstruction. Global Biogeochemical Cycles 9, 289–303.
Froelich, P.N., 1988. Kinetic control of dissolved phosphate in natural rivers and estuaries: a primer on the phosphate buffer mechanism. Limnology and Oceanography 33, 649–668.
Fürsich, F.T., Oschmann, W., 1993. Shell beds as tool in facies analysis: the Jurassic of Kachchh, western India. Journal of the Geological Society of London 150, 169–185.
Fürsich, F.T., Oschmann, W., Jaitly, A.K., 1991. Faunal response to transgressive-regressive cycles: example from the Jurassic of Western India. Palaeogeography, Palaeoclimatology, Palaeoecology 85, 149–159.
Fürsich, F.T., Oschmann, W., Singh, I.B., 1992. Hardgrounds, reworked concretion levels and condensed horizons in the Jurassic of Western India: their significance for basin analysis. Journal of the Geological Society of London 149, 313–331.
Fürsich, F.T., Pandey, D.K., Collomon, J.H., 2001. Marker beds in the Jurassic of the Kachchh Basin, western India: their depositional environment and sequence stratigraphic significance. Journal of the Palaeontological Society of India 46, 173–198.
Fürsich, F.T., Callomon, J.H., Pandey, D.K., Jaitly, A.K., 2004. Environments and faunal patterns in the Kachchh rift basin, Western India, during the Jurassic. Rivista Italiana di Paleontologia e Stratigrafia 110, 181–190.
Fürsich, F.T., Singh, I.B., Joachimski, M., Krumm, S., Schlirl, M., Schlirl, S., 2005. Palaeoclimate reconstructions of the Middle Jurassic of Kachchh (Western India): an integrated approach based on palaeoecological, oxygen isotopic and clay mineralogical data. Palaeogeography, Palaeoclimatology, Palaeoecology 217, 289–309.
Gaetani, M., Garzanti, E., 1991. Multicyclic history of the northern history of the northern India continental margin (North Western Himalaya). American Association of Petroleum Geologists Bulletin 75, 1427–1446.
Geiger, M., Schweigert, G., 2005. Toarcian–Kimmeridgian depositional cycles of the south-western Morondava Basin along the rifted continental margin of Madagascar. Facies 52, 85–112.
Heydari, E., 2000. Porosity loss, fluid flow and mass transfer in limestone reservoirs: application to the upper Jurassic Smackover formation. American Association of Petroleum Geologists Bulletin 84, 100–118.
Hoppie, B., Garrison, R.E., 2001. Miocene phosphate accumulation in the Cuyama Basin, southern California. Marine Geology 177, 353–380.
Jacquin, T., Dardeau, G., Durlet, C., de Graciansky, P.C., Hantzpergue, P., 1998. The North Sea cycle: an overview of 2nd-order transgressive/regressive facies cycles in Western Europe. In: de Graciansky, P.C., Hardenbol, J., Jacquin, T., Vail, P.R. (Eds.), Mesozoic and Cenozoic Sequence Stratigraphy of European Basins, vol. 60. SEPM Special Publication, pp. 445–466.
Jaffe, L.A., Peucker-Ehrenbrink, B., Petsch, S.T., 2002. Mobility of rhenium, platinum group elements and organic carbon during black shale weathering. Earth and Planetary Science Letters 198, 339–353.
Johnson, R.A., Wichern, D.W., 1992. Applied Multivariate Statistical Analysis, third ed. Prentice-Hall of India. Pvt. Ltd., New Delhi. 372 pp.
Kakizaki, Y., Kano, A., 2009. Architecture and chemostratigraphy of late Jurassic shallow marine carbonates in NE Japan, western Paleo-Pacific. Sedimentary Geology 214, 49–61.
Kasten, S., Ruehlemann, C., Haese, R.R., Zabel, M., Mulitza, S., Funk, J., Schulz, H.D., 1998. Barium Peaks at Glacial Terminations in Sediments of the Equatorial Atlantic Ocean—Relics of Deglacial Productivity Pulses? Goldschmidt Conference Program Abstract 749–750.
Keller, G., Adatte, T., Stinnesbeck, W., Stüben, D., Kramar, U., Berner, Z., Li, L., Salis Perch-Nielsen, K.V., 1998. The Cretaceous-Tertiary transition on the shallow Saharan platform of southern Tunisia. Geobios 30, 951–975.
Kramar, U., 1997. Advances in energy-dispersive X-ray fluorescence. Journal of Geochemical Exploration 58, 73–80.
Kulkarni, K.G., Borkar, V.D., 2000. Palaeonvironmental Niches towards the close of deposition of Dhosa oolite (early Oxfordian), Kutch, India. Indian Journal of Petroleum Geology 9, 49–58.

Kumar, A., Ratha, D.S., Nandy, P., 1995. Chemical variations in the Tertiary carbonates of southwestern Kutch, Gujarat—a statistical approach. Journal of the Geological Society of India 46, 295–301.
Land, L.S., Behrens, E.W., Frishman, S.A., 1979. The ooids of Baffin Bay, Texas. Journal of Sedimentary Petrology 49, 1269–1278.
Leinfelder, R.R., 1993. A sequence stratigraphic approach to the Upper Jurassic mixed carbonate-siliciclastic succession of the central Lusitanian Basin, Portugal. Profil 5, 119–140.
Li, X., Grant-Mackie, J.A., 1993. Jurassic sedimentary cycles and eustatic sea-level changes in southern Tibet. Palaeogeography, Palaeoclimatology, Palaeoecology 101, 27–48.
Lyle, M., Zahn, R., Prahl, F., Dymond, J., Collier, R., Pisias, N., Suess, E., 1992. Paleoproductivity and carbon burial across the California current: the multitracers transect, 42°N. Paleoceanography 7, 251–272.
Madhavaraju, J., Il Lee, Y., 2009. Geochemistry of the Dalmiapuram Formation of the Uttatur Group (early Cretaceous), Cauvery Basin, southeastern India: Implications on provenance and paleo-redox conditions. Revista Mexicana de Ciencias Geológicas 26, 380–394.
Malmgren, B.A., Haq, B.U., 1982. Assessment of quantitative techniques in paleobiogeogaphy. Marine Micropalaeontology 7, 213–236.
McManus, J., Berelson, W.M., Klinkhammer, G.P., Johnson, K.S., Coale, K.H., Anderson, R.F., Kumar, N., Burdigo, D.J., Hammond, D.E., Brumsack, H.J., McCorkle, D.C., Rushdi, A., 1998. Geochemistry of barium in marine sediments: implications for its use as a paleoproxy. Geochimica et Cosmochimica Acta 62, 3453–3473.
Moore, C.H., 2001. Carbonate Reservoirs: Porosity Evolution and Diagenesis in a Sequence Stratigraphic Framework. Elsevier, Amsterdam. 444 pp.
Munnecke, A., Calner, M., Harper, D.A.T., Servais, T., 2010. Ordovician and Silurian sea-water chemistry, sea-level, and climate: a synopsis. Palaeogeography, Palaeoclimatology, Palaeoecology 296, 389–413.
Murray, R.W., Leinen, M., 1993. Chemical transport to the seafloor of the equatorial Pacific Ocean across a latitudinal transect at 135°W: tracking sedimentary major, trace and rare earth element fluxes at the equator and the intertropical convergence zone. Geochimica et Cosmochimica Acta 57, 4141–4163.
Nayak, B.J., Rajev, S.M., Sahoo, R.K., 1997. R-mode factor analysis and its implications on the geochemistry of Nishikhal manganese deposit, Raiguda District, Orissa. Journal of the Geological Society of India 49, 133–144.
Norris, M.S., Hallam, A., 1995. Facies variations across the Middle–Upper Jurassic boundary in Western Europe and the relationship to sea-level changes. Palaeogeography, Palaeoclimatology, Palaeoecology 116, 189–245.
Oppo, D.W., Fairbanks, R.G., 1989. Carbon isotope composition of tropical surface water during the past 22,000 years. Paleoceanography 4, 333–351.
Pandey, D.K., Sha, J.G., Choudhary, S., 2006. Depositional history of the early part of the Jurassic succession on the Rajasthan shelf, western India. Progress in Natural Science 16, 176–185.
Pandey, D.K., Fürsich, F.T., Sha, J.G., 2009. Interbasinal marker intervals – a case study from the Jurassic basins of Kachchh and Jaisalmer, Western India. Science in China Series D: Earth Sciences 52, 1924–1931.
Pearce, T.J., Jarvis, I., 1991. Applications of geochemical data to modeling sediment dispersal patterns in distal turbidites: late Quaternary of the Madeira abyssal plain. Journal of Sedimentary Petrology 62, 1112–1129.
Pearce, T.J., Besly, B.M., Wray, D.S., Wright, D.K., 1999. Chemostratigraphy: a method to improve interwell correlation in barren sequences — a case study using onshore Duckmantian/Stephanian sequences (West Midlands, U.K.). Sedimentary Geology 124, 197–220.
Peryt, T.M., 1983. Coated Grains. Springer-Verlag, Berlin. 655 pp.
Plank, T., Langmuir, C.H., 1998. The chemical composition of subducting sediment and its consequences for the crust and mantle. Chemical Geology 145, 325–394.
Pomar, L., Hallock, P., 2008. Carbonate factories: a conundrum in sedimentary geology. Earth Science Reviews 87, 134–169.
Pujol, F., Berner, Z., Stüben, D., 2006. Palaeoenvironmental changes at the Frasnian/Famennian boundary in key European sections: chemostratigraphic constraints. Palaeogeography, Palaeoclimatology, Palaeoecology 240, 120–145.

Ramkumar, M., 1999. Role of chemostratigraphic technique in reservoir characterisation and global stratigraphic correlation. Indian Journal of Geochemistry 14, 33–45.

Ramkumar, M., 2001. Sedimentary microenvironments of modern Godavari delta: characterization and statistical discrimination – towards computer assisted environment recognition scheme. Journal of the Geological Society of India 57, 49–63.

Ramkumar, M., Guha, A.K., 2000. Multivariate statistical verification of petrographic and standard microfacies types and lithostratigraphy of Tertiary carbonates of western Kutchch, Gujarat, India: Implications on global stratigraphic correlation and hydrocarbon exploration. Indian Journal of Petroleum Geology 9, 52–74.

Ramkumar, M., Harting, M., Stüben, D., 2005. Barium anomaly preceding K/T boundary: Plausible causes and implications on end Cretaceous events of K/T sections in Cauvery Basin (India), Israel, NE-Mexico and Guatemala. International Journal of Earth Sciences 94, 475–489.

Ramkumar, M., Stüben, D., Berner, Z., 2006. Elemental interrelationships and depositional controls of Barremian-Danian strata of the Cauvery Basin, South India: implications on scales of chemostratigraphic modelling. Indian Journal of Geochemistry 21, 341–367.

Ramkumar, M., Stüben, D., Berner, Z., 2009. Episodic accumulation of phosphorus during Cretaceous in the Cauvery Basin, South India: implications on palaeoclimate, productivity and weathering. Current Science 97, 262–267.

Ramkumar, M., Stüben, D., Berner, Z., Schneider, J., 2010a. $^{87}Sr/^{86}Sr$ anomalies in late Cretaceous-Early Tertiary strata of the Cauvery Basin, South India: constraints on nature and rate of environmental changes across K-T boundary. Journal of Earth Systems Sciences 119, 1–17.

Ramkumar, M., Stüben, D., Berner, Z., 2010b. Hierarchical delineation and multivariate statistical discrimination of chemozones of the Cauvery Basin, South India: implications on Spatio-temporal scales of stratigraphic correlation. Petroleum Science 7, 435–447.

Ramkumar, M., Stüben, D., Berner, Z., 2011. Barremian-Danian chemostratigraphic sequences of the Cauvery Basin, South India: implications on scales of stratigraphic correlation. Gondwana Research 19, 291–309.

Ramkumar, M., Alberti, M., Fürsich, F.T., 2012. Anomalous phosphorous episodes during Callovian-Oxfordian in the Kachchh Basin, northwestern India: Implications on palaeoclimate, productivity and weathering. International Journal of Geochemistry 72, 141–148.

Ramkumar, M., Alberti, M., Fürsich, F.T., Pandey, D.K., 2013. Depositional and diagenetic environments of the Dhosa Oolite Member (Oxfordian), Kachchh Basin, India, based on Petrographic data: Implications on the origin and occurrence of ooids and their correlation with global oolite peak. In: Ramkumar, M., (Ed.). On a Sustainable Future of the Earth's Natural Resources. Springer-Verlag, Heidelberg, pp. 179–230.

Rangel, A., Parra, P., Nino, C., 2000. The La Luna formation: chemostratigraphy and organic facies in the middle Magdelena Basin. Organic Geochemistry 31, 1267–1284.

Raymo, M.E., 1994. The Himalayas, organic carbon burial, and climate in the Miocene. Paleoceanography 9, 399–404.

Raymo, E., Oppo, D.W., Curry, W., 1997. The mid-Pleistocene climate transition: a deepsea carbon isotopic perspective. Paleoceanography 12, 546–559.

Ruttenberg, K.C., 1993. Reassessment of the oceanic residence time of phosphorus. Chemical Geology 107, 405–409.

Sageman, B.B., Murphy, A.E., Werne, J.P., Ver Straetend, C.A., Hollander, D.J., Lyons, T.W., 2003. A tale of shales: the relative roles of production, decomposition, and dilution in the accumulation of organic-rich strata, Middle–Upper Devonian, Appalachian Basin. Chemical Geology 195, 229–273.

Sahagian, D., Pinous, O., Olferiev, A., Zakharov, V., 1996. Eustatic curve for the middle Jurassic–Cretaceous based on Russian platform and Siberian stratigraphy: zonal resolution. American Association of Petroleum Geologists Bulletin 80, 1433–1458.

Sahu, B.K., 1995. Statistical inference from geochemical and petrographic data. In: Proceedings of Recent Researches in Geology of Western India, pp. 59–64.

Saraswati, P.K., Ramesh, R., 1992. Eocene-Oligocene stable isotope stratigraphy of Kutch. Journal of the Geological Society of India 39, 427–432.

Schroeder, J.O., Murray, R.W., Leinen, M., Pflaum, R.C., Janecek, T.R., 1997. Barium in equatorial Pacific carbonate sediment: terrigenous, oxide and biogenic associations. Paleoceanography 12, 125–146.

Sensarma, S., Rajamani, V., Tripathi, J.K., 2008. Petrography and geochemical characteristics of the sediments of the small River Hemavati, Southern India: implications for provenance and weathering processes. Sedimentary Geology 205, 111–125.

Singh, I.B., 1978. Microfacies, petrography and mineralogy of the Tertiary rocks of Guvar nala near Narain Sarovar, Kutch, India and their palaeoecological significance. Journal of the Palaeontological Society of India 21 & 22, 78–95.

Singh, I.B., 1989. Dhosa oolite – a transgressive condensation horizon of Oxfordian age in Kachchh, western India. Journal of the Geological Society of India 34, 152–160.

Spalletti, L.A., Poire, D.G., Schwarz, E., Veiga, G.D., 2001. Sedimentologic and sequence stratigraphic model of a Neocomian marine carbonate-siliciclastic ramp: Nequen Basin, Argentina. American Earth Sciences 14, 609–624.

Strasser, A., Samankassou, E., 2003. Carbonate sedimentation rates today and in the past: Holocene of Florida Bay, Bahamas, and Bermuda versus Upper Jurassic and Lower Cretaceous of the Jura Mountains (Switzerland and France). Geologia Croatica 56, 1–18.

Stüben, D., Kramar, U., Berner, Z., Stinnesbeck, W., Keller, G., Adatte, T., 2002. Trace elements, stable isotopes and clay mineralogy of the Elles II K-T boundary section in Tunisia: indications for sea-level fluctuations and primary productivity. Palaeogeography, Palaeoclimatology, Palaeoecology 178, 321–345.

Tappan, H., 1967. Primary production, isotopes, extinctions and the atmosphere. Palaeogeography, Palaeoclimatology, Palaeoecology 4, 187–210.

Taylor, S.R., McLennan, S.M., 1985. The Continental Crust: Its Composition and Evolution. Oxford, Blackwell. 349 pp.

Torres, M.E., Brumsack, H.J., Bohrmann, G., Emeis, K.C., 1996. Barite fronts in continental margin sediments: a new look at barium remobilization in the zone of sulfate reduction and formation of heavy barites in diagenetic fronts. Chemical Geology 127, 125–139.

Tucker, M.E., Wright, V.P., 1990. Carbonate Sedimentology. Blackwell Science. 482 pp.

Wierzbowski, H., Dembicz, K., Praszkier, T., 2009. Oxygen and carbon isotope composition of Callovian–Lower Oxfordian (Middle–Upper Jurassic) belemnite rostra from central Poland: a record of a Late Callovian global sea-level rise? Palaeogeography, Palaeoclimatology, Palaeoecology 283, 182–194.

Wilkinson, B.H., Owen, R.M., Carroll, A.R., 1985. Submarine hydrothermal weathering, global eustacy and carbonate polymorphism in Phanerozoic marine oolites. Journal of Sedimentary Petrology 55, 171–183.

Wright, J.K., 2007. Wath quarry, Hovingham. In: British Upper Jurassic Stratigraphy (Oxfordian to Kimmeridgian). http://www.jspubs.com/xp.

Zavar, E., 2006. A study of jurassic carbonate hardgrounds and the Sclerbionts who inhabit them, Zalas quarry, Kraków, Poland (Unpublished Master's thesis). College of Wooster, 73 pp.

CHAPTER

FACIES AND CARBON ISOTOPE CHEMOSTRATIGRAPHY OF LOWER JURASSIC CARBONATE DEPOSITS, LUSITANIAN BASIN (PORTUGAL): IMPLICATIONS AND LIMITATIONS TO THE APPLICATION IN SEQUENCE STRATIGRAPHIC STUDIES

13

Ricardo L. Silva[1], Luís V. Duarte[2], María José Comas-Rengifo[3]

[1]Basin and Reservoir Lab, Department of Earth Sciences, Faculty of Sciences, Dalhousie University, Life Science Center, Dalhousie University Halifax, Canada; [2]Departamento de Ciências da Terra, IMAR-CMA/MARE, Faculdade de Ciências e Tecnologia, Universidade de Coimbra, Largo Marquês de Pombal, Coimbra, Portugal; [3]Departamento de Paleontología, Facultad de Ciencias Geológicas, Universidad Complutense de Madrid, Madrid, Spain

13.1 INTRODUCTION

Sequence stratigraphy has been extensively used by academia and industry. It is a valuable tool in Basin analysis and numeric modeling of the several parameters that govern sedimentation, and resource availability and its spatial distribution (e.g., Howell and Aitken, 1996; Coe, 2003; Brookfield, 2004; Alan and Alan, 2005; Catuneanu, 2006; Miall, 2008, 2010; Hantschel and Kauerauf, 2009; Embry, 2009). Carbonates share common properties with clastics systems, but several key differences should be kept in mind when dealing with the application of the sequence stratigraphic method. Even though both systems respond to changes in base level and are divided by similar surfaces, carbonate accumulation is regarded as "in situ production" whereas the clastics are transported to their depositional resting place. In carbonate environments, the main parameter controlling the response to sea-level changes is the organisms (e.g., Schlager, 2005). The carbonate factory depends on the type of carbonate producers which, in turn, depends mainly on physiography and environment. These parameters control photosynthesis rates, depth, and proximity to the sea surface, for example. More importantly, carbonate producers react dynamically to environmental changes having, under favorable conditions, the ability to outpace creation of accommodation space or to respond actively to base level lowering (Schlager, 2005). As a result from the interaction of the aforementioned processes, the carbonate chemistry, facies, and fabrics are indicative of sea-level position, water chemistry, etc. (e.g., Folk, 1959; Dunham, 1962; Embry and Klovan, 1971; Wright, 1992; Flügel, 2010). A refined sequence stratigraphy model for carbonate systems has to take into account all the processes conditioning sedimentation and success depends upon the information provided by several disciplines related to

Chemostratigraphy. http://dx.doi.org/10.1016/B978-0-12-419968-2.00013-3

Sedimentary Geology (e.g., Catuneanu, 2006; Miall, 2008, 2010). Stable carbon isotope chemostratigraphy is a tool with a high potential to be included in the sequence stratigraphic method applied to carbonates, mostly because it can be used to characterize and/or to trace changes in the various carbon reservoirs, from continental environments and atmosphere to oceans and organisms (Renard et al., 2008; Weissert et al., 2008). In marine environments, and particularly in epeiric sea settings, these processes are governed by the complex interplay of local (different carbonate producers, transgressive–regressive (T-R) cycles) and/or global (worldwide preservation of organic matter, variation of continental weathering, input of volcanogenic light CO_2) mechanisms (e.g., Schlanger and Jenkyns, 1976; Jenkyns and Clayton, 1986; Wefer and Berger, 1991; Marshall, 1992; Patterson and Walter, 1994; Holmden et al., 1998; Weissert et al., 1998; Kump and Arthur, 1999; Hesselbo et al., 2000; Immenhauser et al., 2003; Swart and Eberli, 2005; Panchuk et al., 2006; Duarte et al., 2007; Suan et al., 2010; Pittet et al., 2014).

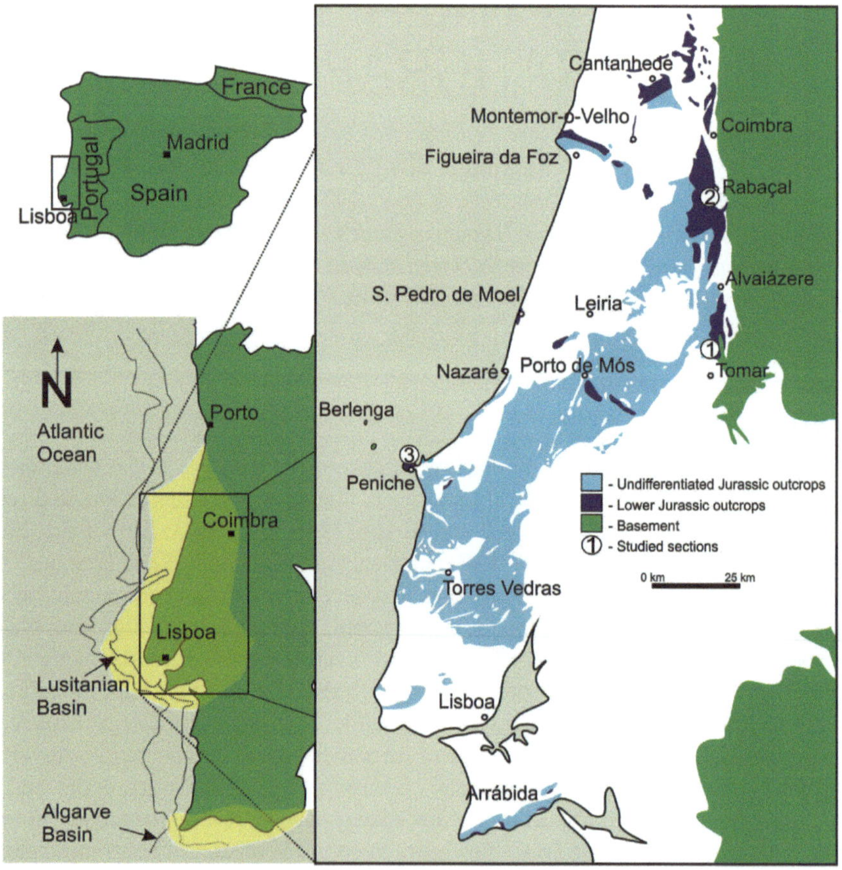

FIGURE 1

Geological map of the Jurassic of the Lusitanian Basin and location of the main studied Pliensbachian outcrops (other sections were locally studied, e.g., S. Pedro de Moel, Figueira da Foz or Coimbra). (1) Tomar; (2) Rabaçal; (3) Peniche.

The on- and offshore sedimentary basins belonging to the Atlantic Conjugate Margins are currently the focus of intense exploration concerning the assessment of their potential for oil and gas. The Lusitanian Basin (Portugal) is regarded either as a target for exploration or, due to the remarkable exposure and high quality of its outcrops, as a starting point to the comprehension of the less-known offshore basins located along the Western European margin, Eastern Canada (Nova Scotia, Newfoundland and Labrador offshore), and North Africa (Morocco), where existence and maturity of source rock is one of the key risks. Data from the Lusitanian Basin and comparison with the available data from offshore areas may hold important clues to the identification and/or understanding of a possible Lower Jurassic source rock interval of regional relevance.

One of the main intervals with source rock potential of the Lusitanian Basin corresponds to the Sinemurian–Pliensbachian, although mainly constrained to two temporal locations: the Polvoeira Member of the Água de Madeiros Formation (Late Sinemurian, Duarte et al., 2012) and the marly limestones with organic-rich facies member of the Vale das Fontes Formation (Pliensbachian, Duarte et al., 2010). Based on the study of three classical sections from the Lusitanian Basin, Portugal (Tomar, Rabaçal, and Peniche, Fig. 1), the objectives of this work are: (1) to present a high-resolution basinwide stratigraphic correlation of the Pliensbachian (Lower Jurassic) organic-rich interval; (2) to establish a third-order sequence stratigraphic scheme; and (3) to compare the sequence stratigraphic scheme and carbon isotopic record of the Lusitanian Basin with its neighboring counterparts.

13.2 GEOLOGICAL BACKGROUND

The Lusitanian Basin is a small, narrow north–south elongated basin, located on the occidental side of the Iberian Massif and bounded to the west by several basement horsts. This basin is one of the several rift-related Peri-Tethyan Mesozoic basins whose origin is related with the opening of the Atlantic Ocean (Fig. 2) (e.g., Wilson et al., 1989; Hiscott et al., 1990. Pinheiro, 1996; Stapel et al., 1996; Rasmussen et al., 1998; Alves et al., 2002, 2003, 2006, 2009).

The works conducted over the past and beginning of this century concerning the Lower Jurassic carbonate series of the Lusitanian Basin are largely related to sedimentology and paleogeography (e.g., Wilson, 1988; Soares et al., 1993; Duarte, 1997, 2007; Azerêdo et al., 2003; Duarte et al., 2004), biostratigraphy (e.g., Mouterde, 1955, 1967; Mouterde and Rocha, 1981; Mouterde et al., 1983; Phelps, 1985; Dommergues, 1987; Boomer et al., 1998; N'zaba-Makaya et al., 2003; Dommergues et al., 2004a,b; among others), and lithostratigraphy (Duarte and Soares, 2002; Azerêdo, 2007). In recent years, supported by the preceding studies, research is more diverse, including the refinement of the ammonite and calcareous nannofossils biostratigraphy (e.g., Mouterde et al., 2007; Oliveira et al., 2007; Reggiani et al., 2010a,b; Comas-Rengifo et al., 2013; Matiolli et al., 2013), sequence analysis (Duarte et al., 2004, 2007, 2010; Duarte, 2007), paleoenvironmental interpretation based on several fossil groups (e.g., Azerêdo et al., 2010; Reggiani et al., 2010a,b; Silva et al., 2012) and sedimentary characterization based on total organic carbon (TOC) and Rock-Eval pyrolysis (Oliveira et al., 2006; Silva et al., 2006, 2007; Duarte et al., 2010, 2011, 2012, 2013), outcrop gamma-ray spectrometry (Correia et al., 2012), elemental (Silva et al., 2009) or isotope geochemistry (whole rock (Oliveira et al., 2006; Hesselbo et al., 2007; Silva et al., 2011b; Pittet et al., 2014), kerogen and fossil wood (Suan et al., 2010; Silva et al., 2011b, 2013) or biogenic calcite (Oliveira et al., 2009; Suan et al., 2008, 2010)),

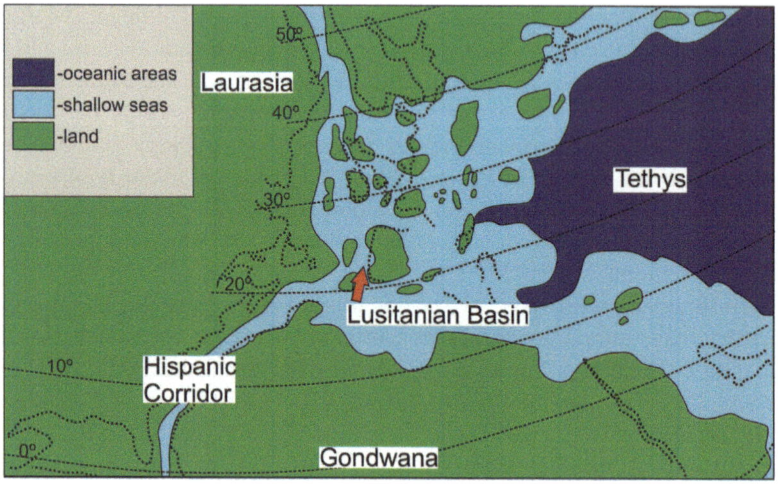

FIGURE 2

Western Tethys paleogeography and location of the Lusitanian Basin during the Late Early Jurassic (modified and simplified from Bassoulet et al., 1993).

palynofacies (e.g., Ferreira et al., 2010; Silva et al., 2011a; Duarte et al., 2012; Silva et al., 2012; Poças Ribeiro et al., 2013), biomarkers (Silva et al., 2010; Mendonça Filho et al., 2013), and biogeochemistry (Silva et al., 2012).

The studied hemipelagic series of the Pliensbachian age (Ibex (Luridum Subchronozone)–Margaritatus (base of Gibbosus Subchronozone) chronozones) correspond to the Marly limestones with organic-rich facies member (MLOF mb) of the Vale das Fontes Formation and is included in the Pliensbachian T-R facies cycle, the first second-order flooding event recognized at a basinal scale (Fig. 3) (see Duarte et al., 2010; Silva et al., 2011b and references therein) contained within in the Upper Triassic–Callovian first-order cycle (e.g., Wilson et al., 1989; Soares et al., 1993; Duarte, 1997; Azerêdo et al., 2002, 2003; Duarte et al., 2004). The MLOF mb, when compared with the under- and overlying units (see Fig. 3 and Duarte and Soares, 2002; Duarte et al., 2010), is characterized by the increasing marly character of the series and by the occurrence of several organic-rich facies, particularly well developed in the western, and distal hemipelagic sectors of the basin. Several well-defined black shales (*s.l.*) are observed throughout the studied sections, where TOC values reach up to 26.3 wt% (see Duarte et al., 2010). This unit is composed of organic-rich marl-limestone hemipelagic alternations with abundant benthonic and nektonic macrofauna, deposited on a northwesterly dipping, low-energy marine carbonate ramp (Fig. 3).

13.3 MATERIALS AND METHODS

This study is based on the detailed field observations and carbon and oxygen isotopic records of the Lower Jurassic reference sections of Tomar, Rabaçal, and Peniche from the Lusitanian Basin, Portugal (Figs. 4–10). In addition to the sedimentological (detailed logs, X-ray diffraction, thin section

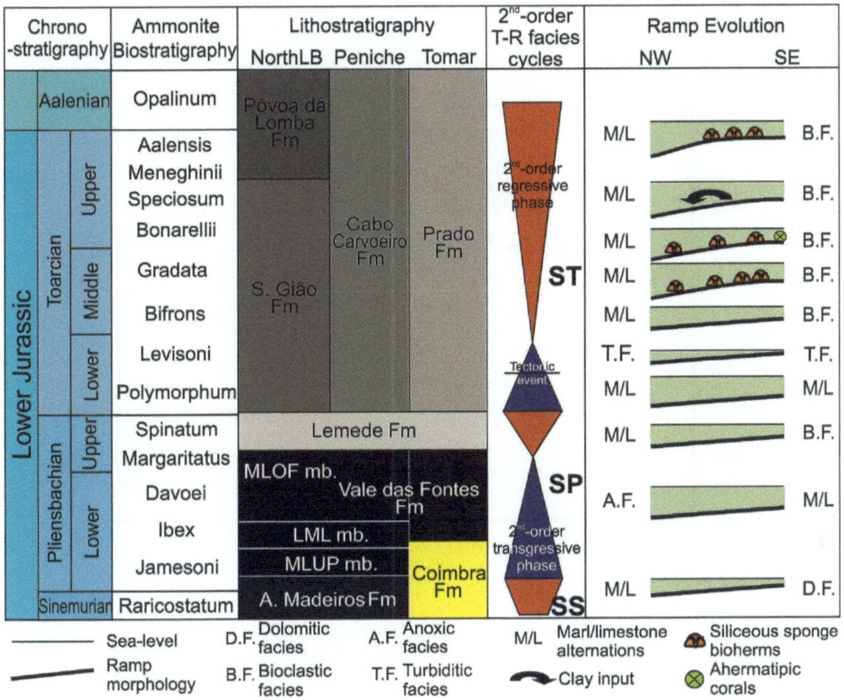

FIGURE 3

Stratigraphic chart for the Upper Sinemurian–Lower Aalenian succession in most of the Lusitanian Basin (modified from Duarte et al., 2011; Duarte, 2007). Fm-Formation; mb-member; MLOF mb-Marly limestones with organic-rich facies member; LML mb-Lumpy marls and limestones member; MLUP mb-Marls and limestones with *Uptonia* and *Pentacrinus* member. SS-Sinemurian sequence; SP-Pliensbachian sequence; ST-Toarcian sequence.

analysis, etc.) and paleontological (predominantly ammonites) analyses, $\delta^{13}C$ and $\delta^{18}O$ have been determined in 213 carbonate samples (19 at Tomar (Fig. 5), 47 at Rabaçal (Fig. 7), and 147 at Peniche (Fig. 9)) collected from the field using a battery-powered hand-drill. The data set corresponding to the Peniche section was presented in Silva et al. (2011b). All the analyses were performed using a Thermo Electron Delta V Advantage mass spectrometer with an automated carbonate preparation device (Gas-Bench III) at the Cardiff University stable isotope laboratory. Stable isotope results were calibrated to the Vienna Pee Dee Belemnite (VPDB) scale by international standard NBS19. The analytical precision is better than ±0.1‰ for both isotopic ratios. The $\delta^{13}C$ variation interval is representative of normal marine values, whereas, the low $\delta^{18}O$ values indicate some degree of diagenetic overprinting, most likely from shallow marine burial diagenesis (see Silva et al., 2011b, for a more detailed discussion). The range of variation of these two variables, within what is considered normal for this kind of data, and the lack of correlation between $\delta^{13}C$ and $\delta^{18}O$ (Fig. 4) suggest that the carbon isotopic records of the studied sections are a reflection of paleoenvironmental variations contemporaneous with sedimentation, though partially obscured by diagenesis.

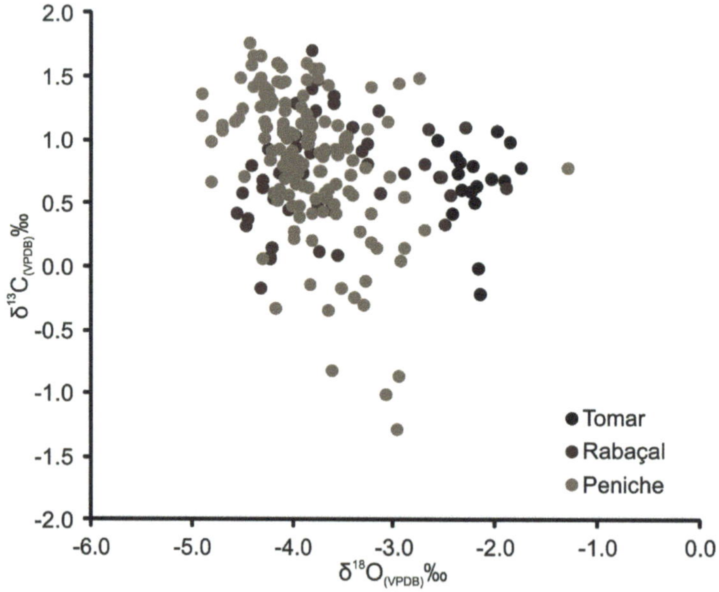

FIGURE 4

$\delta^{13}C$ versus $\delta^{18}O$ cross plot from the Tomar, Rabaçal, and Peniche sections, Lusitanian Basin (Portugal).

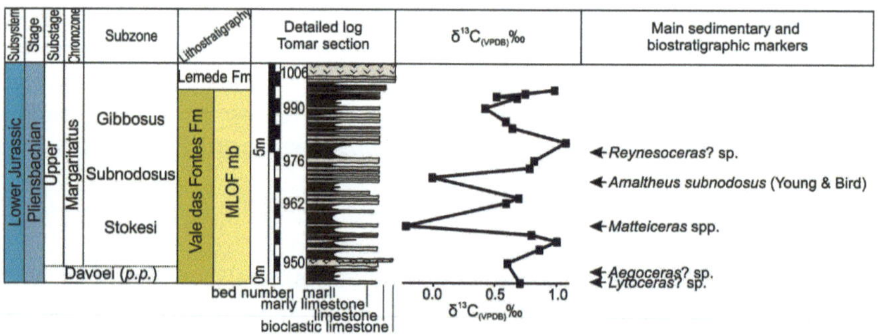

FIGURE 5

Detailed log and $\delta^{13}C$ variation of the top of Vale das Fontes and Lemede formations in the Tomar (Vale Venteiro) section, Lusitanian Basin (Portugal).

13.4 RESULTS: STUDIED SECTIONS, STRATIGRAPHIC IMPROVEMENTS, AND ISOTOPE DATA

13.4.1 TOMAR

The Tomar section is located at Vale Venteiro (UTM coordinates 29 S, 551,190 m; 4,391,144 m), around 7 km north of Tomar (Fig. 1). Mouterde et al. (1971) presented a review of this area and established the ammonite biostratigraphic scheme used in more recent works (e.g., Brunel et al., 1998; Duarte and

FIGURE 6

Field details of the Vale das Fontes and Lemede formations at Tomar (Vale Venteiro section). (A) Base of the studied section; (B) middle portion of the studied section; (C) limit between the Vale das Fontes and Lemede formation; (D) grainstone level belonging to the base of the Lemede formation.

Soares, 2002; Duarte, 2007; Duarte et al., 2010; Reggiani et al., 2010a; Suan et al., 2010). Duarte and Soares (2002), contrarily of the overall situation of the northern part of the basin, recognized only the Vale das Fontes Formation without discriminating its members.

This study analyzed the last 8 m of the Vale das Fontes Formation and the first meter of the Lemede Formation (Fig. 5). The base of the studied part of the section starts at an extremely fossiliferous and bioturbated yellowish marly limestone interval with abundant bivalves, which is overlain by a thick (~2 m) greenish marly package with a few marly limestone beds, where macrofauna is usually scarce (Fig. 6(A)). Afterward, and until the top of the Vale das Fontes Formation, the limestones beds (dominantly mud-wackestone with rare lenticular bioclastic pack-grainstone levels) dominate over the marls (Fig. 6(B)). One bed presents a morphology that resembles the carbonate buildups locally observed in proximal settings of the basin (i.e., Rabaçal). Within this interval, benthic macrofauna (gastropods, bivalves, brachiopods, echinoderms, etc.) and belemnites are abundant. Ammonites are scarce and usually poorly preserved. The base of the Lemede Formation is marked by the sharp increase of the bioclastic character of the series, referred here as a thick fossiliferous pack-grainstone carbonate bed (Fig. 6(C) and (D)).

This section is poorly aged-constrained due to its general scarce ammonite record. Several poorly preserved *Aegoceras*? sp. fragments were recovered at the base of this section. It is not possible to discard that reworking could, locally, result in a younger age for the occurrence of the referred ammonites. For the sake of honesty, it is considered that the studied section encompasses part of the Upper Pliensbachian and that the base may include the Davoei Chronozone (*p.p.*). Reggiani et al. (2010a) also

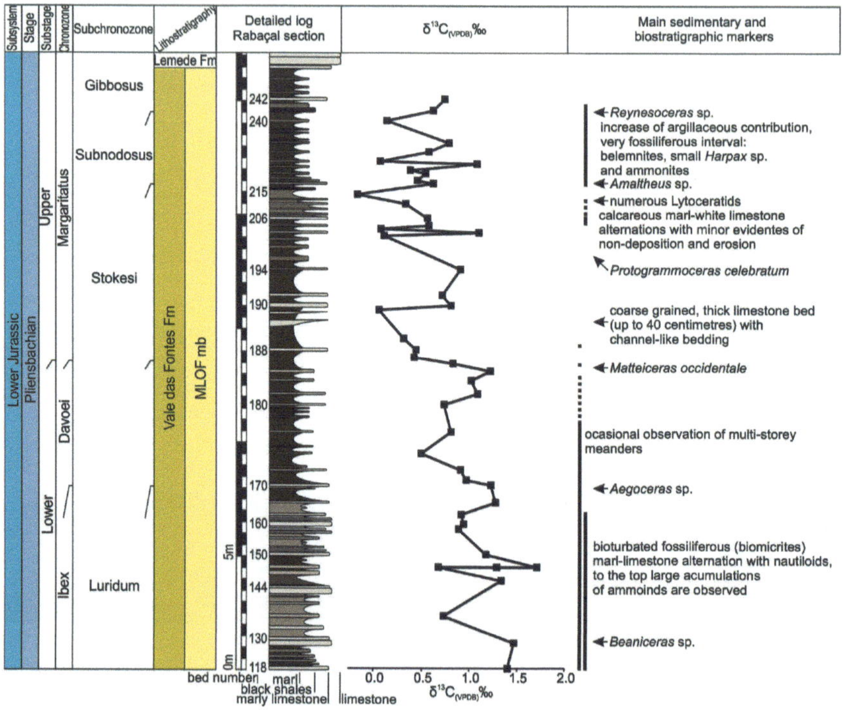

FIGURE 7

Detailed log and δ13C variation of the marly limestones with organic-rich facies member (MLOF mb) and base of Lemede Formation in the Rabaçal section, Lusitanian Basin (Portugal).

established a calcareous nannofossil biozonation, and recognized the boundary between NJ4a and NJ4b (last occurrence of *Paranthropus robustus*) close to the base of the studied part of the series.

In the Tomar section the $\delta^{13}C$ varies from −0.22 to 1.07 ‰. Although there is no apparent vertical trend, the $\delta^{13}C$ presents some variation, presenting maximum values in the Stokesi and Subnodosus subchronozones (Fig. 5).

13.4.2 RABAÇAL

The Rabaçal section (UTM coordinates 29 T, 547,477 m; 4,429,283 m) is located at about 1.8 km southeast of Rabaçal and 4.5 km south of the classical Maria Pares reference section (see, for example, Mouterde et al., 1964–65; Cubaynes et al., 1988; Duarte, 1997, 2007; Boomer et al., 1998; Dommergues et al., 2002) (Figs. 1 and 7). The studied part of this section, ranging about 27 m in thickness, encompasses almost the entire MLOF mb and the base of the Lemede Formation.

Here, the overall tendency is the decrease in the limestone content of the series until the base of the Lemede Formation. The MLOF starts with a succession (about 6.5 m) of several bioturbated fossiliferous (biomicrites) marl-limestone packages (the so-called *calcaires en rangs de pavés*, Cubaynes et al., 1988) (Fig. 8(A) and (B)), with nautiloids, gradually changing into dominant marly lithofacies

FIGURE 8

Field details of the MLOF mb (Vale das Fontes Formation) and Lemede Formation at Rabaçal. (A) Base of the studied section; (B) middle portion of the studied section; (C) upper portion of the studied section (D) limit between the Vale das Fontes and the Lemede formations as observed at the Rabaçal village (different bed numbering).

toward the Lower/Upper Pliensbachian boundary. Benthonic and nektonic macrofauna (e.g., gastropods, bivalves, brachiopods, echinoderms, belemnites, ammonoids) are abundant. Ammonoids are often pyritized. Bioturbation is also significant and occasionally includes multistorey meanders resembling *Spirorhapne* sp. (a graphoglyptid ichnogenus, see Seilacher, 2007). A thick coarse grained limestone bed (up to 40 cm) with channel-like bedding is observed at this location. Toward the top of the section, a small (approximately 1.5 m) pack of dominantly calcareous marl–white limestone alternations (Fig. 8(C)), capped by a limestone containing numerous Lytoceratids, is observed. These limestones present evidences of nondeposition periods. Stratigraphic surfaces are sharp and irregular and/or slightly ferruginous. The fauna also presents evidence of resedimentation and taphonomic reworking (see Fernández-López, 2000), such as bioerosion and incrustation (especially visible in the large ammonoids) and ferruginous coatings. Afterward, the series becomes very argillaceous and fossiliferous (whole fossils or bioclasts) (Fig. 8(C)), sometimes forming extensive pavements. This interval is remarkable by the ubiquitous presence of belemnites, small *Harpax* sp., and ammonites, the latter indicating a Subnodosus and base of Gibbosus subchronozones for this interval. Sparsely, some very small carbonate buildups are observed. The organic-rich facies are present throughout the section, but only a few levels are regarded as black shales (*sensu* Duarte et al., 2010; Silva et al., 2012). The transition to the Lemede Formation is rather sharp (Fig. 8(D)). At this location, the base of this unit is composed of thick limestones (bioclastic) with minor marls.

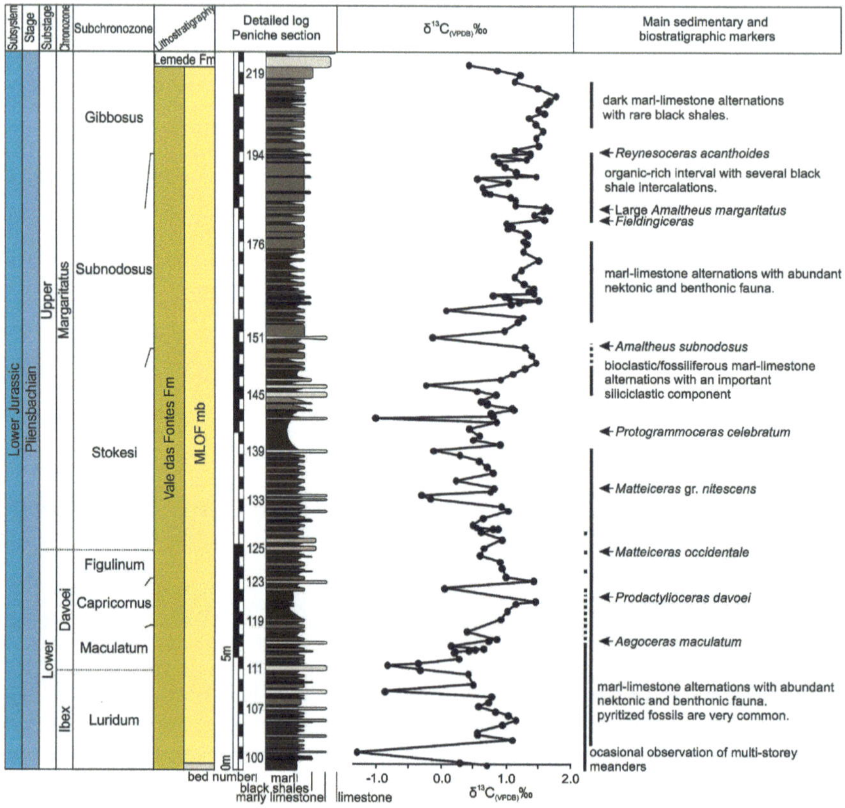

FIGURE 9

Detailed log and $\delta^{13}C$ variation of the marly limestones with organic-rich facies member (MLOF mb) and base of Lemede Formation in the Peniche reference section (modified from Silva et al., 2011b), Lusitanian Basin (Portugal).

The succession of ammonite associations allows a better biostratigraphic control when compared with Tomar (Fig. 7). The new ammonite data allow to recognize the Ibex (Luridum Subchronozone?), Davoei and Margaritatus chronozones, and some of their subchronozones, with some incertitude on boundary positions. For this interval, comparison with previously published biostratigraphic data (e.g., Mouterde et al., 1964–1965; Dommergues et al., 2002) is difficult because thickness of the MLOF mb at this location seems to be highly variable. The Margaritatus/Spinatum chronozones boundary is located within the Lemede Formation, with the first *Pleuroceras solare* (Phillips) being observed around 7 m above the base of this unit (Fonte Coberta section). This observation is in good agreement with the data presented by Mouterde et al. (1964–1965) for the Maria Pares reference section.

In the Rabaçal section, $\delta^{13}C$ varies from −0.17 to 1.71‰. A negative trend from the base of the section until the Stokesi/Subnodosus chronozones boundary (−0.17‰) is observed. Afterward, and although presenting some variation, the determined $\delta^{13}C$ tends to be more positive (Fig. 7).

FIGURE 10

Field details of the marly limestones with organic-rich facies member (MLOF mb; Vale das Fontes Formation) and Lemede Formation at Peniche. (A) *Chondrites* sp. in a organic-rich interval; (B) base of the studied section; (C) middle portion of the studied section; (D) limit between the Vale das Fontes and Lemede formations. ((A) and (C) photos by Rui Ferreira).

13.4.3 PENICHE

At Peniche, the MLOF mb crops out in the Portinho da Areia do Norte beach (UTM coördinates 29 S, 467,389 m; 4,357,730 m) (Fig. 1). Here, it is represented by approximately 30 m of marl–limestone alternations, locally with abundant nektonic (belemnites and ammonites) and benthonic (brachiopods, bivalves, crinoids, and gastropods) macrofossils (Fig. 9). This section is characterized by the pervasive occurrence of organic-rich facies, from dark marls and marly limestones, often bioturbated (mainly *Phymatoderma* and *Chondrites* ichogenera) (Fig. 10(A)), to true black shales (see Duarte and Soares, 2002; Oliveira et al., 2006; Duarte et al., 2010; Silva et al., 2011b, 2012).

The first 4.5 m of the series, corresponding to the Ibex Chronozone, are composed of regular marl–limestone alternations grading to a more irregular marl-dominated interval (around 5–6 m) dated from the Davoei Chronozone (Fig. 10(B)). The limit with the lumpy marls and limestones mb is sharp and marked by the disappearance of the lumpy facies—subspherical micritic grumose concretions, showing some cryptalgal oncolite structures (Dromart and Elmi, 1986; Elmi et al., 1988; Fernández-López et al., 2000; Duarte and Soares, 2002; Duarte et al., 2010). Organic-rich facies is abundant and usually located at the interface between marls and limestones. Pyrite is very abundant and can be found dispersed, as nodule-like forms or as fossil (including ichnofossils) moulds. The Upper Pliensbachian is characterized by more regular marl–limestone alternations, persisting until the limit with the Lemede

Formation. Although rarely, it is possible to observe small lenticular coarse bioclastic beds (this feature is observed throughout the section, but it seems clearer in this interval). Despite the regularity of the marl–limestone couplets and black shales occurrences, a minor variation in the lithofacies is observed, approximately at 17 m thickness mark. This particular variation corresponds to two well-defined black shale layers, passing upward to a more calcareous interval, composed of three mud-wackestone limestone beds. This interval corresponds to the Margaritatus Chronozone (Stokesi Subchronozone) (Fig. 10(C)). The marly interval is very rich in bioclasts and contains a significant terrigenous and coarser bioclastic fraction, being dated from the Margaritatus Chronozone, uppermost Stokesi—base of Subnodosus (marked by the first occurrence of *Amaltheus subnodosus* (Young and Bird)) subchronozones. At the 25 m mark, the development of a thick limestone–marl–black shale package is observed. The MLOF mb is sharply overlain by the Lemede Formation, which is composed of centimeter marl–decimeter limestone bioturbated alternations, very rich in belemnites, ammonoids, bivalves, and brachiopods (Fig. 10(D)).

The Lower Jurassic of Peniche is paleontologically very rich, both in nektonic (ammonites and belemnites) and benthonic (brachiopods, bivalves, gastropods, crinoids, echinoids, and rare coral debris, etc.) macrofauna. The ammonites occurring throughout the Upper Sinemurian–Middle Toarcian (Gradata Chronozone) (Mouterde, 1955; Phelps, 1985; Dommergues, 1987), allow a detailed biostratigraphic zonation (e.g., Fig. 10(B)), reinforced by additional new data from recent ammonite collection campaigns (see, for example, Silva et al., 2011b) and also by a comprehensive calcareous nannofossil biostratigraphic chart for the Pliensbachian–Toarcian (Perilli and Duarte, 2006; Oliveira et al., 2007; Reggiani et al., 2010a,b; Mattioli et al., 2013). Newly collected bed-by-bed ammonite data are, at a broader scale, in good agreement with the biozonation charts established for the Pliensbachian of the Lusitanian Basin (e.g., Phelps, 1985; Dommergues et al., 1997). These data also allowed improving of some biostratigraphic boundaries. The main improvement is related with the age of the Lemede Formation, which was previously assigned to the Spinatum Chronozone (e.g., Mouterde, 1955; Duarte and Soares, 2002). The new high-resolution biostratigraphic data revealed that the base of this formation corresponds to the Margaritatus Chronozone, uppermost Gibbosus Subchronozone, with a distinct level of *Arieticeras* gr. *algovianum* (Oppel) found approximately around 3 m above the base of this unit. This probably corresponds to the same level of *Arieticeras* sp. mentioned in Mouterde et al. (2007) for the Brenha (Figueira da Foz) section, nowadays not assessable. Mouterde and Rocha (1988) had already suggested that the Margaritatus–Spinatum chronozones boundary was probably located within the Lemede Formation while acknowledging the highly variable nature of this biostratigraphic boundary and its difficult recognition due to the lack of data from the base of the unit.

The $\delta^{13}C$ in the carbonates of the Peniche section varies between −1.29 and 1.76‰ (Silva et al., 2011b) (Fig. 9). The $\delta^{13}C$ evolves from about 0 to 1‰ at the base of the section and then decreases toward the Ibex/Davoei chronozones boundary, reaching −0.86‰. Upward, $\delta^{13}C$ values increase until reaching a maximum of 1.44‰ in the middle part of the Davoei Chronozone. The upper Davoei–lowermost Margaritatus (Stokesi Subchronozone) isotopic record is characterized by a slight decrease of about 0.6‰. Starting at the uppermost Stokesi Subchronozone, an increasing trend is recorded upward, where $\delta^{13}C$ values reach a maximum of 1.66‰ at the start of the Subnodosus Subchronozone. This trend is interrupted by a drop of around 1‰ just prior to the Subnodosus/Gibbosus organic-rich interval. The MLOF mb and Lemede Formation boundary is marked by a sharp decrease of about 1.6‰, and then by the return to more positive values, about 1.3‰.

13.5 DEPOSITIONAL ENVIRONMENT

Based on sedimentological criteria, it is possible to distinguish three sedimentation domains in the Pliensbachian of Lusitanian Basin, related to the different environmental subdivisions predicted for carbonate ramp systems (see Burchette and Wright, 1992 and references therein) (Fig. 11). Westward, corresponding to Peniche sector (including also S. Pedro de Moel and Figueira da Foz sections) the main feature of the MLOF mb is its richness in organic matter (e.g., Duarte and Soares, 2002). Ammonites and belemnites are abundant and benthic macrofauna, mainly brachiopods, bivalves, and echinoderms occur in this domain and correspond to outer ramp/basin environments. Paleobathimetric estimates based in microfossil data suggest that water column thickness averaged around 100–200 m (e.g., N'zaba-Makaya et al., 2003; Reggiani et al., 2010a). The central-eastern domain, at Rabaçal, is distinguished by the significant occurrence of nautiloids (macro- and, sometimes, microshells), the widespread presence of usually small mud-wacke-packstone carbonate buildups (associated with nautiloids at the base or with large Lytoceratids higher in the section) and occurrence of bones of larger animals (ichthyosaurs?). Locally, rocks of organic-rich facies are observed, although these are not expressive when compared with the western domain. This sector corresponds to outer/mid-ramp environments. The Tomar sector, in the south-eastern domain, represents the shallowest of the studied environments. Ostracod data from N'zaba-Makaya et al. (2003) suggest that here the minimum thickness of the water was more than 50 m. It contrasts with the other two domains by the lack of organic-rich facies, the increasing limestone and bioclastic character of the series and richness in benthic macrofauna (brachiopods and bivalves). This corresponds mainly to mid-ramp environments. Similar environment was assigned to the Pliensbachian series exposed in outcrops of the Arrábida mountain range, in the south of the Lusitanian Basin (Manuppella and Azerêdo, 1996; Azerêdo et al., 2003).

From all these observations and depositional environmental inferences, it is clear that the Lower–Upper Pliensbachian series are incomplete, since the most proximal environments are not observable

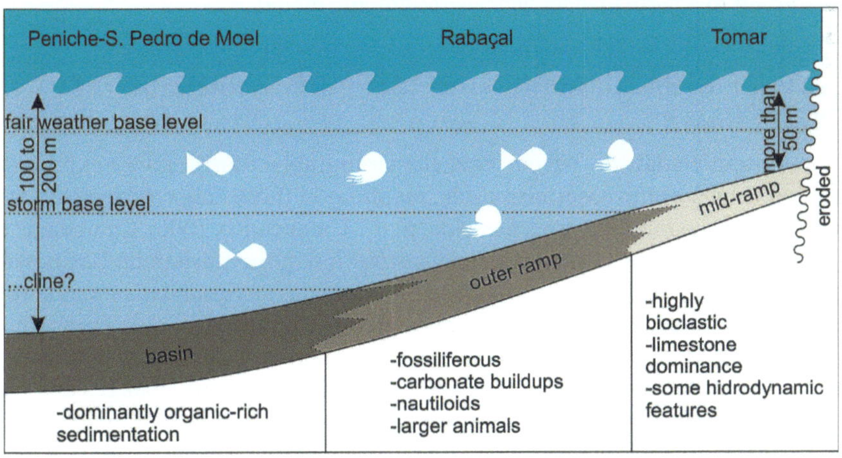

FIGURE 11

Schematic diagram showing inferred paleoenvironmental settings (homoclinal ramp) during deposition of the studied Pliensbachian carbonate successions.

cropping out anywhere (Fig. 11). All the deposits from the proximal environments located eastward of the nowadays limit of the Lusitanian Basin have been eroded during the evolution of the Iberian Massif (Porto-Tomar fault, e.g., Wilson, 1988; Azerêdo et al., 2003 and references therein). The Lower Jurassic outcrops of the western Iberian margin (Fig. 1), corresponding to the Arrábida mountain range (southernmost outcropping limit of the Lusitanian Basin) and Santiago do Cacém (Santiago do Cacém subbasin *sensu* Azerêdo et al., 2003 and references therein) or Alentejo Basin (see also Pereira and Alves, 2012), present a marked difference in facies development. The former shows a tendency to deepen northward (e.g., Azerêdo et al., 2003), while the latter records an opposite deepening trend, from north to south (Inverno et al., 1993; Stapel et al., 1996). This evidence suggests the existence of a structural high between the Lusitanian Basin and the Santiago do Cacém subbasin or Alentejo Basin (e.g., Stapel et al., 1996).

13.6 SEQUENCE STRATIGRAPHY: THIRD-ORDER SEQUENCES AND CARBON ISOTOPE STRATIGRAPHY

A second-order T-R facies cycle scheme for the Sinemurian–Toarcian interval has been already established for the Lusitanian Basin (see Duarte et al., 2004, 2010, Fig. 3) but, although fourth- to fifth-order cycles have also been defined for a part of the Pliensbachian series (Fernández-López et al., 2000), the recognition third-order sequences (*sensu* Jacquin and de Graciansky et al., 1998) remained elusive up to now.

In order to apply the sequence stratigraphy to the studied series, the conceptual standard workflow of Catuneanu (2006) and Catuneanu et al. (2009, 2010, 2011) was followed, starting with the definition of a type section, recognition of sequence stratigraphic surfaces and identification of genetic units (systems tracts). These three steps constituted the backbone of the model-dependent analysis, in which a genetic significance is assigned to sequence stratigraphic surfaces, and changing them to sequence boundaries, according to a particular sequence stratigraphic model.

13.6.1 SEQUENCE STRATIGRAPHIC SURFACES AND STRATIGRAPHIC GENETIC UNITS

In sequence stratigraphic analysis, seven sequence stratigraphic surfaces are expected to occur in relation to the four events of the base-level cycle, resulting in three major systems tracts, i.e., forced regression, normal regression, and transgression (e.g., Catuneanu, 2006; Catuneanu et al., 2009, 2010, 2011). The specificity of the Meso-Cenozoic evolution of the Lusitanian Basin (which resulted in the absence of proximal marine, transitional and continental environments records for the Pliensbachian interval), coupled with the fact that only outcrop data are available (hence no large-scale stratal terminations or geometric features can be inferred), makes it difficult to confidently assign the observed sedimentological features to base-level or shoreline trajectory changes and therefore, to a particular shoreline-related system tracts (*sensu* Catuneanu et al., 2011). This is also problematic because offshore depositional trends may uncouple from shoreline environments (e.g., Bruchette and Wright, 1992; Tucker et al., 1993; Einsele, 2000; Coe, 2003). Thus, and following the recommendation of Catuneanu et al. (2009), this work was based on the definition and observation of two unconventional systems tracts (this nomenclature is preferred over the shoreline independent systems tracts presented in Catuneanu et al. (2011)) and two sequence stratigraphic surfaces, made up by the arrangement of main discernible architectural elements of the studied series—the thickness of limestone and marl beds. The unconventional systems tracts are named decreasing limestone bed thickness (DLBT) and increasing limestone bed thickness (ILBT) systems tracts and the bounding sequence stratigraphic surfaces are defined as trend inflection surfaces (TIS) type 1(at the base of ILBT) or type 2 (at the base of DLBT) (Fig. 12). Hierarchically, these unconventional sequences correspond to third-order sequences, contained in the second-order Pliensbachian T-R facies cycle (see Duarte et al., 2010).

The definition of these unconventional systems tracts and sequence stratigraphic surfaces (basically outcrop bedding rhythms) was based on the assumption that the weathering profile reflects a variation of carbonate content due to production, dilution, and early diagenesis (see Einsele et al., 1991; Einsele, 2000, and references therein for more details), and that the carbonate content threshold that marks the weathering boundary in any section is similar at outcrop scale, although it may vary between locations.

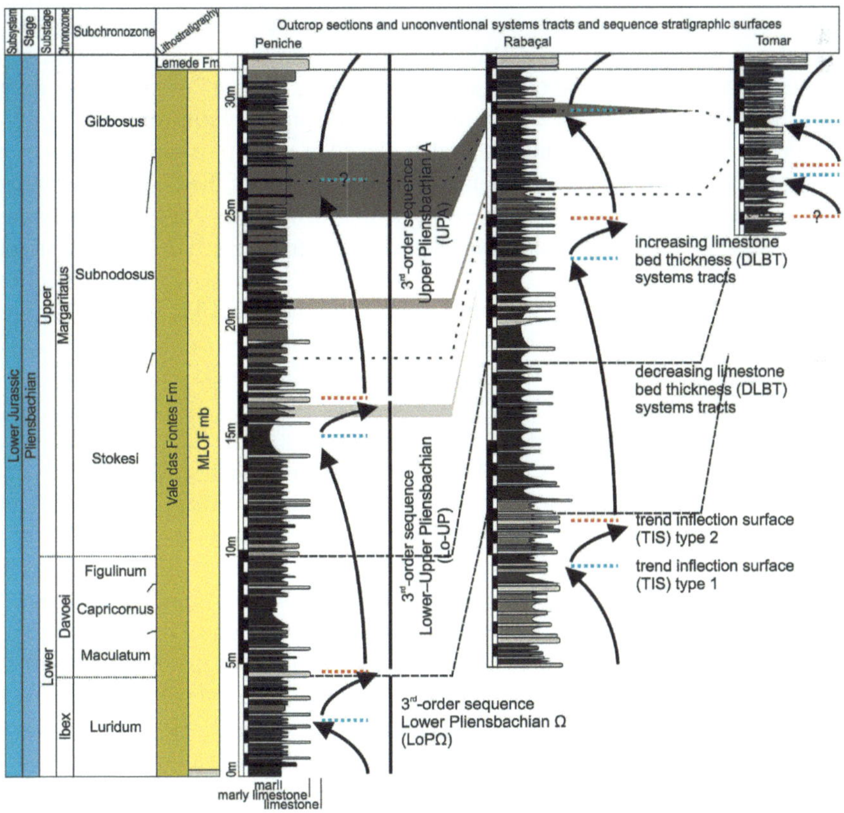

FIGURE 12

High-resolution correlation and inferred unconventional systems tracts and sequence stratigraphic surfaces in S. Pedro de Moel, Peniche, Rabaçal and Tomar sections.

13.6.2 DEFINITION OF A SEQUENCE STRATIGRAPHY TYPE SECTION AND THIRD-ORDER SEQUENCES

The sequence stratigraphy type section corresponds to the Rabaçal section. Three third-order unconventional sequences were recognized in this section and correlated throughout the basin using the available biostratigraphic data and black shales occurrences. The three sequences correspond to the third-order sequence Lower Pliensbachian Ω (LoPΩ), the third-order sequence Lower–Upper Pliensbachian (Lo-UP) and the third-order sequence Upper Pliensbachian A (UPA) (Fig. 12).

13.6.2.1 Third-Order Sequence LoPΩ

The LoPΩ sequence corresponds to the initial part of the MLOF mb. Its base is not defined. In the Rabaçal type section, the recognized part of this sequence is about 6.5 m thick (Fig. 8(A)). The DLBT systems tract (about 4.5 m) is characterized by an important carbonate (micrite and bioclastic) contribution and sporadic occurrence of black shales. The series has a clear pattern of decreasing limestone contribution, with the marl-limestone couplets usually constituting 5/4:1 bundles. The top of the system tracts is placed within an organic-rich argillaceous interval, corresponding to the TIS type 1. Laterally, this unit presents some variation, with the black shales being more expressive at S. Pedro de Moel. The ILBT systems tract is materialized in the Rabaçal type section by a clear increment in the number, thickness, and bioclastic character of the limestone beds. The TIS type 2 corresponds roughly to Ibex/Davoei chronozones boundary interval and to the top of the thickest limestone bed of this interval. At Peniche and S. Pedro de Moel, this boundary is also placed at the top of the thickest limestone bed in this interval (Fig. 10(B)) whose base corresponds to the lower part of the Davoei Chronozone (Phelps, 1985). The carbon isotopic record shows an increase in $\delta^{13}C$ toward the TIS type 1, reaching around 1.5 and 1.0‰ at Rabaçal and Peniche respectively. The development of the ILBT systems tract is followed by a decrease in $\delta^{13}C$.

The recognized portion of the LoPΩ sequence spans from the Ibex Chronozone (Luridum Subchronozone)—extreme base of the Davoei Chronozone (Phelps, 1985). According to the 2012 geological time scale (Gradstein et al., 2012), the LoPΩ is at least 170k years long.

13.6.2.2 Third-Order Sequence Lo-UP

Of all the sequences defined in this work, the Lo-UP sequence (with about 14 m at Rabaçal) is the best constrained. At Rabaçal, the DLBT systems tract rests over sequence LoPΩ (diastem?) and its base is very similar throughout the basin. The base of this unit corresponds to a major influx of argillaceous materials and high accumulation of shells (e.g., *Aegoceras* sp.), some evidencing minor transport and exposure features. The good basinal development of this sequence, the black shales and the biostratigraphic data allow correlating this sequence at bed scale. The mid-part of the series, apart from a very minor variation in siliciclastic and bioclast contents, is very monotonous, hampering the recognition of staking patterns or sequence boundaries and the clear definition of the TIS type 1. At Rabaçal, the top of this sequence, corresponding to the upper part of the ILBT systems tract, is well expressed in the terrain morphology. The monotonous marly sedimentation gradually gains a more calcareous, fossiliferous, and bioclastic character. Here, evidences of prolonged exposure (e.g., ferruginous surfaces), high bioturbation, and shell accumulations are observed (see previous section). The TIS type 2 is placed within this calcareous package, bed 206 of the Rabaçal section (Fig. 8(C)). The $\delta^{13}C$ record of this sequence presents a higher variability than the previous one. In all the sections, the initiation of the ILBT systems tract is mirrored by a sharp increase in $\delta^{13}C$ in the base and middle portion of the section belonging to the Davoei Chronozone. During the middle/upper part of the Davoei Chronozone, $\delta^{13}C$ presents a decreasing trend, although the ILBT systems tract tends to present somewhat higher values.

The ammonite data indicates an age ranging from the extreme base of the Davoei Chronozone to the Margaritatus Chronozone, Stokesi Subchronozone. The Lo-UP is about 1900k years long (Gradstein et al., 2012). Collectively, the faunas indicate a more or less permanent opening toward north boreal domains, although some Tethyan influence is sporadically recognized (Dommergues et al., 2009).

13.6.2.3 Third-Order Sequence UPA

The UPA sequence corresponds to the top of the MLOF mb and base of the Lemede Formation (Fig. 8(C) and (D)). The top of the unit is not defined in this work, but is considered to be located within the Lemede Formation. At Rabaçal, the DLBT systems tract corresponds to the sediments deposited until the thickest black shale interval (see previous subsections), which represents the maximum flooding interval of the Pliensbachian second-order T-R facies cycle. The TIS type 1 is placed in this interval, however not in a definitive position. The base of the DLBT systems tract, as in the previous ones, is marked by a major influx of terrigenous materials and a high accumulation of fossils (see previous section). The ILBT systems tract corresponds to the return of limestone dominated sedimentation over the whole of the basin, which is materialized, in part, by the Lemede Formation. At its base, the $\delta^{13}C$ record of this sequence shows similarity with the pattern of the Lo-UP. The initiation of the DLBT systems tract is followed by an abrupt positive trend in $\delta^{13}C$. This positive trend is interrupted by a sharp negative excursion at the limit between the Subnodosus and Gibbosus chronozones, observable basinwide (see Silva et al., 2011b for a detailed explanation). The ILBT systems tract $\delta^{13}C$ is characterized by the continuation of the positive trend. At Peniche, the limit between the MLOF mb and the Lemede Formation is marked by a sharp negative excursion.

The UPA is more than 1500k years long (Gradstein et al., 2012).

13.6.3 MODEL-DEPENDENT ANALYSIS

In the definition of the second-order T-R facies cycle for the Lusitanian Basin, Duarte et al. (2010) adopted the model proposed by Jacquin and de Graciansky (1998). According to the later authors, the T-R facies cycles are composed of four types of third-order depositional sequences: infilling and forestepping during regressive phases and aggrading and backstepping during transgressive phases. These units are defined by the shelfal accommodation changes and bounded by unconformities or correlative conformities induced by third- or fourth-order relative sea-level falls (Mitchum, 1977; Van Wagoner et al., 1987). However, and as in the current case study, such approach is limited by the difficult recognition of correlative conformities based only in outcrop data (see, for example, Embry, 2009). Much debate has been generated around this concept over the recent years (see Catuneanu, 2006; Catuneanu et al., 2009, 2010, 2011) and the current agreement is that the correlative conformities can only be recognized by the integration of seismic and outcrop data (e.g., Catuneanu et al., 2009). The fact that correlative conformities cannot be confidentially defined inhibits the depositional (Mitchum et al., 1977 and consequent derivations) and genetic (Frazier, 1974; Galloway, 1989) sequence models from being used.

In a very simplistic way, the DLBT and the ILBT can be regarded as corresponding to fining- and coarsening-upward successions, respectively. Using such reasoning, the DLBT represents a transgressive unit and ILBT a regressive unit. The reasoning is that a coarsening-upward succession (increasing limestone and bioclastic character) reflects a shift of the current location to higher energy settings, subjected to shallowing and progradation, while fining-upward successions indicate a decline in the environmental energy, similarly to a deepening, retrogradational event, i.e., Walter law of facies (e.g., Posamentier and Allen, 1999; Einsele, 2000; Coe, 2003; Schlager, 2005). This way, TIS type 1 can be regarded as the maximum flooding surface and the TIS type 2 to the maximum regressive surface/transgressive surface (see Catuneanu et al., 2011). This simple division is in good agreement with observed sedimentological data, where the TIS type 1 is associated with the finest and organic-rich sedimentation and the TIS type 2 is associated with shallow-water exposure features and the earliest signs of transgression (Fig. 12). Based on the aforementioned, the T-R sequence model (Johnson and Murphy, 1984; Embry and Johannessen, 1992; Embry, 1993, 2009) is a good fit to a model-dependent analysis of the unconventional system tracts, where the DLBT is equivalent to the transgressive systems tract and the ILBT to the regressive systems tract (Fig. 13). These inferences are in good agreement with the foraminifera (Rey et al., 2000), ostracods (N'zaba-Makaya et al., 2003), and gamma-ray outcrop data (Correia et al., 2012).

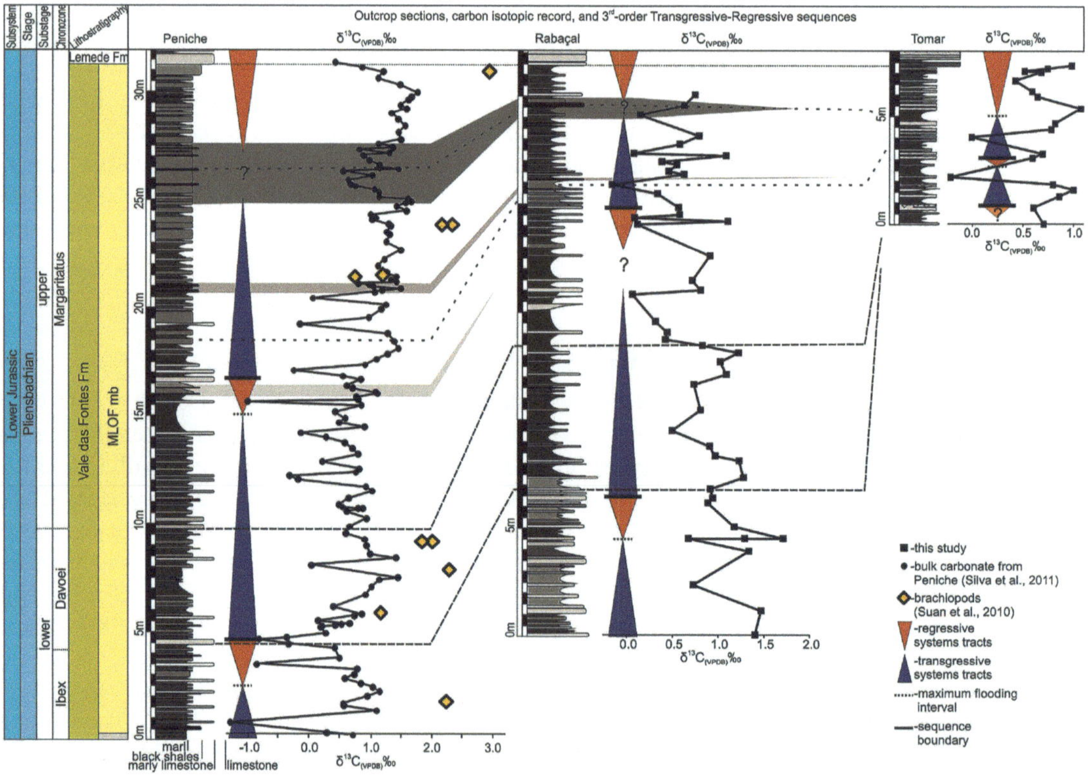

FIGURE 13

High-resolution correlation, carbon isotopic record, and Transgressive–Regressive sequences in S. Pedro de Moel, Peniche, Rabaçal, and Tomar sections.

13.6.4 CARBON ISOTOPE STRATIGRAPHY AND RELATION WITH SECOND- AND THIRD-ORDER SEQUENCE STRATIGRAPHY

It is recognized that $\delta^{13}C$ excursions can be caused by changes in carbon cycling associated with rapid T-R cycles and increased burial of organic carbon in transgressive deposits (e.g., Schlanger and Jenkyns, 1976; Jenkyns and Clayton, 1986; Wefer and Berger, 1991; Marshall, 1992; Patterson and Walter, 1994; Mitchell et al., 1996; Holmden et al., 1998; Weissert et al., 1998; Kump and Arthur, 1999; Hesselbo et al., 2000; Stephens and Summer, 2003; Immenhauser et al., 2003; Swart and Eberli, 2005; Panchuk et al., 2006; Bergström et al., 2010; Sakuma et al., 2012; Huck et al., 2013). The main factors and processes that affect the $^{13}C/^{12}C$ ratio are the equilibrium between oceanic dissolved inorganic carbon and atmospheric CO_2, photosynthesis, and respiration/remineralisation/preservation of organic matter. Burial of isotopically light organic carbon (consequence of the photosynthesis fractionation effect) generates a relative enrichment in ^{13}C of the inorganic carbon dissolved in seawater and therefore an increase in the $\delta^{13}C$ of marine carbonates (e.g., Mitchell et al., 1996; Newton and Bottrell, 2007; Emerson and Hedges, 2008; Silva et al., 2011b). Rosales et al. (2006) suggested a link between second-order relative sea-level changes in the Basque–Cantabrian basin and variations in seawater geochemistry during the Lower Jurassic. The authors noted a coincidence between transgressions and increasing $\delta^{13}C$, TOC and Hydrogen Index (HI) (and vice versa for regressions). The secular $\delta^{13}C$ trend of the Lusitanian Basin is concordant with the Pliensbachian T-R facies cycle, where it is observed with an overall positive excursion in carbon isotopic record (either in rock or brachiopods) with transgression, reaching its maximum just above the maximum flooding interval and enhanced organic matter preservation (Oliveira et al., 2006; Duarte et al., 2010; Suan et al., 2010; Silva et al., 2011b) (Figs. 3 and 13).

The transient positive excursions observed at the onset of transgression (DLBT system tract) most likely resulted from the influx of oceanic waters and opening of the environmental conditions. Immenhauser et al. (2003) suggested that enhanced circulation of platform-top water masses and reduced influence of meteoric diagenesis during high-amplitude transgressive events would result in higher (more marine) $\delta^{13}C$ relatively to the overall lower background. During the deposition of the Lo-UP sequence, and after the initial transient positive $\delta^{13}C$ trend, spatial increase in productivity areas rapidly led to seawater "ageing" (Holmden et al., 1998), where oxidation of organic matter to CO_2 (which then forms dominantly bicarbonate in the seawater) resulted in carbonate materials with low $\delta^{13}C$ values. This interval is relatively leaner in organic matter when compared with the equivalent system tracts of the other sequences. A sharp $\delta^{13}C$ negative excursion is observed just before the Pliensbachian T-R facies cycle (second-order) maximum flooding interval, traceable through the entire basin (Fig. 13). Silva et al. (2011) suggested that the drop in marine $\delta^{13}C$ resulted from the release into the marine reservoir of the ^{12}C previously accumulated in the sea-floor sediments during a period of enhanced preservation of organic matter, probably due to the stratification (thermal?) of the water column. Overturn of the nutrient enriched bottom waters may have become periodic (alternating with intervals of enhanced organic matter preservation), resulting in a greater nutrient availability in the upper part of the water column, as indicated by the high abundance of placoliths (Reggiani et al., 2010a,b and references therein).

13.6.5 COMPARISON WITH THE BOREAL AND TETHYAN DOMAINS

Regardless of the nomenclature, if all the available sequence stratigraphic data are "normalized" for the sequence stratigraphic scheme used for the Lusitanian Basin (i.e., T-R sequence model, Jonhson and Murphy, 1984; Embry and Johannessen, 1992; Embry, 1993, 2009), it seems clear that several "events" can be traced at a European scale (Fig. 14), as already evidenced by de Graciansky et al. (1998). At a greater detail, it is seen that the Early/Late Pliensbachian boundary marks a major change in the sequence stratigraphic architecture of the several European basins. Following the trend evidenced by the paleobiogeographical data (see, for example, Dommergues et al., 2009; Dera et al., 2010), the Upper Pliensbachian series of several locations are much more similar, in terms of sequence stratigraphy, than the Lower Pliensbachian ones (e.g., Hardenbol et al., 1998; Jacquin and de Graciansky, 1998; Quesada et al., 2005; Hesselbo, 2008). During the Late Pliensbachian, it is possible to infer multiple flooding events that can be traced at a regional scale (Fig. 14). The main events recorded in this study for the Lusitanian Basin correspond to the upper Stokesi Subchronozone, Subnodosus Subchronozone, and Subnodosus/Gibbosus subchronozones boundary. The upper Stokesi Subchronozone maximum flooding interval can be traced over most of the Tethyan and Boreal domains, marking a phase of extensive interconnection of the European epicontinental seas, well expressed in the northward expansion of Tethyan faunas, inclusively *Protogrammoceras celebratum*, the first Tethyan/Boreal common horizon index ammonite from the Pliensbachian ammonite biozonation (see Dommergues et al., 1997). The Subnodosus Subchronozone flooding event is subtly recorded in the Lusitanian Basin as a fourth- to fifth-order sequence, although concomitant with deposition of black shales. However, in de Graciansky et al. (1998) this is a major third-order maximum flooding interval. The Subnodosus/Gibbosus subchronozones boundary, although corresponding to a second-order maximum flooding interval of the Pliensbachian T-R facies cycle in the Lusitanian Basin, is poorly recorded elsewhere. This event marks the rapid switch from a northward expansion of southern ammonite faunas (notice that *Reynesoceras ragazzoni* is the subchronozone index ammonite for both Tethyan and Boreal domains (Dommergues et al., 1997)) to a domination of boreal ammonites in the Lusitanian Basin record (see references above). The short-lived northward expansion of southern ammonite faunas is traceable up to SW Germany (Schweigert, 2005). With the exception of the Cleveland Basin (Hesselbo, 2008), it seems that this

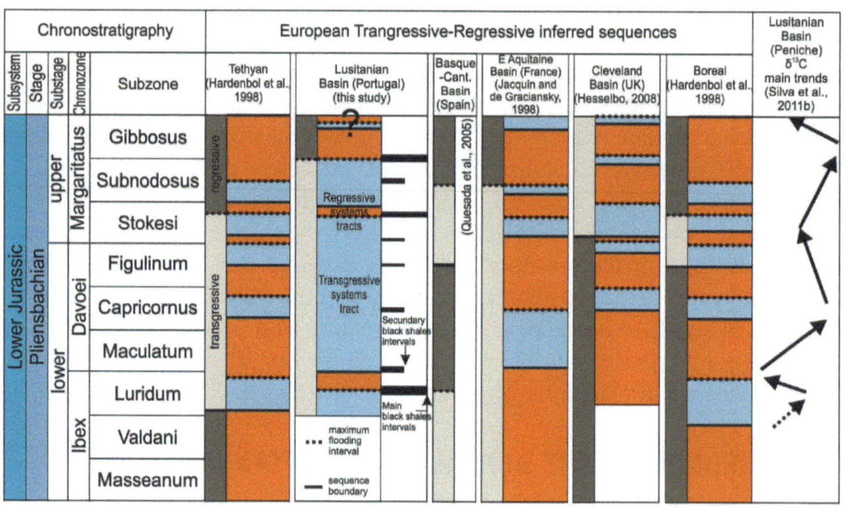

FIGURE 14

Comparison between the different third-order Transgressive–Regressive sequences of several European locations and carbon isotopic record from the Lusitanian Basin (Peniche section, Portugal).

flooding interval does not have a high-order counterpart. In the Lusitanian, Basque–Cantabrian and Asturian basins, these intervals correspond to the deposition of organic-rich facies (e.g., Borrego et al., 1996; Queseda et al., 1997; Duarte et al., 2010; Silva et al., 2011b, 2012 and references therein).

Overall, the major trend of the $\delta^{13}C$ record in the Peniche section is concordant with those in time-equivalent series from other locations around the world (see also Jenkyns et al., 2002), namely from southern Switzerland and Italy (Jenkyns and Clayton, 1986; Marino and Santantonio, 2010; Morettini et al., 2002), United Kingdom (Bailey et al., 2003; Katz et al., 2005; McArthur et al., 2000; van de Schootbrugge et al., 2005), Spain (Rosales et al., 2001, 2004a,b, 2006), Germany (Bailey et al., 2003), France (Dera et al., 2009; van de Schootbrugge et al., 2010), and Argentina (Valencio et al., 2005), implying that the key regional paleoenvironmental constrains governed the sedimentation and the carbon cycle. It seems clear that OM productivity and preservation were enhanced during transgressive episodes (Hallam, 1981), affecting the carbon cycle (e.g., Schlanger and Jenkyns, 1976; Jenkyns, 1988; Suan et al., 2010). Organic-rich intervals are known from the western part of Europe, namely Portugal (Duarte et al., 2010), Northern Spain (e.g., Borrego et al., 1996; Quesada et al., 1997, 2005; Rosales et al., 2006; Comas-Rengifo and Goy, 2010), France (Bessereau et al., 1995; Léonide et al., 2007; van de Schootbrugge et al., 2010) and Austria (Kodina et al., 1988), and also from Siberia, north-eastern Russia, Canadian Arctic and Arctic platform of Alaska (see Nikitenko and Mickey, 2004). Other (upper?) poorly age constrained Pliensbachian organic-rich deposits are known elsewhere, from several North European locations (Fleet et al., 1987), Canada (western coast, Riediger, 2002; Asgar-Deen et al., 2003) to Australia (He, 2002).

13.7 CONCLUSIONS

The main results of this paper are summarized herein:

New, detailed sedimentological, biostratigraphic, and carbon and oxygen stable isotopes data from the main Pliensbachian outcrops of the Lusitanian Basin enabled the high-resolution correlation of the studied series and the definition of a third-order sequence stratigraphic scheme.

In the MLOF mb of the Vale das Fontes Formation three unconventional sequences were defined for the Pliensbachian time interval (upper Ibex–Margaritatus zones). The unconventional systems tracts are named DLBT and ILBT systems tracts and the bounding sequence stratigraphic surfaces are defined as TIS type 1 (at the base of ILBT) or type 2 (at the base of DLBT).

These sequences correspond to the third-order sequence LoPΩ, the third-order sequence Lo-UP and the third-order sequence UPA.

The secular $\delta^{13}C$ trend of the Lusitanian Basin is concordant with the (second-order) Pliensbachian T-R facies cycle, where it is observed with an overall increase in carbon isotopic record with transgression, reaching its maximum just above the maximum flooding interval and enhanced organic matter preservation. When compared with the defined third-order sequences, a transient $\delta^{13}C$ increase during the onset of transgression is observed. Higher frequency variations in the carbon isotopic record can possibly be traced along the offshore studied sections and are a reflection of local controls (which may eventually outpace regional variation).

The newly defined sequences in the Lusitanian Basin for the MLOF mb seem to be linked to, or at least reflected in, several neighboring basins.

This paper suggests that the Pliensbachian European epeiric seas share a more similar story than previously admitted.

ACKNOWLEDGMENTS

Ricardo L. Silva thanks the FCT-MCTES for funding through a PhD fellowship (SFRH/BD/47362/2008), ExxonMobil Canada Properties (on behalf of the Sable Offshore Energy Project) for financial support through the Petroleum Geoscience Research Grant to Dalhousie University (Basin and Reservoir Lab), and the Industry partners of the Consortium *Source Rock and Geochemistry of the Central Atlantic Margins*. We thank Ana C. Azerêdo and Grant Wach for the insightful discussions about several of the studied topics, Selma Lima for her valuable help in the field work, and Julia Becker from the School of Earth & Ocean Sciences (Cardiff University) for technical support in the carbon and oxygen isotopes analysis. We also acknowledge Muthuvairavasamy Ramkumar (editor) and anonymous referees for the very helpful comments on an earlier version of this manuscript. This work has been partially supported by project PTDC/CTE-GIX/098968/2008—"High resolution stratigraphy of the Lower Jurassic organic-rich marine series in the Lusitanian Basin" (FCT-Portugal and COMPETE-FEDER).

REFERENCES

Allan, P.A., Allan, J.R., 2005. Basin Analysis. Principles and Applications, second ed. Blackwell Publishing.

Alves, T.M., Gawthorpe, R.L., Hunt, D.W., Monteiro, J.H., 2002. Jurassic tectono-sedimentary evolution of the Northern Lusitanian Basin (offshore Portugal). Marine and Petroleum Geology 19, 727–754.

Alves, T.M., Manuppella, G., Gawthorpe, R.L., Hunt, D.W., Monteiro, J.H., 2003. The depositional evolution of diapir- and fault-bounded rift basins: examples from the Lusitanian Basin of West Iberia. Sedimentary Geology 162, 273–303.

Alves, T.M., Moita, C., Cunha, T., Ullnaess, M., Myklebust, R., Monteiro, J.H., Manuppella, G., 2009. Diachronous evolution of Late Jurassic–Cretaceous continental rifting in the northeast Atlantic (west Iberian margin). Tectonics 28, TC4003. http://dx.doi.org/10.1029/2008TC002337.

Alves, T.M., Moita, C., Sandnes, F., Cunha, T., Monteiro, J.H., Pinheiro, L.M., 2006. Mesozoic–Cenozoic evolution of North Atlantic continental slope basins: the Peniche basin, western Iberian margin. AAPG Bulletin 90, 31–60.

Asgar-Deen, M., Hall, R., Craig, J., Riediger, C., 2003. New biostratigraphic data from the Lower Jurassic Fernie Formation in the subsurface of west-central Alberta and their stratigraphic implications. Canadian Journal of Earth Sciences 40, 45–63.

Azerêdo, A.C., 2007. Formalização da litostratigrafia do Jurássico Inferior e Médio do Maciço Calcário Estremenho (Bacia Lusitânica). Comunicações Geológicas 94, 29–51.

Azerêdo, A.C., Duarte, L.V., Henriques, M.H., Manuppella, G., 2003. Da dinâmica continental no Triásico aos mares do Jurássico Inferior e Médio. Cadernos de Geologia de Portugal. Instituto Geológico e Mineiro.

Azerêdo, A.C., Silva, R.L., Duarte, L.V., Cabral, M.C., 2010. Subtidal stromatolites from the Sinemurian of the Lusitanian Basin (Portugal). Facies 56, 211–230.

Azerêdo, A.C., Wright, V.P., Ramalho, M.M., 2002. The Middle–Late Jurassic forced regression and disconformity in central Portugal: eustatic, tectonic and climatic effects on a carbonate ramp system. Sedimentology 49, 1339–1370.

Bailey, T.R., Rosenthal, Y., Mcarthur, J.M., Van De Schootbrugge, B., Thirlwall, M.F., 2003. Paleoceanographic changes of the Late Pliensbachian–Early Toarcian interval: a possible link to the genesis of an Oceanic Anoxic Event. Earth Planetary Science Letters 212, 307–320.

Bassoulet, J.-P., Elmi, S., Poisson, A., Cecca, F., Bellion, Y., Guiraud, R., Baudin, F., 1993. Middle Toarcian (184–182 Ma). In: Dercourt, J., Ricou, L.E., Vrielynck, B. (Eds.), Atlas Tethys Paleoenvironmental Maps. BEICIP-FRANLAB, Rueil-Malmaison, pp. 63–80.

Bergström, S.M., Young, S., Schmitz, B., 2010. Katian (Upper Ordovician) $\delta^{13}C$ chemostratigraphy and sequence stratigraphy in the United States and Baltoscandia: a regional comparison. Palaeogeography, Palaeoclimatology, Palaeoecology 296, 217–234.

Bessereau, G., Guillocheau, F., Huc, A.-Y., 1995. Source rock occurrence in a sequence stratigraphic framework: the example of the Lias of the Paris Basin. In: Huc, A.-Y. (Ed.), Paleogeography, Paleoclimate, and Source Rocks. AAPG Studies in Geology, 40. AAPG, Tulsa, pp. 273–301.

Boomer, I., Ainsworth, N.R., Exton, J., 1998. A re-examination of the Pliensbachian and Toarcian Ostracoda of Zambujal, west-central Portugal. Journal of Micropalaeontology 17, 1–14.

Borrego, A.G., Hagemann, H.W., Blanco, C.G., Valenzuela, M., Suárez De Centi, C., 1996. The Pliensbachian (Early Jurassic) "anoxic" event in Asturias, northern Spain: Santa Mera Member, Rodiles Formation. Organic Geochemistry 25 (5–7), 295–309.

Brookfield, M.E., 2004. Principles of Stratigraphy. Blackwell Publishing.

Burchette, T.P., Wright, V.P., 1992. Carbonate ramp depositional systems. In: Sellwood, B.W. (Ed.), Ramps and Reefs. Sedimentary Geology, 79, pp. 3–57.

Brunel, F., Bonnet, L., Ruget, C., Rey, J., Mouterde, R., Rocha, R., 1998. Variations des associations de Foraminifères dans le Domérien du bassin lusitanien en relation avec les flutuations de l'environment. Comunicações do Instituto Geológico e Mineiro 84, A49–A52.

Catuneanu, O., 2006. Principles of Sequence Stratigraphy. Elsevier B.V.

Catuneanu, O., Abreu, V., Bhattacharya, J.P., Blum, M.D., Dalrymple, R.W., Eriksson, P.G., Fielding, C.R., Fisher, W.L., Galloway, W.E., Gibling, M.R., Giles, K.A., Holbrook, J.M., Jordan, R., Kendall, C.G.StC., Macurda, B., Martinsen, O.J., Miall, A.D., Neal, J.E., Nummedal, D., Pomar, L., Posamentier, H.W., Pratt, B.R., Sarg, J.F., Shanley, K.W., Steel, R.J., Strasser, A., Tucker, M.E., Winker, C., 2009. Towards the standardization of sequence stratigraphy. Earth-Science Reviews 92, 1–33.

Catuneanu, O., Bhattacharya, J.P., Blum, M.D., Dalrymple, R.W., Eriksson, P.G., Fielding, C.R., Fisher, W.L., Galloway, W.E., Gianolla, P., Gibling, M.R., Giles, K.A., Holbrook, J.M., Jordan, R., Kendall, C.G.StC., Macurda, B., Martinsen, O.J., Miall, A.D., Nummedal, D., Posamentier, H.W., Pratt, B.R., Shanley, K.W., Steel, R.J., Strasser, A., Tucker, M.E., 2010. Sequence stratigraphy: common ground after three decades of development. First Break 28, 21–34.

Catuneanu, O., Galloway, W.E., Kendall, C.G.StC., Miall, A.D., Posamentier, H.W., Strasser, A., 2011. Sequence stratigraphy: methodology and nomenclature. Newsletters on Stratigraphy 44, 173–245.

Coe, A.L., 2003. The Sedimentary Record of Sea-Level Change. Cambridge University Press.

Comas-Rengifo, M.J., Duarte, L.V., Goy, A., Paredes, R., Silva, R.L., 2013. El Sinemuriense Superior (cronozonas Oxynotum y Raricostatum) en la región de S. Pedro de Moel (Cuenca Lusitánica, Portugal). In: Duarte, L.V., Silva, R.L., Azerêdo, A.C. (Eds.), Fácies carbonatadas ricas em matéria orgânica do Jurássico da Bacia Lusitânica. Novos contributos paleontológicos, sedimentológicos e geoquímicos. Comunicações Geológicas, 100, volume especial I, pp. 15–19.

Comas-Rengifo, M.J., Goy, A., 2010. Caracterización biocronoestratigráfica del Sinemuriense Superior y el Pliensbachiense entre los afloramientos de la Playa Vega y Lastres (Asturias). In: García-Ramos, J.C. (Ed.), Las sucesiones margo-calcáreas marinas del Jurásico Inferior y las series fluviales del Jurásico Superior. Acantilados de la playa de Vega (Ribadesella). Guía de campo (excursión A), V Congreso del Jurásico de España, Servitec, Oviedo, pp. 10–18.

Correia, G.G., Duarte, L.V., Pereira, A., Silva, R.L., 2012. Outcrop gamma-ray spectrometry: applications to the Sinemurian–Pliensbachian organic-rich facies of the Lusitanian Basin (Portugal). Journal of Iberian Geology 38, 373–388.

Cubaynes, R., Marques, J.F., Ruget, C., 1988. L'ecozone à *cladocrinus* dans le Carixien-Domérien de la coupe de Maria Pares (Rabaçal, Condeixa, Portugal). In: Rocha, R.B., Soares, A.F. (Eds.), 2nd International Symposium on Jurassic Stratigraphy, International Subcommission on Jurassic Stratigraphy, Instituto Nacional de Investigação Científica. Centro de Estratigrafia e Paleobiologia da Universidade Nova de Lisboa: Centro de Geociências da Universidade de Coimbra, pp. 677–686.

Dera, G., Neige, P., Dommergues, J.-L., Fara, E., Laffont, R., Pellenard, P., 2010. High-resolution dynamics of Early Jurassic marine extinctions: the case of Pliensbachian–Toarcian ammonites (Cephalopoda). Journal of the Geological Society 167, 21–33.

Dera, G., Pucéat, E., Pellenard, P., Neige, P., Delsate, D., Joachimski, M.M., Reisberg, L., Martinez, M., 2009. Water mass exchange and variations in seawater temperature in the NW Tethys during the Early Jurassic: evidence from neodymium and oxygen isotopes of fish teeth and belemnites. Earth Planetary Science Letters 286, 198–207.

De Graciansky, P.C., Jacquin, T., Hesselbo, S.P., 1998. The Ligurian Cycle: an overview of Lower Jurassic 2nd-order transgressive/regressive facies cycles in western Europe. In: de Graciansky, P.C., Hardenbol, J., Jacquin, T., Vail, P.R. (Eds.), Mesozoic and Cenozoic Sequence Stratigraphy of European Basins. SEPM Special Publication, 60, pp. 467–479.

Dommergues, J.-L., 1987. L'évolution chez les Ammonitina du Lias Moyen (Carixien, Domerien basal) en Europe occidentale. Documents des Laboratoires de Géologie Lyon 98, 1–297.

Dommergues, J.-L., Fara, E., Meister, C., 2009. Ammonite diversity and its palaeobiogeographical structure during the early Pliensbachian (Jurassic) in the western Tethys and adjacent areas. Palaeogeography, Palaeoclimatology, Palaeoecology 280, 64–77.

Dommergues, J.-L., Meister, C., Mouterde, R., 1997. Pliensbachien. In: Cariou, E., Hantzpergue, P. (Eds.), Groupe français d'étude du Jurassique: Biostratigraphie du Jurassique ouest-européen et méditerranéen: zonations paralleles et distribution des invertébrés et microfossiles Bulletin des Centres de Recherche Exploration-Production Elf-Aquitaine, Mémoire, 17, pp. 15–23.

Dommergues, J.-L., Meister, C., Mouterde, R., 2002. *Fuciniceras paradoxus* nov. sp. (Harpoceratinae, Ammonitina) from the Domerian of Portugal. Remark about the taxonomic meaning of a paradoxical set of features. Geobios 35, 457–468.

Dommergues, J.-L., Meister, C., Neige, P., 2004a. *Fuciniceras capilliolusitanicum* nov. sp., une nouvelle espèce d'ammonite du Domérien lusitanien (Pliensbachien, Portugal). Revue de Paléobiologie 23, 373–386.

Dommergues, J.-L., Meister, C., Neige, P., Rocha, R.B., 2004b. Endemic Sinemurian (Early Jurassic) ammonites from the Lusitanian Basin (Portugal). Revue de Paléobiologie 23, 529–549.

Dromart, G., Elmi, S., 1986. Développement de structures cryptalgaires en domaine pélagique au cours de l'ouverture des bassins jurassiques (Atlantique central, Téthys occidentale). Comptes rendus de l'Académie des sciences de Paris, Série 2 (303), 311–316.

Duarte, L.V., 1997. Facies analysis and sequential evolution of the Toarcian-Lower Aalenian series in the Lusitanian Basin (Portugal). Comunicações do Instituto Geológico e Mineiro 83, 65–94.

Duarte, L.V., 2007. Lithostratigraphy, sequence stratigraphy and depositional setting of the Pliensbachian and Toarcian series in the Lusitanian Basin (Portugal). In: Rocha, R.B. (Ed.), The Peniche Section (Portugal). Contributions to the Definition of the Toarcian GSSP. International Subcommission on Jurassic Stratigraphy, pp. 17–23.

Duarte, L.V., Oliveira, L.C., Rodrigues, R., 2007. Carbon isotopes as a sequence stratigraphic tool: examples from the Lower and Middle Toarcian marly limestones of Portugal. Boletín Geológico y Minero 118, 3–18.

Duarte, L.V., Silva, R.L., Mendonça Filho, J.G., 2011. The Lower Jurassic of the west coast of Portugal: stratigraphy and organic matter in carbonate sedimentation. In: Flores, D., Marques, M. (Eds.), 63rd Annual Meeting of the International Committee for Coal and Organic Petrology - Field Trip Guide Book, Memórias N° 17, p. 42.

Duarte, L.V., Silva, R.L., Mendonça Filho, J.G., 2013. Variação do COT e Pirólise Rock-Eval do Jurássico Inferior da região de S. Pedro de Moel (Portugal). Potencial de geração de hidrocarbonetos. In: Duarte, L.V., Silva, R.L., Azerêdo, A.C. (Eds.), Fácies carbonatadas ricas em matéria orgânica do Jurássico da Bacia Lusitânica. Novos contributos paleontológicos, sedimentológicos e geoquímicos. Comunicações Geológicas, 100, pp. 107–111 volume especial I.

Duarte, L.V., Silva, R.L., Mendonça, Filho, J.G., Poças Ribeiro, N., Chagas, R.B.A., 2012. High-resolution stratigraphy, Palynofacies and source rock potential of the Água de Madeiros Formation (Upper Sinemurian) of the Lusitanian Basin, Portugal. Journal of Petroleum Geology 35, 105–126.

Duarte, L.V., Silva, R.L., Oliveira, L.C.V., Comas-Rengifo, M.J., Silva, F., 2010. Organic-rich facies in the Sinemurian and Pliensbachian of the Lusitanian Basin, Portugal: Total Organic Carbon distribution and relation to transgressive-regressive facies cycles. Geologica Acta 8, 325–340.

Duarte, L.V., Soares, A.F., 2002. Litostratigrafia das séries margo-calcárias do Jurássico Inferior da Bacia Lusitânica (Portugal). Comunicações do Instituto Geológico e Mineiro 89, 135–154.

Duarte, L.V., Wright, V.P., López, S.F., Elmi, S., Krautter, M., Azerêdo, A.C., Henriques, M.H., Rodrigues, R., Perilli, N., 2004. Early Jurassic carbonate evolution in the Lusitanian Basin (Portugal): facies, sequence stratigraphy and cyclicity. In: Duarte, L.V., Henriques, M.H. (Eds.), Carboniferous and Jurassic Carbonate Platforms of Iberia. 23rd IAS Meeting of Sedimentology Field Trip Guide Book, Coimbra, vol. 1, pp. 45–71.

Dunham, R.J., 1962. Classification of carbonate rocks according to depositional texture. In: Ham, W.E. (Ed.), Classification of Carbonate Rocks: American Association of Petroleum Geologists Memoir, pp. 108–121.

Einsele, G., 2000. Sedimentary Basins: Evolution, Facies, and Sediment Budget, second ed. Springer-Verlag.

Einsele, G., Ricken, W., Seilacher, A. (Eds.), 1991. Cycles and Events in Stratigraphy. Springer-Verlag.

Elmi, S., Rocha, R.B., Mouterde, R., 1988. Sediméntation pelagique et encroutements crytalgaires: les calcaires grumuleux du Carixien portugais. Ciências da Terra 9, 69–90.

Embry, A.F., 1993. Transgressive-regressive (T-R) sequence analysis of the Jurassic succession of the Sverdrup Basin, Canadian Arctic Archipelago. Canadian Journal of Earth Sciences 30, 301–320.

Embry, A.F., 2009. Practical Sequence Stratigraphy. Canadian Society of Petroleum Geologists. Online at www.cspg.org.

Embry, A.F., Johannessen, E.P., 1992. T-R sequence stratigraphy, facies analysis and reservoir distribution in the uppermost Triassic-Lower Jurassic succession, western Sverdrup Basin, Arctic Canada. In: Vorren, T.O., Bergsager, E., Dahl-Stamnes, O.A., Holter, E., Johansen, B., Lie, E., Lund, T.B. (Eds.), Arctic Geology and Petroleum Potential. Norwegian Petroleum Society (NPF), Special Publication, 2, pp. 121–146.

Embry, A.F., Klovan, J.E., 1971. A Late Devonian reef tract on Northeastern Banks Island, NWT. Canadian Petroleum Geology Bulletin 19, 730–781.

Emerson, S.R., Hedges, J.I., 2008. Chemical Oceanography and the Marine Carbon Cycle. Cambridge University Press, New York.

Fernández-López, S.R., 2000. Temas de Tafonomía. Departamento de Paleontología Universidad Complutense de Madrid.

Fernández-López, S., Duarte, L.V., Henriques, M.H., 2000. Ammonites from lumpy limestones in the Lower Pliensbachian of Portugal: taphonomic analysis and palaeoenvironmental implications. Revista de la Sociedad Geológica de España 13, 3–15.

Ferreira, R., Mendonça Filho, J.G., Duarte, L.V., Silva, R.L., Mendonça, J.O., 2010. Hydrocarbon generation potential of the Pliensbachian organic-rich series of Peniche (Lusitanian Basin, Portugal): an organopetrographic and thermal maturation assessment integrated analysis. In: Pena Dos Reis, R., Pimentel, N. (Eds.), II Central & North Atlantic Conjugate Margins Conference. Extended Abstracts, Volume III, pp. 116–120 Lisbon.

Fleet, A.J., Clayton, C.J., Jenkyns, H.C., Parkinson, D.N., 1987. Liassic source rock deposition in western Europe. In: Brooks, J., Glennie, K. (Eds.), Petroleum Geology of North West Europe. Graham & Trotman, London, pp. 59–70.

Flügel, E., 2010. Microfacies of Carbonate Rocks. Analysis, Interpretation and Application, second ed. Springer-Verlag.

Folk, R.L., 1959. Practical petrographic classification of limestones. American Association of Petroleum Geologists Bulletin 43, 1–38.

Frazier, D.E., 1974. Depositional Episodes: Their Relationship to the Quaternary Stratigraphic Framework in the Northwestern Portion of the Gulf Basin. Geological Circular, 4. University of Texas at Austin, Bureau of Economic Geology (1).

Galloway, W.E., 1989. Genetic stratigraphic sequences in basin analysis, I. Architecture and genesis of flooding-surface bounded depositional units. AAPG Bulletin 73, 125–142.

Gradstein, F.M., Ogg, J.G., Schmitz, M.D., Ogg, G.M., 2012. The Geologic Time Scale 2012, 2 volumes, Elsevier.

Hallam, A., 1981. A revised sea-level curve for the early Jurassic. Journal of the Geological Society 138, 735–743.

Hantschel, T., Kauerauf, A.I., 2009. Fundamentals of Basin and Petroleum Systems Modeling. Springer-Verlag.

Hardenbol, J., Thierry, J., Farley, M.B., Jacquin, T., De Graciansky, P.C., Vail, P., 1998. Mesozoic and Cenozoic sequence chronostratigraphic framework of european basins. In: Graciansky, P.C., Hardenbol, J., Jacquin, T., Vail, P. (Eds.), Mesozoic and Cenozoic Sequence Stratigraphy of European Basins. SEPM Special Publication, 60, pp. 3–13 charts 1–8.

He, S., 2002. Thermal history and deep overpressure modelling in the Northern Carnarvon Basin, North West Shelf, Australia (Unpublished Ph.D. thesis). Curtin University of Technology, Department of Applied Geology.

Hesselbo, S.P., 2008. Sequence stratigraphy and inferred relative sea-level change from the onshore British Jurassic. Proceedings of the Geologists' Association 119, 19–34.

Hesselbo, S.P., Gröcke, D.R., Jenkyns, H.C., Bjerrum, C.J., Farrimond, P., Morgans Bell, H.S., Green, O.R., 2000. Massive dissociation of gas hydrates during a Jurassic oceanic anoxic event. Nature 406, 392–395.

Hesselbo, S.P., Jenkyns, H.C., Duarte, L.V., Oliveira, L.C.V., 2007. Carbon-isotope record of the Early Jurassic (Toarcian) Oceanic Anoxic Event from fossil wood and marine carbonate (Lusitanian Basin, Portugal). Earth and Planetary Science Letters 253, 455–470.

Hiscott, R.N., Wilson, R.C.L., Gradstein, F.M., Pujalte, V., García-Mondéjar, J., Boudreau, R.R., Wishart, H.A., 1990. Comparative stratigraphy and subsidence history of Mesozoic rift basins of North Atlantic. AAPG Bulletin 74, 60–76.

Holmden, C., Creaser, R.A., Muehlenbachs, K., Leslie, S.A., Bergström, S.M., 1998. Isotopic evidence for geochemical decoupling between ancient epeiric seas and bordering oceans: Implications for secular curves. Geology 26, 567–570.

Howell, J.A., Aitken, J.F., 1996. High resolution sequence Stratigraphy: Innovations and applications. Geological Society, London, Special Publications 104.

Huck, S., Heimhofer, U., Immenhauser, A., Weissert, H., 2013. Carbon-isotope stratigraphy of Early Cretaceous (urgonian) shoal-water deposits: diachronous changes in carbonate-platform production in the north-western Tethys. Sedimentary Geology 290, 157–174.

Immenhauser, A., Della Porta, G., Kenter, J.A.M., Bahamonde, J.R., 2003. An alternative model for positive shifts in shallow-marine carbonate $\delta^{13}C$ and $\delta^{18}O$. Sedimentology 50, 953–959.

Inverno, C.M.C., Manupella, G., Zbyszewski, G., Pais, J., Ribeiro, M.L., 1993. Notícia explicativa da folha 42 C, Santiago do Cacém, 1:50.000. Serviços Geologicos de Portugal, Lisbon. 75 pp.

Jacquin, T., de Graciansky, P.C., 1998. Transgressive-regressive (second order) facies cycles: the effects of tectonoeustacy. In: de Graciansky, P.C., Hardenbol, J., Jacquin, T., Vail, P. (Eds.), Mesozoic and Cenozoic Sequence Stratigraphy of European Basins. SEPM Special Publication, 60, pp. 445–466.

Jenkyns, H.C., 1988. The Early Toarcian (Jurassic) anoxic event: stratigraphic, sedimentary and geochemical evidence. American Journal of Science 288, 101–151.

Jenkyns, H.C., Clayton, C.J., 1986. Black shales and carbon isotopes in pelagic sediments from the Tethyan Lower Jurassic. Sedimentology 33, 87–106.

Jenkyns, H.C., Jones, C.E., Gröcke, D.R., Hesselbo, S.P., Parkinson, D.N., 2002. Chemostratigraphy of the Jurassic System: applications, limitations and implications for palaeoceanography. Journal of the Geological Society 159, 351–378.

Johnson, J.G., Murphy, M.A., 1984. Time-rock model for Siluro-Devonian continental shelf, western United States. Geological Society of America Bulletin 95, 1349–1359.

Katz, M.E., Wright, J.D., Miller, K.D., Cramer, B.S., Fennel, K., Falkowski, P.G., 2005. Biological overprint of the geological carbon cycle. Marine Geology 217, 323–338.

Kodina, L.A., Bogatcheva, M.P., Lobitzer, H., 1988. An organic geochemical study of austrian bituminous rocks. Jahrbuch Der Geologischen Bundesanstalt 291–300 Band 131, Heft 2.

Kump, L.R., Arthur, M.A., 1999. Interpreting carbon-isotope excursions: carbonates and organic matter. Chemical Geology 161, 181–198.

Léonide, P., Floquet, M., Villier, L., 2007. Interaction of tectonics, eustasy, climate and carbonate production on the sedimentary evolution of an early/middle Jurassic extensional basin (southern provence sub-basin, SE France). Basin Research 19, 125–152.

Manuppella, G., Azerêdo, A.C., 1996. Contribuição para o conhecimento da geologia da região de Sesimbra. Comunicações Instituto Geológico e Mineiro 82, 37–50.

Marino, M., Santantonio, M., 2010. Understanding the geological record of carbonate platform drowning across rifted Tethyan margins: examples from the Lower Jurassic of the Apennines and Sicily (Italy). Sedimentary Geology 225, 116–137.

Marshall, J.D., 1992. Climatic and oceanographic isotopic signals from the carbonate rock record and their preservation. Geological Magazine 129, 143–160.

Mattioli, E., Plancq, J., Boussaha, M., Duarte, L.V., Pittet, B., 2013. Calcareous nannofossil biostratigraphy: new data from the Lower Jurassic of the Lusitanian Basin. In: Duarte, L.V., Silva, R.L., Azerêdo, A.C. (Eds.), Fácies carbonatadas ricas em matéria orgânica do Jurássico da Bacia Lusitânica. Novos contributos paleontológicos, sedimentológicos e geoquímicos. Comunicações Geológicas, 100, pp. 69–76 volume especial I.

Mendonça Filho, J.G., Silva, T.F., Silva, R.L., Duarte, L.V., Poças Ribeiro, N., 2013. Distribuição dos biomarcadores hopanóides e sua relação com a matéria orgânica nas séries jurássicas de S. Pedro de Moel, Bacia Lusitânica (Portugal). In: Duarte, L.V., Silva, R.L., Azerêdo, A.C. (Eds.), Fácies carbonatadas ricas em matéria orgânica do Jurássico da Bacia Lusitânica. Novos contributos paleontológicos, sedimentológicos e geoquímicos. Comunicações Geológicas, 100, pp. 101–105 volume especial I.

McArthur, J.M., Donovan, D.T., Thirlwall, M.F., Fouke, B.W., Mattey, D., 2000. Strontium isotope profile of the early toarcian (jurassic) oceanic anoxic event, the duration of ammonite biozones, and belemnite palaeotemperatures. Earth Planetary Science Letters 179, 269–285.

Miall, A.D. (Ed.), 2008. Sedimentary basins of the world. The sedimentary basins of the United States and Canada. Elsevier B.V.

Miall, A.D., 2010. The Geology of Stratigraphic Sequences. Springer-Verlag.

Mitchell, S.F., Paul, C.R.C., Gale, A.S., 1996. Carbon isotopes and sequence stratigraphy. Geological Society, London, Special Publications 104, 11–24.

Mitchum Jr., R.M., 1977. Seismic stratigraphy and global changes of sea level, part 11: glossary of terms used in seismic stratigraphy. In: Payton, C.E. (Ed.), Seismic Stratigraphy – Applications to Hydrocarbon Exploration American Association of Petroleum Geologists Memoir, 26, pp. 205–212.

Mitchum Jr., R.M., Vail, P.R., Thompson III, S., 1977. Seismic stratigraphy and global changes of sea-level, part 2: the depositional sequence as a basic unit for stratigraphic analysis. In: Payton, C.E. (Ed.), Seismic Stratigraphy – Applications to Hydrocarbon Exploration. American Association of Petroleum Geologists Memoir, 26, pp. 53–62.

Morettini, E., Santantonio, M., Bartolini, A., Cecca, F., Baumgartner, P.O., Hunziker, J.C., 2002. Carbon isotope stratigraphy and carbonate production during the Early–Middle jurassic: examples from the Umbria–Marche–Sabina Apennines (central Italy). Palaeogeography, Palaeoclimatology, Palaeoecology 184, 251–273.

Mouterde, R., 1955. Le Lias de Peniche. Comunicações dos Serviços Geológicos de Portugal 36, 5–33.

Mouterde, R., 1967. Le Lias du Portugal. Vue d'ensemble et division en Zones. Comunicações dos Serviços Geológicos de Portugal 52, 209–226.

Mouterde, R., Dommergues, J.-L., Rocha, R.B., 1983. Atlas des fossiles caractéristiques du Lias portugais. II-Carixien. Ciências da Terra 7, 187–254.

Mouterde, R., Dommergues, J.-L., Meister, C., Rocha, R.B., 2007. Atlas des fossiles caractéristiques du Lias portugais. III a) Domérien (Ammonites). Ciências da Terra 16, 67–111.

Mouterde, R., Rocha, R.B., 1981. Atlas des fossiles caractéristiques du Lias portugais. I. - Lias inférieur. Ciências da Terra 6, 49–76.

Mouterde, R., Rocha, R.B., 1988. Essai de biozonation du Domérien portugais. Ciências da Terra 3, 66–68.

Mouterde, R., Rocha, R.B., Ruget, C., 1971. Le Lias moyen et supérieur de la région de Tomar. Comunicações dos Serviços Geológicos de Portugal 55, 55–86.

Mouterde, R., Ruget, C., Moitinho De Almeida, F., 1964–65. Coupe du Lias au Sud de Condeixa. Comunicações dos Serviços Geológicos de Portugal 48, 61–91.

Newton, B., Bottrell, S., 2007. Stable isotopes of carbon and sulphur as indicators of environmental change: past and present. Journal of the Geological Society 164, 691–708.

Nikitenko, B.L., Mickey, M.B., 2004. Foraminifera and ostracodes across the Pliensbachian-Toarcian boundary in the Arctic Realm (stratigraphy, palaeobiogeography and biofacies). In: Beaudoin, A.B., Head, M.J. (Eds.), The Palynology and Micropalaeontology of Boundaries. Geological Society Special Publication N°230. The Geological Socyety, London, pp. 137–174.

N'zaba-Makaya, O., Andreu, B., Brunel, F., Mouterde, R., Rey, J., Rocha, R., 2003. Biostratigraphie et paléoécologie des puplements d'ostracodes dans le Domérien du Bassin Lusitanien, Portugal. Ciências da Terra 15, 21–44.

Oliveira, L.C.V., Duarte, L.V., Lemos, V.B., Comas-Rengifo, M.J., Perilli, N., 2007. Bioestratigrafia de nanofósseis calcários e correlação com as zonas de amonites do Pliensbaquiano-Toarciano basal (Jurássico inferior) de Peniche (Bacia Lusitânica, Portugal). In: Carvalho, I.S., Cassab, R.C.T., Schwanke, C., Carvalho, M.A., Fernandes, A.C.S., Rodrigues, M.A.C., Carvalho, M.S.S., Arai, M., Oliveira, M.E.Q. (Eds.), Paleontologia: Cenários de Vida, Proceedings of XIX Congresso Brasileiro de Paleontologia, Búzios (Brazil), Editora Interciência, pp. 411–420.

Oliveira, L.C., Duarte, L.V., Silva, R.L., Rodrigues, R., 2009. Belemnite $\delta^{18}O$ and $\delta^{13}C$ record of the Lusitanian Basin Pliensbachian carbonate series (Portugal). Geochimica and Cosmochimica Acta 73 (13S), A969.

Oliveira, L.C.V., Rodrigues, R., Duarte, L.V., Lemos, V., 2006. Avaliação do potencial gerador de petróleo e interpretação paleoambiental com base em biomarcadores e isótopos estáveis do carbono da seção Pliensbaquiano-Toarciano inferior (Jurássico inferior) da região de Peniche (Bacia Lusitânica, Portugal). Boletim de Geociências da Petrobras 14 (2), 207–234.

Panchuk, K.M., Holmden, C.E., Leslie, S.A., 2006. Local controls on carbon cycling in the Ordovician midcontinent region of North America, with implication for carbon isotope secular curves. Journal of Sedimentary Research 76, 200–211.

Patterson, W.P., Walter, L.M., 1994. Depletion of ^{13}C in seawater ΣCO_2 on modern carbonate platforms: significance for the carbon isotopic record of carbonates. Geology 22, 885–888.

Pereira, R., Alves, T.M., 2012. Tectono-stratigraphic signature of multiphased rifting on divergent margins (deep-offshore southwest Iberia, North Atlantic). Tectonics 31, TC4001.

Phelps, M., 1985. A refined ammonite biostratigraphy for the Middle and Upper Carixian (Ibex and Davoei zones, Lower Jurassic) in North-West Europe and stratigraphical details of the Carixian-Domerian boundary. Geobios 18, 321–362.

Perilli, N., Duarte, L.V., 2006. Toarcian nannobiohorizons from the Lusitanian Basin (Portugal) and their calibration against ammonite zones. Rivista Italiana di Paleontologia e Stratigrafia 112, 417–434.

Pinheiro, L.M., Wilson, R.C.L., Pena dos Reis, R., Whitmarsh, R.B., Ribeiro, A., 1996. The western Iberia margin: a geophysical and geological overview. In: Whitmarsh, R.B., Sawyer, D.S., Klaus, A., Masson, D.G. (Eds.), Proceedings of the Ocean drilling Program. Scientific Results, 149, pp. 3–23.

Pittet, B., Suan, G., Lenoir, F., Duarte, L.V., Mattioli, E., 2014. Carbon isotope evidence for sedimentary discontinuities in the lower Toarcian of the Lusitanian Basin (Portugal): sea level change at the onset of the oceanic anoxic event. Sedimentary Geology 303, 1–14.

Poças Ribeiro, N., Mendonça Filho, J.G., Duarte, L.V., Silva, R.L., Mendonça, J.O., Silva, T.F., 2013. Palynofacies and organic geochemistry of the Sinemurian carbonate deposits in the western Lusitanian Basin (Portugal): coimbra and Água de Madeiros Formations. International Journal of Coal Geology 111, 37–52.

Posamentier, H.W., Allen, G.P., 1999. Siliciclastic sequence stratigraphy: concepts and applications. SEPM Concepts in Sedimentology and Paleontology 7, 210.

Quesada, S., Dorronsoro, C., Robles, S., Chaler, R., Grimalt, J.O., 1997. Geochemical correlation of oil from the Ayoluengo field to Liassic black shale units in the southwestern Basque-Cantabrian Basin (northern Spain). Organic Geochemisty 27, 25–40.

Quesada, S., Robles, S., Rosales, I., 2005. Depositional architecture and transgressive–regressive cycles within Liassic backstepping carbonate ramps in the Basque–Cantabrian basin, northern Spain. Journal of the Geological Society 162, 531–548.

Rasmussen, E.S., Lomholt, S., Andersen, C., Vejbæk, O.V., 1998. Aspects of the structural evolution of the Lusitanian Basin in Portugal and the shelf and slope area offshore Portugal. Tectonophysics 300, 199–225.

Renard, M., Corbin, J.C., Daux, V., Emmanuel, L., Baudin, F., Tamburini, F., 2008. Chapter. 3: chemostratigraphy. In: Rey, J., Galeotti, S. (Eds.), Stratigraphy: Terminology and Practice. Editions TECHNIP, pp. 41–52.

Rey, J., Bonnet, L., Brunel, F., Ruget, C., Cubaynes, R., Mouterde, R., Bordalo Da Rocha, R., 2000. Comparaison des associations de foraminifères dans le Domérien des bassins aquitain (France) et lusitanien (Portugal). Geobios 33, 135–144.

Reggiani, L., Mattioli, E., Pittet, B., 2010a. Spatial distribution of Late Pliensbachian (Early Jurassic) calcareous nannofossils within the Lusitanian basin (Portugal). Geobios 43, 87–97.

Reggiani, L., Mattioli, E., Pittet, B., Duarte, L.V., Veiga De Oliveira, L.C., Comas-Rengifo, M.J., 2010b. Pliensbachian (Early Jurassic) calcareous nannofossils from the Peniche section (Lusitanian Basin, Portugal): a clue for palaeoenvironmental reconstructions. Marine Micropaleontology 75 (1–4), 1–16.

Riediger, C.L., 2002. Hydrocarbon source rock potential and comments on correlation of the Lower Jurassic Poker Chip Shale, west-central Alberta. Bulletin of Canadian Petroleum Geology 50 (2), 263–276.

Rosales, I., Quesada, S., Robles, S., 2001. Primary and diagenetic isotopic signal in fossils and hemipelagic carbonates: the Lower Jurassic of northern Spain. Sedimentology 48, 1149–1169.

Rosales, I., Quesada, S., Robles, S., 2004a. Paleotemperature variations of Early Jurassic seawater recorded in geochemical trends of belemnites from the Basque–Cantabrian basin, northern Spain. Palaeogeography, Palaeoclimatology, Palaeoecology 203, 253–275.

Rosales, I., Robles, S., Quesada, S., 2004b. Elemental and oxygen isotope composition of Early Jurassic belemnites: salinity vs temperature signals. Journal of Sedimentary Research 74, 343–355.

Rosales, I., Quesada, S., Robles, S., 2006. Geochemical arguments for identifying second-order sea-level changes in hemipelagic carbonate ramp deposits. Terra Nova 18, 233–240.

Sakuma, H., Tada, R., Ikeda, M., Kashiyama, Y., Ohkouchi, N., Ogawa, N.O., Watanabe, S., Tajika, E., Yamamoto, S., 2012. High-resolution lithostratigraphy and organic carbon isotope stratigraphy of the Lower Triassic pelagic sequence in central Japan. Island Arc 21 (2), 79–100.

Schlager, W., 2005. Carbonate Sedimentology and Sequence Stratigraphy, SEPM Concepts in Sedimentology and Paleontology 08. SEPM Foundation Inc.

Schlanger, S.O., Jenkyns, H.C., 1976. Cretaceous oceanic anoxic events: causes and consequences. Geology en Mijnbouw 55 (3–4), 179–184.

Schweigert, G., 2005. The occurrence of the Tethyan ammonite genus *Meneghiniceras* (Phylloceratina: juraphyllitidae) in the Upper Pliensbachian of SW Germany. Stuttgarter Beiträge zur Naturkunde Serie B (Geologie und Paläontologie) 356 (15 S).

Seilacher, A., 2007. Trace Fossil Analysis. Springer-Verlag.

Silva, R.L., Duarte, L.V., Azerêdo, A.C., Mendonça Filho, J.G., 2011a. Contributos sedimentológicos e geoquímicos para a modelação dos sistemas petrolíferos portugueses: as séries jurássicas ricas em matéria orgânica da Bacia Lusitânica (Portugal). Modelação de Sistemas Geológicos, Laboratório de Radioactividade Natural da Universidade de Coimbra. pp. 383–391.

Silva, R.L., Duarte, L.V., Comas-Rengifo, M.J., Mendonça Filho, J.G., Azerêdo, A.C., 2011b. Update of the carbon and oxygen isotopic records of the Early–Late Pliensbachian (Early Jurassic, ~187 Ma): insights from the organic-rich hemipelagic series of the Lusitanian Basin (Portugal). Chemical Geology 283, 177–184.

Silva, R.L., Duarte, L.V., Mendonça Filho, J.,G., 2013. Optical and geochemical characterization of Upper Sinemurian (Lower Jurassic) fossil wood from the Lusitanian Basin (Portugal). Geochemical Journal 47, 489–498.

Silva, R.L., Mendonça Filho, J.G., Duarte, L.V., Comas-Rengifo, M.J., Azerêdo, A.C., Ferreira, R., 2010. Organic-rich facies of the top Ibex–Margaritatus zones (Pliensbachian) of the Lusitanian Basin (Portugal): TOC and biomarkers variation. Geochimica et Cosmochimica Acta 74 (12–S1), A962.

Silva, R.L., Mendonça Filho, J.G., Da Silva, F.S., Duarte, L.V., Silva, T.F., Ferreira, R., Azerêdo, A.C., 2012. Can biogeochemistry aid in the palaeoenvironmental/early diagenesis reconstruction of the ~187 Ma (Pliensbachian) organic-rich hemipelagic series of the Lusitanian Basin (Portugal)? Bulletin of Geosciences 87, 373–382.

Silva, R.L., Duarte, L.V., Oliveira, L.C.V., 2009. Compositional classification of sedimentary rocks based in geochemical indicators: the case of the Upper Pliensbachian–Lower Toarcian of Peniche (Portugal) (translated from Portuguese). Extended abstracts of the VII Congreso Ibérico-X Congreso Nacional de Geoquímica. Biblioteca Electrónica de la Excma. Diputación Provincial de Soria. 17, 118–127.

Silva, F., Duarte, L.V., Oliveira, L.C.V., Comas-Rengifo, M.J., Rodrigues, R., 2006. A Formação de Vale das Fontes no sector norte da Bacia Lusitânica (Portugal): caracterização e avaliação preliminar da distribuição do Carbono Orgânico Total. In: Mirão, J., Balbino, A. (Eds.), Actas do VII Congresso Nacional de Geologia, vol. II, pp. 669–672 Évora.

Silva, F., Duarte, L.V., Oliveira, L.C.V., Rodrigues, R., Comas-Rengifo, M.J., 2007. Caracterização do Carbono Orgânico Total e pirólise *Rock-Eval* no intervalo Sinemuriano superior-Pliensbaquiano do sector norte da Bacia Lusitânica (Portugal). In: Gomes, E.P. and Alencoão, A.M., (Coords.), Actas do VI Congresso Ibérico de Geoquímica. Vila Real, pp. 564–567.

Soares, A.F., Rocha, R.B., Elmi, S., Henriques, M.H., Mouterde, R., Almeras, Y., Ruget, C., Marques, J., Duarte, L.V., Carapito, C., Kullberg, J.C., 1993. Le sous-bassin nord-lusitanien (Portugal) du Trias au Jurassique moyen: histoire d'un "rift avorté". Comptes rendus de l'Académie des sciences de Paris 317, 1659–1666.

Stapel, G., Cloetingh, S., Pronk, B., 1996. Quantitative subsidence analysis of the Mesozoic evolution of the Lusitanian Basin (western Iberian margin). Tectonophysics 266, 493–507.

Stephens, N.P., Summer, D.Y., 2003. Late Devonian carbon isotope stratigraphy and sea level fluctuations, canning basin, Western Australia. Palaeogeography, Palaeoclimatology, Palaeoecology 191, 203–219.

Suan, G., Mattioli, E., Pittet, B., Lecuyer, C., Sucheras-Marx, B., Duarte, L.V., Philippe, M., Reggiani, L., Martineau, F., 2010. Secular environmental precursors to Early Toarcian (Jurassic) extreme climate changes. Earth and Planetary Science Letters 290, 448–458.

Suan, G., Mattioli, E., Pittet, B., Mailliot, B., Lecuyer, C., 2008. Evidence for major environmental perturbation prior to and during the Toarcian (Early Jurassic) oceanic anoxic event from the Lusitanian Basin, Portugal. Paleoceanography 23, PA1201.

Swart, P.K., Eberli, G., 2005. The nature of the $\delta^{13}C$ of periplatform sediments: implications for stratigraphy and the global carbon cycle. Sedimentary Geology 175, 115–129.

Tucker, M.E., Calvet, F., Hunt, D., 1993. Sequence stratigraphy of carbonate ramps: systems tracts, models and application to the Muschelkalk carbonate platforms of eastern Spain. In: Posamentier, H.W., Summerhayes, C.P., Haq, B.U., Allen, G.P. (Eds.), Sequence Stratigraphy and Facies Associations International Association of Sedimentologists, Special Publication, 18, pp. 397–415.

Valencio, S.A., Cagnoni, M.C., Ramos, A.M., Riccardi, A.C., Panarello, H.O., 2005. Chemostratigraphy of the Pliensbachian, Puesto Araya Formation (Neuquén Basin, Argentina). Geologica Acta 3 (2), 147–154.

Van De Schootbrugge, B., Bailey, T.R., Rosenthal, Y., Katz, M.E., Wright, J.D., Miller, K.G., Feist-Burkhardt, S., Falkowski, P.G., 2005. Early Jurassic climate change and the radiation of organic walled phytoplankton in the Tethys Ocean. Paleobiology 31 (1), 73–97.

Van De Schootbrugge, B., Harazim, D., Sorichter, D., Oschmann, W., Fiebig, J., Püttmann, W., Peinl, M., Zanella, F., Teichert, B., Hoffmann, J., Stadnitskaia, A., Rosenthal, Y., 2010. The enigmatic ichnofossil *Tisoa siphonalis* and widespread authigenic seep carbonate formation during the Late Pliensbachian in southern France. Biogeosciences Discussions 7, 4181–4222.

Van Wagoner, J.C., Mitchum, R.M., Posamentier, H.W., Vail, P.R., 1987. An overview of sequence stratigraphy and key definitions. In: Bally, A.W. (Ed.), Atlas of Seismic Stratigraphy, Volume 1AAPG Studies in Geology, 27, pp. 11–14.
Wefer, G., Berger, W.H., 1991. Isotope paleontology: growth and composition of extant calcareous species. Marine Geology 100, 207–248.
Weissert, H., Joachimski, M., Sarnthein, M., 2008. Chemostratigraphy. Newsletters on Stratigraphy 42 (3), 145–179.
Weissert, H., Lini, A., Föllmi, K., Kuhn, O., 1998. Correlation of Early Cretaceous carbon isotope stratigraphy and platform drowning events: a possible link? Palaeogeography, Palaeoclimatology, Palaeoecology 137, 189–203.
Wilson, R.C.L., 1988. Mesozoic development of the Lusitanian Basin, Portugal. Revista de la Sociedad Geológica de España 1, 393–407.
Wilson, R.C.L., Hiscott, R.N., Willis, M.G., Gradstein, F.M., 1989. The Lusitanian Basin of west-central Portugal: Mesozoic and Tertiary tectonic, stratigraphic and subsidence history. In: Tankard, A.J., Balkwill, H.R. (Eds.), Extensional Tectonics and Stratigraphy of the North Atlantic Margins AAPG Memoir, 46, pp. 341–361.
Wright, V.P., 1992. A revised classification of limestones. Sedimentary Geology 76, 177–185.

CHAPTER

CHEMOSTRATIGRAPHY OF THE PERMIAN–TRIASSIC STRATA OF THE OFFSHORE PERSIAN GULF, IRAN

14

Vahid Tavakoli

School of Geology, College of Science, University of Tehran, Tehran, Iran

14.1 INTRODUCTION

The Late Permian–Early Triassic transition is generally referred to as a global greenhouse interval (e.g., Fluteau et al., 2001; Woods, 2005; Schneider et al., 2006). It is characterized by the widespread deposition of carbonates and evaporites in the Middle East and negative carbon stable-isotope excursions linked to the end-Permian mass extinction. Several hypotheses have been suggested to explain this catastrophic event. The most commonly accepted theories include oceanic anoxia, changes in seawater chemistry, volcanism, asteroid/comet impact, methane clathrate dissociation events, and extreme climate change (Stanley, 1988; Ellis and Schramm, 1995; Renneet al., 1995; Knoll et al., 1996; Isozaki, 1997; Bowring et al., 1998; Retallack et al., 1998; Becker et al., 2001; Erwin et al., 2002; Wignall and Twitchett, 2002; Wit et al., 2002; Heydari et al., 2003; Kidder and Worsley, 2004; Kump et al., 2005; Retallak and Jahren, 2008; Tavakoli and Rahimpour-Bonab, 2012).

Regardless of any of the hypotheses, the ocean chemistry has been changed during that time. These changes were reflected in the geochemistry of the sediments. Geochemical data may point to the nature of original seawater and paleoclimate in various carbonate environments (e.g., Walgenwitz et al., 1992; Emmanuel et al., 1999). So, Permian–Triassic boundary (PTB) sections with carbonate lithology are unique for investigating the end-Permian mass extinction because of their potential for recording changes in seawater geochemistry (Grotzinger and Knoll, 1995; Lehrmann et al., 2003; Ehrenberg et al., 2008). Chemostratigraphy of these strata can reveal the paleoceanic conditions at the time of deposition and may aid in high-resolution correlation using the results of isotopic and elemental analyses. Anomalous diagenetic overprints and possible siliciclastic influx must be differentiated from such evidences (Ehrenberg et al., 2008; Rahimpour-Bonab et al., 2009; Tavakoli et al., 2011). Most of the chemostratigraphic researches have been focused on the nature of changes in isotopic or elemental concentrations or the cause of extinction (e.g., Heydari et al., 2008; Liu et al., 2013; Heydari and Hassanzadeh, 2003; Wang et al., 2007; Heydari et al., 2000; Twichett et al., 2001). The results of the previous studies (*op cit.*) show that the end-Permian extinction is associated with distinct changes in the C and Sr isotopic ratios of seawater. Although fewer studies considered the elemental concentrations of sediments at the Permian–Triassic transitions (e.g., Dolenec et al., 2001; Kaiho et al., 2006; Algeo et al., 2007), they focused only on the redox conditions at the time of deposition as the cause of extinction. Thus, chemostratigraphic application in documenting the paleoceanic conditions at the time of deposition is yet to be attempted.

Chemostratigraphy. http://dx.doi.org/10.1016/B978-0-12-419968-2.00014-5

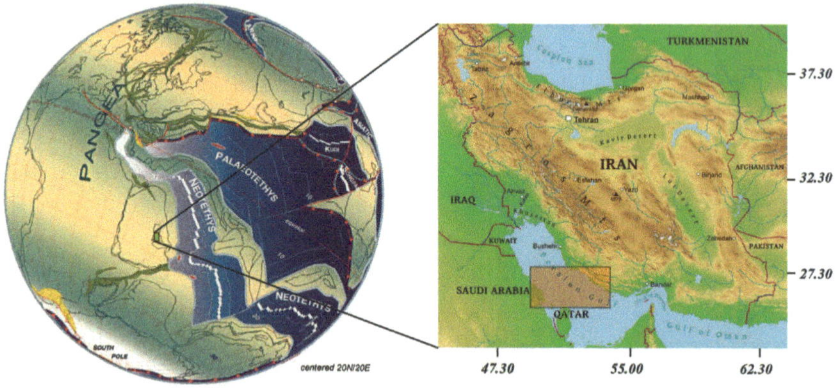

FIGURE 1

Study area at the time of deposition (left) and at the present time (right). The study area is highlighted in both pictures (the Permian–Triassic global map from Stampfli and Borel, 2002).

The PTB sections studied here are located in the offshore Persian Gulf, Iran (Fig. 1). The Persian Gulf PTB sections are significant in terms of the occurrence of organosedimentary thrombolites that have been reported from the other PTB sections around the world (Sano and Nakashima., 1997; Lehrmann, 1999; Kershaw et al., 1999, 2002; Wignall and Twitchett, 2002; Ezaki et al., 2003; Hips and Haas, 2006; Pruss et al., 2006). These thrombolites have been generally described as microbialite buildups that were formed by a result of a benthic microbial community (Burne and Moore, 1987). Microbialites exhibit unique clotted accretionary textures that are rarely found in post-Cambrian strata, and may have resulted from the distinct ecological and geochemical conditions of the post-PTB mass extinction. This marker bed could be used as a good indicator for Early Triassic time correlation through the area.

In this paper, high-resolution carbon, oxygen, and strontium isotope ratios as well as elemental analysis (Fe, Mg, Sr, Na, and Th/U ratios) of the three wells in the Persian Gulf are presented, which provide new insights to calibrate the PTB from geochemical aspects. Elemental concentrations are introduced here for the first time in this area. These new data are important to understand the Late Permian and Early Triassic major paleoceanographic conditions in this region. Also, this study provides a new approach for calibration and correlation of the Upper Permian to Lower Triassic strata if the section does not have enough biostratigraphic data for absolute age dating. These data fill the gap in Permian–Triassic ocean geochemistry in this part of the world. Based on the chemostratigraphic data as well as sedimentological and biological characteristics, the exact horizon of the PTB is located, and the sequence is compared to other PTB sections around the world. Discussion will be extended to find the significance of the geochemical data that provides an insight into paleoceanographic conditions during the greatest crisis in the history of life.

14.2 GEOLOGICAL SETTING AND STRATIGRAPHY

The area concerned forms an NW–SE trending linear belt, which is located along the Iran–Qatar political border in the Persian Gulf (Fig. 1). The area is known for huge South Pars and other hydrocarbon-producing fields in offshore Persian Gulf, which is situated in the Arabian Peninsula. The basin contains a very thick sequence of sediments accumulated from the Late Proterozoic to the Holocene (Beydoun, 1991; Alsharhan and Nairn, 1997; Edgell, 1996; Konert et al., 2001; Sharland et al.,. 2001; Ziegler, 2001). In terms of tectonic setting, the studied region is located between the Gotnia Trough in the northwest, Ghawar High in the west, Arabian Basin in the south, and Rub-Al-Khali in the east.

The geological history of the basin is characterized by uniform clastic continental and shallow marine sedimentation on a stable passive margin setting during most of the Paleozoic (Silurian shale and Faraghun Formation in Iran). The Permian–Triassic transition is characterized by the deposition of carbonate–evaporite series on a shallow homoclinal ramp named the Kangan and Dalan formations in Iran that are time equivalent of the Khuff Formation in other regions around the Persian Gulf. The upper Dalan and Kangan formations studied here are divided into K4 to K1 reservoir units from the bottom to the top, respectively. The PTB lies between the K3 and K2 units. The deposition continued to the Paleogene on a broad, stable platform where the deposition of mainly shallow-water carbonates with minor anhydrites and shales occurred (Perotti et al., 2011). The Zagros Orogeny formed as a consequence of the Arabia/Eurasia collision at about 35 Ma (Mouthereau et al., 2012), and then the northeastern part of the basin was undergoing a shortening as a consequence of this collision.

From a stratigraphic point of view, the Kangan and Dalan formations were deposited during the Late Permian–Early Triassic time. In the studied region, the Early Permian time is characterized with the deposition of Faraghun siliciclastic Formation, which in turn continued with the deposition of the limestone of the Dalan Formation (lower Dalan) and capped by the anhydritic Nar member. The upper Dalan and overlying Kangan Formation are composed of limestone and dolomite. Deposition of the Kangan Formation was terminated with the accumulation of shale and evaporites of the Dashtak Formation.

Sedimentology and diagenesis of the studied formations in the Persian Gulf area were the subjects of many previous publications (e.g., Alsharhan, 1993; Al-Jallal,1994; Alsharhan and Nairn, 1997; Esrafili-Dizaji and Rahimpour-Bonab, 2009; Tavakoli et al., 2011). These authors interpreted the deposition in a ramp environment with less than a 30-m water depth. The diagenetic environments include marine and meteoric in the studied formations. The main diagenetic effects are dolomitization and dissolution that enhanced the porosity and permeability within these strata; both of them are the results of early diagenesis.

14.3 MATERIALS AND METHODS

Three wells have been selected for this study covering the Late Permian to Early-Middle Triassic time in the studied region. The cores were studied from a macroscopic point of view for detecting any lithological changes, stratal surfaces, sedimentary structures, and other sedimentological characteristics. Standard thin sections were prepared for rock samples collected at every 30 cm from approximately 1200 m of the core. The thin sections were stained with Alizarin Red-S, and examined under a petrographic microscope. A vehicle-mounted spectral gamma-ray probe sensor has measured the energy of each gamma ray emitted from U and Th elements in the wells, and the concentrations of the two radioelements were determined.

C and O isotope data were available for two wells from a previous study (Rahimpour-Bonab et al., 2009). Stable isotopic data were generated for one new well in the present study. CO_2 gas was extracted from powdered carbonate samples by reacting 10–20 mg of the sample with an appropriate acid, and the C and O isotope ratio was measured. The samples were dissolved, and Sr was isolated using conventional Sr separation techniques. Subsequently, the samples were analyzed using high-resolution multicollector inductively coupled plasma mass spectrometry techniques. Elemental concentration was measured using simultaneous major to ultratrace elemental analysis by fusion dissolution and inductively coupled plasma mass spectrometry techniques (Madinabeitia et al., 2008).

14.4 RESULTS

A total of 192 carbonate rock samples from the Kangan and upper Dalan formations were analyzed for their major and trace element concentrations, and C and O as well as Sr isotopic compositions with the aim of identifying chemostratigraphic trends and characterizing the contemporaneous seawater geochemistry.

14.4.1 Sr ISOTOPES

Early studies proved the applicability of Sr as a trace element in the analysis of carbonate diagenetic systems (Holland et al., 1964; Kinsman, 1969). Subsequently, the study of Sr isotopes from ancient marine carbonates had illustrated the presence of significant variations in seawater $^{87}Sr/^{86}Sr$ through time. The Sr isotopic ratio has been used as a powerful tool for the reconstruction of primary conditions and secondary changes in carbonate strata of Permian–Triassic time around the world (Korte et al., 2003 and references therein). In the studied wells, there is a gradual increase from the lower part of the upper Dalan Formation toward the upper part of the Kangan Formation. Between these two end points, the Sr ratio changes from 0.7072 in the upper Dalan up to 0.7080 in the Kangan Formation. In the PTB, there is an abrupt increase in the $^{87}Sr/^{86}Sr$ ratio (Fig. 2). Such an increase has been reported from the other studied sections in the Persian Gulf previously (Tavakoli and Rahimpour-Bonab, 2012). Confirming the mentioned work, our new data show an increasing trend at about 20 m below the PTB toward it. Such an increase could be due to the primary positive excursion or secondary diagenesis effects.

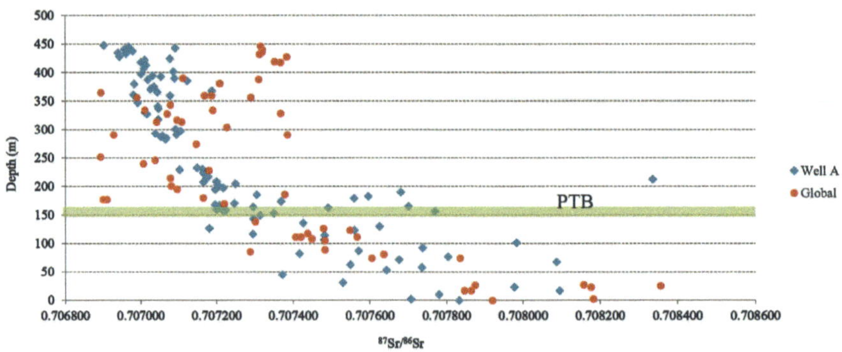

FIGURE 2

Sr isotope ratio in well A and global Sr change at the same time. High Sr at the Permian–Triassic Boundary (PTB) at the studied strata is visible.

Global Sr data from Korte et al. (2003, 2006).

14.4.2 CHARACTERISTICS OF δ13C DATA

The bulk carbonate δ13C ranges from −1.85‰ to 7.24‰ (average = 2.79), and the curve shows a gradual decrease from the bottom of the section to the PTB. In well A, there is a sharp increase at about 50 m below the boundary from −1.7‰ to 7.2‰. Although subdued, this pattern is also visible in well B. In the studied sections, the δ13C profiles exhibit a negative excursion in this boundary. There is an increasing trend after the PTB toward the top of the Kangan Formation. There are two more later excursions and these are detectable in all the three wells (Fig. 3).

The prominent negative δ13C excursion at the PTB has been reported from numerous other sections around the world (e.g., Magaritz et al., 1988; Baud et al., 1989; Holser et al., 1989; Korte et al., 2004a,b; Corsetti et al., 2005; Holser et al., 2007) and was also reported from the Persian Gulf region (Tavakoli

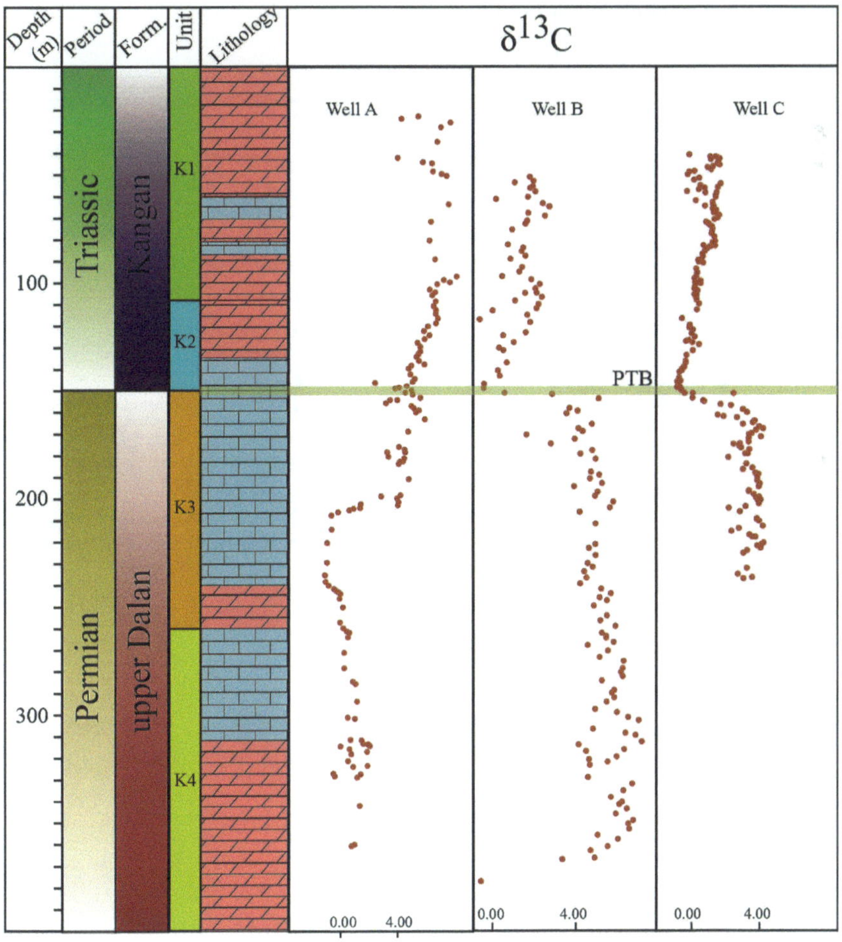

FIGURE 3

Excursion of δ13C data in the three studied wells.

and Rahimpour-Bonab, 2012). In the studied sections, the $\delta^{13}C$ decreased down to about −1.4‰ in this event, which is comparable with that in the other sections of the Permian–Triassic transitions of the world. In other wells, there is also the decreasing trend, but the lowest value is different. There is also a positive excursion below the PTB in all the wells at about 10 m below the boundary. However, this increase is not very sharp in all the studied wells.

14.4.3 CHARACTERISTICS OF $\delta^{18}O$ DATA

The $\delta^{18}O$ values show a gradual increase from the lower part of the upper Dalan Formation toward the PTB. Similar to the $\delta^{13}C$ values, there is also a negative excursion of the $\delta^{18}O$ profile in all the studied wells at the PTB. After the boundary, the values are nearly constant until the end of the Kangan Formation deposits. In the upper part of the Kangan Formation, $\delta^{18}O$ increases moderately up to −1‰ at a horizon 80 m below the Kangan–Aghar boundary (Fig. 4).

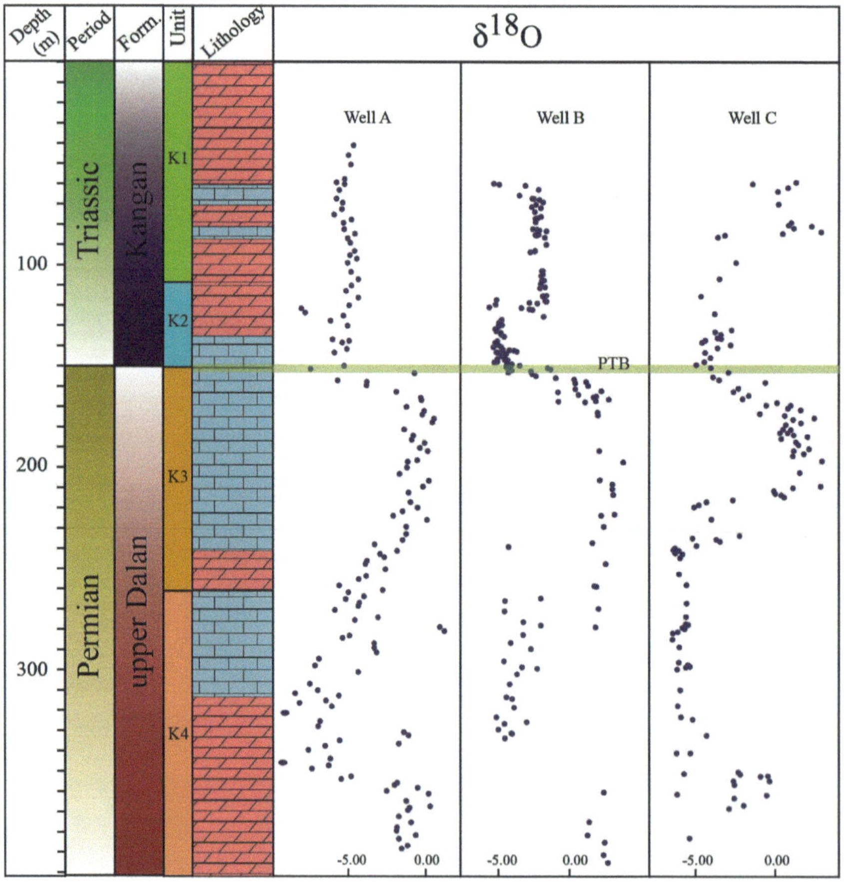

FIGURE 4

Excursion of $\delta^{18}O$ data in the three studied wells.

In contrast with the $\delta^{13}C$, the negative excursion of the $\delta^{18}O$ is not globally traceable (Heydari et al., 2001; Son et al., 2007). The Permian–Triassic sections in various places of the world show different excursions. So, changes in $\delta^{18}O$ could not be related to a global event at this time, and it could not be associated with the oceanic conditions during the Permian–Triassic transition.

14.4.4 Th/U RATIO

The Th/U ratio is a routine wire line log data studied as an indicator of the redox condition of the primary depositional environment (Wignall and Twitchett, 1996; Son et al., 2007; Ehrenberg et al., 2008) in many of the reservoirs in the world including the Persian Gulf area (Tavakoli and Rahimpour-Bonab, 2012). The study of this ratio in the three studied wells (Fig. 5) shows values <2 in almost all the samples located below the PTB. This ratio increases suddenly at the PTB (up to 12) and remains high in almost all the Triassic samples with some variations. Diagenesis could change the U concentration of the sediments, but it is not the case with the sudden changes in the PTB of the studied units because the diagenetic pattern of the Kangan and Dalan formations is almost the same in the Persian Gulf area, so the resultant ratio could not be related to the diagenetic impacts in these formations. Change in uranium concentration could also be interpreted by the change in organic content as high organic content is associated with increased uranium values in many cases (Klinkhammer and Palmer, 1991; Spirakis, 1996; Fiet and Gorin, 2000).

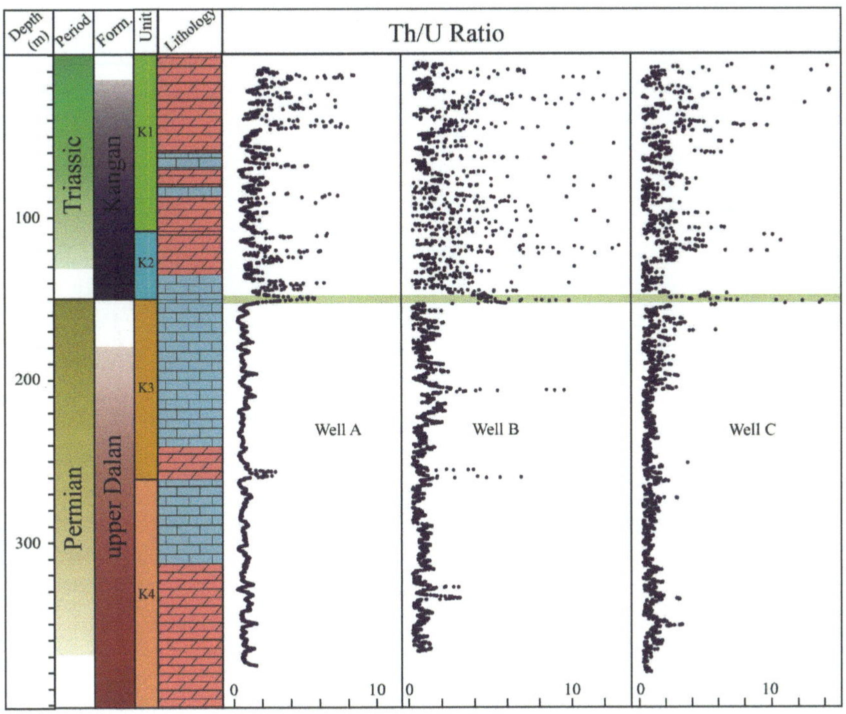

FIGURE 5

Th/U ratio in the three studied wells. This ratio increases dramatically at the Permian–Triassic Boundary and remains high in the Triassic time.

14.4.5 ELEMENTAL ANALYSIS

Sr, Mn, Fe, and Mg are elements that have been classically selected for analyzing carbonate strata in terms of primary depositional environments and diagenesis (Veizer et al., 1971; Veizer and Demovic, 1974; Walgenwitz et al., 1992; Emmanuel et al., 1999). Taking into account both sedimentological data and the diagenetic context, the variations in these elements' content in the studied formations are a good indicator of primary geochemical paleoceanic conditions.

Due to the high dolomitization rate in the Kangan and Dalan formations, the investigation of paleoceanic conditions based on the Mg content is found to be unsuitable. Variations in the Mg content in the studied formations confirm this interpretation (Fig. 6), so Mg is excluded from further discussions.

Previous studies suggested that Fe and Mn contents could show variations of detrital inputs. The Mn concentration is below the detection limit of the used technique in almost all the samples. However, the concentration of Fe is low below the PTB sediments (average 0.15% in well A), but the Fe content increase upward (average 0.24 in well A). The overall low Fe content indicates a low detrital input during the deposition of the Kangan and Dalan formations, so the depletion of Mn content is explained.

The Sr concentrations in the studied wells present some variations. At the lower part of the K4 unit, a high Sr content is noticeable (Fig. 3). The Sr values decrease gently upward (average of 2969.86 ppm

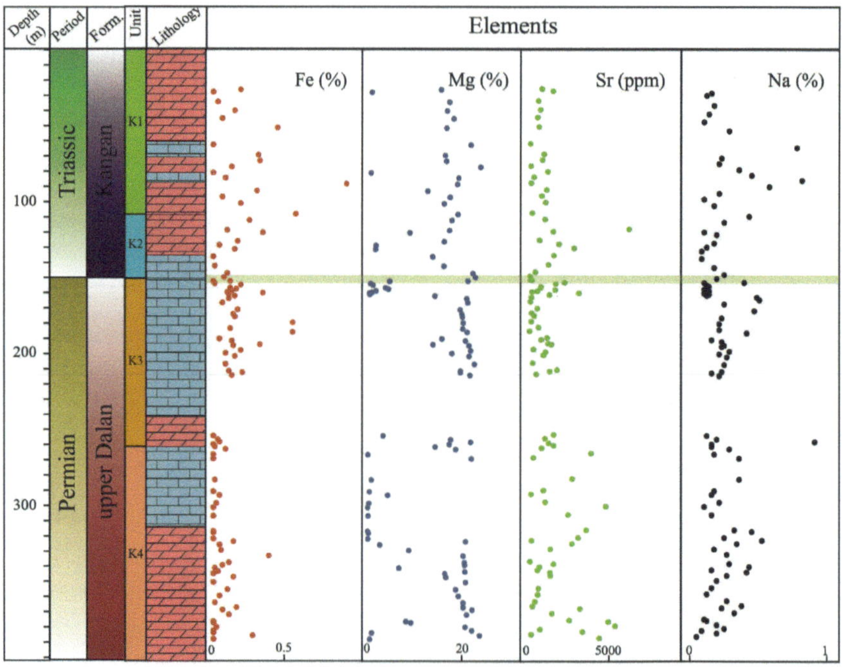

FIGURE 6

Elemental data (Fe, Mg, Sr, and Na) in well A from the Kangan and Dalan formations in the Persian Gulf area.

through the K4 unit). There is another decreasing trend from the middle of the K3 unit toward the PTB. A sharp increase could be seen just below the boundary. The increasing trend of the Sr element continues until 20 m up the Permian–Triassic transition. After the Sr peak at this point, the decreasing trend continues until the end of the studied sections.

14.5 DISCUSSION

Before the use of element content as a potential paleoceanic indicator, the diagenetic context needs to be considered to constrain the possible alterations of the primary oceanic conditions. So, the diagenetic effect on each element or isotope ratio is considered here besides each elemental variation in the studied sections.

14.5.1 Sr DATA

The application of Sr isotope ratios for age dating has been demonstrated by previous researchers (Burke et al., 1982; DePaolo and Ingram, 1985; Hess et al., 1986; DePaolo, 1986; McArthur et al., 2001). This relies on the fact that this ratio in ocean waters is relatively constant worldwide at any one point in time, but the ratio has varied in a known manner throughout geologic history. As there is minor fractionation in this ratio, marine carbonates have the same $^{87}Sr/^{86}Sr$ ratio as the seawater in which they formed. So, the Sr isotopic ratio can be used as a stratigraphic correlation tool and for dating the sediments. Late stage diagenesis can change the Sr isotope ratio (Veizer and Compston, 1974), depending on the $^{87}Sr/^{86}Sr$ ratio of the diagenetic solution. If late diagenesis has modified the original Sr absolute values (as mentioned by Frank and Lohmann, 1996), it affects the samples in their entirety.

The Sr excursions in the Permian–Triassic strata of the Persian Gulf region show the same trends with the global Sr change (Korte et al., 2003, 2006). So, the ocean chemistry in this region was the same as in the other regions in the world. However, the difference is in the PTB that the Persian Gulf data show a sharp increase. As this phenomenon is not global, a worldwide reason could not be considered for this change (Fig. 2). The high isotopic Sr ratio on a rather small-scale thickness in sediments at the PTB may have resulted from the meteoric waters with a high radiogenic ^{87}Sr content. Inference of meteoric diagenesis is also corroborated through depleted $^{18}O/^{16}O$ and the textural features indicative of meteoric diagenesis as evidenced from the petrographic study of the samples collected at and below the PTB.

14.5.2 $\delta^{13}C$ DATA

The negative excursion of the $\delta^{13}C$ isotope ratio in the PTB is a global event, as mentioned earlier. It is assumed that this phenomenon is related to the end-Permian mass extinction event. Many studies suggested that a huge input of ^{12}C-enriched carbon into the ocean system has occurred and may have caused a global environmental change. This is due to the fact that the organic bodies with negative ^{12}C (vital effect) died together and accumulated in the upper Permian strata at this time. The result is a global negative excursion of this isotope ratio in the sediments. The magnitude of this shift is also comparable with the worldwide $\delta^{13}C$ depletion at this time (e.g., Dolenec et al., 2001; Heydari and Hassanzadeh, 2003; Lehrmann et al.,

2003; Korte et al., 2004a,b). The sharp decrease of the $\delta^{13}C$ at the PTB and gradual recovery pattern in Triassic time (Fig. 3) seem to be due to the recovery of biological productivity after the huge mass extinction.

The global warming (Kidder and Worsley, 2004) caused massive methane release from sea-floor gas hydrate reservoirs and may have enhanced the sharp negative $\delta^{13}C$ excursion at the PTB (Erwin, 1993; Krull and Retallack, 2000; Wignall, 2001; Berner, 2002).

14.5.3 $\delta^{18}O$ DATA

Previous study of $\delta^{18}O$ from various sections in the Persian Gulf showed that the oxygen isotope depletion during the Permian–Triassic transition is not a global phenomenon (Tavakoli and Rahimpour-Bonab, 2012). Regarding the diagenetic effect on this boundary after the deposition of Permian strata, the negative excursion of the oxygen isotope could be due to meteoric diagenesis. The meteoric waters with low oxygen isotope ratio affected the sediments and caused lower $\delta^{18}O$. Petrographic evidences support such interpretation (Rahimpour-Bonab et al., 2009). Generally, carbon isotopes are relatively much more resistant from meteoric diagenesis than oxygen isotopes especially in carbonate rocks (Marshall, 1992). So, the depletion of carbon isotope ratio could not be interpreted completely as the result of meteoric diagenesis. This inference is also supported by the global decrease of $\delta^{13}C$ at the PTB. A local increase in $\delta^{18}O$ values at about 30 m below the boundary is due to high local evaporation. This is also evident from the Na content in the sediments, which shows the high rate of evaporation (Fig. 6).

Figure 7 shows the distribution of $\delta^{13}C$ versus $\delta^{18}O$ data in well A. As could be seen in this figure, the Permian seawater oxygen isotopic ratio hovers between −6‰ and 2‰, and carbon isotope ratio

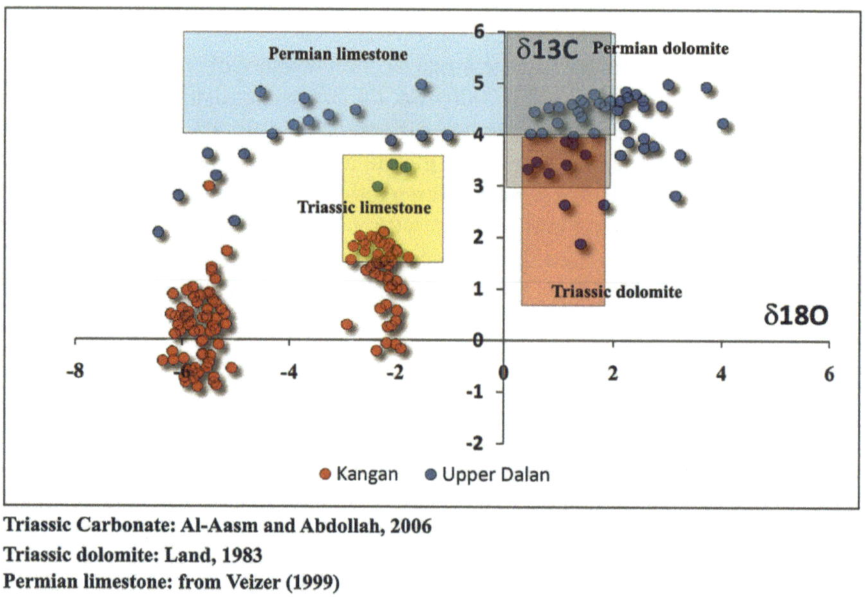

FIGURE 7

Distribution of $\delta^{13}C$ versus $\delta^{18}O$ data for Permian and Triassic samples as well as ocean signature of these isotopes at the time of deposition.

ranges between 3‰ and 6‰. Most of the Permian samples (upper Dalan) have different ratios. The maximum difference could be seen in samples near the PTB (Fig. 8). Comparing the isotope data of the Permian samples with previously proved various diagenetic environments shows that most of the samples experienced a shallow diagenetic realm (Fig. 9).

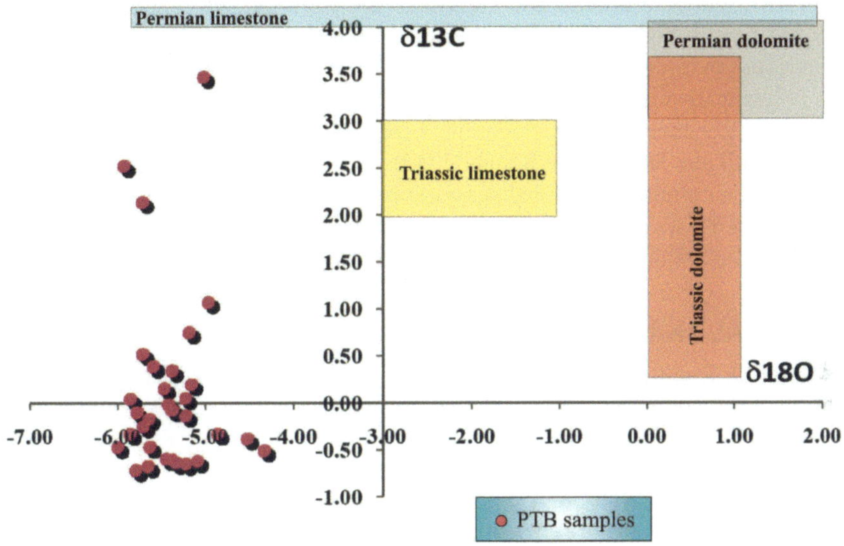

FIGURE 8

The $\delta^{13}C$ versus $\delta^{18}O$ isotope signature for near the Permian–Triassic Boundary (PTB) samples showing their signature change from the original values.

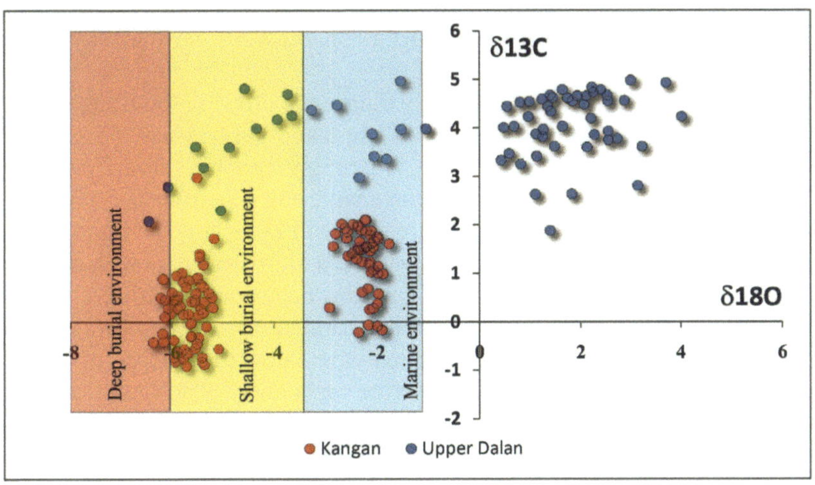

FIGURE 9

Distribution of $\delta^{13}C$ versus $\delta^{18}O$ data with diagenetic realms.

14.5.4 Th/U RATIO CHANGES

The Th/U ratio is a good indicator of environmental redox conditions (Wignall and Twitchett, 1996; Son et al., 2007). This is due to the fact that uranium is reduced to the insoluble uranous (U^{4+}) complex in anoxic conditions and can be preserved in sediments. In oxidizing situations, soluble uranyl (U^{6+}) carbonate is formed, which is mobile and carried in the solution. Such precipitation of uranium in anoxic conditions causes a higher U content in anoxic sediments. In contrast, thorium is almost insoluble (Th^{4+}) in various environmental redox conditions. So, the environment is depleted of oxygen when the Th/U ratio is lower than in oxic conditions. The change in the Th/U ratio in the studied sections could not be due to the change in organic content because there is no direct relationship between these two variables (Tavakoli and Rahimpour-Bonab, 2012).

The inferred anoxic condition of global seawater across the PTB may have affected the extinction of Permian biota, even if it was not the only reason for the extinction, since there were sessile or benthic animals that lived at the bottom of the shallow marine setting.

14.5.5 ELEMENTAL ANALYSIS

The low concentration levels of the Mn and Fe found in the upper Dalan are compatible with initial marine calcite contents and show no freshwater fluid dominated system (Brand and Veizer, 1981; Banner and Hanson, 1990). In contrast, Fe increases dramatically in the Kangan Formation showing the effect of fluvial/detrital input to the environment. Two main sources of Fe in marine carbonates are advocated, namely, the hydrothermal activity of the oceanic ridges (Emmanuel and Renard, 1993; Von Damm, 1995; Corbin et al., 2000) and inputs from the erosion of exposed continental rocks. Hydrothermal influx can be ruled out here because the ridges are a long way from the study area and also the Kangan and Dalan formations were deposited in the inner to midramp shallow marine environments. The presence of a higher shale volume in this part confirms this interpretation. This is also in agreement with the higher stylolite frequency in the Kangan Formation. The entrance of SO_2, Cl, and F to the atmosphere due to volcanic emissions may have caused acidic rains that in turn might have ruined the continental vegetation leading to an increase in the intensity of the weathering (Wignall, 2001).

The dolomitization process is a ubiquitous early diagenetic effect in the Kangan and Dalan formations in the Persian Gulf region. So, the presence of Mg could not be used as a good indicator of the primary depositional conditions. The Mg content is strongly variable through the Permian–Triassic studied interval, but the dolomitic parts generally have more Mg (Fig. 6).

It is generally considered that the sedimentary deposits of low salinity environments contain low Sr, and relatively high Fe and Mn concentrations. In contrast, a high Sr content with relatively low Fe and Mn concentrations is observable in a proximal open marine depositional setting. None of these general considerations are true for the Kangan and upper Dalan formations as Mn is always low and also there is no evidence of an open marine setting in these formations. Despite this apparent view that the highest Sr values correspond to the lowest Fe contents and vice versa, there is no significant linear correlation between Fe and Sr ($r=0.01$) or Fe and Mg (Fig. 10).

It is believed that the Sr concentration in aragonite is more than that in calcite (Tucker, 2001). So, a higher Sr content shows more aragonitic seawater. This is contemporaneous with the more alkaline conditions of the seawater at the time of deposition, as aragonite is less stable compared to calcite, and acidic conditions will shut down the precipitation of this mineral. Heydari et al. (2013) studied the

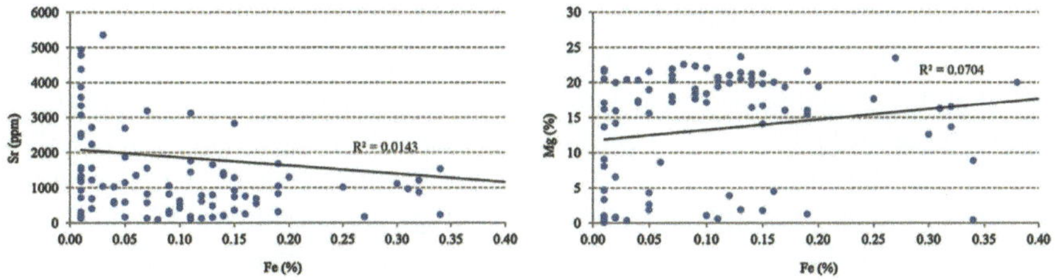

FIGURE 10

Crossplots of Fe versus Sr and Mg show no meaningful trends.

seawater conditions near the PTB in the Shareza section and described the near boundary primary seawater conditions. Their study limits to about 70 m below and 20 m above the PTB, which is comparable to part of this study. The sampling of the early Dzhulfian (equivalent to Wuchiapingian) in this study showed that the aragonite sea was dominated at the Persian Gulf region at that time. The Sr content of the samples and aragonite contribution decrease upward until the PTB. This observation also demonstrates the gradual increase of pH of the depositional environment. The acidic conditions before the PTB have been reported in many previous studies (e.g., Hinojosa et al., 2012; Beauchamp and Grasby, 2012; Heydari et al., 2013).

The sharp increase in the Sr content just below the PTB could be related to a sudden increase in the seawater pH or meteoric water input at the time and after the deposition of the upper Permian strata. As discussed previously, much of the evidence confirms the role of meteoric diagenesis at the PTB in the Persian Gulf region. Other comparable studies (e.g., Heydari et al., 2013) have not reported such an excursion at the PTB, so it could not be related to a global change at this time. So, the increase in the Sr content in the PTB could be the result of exposure-related or very shallow early meteoric diagenesis involving freshwater associated cementation and/or dissolution. If the end-Permian regression is responsible for such observations, a sequence boundary has to be visualized, and a sedimentation gap or at least sediment starvation has to be hypothesized at the end-Permian time in this region. The followed-up (Early Triassic) transgression could have eroded some sedimentological features, so Sr contents could be a good indication of an exposure event. This inference is also supported by the depleted nature of the oxygen isotope. So, there was a regression at the end-Permian time at the Persian Gulf region as confirmed by the geochemical evidences. However, this was for a very brief period.

There is a reverse order after the PTB in the Early Triassic time. Sr has an increasing trend showing the dominance of the aragonitic sea again that favored the restoration of biota. This increase continued from 50 m above the boundary and followed by a decreasing trend and stable depositional environment became dominant until the end of the section.

14.5.6 GLOBAL CORRELATION

Provided that the isotopes and elemental data are representative of the global Permian–Triassic ocean conditions, they have a high potential for accurately correlating sedimentary successions. Such a comparison allows a high-precision chemostratigraphic correlation between the studied sections in this region and with the other Permian–Triassic strata in various parts of the Tethys Ocean.

The Abadeh section of Iran was studied by Heydari et al. (2000), Korte et al. (2004a), Richoz et al. (2010), and Liu et al. (2013). The negative shift reported in the previous sections is documented just below the PTB in all these studies. In the Zal and Shahreza sections in Iran, the uppermost Permian is made up of mud-supported facies rich in pelagic fauna. The carbon isotope signature shows the same trend in all the sections. Heydari et al. (2003) reported the carbonate cement and grain-dominated facies after the PTB that were deposited in shallow marine settings indicating a rapid and major drop in the relative sea level. These authors suggested the domination of oxygenated condition in the boundary interval that did not change significantly at the boundary. However, the present study shows the domination of anoxia at the end-Permian time. Anoxic conditions have been reported from other sections of the world before (Wignall and Hallam, 1992). So, the environmental conditions were different conditions at the PTB in these locations of the Persian Gulf region.

Relatively deep marine sections of the PTB reported from Kashmir, India, located on the southern part of the Tethys on the Gondwana margin also show a negative $\delta^{13}C$ peak in the uppermost Permian, below the biostratigraphic boundary (Baud et al., 1996). The PTB sections in the Bukk Mountains at the western Tethys margin have been deposited in the distal part of a carbonate ramp system (Haas et al., 2007). The sections reported in the present study and the PTB sections located in Hungary and the Meishan (China) show a similarity in biotic decline and the pattern of $\delta^{13}C$ values.

The phenomena discussed above reflect a series of environmental changes that took place during the end-Permian and Early Triassic time in the Persian Gulf region. While some of them can be correlated with the other PTB sections of the world, others are unique to the Persian Gulf region, especially in the studied wells. In addition, the present study shows no remarkable terrestrial input and change in marine carbonate production at the studied sections, in contrast to the other PTB sections of the World. These similarities and differences were brought to light with the help of stratigraphic variations of geochemical changes.

14.6 CONCLUSIONS

The chemostratigraphy of the carbonate–evaporite successions from the Persian Gulf region was determined based on carbon, oxygen, and Sr isotope data as well as Sr, Fe, Mg, and Th/U elemental analysis. The geochemical data in all the three wells show a general concurrent trend in the Late Permian and Early Triassic time, which might be an important feature created in response to the changes in oceanic geochemistry and paleoceanic conditions.

The results showed that there is a general upward decrease in Sr content from the base of K4 toward the PTB in the studied sections. This could be due to the decreasing trend in the pH of the depositional environment or increasing PCO_2 from the early Dzhulfian to the end of Dorashamian (end-Permian). There was a regression event at the end-Permian time as indicated by the geochemical data. The Sr isotope and elemental data as well as oxygen isotope ratio indicate significant fluvial influx during the deposition of the upper few tens of meters of the upper Dalan Formation.

The late Dorashamian is characterized by a gradual positive trend of $\delta^{18}O$ isotope followed by a negative trend that culminated at the negative shift across the PTB of all the studied wells. This negative shift is not related to a global event, so it may not serve as a potential chemostratigraphic marker for

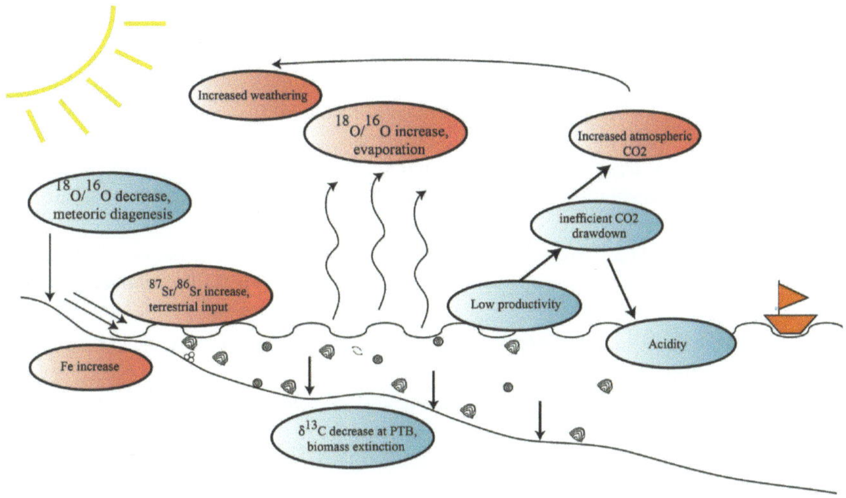

FIGURE 11

Summary and relationships of Permian–Triassic events in the studied intervals. The $\delta^{18}O$ increase happened about the 30 m below the boundary and depleted at the Permian–Triassic Boundary (PTB).

global correlation in this region. The negative shift of $\delta^{13}C$, in contrast, is a global event and could be correlated worldwide. Disappearance of the Permian faunal assemblages is accompanied by a global negative $\delta^{13}C$ peak at the PTB that may be used as a chemostratigraphic marker of this boundary. Subsequently, microbial thrombolites appear that are regionally traceable and may be correlated with peculiar carbonate cement layers and oolite-dominated facies in the other sections. This might have potentially resulted from the enhanced organic carbon burial and subsequent ^{12}C enrichment of the sediments. The gradual positive shift of the carbon isotopes after the PTB is related to the biotic recovery after the end-Permian mass extinction.

Comparison of the Persian Gulf sections with other marine boundary sections in the western to eastern Tethys revealed some similarities in the nature of the geochemistry and paleooceanic conditions of the end-Permian events that include environmental changes (e.g., negative shift in $\delta^{13}C$ values, abrupt Th/U change and subsequent oxidation, postextinction microbialite). However, there are also some features (e.g., oxygen isotope depletion, no change in lithology), which are attributed mainly to the local characteristics of the depositional environment. A summary of the end-Permian events and their relations is presented in Fig. 11.

ACKNOWLEDGMENTS

The vice president of Research and Technology of the University of Tehran provided financial support for this research, for which we are grateful. We also extend our appreciation to the POGC (Pars Oil and Gas Company of Iran) and MAPSA (Abdal Industrial Projects Management Company) for sponsoring, data preparation, and permission to publish this study. The authors thank M. Naderi for her contributions and for fruitful discussions, as well as J. I. Gil Ibarguchi (Spain) for his contribution in the elemental analysis.

REFERENCES

Al-Aasm, I.S., Azmy, K., 1996. Diagenesis and evolution of microporosity of Middle-Upper Devonian Kee Scarp reefs, Norman Wells, N.W.T., Canada: petrologic and chemical evidence. American Association of Petroleum Geologists Bulletin 80, 82–100.

Al-Jallal, A.I., 1994. Stratigraphy, depositional environments and reservoir potential of the Khuff formation in the Arabian Gulf countries. In: Proceedings of the 6th Abu Dhabi International Petroleum Exhibition and Conference, p. 402.

Algeo, T.J., Hannigan, R., Rowe, H., Brookfield, M., Baud, A., Krystyn, L., Ellwood, B.B., 2007. Sequencing events across the Permian–Triassic boundary, Guryul Ravine (Kashmir, India). Palaeogeography, Palaeoclimatology, Palaeoecology 252, 328–346.

Alsharhan, A.S., 1993. Facies and sedimentary environment of the Permian carbonates (Khuff formation) in the United Arab Emirates. Sedimentary Geology 84, 89–99.

Alsharhan, A.S., Nairn, A.E.M., 1997. Sedimentary Basins and Petroleum Geology of the Middle East. Elsevier, Amsterdam. p. 878.

Banner, J.L., Hanson, G.N., 1990. Calculation of simultaneous isotopic and trace element variations during water–rock interaction with applications to carbonate diagenesis. Geochimica et Cosmomchimica Acta 54, 3123–3137.

Baud, A., Atudorei, N.V., Sharp, Z., 1996. Permian and Early Triassic evolution of the Northern Indian margin: carbon isotope and sequence stratigraphy. Geodinamica Acta 9, 57–77.

Baud, A., Margaritz, M., Holser, W.T., 1989. Permian–Triassic of the Tethys: carbon isotope studies. Geologische Rundschau 78, 647–677.

Beauchamp, B., Grasby, S.E., 2012. Permian lysocline shoaling and ocean acidification along NW Pangea led to carbonate eradication and chert expansion. Palaeogeography, Palaeoclimatology, Palaeoecology 350–352, 73–90.

Becker, L., Poreda, R.J., Hunt, A.G., Bunch, T.E., Rampino, M., 2001. Impact event at the Permian–Triassic boundary: evidence from extraterrestrial noble gases in fullerenes. Science 291, 1530–1533.

Berner, R.A., 2002. Examination of hypothesis for the Permo–Triassic boundary extinction by carbon cycle modeling. Proceedings of the National Academy of Sciences 99, 4172–4177.

Beydoun, Z.R., 1991. Arabian plate hydrocarbon geology and potential—a plate tectonic approach. AAPG Studies in Geology 33, 77.

Bowring, S.A., Erwin, D.W., Jin, Y.G., Martin, M.W., Davidek, K., Wang, W., 1998. U/Pb zircon geochronology and temp of the end-Permian mass extinction. Science 280, 1039–1045.

Brand, U., Veizer, J., 1981. Chemical diagenesis of multicomponent carbonate system 2: stable isotopes. Journal of Sedimentary Petrology 51, 987–997.

Burke, W.H., Denison, R.E., Hetherington, E.A., Koepnick, R.B., Nelson, H.F., Oto, J.B., 1982. Variation of seawater $^{87}Sr/^{86}Sr$ throughout Phanerozoic time. Geology 10, 516–519.

Burne, R.V., Moore, L.S., 1987. Microbialites: organosedimentary deposits of benthic microbial communities. Palaios 2, 241–254.

Corbin, J.C., Person, A., Iatzoura, A., Ferre, B., Renard, M., 2000. Manganese in pelagic carbonates: indication of major tectonic events during the geodynamic evolution of a passive continental margin. Palaeogeography, Palaeoclimatology, Palaeoecology 156, 123–138.

Corsetti, F.A., Baud, A., Marenco, P.J., Richoz, S., 2005. Summary of Early Triassic carbon isotope records. Comptes Rendus Palevol 4, 473–486.

DePaolo, D.J., 1986. Detailed record of the Neogene Sr isotopic evolution of seawater from DSDP site 590B. Geology 14, 103–106.

DePaolo, O.J., Ingram, B.L., 1985. High-resolution stratigraphy with strontium isotopes. Science 227, 938–941.

Dolenec, T., Lojen, S., Ramovs, A., 2001. The Permian–Triassic boundary in Western Slovenia (Idrijca Valley section): magnetostratigraphy, stable isotopes, and elemental variations. Chemical Geology 175, 175–190.

Edgell, H.S., 1996. Salt tectonics in the Persian Gulf basin. In: Alsop, G.L., Blundell, D.L., Davison (Eds.), Salt Tectonics, vol. 100. Special Publication of the Geological Society of London, pp. 129–151.

Ehrenberg, S.N., Svana, T.A., Swart, P.K., 2008. Uranium depletion across the Permian–Triassic boundary in Middle East carbonates: signature of oceanic anoxia. American Association of Petroleum Geologists Bulletin 92, 691–707.

Ellis, J., Schramm, D.N., 1995. Could a nearby supernova explosion have caused a mass extinction. Proceedings of the National Academy of Sciences 92, 235–238.

Emmanuel, L., Renard, M., 1993. Carbonate geochemistry (Mn, $\delta^{13}C$, $\delta^{18}O$) of the late Tithonian-Berriasian pelagic limestones of the Vocontian trough (SE France). Bulletin of the Centres Rech Elf Exploration and Production 17, 205–221.

Emmanuel, L., Vincent, B., Renard, M., De Carlo, E.H., 1999. Evolution of Mn contents of Neogene carbonates along the Bahamas transect (ODP Leg 166). Relationship of geochemical data to sea-level change. In: Proceedings of the 5th International Symposium Geochemistry of Earth Surface, Reykjavik, Balkema Rotterdam, pp. 315–318.

Erwin, D.H., 1993. The Great Paleozoic Crisis: Life and Death in the Permian. Columbia University Press, New York. p. 327.

Erwin, D.H., Bowring, S.A., Yugan, J., 2002. End-Permian mass extinction: a review. In: Koeberl, C., MacLeod, K.G. (Eds.), Catastrophic Events and Mass Extinctions: Impacts and beyond, vol. 356. Geological Society of America Special Paper, pp. 363–383.

Esrafili-Dizaji, B., Rahimpour-Bonab, H., 2009. Effects of depositional and diagenetic characteristics on carbonate reservoir quality: a case study from the South Pars Gas field in the Persian Gulf. Petroleum Geoscience 15, 325–344.

Ezaki, Y., Liu, J., Adach, N., 2003. Earliest Triassic microbialite micro- and megastructures in the Huaying area of Sichuan Province, south China: implications for the nature of oceanic conditions after the end-Permian extinction. Palaios 18, 388–402.

Fiet, N., Gorin, G.E., 2000. Gamma-ray spectrometry as a tool for stratigraphic correlations in the carbonate dominated, organic rich, pelagic Albian sediments in central Italy. Eclogae Geologicae Helvetiae 93, 175–181.

Fluteaua, F., Bessea, J., Broutinb, J., Ramstein, G., 2001. The Late Permian climate. What can be inferred from climate modelling concerning Pangea scenarios and Hercynian range altitude? Palaeogeography, Palaeoclimatology, Palaeoecology 167, 39–71.

Frank, T., Lohmann, K.G., 1996. Diagenesis of fibrous magnesian calcite marine cement: implications for the interpretation of d18 and $\delta^{13}C$ values of ancient equivalents. Geochimica et Cosmomchimica Acta 60, 2427–2436.

Grotzinger, J.P., Knoll, A.H., 1995. Anomalous carbonate precipitates: is the Precambrian the key to the Permian? Palaios 10, 578–596.

Haas, J., Demeny, A., Hips, K., Zajzon, N., Weiszburg, T.G., Sudar, M., Palfy, J., 2007. Biotic and environmental changes in the Permian–Triassic boundary interval recorded on a western Tethyan ramp in the Bukk Mountains, Hungary. Global and Planetary Change 55, 136–154.

Hess, J., Bender, M.L., Schilling, J.G., 1986. Sea-water $^{87}Sr/^{86}Sr$ evolution from cretaceous to present—applications to paleoceanography. Science 231, 979–984.

Heydari, E., Arzani, N., Hassanzadeh, J., 2008. Mantle plume: the invisible serial killer—application to the Permian–Triassic boundary mass extinction. Palaeogeography, Palaeoclimatology, Palaeoecology 264, 147–162.

Heydari, E., Arzani, N., Safaei, M., Hassanzadeh, J., 2013. Ocean's response to a changing climate: clues from variations in carbonate mineralogy across the Permian–Triassic boundary of the Shareza section, Iran. Global and Planetary Change 105, 79–90.

Heydari, E., Hassandzadeh, J., Wade, W.J., 2000. Geochemistry of central Tethyan upper Permian and lower Triassic strata, Abadeh region, Iran. Sedimentary Geology 137, 85–99.

Heydari, E., Hassanzadeh, J., 2003. Deevjahi model of the Permian–Triassic boundary mass extinction: a case for gas hydrates as the main cause of biological crisis on Earth. Sedimentary Geology 163, 147–163.

Heydari, E., Hassanzadeh, J., Wade, W.J., Ghazi, A.M., 2003. Permian–Triassic boundary interval in the Abadeh section of Iran with implications for the mass extinction: part 1. Sedimentology. Palaeogeography, Palaeoclimatology, Palaeoecology 193, 405–423.
Heydari, E., Wade, W.J., Hassanzadeh, J., 2001. Diagenetic origin of carbon and oxygen isotope compositions of Permian–Triassic boundary strata. Sedimentary Geology 143, 191–197.
Hinojosa, J.L., Brown, S.T., Chen, J., DePaolo, D.J., Paytan, A., Shen, S., Payne, J.L., 2012. Evidence for end-Permian ocean acidification from calcium isotopes in biogenic apatite. Geology 40, 743–746.
Hips, K., Haas, J., 2006. Calci microbial stromatolites at the Permian–Triassic boundary in a western Tethyan section, Bukk Mountains, Hungary. Sedimentary Geology 185, 239–253.
Holland, H.D., Kirsipu, T.V., Huebner, J.S., Oxburgh, U.M., 1964. On some aspects of the chemical evolution of cave waters. Journal of Geology 72, 36–67.
Holser, W.T., Schonlaub, H.P., Attrep, M., 2007. Evidence for recurrent changes in Lower Triassic oceanic circulation of the Tethys: the $\delta^{13}C$ record from marine sections in Iran. Palaeogeography, Palaeoclimatology, Palaeoecology 252, 355–369.
Holser, W.T., Schonlaub, H.P., Attrep, M., Boeckelmann, K., Klein, P., Magaritz, M., Orth, C.J., Fenninger, A., Jenny, C., Kralik, M., Mauritsch, H., Pak, E., Schramm, J.M., Stattegger, K., Schmoller, R., 1989. A unique geochemical record at the Permian/Triassic boundary. Nature 337, 39–44.
Isozaki, Y., 1997. Permo–Triassic boundary superanoxia and stratified superocean: record from the deep sea. Science 276, 235–238.
Kaiho, K., Chen, Z., Kawahata, H., Kajiwara, Y., Sato, H., 2006. Close-up of the end-Permian mass extinction horizon recorded in the Meishan section, South China: sedimentary, elemental, and biotic characterization and a negative shift of sulfate sulfur isotope ratio. Palaeogeography, Palaeoclimatology, Palaeoecology 239, 396–405.
Kershaw, S., Guo, L., Swift, A., Fan, J., 2002. Microbialites in the Permian–Triassic boundary interval in central China: structure, age, and distribution. Facies 47, 83–90.
Kershaw, S., Zhang, J., Lan, G., 1999. A microbialite carbonate crust at the Permian–Triassic boundary in south China, and its palaeoenvironmental significance. Palaeogeography, Palaeoclimatology, Palaeoecology 146, 1–18.
Kidder, D.L., Worsley, T.R., 2004. Causes and consequences of extreme Permo–Triassic warming to globally equable climate and relation to the Permo–Triassic extinction and recovery. Palaeogeography, Palaeoclimatology, Palaeoecology 203, 207–237.
Kinsman, D.J.J., 1969. Interpretation of Sr concentration in carbonate minerals and rocks. Journal of Sedimentary Petrology 39, 486–508.
Klinkhammer, G.P., Palmer, M.R., 1991. Uranium in the oceans: where it goes and why. Geochimica et Cosmochimica Acta 55, 1799–1806.
Knoll, A.H., Bambach, A.K., Canfield, D.E., Grotzinger, J.P., 1996. Comparative Earth history and Late Permian mass extinction. Science 273, 452–457.
Konert, G., Afif, A.M., AL-Hajari, S.A., Droste, H., 2001. Paleozoic stratigraphy and hydrocarbon habitat of the Arabian Plate. GeoArabia 6, 407–442.
Korte, C., Jasper, T., Kozur, H.W., Veizer, J., 2006. $^{87}Sr/^{86}Sr$ record of Permian seawater. Palaeogeography, Palaeoclimatology, Palaeoecology 240, 89–107.
Korte, C., Kozur, H.W., Bruckschen, P., Veizer, J., 2003. Strontium isotope evolution of Late Permian and Triassic seawater. Geochimica et Cosmochimica Acta 67, 47–62.
Korte, C., Kozur, H.W., Joachimski, M.M., Strauss, H., Veizer, J., Schwark, L., 2004a. Carbon, sulfur, oxygen and strontium isotope records, organic geochemistry and biostratigraphy across the Permian/Triassic boundary in Abadeh, Iran. International Journal of Earth Sciences 93, 565–581.
Korte, C., Kozur, H.W., Mohtat-Aghai, P., 2004b. Dzhulfian to lowermost Triassic $\delta^{13}C$ record at the Permian/Triassic boundary section at Shahreza, Central Iran. Hallesches Jahrbuch der Geowissenschaften Beiheft 18, 73–78.
Krull, E.S., Retallack, G.J., 2000. $\delta^{13}C$ depth profile from paleosols across the Permian–Triassic boundary: evidence for methane release. Geological Society of America Bulletin 112, 1459–1472.

Kump, L.R., Pavlov, A., Arthur, M.A., 2005. Massive release of hydrogen sulfide to the surface ocean and atmosphere during interval of oceanic anoxia. Geology 33, 397–400.

Land, L.S., 1983. The application of stable isotopes to studies of the origin of dolomite and to problems of diagenesis of clastic sediments. In: Arthur (Ed.), Stable Isotopes in Sedimentary Geology SEPM Short Course, vol. 10, pp. 4.1–4.22.

Lehrmann, D.J., 1999. Early Triassic calcimicrobial mounds and biostromes of the Nanpanjiang Basin, South China. Geology 27, 359–362.

Lehrmann, D.J., Payne, J.L., Felix, S.V., Dillett, P.M., Wang, H., Yu, Y., Wei, J., 2003. Permian–Triassic boundary sections from shallow-marine carbonate platforms of the Nanpanjiang Basin, South China: implications for oceanic conditions associated with the end-Permian extinction and its aftermath. Palaios 18, 138–152.

Liu, X., Wang, W., Shen, S., Gorgij, M.N., Ye, F., Zhang, Y., Furuyama, S., Kano, A., Chen, X., 2013. Late Guadalupian to Lopingian (Permian) carbon and strontium isotopic chemostratigraphy in the Abadeh section, central Iran. Gondwana Research 24, 222–232.

Madi, A., Savard, M.M., Bourque, P.A., Chi, G., 2000. Hydrocarbon potential of the Mississippian carbonate platform, Bechar Basin, Algerian Sahara. AAPG Bulletin 84, 266–287.

Madinabeitia, S.G., Lorda, M.E.S., Ibarguchi, J.I.G., 2008. Simultaneous determination of major to ultratrace elements in geological samples by fusion-dissolution and inductively coupled plasma mass spectrometry techniques. Analytica Chimica Acta 625, 117–130.

Magaritz, M., Baer, R., Baud, A., Holser, W.T., 1988. The carbon isotope shift at the Permian/Triassic boundary in the southern Alps is gradual. Nature 331, 337–339.

Marshall, J.D., 1992. Climatic and oceanographic isotopic signals from the carbonate record and their preservation. Geological Magazine 129, 143–160.

McArthur, J.M., Ravenscroft, P., Safiullah, S., Thirlwall, M.F., 2001. Arsenic in groundwater: testing pollution mechanisms for sedimentary aquifers in Bangladesh. Water Resources Research 37, 109–117.

Mouthereau, F., Lacombe, O., Verges, J., 2012. Building the Zagros collisional orogen: timing, strain distribution and the dynamics of Arabia/Eurasia plate convergence. Tectonophysics 532–535, 27–60.

Perotti, C.R., Carruba, S., Rinaldi, M., Bertozzi, G., Feltre, L., Rahimi, M., 2011. The Qatar–South Fars Arch development (Arabian platform, Persian Gulf): insights from Seismic interpretation and analogue modelling. In: Schattner, U. (Ed.), New Frontiers in Tectonic Research—At the Midst of Plate Convergence. InTech. Available from: http://www.intechopen.com/books/new-frontiers-in-tectonic-researchat-the-midst-of-plate-convergence/the-qatar-south-fars-arch-development-arabian-platform-persian-gulfinsights-from-seismic-interpreta (accessed November 2013).

Pruss, S.B., Bottjer, D.J., Corsetti, F.A., Baud, A., 2006. A global marine sedimentary response to the end-Permian mass extinction: examples from southern Turkey and the western United States. Earth-Science Reviews 78, 193–206.

Rahimpour-Bonab, H., Asadi-Eskandar, A., Sonei, R., 2009. Controls of Permian–Triassic boundary over reservoir characteristics of South Pars Gas field, Persian Gulf. Geological Journal 44, 341–364.

Renne, P.R., Zichao, Z., Richards, M.A., Black, M.T., Basu, A.R., 1995. Synchrony and causal relations between Permian–Triassic boundary crises and Siberian flood volcanism. Science 269, 1413–1416.

Retallack, G.J., Jahren, A.H., 2008. Methane release from igneous intrusion of coal during Late Permian extinction events. Journal of Geology 116, 1–20.

Retallack, G.J., Seyedolali, A., Krull, E.S., Holser, W.T., Ambers, C.P., Kyte, F.T., 1998. Search for evidence of impact at the Permian–Triassic boundary in Antarctica and Australia. Geology 26, 979–982.

Richoz, S., Krystyn, L., Baud, A., Brandner, R., Horacek, M., Mohtat-Aghai, P., 2010. Permian–Triassic boundary interval in the Middle East (Iran and N. Oman): progressive environmental change from detailed carbonate carbon isotope marine curve and sedimentary evolution. Journal of Asian Earth Sciences 39, 238–253.

Sano, H., Nakashima, K., 1997. Lowermost Triassic (Griesbachian) microbial bindstone–cementstone facies, south-west Japan. Facies 36, 1–24.

Schneider, J.W., Korner, F., Roscher, M., Kroner, U., 2006. Permian climate development in the northern peri-Tethys area—the Lodève basin, French Massif Central, compared in a European and global context. Palaeogeography, Palaeoclimatology, Palaeoecology 240, 161–183.

Sharland, P.R., Archer, R., Casey, D.M., Davies, R.B., Hall, S.H., Heward, A.P., Horbury, A.D., Simmons, M.D., 2001. Arabian Plate Sequence Stratigraphy. GeoArabia Special Publication, vol. 2,Gulf PetroLink, Manama. p. 372.

Son, T.H., Koeberl, C., Ngocb, N.L., Huyenb, D.T., 2007. The Permian–Triassic boundary sections in northern Vietnam (Nhi Tao and Lung Cam sections): carbon-isotope excursion and elemental variations indicate major anoxic event. Palaeoworld 16, 51–66.

Spirakis, C.S., 1996. The roles of organic matter in the formation of uranium deposits in sedimentary rocks. Ore Geology Reviews 11, 53–69.

Stampfli, G.M., Borel, G.D., 2002. A plate tectonic model for the Paleozoic and Mesozoic constrained by dynamic plate boundaries and restored synthetic oceanic isochrones. Earth and Planetary Science Letters 196, 17–33.

Stanley, S.M., 1988. Paleozoic mass extinctions; shared patterns suggest global cooling as a common cause. American Journal of Science 288, 334–352.

Tavakoli, V., Rahimpour-Bonab, H., Esrafili-Dizaji, B., 2011. Diagenetic controlled reservoir quality of South Pars Gas field, an integrated approach. Comptes Rendus Geoscience 343, 55–71.

Tavakoli, V., Rahimpour-Bonab, H., 2012. Uranium depletion across Permian–Triassic Boundary in Persian Gulf and its implications for paleooceanic conditions. Palaeogeography, Palaeoclimatology, Palaeoecology 350–352, 101–113.

Tucker, M.E., 2001. Sedimentary petrology: An introduction to the origin of sedimentary rocks, third ed. Wiley-Blackwell. p. 272.

Twitchett, R.J., Looy, C.V., Morante, R., Visscher, H., Wignall, P.B., 2001. Rapid and synchronous collapse of marine and terrestrial ecosystems during the end-Permian biotic crisis. Geology 29, 351–354.

Veizer, J., Demovic, R., 1974. Strontium as a tool in facies analysis. Journal of Sedimentary Petrology 44, 93–115.

Veizer, J., Ala, D., Azmy, K., Bruckschen, P., Buhl, D., Bruhn, F., Carden, G.A.F., Diener, A., Ebneth, S., Godderis, Y., Jasper, T., Korte, C., Pawellek, F., Podlaha, O.G., Strauss, H., 1999. $^{87}Sr/^{86}Sr$, $\delta^{13}C$ and $\delta^{18}O$ evolution of Phanerozoic seawater. Chemical Geology 161, 59–88.

Veizer, J., Compston, W., 1974. $^{87}Sr/^{86}Sr$ composition of seawater during the Phanerozoic. Geochimica et Cosmochimica Acta 38, 1461–1484.

Veizer, J., Demovic, R., Turan, J., 1971. Possible use of strontium in sedimentary carbonates rocks as a paleoenvironmental indicator. Sedimentary Geology 5, 5–22.

Von Damm, K.L., 1995. Temporal and compositional diversity in sea-floor hydrothermal fluids. Reviews of Geophysics 33, 1297–1305.

Walgenwitz, F., Le Dluz, A., Eichenseer, H., 1992. Isotope and trace element record of relative sea-level in Albian carbonates from Congo Atlantic margin. In: Kharaka, Y.K., Maest, A.S. (Eds.), 7th Int Symp Water–Rock Interaction, Balkema, Rotterdam, pp. 487–491.

Wang, W., Kano, A., Okumura, T., Ma, Y., Matsumoto, R., Matsuda, N., Ueno, K., Chen, X., Kakuwa, Y., Gharaie, M.H.M., Rahmati, M., 2007. Isotopic chemostratigraphy of the microbialite-bearing Permian–Triassic boundary section in the Zagros Mountains, Iran. Chemical Geology 244, 708–714.

Wignall, P., 2001. Large igneous provinces and mass extinctions. Earth-Science Reviews 53, 1–33.

Wignall, P.B., Hallam, A., 1992. Anoxia as a cause of the Permian/Triassic mass extinction: facies evidence from northern Italy and the western United States. Palaeogeography, Palaeoclimatology, Palaeoecology 93, 21–46.

Wignall, P.B., Twitchett, R.J., 1996. Oceanic anoxia and the end Permian mass extinction. Science 372, 1155–1158.

Wignall, P.B., Twitchett, R.J., 2002. Extent, duration, and nature of the Permian Triassic superanoxia. In: Koeberl, C., MacLeod, K.G. (Eds.), Catastrophic Events and Mass Extinctions: Impacts and beyond, vol. 356. Geological Society of America, Special Publication, pp. 395–423.

Wit, M.J., Ghosh, J.G., Villiers, S., Rakotosolofo, N., Alexander, J., Tripathi, A., Looy, C., 2002. Multiple organic carbon isotope reversals across the Permian–Triassic boundary of terrestrial Gondwana sequences: clues to extinction patterns and delayed ecosystem recovery. Journal of Geology 110, 227–240.

Woods, A.D., 2005. Paleoceanographic and paleoclimatic context of Early Triassic time. Comptes Rendus Palevol 14, 463–472.

Ziegler, M.A., 2001. Late Permian to Holocene Paleofacies Evolution of the Arabian Plate and its Hydrocarbon Occurrences. GeoArabia 6, 445–504.

CHAPTER

THE POSITION OF THE ORDOVICIAN–SILURIAN BOUNDARY IN ESTONIA TESTED BY HIGH-RESOLUTION $\delta^{13}C$ CHEMOSTRATIGRAPHIC CORRELATION

15

Leho Ainsaar[1], Janika Truumees[2], Tõnu Meidla[1]

[1]Department of Geology, University of Tartu, Tartu, Estonia; [2]Institute for Geology and Palaeontology, University of Hamburg, Bundesstrasse, Hamburg, Germany

15.1 INTRODUCTION

The primary approach utilized in defining the boundaries of global chronostratigraphic units in the Phanerozoic has traditionally been biostratigraphy. This approach is used, exclusively or in combination with other methods, by definition of the majority of stage boundaries in the stratigraphic interval from Cambrian to Pliocene. Magnetostratigraphy that has been used for primary definition of the boundaries within the Pliocene and Pleistocene is often applied for older strata as well, in combination with biostratigraphy or other methods.

A new tool of stratigraphic correlation that emerged in the twentieth century is chemostratigraphy. This tool is based on chemical archives in sedimentary rocks and utilizes marine or terrestrial sedimentary records of various geochemical proxies that document changes in oceanography and climate. Several of recorded changes are related to large-scale or global events. The geochemical records of such changes could be used effectively as isochronous stratigraphic markers (e.g., Weissert et al., 2008).

Stable isotope geochemistry is most widely used in chemostratigraphic studies. The basic principles of using changing ratios of stable carbon and oxygen isotopes in stratigraphy and paleoenvironmental analysis were established in 1950s and refined during the next decades. It has also found its utility as markers for definition of the stratigraphic boundaries of global importance.

Good example for chemostratigraphic scale is the stable oxygen isotope stratigraphy developed for late Quaternary (Shackleton and Opdyke, 1977). One of the pioneer works in the stable carbon isotope chemostratigraphy was a study by Scholle and Arthur (1980) who described seven global $\delta^{13}C$ excursions in the Cretaceous system, several of them are associated with important stratigraphic boundaries (including Aptian–Albian, Cenomanian–Turonian, Jurassic–Cretaceous, and Cretaceous–Paleogene). The first formal definition of a chronostratigraphic boundary that integrated the results of stable oxygen isotopic geochemistry was that of the base of Aquitanian stage (Miocene), ratified in 1996 (Steininger et al., 1997). The first application of changes in stable carbon isotopic ratio as the primary marker in

Chemostratigraphy. http://dx.doi.org/10.1016/B978-0-12-419968-2.00015-7

defining a global chronostratigraphic unit was the base of Ypresian (Paleogene), ratified in 2003 (Aubry et al., 2007). The primary definition of the base of Serravallian Stage (Middle Miocene) at the oxygen isotopic event (global cooling episode) Mi3b was ratified in 2007 (Hilgen et al., 2009).

The use of stable isotopic markers in global chronostratigraphy seems to be extending further. The initial boundary definition of the Gelasian Stage in Pliocene (Rio et al., 1998) was improved in course of redefining the base of the Quaternary System, when a reference to the marine oxygen isotope stage 103 was added (Gibbard and Head, 2010). References to possible stable isotopic criteria for definition of boundaries of the chronostratigraphic units are also published for the Chattian, Albian, and Olenekian stages (www.stratigraphy.org), as well as for the Ediacaran System (Knoll et al., 2006). Apart from formal definitions, there are many more stratigraphic boundaries where the stable isotopic changes seem to play an important role or may serve as a secondary marker. The stable isotopic excursions are also mentioned in the context of major extinctions during/across the end Triassic (e.g., Williford et al., 2007), the Permian/Triassic (e.g., Fio et al., 2010), the Late Devonian (e.g., Buggish and Joachimski, 2006), and the latest Ordovician (e.g., Brenchley et al., 1994).

Chemostratigraphic correlation of the sections is particularly important in the intervals of major biotic changes. The shapes of the isotopic curves may reveal gaps in the successions (e.g., Brenchley et al., 2003). The isotopic changes may also help to correlate barren intervals in the sedimentary successions and validate the correlation across biofacies zones. These opportunities of correlation may add new important aspects in deciphering the complex environmental and evolutionary patterns created by major perturbations of the global climate system.

The stratigraphy of the Ordovician/Silurian boundary interval has attracted a great deal of interest among geoscientists in the recent years in different parts of the world. This is due to the increasing success in deciphering the paleoenvironmental signal indicative of cooling and glaciation from this stratigraphic interval and accumulating new data on related biotic changes. This cooling—the Hirnantian glacial event—caused a global biotic crisis and is well expressed in sea level, facies, and faunal changes in tropical shelves of different continents. The environmental and oceanographic changes related to this glaciation are expressed in marine carbonates as changes in the stable carbon isotope composition. The $\delta^{13}C_{carb}$ values of the sediments show a positive excursion (the Hirnantian carbon isotope excursion, HICE), starting from the earliest Hirnantian, rising to a peak of up to 7‰, and falling back to original values in the mid or late Hirnantian, below the Ordovician/Silurian boundary (Marshall et al., 1997; Kump et al., 1999; Brenchley, 2004). A similar trend is shown by the $\delta^{13}C_{org}$ values (Underwood et al., 1997; LaPorte et al., 2009; Fan et al., 2009). The HICE has been widely used in chemostratigraphic correlation of the global Hirnantian Stage in different continents and basins. However, recent detailed bio- and chemostratigrapic studies show that both the position of the base of Hirnantian and base of the Silurian System could not be established without uncertainties or do not fit traditional stratigraphic schemes in several local successions where HICE is documented (Delabroye and Vecoli, 2010).

The Baltoscandian paleobasin offers one of the best studied and most representative Lower Paleozoic successions of the world, with a regional chronostratigraphic framework that is extremely detailed. The Estonian regional stage standard is widely used for the Ordovician succession of Baltoscandia (e.g., Webby et al., 2004). Traditionally, the Porkuni Regional Stage is considered to be a regional time equivalent for the Hirnantian Global Stage, while its boundary with the overlying Juuru Regional Stage is traditionally referred to as the Ordovician/Silurian boundary in the region (Männil and Meidla, 1994; Webby et al., 2004; Nõlvak et al., 2006). This consideration was supported by the general distribution of the shelly fauna and by the stratigraphic position of the HICE that was found to be equivalent to the whole Porkuni Stage in Estonia (Kaljo et al., 2001, 2008).

The base of the Silurian System is now fixed by an international agreement at the level of first appearance of the graptolite species *Akidograptus ascensus* (Rong et al., 2008). As this graptolite is unknown in the carbonate-dominated shelves, like Estonian or US Midcontinent successions, other evidence must be used for establishing the position of the systemic boundary. Additional difficulties in drawing this boundary in non-graptolitiferous shelf successions are due to widespread unconformities of glacioeustatic origin in this interval, and to indications that the change from "Ordovician-type" to "Silurian-type" shelly and conodont faunas may have taken place before, rather than near the Ordovician/Silurian transition (Meidla et al., 2011; Bergström et al., 2012a).

In the present study, we analyzed the $\delta^{13}C$ trends in carbonates of the Ordovician/Silurian boundary interval in four new drillcore sections from the Pandivere area, close to the stratotype of the Porkuni Regional Stage in the central Estonia. Despite the sedimentary gaps, presumably occurring in the Estonian carbonate shelf facies, and difficulties in interpreting the poor faunal record in the critical interval, our results of chemostratigraphic analysis enable to reconsider the position of the systemic boundary in the Baltoscandian regional standard.

15.2 GEOLOGICAL SETTING AND STRATIGRAPHY

The Baltoscandian epicontinental paleobasin existed on the Baltica plate during the early and middle Paleozoic. From Late Ordovician Sandbian Epoch until the end of the Silurian Period, the basin was characterized by extensive tropical carbonate shelves (Nestor and Einasto, 1997; Dronov and Rozhnov, 2007). The Late Ordovician Estonian shelf (northern Estonian confacies by Jaanusson, 1976) is characterized by limestones and dolomites formed below the fair-weather wave base. These shelf carbonates are facially replaced by argillaceous carbonate sediments of the deeper-shelf Livonian Basin, a broad embayment of the Scandinavian Basin extending to southern Estonia and western Latvia (Fig. 1). The facies distribution during the Llandovery Epoch is generally similar, except for the more frequent occurrence of nearshore bioclastic accumulations at different stratigraphic levels.

The Hirnantian and Rhuddanian stratigraphy of the Baltoscandian region is summarized in the Fig. 2. The Porkuni Regional Stage, traditionally considered to be a regional equivalent of the Hirnantian Stage, is represented in northern and central Estonia (Estonian Shelf) by various shallow water limestones and dolomites of the Ärina Formation (Kaljo et al., 1988, 2001). The Ärina Formation is subdivided into five members in the area (Fig. 2). The lower part of the formation, the Röa Member, is characterized by dolomites with sparse skeletal remains of crinoid columnals (Hints and Meidla, 1997; Hints et al., 2000; Kaljo et al., 2001). The middle part of the formation comprises a complex of stromatoporoid-coral patch reefs with interreef rocks. Three types of lithology, described as the Vohilaid, Siuge, and Tõrevere members, have been distinguished in this reef complex although the stratigraphic order of the "members" may vary even in nearby sections. The Vohilaid Member is characterized by variegated skeletal limestones (grainstones, packstones), the Siuge Member by bituminous argillaceous limestones (packstones), and the Tõrevere Member by massive biohermal limestone with interlayers of micritic limestone (Hints and Meidla, 1997; Hints et al., 2000; Kaljo et al., 2001). The upper part of the Ärina Formation, the Kamariku Member, is a 3-m thick (Viiding and Oraspõld, 1972; Oraspõld, 1975) distinctive sandy carbonate bed distributed in a 30-km wide

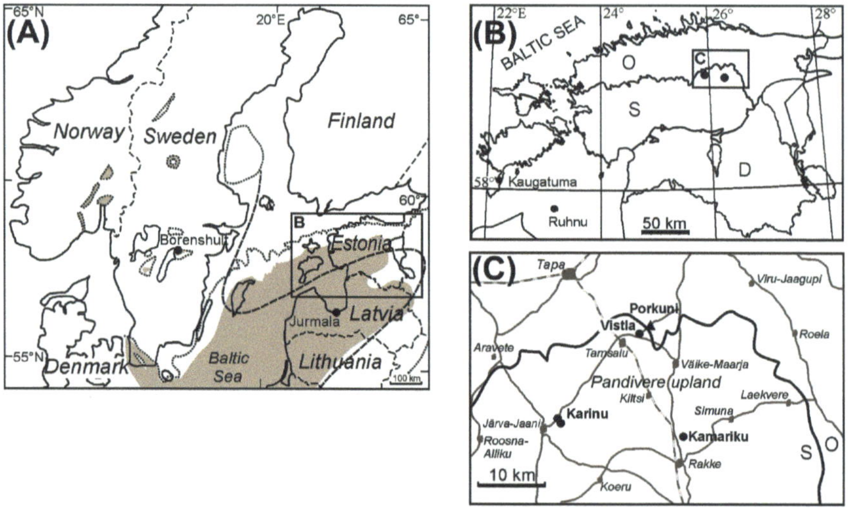

FIGURE 1

Location of the studied sections. (A) Distribution of Silurian (gray area) and Ordovician (bordered by dotted line) sediments in Baltoscandia with boundary between the major Late Ordovician facies belts (dashed line). *(Modified from Jaanusson, (1976)).* Filled dots—drillcores mentioned in the text; triangle—Porkuni quarry section. (B) Location of the study area. (C) Detailed road map of the study area.

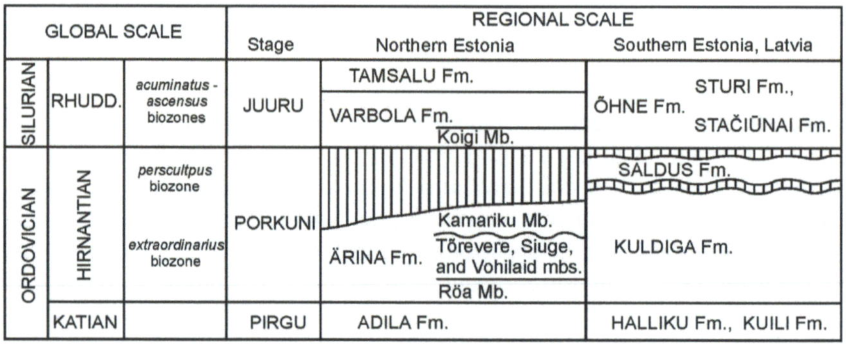

FIGURE 2

The latest Ordovician and earliest Silurian stratigraphy in Estonia showing the traditional correlation with global standard units (Kaljo et al., 2001; Brenchley et al., 2003).

and 200-km long east–west oriented belt in the North Estonia. It is characterized by sandy limestones and quartzose calcareous sandstones with 40–80% of carbonate that are often dolomitized (Viiding and Oraspõld, 1972; Hints et al., 2000). Although the reef bodies are documented mainly in the middle part of the Ärina Formation, occasional patch reefs have been found also in the lowermost part (this study) and in the upper part of the formation (in the Kamariku Member; Hints et al., 2000). Thickness of the Ärina Formation varies between 5 and 10 m in the study area (Hints and Meidla, 1997).

The Llandovery succession in Estonia begins with the Juuru Regional Stage that is in central Estonia represented by the Varbola and Tamsalu formations. The Varbola Formation is characterized by 10–25-m thick complex of micritic and skeletal limestones (wackestones, packstones) with thin intercalations of marlstone (Nestor, 1997). A bed of micritic limestone, 0.1–3-m thick, called the Koigi Member, is widely distributed at the base of the Varbola Formation in central Estonia (Nestor, 1997). The Varbola Formation is overlain by a bank of brachiopod coquinoid limestone, characterized by flood of a pentamerid brachiopod species *Borealis borealis* and called the Tamsalu Formation.

According to the traditional biostratigraphic correlations, the Ärina Formation together with its basinward counterpart, the argillaceous–calcareous Kuldiga Formation, is tentatively attributed to the *Metabolograptus extraordinarius* (formerly referred to as *Normalograptus extraordinarius*—see Melchin et al., 2011) Biozone and the upper Hirnantian *M. persculptus* Biozone (the former *Normalograptus persculptus*—ibid.) is thought to be missing in the central Estonia (Fig. 2). The oolitic and sandy grainstones of the Saldus Formation fill in this sedimentary gap in the southern Estonia and Latvia and may represent the *M. persculptus* Biozone in the area (Kaljo et al., 2001, 2008; Brenchley et al., 2003). The Varbola and Tamsalu formations, transgressively overlying the older strata, are thought to represent the *A. ascensus–Parakidograptus acuminatus* Biozone (Kaljo et al., 2008). However, it is important to mention, that no zonal graptolites have been found in the Porkuni and Juuru regional stages in East Baltic area, except in one lithologically different offshore Baltic drillcore (Kaljo et al., 2008). Therefore, the reinterpretation of chitinozoan data and $\delta^{13}C$ curves allowed to propose a rather different correlation for the end-Ordovician strata in Baltoscandia in some recent publications (Melchin and Holmden, 2006; Fan et al., 2009; Holmden et al., 2013), where the whole Porkuni Regional Stage is attributed to the upper Hirnantian *M. persculptus* Biozone. The problem regarding position of the lower boundary of the Hirnantian Stage in Baltic succession and correlation of the Porkuni Regional Stage with the global standard needs to be solved by means of more detailed chemo- and biostratigraphic studies in stratigraphically complete basinal sections. Unfortunately, sedimentary gaps at the base of the Porkuni Stage in our study area do not allow far-reaching conclusions in this ongoing discussion and this study focuses only on the position of system boundary in the area.

The correlation of the Ordovician/Silurian boundary with the Porkuni/Juuru regional stage boundary in Estonia is usually argued with the presence of the Hirnantian brachiopods, trilobites, and ostracods in the Kuldiga and Saldus formations and records of the brachiopod species *Stricklandia lens prima* from the Varbola Formation (Kaljo et al., 1988; Nestor, 1997). The correlation of the Kuldiga and Ärina formations as lateral facies was proposed already for several decades (e.g., Oraspõld, 1975; Kaljo et al., 1988; Männil and Meidla, 1994) and later on confirmed by chitinozoan biostratigraphy (the ranges of *Spinachitina taugourdourdeaui*) and HICE $\delta^{13}C$ isotope excursion (Hints et al., 2000; Kaljo et al., 2001, 2008). Looking closer at the biostratigraphic evidence in the boundary interval, we see the lack of reliable data from the units immediately below and above the Porkuni/Juuru stage boundary. The sandy carbonates of the Kamariku Member, and the topmost part of the Porkuni Stage in central Estonia, are practically unfossiliferous (Oraspõld, 1975; Kaljo et al., 2001). The micritic limestone of the Koigi Member, the basal bed of the Juuru Stage, is poorly of fossiliferous. It contains occasional brachiopods *Spirigerina (Eospirigerina) porkuniana* and *Onniella* sp., a trilobite *Proetus* sp., the tabulate coral *Catenipora* cf. *gotlandica,* and the stromatoporoid *Clathrodictyon boreale* (Nestor and Kala, 1968; Nestor, 1970). A number of brachiopod and rugose coral fragments recorded in the new drillcores from Karinu, in spite of their rather small diameter (3 cm) suggest that the unit may locally be rather fossiliferous, however, none of the recorded fragmentary specimens could be identified on the species level. A few chitinozoans of wide stratigraphical range (e.g., *Ancyrochitina ancyrea*) have been found also in the Koigi Member in some core sections (Nestor, 1994). Although the distribution range of the brachiopod species *S. lens prima* in the Juuru Stage starts above the Koigi Member (Rubel, 1970), the poor brachiopod association of the Koigi bed is formally attributed to the *Stricklandia* Community (Rubel, 1970), based on the occurrence of the supposedly Silurian species *Spirigerina (Eospirigerina) porkuniana,* and thus considered to be post-Hirnantian in age (Kaljo et al., 1988, 2008).

Thus, the Porkuni/Juuru boundary interval in the central Estonia, and the Kamariku and Koigi members in particular, do not reveal strong biostratigraphic evidence for correlation of the regional stage boundary with the Ordovician/Silurian boundary and both the regional stage and system boundaries are mainly drawn at the level of proposed sedimentary gap in the succession. The macrofossil distribution data display a faunal change in the critical interval (e.g., Nestor et al., 1991). In this chapter, we present and discuss the new $\delta^{13}C$ chemostratigraphic data for testing the correlation in this boundary interval.

The study area, the Pandivere Upland is a slightly elevated area in the central–northern Estonia, where the surface of Paleozoic bedrock is situated at 60–80m higher than the surrounding lowlands. The northern part of the Pandivere Upland is characterized by outcropping area of the Upper Ordovician carbonates, which are covered by the Silurian carbonate rocks in the southern part of the upland. Bedrock deposits are mostly covered by the Quaternary sediments. The uppermost Paleozoic bed in the Pandivere Upland is the pure coquinoid limestone bank of the Tamsalu Formation, Juuru Stage, which is more resistant to the erosion than underlying beds and might be partly responsible for the formation of elevated relief during the Pleistocene glacial erosion. The Ordovician/Silurian boundary beds are outcropping only in few old quarries (including Porkuni). In the southern part of the area, this interval lies up to 50-m below the surface.

15.3 MATERIAL AND METHODS

We analyzed in detail the uppermost Pirgu, Porkuni, and lower Juuru regional stage intervals in four drillcore sections from the southern part of the Pandivere area, drilled by Nordkalk Co. in the year 2008 (Fig. 1). Three of these drillcores penetrate the Karinu and the fourth one the Kamariku limestone deposit, where shelly limestone of the Tamsalu Formation is exploited for the lime industry. The Karinu 3 drill site is situated at 100-m SE of the Karinu 2, and 1-km NE of the Karinu 1, whereas the Kamariku 5 drill site is situated at 20-km east of the Karinu deposit. The core sections were described in detail. Altogether 106 samples were collected from the 20-m thick interval, spaced at 10–30cm and analyzed for carbon and oxygen stable isotope composition in bulk carbonate. For the stable carbon and oxygen isotopic analysis, ca 2g of each rock sample was selected avoiding secondary veins and large crystals, and powdered. This method enables better averaging of analyzed material compared to microdrilling. The powdered sample materials were analyzed by the mass spectrometer Delta V Advantage and (for preparation of gases) GasBench II by Thermo Scientific, using the international standards NBS 18, NBS 19, and LSVEC. The analytical work was conducted in the mass spectrometry laboratory in the Department of Geology, the University of Tartu, Estonia.

15.4 $\delta^{13}C$ CHEMOSTRATIGRAPHY

Formed on the tectonically stable southern slope of the Baltic Shield, the Paleozoic carbonate rocks of Estonia are well preserved in most localities and unaffected by deep burial diagenesis and tectonic dislocations. The limestones, dolomites, and calcareous quartzose sandstones in the studied

Ordovician/Silurian boundary interval have obviously different diagenetic histories. However, the results of this study, together with those of the previous studies on this interval in Estonia (Brenchley et al., 1994, 2003; Kaljo et al., 2001, 2008) show that $\delta^{13}C$ values of the bulk carbonate can generally reflect the predicted HICE stratigraphy in the uppermost Ordovician of this region and are not significantly affected by diagenetic changes, e.g., dolomitization. The $\delta^{18}O$ of bulk carbonate rocks is more sensitive to diagenetic changes and these data are considered unreliable for chemostratigraphic work. In the present study, most of the $\delta^{18}O$ values were found to be ranging between −6 and −2‰ that is common for the Paleozoic carbonate rocks, showing no stratigraphic trend or correlation with lithologies. No clear mixing trends of secondary and primary components (like described, for example, by Metzger and Fike (2013) from the Ordovician carbonates of US Midcontinent) were observed in the $\delta^{13}C/\delta^{18}O$ plots of sample clusters of different lithologies and stratigraphic units.

The pre-HICE $\delta^{13}C$ values are lying in between −0.5 and +1‰, slightly increasing upward in the topmost 4 m of the Adila Fm, Pirgu Regional Stage, in the Karinu core sections (Fig. 3). The Röa Member shows a clear increase in $\delta^{13}C$ values from 0.5 up to 2‰ in the Karinu 3 section and from 2 to 3‰ in the Karinu 2 section. The lithologically distinctive Röa Member in the Karinu 1 section is

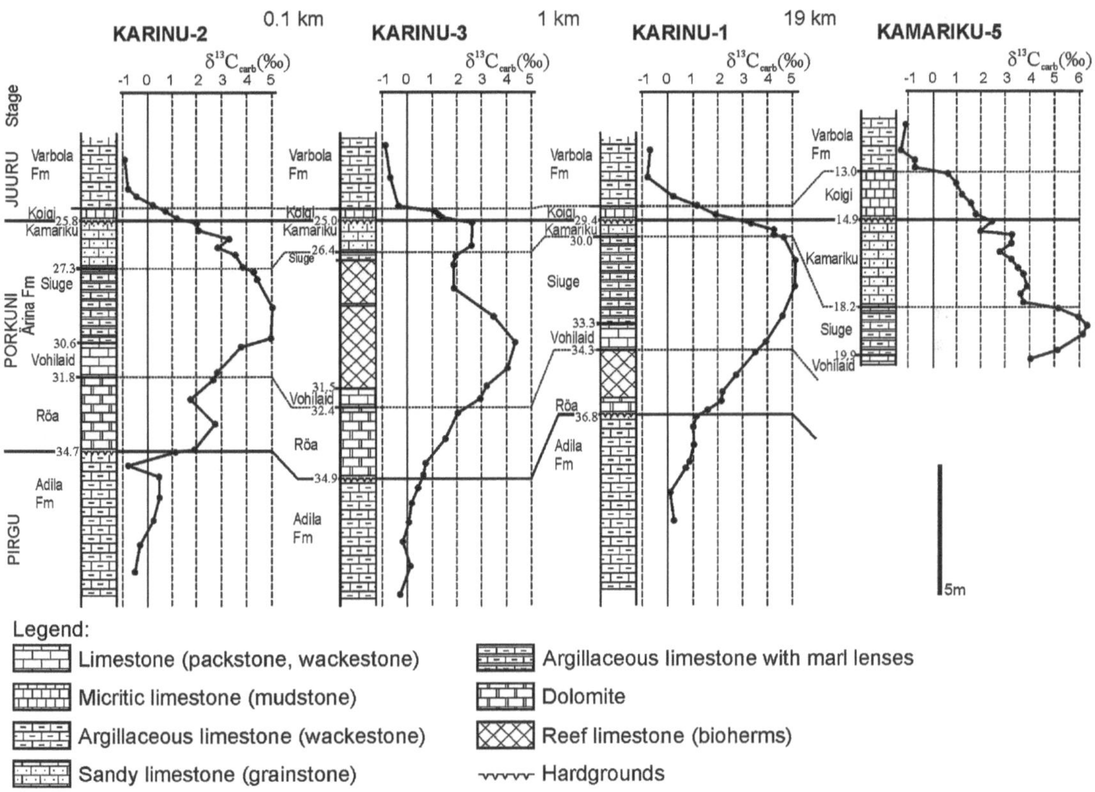

FIGURE 3

The carbonate carbon isotope ($\delta^{13}C_{carb}$) curves of the studied drillcore sections from the Pandivere area.

very thin, having 1‰ higher $\delta^{13}C$ value (+2‰) than the underlying beds and being overlain by a 2-m thick reef body. This rise of the $\delta^{13}C$ curves in the lower part of the Porkuni Stage is traditionally interpreted as the start of HICE (Brenchley et al., 2003; Kaljo et al., 2001, 2008).

The HICE reaches its maximum (+4 to +6‰) in the middle part of the Ärina Formation, in the reef complex. The $\delta^{13}C$ curves form a smooth peak at the top of interval of gradually rising values in the lower part of the formation. The peak is followed by a gradual decline. In all the three sections, (Karinu 1, Karinu 2, and Kamariku 5, see Fig. 3) the maximum $\delta^{13}C$ values are found to be related to the argillaceous limestones of the Siuge Member, but in the Karinu 3 section, they fall into the lower part of the reef body. The upper part of the same reef in this core section shows lower $\delta^{13}C$ values (+2‰) than the upper part of the reefs in other sections, which may refer to some alteration of the carbon isotopic composition during the reef diagenesis or slightly diachronous reef formation. Except for this uppermost reef interval in the Karinu 3 core, the reef bodies in the studied sections (Karinu 1 and Karinu 3) display similar shape of the HICE rise and peak than the interreef sediments in other sections. This may suggest that the reef formation took place approximately at the same speed as the interreef accumulation. The fact that reef formation occurred during the peak glaciation is particularly noteworthy. The reef limestones contain a highly diverse shelly fauna that includes no elements of the classical *Dalmanitina*/*Hirnantia* fauna, although some decrease in diversity has been recorded, compared to the latest pre-Hirnantian (see e.g., Hints and Meidla, 1997; Meidla, 1996a; Kaljo et al., 2011).

The $\delta^{13}C$ values in the Kamariku Member are characterized by a gradual upward decrease from +4 to +2‰ in the Karinu 1, Karinu 2, and Kamariku sections (Fig. 3). Importantly, the falling trend of the isotope curves does not stop or demonstrate sharp changes at the Porkuni/Juuru stage boundary (at the level of suggested major gap). The $\delta^{13}C$ values continue to fall in the basal beds of the Juuru Stage where rocks of the Koigi Member show values decreasing from +2 to +1‰. The lowest $\delta^{13}C$ values, 0 to −1‰, are observed in the overlying beds of the Varbola Formation, whereas top of the Koigi Member marks a more sharp change of the stable isotope values than the base of the Juuru Stage, i.e., the traditional level of Ordovician/Silurian boundary.

15.5 DISCUSSION

The chemostratigraphic zonation based on the comparison of the $\delta^{13}C$ curves is proposed for the Baltoscandian Ordovician succession by Ainsaar et al. (2010). The HICE, corresponding to the Porkuni Regional Stage, was subdivided into two isotope zones (Fig. 4). The Zone BC16 (lower part of the Porkuni Stage) corresponds to the steeply rising limb of the Hirnantian Excursion with an increase of up to 5‰ in the $\delta^{13}C$ values (interval A by Kaljo et al., 2001) in Estonia. The Zone BC17 (upper part of the Porkuni Stage) represents the gently falling limb of the HICE (intervals B and C of Kaljo et al., 2001). The Hirnantian carbon isotope curves from the northern Estonia show usually (e.g., in the Rapla, Hiiumaa, and Saaremaa sections) the HICE excursion that is sharply cut in the top (Ainsaar et al., 2010). This is interpreted as an evidence of glacioeustatically formed stratigraphic gap and the absence of the upper part of the Porkuni Stage, i.e., the Isotopic Zone BC17 (Brenchley et al., 2003; Ainsaar et al., 2010). In some sections of Estonia (e.g., Viljandi), the Zone BC16 is missing because of erosion in the middle Hirnantian while both parts of the HICE are usually represented in the sections of southern Estonia and Latvia (Brenchley et al., 2003; Ainsaar et al., 2010).

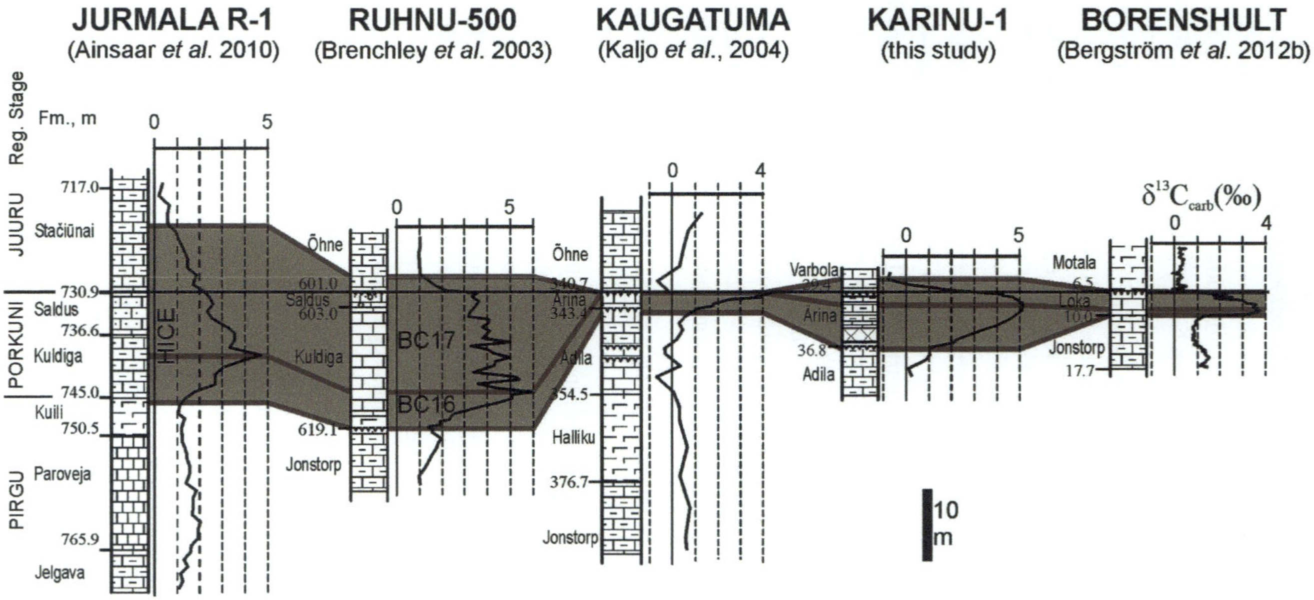

FIGURE 4

Correlation of the Ordovician/Silurian boundary beds in Baltoscandia by $\delta^{13}C_{carb}$ curves.

The carbon isotope zonation modified from Ainsaar et al. (2010). For legend see Fig. 3.

Our detailed curves, supported by the earlier data from the Pandivere area (e.g., Vistla section by Kaljo et al., 2001) show that the upper part of the Porkuni Stage is chemostratigraphically positioned on the falling limb of the HICE event, thus probably belonging to the Isotopic Zone BC17 (Figs 3, 4). This is most clear for the Kamariku Member, which displays a falling trend of $\delta^{13}C$ values that is similar to the falling trend in the uppermost Kuldiga and Saldus formations in the southern Estonia and Latvia. This suggests that the quartz sand containing Kamariku Member may be of late Porkuni age and also of the late Hirnantian age, like another sand-containing unit, the Saldus Formation, distributed in more distal basinal setting. Bergström et al. (2006), and Schmitz and Bergström (2007) labeled two late Hirnantian unconformities as HA and HB, bordering the Saldus Formation from base and top, respectively, in Baltoscandia. Comparison of the carbon isotope curves suggests that the Kamariku Member in the central Estonia may be positioned in the same stratigraphic interval between these two unconformities.

The basal beds of the Juuru Regional Stage, comprising the Koigi Member, form chemostratigraphically obviously a part of the same falling segment of the HICE excursion, the Isotope Zone BC17 (Figs 3, 4). This suggests that the basal part of the Juuru Stage is stratigraphically more complete (or less condensed) in the central Estonian Pandivere area than in other parts of northern Estonia (e.g., in the Rapla or Männamaa sections) or in western Estonia (in Kaugatuma; Kaljo et al., 2004; Ainsaar et al., 2010). The isotope curves of the studied sections in the Pandivere area show some similarities with curves in the Valga and Jurmala core sections from the Livonian Basin (Kaljo et al., 2007, 2008; Ainsaar et al., 2010), where the gradual decrease in the $\delta^{13}C$ values across the system boundary is taken as the evidence for stratigraphic completeness of the Ordovician/Silurian boundary interval in the region.

Analysis of the ostracod fauna in the Jurmala core section, Latvia, suggested that the typical Hirnantian-type fauna (the *Medianella aequa* association and the *Harpabollia* fauna; Meidla, 1996b) is replaced by the new "Silurian-type" ostracod fauna at the lithological boundary between the Saldus and Stačiūnai formations (Meidla et al., 2011). This change at the Porkuni/Juuru stage boundary occurs far below the end of HICE (Meidla et al., 2011). The succession of events is thus similar in the Pandivere area. The strata containing the new "Silurian-type" brachiopod fauna (the *Stricklandia* Community), i.e., the Koigi Member *sensu* Rubel (1970), are stratigraphically positioned on the falling limb of the HICE excursion. This means either, that the HICE excursion in Baltoscandia is partly of Silurian age, or that the position of the system boundary in this region should be positioned higher in the succession than traditionally assumed.

The Hirnantian carbon isotope curves of Baltoscandia have been compared with other regions by several authors (Brenchley et al., 2003; Melchin et al., 2003, in press; Melchin and Holmden, 2006; Kaljo et al., 2007, 2008; Bergström et al., 2006, 2012a,b; Schmitz and Bergström, 2007; Fan et al., 2009; Young et al., 2010; Achab et al., 2011; Gorjan et al., 2012; Holmden et al., 2013). These comparisons are based on the Estonian sections that are thought to be stratigraphically incomplete in the topmost Ordovician interval (e.g., Ruhnu, Kardla; see Brenchley et al., 2003) or have insufficient data from the lowermost Silurian interval (e.g., Valga; see Kaljo et al., 2008). Still, based on the comparison of the carbon isotope curves, few authors have placed the position of the traditional Ordovician/Silurian boundary at the *Conochitina scabra/Spinachitina fragilis* Chitinozoan Biozone boundary under discussion and argued for its higher position in the succession (Melchin et al., 2003; Melchin and Holmden, 2006; Fan et al., 2009; Gorjan et al., 2012).

To draw conclusions about the correlation of the end of HICE and the Ordovician/Silurian boundary, examining the stratigraphically most complete sections, where the glacioeustatic sea level falls had minimum effect to the deposition is necessary. The most important stratigraphic sections of the boundary interval, the Dob's Linn section (Scotland, stratotype for the base of the Silurian) and the Wangjiawan Riverside section (China, stratotype for the base of the Hirnantian) have been studied for organic carbon isotopic composition (Underwood et al., 1997; Chen et al., 2006) and these curves show clearly that relatively high $\delta^{13}C_{org}$ values of the HICE interval fall down to pre-HICE level in the middle of the *M. persculptus* Graptolite Biozone, well below the Ordovician/Silurian boundary (Fig. 5).

The HICE is expressed as roughly synchronous positive peaks in both $\delta^{13}C_{org}$ and $\delta^{13}C_{carb}$ curves as reported from several sections, where both analyses are available (Marshall et al., 1997; LaPorte et al., 2009; Gorjan et al., 2012). Comparing the carbonate carbon isotope curves of the HICE interval from deeper shelf environment of Nevada (Finney et al., 1999; Kump et al., 1999; LaPorte et al., 2009) we see a similar pattern, where the HICE ends below the Ordovician/Silurian boundary. Given cognizance to these data, Bergström et al. (2009) defined the terminal Ordovician stage slice Hi2 as the post-HICE interval of the Hirnantian Stage. Recent studies (Munnecke et al., 2011; Gorjan et al., 2012) on several sections located in the South China place the entire *M. persculptus* Graptolite Biozone into the post-HICE position in the carbonate carbon isotope curves (Fig. 5).

All these suggest that the basal beds of the Juuru Regional Stage with the elevated and upward declining $\delta^{13}C_{carb}$ values in Baltoscandia may be attributed to the upper Hirnantian interval. It is likely that the lower boundary of the Silurian System is located in the lower part of the Varbola Formation in the Pandivere area and within the Stačiūnai Formation in the Livonian Basin (Jurmala). Thus, the Koigi Member in Estonia may be considered to be of the latest Ordovician age (Fig. 6). This also means that the criteria used by previous authors for locating the lower boundary of the Silurian System in the sections of the Baltoscandian region are most likely not valid and this boundary needs to be reconsidered. It also suggests a need for critical reexamination of the stratigraphic correlations of all the non-graptolitiferous Ordovician/Silurian boundary successions.

Bergström et al. (2011, 2012a) recently analyzed the Ordovician/Silurian boundary beds exposed in the southern Ontario and in the Upper Mississippi Valley, North America. Carbon isotopic chemostratigraphic results of these analyses have shown, that the Ordovician/Silurian boundary should be positioned higher in the both successions than it is shown in traditional schemes (Bergström et al., 2011, 2012a). Recent studies in the Anticosti Island, North America, offer detailed chemostratigraphic, sequence stratigraphic, and biostratigraphic data on the Ordovician/Silurian boundary beds (Achab et al., 2011; Desrochers et al., 2010; Delabroye et al., 2011). According to the chemostratigraphic and sea level history interpretations on these beds, the system boundary may be situated several meters higher from the base of the Becscie Formation, which has been traditional position for the base of Silurian (Desrochers et al., 2010; Delabroye et al., 2011). Similar upward correction of the Ordovician/Silurian boundary position in the Subpolar Urals stratigraphical scheme was made by Beznosova et al. (2011) after the isotope chemostratigraphic analysis. These chemostratigraphic corrections of the traditional biostratigraphic schemes in Estonia, Latvia, North America and elsewhere are in agreement with the former suggestion that the major changes in several fauna groups likely preceded the system boundary (Brenchley et al., 2001). This also agrees with the evidence on global warming and sea level rise during the latest Hirnantian (e.g., Finnegan et al., 2011; Hammarlund et al., 2012; Melchin et al.,

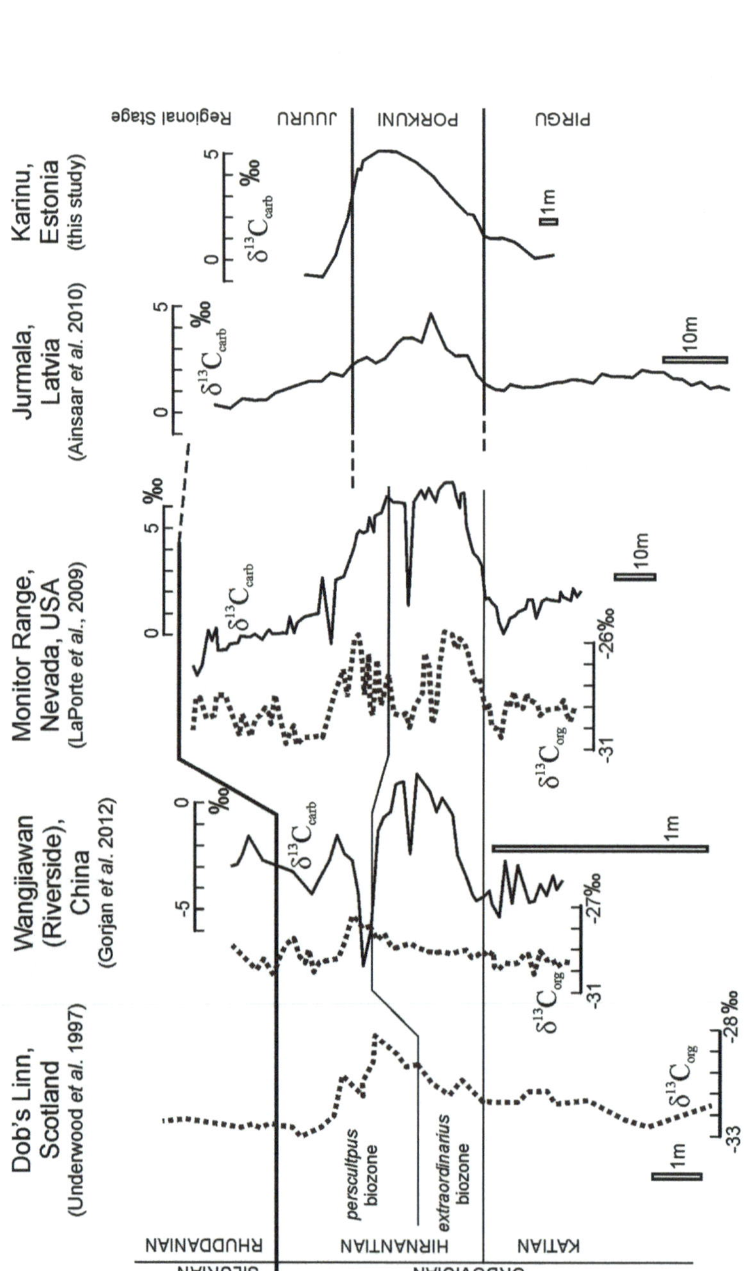

FIGURE 5

Comparison of the stable carbon isotope trends of Baltoscandia with graptolitiferous key sections in other areas.

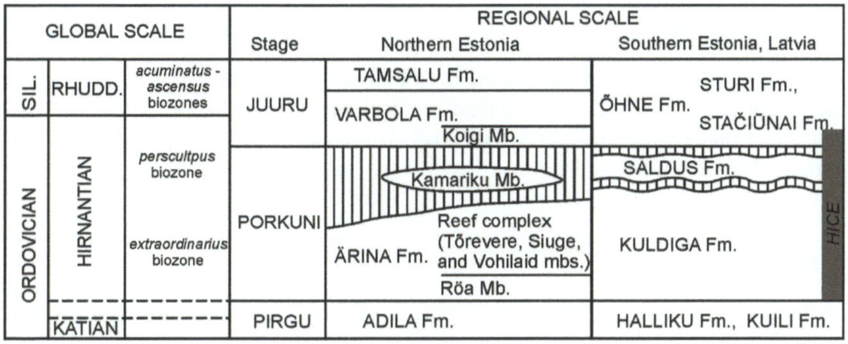

FIGURE 6

Stratigraphic chart showing the proposed correlation of the Baltoscandian successions with global standard.

2013), in the interval that is globally characterized as time of "survival fauna" after the second Hirnantian extinction (Brenchley et al., 2001; Brenchley, 2004). The second phase of the Hirnantian extinction resulted in extremely low diversity ecosystems that only slowly recovered during a considerably long epoch of the latest Hirnantian and Rhuddanian (Brenchley et al., 2001; Brenchley, 2004; Hammarlund et al., 2012).

15.6 CONCLUSIONS

Comparison of the Hirnantian and Rhuddanian $\delta^{13}C$ curves of the new core sections from the Pandivere area, central Estonia, suggests some changes in the traditional stratigraphic correlation scheme of Estonia. The Ärina Formation, Porkuni Regional Stage, has previously been considered as the early Hirnantian while the upper Hirnantian was supposed to be missing in the area and the Varbola Formation was dated as the early Llandovery (e.g., Kaljo et al., 2001). The sandy unfossiliferous Kamariku Member in the uppermost part of the Ärina Formation is located on the falling limb of the HICE and may thus be of late Hirnantian age. It is possible, that the Kamariku Member belongs to the interval between the widespread Hirnantian unconformities HA and HB, like the Saldus Formation in Latvia and southern Estonia (Bergström et al., 2006). The basal part of the Varbola Formation, the Koigi Member, falls into the interval of upward declining $\delta^{13}C$ values in the sections of the Pandivere area and obviously represents the latest part of the HICE. This suggests the age of the Koigi Member to be late Hirnantian and places the Ordovician/Silurian boundary into the lower part of the Varbola Formation.

The second phase of the Hirnantian extinction related to the unconformity HB was followed by a slow faunal recovery (Brenchley et al., 2001). This recovery is characterized by the slow appearance of postextinction biota that obviously included some "Silurian-type" shelly fossil groups that were used in traditional correlation schemes of non-graptolitiferous Ordovician/Silurian boundary successions for positioning the system boundary. The detailed chemostratigraphic comparison of these sections with graptolitiferous successions allowed testing and correcting the position of the Ordovician/Silurian boundary, which is likely be positioned higher in the succession than the traditional placement.

ACKNOWLEDGMENTS

Authors are grateful to Nordkalk Co. and particularly to H. Pihl and R. Harinen for the high-quality drillings which they made available for our research. H. Sepp is thanked for the analytical support. The study was supported by the Estonian Research Council grants IUT20-34 and 8049. This chapter is a contribution to the IGCP project 591 "The Early to Middle Paleozoic Revolution."

REFERENCES

Achab, A., Asselin, E., Desrochers, A., Riva, J.F., Farley, C., 2011. Chitinozoan biostratigraphy of a new Upper Ordovician stratigraphic framework for Anticosti Island, Canada. Geological Society of America Bulletin 123 (1–2), 186–205.

Ainsaar, L., Kaljo, D., Martma, T., Meidla, T., Männik, P., Nõlvak, J., Tinn, O., 2010. Middle and Upper Ordovician carbon isotope chemostratigraphy in Baltoscandia: a correlation standard and clues to environmental history. Palaeogeography, Palaeoclimatology, Palaeoecology 294 (3–4), 189–201.

Aubry, M.-P., Ouda, K., Dupuis, C., Berggren, W.A., Van Couvering, J.A., Ali, J.A., Brinkhuis, H., Gingerich, P.R., Heilmann-Clausen, C., Hooker, J., Kent, D.V., King, C., Knox, R.W., Laga, P., Molina, E., Schmitz, B., Steurbaut, E., Ward, D.R., 2007. The Global Standard Stratotype-section and Point (GSSP) for the base of the Eocene series in the Dababiya section, Egypt. Episodes 30 (4), 271–286.

Bergström, S.M., Chen, X., Gutierrez-Marco, J.C., Dronov, A., 2009. The new chronostratigraphic classification of the Ordovician System and its relations to major regional series and stages and to $\delta^{13}C$ chemostratigraphy. Lethaia 42 (1), 97–107.

Bergström, S.M., Kleffner, M., Schmitz, B., Cramer, B.D., 2011. Revision of the position of the Ordovician-Silurian boundary in southern Ontario: regional chronostratigraphic implications of $\delta^{13}C$ chemostratigraphy of the Manitoulin Formation and associated strata. Canadian Journal of Earth Sciences 48 (11), 1447–1470.

Bergström, S.M., Kleffner, M., Schmitz, B., 2012a. Late Ordovician–Early Silurian $\delta^{13}C$ chemostratigraphy in the Upper Mississippi Valley: implications for chronostratigraphy and depositional interpretations. Earth and Environmental Science Transactions of the Royal Society of Edinburgh 102 (3), 159–178.

Bergström, S.M., Lehnert, O., Calner, M., Joachimski, M.M., 2012b. A new Upper Middle Ordovician-Lower Silurian drillcore standard succession from Borenshult in Östergötland, southern Sweden: 2. Significance of $\delta^{13}C$ chemostratigraphy. GFF 134 (1), 39–63.

Bergström, S.M., Saltzman, M.R., Schmitz, B., 2006. First record of the Hirnantian (Upper Ordovician) $\delta^{13}C$ excursion in the North American Midcontinent and its regional implications. Geological Magazine 143 (5), 657–678.

Beznosova, T.M., Maidl, T.V., Männik, P., Martma, T., 2011. The Ordovician-Silurian boundary on the western slope of the Subpolar Urals. Stratigraphy and Geological Correlation 19 (4), 385–403.

Brenchley, P.J., 2004. End Ordovician glaciation. In: Webby, B.D., Paris, F., Droser, M.L., Percival, I.G. (Eds.), The Great Ordovician Biodiversification Event. Columbia University Press, New York, pp. 81–83.

Brenchley, P.J., Carden, G.A., Hints, L., Kaljo, D., Marshall, J.D., Martma, T., Meidla, T., Nõlvak, J., 2003. High-resolution stable isotope stratigraphy of Upper Ordovician sequences: constraints on the timing of bioevents and environmental changes associated with mass extinction and glaciation. Geological Society of America Bulletin 115 (1), 89–104.

Brenchley, P.J., Marshall, J.D., Carden, G.A.F., Robertson, D.B.R., Long, D.G.F., Meidla, T., Hints, L., Anderson, T.F., 1994. Bathymetric and isotopic evidence for a short-lived Late Ordovician glaciation in a greenhouse period. Geology 22 (4), 295–298.

Brenchley, P.J., Marshall, J.D., Underwood, C.J., 2001. Do all mass extinctions represent an ecological crisis? Evidence from the Late Ordovician. Geological Journal 36 (3–4), 329–340.

Buggisch, W., Joachimski, M.M., 2006. Carbon isotope stratigraphy of the Devonian of central and southern Europe. Palaeogeography, Palaeoclimatology, Palaeoecology 240 (1–2), 68–88.

Chen, X., Rong, J.Y., Fan, J.X., Zhan, R.B., Mitchell, C.E., Harper, D.A.T., Melchin, M.J., Peng, P., Finney, S.C., Wang, X.F., 2006. The Global boundary Stratotype Section and Point (GSSP) for the base of the Hirnantian Stage (the uppermost of the Ordovician System). Episodes 29 (3), 183–196.

Delabroye, A., Munnecke, A., Vecoli, M., Copper, P., Tribovillard, N., Joachimski, M.M., Desrochers, A., Servais, T., 2011. Phytoplankton dynamics across the Ordovician/Silurian boundary at low palaeolatitudes: correlations with carbon isotopic and glacial events. Palaeogeography, Palaeoclimatology, Palaeoecology 312 (1–2), 79–97.

Delabroye, A., Vecoli, M., 2010. The end-Ordovician glaciation and the Hirnantian Stage: a global review and questions about Late Ordovician event stratigraphy. Earth-Science Reviews 98 (3–4), 269–282.

Desrochers, A., Farley, C., Achab, A., Asselin, E., Riva, J.F., 2010. A far-field record of the end Ordovician glaciation: the Ellis Bay Formation, Anticosti Island, eastern Canada. Palaeogeography, Palaeoclimatology, Palaeoecology 296 (3–4), 248–263.

Dronov, A., Rozhnov, S., 2007. Climatic changes in the Baltoscandian basin during the Ordovician: sedimentological and palaeontological aspects. Acta Palaeontologica Sinica 46 (Suppl.), 108–113.

Fan, J., Peng, P., Melchin, M.J., 2009. Carbon isotopes and event stratigraphy near the Ordovician-Silurian boundary, Yichang, South China. Palaeogeography, Palaeoclimatology, Palaeoecology 276 (1–4), 160–169.

Finnegan, S., Bergmann, K., Eiler, J.M., Jones, D.S., Fike, D.A., Eisenman, I., Hughes, N.C., Tripati, A.K., Fischer, W.W., 2011. The magnitude and duration of Late Ordovician-Early Silurian glaciation. Science 331 (6019), 903–906.

Finney, S.C., Berry, W.B.N., Cooper, J.D., Ripperdan, R.L., Sweet, W.C., Jacobson, S.R., Soufiane, A., Achab, A., Noble, P.J., 1999. Late Ordovician mass extinction: a new perspective from stratigraphic sections in central Nevada. Geology 27 (3), 215–218.

Fio, K., Spangenberg, J.E., Vlahović, I., Sremac, J., Velić, I., Mrinjek, E., 2010. Stable isotope and trace element stratigraphy across the Permian–Triassic transition: a redefinition of the boundary in the Velebit Mountain, Croatia. Chemical Geology 278 (1), 38–57.

Gibbard, P.L., Head, M.J., 2010. The newly-ratified definition of the Quaternary System/Period and redefinition of the Pleistocene Series/Epoch, and comparison of proposals advanced prior to formal ratification. Episodes 33 (3), 82–87.

Gorjan, P., Kaiho, K., Fike, D.A., Xu, C., 2012. Carbon- and sulfur-isotope geochemistry of the Hirnantian (Late Ordovician) Wangjiawan (Riverside) section, South China: global correlation and environmental event interpretation. Palaeogeography, Palaeoclimatology, Palaeoecology 337–338 (0), 14–22.

Hammarlund, E.U., Dahl, T.W., Harper, D.A.T., Bond, D.P.G., Nielsen, A.T., Bjerrum, C.J., Schovsbo, N.H., Schönlaub, H.P., Zalasiewicz, J.A., Canfield, D.E., 2012. A sulfidic driver for the end-Ordovician mass extinction. Earth and Planetary Science Letters 331–332, 128–139.

Hilgen, F.J., Abels, H.A., Iaccarino, S., Krijgsman, W., Raffi, I., Sprovieri, R., Turco, E., Zachariasse, W.J., 2009. The global stratotype section and point (GSSP) of the serravallian stage (Middle Miocene). Episodes 32 (3), 152–166.

Hints, L., Meidla, T., 1997. Porkuni Stage. In: Raukas, A., Teedumäe, A. (Eds.), Geology and Mineral Resources of Estonia. Estonian Academy Publishers, Tallinn, pp. 85–88.

Hints, L., Oraspõld, A., Kaljo, D., 2000. Stratotype of the Porkuni Stage with comments on the Röa member (Uppermost Ordovician, Estonia). Proceedings of the Estonian Academy of Sciences. Geology 49 (3), 177–199.

Holmden, C., Mitchell, C.E., LaPorte, D.F., Patterson, W.P., Melchin, M.J., Finney, S.C., 2013. Nd isotope records of late Ordovician sea-level change—Implications for glaciation frequency and global stratigraphic correlation. Palaeogeography, Palaeoclimatology, Palaeoecology 386, 131–144.

Jaanusson, V., 1976. Faunal dynamics in the Middle Ordovician (Viruan) of Balto-Scandia. In: Bassett, M.G. (Ed.), The Ordovician System: Proceedings of a Palaeontological Association Symposium, Birmingham, September 1974. University of Wales Press, Cardiff, pp. 301–326.

Kaljo, D., Hints, L., Hints, O., Männik, P., Martma, T., Nõlvak, J., 2011. Katian prelude to the Hirnantian (Late Ordovician) mass extinction: a Baltic perspective. Geological Journal 46 (5), 464–477.

Kaljo, D., Hints, L., Martma, T., Nõlvak, J., 2001. Carbon isotope stratigraphy in the latest Ordovician of Estonia. Chemical Geology 175 (1–2), 49–59.

Kaljo, D., Hints, L., Martma, T., Nõlvak, J., Oraspõld, A., 2004. Late Ordovician carbon isotope trend in Estonia, its significance in stratigraphy and environmental analysis. Palaeogeography, Palaeoclimatology. Palaeoecology 210 (2–4), 165–185.

Kaljo, D., Hints, L., Männik, P., Nõlvak, J., 2008. The succession of Hirnantian events based on data from Baltica: brachiopods, chitinozoans, conodonts, and carbon isotopes. Estonian Journal of Earth Sciences 57 (4), 197–218.

Kaljo, D., Martma, T., Saadre, T., 2007. Post-Hunnebergian Ordovician carbon isotope trend in Baltoscandia, its environmental implications and some similarities with that of Nevada. Palaeogeography, Palaeoclimatology, Palaeoecology 245 (1–2), 138–155.

Kaljo, D., Nestor, H., Põlma, L., 1988. East Baltic region. British Museum (Natural History) Bulletin (Geology) 43, 85–91.

Knoll, A.H., Walter, M.R., Narbonne, G.M., Christie-Blick, N., 2006. The Ediacaran Period: a new addition to the geologic time scale. Lethaia 39 (1), 13–30.

Kump, L.R., Arthur, M.A., Patzkowsky, M.E., Gibbs, M.T., Pinkus, D.S., Sheehan, P.M., 1999. A weathering hypothesis for glaciation at high atmospheric pCO_2 during the Late Ordovician. Palaeogeography, Palaeoclimatology, Palaeoecology 152 (1–2), 173–187.

LaPorte, D.F., Holmden, C., Patterson, W.P., Loxton, J.D., Melchin, M.J., Mitchell, C.E., Finney, S.C., Sheets, H.D., 2009. Local and global perspectives on carbon and nitrogen cycling during the Hirnantian glaciation. Palaeogeography, Palaeoclimatology, Palaeoecology 276 (1–4), 182–195.

Marshall, J.D., Brenchley, P.J., Mason, P., Wolff, G.A., Astini, R.A., Hints, L., Meidla, T., 1997. Global carbon isotopic events associated with mass extinction and glaciation in the Late Ordovician. Palaeogeography, Palaeoclimatology, Palaeoecology 132 (1–4), 195–210.

Meidla, T., 1996a. Late Ordovician ostracodes of Estonia. Fossilia Baltica 2, 1–222.

Meidla, T., 1996b. Latest Ordovician ostracodes of Baltoscandia. In: Stouge, S. (Ed.), WOGOGOB-94 Symposium. Working Group of Ordovician Geology of Baltoscandia, Bornholm 94, 98, pp. 63–68 Geological Survey of Denmark and Greenland Report.

Meidla, T., Ainsaar, L., Truuver, K., 2011. Ostracods in Baltoscandia through the Hirnantian crises. In: Gutierrez-Marco, J.C., Rabano, I., Garcia-Bellido, D. (Eds.), Ordovician of the World. Cuadernos del Museo Geominero, 14. Instituto Geologico y Minero de España, Madrid, pp. 353–357.

Melchin, M.J., Holmden, C., 2006. Carbon isotope chemostratigraphy in Arctic Canada: sea-level forcing of carbonate platform weathering and implications for Hirnantian global correlation. Palaeogeography, Palaeoclimatology, Palaeoecology 234 (2–4), 186–200.

Melchin, M.J., Holmden, C., Williams, S.H., 2003. Correlation of graptolite biozones, chitinozoan biozones, and carbon isotope curves through the Hirnantian. In: Albanesi, G.L., Beresi, M.S., Peralta, S.H. (Eds.), Ordovician from the Andes, 17. INSUGEO, Serie de Correlacion Geologica, Tucuman, pp. 101–104 .

Melchin, M.J., Mitchell, C.E., Holmden, C., Štorch, P., 2013. Environmental changes in the Late Ordovician–early Silurian: review and new insights from black shales and nitrogen isotopes. Geological Society of America Bulletin 125 (11–12), 1635–4670.

Melchin, M.J., Mitchell, C.E., Naczk-Cameron, A., Fan, J.X., Loxton, J., 2011. Phylogeny and adaptive radiation of the Neograptina (Graptoloida) during the Hirnantian mass extinction and Silurian recovery. Proceedings of the Yorkshire Geological Society 58 (4), 281–309.

Metzger, J.G., Fike, D.A., 2013. Techniques for assessing spatial heterogeneity of carbonate $\delta^{13}C$ values: implications for craton-wide isotope gradients. Sedimentology 60 (6), 1405–1431.

Munnecke, A., Zhang, Y., Liu, X., Cheng, J., 2011. Stable carbon isotope stratigraphy in the Ordovician of South China. Palaeogeography, Palaeoclimatology, Palaeoecology 307 (1–4), 17–43.

Männil, R., Meidla, T., 1994. The Ordovician System of the East European platform (Estonia, Latvia, Lithuania, Byelorussia, parts of Russia, Ukraine and Moldova). In: Webby, B.D., Ross, R.J., Zhen, Y.Y. (Eds.), The Ordovician System of the East European Platform and Tuva (Southeastern Russia): Correlation Charts and Explanatory Notes, 28A. IUGS Publication, pp. 1–52.

Nestor, H., 1970. Yuuruskiy gorizont [Juuru stage]. In: Kaljo, D. (Ed.), Silur Estonii [Silurian of Estonia]. Valgus, Tallinn, pp. 204–221 (in Russian with English summary).

Nestor, H., 1997. Silurian. In: Raukas, A., Teedumäe, A. (Eds.), Geology and Mineral Resources of Estonia. Estonian Academy Publishers, Tallinn, pp. 89–106.

Nestor, H., Einasto, R., 1997. Ordovician and Silurian carbonate sedimentation basin. In: Raukas, A., Teedumäe, A. (Eds.), Geology and Mineral Resources of Estonia. Estonian Academy Publishers, Tallinn, pp. 192–204.

Nestor, H., Kala, E., 1968. Reviziya stratigrafii nizov silura Severnoi Pribaltiki (A stratigraphic revision of the lowermost Silurian in the Northern East-Baltic area). In: Grigelis, A. (Ed.), Stratigraphy of the Baltic Lower Paleozoic and its correlation with other areas. Mintis, Vilnius, pp. 188–214 (in Russian with English summary).

Nestor, H., Klaamann, E., Meidla, T., Männik, P., Männil, R., Nestor, V., Nõlvak, J., Rubel, M., Sarv, L., Hints, L., 1991. Dinamika fauny v Baltiiskom basseine na granitse ordovika i silura (Faunal dynamics in the Baltic Basin on the Ordovician/Silurian boundary). In: Kaljo, D., Modzalevskaya, T.L., Bogdanova, T.N. (Eds.), Major Biological Events in Earth History. Transactions of the XXXII Session of All-Union Paleontological Society. Academy of Sciences of Estonia, Tallinn, pp. 79–86 (in Russian).

Nestor, V., 1994. Early Silurian chitinozoans of Estonia and North Latvia. Academia 4, 1–163.

Nõlvak, J., Hints, O., Männik, P., 2006. Ordovician timescale in Estonia: recent developments. Proceedings of the Estonian Academy of Sciences Geology 55 (2), 95–108.

Oraspõld, A., 1975. Litologiya porkuniskogo gorizonta v Estonii (On the lithology of the Porkuni Stage in Estonia). Acta et Commentationes Universitatis Tartuensis 359, 33–75 (in Russian with English summary).

Rio, D., Sprovieri, R., Castradori, D., Di Stefano, E., 1998. The Gelasian Stage (Upper Pliocene): a new unit of the global standard chronostratigraphic scale. Episodes 21 (2), 82–87.

Rong, J.Y., Melchin, M., Williams, S.H., Koren, T.N., Verniers, J., 2008. Report of the restudy of the defined global stratotype of the base of the Silurian system. Episodes 31 (3), 315–318.

Rubel, M., 1970. On the distribution of brachiopods in the lowermost Llandovery of Estonia. Eesti NSV Teaduste Akadeemia Toimetised Keemia Geoloogia 19 (1), 69–79.

Schmitz, B., Bergström, S.M., 2007. Chemostratigraphy in the Swedish Upper Ordovician: regional significance of the Hirnantian $\delta^{13}C$ excursion (HICE) in the Boda Limestone of the Siljan region. GFF 129, 133–140.

Scholle, P.A., Arthur, M.A., 1980. Carbon isotope fluctuations in Cretaceous pelagic limestones: potential stratigraphic and petroleum exploration tool. American Association of Petroleum Geologists Bulletin 64 (1), 67–87.

Shackleton, N.J., Opdyke, N.D., 1977. Oxygen isotope and paleomagnetic evidence for early northern hemisphere glaciation. Nature 270, 216–219.

Steininger, F.F., Aubry, M.P., Berggren, W.A., Biolzi, M., Borsetti, A.M., Cartlidge, J.E., Cati, F., Corfield, R., Gelati, R., Iaccarino, S., Napoleone, C., Ottner, F., Rogl, F., Roetzel, R., Spezzaferri, S., Tateo, F., Villa, G., Zevenboom, D., 1997. The Global Stratotype Section and Point (GSSP) for the base of the Neogene. Episodes 20 (1), 23–28.

Underwood, C.J., Crowley, S.F., Marshall, J.D., Brenchley, P.J., 1997. High-resolution carbon isotope stratigraphy of the basal Silurian stratotype (Dob's Linn, Scotland) and its global correlation. Journal of the Geological Society 154, 709–718.

Viiding, H., Oraspõld, A., 1972. Litologiya i mineralogiya kamarikuskoy pachki porkuniskogo gorizonta (Lithology and mineralogy of the Kamariku Member of the Porkuni Stage). Eesti NSV Teaduste Akadeemia Toimetised Keemia Geoloogia 21 (3), 245–258 (in Russian with English Summary).

Webby, B.D., Cooper, R.A., Bergström, S.M., Paris, F., 2004. Stratigraphic framework and time slices. In: Webby, B.D., Paris, F., Droser, M.L., Percival, I.G. (Eds.), The Great Ordovician Biodiversification Event. Columbia University Press, New York, pp. 41–47.

Weissert, H., Joachimski, M., Sarnthein, M., 2008. Chemostratigraphy. Newsletters on Stratigraphy 42 (3), 145–179.

Williford, K.H., Ward, P.D., Garrison, G.H., Buick, R., 2007. An extended stable organic carbon isotope record across the Triassic-Jurassic boundary in the Queen Charlotte Islands, British Columbia, Canada. Palaeogeography, Palaeoclimatology, Palaeoecology 244 (1–4), 290–296.

Young, S.A., Saltzman, M.R., Ausich, W.I., Desrochers, A., Kaljo, D., 2010. Did changes in atmospheric CO_2 coincide with latest Ordovician glacial-interglacial cycles? Palaeogeography, Palaeoclimatology, Palaeoecology 296 (3–4), 376–388.

CHAPTER

16 STABLE ISOTOPE STRATIGRAPHY: CORRELATIONS AND IMPLICATIONS FOR HYDROCARBON MICROSEEPAGE AND PROSPECTING

Devleena Mani, D.J. Patil, A.M. Dayal
CSIR-National Geophysical Research Institute, Hyderabad, India

16.1 INTRODUCTION

The source of petroleum has long been known as the organic matter deposited along with the sediments in a basin through geological times. The evolutionary pathway of the sedimentary organic matter leading to prolific hydrocarbon source rocks depends upon the (1) environmental conditions—which support biologic activities resulting in the production of large quantities of oil/gas prone organic matter, (2) depositional conditions—which concentrate the organic matter via burial, and (3) postdepositional conditions—which permit the preservation and maturation of organic matter under the influences of temperature and pressure to liquid or gaseous hydrocarbons (Hunt, 1996; Tissot and Welte, 1984; McCarthy et al., 2011) (Fig. 1). Apart from the geological controls on the extent and mechanism of organic carbon burial, useful perspectives are derived from the global carbon (C), oxygen (O), and sulfur (S) cycles (Berner, 1999, 2006; Hedges and Keil, 1995; Holser et al., 1988; Des Marais et al., 1992; Karhu and Holland, 1996). Burial of organic matter and pyrite is the only source of atmospheric oxygen and it is intimately coupled with the global cycles of C, S, and O over geologic time (Berner, 1982; Hedges and Keil, 1995). Substantial insights to the processes pertaining to hydrocarbon accumulation, generation, and preservation, can be obtained through the stable isotopic and molecular records preserved in the sedimentary sequences. These records are associated with the marine and terrestrial organic input, redox condition, depositional, and paleoenvironment, history of sedimentary cycle, erosion, and oceanic circulation (Arthur et al., 1985; Kump and Arthur, 1999; Berner, 2003; Holser et al., 1996; Klemme and Ulmishek, 1991; Kaufman et al., 1991; Jenkyns, 1988, 2010; Schidlowski, 1987; Veizer et al., 1999 and several others). Specifically, the stable isotopes of carbon and oxygen hold immense significance in event stratigraphy, paleoceanography, and paleotemperature studies (Veizer and Hoefs, 1976), whereas the molecular stratigraphy focuses on the origin and fate of sedimentary organic matter, reflecting various aspects of depositional environment and early life (Meyers, 1994; Brassell et al., 1986).

Chemostratigraphy. http://dx.doi.org/10.1016/B978-0-12-419968-2.00016-9

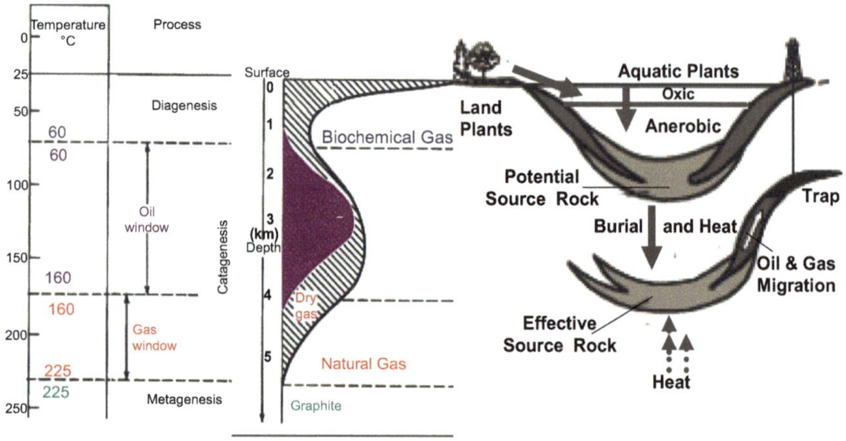

FIGURE 1

Generalized evolution of sedimentary organic matter upon increasing burial and temperature.

Mani et al. (2013).

A synoptic view of the carbon cycle indicates the manifestation of carbon as organic and inorganic species in the sedimentary rocks (Fig. 2; after Schidlowsky, 1987; Hunt, 1972). Organic form is the reduced carbon formed by biological processes over geologic time. Inorganic carbon is the oxidized carbon, primarily in the form of carbonate. Of the total organic carbon (TOC) present in the sediments, ~10% is the soluble organic matter or bitumen and remaining (~90%) is kerogen, the complex, macromolecular, insoluble geopolymer. Inorganic carbon is present largely in the form of carbonate, bicarbonate, dissolved inorganic carbon, and atmospheric CO_2. The transformation of inorganic carbon into living matter entails a marked bias, as a result of kinetic isotope effect, in favor of light isotopes (^{12}C), with heavy species (^{13}C) retained in the inorganic reservoir (Schidlowski, 1987). This isotopic discrimination leads to the preferential accumulation of ^{12}C in all forms of biogenic (reduced) carbon as compared with the inorganic (oxidized) carbon pool of surficial environment, mainly as the atmospheric carbon dioxide and dissolved marine carbonate (Schidlowski, 1987, 1988). When the biogenic materials and carbonates are incorporated in newly formed sediments, the kinetic isotope effect associated with the autotrophic carbon fixation is propagated from the surficial exchange reservoir into the rock section of the carbon cycle (Schidlowski, 1987). Over geological time, this effect has ultimately brought about a conspicuous isotope disproportionation of earth's primordial carbon into a light and heavy crustal carbon reservoir (Schidlowski, 1987).

The preservation of sedimentary carbon allows a comparable amount of photosynthetically produced oxygen to accumulate in the atmosphere (Hedges and Keil, 1995; Schidlowski et al., 1975). The main sink for the oxygen generated by the sediment burial is thought to be the oxidation of kerogen and pyrite in shales during continental weathering (Berner, 1989; Hedges and Keil, 1995). Thus, the global C and O cycles dynamically influence and control each other through multitudes of interactions between the carbon dioxide, bicarbonate/carbonate system and associated organic carbon (Fig. 2). The isotopic signatures of carbon and oxygen in the stratigraphically controlled sedimentary rocks represent the isotopic differences established in response to earth processes in the past and are useful in

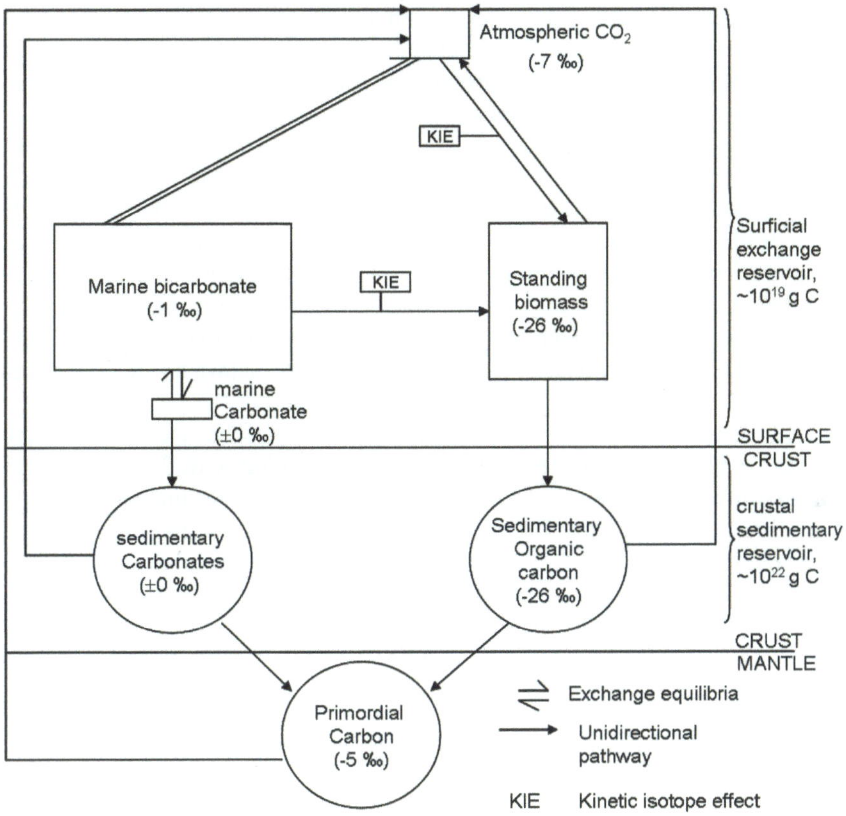

FIGURE 2

Isotopic carbon fractionation between the surficial and crustal reservoirs.

After Schidlowski (1987).

understanding the depositional/diagenetic conditions of the source rocks. The isotopic variations preserved in the sedimentary carbonates and shales provide information on the productivity and gross carbon storage leading to the development of prolific source rocks in sedimentary basins. Significant short-term carbon and oxygen isotope fluctuations have been used as potential stratigraphic and exploration tool to map the stratigraphic distribution of potential petroleum source beds in oceanic basins and continental margins (Scholle and Arthur, 1980).

16.2 CARBON ISOTOPE EVENT STRATIGRAPHY

Chemostratigraphy uses chemical fingerprints stored in sediments and rocks for stratigraphic correlation. The stable isotope signatures of the sedimentary inorganic and organic matter are among the most powerful proxies used in chemostratigraphy (Weissert et al., 2008). Considering the photosynthesis (Calvin cycle) as principal contrivance for biologically mediated reduction of CO_2, the organic carbon covers the $\delta^{13}C$ range between −23‰ and −34‰, with an average close to −27‰ (Schidlowski, 1987). $\delta^{13}C$ signatures of the marine carbonate have remained close to zero per mil. Pertubation in the global biogeochemical cycles, particularly that of carbon, in the form of positive and negative excursions has occurred in the earth's geologic history. The long-term C-isotope trend for carbonates has been punctuated by sudden shifts over short time intervals and have been used as time markers for stratigraphic correlations leading to the widespread use of chemostratigraphic correlations (Knoll et al., 1986; Knoll and Walter, 1992). Extreme climatic conditions such as those during the Proterozoic and Mesozoic, led to the drawdown of global atmospheric carbon inventory by increased carbon and pyrite burial in the anoxic sediments on ocean floor (Lambert et al., 1987; Garrels and Lerman, 1981; Jarvis et al., 2006). During the periods of higher accumulation of organic carbon in sediments, which preferentially removes the isotopically light ^{12}C via autotrophic carbon fixation, the marine bicarbonate ions become $\delta^{13}C$ enriched and hence the carbonate precipitated from such seawater show enrichment in $\delta^{13}C$ values.

The associated isotopic fractionation is explicated by the fact that isotopic compositions of the organic carbon (C_{org}) and the carbonate carbon (C_{carb}) are coupled with their relative proportions in the exogenic reservoir by the constraints of an isotope mass balance (Schidlowski, 1987). During the primary carbon ($\delta^{13}C_{prim} = -5‰$) flux, the ocean–atmosphere system shows partitioning between C_{org} and C_{carb} with a fractionation of −25‰. It results in the formation of biogenic carbon, with $\delta^{13}C_{carb}$ values around 0‰ which in turn would indicate a ratio $C_{org}/C_{carb} = 1/4$ in surficial exchange reservoir (Schidlowski, 1987). This follows from the relation.

$$\delta^{13}C_{prim} = R\,\delta^{13}C_{org} + (1 - R)\ \delta^{13}C_{carb}$$

rendering $R = C_{org}/(C_{org} + C_{carb}) = 0.2$ and consequently, $C_{org}/C_{carb} = 0.2/0.8 = 1/4$ for above parameters. A steady state complying with the above is established in a geologically short time interval (Schidlowski, 1987). Accordingly, any shift of $\delta^{13}C_{carb}$ from the primordial value of −5‰ in positive direction depicts that the total carbon removed through the surficial exchange reservoir has been converted into organic carbon.

Berner (2003) compared the global organic burial rate during the Phanerozoic eon with the major oil and gas source rocks depositional intervals (Fig. 3; after Berner, 2003). The burial curve was calculated using the variable isotope fractionation during photosynthesis, data on $\delta^{13}C$ for the late Permian period and data on the rapid recycling of carbon in younger rocks. The organic carbon burial curve shows a broad maximum peak centered around 300 Ma (Permo-Carboniferous). Prominent minima occurs before ~470 Ma (early Paleozoic) and around 220–180 Ma (Triassic–Early Jurassic), whereas secondary maxima are at 440–410 Ma (Silurian) and 90–120 Ma (Middle Cretaceous) (Berner, 2003). The times of maximum model-calculated organic carbon burial coincide with independent estimates of the times of major oil source rock deposition suggesting the linkage between organic burial and hydrocarbon preservation. The relative areas of the six principal source-rock periods (Klemme and Ulmishek, 1991) expressed as a percentage of the total for all areas coincide in time with the maxima in the calculated carbon burial curve. Also, the minima in the carbon–burial curve coincide with periods of very low occurrence of source rocks for oil and coal. This agreement suggests that a major factor in oil generation is the increased global deposition of organic matter (Berner, 2003).

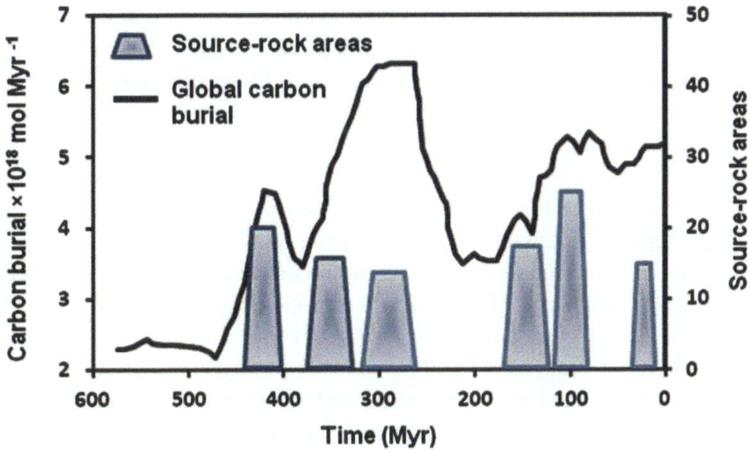

FIGURE 3

Organic carbon burial rate estimated from the $\delta^{13}C$ record of carbonates overlain with peaks in times of occurrence of major oil source rocks.

After Berner (2003).

The prolific Infracambrian and Mesozoic petroleum systems around the globe are characterized by the depositions of organic rich source rocks (Bois et al., 1982; Cozzi et al., 2012; Craig et al., 2012). These periods are also characterized by marked increase in carbonate carbon (C_{carb}) and organic carbon (C_{org}) budget (Ronov, 1958, 1982). In general, increases in the $\delta^{13}C$ of carbonates are considered to be indicative of increased burial and preservation of isotopically negative organic carbon; whereas, the decreases in the $\delta^{13}C$ suggest transfer of carbon from the organic into the inorganic reservoir. Hence, variation in the $\delta^{13}C$ of carbonate material serves as a proxy for the amount of buried organic carbon (Holser et al., 1996). Therefore, systematic studies of carbonate sequences with respect to their stratigraphic positions demarcate the time intervals of higher organic carbon burial/productivity and provide insights into their depositional, diagenetic, and paleoenvironmental conditions. In addition, the oxygen isotopic compositions of the carbonates can shed light towards evaluation of depositional/diagenetic conditions (Hoefs, 1996; Knoll et al., 1986; Scholle et al., 1991; Holser et al., 1996; Veizer et al., 1999; Derry et al., 1994).

16.3 CHEMOSTRATIGRAPHIC CORRELATIONS FOR PETROLEUM PROSPECTING

Several workers have described the stratigraphic variation of geochemical parameters, specifically $\delta^{13}C$ and $\delta^{18}O$ as a correlation tool and as an index of changing palaeoceanographical conditions, which have significant implications on the source rock deposition and development. Variation of oxygen isotope composition in planktonic foraminifera studied from the piston cores recovered from the Caribbean was demonstrated by Emiliani (1955), who interpreted the oscillations of the curve as a paleotemperature curve recording seven glacial–interglacial cycles. Improved understanding of oxygen isotope geochemistry and associated fractionation processes in the water cycle have shown that the oxygen isotope variations through time record changes in both temperature and isotopic composition of seawater, and the control excercised by the ice volume variations, thus producing a signal of global sea level variations (Shackleton and Opdyke, 1973, 1977; Shackleton et al., 1984; Shackleton, 1987; Weissert et al., 2008).

The diagenetic processes alter the oxygen isotope ratios of the deeply buried marine sediments, whereas the lithified pelagic sedimentary successions store reliable information on carbon isotope stratigraphy (Weissert et al., 2008). Thus, carbon isotopic records have been used widely to decipher the ancient climatic conditions. The Valanginian carbon isotope excursion records the first major carbon cycle perturbation in the Early Cretaceous. Hennig (2003) traced the carbon isotope excursion of the Valanginian along a northwest–southeast trending transect through the Alpine Tethys Ocean. The carbon isotope excursion spanning the Late Valanginian and the earliest Hauterivian has been identified in pelagic successions along Southern Alps of northern Italy, in other oceans (Erba et al., 2004; Weissert and Erba, 2004), neritic environments (Gréselle, 2007) and continental successions (Gröcke et al., 2005). The Valanginian–Hauterivian carbon isotope anomaly serves as a proxy of global carbon cycling and also as an excellent stratigraphic marker, which coincides with the deposition of Black shale globally. Spatial and temporal distributions of Black shales in several sedimentary basins across the globe have resulted in commercial petroleum deposits.

In another significant work by Scholle and Arthur (1980), using more than 1000 closely spaced samples of Cretaceous Pelagic Limestones, seven major $\delta^{13}C$ excursions were correlated from section to section. The most important "heavy events" occur near the Aptian–Albian and Cenomanian–Turonian boundaries, whereas "light events" are near the Jurassic–Cretaceous, Albian–Cenomanian, Turonian–Coniacian, and Cretaceous–Tertiary boundaries (Scholle and Arthur, 1980).

Possible Late Middle Ordovician organic carbon isotope excursion has been shown in Ordovician oils and hydrocarbon source rocks, from Mid-Continent and East-Central United States (Hatch et al., 1987). The wide ranges in $\delta^{13}C$ for oils and rock extracts reflect a major, positive excursion(s) (6–9‰) in organic matter $\delta^{13}C$ in late Middle Ordovician rocks. This excursion has at least a regional significance in that it can be documented in sections located 770 km apart in south-central Kansas and eastern Iowa (Hatch et al., 1987).

Carbon isotope studies have been intensively used to unravel the changes in the Paleozoic global carbon cycle and for chemostratigraphic correlation of sections from various palaeocontinents. The Paleozoic $\delta^{13}C$ excursions are significantly larger than the excursions observed in the Mesozoic and Cenozoic, but +10‰ lower than the excursions observed in Neoproterozoic (Johnston et al., 2012). Prominent positive excursions in $\delta^{13}C$ with amplitudes of up 6‰ are observed in the Upper Cambrian, Late Ordovician to Silurian, and Late Devonian to Mississippian carbonates (Saltzman and Young, 2005). The Devonian whole rock carbon isotope record (Buggisch and Mann, 2004; Saltzman and Young, 2005; Buggisch and Joachimski, 2006) shows several positive excursions, which in most cases coincide with sea-level changes, Black shale deposition and in part with bio-events. The association of isotopic variations with gross carbon storage leading to development of organic rich source rocks is potentially useful for evaluation of precise timings and magnitude of preservation of organic matter in sedimentary basins. Such studies may aid in estimating potential hydrocarbon resources in largely unexplored oceanic basins or along continental margins (Scholle and Arthur, 1980).

16.4 METHODOLOGY FOR SAMPLE SELECTION AND ANALYSIS

16.4.1 POSTDEPOSITIONAL ALTERATION

Postdepositional overprints in sedimentary rocks such as diagenesis and metamorphism may affect the primary isotopic signatures of the organic and inorganic carbon species (Arthur, 2009; Schidlowski, 1987; Veizer and Fritz, 1976). Sedimentary carbonates commonly preserve the isotopic composition of their parent carbonates within ±1‰ of the original value, which in turn is inherited from a bicarbonate precursor. The isotopic changes during burial and subsequent diagenetic reconstitution of sedimentary organic carbon are more pronounced, spanning the range from a few to several part per million over the maturation pathway of these substances in the sediment (Schidlowski, 1987). The preferential loss of isotopically light lipids, hydrocarbons and functional groups during the thermal maturation leads to enrichment relative to the precursor materials (Schidlowski, 1987). Sizable alterations in $^{13}C/^{12}C$ in organic and carbonate carbon occur when subjected to metamorphism. Siliceous carbonates are bound to undergo a sequence of decarbonation reactions in which primary carbonate constituents (calcite, dolomite) react with silica to give Ca–Mg silicates with a concomitant liberation of CO_2 and the residual carbonates are ^{12}C enriched (Schidlowski, 1987). Therefore, the careful examination of samples, to avoid oxidative weathering and alterations, should be a prime consideration when selecting the samples for the isotopic ratio determination. The unaltered or the least altered samples should be chosen, as only these can provide the isotopic signatures close to the original or the primary composition. Petrographic, cathodoluminescence, and trace element concentration investigations can help in the selected of most well-preserved portions of the rock for the geochemical analysis (Schidlowski, 1987).

16.4.2 ANALYTICAL PROCEDURE

The isotopic composition of carbon is expressed in units of the abundance ratio of carbon relative to a standard by the $\delta^{13}C$ parameter. $\delta^{13}C$ is defined using the following equation to calculate the ratio difference (δ) in parts per 1000 (parts per million, or ‰):

$$\delta^{13}C = \left[\frac{\frac{^{13}C}{^{12}C}_{(\text{sample})} - \frac{^{13}C}{^{12}C}_{(\text{standard})}}{\frac{^{13}C}{^{12}C}_{(\text{standard})}} \right] \times 1000$$

where, $^{13}C/^{12}C$ is the atomic abundance ratio of heavy to light isotope of carbon in the sample and standard (Hoefs, 1996). The standard that has been most widely used is the belemnite from Peedee Formation (PDB) in South Carolina. The δ values can be positive indicating enrichment in the ^{13}C relative to standard or negative indicating depletion of the heavier isotope. For the oxygen isotopes, Vienna Standard Mean Ocean Water (VSMOW) is used as the standard and the relationship between sample and standard expressed in parts per million is as below:

$$\delta^{18}O = \left[\frac{\frac{^{18}O}{^{16}O}_{(\text{sample})} - \frac{^{18}O}{^{16}O}_{(\text{standard})}}{\frac{^{18}O}{^{16}O}_{(\text{standard})}} \right] \times 1000$$

The carbonate and organic carbon isotope signatures are determined using the isotope ratio monitoring spectrometer (IRMS). Gas Bench IRMS is used for the C and O isotope ratio determinations on carbonates. Carbon and oxygen measurements are carried out on bulk carbonate samples by treating them with phosphoric acid which decomposes the carbonates and the CO_2 is evolved. Elemental Analyzer IRMS is used for the determination of isotopic composition of coexisting organic matter in carbonate rock samples. The technique used for the determination of organic carbon is based on the quantitative dynamic flash combustion method. The acid treated decarbonated samples are combusted in quartz tube maintained at 1020 °C (combustion reactor) and completely oxidized to CO_2.

The gaseous samples, in the form of CO_2 obtained after chemical treatment of geological samples, are introduced via needle valve into the ion source. A hot filament in the ion source generates beam of electrons which ionizes the molecules of CO_2. The ions are collimated into a fine focused beam by electrostatic lenses and accelerated into the flight tube, where they encounter a strong magnetic field created by the electromagnet. The mass separation is done on the basis of mass to charge (m/z) ratio. The ions are detected by an array of Faraday cups, the output from which is used to calculate the final stable isotope ratio (Fig. 4; after Hoefs, 1996). This is calculated relative to the standard of known isotopic composition and expressed using the dimensionless "per mil" notation.

16.5 C AND O ISOTOPE ANOMALIES AND NEAR SURFACE HYDROCARBON MANIFESTATIONS

The surface manifestations of light gaseous hydrocarbons are based on the concept of the microseepage of oil and gas and form the basis of one of the widely used geochemical prospecting technique, particularly in the unexplored basins. The geochemical approach for the hydrocarbon prospecting assumes that the sediment layers in a basin including the caprocks of petroleum reservoirs do not have perfect seals and are permeable to the subsurface hydrocarbons. Light (primarily methane-C1; ethane-C2; propane-C3; butane-C4; and pentane-C5) and heavy (volatile and semivolatile C10+) hydrocarbon molecules tend to seep through natural geological structures namely, faults and fractures and through processes like diffusion, buoyancy of microbubble etc. either from the source rocks or from the oil/gas reservoirs (Mani et al., 2011a, 2012; Jones and Drozd, 1983; Horvitz, 1985; Klusman, 1993; Schumacher, 2003; Tedesco, 1995). The migration of hydrocarbons in low concentrations towards the surface or the invisible seepage causes chemically detectable alterations in the near-surface soils and sediments (hydrocarbon-induced alterations of soils and sediments) (Fig. 4; after Schumacher, 2003). A detailed geochemical examination of shallow soils and sediments in a sedimentary basin, with the use of sensitive analytical techniques is capable of detecting the hydrocarbons that have migrated from the deep seated reservoirs. The quality and quantity of the light hydrocarbon gases (C_1–C_4) are then determined using a gas chromatograph coupled to a flame ionization detector, and the source of these gases is studied using the stable carbon isotopic ratios measured on a gas chromatograph–combustion IRMS (Stahl et al., 1981). The magnitudes of the hydrocarbon gases observed in the near-surface soils are controlled by regional geology, tectonic activity, and fault/lineament pattern (Mani et al., 2011b) and also by the seepage processes listed earlier.

An example from the Proterozoic Vindhyan Basin shows an association of chemostratigraphic carbon and oxygen isotope anomalies with near surface manifestation of elevated concentrations of light gaseous hydrocarbons. The study shows that the relative enrichments of C_{org} and C_{carb} in the carbonate rocks of the basin are coincident with the occurrence of thermally generated hydrocarbon in the basins.

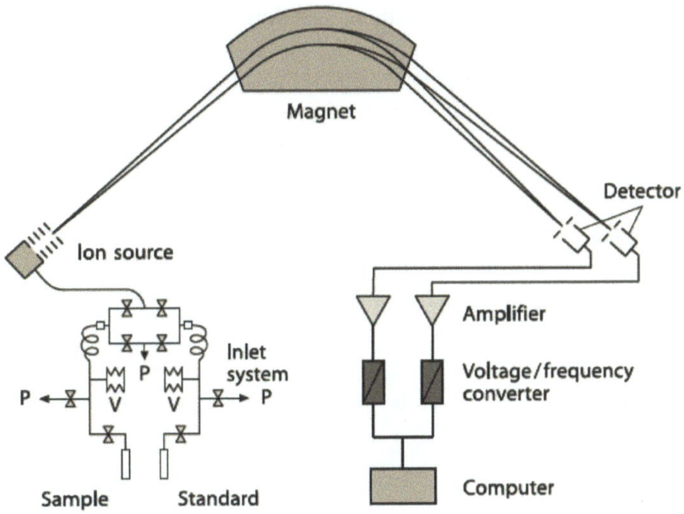

FIGURE 4

A generalized schematic of isotope ratio monitoring spectrometer.

After Hoefs (1996).

Similar associations have been observed from the Kutch Basin (Patil et al., 2011; Srinivas Rao et al., 2013) and Cuddapah Basin (Patil et al., 2002; Kalpana et al., 2010).

The carbon isotopic excursions leading to organic carbon burial and preservation, as evident from the isotopic signatures of the carbonate rocks from these basins, possibly indicate the occurrences of potential source rocks at depth, albeit noncommercial as of present, which in turn formed the source of the light gaseous hydrocarbons observed in the near-surface soils. Such associations of microseepage of hydrocarbons from the potential subsurface source rocks and organic rich source beds are extremely useful in reconnaissance exploratory studies for the delineation of priority areas planned for advanced resource evaluations.

16.5.1 VINDHYAN BASIN

Findings of commercial quantities of oil at Baghewala-1 in the Bikaner–Nagpur Basin of Rajasthan encouraged high resolution studies in the Indian Proterozoic basins, among which the Vindhyan comprises a large area, and remaines relatively unexplored for hydrocarbon resource potential. The Vindhyans are considered to be analogous in terms of their petroleum geology with the other Proterozoic sedimentary basins of the world which contain commercial quantities of hydrocarbons (Ram et al., 1996). The favorable tectonics and entrapment conditions corroborate further to the hydrocarbon potential of Vindhyan, making it a geologically prospective Category III sedimentary basin of India (DGH, 2012). The basin consists of more than 5000 m thick sequence of sandstones, shales, and limestones, which are characterized by the presence rich fossil assemblage (Bengtson et al., 2009; Banerjee et al., 1992, 2006, 2007; Kumar, 2001; Sharma et al., 2009; Sharma and Shukla, 2009, and others). Documented occurrence of the well-preserved algal/bacterial rich organic matter in this basin could be a suitable source for the generation of liquid hydrocarbons (Hunt, 1996; Tissot and Welte, 1984).

The Vindhyan strata are exposed in Rajasthan, Bundelkhand, and Son valley. The inorganic and organic carbon excursions have been reported from the Vindhyan Basin by several workers (Chakraborty et al., 2002; Kumar et al., 2002; Jyotiranjan et al., 2003; Sarkar et al., 2003; Krishnamurthy et al., 1987). Kumar et al. (2002) from their study on carbonates from the central Vindhyan, eastern Son Valley and Rajasthan sections inferred that the carbon and oxygen isotope compositions of the rocks can be correlated with the isotope evolution curves of marine carbonates of the late Proterozoic (Fig. 5; after

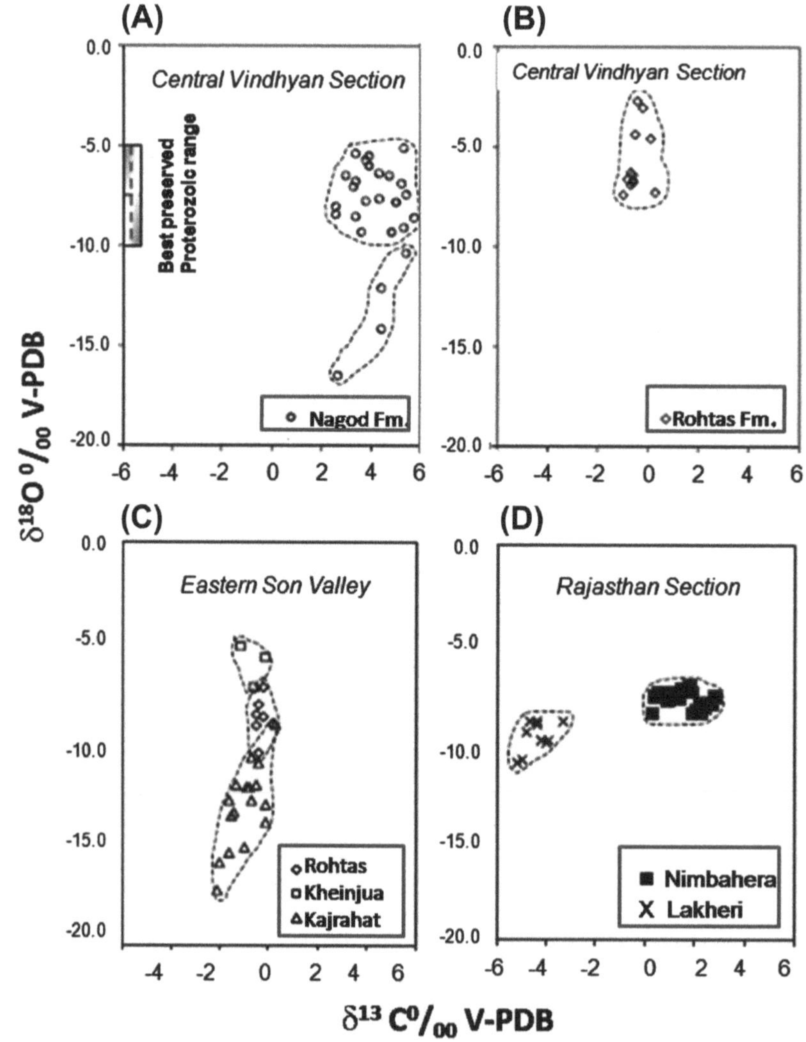

FIGURE 5

Carbon and oxygen isotope compositions from the sedimentary carbonates of Vindhyan Basin, Central India. **(A)** Nagod Formation, central Vindhyan section; **(B)** Rohtas Formation, central Vindhyan section; **(C)** Kajrahat, Kheinjua and Rohtas Formations, eastern Son Valley section; and **(D)** Lakheri and Nimbahera Formations, Rajasthan section.

After Kumar et al. (2002).

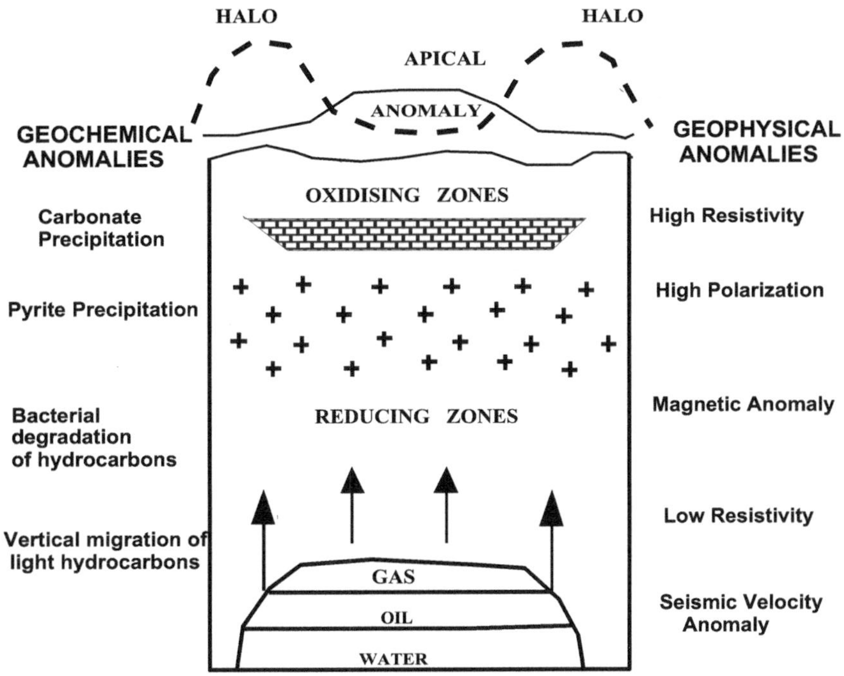

FIGURE 6

Geochemical model for hydrocarbon-induced alteration of soils and sediments.

After Schumacher (1996).

Kumar et al., 2002). The carbonate rocks of the Lower Vindhyan Supergroup from eastern Son Valley and central Vindhyan sections show $\delta^{13}C$ values of ~0‰ (V-PDB) and those from Rajasthan section are enriched up to +2.8‰ (Kumar et al., 2002). In contrast, the carbonate rocks of the Upper Vindhyan succession record both positive and negative shifts in $\delta^{13}C$. In the central Vindhyan section, the carbonates exhibit positive $\delta^{13}C$ values up to +5.7‰ and those from Rajasthan show negative values down to −5.2‰ (Kumar et al., 2002). The $\delta^{18}O$ values of most of the carbonate rocks from the Vindhyan Supergroup show a narrow range between −10‰ and −5‰ (V-PDB) and are similar to the best preserved ^{18}O compositions of the Proterozoic carbonate rocks (Kumar et al., 2002).

Surface geochemical prospecting studies have been reported from the Chambal Valley in the western part of Vindhyan Basin on a regional scale (Dayal et al., 2014; Ram, 2012; Kumar et al., 2006; Ojha, 2012; Unpublished Technical Report, NGRI, 2003). The adsorbed soil gas data on 350 soil samples have shown varied concentrations of light gaseous hydrocarbon methane to butane (CH_4, C_2H_6, C_3H_8, iC_4H_{10}, and $nC_4\ H_{10}$) ranging between 1 and 2547 ppb, 1 and 558 ppb, 1 and 181 ppb, 1 and 37 ppb, and 1 and 32 ppb, respectively (Kumar et al., 2006). The results show that the region around Baran–Jhalawar–Bhanpur–Garot is characterized by a high concentration of methane (>100 ppb) and the higher hydrocarbons ($\Sigma C2+$; sum of ethane, propane, and butane) is >40 ppb (Fig. 7; after Ojha, 2012). The correlation between the hydrocarbon variables C_1, C_2, C_3, and C_4 is linear, which suggests that the hydrocarbons have similar sources, likely to be thermogenic and are not affected by secondary alterations. Bimodal and polymodal distribution patterns, as seen in the frequency distribution plots of

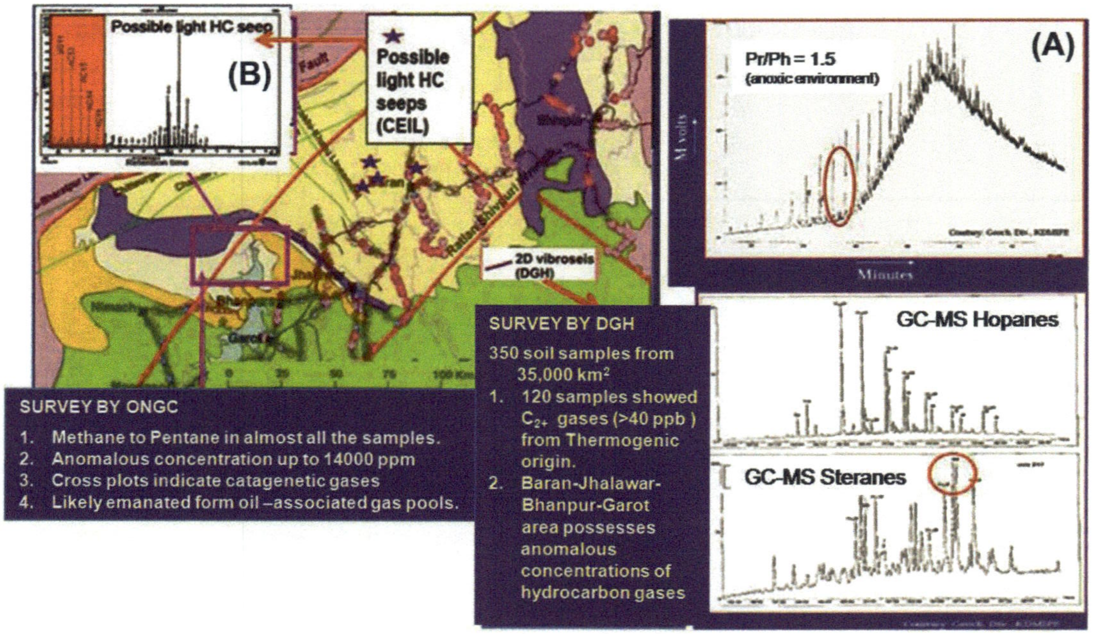

FIGURE 7

Light gaseous hydrocarbon anomalies in the western part Vindhyan Basin, Central India. **(A)** Gas chromatogram of the extracted organic matter indicating a non-biodegradable marine origin for oil and mass spectra indicating algal/carbonate sourced oil generated within anoxic conditions. **(B)** Area surveyed by DGH, ONGC and CEIL exhibiting light hydrocarbon seeps.

After Ram (2012).

these hydrocarbon components suggest that the adsorbed hydrocarbon data from the region is characterized by the presence of different populations with background and anomalous levels (Kumar et al., 2006; Ojha 2012; Unpublished Technical Report, NGRI, 2003).

The concentrations of C_1, C_2, C_3, iC_4, and nC4 desorbed from the soils of eastern Vindhyan Basin (Son Valley) range between 0 and 186 ppb, 0 and 4 ppb, 0 and 5 ppb, and 0 and1 ppb, respectively (Dayal and Raju, 2010; Unpublished Technical Report, NGRI, 2009; Ojha, 2012). The presence of light gaseous hydrocarbon in the studied samples shows that the methane concentration varies from moderate to low levels while the ethane and propane vary from low to zero in concentrations (Fig. 8; after Ojha, 2012). The Pixler plot (Pixler, 1969) indicates that the soil samples lie in oil and gas zone. The carbon isotopic compositions of methane ($\delta^{13}C_1$), ethane ($\delta^{13}C_2$), and propane ($\delta^{13}C_3$) showed the thermogenic source for these light hydrocarbon gases. The carbon isotope values for methane ($\delta^{13}C_1$) vary between −45.7‰ and −25.24‰ (PDB) (Dayal and Raju, 2010; Unpublished Technical Report, NGRI, 2009; Ojha, 2012). Ethane carbon isotopes for the soil samples vary between −35.3‰ and −20.07‰.

Stratigraphic sections of the Vindhyan Basin from where the carbonates were sampled include the (1) Jabera–Damoh–Panna–Satna–Maihar–Badanpur (Central Vindhyan section); (2) Chopan–Robertsganj (Eastern Son Valley section) Bundi–Kota–Jhalawar (Rajasthan section) section). These areas exhibit the occurrences of light gaseous hydrocarbon anomalies. Accumulation and preservation of organic matter in the Vindhyan Basin has been reported by several workers. Mathur (1983) suggested that the sedimentary organic matter in Vindhyan Basin is capable of generating dry gas. Kajrahat Formation, the Rampur Shale and the Bijaigarh Shale have >1% TOC and this organic matter contains Type I and II kerogens (Ram, 2012). The organic matter rich shales can form good source rocks under suitable depth and temperature conditions. Stromatolites are the main constituents of the Kajrahat, Rohtas, and Bhander limestones. The organic facies in the Lower Vindhyan sequences are buried to a depth of more than 3000 m and are likely to be mature in the Son Valley area (Ojha, 2012). About 4500–5000 m thickness of Upper Vindhyan sediments are expected in the Chambal Valley. The Lower Vindhyan shales, particularly, Hinota and Pulkova shales have organic carbon contents of up to 3.89% and 1.86%, respectively (DGH, 2012). Jabera-1 well drilled in Son Valley showed the gas (methane, ethane, propane, butane and pentane) flow at 4000 m^3/day from the Charkaria succession (Fuloria, 1996; Ojha, 2012). The gas indices comprising of butane and pentane ratios suggest a mature source for the hydrocarbon gases. Also, seepage of thermogenic gases, methane through butane, has been reported from the Sagar District in Madhya Pradesh (Prasanna et al., 2010). Exploratory drilling has not yet been initiated in the western part of Vindhyan Basin. Composition of gas seeps from the Chambal Valley show unaltered oil of marine origin with the presence of hopane and sterane biomarkers (Fig. 6). These suggest anoxic depositional settings with algal/carbonate source contributions (Ojha, 2012). Thus, organic carbon burial, as indicated by the stable isotope enrichments of carbonates in the Vindhyan Basin is corroborated by the occurrence of organic rich shales in the region, which could be potential hydrocarbon source rocks in this basin.

16.5.2 KUTCH AND CUDDAPAH BASINS

Patil et al. (2011) reported the carbon, oxygen, and strontium compositions of sedimentary carbonates from the Mesozoic Kutch Basin and have shown the best preserved $\delta^{13}C$ values to be in agreement with the global organic carbon burial curve. Surface geochemical hydrocarbon prospecting studies from the Kutch–Saurashtra basin indicate the occurrence of adsorbed light hydrocarbon (methane to butane)

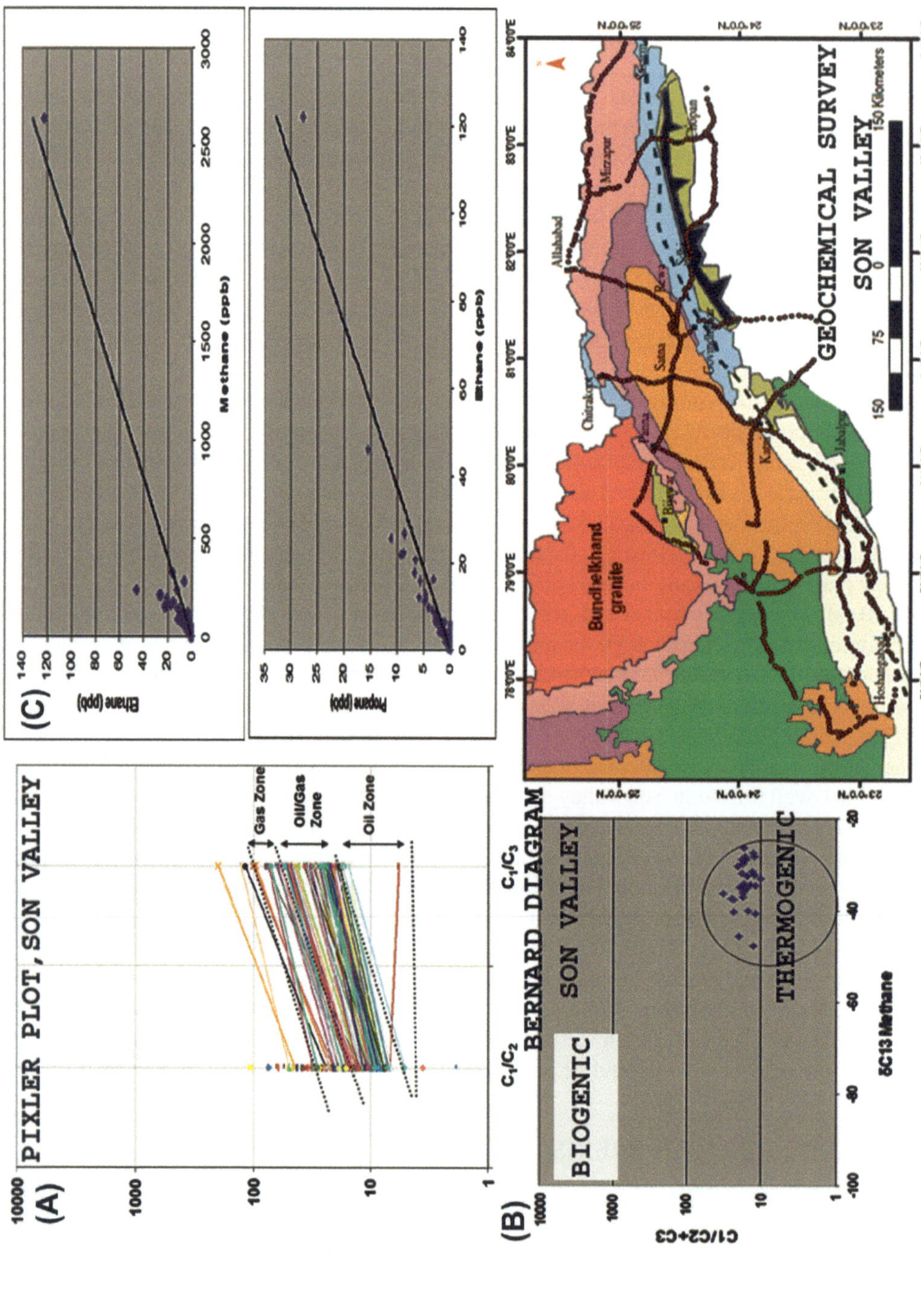

FIGURE 8

Light gaseous hydrocarbon anomalies in the eastern part of Vindhyan Basin, Central India. **(A)** Pixler plot showing samples falling within hydrocarbon window; **(B)** Bernard diagram showing thermogenic origin of the hydrocarbons; **(C)** Correlation plots of methane and ethane showing genetically related hydrocarbon.

After Ojha (2012).

anomalies (Patil et al., 2013; Srinivasarao et al., 2013), suggesting the generation of gaseous hydrocarbons in the basin. In a study of similar kind, the stable isotope systematics of C, O, and Sr from the Cuddapah Basin (Patil et al., 2002), where the Proterozoic strata has one of the maximum thickness of sedimentary successions compared to other analogous Indian basins, suggests the positive $\delta^{13}C$ values for the Koilkuntla limestones. Gross organic carbon preservation in the basin in due course of geological time possibly led to the development of light hydrocarbon anomalies (Kalpana et al., 2010) which in turn are observed in the near-surface soils of the Cuddapah Basin.

16.6 CONCLUSION

The stable carbon isotopic data from the sedimentary carbonates and shales indicate the net organic carbon burial that took place during different geological time periods on the earth. Under given conditions of temperature, pressure and tectonic activities, digenesis and compaction of organic rich sediments lead to the formation of oil and gas in a basin. Any deviation of organic carbon productivity/burial leads to consequent changes in the $\delta^{13}C$ signatures of marine carbonates, which in turn, serves as a proxy for productivity/burial. Examples from the Proterozoic Vindhyan, Cuddapah, and Mesozoic Kutch sedimentary basins of India indicate the stable carbon and oxygen isotope events to be related with the light gaseous hydrocarbon microseepge, indicating generation of hydrocarbons in these basins.

Thus, carbon isotopic excursions leading to organic carbon burial and preservation provide useful insights on potential source rocks at depth. The thermogenic light gaseous hydrocarbons observed in the near-surface soils are possibly generated from such source beds and migrated towards near surface shallow environment where they are retained in the carbonate and clay matrices of the soil. Such associations of isotopic anomalies and microseepage of hydrocarbons from the subsurface source rocks are extremely useful indicators to prioritize exploration activities in the frontier basins where limited information is available. Reconnaissance and high-resolution exploratory studies of these types can aid in the mapping of hydrocarbon rich source beds and areas.

ACKNOWLEDGMENTS

The Oil Industry Development Board is acknowledged for granting the financial support in setting up of National Facility of Hydrocarbon Prospecting at NGRI. The Director, NGRI is acknowledged for permitting the publication of this research work. DM acknowledges CSIR for the Senior Research Associate ship. The authors thank the anonymous reviewers for constructive suggestions towards improvement of manuscript.

REFERENCES

Arthur, M.A., 2009. Carbonate rocks deconstructed. Nature 460, 6.

Arthur, M.A., Dean, W.E., Claypool, G.E., 1985. Anomalous ^{13}C enrichment in the modern marine organic carbon. Nature 315, 216–218.

Banerjee, S., Dutta, S., Paikaray, S., Mann, U., 2006. Stratigraphy, sedimentology and bulk organic geochemistry of black shales from the ProterozoicVindhyan Supergroup Central India. Journal of Earth System Science 115 (1), 37–47.

Banerjee, D.M., Deb, M., Strauss, H., 1992. Organic carbon isotopic composition of Proterozoic sedimentary rocks from India: preliminary results. In: Schidlowski, et al. (Ed.), Early Organic Evolution: Implications for Mineral and Energy Resources. Springer, pp. 232–240.

Banerjee, S., Bhattacharya, S.K., Sarkar, S., 2007. Carbon and oxygen isotopic variations in peritidal stromatolite cycles, Paleoproterozoic Kajrahat Limestone, Vindhyan basin of Central India. Journal of Asian Earth Sciences 29, 823–831.

Bengtson, S., Belivanovaa, V., Rasmussen, B., Whitehouse, M., 2009. The controversial "Cambrian" fossils of the Vindhyan are real but more than a billion years older. Proceedings of National Academy of Science 106 (19), 7729–7734.

Berner, R.A., 1982. Burial of organic carbon and pyrite sulfur in the modern ocean: its geochemical and environmental significance. American Journal of Science 282, 451–473.

Berner, R.A., 1989. Biogeochemical cycles of carbon and sulfur and their effect on atmospheric oxygen over Phanerozoic time. Global and Planetary Change 75, 97–122.

Berner, R.A., 1999. Atmospheric oxygen over Phanerozoic time. Proceedings of the National Academy of Sciences 96, 10955–10957.

Berner, R.A., 2006. GEOCARBSULF: a combined model for Phanerozoic atmospheric (isotopes/sediment cycles/natural oxygen reservoirs). Proceedings of the National Academy of Sciences 78 (8), 4652–4656.

Berner, R.A., 2003. The long-term carbon cycle, fossil fuels and atmospheric composition. Nature 426, 323–326.

Bois, C., Bouche, P., Palet, R., 1982. Global geological history, distribution of hydrocarbon reserves. American Association of Petroleum Geologists Bulletin 66, 1248–1270.

Brassell, S.C., Eglinton, G., Marlowe, I.T., Pflaumann, U., Sarnthein, M., 1986. Molecular stratigraphy: a new tool for climatic assessment. Nature 320, 129–133.

Buggisch, W., Joachimski, M.M., 2006. Carbon isotope stratigraphy of the Devonian of Central and Southern Europe. Palaeogeography, Palaeoclimatology, Palaeoecology 240, 68–88.

Buggisch, W., Mann, U., 2004. Carbon isotope stratigraphy of Lochkovian to Eifelian limestones from the Devonian of central and southern Europe. International Journal of Earth Sciences 93, 521–541.

Chakraborty, P.P., Sarkar, A., Bhattacharya, S.K., Sanyal, P., 2002. Isotopic and sedimentological clues to productivity change in Late Riphean Sea: a case study from two intracratonic basins of India. Proceedings of the Indian Academy of Sciences (Earth and Planetary Science) 111 (4), 379–390.

Cozzi, A., Rea, G., Craig, J., 2012. From global geology to hydrocarbon exploration: Pakistan and Oman Ediacaran-Early Cambrian petroleum plays of India. Geological Society, London, Special Publications 366, 131–162.

Craig, J., Biffi, U., Galimberti, R.F., Ghori, K.A.R., Gorter, J.D., Hakhoo, N., Le Heron, D.P., Thurowe, J., Vecoli, M., 2012. The palaeobiology and geochemistry of Precambrian hydrocarbon source rocks. Marine and Petroleum Geology 40, 1–47.

Dayal, A.M.; Mani, D.; Madhavi, T.; Kavitha, S.; Kalpana, M.S.; Patil, D.J.; Sharma, M., 2014. Organic geochemistry of the Vindhyan sediments: Implications for hydrocarbons. Journal of Asian Earth Sciences 91, 329–338.

Dayal, A.M., Raju, S.V., 2010. Discovery of Hydrocarbon in Proterozoic Vindhyan Basin, East India Using Geochemical Signatures. (Goldschmidt Conference Abstracts, A214). Retrieved from: http://goldschmidt.info/2010/abstracts/finalPDFs/A214.pdf downloaded on June 12, 2013.

Des Marais, D.J., Strauss, H., Summons Roger, E., Hays, J.M., 1992. Carbon isotope evidence for step wise oxidation of the Proterozoic environment. Nature 359, 605–609.

Derry, L.A., Brasier, M.D., Corfield, R.M., Rozanov, A.Y., Zhuravlev, A.Y., 1994. Sr and C isotopes in Lower Cambrian carbonates from the Siberian craton; a paleoenvironmental record during the 'Cambrian explosion'. Earth and Planetary Science Letters 128 (3–4), 671–681.

DGH, 2012. Directorate General of Hydrocarbons, India. Retrieved from: www.dghindia.org downloaded on December 16, 2012.

Emiliani, C., 1955. Pleistocene temperatures. Journal of Geology 63, 538–578.

Erba, E., Bartolini, A., Larson, R.L., 2004. Valanginian Weissert oceanic anoxic event. Geology 32, 149–152.

Fuloria, R.C., 1996. Geology and hydrocarbon prospects of the Vindhyan sediments in Ganga Valley. In: Bhattacharyaa, A. (Ed.), Recent Advances in Vindhyan Geology, vol. 36. Geological Society of India Bangalore, Memoir, p. 235.

Garrels, R.M., Lerman, A., 1981. Phanerozoic cycles of sedimentary carbon and sulfur. Geology 24, 867–870.

Gréselle, B., 2007. Impact des variations paleoclimatiques sur la sedimentation carbonatée au Valanginien. Thèse Université Lyon 1 – Claude Bernard, Part I, 154 p.

Grocke, D.R., Price, G.D., Robinson, S.A., Baraboshkin, E.Y., Mutterlose, J., Ruffel, A.H., 2005. The upper Valanginian (Early Cretaceous) positive carbon-isotope event recorded in terrestrial plants. Earth and Planetary Science Letters 240, 495–509.

Hatch, J.R., Jacobson, S.R., Witzke, B.J., Risatti, B.,J., Anders, D.E., Watney, L.W., Newell, D.K., Vuletich, A.K., 1987. Possible late middle ordovician organic carbon isotope excursion: evidence from ordovician oils and hydrocarbon source rocks, Mid-Continent and East-Central United States. AAPG Bulletin 71, 1342–1354.

Hedges, J.I., Keil, R.G., 1995. Sedimentary organic matter preservation: an assessment and speculative synthesis. Marine Chemistry 49, 81–115.

Hennig, S., 2003. Geochemical and Sedimentological Evidence for Environmental Changes in the Valanginian (Early Cretaceous) of the Tethys Region, Dissertation Nr. 15106, ETH Zurich, 185 pp.

Hoefs, J., 1996. Stable Isotope Geochemistry, fourth ed. Springer-Verlag, Heidelberg.

Holser, W.T., Schidlowsky, M., Mackenzie, F.T., Maynard, J.B., 1988. In: Gregor, B.C., Garrels, R.M., Mackenzie, F.T. (Eds.), Geochemical Cycles of Carbon and Sulfur.

Holser, W.T., Magaritz, M., Ripperdan, R.L., 1996. Global isotopic events. In: Walliser, O.H. (Ed.), Global Events and Event Stratigraphy. Springer-Verlag, Heidelberg, pp. 63–88.

Horvitz, L., August 30, 1985. Geochemical exploration for petroleum. Science 229 (4716), 821–827.

Hunt, J.M., 1996. Petroleum Geology and Geochemistry. W. H. Freeman and Company, San Francisco. 617.

Hunt, J.M., 1972. Distribution of Carbon in the crust of earth. AAPG Bulletin 56, 2273–2277.

Jarvis, I., Gale, A.S., Jenkyns, H.C., Pearce, M.A., 2006. Secular variation in Late Cretaceous carbon isotopes: a new $\delta^{13}C$ carbonate reference curve for the Cenomanian–Campanian (99.6–70.6 Ma). Geological Magazine 143 (5), 561–608.

Jenkyns, H.C., 1988. The Early Toarcian (Jurassic) anoxic event: stratigraphy, sedimentary and geochemical evidence. American Journal of Science 288, 101–151.

Jenkyns, H.C., 2010. Geochemistry of oceanic anoxic events. Geochemistry, Geophysics, Geosystems 11 (3). http://dx.doi.org/10.1029/2009GC002788.

Johnston, D.T., Macdonald, F.A., Gill, B.C., Hoffman, P.F., Schrag, D.P., 2012. Uncovering the Neoproterozoic carbon cycle. Nature 483, 320–324.

Jones, V.T., Drozd, R.J., 1983. Predictions of oil or gas potential by near-surface geochemistry. American Association of Petroleum Geologists Bulletin 67, 932–952.

Jyotiranjan, S.R., Veizer, J., Davis, W.J., 2003. C, O, Sr and Pb isotope systematics of carbonate sequences of the Vindhyan Supergroup, India: age, diagenesis, correlations and implications for global events. Precambrian Research 121, 103–140.

Kalpana, G., Madhavi, T., Patil, D.J., Dayal, A.M., Raju, S.V., 2010. Light gaseous hydrocarbon anomalies in the near surface soils of Proterozoic Cuddapah Basin: implications for hydrocarbon prospects. Journal of Petroleum Science and Engineering 73 (1), 161–170.

Karhu, J.A., Holland, H.D., 1996. Carbon isotopes and the rise of atmospheric oxygen. Geology 24 (10), 867–870.

Kaufman, A.J., Hayes, J.M., Knoll, A.H., Germs, G.J.B., 1991. Isotopic compositions of carbonates and organic carbon from upper Proterozoic successions in Namibia: stratigraphic variation and the effects of diagenesis and metamorphism. Precambrian Research 49, 301–327.

Klemme, H.D., Ulmishek, G.F., 1991. Effective petroleum source rocks of the world: stratigraphic distribution and controlling depositional factors. Bulletin American Association of Petroleum Geologist 75, 1809–1851.

Klusman, R.W., 1993. Soil Gas and Related Methods for Natural Resource Exploration. John Wiley & Sons, New York. 483 pp.

Knoll, A.H., Hayes, J.M., Kaufman, A.J., Swett, F., Lambert, T.B., 1986. Secular variations in carbon isotope ratios from upper Proterozoic successions of Svalbard and East Greenland. Nature 321, 832–838.

Knoll, A.H., Walter, M.R., 1992. Latest Proterozoic stratigraphy and earth history. Nature 356, 673–674.

Krishnamurthy, R.V., Bhattacharya, S.K., Mathur, S.M., 1987. Carbon isotope ratio of coaly matter from the basal part of the Proterozoic Vindhyan Supergroup. Journal of Geological Society of India 27, 119–120.

Kumar, B., Das Sharma, S., Sreenivas, B., Dayal, A.M., Rao, M.N., Dubey, N., Chawla, B.R., 2002. Carbon, oxygen and strontium isotope geochemistry of Proterozoic carbonate rocks of the Vindhyan Basin, Central India. Precambrian Research 113, 43–63.

Kumar, B., Raju, S.V., Patil, D.J., Kalpana, G., Vishnu Vardhan, C., 2006. Integrating Surface Geochemical Expressions with Geological, Geophysical and Remote Sensing Data: Case Histories from Selected Proterozoic Basins of India. http://www.searchanddiscovery.com/documents/2006/06088houston_abs/abstracts/kumar.htm.

Kumar, S., 2001. Mesoproterozoic megafossil *Chuaria–Tawuia* association may represent parts of a multicellular plant, Vindhyan Supergroup, Central India. Precambrian Research 106, 187–211.

Kump, L.R., Arthur, M.A., 1999. Interpreting carbon-isotope excursions: carbonates and organic matter. Chemical Geology 161, 181–198.

Lambert, I.B., Walter, M.R., Wenlong, Z., Songnian, L., Guogan, M., 1987. Palaeo environmental and carbon isotope stratigraphy of upper Proterozoic carbonates of the Yangtze platform. Nature 325 (8), 140–142.

Mani, D., Patil, D.J., Dayal, A.M., 2011a. Stable carbon isotope geochemistry of near surface adsorbed alkane gases in Saurashtra Basin. Chemical Geology 280 (1–2), 144–153.

Mani, D., Dayal, A.M., Rasheed, M.A., Satish Kumar, T., Rao, T.G., Balaram, V., 2011b. Soil iodine determination in Deccan Syneclise basin of India: implications for near surface geochemical hydrocarbon prospecting. Natural Resources Research 20 (1), 75–88.

Mani, D., Patil, D.J., Kalpana, M.S., Dayal, A.M., 2012. Evaluation of hydrocarbon prospects using surface geochemical data with constraints from geological and geophysical observations in Jamnagar area, Saurashtra Basin, India. Journal of Petroleum Geology 35 (1), 67–84.

Mani, D., Patil, D.J., Dayal, A.M., 2013. Evolution of sedimentary organic matter and formation of oil and gas: a geochemical approach. In: International Seminar on Earth Science for Society, 7–8 December 2012. CSIR-National Geophysical Research Institute, Hyderabad, India.

Mathur, S.M., 1983. A reappraisal of trace fossils described by Vredenburg (1908) and Beer (1919) in rocks of the Vindhyan Supergroup. Records of Geological Survey of India 113 (2), 111–113.

McCarthy, K., Niemann, M., Palmowski, D., Peters, K., Stankiewicz, A., 2011. Basic petroleum geochemistry for source rock evaluation. Oilfield Review 23, 2.

Meyers, P.A., 1994. Preservation of elemental and isotopic source identification of sedimentary organic matter. Chemical Geology 114, 289–302.

Ojha, P.S., 2012. Precambrian sedimentary basins of India: an appraisal of their petroleum potential. In: Bhat, G., Craig, J., Thurow, J.W., Thusu, B., Cozzi, A. (Eds.), Geology and Hydrocarbon Potential of Neoproterozoic–Cambrian Basins in Asia Geological Society, London, Special Publications 366. http://dx.doi.org/10.1144/sp366.11.

Patil, D.J., Mani, D., Madhavi, T., Sudarshan, V., Srikarni, C., Kalpana, M.S., Sreenivas, B., Dayal, A.M., 2013. Near surface hydrocarbon prospecting in Mesozoic Kutch sedimentary basin, Gujarat, Western India – a reconnaissance study using geochemical and isotopic approach. Journal of Petroleum Science and Engineering 108, 393–403.

Patil, D.J., Sreenivas, B., Srikarni, C., Ramamurthy, P.B., Babu, E.V.S.S.K., Gopal, V., Dayal, A.M., 2011. C, O, Sr isotope compositions of sediments of the Mesozoic Kutch Basin, NW India. In: Goldschmidt Conference Abstracts P1606.

Patil, D.J., Sharma, S.D., Kumar, B., Dayal, A.M., Shukla, M., 2002. Carbon, oxygen and strontium isotope geochemistry of carbonate rocks from Kurnool Group, Southern India. Journal of the Geological Society of India 60 (6), 615–622.

Pixler, B.O., 1969. Formation evaluation by analysis of hydrocarbon ratios. Journal of Petroleum Technology 21, 665–670.

Prasanna, M.V., Rasheed, M.A., Madhavi, T., Kalpana, G., Patil, D.J., Dayal, A.M., 2010. Light gaseous hydrocarbon anomalies in the near surface soils of Sagar district, Vindhyan basin, India. Current Science 99 (11), 1586–1590.

Ram, J., 2012. Neoproterozoic successions in peninsular India and their hydrocarbon prospectivity. Geological Society, London, Special Publications 366. http://dx.doi.org/10.1144/sp366.4.

Ram, J., Shukla, S.N., Pramanik, A.G., Verma, B.K., Chandra, G., Murthy, M.S.N., 1996. Recent investigation in Vindhyan Basin: implications for the basin tectonics. Memoir of the Geological Society of India 36, 267–286.

Ronov, A.B., 1958. Organic carbon in sedimentary rocks (in relation to the presence of petroleum) (a translation of Geokhimiya) Geochemistry 5, 510–536.

Ronov, A.B., 1982. Earth's sedimentary shell (quantitative patterns of its structure, composition and evolution). The 20th Vernadsky Lecture, March 12, 1978. International Geology Review 24, 1365–1388.

Saltzman, M.R., Young, S.A., 2005. Long-lived glaciation in the late Ordovician? Isotopic and sequence-stratigraphic evidence from Western Laurentia. Geology 33, 109–112.

Sarkar, A., Sarangi, S., Ebihara, M., Bhattacharya, S.K., Ray, A.K., 2003. Carbonate geochemistry across the Eocene/Oligocene boundary of Kutch, Western India: implications to oceanic O_2-poor condition and foraminiferal extinction. Chemical Geology 201, 281–293.

Schidlowski, M., 1987. Application of stable carbon isotopes to early biochemical evolution on Earth. Annual Review of Earth and Planetary Sciences 15, 47–72.

Schidlowski, M., 1988. A 3,800-million-year isotopic record of life form carbon in sedimentary rocks. Nature 333, 313–318.

Scholle, P.A., Stemmerik, L., Ulmer, D.S., 1991. Diagenetic history and hydrocarbon potential of upper Permian carbonate buildups, Wegener Holvo area, Tamerson Land Basin, East Greenland. Bulletin American Association of Petroleum Geology 75, 701–725.

Scholle, P.A., Arthur, M.A., 1980. Carbon isotope fluctuations in Cretaceous pelagic limestones: potential stratigraphic and petroleum exploration tool. Bulletin American Association of Petroleum Geology 64, 67–87.

Schumacher, D., 1996. Hydrocarbon induced alteration of soils and sediments. In: Schumacher, D., Abrams, M.A. (Eds.), Hydrocarbon Migration and Its Near-Surface Expression, vol. 66. American Association of Petroleum Geologist Memoir, Tulsa, Oklahoma, pp. 71–89.

Schumacher, D.D., December 8, 2003. Geochemical Exploration for Oil and Gas. An Exploration Short Course at International Symposium on Recent Trends in Surface Geochemical Prospecting and Risk Reduction for Hydrocarbon Exploration and Development. National Geophysical Research Institute, Hyderabad.

Schidlowski, M., Eichmann, R., Junge, C.E., 1975. Precambrian sedimentary carbonates. Carbon and oxygen isotope geochemistry and implications for the terrestrial oxygen budget. Precambrian Research 2, 1–69.

Shackleton, N.J., Opdyke, N., 1973. Oxygen isotope and paleomagnetic stratigraphy of Equatorial Pacific core V28–238: oxygen isotope temperatures and ice volumes on a 105 year and 106 year scale. Quaternary Research 3, 39–55.

Shackleton, N.J., Opdyke, N., 1977. Oxygen isotope and paleomagnetic evidence for early northern hemisphere glaciation. Nature 270, 216–219.

Shackleton, N.J., 1987. Oxygen isotopes, ice volume and sea level. Quartenary Science Reviews 6, 183–190.

Shackleton, N.J., Backman, J., Zimmerman, H., Kent, D.V., Hall, M.A., Roberts, D.G., Schnitker, D., Baldauf, J.G., Desprairies, A., Homrighausen, R., Huddlestun, P., Keene, J.B., Kaltenback, A.J., Krumsiek, K.A.O., Morton, A.C., Murray, J.W., Westbergsmith, J., 1984. Oxygen isotope calibration of the onset of ice-rafting and history of glaciation in the North-Atlantic region. Nature 307, 620–623.

Sharma, M., Mishra, S., Dutta, S., Banerjee, S., Shukla, 2009. On the affinity of *Chuaria–Tawuia* complex: a multidisciplinary study. Precambrian Research 173 (1–4), 123–136.

Sharma, M., Shukla, Y., 2009. Taxonomy and affinity of early Mesoproterozoic megascopic helically coiled and related fossils from the Rohtas Formation, the Vindhyan Supergroup, India. Precambrian Research 173, 105–122.

Srinivasa Rao, P.L., Madhavi, T., Srinu, D., Kalpana, M.S., Patil, D.J., Dayal, A.M., 2013. Geochemical assessment of light gaseous hydrocarbons in near-surface soils of Kutch–Saurashtra: implication for hydrocarbon prospects. Journal of Earth System Science 122 (1), 55–63.

Stahl, W., Faber, E., Carey, B.D., Kirksey, D.L., 1981. Near surface evidence of migration of natural gas from deep reservoirs and source rocks. American Association of Petroleum Geology Bulletin 65, 1543–1550.

Tedesco, S.A., 1995. Surface Geochemistry in Petroleum Exploration. Chapman & Hall, USA. 205 pp.

Tissot, B., Welte, D.H., 1984. Petroleum Formation and Occurrence, second ed. Springer Verlag, Heidelberg.

Unpublished Tech Report, NGRI, 2003. Adsorbed soil gas surveys for hydrocarbon research and exploration in western part of Vindhyan Basin (Chambal Valley). In: NGRI Tech. Report No 2002-Exp-361.

Unpublished Tech Report, NGRI, 2009. Surface geochemical survey for oil and Gas exploration in East Vindhyan Basin. In: NGRI Tech. Report No 2009-Exp-693.

Veizer, J., Hoefs, J., 1976. The nature of O^{18}/O^{16} and C^{13}/C^{12} secular trends in sedimentary carbonate rocks. Geochimica et Cosmochimica Acta 40, 1387–1395.

Veizer, J.N., Fritz, P., 1976. Possible controls of post depositional alteration in oxygen paleotemperature determinations. Earth and Planetary Science Letters 33, 255–260.

Veizer, J., Ala, D., Azmy, K., Bruckschen, P., Buhl, D., Giles, F.B., Carden, A.F., Diener, A., Ebneth, S., Gdderies, Y., Jasper, T., Korte, C., Pawellek, F., Podlaha, O.G., Strauss, H., 1999. $^{87}Sr/^{86}Sr$, $\delta^{13}C$ and $\delta^{18}O$ evolution of Phanerozoic seawater. Chemical Geology 161, 59–88.

Weissert, H., Joachimski, M., Sarnthein, M., 2008. Chemostratigraphy. Newsletters on Stratigraphy 42 (3), 145–179.

Weissert, H., Erba, E., 2004. Volcanism, CO_2 and palaeoclimate: a Late Jurassic-Early Cretaceous carbon and oxygen isotope record. Journal of the Geological Society 161, 695–702.

CHAPTER

CHEMOSTRATIGRAPHY OF NEOPROTEROZOIC BANDED IRON FORMATION (BIF): TYPES, AGE AND ORIGIN

17

Claudio Gaucher[1], Alcides N. Sial[2], Robert Frei[3]

[1]*Departamento de Geología, Facultad de Ciencias, Universidad de la República, Montevideo, Uruguay;*
[2]*NEG-LABISE, Department of Geology, Federal University of Pernambuco, Recife, Pernambuco, Brazil;*
[3]*Department of Geosciences and Natural Resource Management, Geology Section, University of Copenhagen and Nordic Center for Earth Evolution (NordCEE), Copenhagen, Denmark*

17.1 INTRODUCTION

The Banded iron formations (BIFs) are nonactualistic rocks deposited in the Archean, Paleoproterozoic, and Neoproterozoic. They are characterized by rhythmic alternations of iron-rich layers ("bands") and chert layers. Based on their iron mineralogy, BIFs can be classified as oxide-, carbonate-, silicate-, or sulfide-facies (James, 1954), among which, the oxide-facies is most important. Based on depositional and geotectonic setting, Gross (1980) distinguished between Algoma- and Lake Superior-type BIFs. Accordingly, the Lake Superior-type BIFs were deposited on stable continental platforms, and Algoma-type BIFs are associated with volcanic rocks and were deposited in arc or rift settings. A third BIF type, exclusively Neoproterozoic in age, is the Rapitan type (i.e., Beukes and Klein, 1992), and is characterized by a glacially influenced depositional environment.

Geochemistry has played a central role in unraveling the origin of BIFs, and many proxies such as rare earth element (REE) concentrations, have been widely used (e.g., Kato et al., 2006). However, chemostratigraphic studies are scarce, probably because traditional stable and radiogenic isotope studies such as $\delta^{13}C$, $\delta^{18}O$, $\delta^{34}S$ and $^{87}Sr/^{86}Sr$ are impossible to apply for the more widespread oxide-facies BIF. Isotope chemostratigraphy has played a significant role in the study of Neoproterozoic successions since the pioneer reports by Knoll et al. (1986) and Kaufman et al. (1991). The establishment of the "snowball Earth" hypothesis was based mainly on chemostratigraphic and paleomagnetic evidences (Hoffman et al., 1998). Curves of global secular variations of $\delta^{13}C$ and $^{87}Sr/^{86}Sr$ in the Neoproterozoic (e.g., Jacobsen and Kaufman, 1999; Melezhik et al., 2001; Halverson et al., 2010) have been important tools for correlation and geochronology, mostly in combination with biostratigraphy and U–Pb radiochronology. The caveat of all these studies is that, for analytical reasons, they can only be applied to the carbonate successions or, more rarely, to the organic-rich shales (e.g., Johnston et al., 2010), but not to the oxide-facies Neoproterozoic iron formations.

The development of new techniques, such as Fe speciation, Cr and Fe isotopes ($\delta^{53}Cr$, $\delta^{56}Fe$), and Mo concentrations and isotopes ($\delta^{98}Mo$) opened new horizons and enabled thorough chemostratigraphic

Chemostratigraphy. http://dx.doi.org/10.1016/B978-0-12-419968-2.00017-0

studies of thick BIF successions (Canfield et al., 2008; Frei et al., 2009; Halverson et al., 2011; Baldwin et al., 2013). Hence, for the first time, the study of secular variations within the iron formations became possible, enabling insights into the basin evolution and changes in the redox conditions. Studies using REE concentrations changed from only reporting REE distributions for whole iron formations (e.g., Graf et al., 1994; Kato et al., 2006) to reporting stratigraphic variations of certain REE parameters, such as Ce and Eu anomalies (e.g., Stern et al., 2013). This development had added the time dimension to fundamental parameters such as the importance of hydrothermal processes, redox conditions and sources of iron, which need not be constant troughout deposition of individual iron formations.

The new techniques, coupled with a growing interest on the Neoproterozoic iron formations, show a more complex picture than previously thought. We review here the occurrence of Rapitan, Algoma, and Lake Superior BIF types in the Neoproterozoic, and present recent chemostratigraphic data that enable their characterization.

17.2 AGE OF NEOPROTEROZOIC BIFS

It has been recently proposed that all the Neoproterozoic iron formations were deposited during the middle Cryogenian ("Sturtian") glaciation at c. 715 Ma (Macdonald et al., 2010a; Cox et al., 2013). This is certainly the case for BIF of the Rapitan Group (Sayunei Formation, Canada), which is constrained by an U–Pb age of a felsic tuff of 716.5 ± 0.24 Ma (Macdonald et al., 2010b). Similarly, glacially influenced BIF of the Fulu Formation and associated diamictites (Jiangkou Group, South China) directly overlie a tuffaceous siltstone that yielded an U–Pb age of 725 ± 10 Ma. The Chuos Formation of the Otavi Group (NW-Namibia) includes BIF associated to glacial diamictites, which rests atop a porphyry lava dated U-Pb at 746 ± 2 Ma (Hoffman et al., 1996). This, combined with the fact that there are more than 500 m of intervening carbonates between the BIF and the end-Cryogenian Ghaub Formation, dated at 636 ± 1 Ma (Hoffmann et al., 2004), makes an age around 715 Ma for the Chuos Formation unlikely, but it cannot be completely ruled out. The Holowilena Ironstone of the Flinders Ranges, South Australia, is an interesting example that illustrates the problems of global correlations of Neoproterozoic ice ages. Fanning and Link (2008) reported U–Pb SHRIMP zircon age of 660 ± 5 Ma for the volcaniclastic intercalations in the partially glaciogenic lower Wilyerpa Formation, which is part of the Sturtian type section. The Holowilena Ironstone occurs some 250 m stratigraphically beneath the Wilyerpa Formation (Le Heron et al., 2011). Thus, although the Holowilena Ironstone is undoubtedly Sturian, because it occurs at the Sturtian type section in Australia, it is probably much younger than the Rapitan glacials.

BIF occurring at Wadi Karim, Wadi El Dabbah, and Um Anab in Egypt (Basta et al., 2011, Fig. 1) and the correlative Sawawin BIF in northwestern Saudi Arabia are constrained between underlying metavolcanics dated U–Pb at 759 ± 17 Ma and intrusive rocks yielding U–Pb age of 710 ± 5 Ma (Stern et al., 2013, and references therein). Therefore, these BIFs are distinctively older than the Rapitan iron formation, and they are not related to glacial processes (Basta et al., 2011). The 10-m-thick Erzin BIF in southern Tuva (Russian Federation) and northern Mongolia occurs within the Mugur Formation, and rests on the top of the metavolcanics of the same unit dated U–Pb at 767 ± 15 Ma (Ilyin, 2009). The Shilu Group in South China (Fig. 1) hosts 50 to 400-m-thick, high-grade iron formation deposited in the Tonian between 960 and 830 Ma, as shown, respectively, by U–Pb ages of detrital zircons and the age of metamorphic overprint (Xu

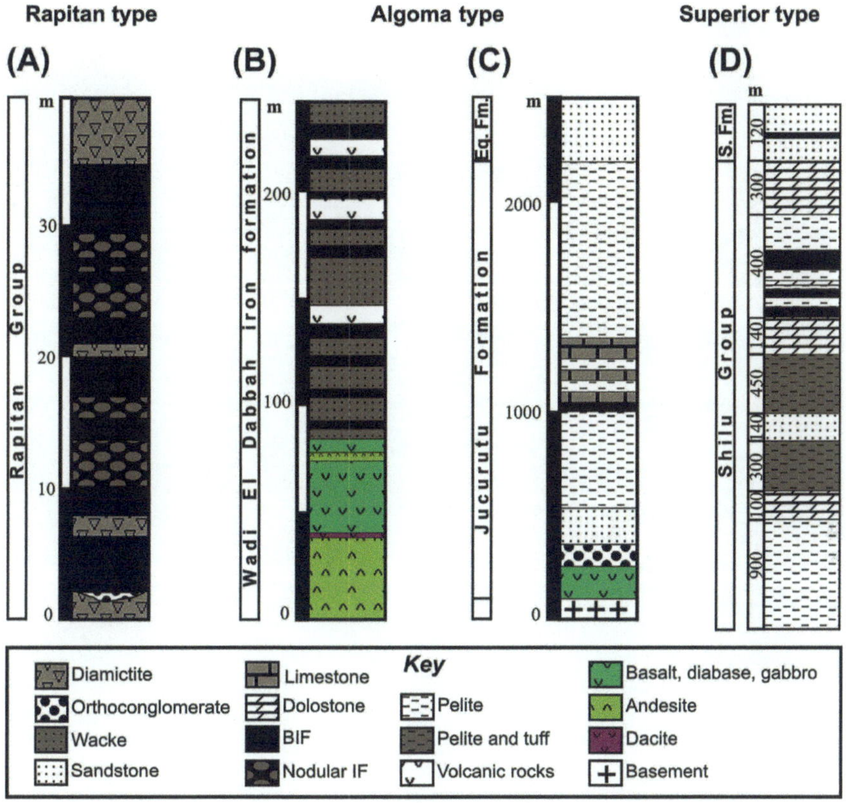

FIGURE 1

Lithostratigraphy and facies association of Neoproterozoic iron formations. (A) Iron formation at the boundary between the Suyunei and Shezal formations, Rapitan Group, Canada *(From Baldwin et al. (2013).).* (B) Wadi El Dabbah iron formation, Egypt *(From Ali et al. (2009).).* (C) Jucurutu Formation, Seridó Belt, NE Brazil *(From Van Schmus et al. (2003), Nascimento et al. (2007).).* Eq. Fm.: Equador Formation. (D) Shilu Group, South China *(From Xu et al. (2013b).).* Indicated are maximum thicknesses observed. S. Fm.: Shihuiding Formation.

et al., 2013a). Moreover, the Shilu BIFs are unrelated to the glacial processes and are considered to be Lake Superior type by Xu et al. (2013b; Fig. 1).

Several occurrences of Neoproterozoic BIFs are demonstrably younger than the middle Cryogenian glaciation, including the largest one, the >300-m-thick BIF and associated manganese formations of the Jacadigo Group (Brazil) and its correlative Boqui Group (Bolivia). The glacially influenced BIF (Figs 2(A)–(B) and 3(C)) yielded youngest U–Pb detrital zircon age from interbedded sandstones of 695 ± 17 Ma (Døssing et al., 2010). Ar–Ar dating on cryptomelane within the Mn ores reveals a minimum depositional age of 587 ± 7 Ma (Piacentini et al., 2013), which may be interpreted as the age of early diagenesis. Considering that the metamorphic braunite yielded Ar–Ar ages between 547 and 513 Ma (Piacentini et al., 2013), which matches with the age of the Pan African-Brasiliano orogeny (587 ± 7 Ma) in the region (McGee et al., 2012), it could well approximate the depositional age and is within the error of the

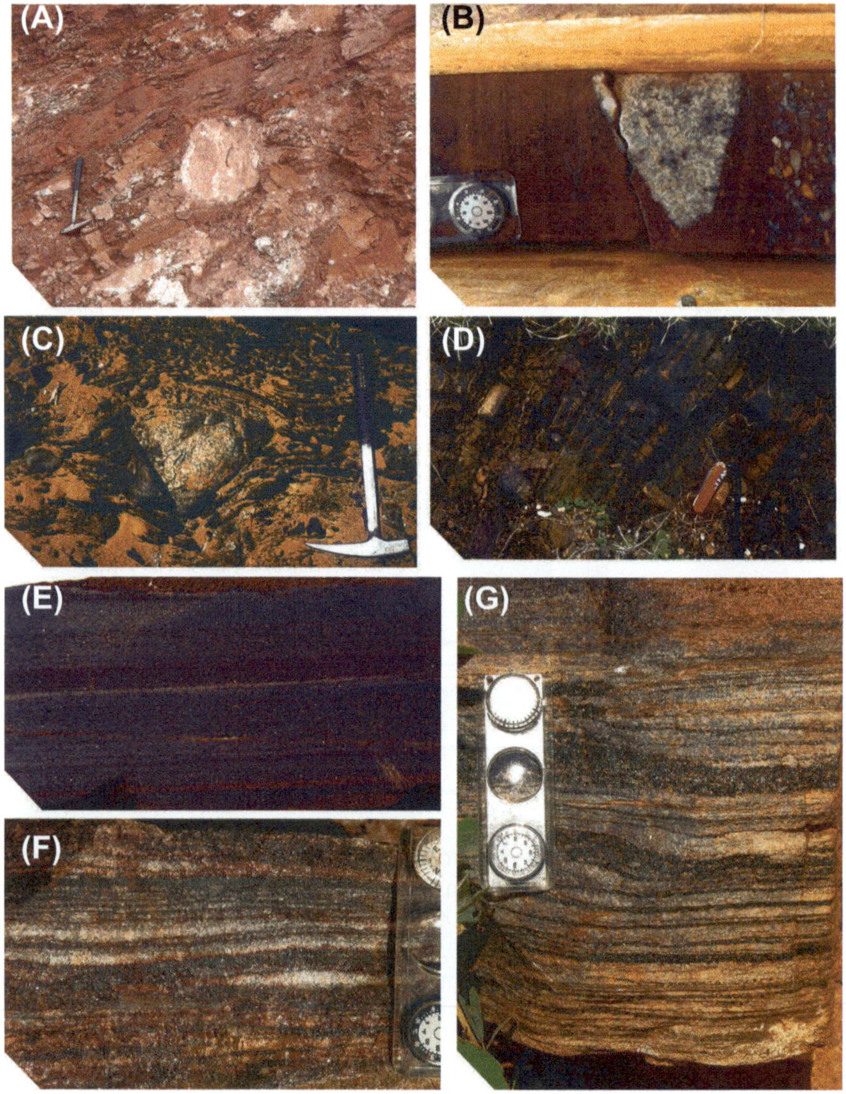

FIGURE 2

Outcrop and polished slab photographs of Neoproterozoic banded iron formation (BIF). (A) Elongate gneiss dropstone with long axis at high angle to bedding in BIF of the Jacadigo Group (W Brazil). (B) Granite dropstone in pure BIF of the Jacadigo Group. Note the separation between the dropstone and the gritty layer below. (C) Basement dropstone in BIF of the Jakkalsberg Member, Numees Formation (Namibia). (D) BIF of the Yerbal Formation (Uruguay) as exposed in a shallow exploration trench. (E) Polished slab of magnetite + hematite BIF of the Yerbal Formation, showing thin chert bands *(From Gaucher et al. (2004).)*. Field of view is 9 cm wide. (F) BIF of the Jucurutu Formation, Seridó Belt, at Riacho Fundo section (NE Brazil). (G) Same unit as previous (Bonito Mine section), but showing thicker and more frequent chert bands. Scale in B, F, and G is 8 cm long.

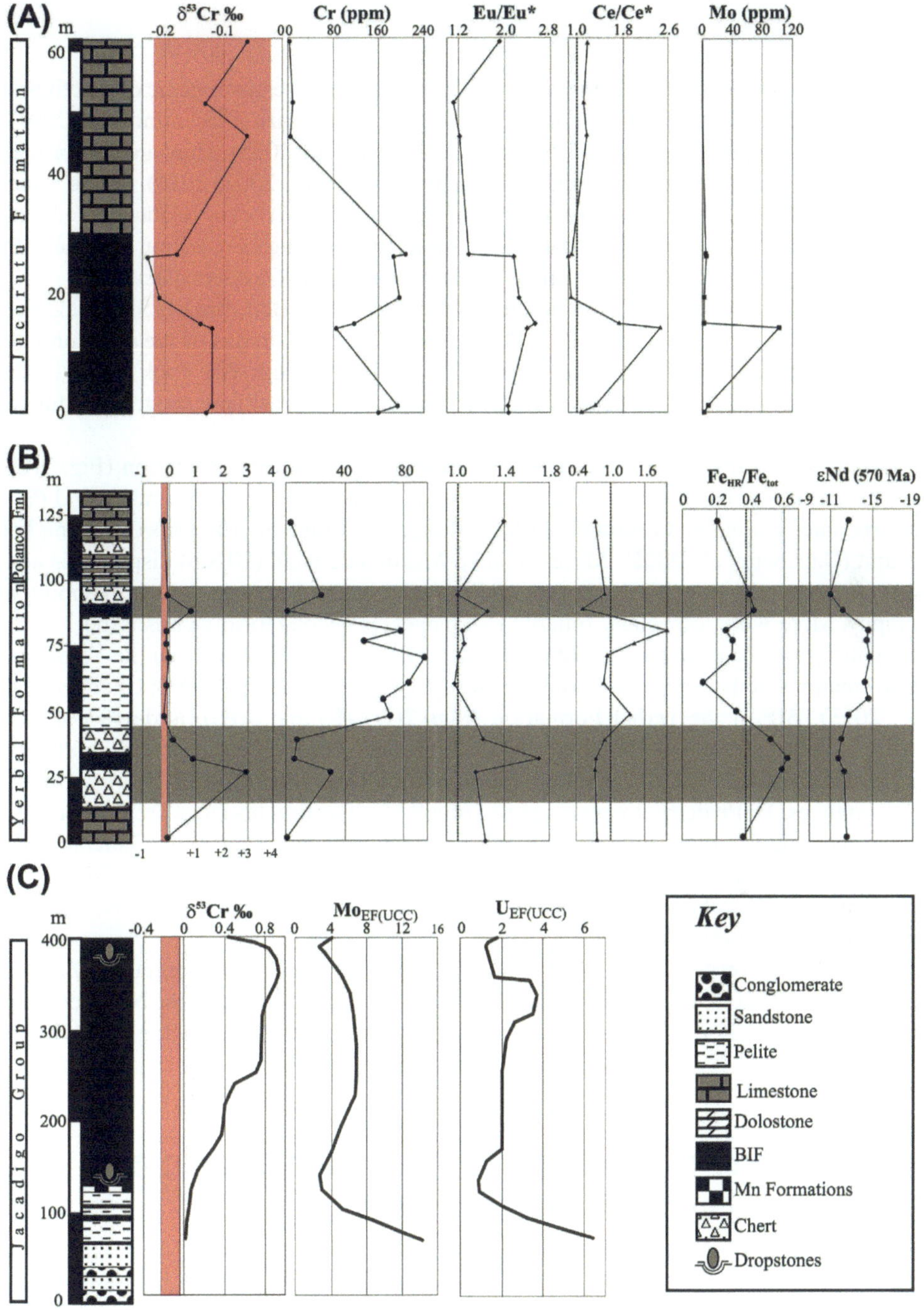

FIGURE 3

Chemostratigraphy of Neoproterozoic iron formations from South America. (A) Algoma-type banded iron formation (BIF), Jucurutu Formation, Seridó Belt (NE Brazil; Sial et al., submitted). Note high Eu/Eu* and Cr concentrations, and $\delta^{53}Cr$ entirely within the high-temperature field (colored area). (B) Lower Arroyo del Soldado Group, Uruguay (Frei et al., 2013). Shaded areas mark chemical sediments characterized by high $\delta^{53}Cr$, Eu/Eu*, and FeHR/FeTot, and low εNd. Dashed line in Fe speciation column marks boundary between oxic and anoxic conditions. (C) Jacadigo Group (W Brazil, Døssing et al., 2010). Note secular variations in $\delta^{53}Cr$ within the iron formation.

Gaskiers Glaciation (582.4±0.5 Ma; Bowring et al., 2003). Correlative glaciogenic diamictites interbedded with BIF of the Puga Formation in the southern Paraguay Belt (Brazil, Piacentini et al., 2007) yielded U–Pb age of detrital zircons as young as 709±6 Ma (Babinski et al., 2013). They are conformably overlain by the cap carbonates of the Corumbá Group (Boggiani et al., 2003, 2010), above which, well-preserved Ediacaran acritarchs and metazoans (*Cloudina*, *Corumbella*) occur (Gaucher et al., 2003). All these data suggest that deposition of the Jacadigo Group and the Puga Formation (southern Paraguay Belt) took place in the end-Cryogenian or, more probably, during the Gaskiers glaciation.

BIF also occurs in the Jacurutu Formation located in the Seridó Belt, Brazil (Sial et al., submitted; Fig. 1). Van Schmus et al. (2003) reported U–Pb age for the detrital zircons of the same unit at its type area as young as 634±13 Ma. Therefore, deposition of the Jucurutu BIF took place in the lower Ediacaran, between 634 Ma and the peak of metamorphism in the Brasiliano orogeny in this region (c. 600–585 Ma; Hollanda et al., 2010).

Dropstone-bearing BIFs of the Jakkalsberg Member of the Numees Formation (Fig. 2(C)) in the Gariep Belt (Namibia) were assigned to the Ediacaran on the basis of Sr isotope ratios and Pb–Pb ages of its cap carbonate (Fölling and Frimmel, 2002) and organic-walled microfossils occurring below and above the unit (Gaucher et al., 2005). More recently, Macdonald et al. (2010b) dispute this age assignment and prefer a correlation with the Rapitan BIF, on the basis of new geological mapping and chemostratigraphy. More work, especially reliable U–Pb ages, is needed to solve this issue.

Arguably the youngest examples of Neoproterozoic BIFs were reported from the Yerbal and Cerro Espuelitas formations of the Arroyo del Soldado Group, Uruguay (Gaucher, 2000; Gaucher et al., 2004; Frei et al., 2013). BIF of the Yerbal Formation (Figs 2(D)–(E) and 3(B)), is devoid of any glacial features, occurs 800 m above the sandstones for which Blanco et al. (2009) reported U–Pb detrital zircon age as young as 664±14 Ma. Equivalent, more proximal sandstones of the Barriga Negra Formation yielded youngest U–Pb detrital zircon age of 566±8 Ma. The iron horizons are further interbedded with siltstones bearing *Cloudina riemkeae*, an index fossil for the late Ediacaran (Gaucher and Sprechmann, 1999; Gaucher et al., 2003; Gaucher and Poiré, 2009). The magnetite-bearing BIFs of the Cerro Espuelitas Formation (Gaucher, 2000) occur at stratigraphically two formations above the Yerbal Formation, and are thus of latest Ediacaran age.

In summary, the Neoproterozoic BIFs, pre- and postdate the middle Cryogenian, c. 715 Ma glaciation, and occur in the Tonian, Cryogenian, and Ediacaran. Interestingly, not all the Neoproterozoic BIFs were deposited in glacially influenced settings. In fact, as we demonstrate in the following sections, examples of the three BIF types have been described from the Neoproterozoic successions.

17.3 DEPOSITIONAL ENVIRONMENT OF NEOPROTEROZOIC BIFS

17.3.1 SEDIMENTOLOGY AND FACIES ASSOCIATION

17.3.1.1 Glacially Influenced Iron Formations

Several Neoproterozoic iron formations were deposited in glacially influenced environments, and they are assigned to the "Rapitan-type" (Beukes and Klein, 1992). Apart from well-documented glaciomarine BIF of the Rapitan Group (Baldwin et al., 2012) other analogues show clear indications of having been influenced by glaciomarine environment. Basement dropstones occur in BIF of the Jakkalsberg Member of the Numees Formation (Fig. 2(C)), southern Namibia (Frimmel, 2008). Similarly, the Chuos Formation in northern Namibia comprises glacial diamictite with predominantly basement clasts, including some striated, and associated thin horizons of BIF (Hoffman and Halverson, 2008;

Miller, 2008). A well-developed cap carbonate yielding negative $\delta^{13}C$ values (Rasth of Formation) rests on top of the Chuos Formation (Hoffman and Schrag, 2002). Thus, BIFs of the Chuos Formation were deposited in unambiguously glaciomarine environment and can be assigned to the Rapitan type.

BIFs of the Jacadigo Group (Banda Alta or Santa Cruz Formation) and Puga Formation bear convincing evidences of a glacial environment. The Puga Formation includes both associated diamictites and dropstones in the BIF (Piacentini et al., 2007). In the Jacadigo Group, both planar and elongate, outsized clasts occur (Fig. 2(A)–(B)). Planar clasts tend to lie parallel to bedding, but elongated clasts often show major axes perpendicular to bedding (i.e., bullet clast). A 6 m long granite bullet-clast was illustrated by Möller (2004) from the top of the Banda Alta Formation (Fig. 3(C)). Attempts to explain these outsized clasts as nonglacial in origin, for example as the result of downslope sliding and/or mass flows (e.g., Freitas et al., 2011) are not convincing. Downslope sliding cannot account for the orientation of bullet clasts. The thin, sandy layers occasionally accompanying outsized clasts (Freitas et al., 2011; Trompette et al., 1998) do not contradict a glacial origin for the latter, because they can also be derived from melting icebergs. Granite dropstones weighing more than 100 tons in chemical sediments such as BIF and MnF are possibly the most convincing evidences of glaciation so far reported from the Neoproterozoic deposits.

Thus, a glaciomarine setting is ascribable for a number of Neoproterozoic iron formations, including the thickest ones (e.g., Jacadigo Group, Rapitan Group). Interestingly, most of these occurrences were deposited in rift basins. This has led to the development of models that dispute the importance of glaciation and emphasize the role of rifting and hydrothermal activity (e.g., Young, 2002; Eyles and Januszczak, 2004; Freitas et al., 2011).

17.3.1.2 BIF Associated with Volcanic Rocks

Neoproterozoic iron formations in Egypt and their correlative units in northwestern Saudi Arabia, up to 90 m thick, are interbedded with wackes, tuffs, lapilli and subordinate volcanic flows (basalts, andesites, dacites; Basta et al., 2011; Stern et al., 2013, Fig. 1). The geochemical affinities of the volcanic rocks point to an arc or back arc setting (Ali et al., 2009). εNd (750 Ma) of these rocks is invariably positive, and ranges between +5 and +9 (Ali et al., 2009), showing their juvenile nature.

The 30 m thick BIF (Figs 1(C) and 3(A)) of the Jucurutu Formation (Brazil), is also associated with mafic metavolcanics and metabasalts, but it is overlain by carbonates characterized by negative $\delta^{13}C$ values (Nascimento et al., 2007; Sial et al., submitted). However, these carbonates do not exhibit the typical features of cap carbonates (e.g., Hoffman and Schrag, 2002). Furthermore, the glacial features are absent in the thinly banded BIF (Sial et al., submitted; Fig. 2(F)–(G)). Diamictites with faceted clasts occur elsewhere in the Seridó Belt, but is not associated with the BIF (Legrand et al., 2008). Van Schmus et al. (2003) proposed, on the basis of detrital zircon ages, the nature of the basement and lithostratigraphy, that the Jucurutu Formation was deposited in a rift basin as the result of late Neoproterozoic extension of a pre-existing continental basement, with the development of small marine basins (Fig. 4). This may explain the differences, such as the occurrence of carbonates, compared to BIFs deposited in an arc environment, such as the Egyptian and Saudi Arabian BIF described above.

These examples serve to illustrate that Algoma-type BIFs occur in Neoproterozoic successions, and that they were both associated to compressional and extensional geotectonic settings, as also known for older BIFs (Gross, 1980). As we will discuss in the following sections, these Algoma-type Neoproterozoic BIFs bear distinct geochemical and chemostratigraphic features that allow a distinction from the glacially influenced iron formations.

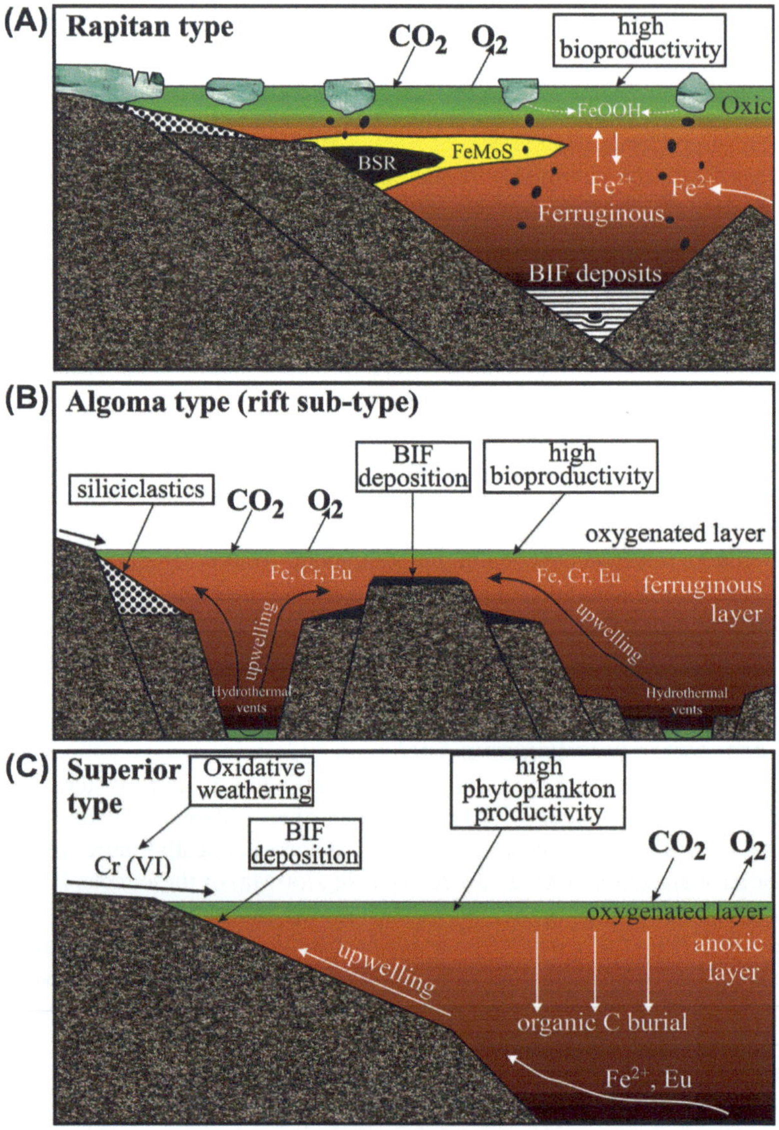

FIGURE 4

Depositional models for the three types of Neoproterozoic iron formations discussed in this work. (A) Deposition of Rapitan type banded iron formation (BIF) during glacial retreat, exemplified by the type occurrence in the Rapitan Group *(Modified from Baldwin et al. (2012).)*. BSR: bacterial sulfate reduction. (B) Depositional model for Algoma type BIF of the Jucurutu Formation (NE Brazil) according to Sial et al. (submitted). (C) Depositional environment of Lake Superior type BIF of the Yerbal Formation, Arroyo del Soldado Group *(From Frei et al. (2013), and references therein.)*.

17.3.1.3 BIF Associated with Platform Deposits but without Glacial Features

These are possibly the most neglected Neoproterozoic iron formations, despite being remarkably thick and laterally persistent. They share three characteristics: (1) they do not show glacial features or associated glacial deposits, (2) they are not associated with volcanic rocks, and (3) they occur in passive margin successions characterized by thick carbonate and shale deposits.

A well-documented example is the Yerbal Formation of the Arroyo del Soldado Group (Uruguay; Fig. 3(B)). Up to 50-m-thick, oxide-facies BIF, composed of alternating magnetite + hematite and chert bands (Fig. 2(D)–(E)), occurs at the top of the unit, just beneath the contact with the overlying carbonates of the Polanco Formation (Gaucher et al., 2004; Frei et al., 2013, Fig. 3(B)). BIFs are interbedded with siltstones, cherts, and dolostones, all of which show considerable lateral continuity (Gaucher, 2000; Gaucher et al., 2004). No volcanic or pyroclastic rocks occur in the whole of the 5-km-thick Arroyo del Soldado Group. Detrital zircon ages of the Yerbal Formation are mostly Paleoproterozoic, with subordinate Archean and Mesoproterozoic populations (Gaucher et al., 2008; Blanco et al., 2009). Thus, both facies association and provenance show that the geotectonic setting of the Yerbal Formation was a stable platform developed at the margin of an old craton (Río de la Plata Craton, Gaucher et al., 2003, 2008). The same setting is inferred for the younger BIFs of the Cerro Espuelitas Formation (Gaucher et al., 1996; Gaucher, 2000). They are predominantly magnetitic and are associated with thick, gray and black shales and chert. A likely metamorphic (amphibolite facies) equivalent of the Arroyo del Soldado BIF is represented by the Marco de los Reyes Formation, which crops out at the eastern edge of the Río de la Plata Craton (Chiglino et al., 2010). This unit hosts 40 m thick oxide-facies (magnetite) BIF having a mean composition of 64% Fe_2O_3 and 33% SiO_2 (Bossi and Navarro, 1991). This BIF is associated with cherts, carbonates, and metapelites. $^{87}Sr/^{86}Sr$ values of the high-Sr limestones range between 0.7070 and 0.7080 and $\delta^{13}C$ between and −3‰ and 4.4‰ VPDB (Chiglino et al., 2010), matching corresponding values of the Polanco Formation (Gaucher et al., 2004; Frei et al., 2011).

A shelf setting has also been proposed for the Tonian BIF of the Shilu Group (Hainan Province, South China). Four BIF horizons are interbedded with dolostones, pelites, and quartz-arenites (Fig. 1(D)) and the whole package has been metamorphosed in the upper greenschist facies (Xu et al., 2013a,b). Thickness of the BIF and intervening strata ranges between 50 and 400 m. No evidence of contemporaneous glaciation or volcanism occurs in the Shilu District which, together with the facies association, favor an assignment to the Lake Superior type (Xu et al., 2013b). A similar depositional setting was proposed also for BIFs of the Neoproterozoic Jingtieshan Formation (Gansu Province, North China) on the basis of their facies association (Sun et al., 1998).

Thus, despite being much smaller than their older counterparts, Lake Superior-type BIFs do occur in the Tonian and Ediacaran. This clearly shows that BIF deposition was not dependent on glacial conditions, a fact that can already be inferred from the ferruginous nature of the Neoproterozoic ocean (Canfield et al., 2008).

17.3.2 GEOCHEMISTRY AND CHEMOSTRATIGRAPHY

17.3.2.1 Rapitan-Type Neoproterozoic BIF

Rare earth elements (REE) are used to infer paleo-redox conditions and to elucidate the primary source of iron in the BIFs (hydrothermal vs. weathering, e.g., Kato et al., 2006). REE + Y data reported by Klein (2005), Halverson et al. (2011) and Baldwin et al. (2012) for the glacially influenced Rapitan iron formation (Canada) display a typical, LREE-depleted pattern and the absence of Eu or Ce anomalies. For the Jacadigo Group, Klein and Ladeira (2004) also report the absence of Eu or Ce anomalies. This suggests that the primary iron source in the glacially influenced BIFs may have been glacially sourced iron oxyhydroxide nanoparticles that were subsequently bacterially reduced under ferruginous conditions, supplemented with distal hydrothermal sources from the deep ocean (Baldwin et al., 2012). In the particular case of the Jacadigo Group, REE patterns, unlike the Rapitan BIFs, are LREE-enriched, resembling modern estuarine or coastal waters (Graf et al., 1994), maybe due to the influx of meltwater.

Other trace elements used in Rapitan-type BIF chemostratigraphy include Mo, U, and W. These elements, when normalized against the Mud from Queensland (Kamber et al., 2005) exhibit secular variations in the Rapitan BIF, with enrichments at the base and top of the unit and low values in most of the BIF interval (Baldwin et al., 2012). This stratigraphic distribution indicates reducing conditions throughout the unit, with sulfidic (euxinic) conditions at the base and top of the iron formation, as indicated especially by Mo (Baldwin et al., 2012). These conclusions were corroborated more recently by Mo isotopes (δ^{98}Mo, Baldwin et al., 2013). Similar low values of U and Mo were found by Døssing et al. (2010) throughout the glacially influenced BIF of the Jacadigo Group (Fig. 3(C)), with significant U and Mo enrichment only at the base (Córrego das Pedras Formation) but not at the top. Thus, it becomes clear that conditions during the deposition of these examples were characterized by ferruginous bottom waters and low sulfate availability.

Iron isotopes are obvious candidates for BIF chemostratigraphy, but the interpretation of isotopic signals is often complex due to the involvement of many processes, biotic and abiotic, that fractionate iron. Halverson et al. (2011) reported an increasing trend of δ^{57}Fe, from −0.7‰ to 1.2‰ IRMM-14, for BIF of the Sayunei Formation (Rapitan Group), which they interpreted as the reflection of rising sea-level and an isotopic gradient across the chemocline. Halverson et al. (2007) reported a similar δ^{57}Fe curve for the glacially influenced Holowilena Ironstone, but with significantly larger amplitude (−0.8‰ to 3.1‰ IRMM-14).

Chromium isotopes have become a useful tool for the study of BIF, carbonates, and other chemogenic rocks, essentially reflecting: (1) the degree of oxidative weathering on the continents (and thus atmospheric oxygenation), and (2) the relative input of oxidized, land-derived, fractionated Cr (VI) versus hydrothermal, unfractionated Cr (III) from vents (Frei et al., 2009). Whereas, the older Archean BIFs show mostly unfractionated, negative δ^{53}Cr values around −0.15‰ $_{\text{SRM 979}}$, the Neoarchean and Paleoproterozoic BIFs exhibit positive values up to +0.5‰ $_{\text{SRM 979}}$, depicting the early atmospheric oxygenation (Frei et al., 2009). Cr-isotope data for Neoproterozoic units include the Rapitan BIF, which yielded positive values between 0.90‰ and 0.96‰ (Frei et al., 2009). In an attempt to study the secular variations of δ^{53}Cr, Døssing et al. (2010) applied this method to the Jacadigo Group. BIF of the Banda Alta/Santa Cruz Formation yielded positive δ^{53}Cr values rising from 0.1‰ at the base to a peak of 0.9‰ in the upper part (Fig. 3(C)), slightly decreasing to c. 0.6‰ at the top (Døssing et al., 2010). The Fe-rich glaciogenic diamictites and associated BIF of the correlative Puga Formation also yielded positive δ^{53}Cr values from +0.1‰ to +0.4‰. Interesting in this regard is the parallelism of δ^{53}Cr and δ^{13}C found in the Neoproterozoic carbonates (Frei et al., 2011), which allow the use of δ^{53}Cr in BIF as a proxy of δ^{13}C of contemporaneous seawater. The coupling of both isotope systems is probably through photosynthetic activity, that affected both carbon burial and oxygen production (Frei et al., 2013).

17.3.2.2 Algoma-Type Neoproterozoic BIF

REE+Y patterns were presented by Basta et al. (2011) and Stern et al. (2013) for Algoma-type BIFs of Egypt and Saudi Arabia. They are LREE-depleted (normalized to the Post-Archean Australian Shale, PAAS), exhibit positive Eu anomalies (Eu/Eu*) of up to 1.41 (mean 1.13), and true negative Ce anomalies (Ce/Ce*) down to 0.7 (mean 0.9). The latter characteristics are in strong contrast with Rapitan-type BIF that do not show Eu or Ce anomalies. When plotted against the stratigraphy, Eu/Eu* and Ce/Ce*, allow the identification of the intervals of elevated hydrothermal input (Stern et al., 2013). Significantly, these intervals are also marked by higher εNd(t) values, which reach +6 in the Sawawin BIF of Saudi Arabia (Stern et al., 2013). Similar, albeit more extreme, REE+Y patterns were found by Sial et al. (submitted) for Algoma-type BIFs of the Jucurutu Formation (Seridó Belt, Brazil). They are LREE-depleted and characterized by strongly positive Eu/Eu* anomalies (up to 3.1) and true positive Ce/Ce* anomalies up to 1.3 (Fig. 3(A); Bau and Dulski, 1996). Strong hydrothermal input related to rifting might have promoted the anoxic (ferruginous) conditions, as shown by positive Ce anomalies.

The proximity to hydrothermal vents is also shown by very high Cr contents of the Jucurutu BIFs (85–300, eventually 400 ppm) and by their δ^{53}Cr values that vary mostly between −0.30‰ and −0.12‰ $_{SRM\ 979}$ (Sial et al., submitted; Fig. 3(A)), and within the field of magmatic values (Schoenberg et al., 2008). A negligible detrital component is shown by the occurrences of low Sc and Zr concentrations in the BIF (average Sc = 2.2 ppm, average Zr = 1.2 ppm). Less negative δ^{53}Cr values tend to appear at the base of the iron formation, becoming more negative up section (Fig. 3(A)). Overlying carbonates are characterized by equally negative to slightly fractionated δ^{53}Cr (−0.19‰ to −0.05‰) and "normal" Cr concentrations between 1 and 10 ppm (Fig. 3(A)). Interestingly, corresponding δ^{13}C values are very negative, around −7‰ VPDB, in the lower 20 m of carbonates (sitting atop BIF), passing into positive values up to 10‰ VPDB in the up section (Sial et al., submitted).

An arc setting was proposed for Algoma-type BIFs in Egypt and Saudi Arabia on the basis of the type, geochemical affinities and Nd isotopes of the accompanying volcanic rocks (Ali et al., 2009, Fig. 1). On the other hand, the Jucurutu BIFs were probably deposited in a starved rift basin (Fig. 4(B)), as shown by their low terrigenous contents, geochemistry and association with overlying carbonates (Sial et al., submitted). In his original definition, Gross (1980) erected the Algoma type to include BIFs that were deposited "along volcanic arcs or rift zones". Thus, both of these Algoma subtypes occur in the Neoproterozoic.

17.3.2.3 Lake Superior Type Neoproterozoic BIF

Frei et al. (2009, 2013) reported REE+Y data for BIF of the Arroyo del Soldado Group (Fig. 3(B)). BIFs of the Yerbal Formation are LREE-depleted and characterized by positive Eu/Eu* anomalies (1.15–1.71) and negative Ce/Ce* anomalies ranging from 0.63 to 0.79 (data from Frei et al., 2009). Fe speciation (Fe_{HR}/Fe_{Tot}) values range from 0.5 to 0.7 for the BIF and associated cherts (Fig. 3(B); Frei et al., 2013) and thus indicate anoxic, ferruginous conditions (Canfield et al., 2008). However, interbedded shales and carbonates yielded Fe_{HR}/Fe_{Tot} values far below the threshold value of 0.38 (Canfield et al., 2008; Frei et al., 2013, Fig. 3(B)), showing that predominantly oxygenated conditions alternated with anoxic events, as do the negative Ce anomalies mentioned above.

The εNd(t) values of the entire Yerbal Formation range between −19 and −5 (Fig. 3(B)), with Tdm values between 2.4 and 2.0 Ga that match with the source rocks of the Río de la Plata Craton (Frei et al., 2013). Understandably, the lowest εNd(t) values (−5 to −12) correspond to the BIFs and cherts, because they are pure chemical precipitates with little siliciclastic input. These values confirm the inference of stable platform setting for the Arroyo del Soldado Group (Fig. 4(C)), deduced from the facies association and detrital zircon U–Pb ages (see above).

The δ^{53}Cr values for BIF of the Arroyo del Soldado Group are strongly fractionated (Fig. 3(B)), yielding values between 0.7‰ and 5.0‰ $_{SRM\ 979}$ (Frei et al., 2009, 2013). The latter are the most positive δ^{53}Cr values reported for BIFs up to date (Frei et al., 2009), which is logical considering the young age of these iron formations and the progressive oxygenation of the surface environments leading up to the

Ediacaran–Cambrian boundary. Moreover, in Rapitan type BIF, rapid denudation and cold climate in the hinterland allows for less intense chemical weathering and could explain the lower $\delta^{53}Cr$ compared to the Neoproterozoic Lake Superior type (see Fig. 4), which were deposited on deeply weathered, old cratons.

The Lake Superior-type BIF of the Shilu Group (China) are also characterized by LREE-depletion (PAAS-normalized), and positive Eu/Eu* (up to 11) and Ce/Ce* (1.0–2.4) anomalies (Xu et al., 2013b). Corresponding εNd(t) values for the BIF range between −7 and −5 (average=−6) and Tdm 1.9 to 2.1 Ga (Xu et al., 2013b). Thus, the Shilu BIFs also require an old cratonic region as source area, and represent a marine shelf setting, in line with the facies association observed. The only difference with the BIF of the Arroyo del Soldado Group is the more anoxic setting for the Chinese occurrence, which is also significantly older (Tonian).

In summary, it could be stated that the REE geochemical characteristics (e.g., Eu and Ce anomalies) of the Chinese and Uruguayan Superior type Neoproterozoic BIFs are intermediate between the Rapitan type and the Algoma type Neoproterozoic BIF. On the other hand, they show the most positive $\delta^{53}Cr$ values due to a more open marine setting of these iron formations (Fig. 4) and the younger age of the Yerbal Formation.

17.4 DISCUSSION

It becomes clear from the data reviewed in the preceding sections that glaciation was not a pre-requisite for deposition of BIFs in the Neoproterozoic, as initially thought (e.g., Beukes and Klein, 1992). Contrary to the recent assumptions (Macdonald et al., 2010a; Cox et al., 2013) the age distribution of Neoproterozoic iron formations actually span the whole Era, from the Tonian to the late Ediacaran, and are not restricted to the Sturtian glaciation or even the Cryogenian. While it is true that many Neoproterozoic BIFs, including some of the largest, were deposited in glacially influenced settings (e.g., Jacadigo Group, Rapitan Group; Fig. 4(A)), many examples bear no indications of a glacial environment (Gaucher et al., 2004; Frei et al., 2013; Stern et al., 2013; Xu et al., 2013b; Sial et al., submitted). Instead, the prevalence of rifting and hydrothermal activity were important for the development of both glacial and nonglacial BIFs (Fig. 4), possibly due to the protracted fragmentation of Rodinia and the associated mantle superplumes (Li et al., 2008). Isley and Abbott (1999) already postulated a link between superplume events and BIF deposition in the late Archean and Paleoproterozoic. Ferruginous conditions were likely fueled by enhanced mid-ocean ridge hydrothermal circulation (Canfield et al., 2008). Both the increased length of mid-ocean ridges and enhanced heat flux during Rodinia breakup might have promoted the elevated hydrothermal activity (e.g., Gaucher et al., 2009), as is shown by extremely low $^{87}Sr/^{86}Sr$ seawater ratios in the Tonian and Cryogenian (Melezhik et al., 2001; Halverson et al., 2007). The re-emergence of BIF in the Neoproterozoic after nearly 1 Gyr of nondeposition has more to do with the return to ferruginous conditions in the oceans than with near-global glaciation. Primary productivity must also be taken into account (Fig. 4), because it was the source of the large amounts of oxygen required to precipitate Neoproterozoic BIFs (e.g., Gaucher et al., 2004; Frei et al., 2013). In a basin covered by thick ice, dissolved oxygen would be rapidly exhausted due to the inhibition of photosynthesis. Thus, in glacial settings, most models require deposition of BIFs after ice retreat (e.g., Baldwin et al., 2012, Fig. 4(A)). Even in glacially influenced iron formations, there were no continued ice cover, and sea ice-free conditions are recorded at some levels (Le Heron et al., 2011). Finally, the progressive oxygen buildup in surface environments determine evolving geochemical and isotopic signatures for the Neoproterozoic BIFs (e.g., Frei et al., 2009), which, in the future may be used as a chemostratigraphic tool for correlation and geochronology.

17.5 CONCLUSIONS

Neoproterozoic BIFs occur in Tonian, Cryogenian, and Ediacaran successions, and are not restricted to the middle Cryogenian (c. 715 Ma) glaciation. Many Neoproterozoic BIFs (North and South America, southern Africa, and Australia) were deposited in glacially influenced settings, and are traditionally assigned to the Rapitan type. Other occurrences in South America, Africa, and Asia, however, bear no indications of glacial processes and can be assigned to the Algoma and Lake Superior types.

Chemostratigraphy is a powerful tool for unraveling the depositional setting of Neoproterozoic BIFs. Some of the most useful methods are REE distribution, especially the Eu and Ce normalized concentrations, iron speciation, Nd and Cr isotopes ($\delta^{53}Cr$). Whereas Rapitan type BIFs exhibit no Eu or Ce anomalies, Algoma type Neoproterozoic BIF show both. Positive $\delta^{53}Cr$ values characterize glacially influenced BIFs, and differentiates them from nonfractionated, mantle-like $\delta^{53}Cr$ values of Algoma type BIFs. The latter are also characterized by high Cr concentrations. The most positive $\delta^{53}Cr$ values occur in the open-shelf, Lake Superior type BIFs, especially those of Ediacaran age. Given the consistent secular variations exhibited by $\delta^{53}Cr$, it is envisaged that it may be used for correlation and geochronology once the available database grows sufficiently.

It is demonstrated clearly that glaciation was not the main mechanism responsible for deposition of BIFs in the Neoproterozoic. Hydrothermal activity related to the rifting of Rodinia probably played the key role in their reappearance after the Mesoproterozoic gap, and promoted a return to ferruginous conditions after the demise of Mesoproterozoic euxinic oceans.

ACKNOWLEDGMENTS

Thanks for funding and support are due to the Comisión Sectorial de Investigación Científica (CSIC, Uruguay) to C.G. Financial support through The Danish Agency for Science, Technology and Innovation grant nr. 11-103378 to RF and through the Danish National Research Foundation's center of excellence NordCEE (DNRF grant number DNRF53) is highly appreciated. This study was partially supported by grants to ANS (CNPq, grants n. 470399/2008 and 472842/2010-2 and FACEPE APQ 0727-1.07/08; APQ-1059-9.05/12). This is the NEG-LABISE scientific contribution n. 270. We thank the constructive suggestions of the anonymous reviewer.

REFERENCES

Ali, K.A., Stern, R.J., Manton, W.I., Kimura, J.-I., Khamees, H.A., 2009. Geochemistry Nd isotopes and U–Pb SHRIMP zircon dating of Neoproterozoic volcanic rocks from the Central Eastern Desert of Egypt: new insights into the ~750 Ma crust-forming event. Precambrian Research 171, 1–22.

Babinski, M., Boggiani, P.C., Trindade, R.I.F., Fanning, C.M., 2013. Detrital zircon ages and geochronological constraints on the Neoproterozoic Puga diamictites and associated BIFs in the southern Paraguay Belt, Brazil. Gondwana Research 23, 988–997.

Baldwin, G.J., Turner, E.C., Kamber, B.S., 2012. A new depositional model for glaciogenic Neoproterozoic iron formation: insights from the chemostratigraphy and basin configuration of the Rapitan iron formation. Canadian Journal of Earth Sciences 49, 455–476.

Baldwin, G.J., Nägler, T.F., Greber, N.D., Turner, E.C., Kamber, B.S., 2013. Mo isotopic composition of the mid-Neoproterozoic ocean: an iron formation perspective. Precambrian Research 230, 168–178.

Basta, F.F., Maurice, A.E., Fontboté, L., Favarger, P.-Y., 2011. Petrology and geochemistry of the banded iron formation (BIF) of Wadi Karim and Um Anab, Eastern Desert, Egypt: Implications for the origin of Neoproterozoic BIF. Precambrian Research 187, 277–292.

Bau, M., Dulski, 1996. Distribution of yttrium and rare earth elements in the Penge and Kuruman Iron Formation, Transval Supergroup, South Africa. Precambrian Research 79, 37–55.

Blanco, G., Rajesh, H.M., Gaucher, C., Germs, G.J.B., Chemale Jr., F., 2009. Provenance of the Arroyo del Soldado Group (Ediacaran to Cambrian, Uruguay): Implications for the paleogeographic evolution of southwestern Gondwana. Precambrian Research 171, 57–73.

Beukes, N.J., Klein, C., 1992. Models for iron-formation deposition. In: Schopf, J.W., Klein, C. (Eds.), The Proterozoic Biosphere—A Multidisciplinary Study. Cambridge University Press, Cambridge, pp. 147–151.

Boggiani, P.C., Ferreira, V.P., Sial, A.N., Babinski, M., Trindade, R.I.F., Aceñolaza, G., Toselli, A.J., Parada, M.A., 2003. The cap carbonate of the Puga Hill (Central South America) in the context of the post-Varanger Galciation. In: IV South American Symposium on Isotope Geology, Short Papers, Salvador, pp. 324–327.

Boggiani, P.C., Gaucher, C., Sial, A.N., Babinski, M., Simon, C.M., Riccomini, C., Ferreira, V.P., Fairchild, T.R., 2010. Chemostratigraphy of the Tamengo Formation (Corumbá Group, Brazil): a contribution to the calibration of the Ediacaran carbon-isotope curve. Precambrian Research 182, 382–401.

Bossi, J., Navarro, R., 1991. Geología del Uruguay. Universidad de la República, Montevideo. pp. 1–970.

Bowring, S.A., Myrow, P., Landing, E., Ramezani, J., Grotzinger, J.P., 2003. Geochronological constraints on terminal Neoproterozoic events and the rise of metazoans. In: European Geophysical Union Annual Meeting, Nice 2003. Geophysical Research Abstracts, vol. 5, p. 13219.

Canfield, D.E., Poulton, S.W., Knoll, A.H., Narbonne, G.M., Ross, G., Goldberg, T., Strauss, H., 2008. Ferruginous conditions dominated later Neoproterozoic deep-water chemistry. Science 321, 949–952.

Chiglino, L., Gaucher, C., Sial, A.N., Bossi, J., Ferreira, V.P., Pimentel, M.M., 2010. Chemostratigraphy of Mesoproterozoic and Neoproterozoic carbonates of the Nico Pérez Terrane, Río de la Plata Craton, Uruguay. Precambrian Research 182, 313–336.

Cox, G.M., Halverson, G.P., Minarik, W.G., Le Heron, D.P., Macdonald, F.A., Bellefroid, E.J., Strauss, J.V., 2013. Neoproterozoic iron formation: An evaluation of its temporal, environmental and tectonic significance. Chemical Geology 362, 232–249.

Døssing, L.N., Gaucher, C., Boggiani, P.C., Frei, R., 2010. Stable Chromium Isotopes as Tracer of Changes in Weathering Processes and Redox State of the Ocean during Neoproterozoic Glaciations. American Geophysical Union, Fall Meeting, San Francisco.

Eyles, N., Januszczak, N., 2004. "Zipper-rift": a tectonic model for Neoproterozoic glaciations during the breakup of Rodinia after 750 Ma. Earth-Science Reviews 65, 1–73.

Fanning, C.M., Link, P.K., 2008. Age constraints for the sturtian glaciation: data from the Adelaide Geosyncline, south Australia and Pocatello formation, Idaho, USA. In: Gallagher, S.J., Wallace, M.W. (Eds.), Neoproterozoic Extreme Climates and the Origin of Early Life, Selwyn Symposium of the GSA Victoria DivisionGeological Society of Australia Extended Abstracts, vol. 91, pp. 57–62.

Fölling, P.G., Frimmel, H.E., 2002. Chemostratigraphic correlation of carbonate successions in the Gariep and Saldania Belts, Namibia and South Africa. Basin Research 14, 69–88.

Frei, R., Gaucher, C., Poulton, S.W., Canfield, D.E., 2009. Fluctuations in Precambrain atmospheric oxygenation recorded by chromium isotopes. Nature 461, 250–253.

Frei, R., Gaucher, C., Døssing, L.N., Sial, A.N., 2011. Chromium isotopes in carbonates—a tracer for climate change and for reconstructing the redox state of ancient seawater. Earth and Planetary Science Letters 312, 114–125.

Frei, R., Gaucher, C., Stolper, D., Canfield, D.E., 2013. Fluctuations in late Neoproterozoic atmospheric oxidation—Cr isotope chemostratigraphy and iron speciation of the late Ediacaran lower Arroyo del Soldado Group (Uruguay). Gondwana Research 23, 797–811.

Freitas, B.T., Warren, L.V., Boggiani, P.C., De Almeida, R.P., Piacentini, T., 2011. Tectono-sedimentary evolution of the Neoproterozoic BIF-bearing Jacadigo Group, SW-Brazil. Sedimentary Geology 238, 48–70.

Frimmel, H.E., 2008. The gariep Belt. In: Miller, R.M. (Ed.), The Geology of Namibia. Neoproterozoic to Lower Palaeozoic, vol. 2. Geological Survey of Namibia, Windhoek, pp. 14-1–14-39.

Gaucher, C., 2000. Sedimentology, palaeontology and stratigraphy of the Arroyo del Soldado Group (Vendian to Cambrian, Uruguay). Beringeria 26, 1–120.

Gaucher, C., Sprechmann, P., Schipilov, A., 1996. Upper and Middle Proterozoic fossiliferous sedimentary sequences of the Nico Pérez Terrane of Uruguay: Litho-stratigraphic units, paleontology, depositional environments and correlations. Neues Jahrbuch für Geologie und Paläontologie, Abhandlungen 199, 339–367.

Gaucher, C., Sprechmann, P., 1999. Upper Vendian skeletal fauna of the Arroyo del Soldado Group, Uruguay. Beringeria 23, 55–91.

Gaucher, C., Boggiani, P.C., Sprechmann, P., Sial, A.N., Fairchild, T., 2003. Integrated correlation of the Vendian to Cambrian Arroyo del Soldado and Corumbá Groups (Uruguay and Brazil): palaeogeographic, palaeoclimatic and palaeobiologic implications. Precambrian Research 120, 241–278.

Gaucher, C., Sial, A.N., Blanco, G., Sprechmann, P., 2004. Chemostratigraphy of the lower Arroyo del Soldado Group (Vendian, Uruguay) and palaeoclimatic implications. Gondwana Research 7, 715–730.

Gaucher, C., Frimmel, H.E., Germs, G.J.B., 2005. Organic-walled microfossils and biostratigraphy of the upper Port Nolloth Group (Namibia): implications for the latest Neoproterozoic glaciations. Geological Magazine 142, 539–559.

Gaucher, C., Finney, S.C., Poiré, D.G., Valencia, V.A., Grove, M., Blanco, G., Pamoukaghlián, K., Gómez Peral, L., 2008. Detrital zircon ages of Neoproterozoic sedimentary successions in Uruguay and Argentina: insights into the geological evolution of the Río de la Plata Craton. Precambrian Research 167, 150–170.

Gaucher, C., Poiré, D., 2009. Biostratigraphy. Neoproterozoic-Cambrian evolution of the Río de la Plata Palaeocontinent. In: Gaucher, C., Sial, A.N., Halverson, G.P., Frimmel, H.E. (Eds.), Neoproterozoic-cambrian Tectonics, Global Change and Evolution: A Focus on Southwestern Gondwana. Developments in Precambrian Geology, vol. 16. Elsevier, pp. 103–114.

Gaucher, C., Frimmel, H.E., Germs, G.J.B., 2009. Tectonic events and palaeogeographic evolution of southwestern Gondwana in the Neoproterozoic and Cambrian. In: Gaucher, C., Sial, A.N., Halverson, G.P., Frimmel, H.E. (Eds.), Neoproterozoic-cambrian Tectonics, Global Change and Evolution: A Focus on Southwestern Gondwana. Developments in Precambrian Geology, vol. 16. Elsevier, pp. 295–316.

Graf, J.L., Oconnor, E.A., Vanleeuwen, P., 1994. Rare earth element evidence of origin and depositional environment of Late Proterozoic ironstone beds and manganese-oxide deposits, SW Brazil and SE Bolivia. Journal of South American Earth Sciences 7, 115–133.

Gross, G.A., 1980. A classification of iron formations based on depositional environments. Canadian Mineralogist 18, 215–222.

Halverson, G.P., Poitrasson, F., Hoffman, P.F., Nédélec, A., Montel, J.-M., 2007. Iron isotope composition of Neoproterozoic syn-glacial iron-formation: implications for sources and cycling of iron in anoxic oceans. In: 7th International Symposium on Applied Isotope Geochemistry, Stellenbosch, pp. 63–64.

Halverson, G.P., Wade, B.P., Hurtgen, M.T., Barovich, K.M., 2010. Neoproterozoic chemostratigraphy. Precambrian Research 182, 337–350.

Halverson, G.P., Poitrasson, F., Hoffman, P.F., Nédélec, A., Montel, J.-M., Kirby, J., 2011. Fe isotope and trace element geochemistry of the Neoproterozoic syn-glacial Rapitan iron formation. Earth and Planetary Science Letters 309, 100–112.

Hoffman, P.F., Hawkins, D.P., Isachsen, C.E., Bowring, S.A., 1996. Precise U–Pb zircon ages for early Damaran magmatism in the Summas Mountains and Welwitschia inlier, northern Damara belt, Namibia. Communications of the Geological Survey of Namibia 11, 47–52.

Hoffman, P.F., Kaufman, A.J., Halverson, G.P., Schrag, D.P., 1998. A Neoproterozoic snowball Earth. Science 281, 1342–1346.

Hoffman, P., Schrag, D.P., 2002. The snowball Earth hypothesis: testing the limits of global change. Terra Nova 14, 129–155.

Hoffman, P.F., Halverson, G.P., 2008. The Otavi Group of the Northern Platform and the Northern Margin Zone. In: Miller, R.M. (Ed.), The Geology of Namibia. Neoproterozoic to Lower Palaeozoic, vol. 2. Geological Survey of Namibia, Windhoek, pp. 13-69–13-136.

Hoffmann, K.H., Condon, D.J., Bowring, S.A., Crowley, J.L., 2004. U–Pb zircon date from the Neoproterozoic Ghaub Formation, Namibia: constraints on Marinoan glaciation. Geology 32, 817–820.

Hollanda, M.H., Archanjo, C.J., Souza, L.C., 2010. Historia de provenieência do Grupo Seridó (Província Borborema), com base em dados isotópicos. In: Congresso Brasileiro de Geologia, Belém, (CD-ROM).

Ilyin, A.V., 2009. Neoproterozoic banded iron formations. Lithology and Mineral Resources 44, 78–86.

Isley, A.E., Abbott, D.H., 1999. Plume-related mafic volcanism and the deposition of banded iron formation. Journal of Geophysical Research 104, 15461–15477.

Jacobsen, S.B., Kaufman, A.J., 1999. The Sr, C and O isotopic evolution of Neoproterozoic seawater. Chemical Geology 161, 37–57.

James, H.L., 1954. Sedimentary facies of iron-formation. Economic Geology 49, 235–293.

Johnston, D.T., Poulton, S.W., Dehler, C., Porter, S., Husson, J., Canfield, D.E., Knoll, A.H., 2010. An emerging picture of Neoproterozoic ocean chemistry: insights from the Chuar Group, Grand Canyon, USA. Earth and Planetary Science Letters 290, 64–73.

Kamber, B.S., Greig, A., Collerson, K.D., 2005. A new estimate for the composition of weathered young upper continental crust from alluvial sediments, Queensland, Australia. Geochimica et Cosmochimica Acta 69, 1041–1058.

Kato, Y., Yamaguchi, K.E., Ohmoto, H., 2006. Rare earth elements in Precambrian banded iron formation: secular changes of Ce and Eu anomalies and evolution of atmosphere oxygen. Geological Society of America, Memoir 198, 269–280.

Kaufman, A.J., Hayes, J.M., Knoll, A.H., Germs, G.J.B., 1991. Isotopic compositions of carbonates and organic carbon from Upper Proterozoic successions in Namibia: stratigraphic variation and the effects of diagenesis and metamorphism. Precambrian Research 49, 301–327.

Klein, C., 2005. Some Precambrian banded iron-formation (BIFs) from around the world: their age, geologic setting, mineralogy, metamorphism, geochemistry, and origin. American Mineralogist 90, 1473–1499.

Klein, C., Ladeira, E.A., 2004. Geochemistry and mineralogy of Neoproterozoic of banded iron formations and some selected, siliceous manganese formations from the Urucum district, Mato Grosso do Sul, Brazil. Economic Geology 99, 1233–1244.

Knoll, A.H., Hayes, J.M., Kaufman, A.J., Swett, K., Lambert, I.B., 1986. Secular variation in carbon isotope ratios from Upper Proterozoic successions of Svalbard and east Greenland. Nature 321, 832–837.

Legrand, M., Martins de Sá, J., Santos, L., 2008. Tilito Neoproterozoico intercalado nas meta-supracrustais da Faixa Seridó, Província Borborema, Nordeste do Brasil. In: 44° Congresso Brasileiro de Geologia (CD-ROM).

Le Heron, D.P., Cox, G., Trundley, A., Collins, A., 2011. Sea ice-free conditions during the Sturtian glaciation (early Cryogenian), South Australia. Geology 39, 31–34.

Li, Z.X., Bogdanova, S.V., Collins, A.S., Davidson, A., De Waele, B., Ernst, R.E., Fitzsimons, I.C.W., Fuck, R.A., Gladkochub, D.P., Jacobs, J., Karlstrom, K.E., Lu, S., Natapov, L.M., Pease, V., Pisarevsky, S.A., Thrane, K., Vernikovsky, V., 2008. Assembly, configuration, and break-up history of Rodinia: a synthesis. Precambrian Research 160, 179–210.

Macdonald, F.A., Strauss, J.V., Rose, C.V., Dudas, F.O., Schrag, D.P., 2010a. Stratigraphy of the Port Nolloth Group of Namibia and South Africa and implications for the age of Neoproterozoic iron formations. American Journal of Science 310, 862–888.

Macdonald, F.A., Schmitz, M.D., Crowley, J.L., Roots, C.F., Jones, D.S., Maloof, A.C., Strauss, J.V., Cohen, P.A., Johnston, D.T., Schrag, D.P., 2010b. Calibrating the Cryogenian. Science 327, 1241–1243.

McGee, B., Collins, A.S., Trindade, R.I.F., 2012. G'day Gondwana—the final accretion of a supercontinent: U–Pb ages from the post-orogenic São Vicente Granite, northern Paraguay Belt, Brazil. Precambrian Research 21, 316–322.

Melezhik, V.A., Gorokov, I.M., Kuznetsov, A.B., Fallick, A.E., 2001. Chemostratigraphy of Neoproterozoic carbonates: implications for "blind dating". Terra Nova 13, 1–11.

Miller, R.M., 2008. Swakop Group. In: Miller, R.M. (Ed.), The Geology of Namibia. Neoproterozoic to Lower Palaeozoic, vol. 2. Geological Survey of Namibia, Windhoek, pp. 13-136–13-178.

Möller, J.C., 2004. A geologia neoproterozóica da região de Corumbá, os recursos minerais de ferro e informações gerais sobre a MCR, Guia da Mineração Corumbaense Reunida. Corumbá 1–25.

Nascimento, R.S.C., Sial, A.N., Pimentel, M.M., 2007. C- and Sr-isotope systematics applied to Neoproterozoic marbles of the Seridó belt, northeastern Brazil. Chemical Geology 237, 209–228.

Piacentini, T., Boggiani, P.C., Kazuo Yamamoto, J., Freitas, B.T., Campanha, G.A.C., 2007. Formação ferrífera associada à sedimentação glaciogênica da Formação Puga (Marinoano) na Serra da Bodoquena. MS. Revista Brasileira de Geociências 37, 530–541.

Piacentini, T., Vasconcelos, P.M., Farley, K.A., 2013. $^{40}Ar/^{39}Ar$ constraints on the age and thermal history of the Urucum Neoproterozoic banded iron-formation, Brazil. Precambrian Research 228, 48–62.

Schoenberg, R., Zink, S., Staubwasser, M., von Blanckenburg, F., 2008. The stable Cr isotope inventory of solid Earth reservoirs determined by double spike MC-ICP-MS. Chemical Geology 249, 294–306.

Sial, A.N., Campos, M.S., Gaucher, C., Frei, R., Ferreira, V.P., Nascimento, R.C., Pimentel, M.M., Pereira, N.S., Rodler, A., (submitted). Algoma-type Neoproterozoic BIFs and related marbles in the Seridó Belt (NE Brazil): REE, C, O, Cr and Sr isotope evidence. Journal of South American Earth Sciences.

Stern, R.J., Mukherjee, S.K., Miller, N.R., Ali, K., Johnson, P.R., 2013. ~750 Ma banded iron formation from the Arabian-Nubian Shield—implications for understanding Neoproterozoic tectonics, volcanism, and climate change. Precambrian Research 239, 79–94.

Sun, H., Wu, J., Yu, P., Li, J., 1998. Geology, geochemistry and sulfur isotope composition of the Late Proterozoic Jingtieshan (Superior-type) hematite–jasper–barite iron ore deposits associated with stratabound Cu mineralization in the Gansu Province, China. Mineralium Deposita 34, 102–112.

Trompette, R., De Alvarenga, C.J.S., Walde, D., 1998. Geological evolution of the Neoproterozoic Corumba graben system (Brazil). Depositional context of the stratified Fe and Mn ores of the Jacadigo group. Journal of South American Earth Science 11, 587–597.

Van Schmus, W.R., Neves, B.B.B., Williams, I.S., Hackspacher, P.C., Fetter, A.H., Dantas, E.L., Babinski, M., 2003. The Seridó Group of NE Brazil, a late Neoproterozoic pre-to syn-collisional basin in West Gondwana: insights from SHRIMP U–Pb detritical zircon ages and Sm–Nd crustal residence (TDM) ages. Precambrian Research 127, 287–327.

Xu, D.R., Wang, Z.L., Cai, J.H., Wu, C.J., Bakun-Czubarow, N., Wang, L., Chen, H., Baker, M., Kusiak, A., 2013a. Geological characteristics and metallogenesis of the Shilu Fe-ore deposit in Hainan Province, South China. Ore Geology Reviews 53, 318–342.

Xu, D.R., Wang, Z.L., Chen, H.Y., Hollings, P., Jansen, N.H., Zhang, Z.C., Wu, C.J., 2013b. Petrography and geochemistry of the Shilu Fe–Co–Cu ore district, South China: Implications for the origin of a Neoproterozoic BIF system. Ore Geology Reviews 57, 322–350.

Young, G.M., 2002. Stratigraphic and tectonic settings of Proterozoic glaciogenic rocks and banded iron-formations: relevance to the snowball Earth debate. Journal of African Earth Sciences 35, 451–466.

CHAPTER

CHEMOSTRATIGRAPHY OF NEOPROTEROZOIC CARBONATE DEPOSITS OF THE TUVA–MONGOLIAN AND DZABKHAN CONTINENTAL BLOCKS: CONSTRAINTS ON THE AGE, GLACIATION AND SEDIMENTATION

18

Irina Vishnevskaya[1,2], Elena Letnikova[1,2], Natalia Pisareva[1,2], Artem Proshenkin[1,2]
[1]*Sobolev Institute of Geology and Mineralogy, Siberian Branch of the Russian Academy of Sciences, Novosibirsk, Russia;* [2]*Geology-Geophysics Department of Novosibirsk State University, Novosibirsk, Russia*

18.1 INTRODUCTION

The Tuva–Mongolian and Dzabkhan continental blocks are the largest Precambrian blocks in the Central Asian fold belt (CAFB) structures (Fig. 1). Tuva–Mongolian block is separated from Dzabkhan by the Khangai fault on the south. Dzabkhan block is detached from Sangilen through ophiolite lake zone on the south east and by Bayankhongor ophiolite zone from north west. In spite of the fact that the Neoproterozoic sediments of these two blocks are well documented for their lithology and biostratigraphy, stratigraphic correlation between the carbonates of these two blocks is not yet clear. In addition, information is not available on their depositional history and synchroneity due to the poorly constrained nature of organic remains and paucity of marker horizons. These limitations have encouraged the authors to venture into the documentation of chemostratigraphic traits for establishing precise correlation and age bracketing the depositional history through isotope stratigraphy.

18.2 GEOLOGIC SETTING

The basement of the Tuva–Mongolian block is represented by Archean—early Proterozoic crystalline rocks of Gargan block (Fig. 2). Gross isochrone age for this block was determined to be 3240 ± 57 Ma (Aktanov et al., 1991). Plagiogneisses of this block were influenced by high grade metamorphism at 2.66 Ga (U–Pb dating by Zircon) (Kovach et al., 2004). Sediments of Irkut Formation (up to 590 m thick) overlie the crystalline rocks of the Gargan block with stratigraphic unconformity. The basal part of the formation consists of thin layer (10 m) of thin-pebble conglomerates, gravelites and sandstones

Chemostratigraphy. http://dx.doi.org/10.1016/B978-0-12-419968-2.00018-2

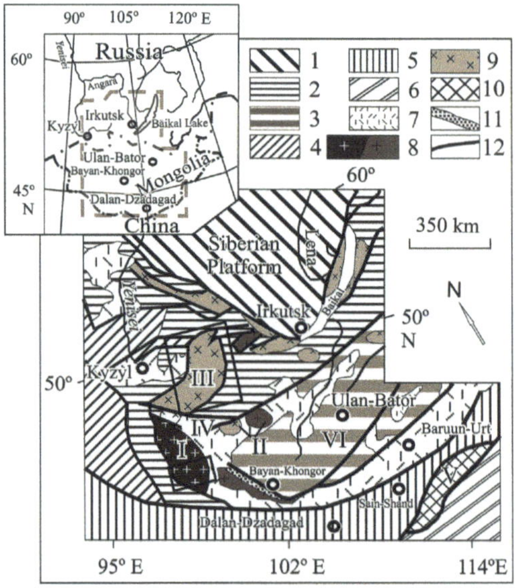

FIGURE 1

Crystalline complexes in the eastern segment of the Central Asian fold belt (CAFB) *(Compiled after Kozakov et al. (2007), Ovchinnikova et al. (2012).)*: (1) ancient platforms; (2–10) CAFB structures: (2) early Caledonides, (3) upper Paleozoic covers, (4) late Caledonides, (5) Hercynides, (6) Indosinides, (7) Late Paleozoic–Mesozoic volcanoplutonic belts; (8–10) fragments of continental massifs: (8a) with the undivided Early and Late Paleozoic basement, (8b) with the Early Precambrian basement, (9) with the Early Baikalian basement, (10) with the Grenvillian basement, (11) South Khangai metamorphic belt, (12) principal tectonic boundaries. Roman numerals: (I) Dzabkhan block, (II) Tarbagatai block, (III) Tuva–Mongolian block, (IV) Songino block, (V) South Gobi block, (VI) Khangai block overlain by Paleozoic–Mesozoic sequences. Rectangular—maps of Tuva–Mongolian block on Fig. 2 and Dzabkhan block on Fig. 3.

which contain basement clasts. The formation is mainly represented by bituminous dolomites and limestones with rare layers of chlorite–mica shale. Carbonate deposits contain structures of columnar stromatolites *Conophyton garganicus, Conophyton cylindricus and Collumnacollenia giga* (Semikhatov and Serebryakov, 1967), which are typical for Early Neoproterozoic sediments formed in the open shelf areas below the storm wave base. During middle Neoproterozoic, the shelf carbonate sedimentation was interrupted by tectonic activation of ancient Precambrian structure and active continental margin was formed resulting in the formation of Oka (and its analogue Hugein) accretion prism. Rocks of the Oka Formation are interrupted by synsedimentary trolleite sills. Age of this time episode determined by U–Pb method on Discordia is 753 ± 16 Ma (Kuzmichev, 2004).

During the Pre-Ediacaran collision, sediments of Gargan blocks including early Neoproterozoic carbonate cover and its middle Neoproterozoic active continental margin were obducted by the sediments of Djungur ophiolite complex, largely overlying basement and cover sediments of the Gargan block. Age of the ophiolite plagiogranites, forming lenticular veins in gabbro and lamellar complexes, determined by U–Pb and Pb–Pb methods on zircons, is 1010–1020 Ma (Khain et al., 1999). Sedimentation of Gargan block cover and its obduction by ophiolite allochton occurred till 790 Ma ago. Age of

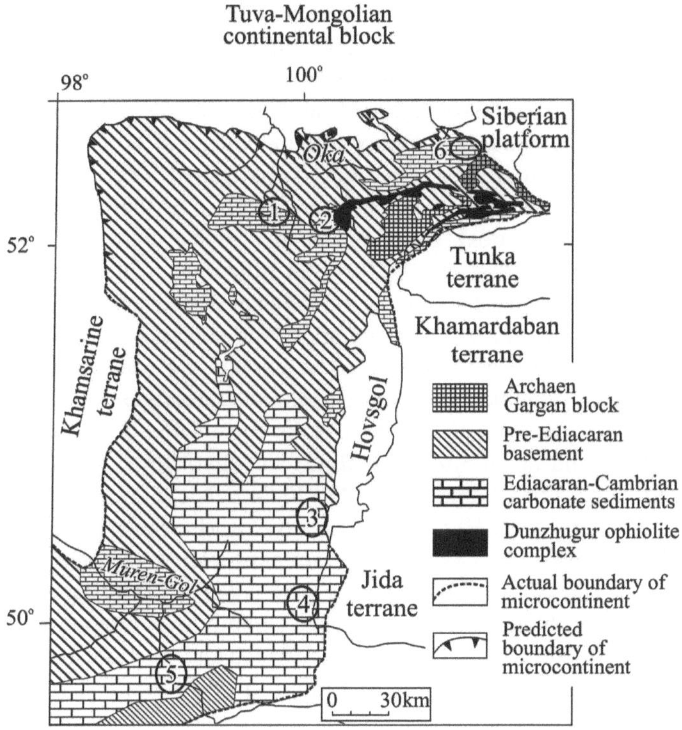

FIGURE 2

Sketch map of the studied sections of carbonate sediments in the Tuva–Mongolian microcontinent (tectonic structure, simplified after (Kuzmichev, 2004)). Numbers in the circles are the studded sections: (1) Bokson Group at the confluence of the Urdo-Bokson and Khoito-Bokson Rivers, (2) Zabit Formation on the Ukhagol River, (3) Hovsgol Group on the Hovsgol phosphorite deposit, (4) Hovsgol Group on the Burenkhan phosphorite deposit, (5) Muren Formation, and (6) Araoshei Formation.

the tonalites were determined by Rb–Sr method (812 ± 18 Ma) and also by U–Pb method in zircons (785 ± 11 Ma) (Kuzmichev et al., 2000). This complex structure is overlain by Ediacaran–Cambrian cover rocks with unconformity. Sediments of this cover are spread along the wide territory in Northern Mongolia and eastern Sayan Mountains and are represented by sediments of the Bokson and Hovsgol groups and of Muren and Araoshei Formations. These sedimentary complexes are studied lithologically in the areas where phosphorite and bauxite deposits occur in the tectonic structure (Osokin, 1999). The sedimentation coincidence of these sediments has not been established for a long time (Vishnevskaya and Letnikova, 2013). Fossil data helped determine the age of these sedimentary complexes to be of late Neoproterozoic. Studies of rare earth and trace elements in carbonate rocks showed that they were formed for a long period of time in the open shelf environment with little to no influence of terrigenous material from continent and complete absence of volcanogenous and hydrothermal material (Letnikova and Geletii, 2005). So there is a chance that these sediments recorded the isotopic composition of oceanic waters on the time when they were formed and there is a possibility to determine their sedimentation interval using Sr-isotopic stratigraphic methods.

The Dzabkhan block is situated below the Tuva–Mongolian microcontinent. Dzabkhan block is not studied so well due to the accessibility and logistic constraints. Dzabkhan block is composed of Baydrik pre-Cambrian block (Fig. 3), represented by Archean and Proterozoic crystalline basement rocks—Baydargin complex with model age of 3100–3200 Ma and granulite metamorphism age of 2664 ± 15 Ma (U–Pb method in zircons) and Bumbuger complex with model age of 3100 Ma and granulite metamorphism age of 2364 ± 6 Ma (U–Pb method in zircons) (Kozakov et al., 2007).

The Mezoproterozoic crystalline basement rocks are overlain by the Uldzitgol Formation deposits (sandstones, limestones, and dolomites). Basement of this Mezoproterozoic cover is formed by pre-Mezoproterozoic complexes (Kozakov et al., 2007). In Ediacaran–Cambrian block, in the range of Dzabkhan continental shelf, the marine sediments of Tsagaan Oloom and Bayan Gol Formations were deposited. These sediments lie with unconformity above the volcano-sedimentary Dzabkhan Formation rocks (Neoproterozoic). These volcanogenous sediments are formed mostly by acid and to a lesser extent by basic effusive rocks with some amount of terrigenous and carbonate deposits. Dzabkhan volcanic deposits are represented by basalts (up to 46.3% SiO_2 and 3–3.5% alkaline), dacite (up to 68.49% SiO_2 and up to 3.92% alkaline) and rhyolites (up to 74.49% SiO_2 and up to 5.25% alkaline). All these are remarkable for $Na_2O > K_2O$. Felsic and mafic rocks are enriched with light-group rare earth elements (LREE) ((La/Yb)n = 4.6–7.8, (La/Sm)n = 2.4–4) and huge ion lithophile elements (Cs, Rb, and Ba). The heavy-group rare earth elements contain values in mafic rocks are less than 100 per chondrite. These values though grow steadily from basalts to rhyolites and reach 150–240 per chondrite, all Dzabkhan volcanic deposits contain low amounts of Nb, Ta, Hf, and Zr in comparison to LREE, Ba, Sr, and Rb. Minimum values of Nb and Ta show that there were subduction environments. The age of volcanic sediments according to U–Pb method in zircons is 770–805 Ma (Levachova et al., 2010).

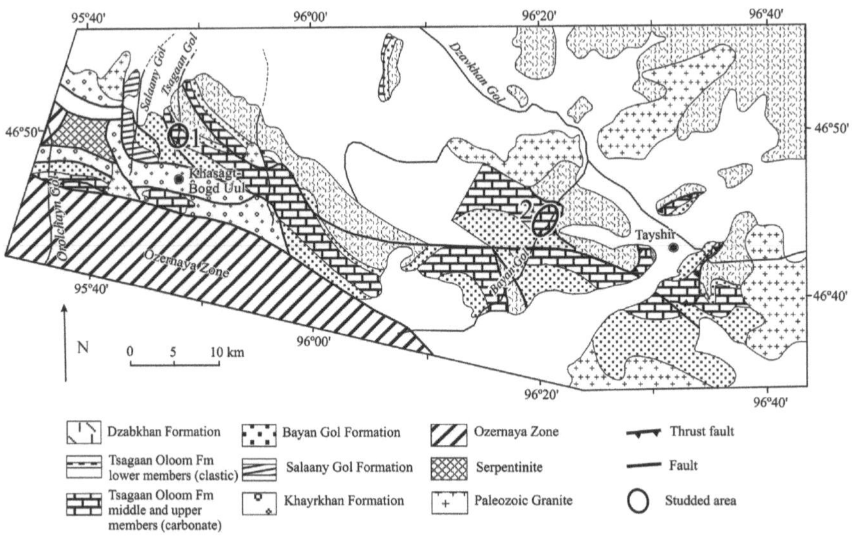

FIGURE 3

Simplified geological map of the Dzabkhan block. Numbers in the circles are the studded sections: (1) Section N1, (2) Section N2.

During the Ediacaran–Cambrian time in the range of Dzabkhan block, shelf sedimentation took place and the sedimentary records are represented by the Tsagaan Oloom and Bayan Gol Formations. These sediments rest over the volcanogeneous rocks with an angular unconformity. Carbonates of the Tsagaan Oloom and Bayan Gol groups are exposed along the banks of the Tsagaan Oloom and Bayan Gol rivers and in the dry channels of Dolon-Tognoni-Hundey. There is a layer of glacial deposits at the basement of Tsagaan Oloom Formation. The U–Pb and Pb–Pb methods were used for determining age of cap limestones of the tillite (Ovchinnikova et al., 2012). The weighted average age is 632 ± 14 Ma (MSWD = 0.11, probability 0.74). Consequently, the lower boundary of the Tsagaan Oloom Formation corresponds to the beginning of the Ediacaran. Dzabkhan tillites are correlative with glacial sediments of the Marinoan Epoch. The low $^{238}U/^{204}Pb$ and $^{232}Th/^{238}U$ ratios observed in initial Pb sources of limestones from the Tsagaan Oloom Formation indicate that the Dzabkhan paleobasin received at its early development stages a bulk of material from eroded upper middle Neoproterozoic juvenile rocks (Ovchinnikova et al., 2012).

18.2.1 STRATIGRAPHY OF THE TUVA–MONGOLIAN BLOCK

The most studied object within the Ediacaran–Cambrian cover of the Tuva–Mongolian block is the Bokson Group (Fig. 2, sections 1 and 2). It is divided into five formations (from the bottom upward): Zabit, Tabinzurta, Khuzhirtai, Nurgata, and Khyuten (Fig. 4). The first two mainly consist of dolomites, and the others of limestones. The lower boundary of the group is disputable; in the stratotype locality, it passes along the roof of the variegated siltstone unit of the Khushatai Formation (Kuzmichev, 2004). At the base of the Zabit Formation, there is a mixtite bed, assigned to tillite by some researchers (Osokin and Tyzhinov, 1998). The group is conformably overlain by the sedimentary rocks of the Mangatgol Formation (upper Cambrian–lower Ordovician). Deposits of this formation are overlain by rocks from Mantagol Formation (upper Cambrian–lower Ordovician).

In the stratotype locality, the Zabit Formation (up to 1000 m) is marked by facies and thickness variability. The upper boundary of the formation is distinct only in the Bokson River basin. A bauxite-bearing bed, which indicates a sedimentation gap, is distinguished here at its top. At the confluence of the Urdo-Bokson and Khoito-Bokson rivers (Fig. 2, section 1), the formation consists of dolomites with rare limestone interbeds as well as phosphate-containing dolomites (50–100 m). The formation is saturated with microphytolites (Vesicularites, Glebosites, Nubecularites, Radiosus). Also, stromatolites (Semikhatov and Serebryakov, 1967), Claudina tubes (Kheraskova and Samygin, 1992), and Ediacaran fauna (Dobretsov, 1988) were found to occur. The section of this formation is more complete near the head of the Ukhagol River (Fig. 2, section 2). Boulder silty pelites and carbonaceous limestones here make up unit 1, and the overlying dolomites are divided into units 2, 3, and 4 (Khomentovsky et al., 1985). It is the presence of limestones at the formation base that makes this section more suitable for chemostratigraphic studies than that in the stratotype locality.

The Tabinzurta Formation (up to 1300 m) has many facies features in common with the Zabit Formation. It is dominated by microphytolitic (Osagia, Volvatella, Vesicularites, Nubecularites, Glebosites) and stromatolitic (*Linella, Stratifera*) dolomites. The upper part of the formation contains Renalcis algae and Early Cambrian archeocyaths (Dobretsov, 1988). This formation has no clear boundary with the overlying Khuzhirtai Formation, because the latter is often also dolomitic and only the predominance of limestones makes their differentiation possible.

The Khuzhirtai Formation (up to 500 m) is composed of archeocyath-algal calcareous bioherms, separated by beds of dolomites, carbonate breccias, and limestones with trilobite and brachiopod

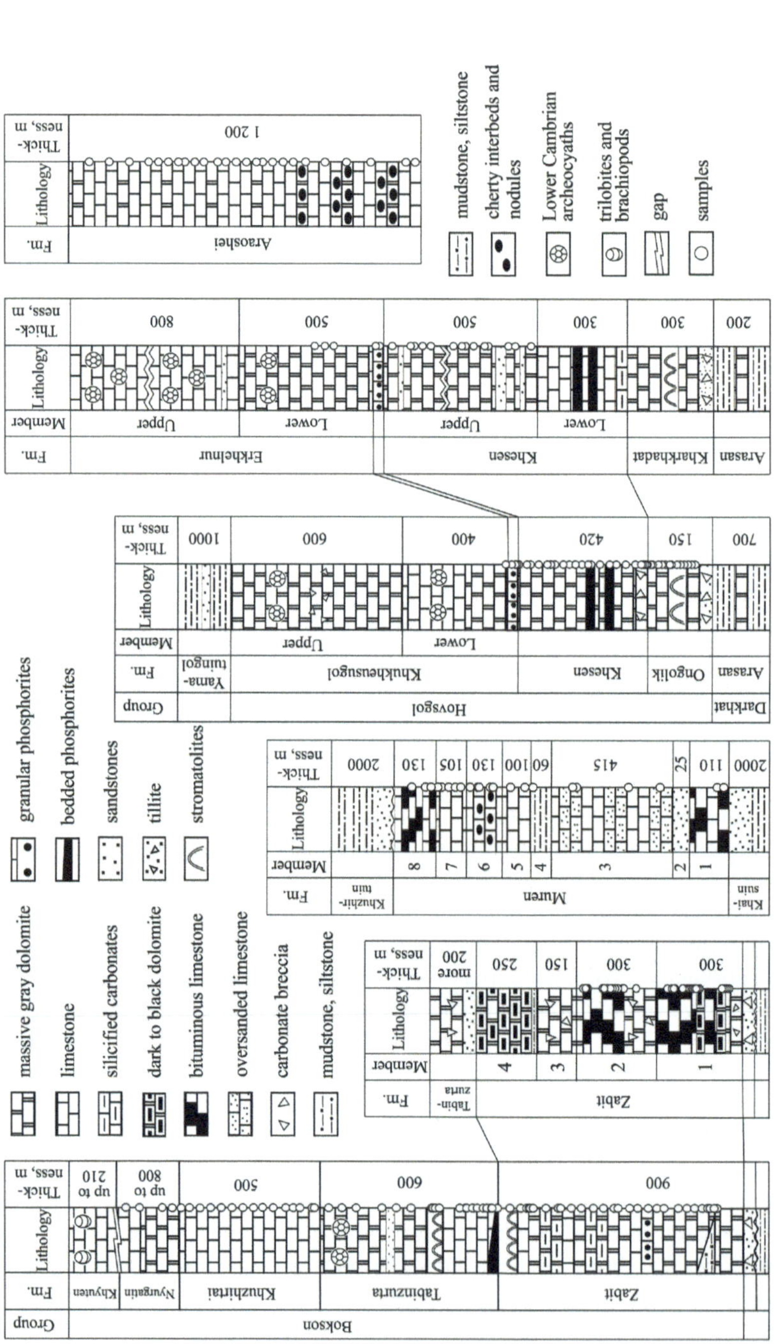

FIGURE 4

Stratigraphic column of the Ediacaran–Cambrian Tuva–Mongolian carbonate sediments.

remnants (Kheraskova and Samygin, 1992). The Nurgata (up to 800m) and Khyuten (up to 210m) formations, unlike the units described above, consist almost exclusively of laminated clayey and massive limestones. The Khyuten Formation contains trilobite and brachiopod remnants. No distinct contact between these formations is observed, but the contact is assumed.

Data on the distribution of stable (C–O) isotopes were obtained previously for the Bokson Group (Pokrovskii et al., 1999). The Ediacaran age of the Zabit Formation was confirmed by correlation of C isotope and O isotope curves for the Bokson Group with those for Ediacaran–Cambrian type sections of the Siberian Platform and western Mongolia. The lower part (75–100 m) of the Tabinzurta Formation also belongs to the Ediacaran, and the rest, to the Early Cambrian. This was confirmed by Sr isotope chemostratigraphy of a pilot study of carbonate sediments from the Bokson Group (Vishnevskaya and Letnikova, 2013).

The carbonate sediments of the Hovsgol Group were studied at two phosphorite deposits in northern Mongolia: Hovsgol (Fig. 2, section 3) and Burenkhan (Fig. 2, section 4). There are several ways of dividing this group into formations (Eganov et al., 1999; Il'in, 1973; Osokin, 1999). We use the subdivision proposed by Osokin (1999). However, since an accurate correlation of the sediments with the above formations at these two deposits is complicated by later tectonic deformations, the group was divided into the lower (below-phosphate) and upper (above-phosphate) parts in the subsequent studies.

The sediments of the Hovsgol Group (Fig. 4), with a sedimentation intermission, overlie the terrigenous rocks of the Darkhat Group, which occur as a tillite bed (20–50m thick) in the upper part of the section (Osokin and Tyzhinov, 1998). At the base of the Hovsgol Group, black dolomites are interbedded with dolomitized limestones, including oncolitic ones. A 120-m-thick phosphorite bed is observed in the middle part of the section. The phosphorites are conformably overlain by gray dolomites and dolomitized limestones. The dolomites are followed toward top to black and dark gray limestones, which are often clayey. The entire phosphate part of the section contains Botomian (lower Cambrian) trilobites and archeocyaths.

In the south western part of the Tuva–Mongolian continental block, the carbonate deposits of the Muren (Agaringol) Formation (Fig. 2, section 5) occur. They overlie terrigenous rocks of the Haisuin Formation (Fig. 4). Sediments of the Muren Formation are abundant (Gibsher and Bat-Ireedui, 1990). Lower part of the Muren Formation consists of fine and medium grained limestone beds and lenses with low amounts of siliceous shale. The next part of the Muren Formation is dominated by bituminous limestones. The presence of carbonaceous matter and thin laminations in the limestones indicates a relatively deep-water depositional environment (below the wave base). The upper part of the section contains organogenic algal limestones of uncertain taxonomy. The age of the sediments of the Muren Formation is uncertain for lack of finds of index biota. The Ediacaran age of the formation was obtained by Sr isotope (Vishnevskaya and Letnikova, 2013). The total thickness of the Muren Formation in this locality is 1075m.

The Cambrian part of the Tuva–Mongolian block cover is represented by the sediments of Araoshei Formation (Fig. 2, section 6), bounded at the bottom with tectonic contacts and is arguably the superstructure to Gorluk Formation. Outcrops of this formation can be found in the eastern Sayan area. Assumed basement of the section (Boos, 1991) is enriched with siliceous material in the form of small (first millimeters) and huge (20–30cm) concretions and lenticular bodies of up to 1 m size. Sediments of Araoshei Formation are overlain by tectonic cover remains of the mélange zone in its basement (Letnikova and Geletii, 2005).

18.2.2 STRATIGRAPHY OF THE DZABKHAN BLOCK

The contact between Tsagaan Oloom and Dzabkhan Formations in the tract Dolon-Tognoni-Hundei has tectonic nature because various parts of stratigraphic sequence of Tsagaan Oloom Formation set against the contact zone. At the same time we can see in a set of places, most clearly on the left flank of River Bayan Gol and dry channel of the river (Fig. 3, section 2) that Dzabkhan effusive rocks stratigraphically contact with basal pack of Tsagaan Oloom Formation (Khomentovsky and Gibsher, 1996).

The Tsagaan Oloom Formation is represented by more than 1500 m of carbonate deposits. It is divided into two members: lower and upper (Fig. 5). Lower one is 400–800 m thick, composed mainly by thick layered dolomites, and contain frequent oncolite and breccia. In the basement of lower member, 2–300 m thick tillite, composed of terrigenous rockclasts of boulder gravelites and sandstones with argillite layers (Brazier et al., 1996; Macdonald et al., 2009) is observed. It is overlain by nonmassive (20–30 m) bituminous thin layered limestone, which grades gradually to thick limestone–dolomite strata (600–650 m). In the upper part there is a separate horizon of stromatolites dolomite (5–50 m). It is typical for Dzabkhan basin and is associated with phosphorite and is its underlying marker, while the horizon itself is the bottom of the upper member. The upper member is characterized by complex rhythmic alternations of aleurolite, black layered limestone, sandstone and clayey limestones whose

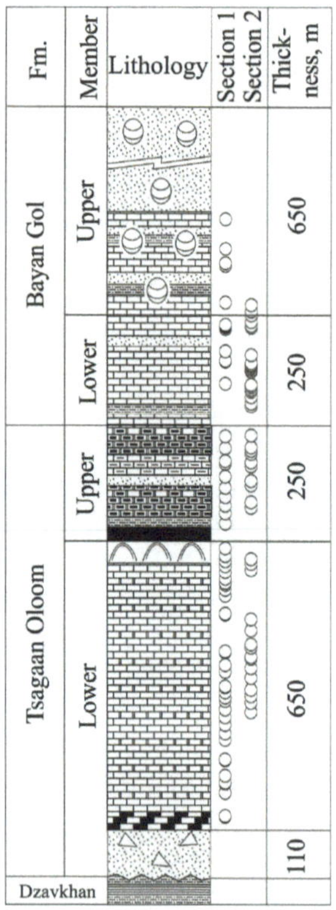

FIGURE 5

Stratigraphic column of the Ediacaran–Cambrian Dzabkhan carbonate sediments (See Fig. 4 for the legend).

thicknesses vary from 15 to 100 m. A variety of fauna including ichnofauna, Ediacaran type fauna (*Medusonesm, Polipes, Helminthopsises*) are present in the sandstones and shales. The boundary between Ediacaran and Cambrian deposits inside the Tsagaan Oloom Formation is bonded (due to fossil remains complexes) to first thick layer of terrigenous rocks (approximately 1150 m from the formation basement) (Osokin, 1999). Average value of 632 ± 14 Ma (SD = 0.11, probability fit 0.74) gives direct geochronological characteristic of cap dolomite of Tsagaan Oloom Formation tillite. These results make possible to limit the sedimentation datum of Dzabkhan block carbonate cover and correlate the whole sedimentary sequence of the Tsagaan Oloom Formation with Ediacarian, and also with corresponding glacial sediments to the Marinoan glaciation (Ovchinnikova et al., 2012).

The Tsagaan Oloom Formation is overlain with gradual transition (thin alternation of limestones and sandstones) by the Bayan Gol Formation (950 m). It is represented mainly by pink and yellow sandstones (Fig. 5). Its lower part has terrigenous carbonate composition and can be counted as transitional interval between the carbonate and terrigenous depositional systems. This also evidences that the sedimentation here was not stratigraphically interrupted. This Bayan Gol Formation can be divided into two parts. Lower part of the Bayan Gol Formation (250 m) is composed of glaukonite aleuro sandstones, argillites, aleurolites, and limestones. The upper part is represented by dolomites, limestones, sandstones, aleurolites and argillites. This formation contains remains of algae, hyolithelminthids, tommotiids, sachitides and brachiopods typical for the time interval from Tommotian to Atdabon. Excellent ichnofossil assemblage consisting of *Planolites isp., Cochlichnus isp. and Didymaulichnus isp., Helminthoida isp.*, group of "arthropods traces," *Rusophycus isp.—Diplichnites isp.—Monomorphichnus isp., Palaeophycus isp.*, cf. *Plagiogmus isp.*, cf. *Taphrhelmithopsis isp.*, various ichnospecies *Treptichnus ispp.*, cf. *Hormosiroidea isp.* are reported from the Bayan Gol Formation (Ragozina et al., 2009). Sediments of Bayan Gol Formation are overlain by Salani Gol Formation (approximately 600 m) represented by terrigenous sediments with huge boulder conglomerate layer (up to 1.5 m) in the basement and huge (up to 320 m) archaeocyath-algae limestone reefs in the upper part of the section.

18.3 METHODS OF INVESTIGATION

The study involved field work for documenting lithological characteristics and contact relationships and collection of rock samples with minimal diagenetic alteration. Pristine nature of the rock samples was further affirmed through petrographic study of thin sections prepared from the rock specimens. It was followed by analysis of Ca, Mg, Mn, Fe, and Sr contents using atomic-absorption spectrometric method. With these data, and with the ratios of these elements, the amount of post sedimentary alterations in rocks (Brand and Veizer, 1980; Banner et al., 1988; Banner and Hanson, 1990; Jacobsen and Kaufman, 1999; Sawaki et al., 2008, 2010; Li et al., 2011, Kuznetsov et al., 2012) was determined. Based on these, we have selected rock samples with least diagenetic alteration for further analyses. Sr is then extracted from the least altered samples on chromatographic columns for $^{87}Sr/^{86}Sr$ ratio determination by mass-spectrometer equipments MI 1201T (Novosibirsk, IGM SB RAS) and Triton TI (IPGG RAS, Saint-Petersburg). Accuracy of Sr-isotopic measurements was checked by replicate measurements of isotopic standards from All-Russia D.I. Mendeleyev Scientific and Research Institute for Metrology ($^{87}Sr/^{86}Sr$ 0.70800 ± 0.00007, $2\sigma_{ave}$, n = 15) and SRM-987 ($^{87}Sr/^{86}Sr$ 0.71026 ± 0.00006, $2\sigma_{ave}$, n = 8) in each sample group. Isotopic compositions of O and C in carbonates were studied using mass spectrometer Finnigan MAT—253 and sample preparation line—Gas Bench II (IGM SB RAN, Novosibirsk). C and O isotopic compositions were measured with running mass spectrometry methodology in constant helium flow. Accuracy of measurements was controlled by international (NBS19 $\delta^{13}C=+1.9‰$, $\delta^{18}O=+28.6‰$), Russian (DVGI $\delta^{13}C=+1.2‰$, $\delta^{18}O=+32.7‰$) and laboratory (Ca770) standards and are 0.1‰ for $\delta^{13}C$ and $\delta^{18}O$ values. All the values are given in per mil (‰), $\delta^{13}C$ counted on standard PDB, $\delta^{18}O$ counted on standard SMOW (Standard Mean Ocean Water).

18.4 RESULTS

18.4.1 TUVA–MONGOLIAN BLOCK CARBONATE DEPOSITS

18.4.1.1 Basal of Bokson Group

Basal horizons of this formation are represented in Zabit Formation section along the River Ukhagol. The results of atomic-absorption analysis (A.B. Kuznetsov data) show most of the rocks from this section to be dolomites (Mg/Ca = 0.54–0.64) (Table 1), with low amount of limestones (Mg/Ca = 0.02–0.17). Samples with the lowest amounts of silica-clastic material (0.2–5 wt%) are in the upper parts of the section (250–600 m from the top). In other samples, terrigenous part exceeds 10 wt%. Higher concentrations of Mn and Fe are found in the middle part of the section (layer with thickness about 30 m)—1000–2900 ppm and 600–960 ppm. The upper part of the section contains Mn varying from 430 to 1270 ppm and Fe—2200–4700 ppm (Fig. 6). Highest concentrations of Fe (6000–8000 ppm) are found in samples with large amounts of terrigenous admixture (13–26 wt%). Sr concentrations in these rocks are relatively low. In the samples with the lowest amounts of silica-clastic material, it is 80–120 ppm. Minimal Mn/Sr (0.02–0.1) and Fe/Sr (0.25–3.1) ratios are observed in the lower part of the section (120–150 m from bottom) (Table 1). Most of these samples satisfy the parameters accepted for Sr-isotopic chemostratigraphy (Mn/Sr ≤ 6, Fe/Sr ≤ 15 for dolomites, Mn/Sr ≤ 4, Fe/Sr ≤ 10 for limestones) (Podkovyrov et al., 1998). Five samples completely satisfying the criteria and nine samples almost satisfying the criteria (Mn/Sr 0.34–4.2, Fe/Sr 6.1–29.9) were selected for further analyses. $\delta^{18}O_{SMOW}$ values for all the samples vary from +20.8‰ to +28.8‰ (Fig. 5) suggestive of insignificant secondary transformations. Only in the sample SB4-1 near the bottom of the section $\delta^{18}O_{SMOW}=+15.2‰$ is quite low (Table 1).

18.4.1.2 Section on the Urdo-Bokson River and Khoito-Bokson Confluence (Stratotype)

In the mixtite horizon underlying the bottom of Bokson Group there is a pack (samples K5/158–K5/166; Table 1) represented by alternation of limestones (Mg/Ca = 0.001–0.016) and dolomites (Mg/Ca ~ 0.55). They contain from 2 to 98 mass % of terrigenous material. Mn contents in these sediments vary from 10 to 240 ppm in samples with low amount of aluminosilicate and up to 1000 ppm in samples with huge amounts of terrigenous material. Fe is spread unequally from 380 to 6300 ppm (Fig. 6; Table 1). Sr concentrations vary from 50 to 100 ppm and in some rare cases reach as much as 2050 ppm. All the samples with low amount of terrigenous material are found to be suitable for C-isotopic chemostratigraphy but only one limestone sample was selected for Sr-isotopic composition measurements (Mn/Sr = 0.05, Fe/Sr = 0.1).

18.4.1.3 Zabit Formation

Dolomites (Mg/Ca ~ 0.55) predominate in this formation. In general, the rocks contain 2–10 wt% of terrigenous material. Only few samples contain up to 17 wt% of terrigenous material and the highest

Table 1 Rb–Sr and Geochemical Characteristics of the Ediacaran–Cambrian Carbonate Rocks of the Tuva–Mongolian Continental Block

Sample	Position of Sample *	SA†, %	Content, ppm						Mn/Sr	Fe/Sr	Mg/Ca	δ¹³C vPDB, ‰	δ¹⁸O vSMOW, ‰	⁸⁷Rb/⁸⁶Sr	⁸⁷Sr/⁸⁶Sr Measured	⁸⁷Sr/⁸⁶Sr Initial
			Mn	Fe	Sr	Rb‡	Mg	Ca								
Bokson Group. Urdo-Bokson–Khoito-Bokson Confluence																
Pre-Bokson Sediments (Submixtite Bed)																
K5/158	–45	15.9	10	530	50	–	125,000	221,000	0.200	11	0.566	–	–	–	–	–
K5/159	–40	98.5	2130	520,130	0	–	50,000	93,000	0	0.0	0.538	–	–	–	–	–
K5/161	–35	24.7	10	970	1860	–	6000	387,000	0.005	0.52	0.016	–	–	–	–	–
K5/162	–30	6.4	10	210	2050	0.02	300	398,000	0.005	0.10	0.001	–	–	0.00003	0.70832	0.70832
K5/163	–25	16.3	920	7690	60	–	117,000	211,000	15	128	0.555	4.7	27.9	–	–	–
K5/222	–25	3.6	90	2880	50	–	123,000	223,000	1.8	58	0.552	2.0	27.9	–	–	–
K5/223	–20	4.5	60	3440	50	–	120,000	226,000	1.2	69	0.531	1.6	29.6	–	–	–
K5/164	–15	10.1	1070	5020	100	–	103,000	186,000	11	50	0.554	4.9	27.8	–	–	–
K5/165	–10	2.8	240	380	40	–	150,000	218,000	6	10	0.688	–	–	–	–	–
K5/166	–4	17.4	540	6330	80	–	113,000	211,000	7	79	0.536	4.6	29.8	–	–	–
Zabit Formation																
K5/167	15	5.5	1150	1990	140	–	109,000	196,000	8	14	0.556	–0.6	25.4	–	–	–
K5/168	30	5.3	930	1280	210	–	115,000	221,000	4.4	6	0.520	–0.1	26.3	–	–	–
K5/169	60	5.5	270	520	50	–	113,000	206,000	5.4	10	0.549	–2.1	25.0	–	–	–
K5/170	90	2.3	450	450	60	–	116,000	209,000	8	8	0.555	–2.0	24.7	–	–	–
K5/171	120	4.2	280	400	50	–	122,000	219,000	6	8	0.557	–2.4	26.2	–	–	–
K5/172	170	1.9	660	530	30	–	121,000	215,000	22	18	0.563	0.2	28.4	–	–	–
K5/173	200	29.3	100	750	30	–	119,000	213,000	3.3	25	0.559	–0.7	27.4	–	–	–
K5/174	230	3.8	130	770	30	–	124,000	224,000	4.3	26	0.554	0.3	30.1	–	–	–
K5/179	430	2.1	160	480	170	0.08	30,500	318,000	0.94	2.8	0.096	–2.9	22.4	0.00150	0.70819	0.70817
K5/180	470	5.4	460	430	20	–	125,000	220,000	23	22	0.568	1.6	30.9	–	–	–
K5/181	510	3.0	90	660	40	–	122,000	218,000	2.3	17	0.560	–0.3	28.8	–	–	–
K5/182	780	19.1	410	8730	50	–	122,000	219,000	8.2	175	0.557	0.2	28.7	–	–	–
K5/183	795	4.4	100	6360	60	–	120,000	221,000	1.7	106	0.543	0.3	28.3	–	–	–

Continued

Table 1 Rb–Sr and Geochemical Characteristics of the Ediacaran–Cambrian Carbonate Rocks of the Tuva–Mongolian Continental Block—cont'd

Sample	Position of Sample *	SA†, %	Content, ppm						Mn/Sr	Fe/Sr	Mg/Ca	$\delta^{13}C$ vPDB, ‰	$\delta^{18}O$ vSMOW, ‰	$^{87}Rb/^{86}Sr$	$^{87}Sr/^{86}Sr$ Measured	$^{87}Sr/^{86}Sr$ Initial
			Mn	Fe	Sr	Rb‡	Mg	Ca								
Tabinzurta Formation																
K5/189	850	1.2	160	1500	52	–	116,000	233,000	3.1	29	0.498	–0.3	28.1	–	–	–
K5/191	870	2.3	10	470	40	–	123,000	218,000	0.25	12	0.564	–0.8	24.7	–	–	–
K5/194	1000	16.2	30	1000	80	–	122,000	203,000	0.35	13	0.601	–3.5	21.0	–	–	–
K5/196	1025	60.1	40	6400	150	–	113,000	211,000	0.25	43	0.536	–4.0	25.5	–	–	–
K5/197	1100	4.4	10	380	82	0.10	120,000	228,000	0.16	4.6	0.526	–2.0	26.0	0.00334	0.70825	0.70822
K5/198	1150	2.1	10	370	86	0.04	119,000	235,000	0.10	4.3	0.506	–1.6	25.8	0.00103	0.70821	0.70820
K5/200	1230	98.9	550	22,560	120	–	102,000	212,000	4.6	188	0.481	–	–	–	–	–
K5/201	1300	3.6	15	640	100	–	118,000	215,000	0.150	6	0.549	–2.8	25.7	–	–	–
K5/202	1350	1.0	10	600	76	–	110,000	210,000	0.132	8	0.524	–1.9	26.3	–	–	–
K5/203	1390	6.8	30	770	62	–	119,000	229,000	0.47	12	0.520	–	–	–	–	–
K5/204	1400	8.7	40	950	66	–	121,000	211,000	0.62	14	0.573	–	–	–	–	–
K5/205	1420	18.8	30	1800	68	–	123,000	209,000	0.44	26	0.589	–	–	–	–	–
K5/206	1450	2.5	25	1000	93	–	113,000	194,000	0.280	11	0.582	–	–	–	–	–
K5/154	1750	13.2	10	600	80	–	121,000	224,000	0.125	7.5	0.540	–	–	–	–	–
Khuzhirtai Formation																
K5/155	1760	3.6	10	390	70	–	122,000	216,000	0.143	5.6	0.565	–	–	–	–	–
K5/156	1770	8.2	20	410	90	0.02	122,000	2,159,000	0.222	4.6	0.057	–	–	0.00079	0.70848	0.70847
K5/157	1780	3.3	10	300	90	0.03	116,000	208,000	0.111	3.3	0.558	–	–	0.00090	0.70845	0.70844
K5/209	1970	1.3	25	1100	1200	–	112,000	197,000	0.020	0.92	0.569	–0.4	25.0	–	–	–
K5/212	2000	0.7	20	1400	240	0.06	107,000	213,000	0.092	5.8	0.502	–0.1	24.6	0.00085	0.70849	0.70848
Nurgata Formation																
K5/216	2400	0.5	35	540	130	0.09	106,000	225,000	0.262	4.2	0.471	–	–	0.00194	0.70854	0.70852
K5/219	2420	0.9	25	900	230	0.10	99,000	243,000	0.113	3.9	0.407	–0.9	25.4	0.00143	0.70855	0.70854
K5/150	2500	2.4	20	380	140	0.06	52,000	314,000	0.143	2.7	0.166	–	–	0.00113	0.70859	0.70858
K5/151	2520	2.6	30	80	80	0.10	122,000	222,000	0.375	1.0	0.550	–	–	0.00351	0.70833	0.70830
K5/152	2570	2.7	30	770	120	–	55,000	298,000	0.250	6.4	0.185	–	–	–	–	–
K5/153	2600	2.3	10	180	550	0.05	11,000	365,000	0.018	0.33	0.030	–	–	0.00030	0.70873	0.70873

Table 1 Rb–Sr and Geochemical Characteristics of the Ediacaran–Cambrian Carbonate Rocks of the Tuva–Mongolian Continental Block—cont'd

Sample	Position of Sample *	SA†, %	Content, ppm						Mn/Sr	Fe/Sr	Mg/Ca	$\delta^{13}C$ vPDB, ‰	$\delta^{18}O$ vSMOW, ‰	$^{87}Rb/^{86}Sr$	$^{87}Sr/^{86}Sr$ Measured	$^{87}Sr/^{86}Sr$ Initial
			Mn	Fe	Sr	Rb‡	Mg	Ca								
Zabit Formation. Ukhagol River																
SB4-1	–22	4.8	420	380	180	–	–	–	2.3	2.1	0.080	5.7	15.2	–	–	–
SB4-2	–20	3.4	360	240	180	0.29	–	–	2.0	1.3	0.060	5.9	20.8	0.00443	0.70756	0.70752
SB4-4	2	10.1	170	3400	270	–	–	–	0.63	13	0.450	3.3	28.2	–	–	–
SB4-5	5	27	480	17,800	300	–	–	–	1.6	59	0.540	−1.8	23.1	–	–	–
SB4-6	10	17.2	150	810	130	0.12	–	–	1.2	6	0.650	0.2	26.6	0.00229	0.70735	0.70733
SB4-7	15	38.5	130	600	180	0.14	–	–	0.72	3.3	0.640	2.4	28.7	0.00237	0.70727	0.70725
SB4-8	120	28.4	70	3400	2700	0.13	–	–	0.026	1.3	0.040	4.4	23.5	0.00012	0.70755	0.70755
SB4-9	130	22.2	110	4300	1400	–	–	–	0.079	3.1	0.170	7.3	25.6	–	–	–
SB4-10	135	18.8	100	4400	1900	0.02	–	–	0.053	2.3	0.140	7.7	25.5	0.00003	0.70732	0.70732
SB4-11	140	9.7	90	4100	1800	0.02	–	–	0.050	2.3	0.170	7.7	25.3	0.00003	0.70733	0.70733
SB4-12	145	7.4	20	30	1000	–	–	–	0.020	0.03	0.020	6.9	24.2	–	–	–
SB4-13	150	8.1	50	120	490	0.03	–	–	0.102	0.24	0.030	–5.2	23.1	0.00017	0.70728	0.70728
SB4-14	160	16.5	450	1100	90	–	–	–	5.0	12	0.630	0.2	25.4	–	–	–
SB4-15	162	22.8	2900	960	60	–	–	–	48	16	0.610	0.1	26.2	–	–	–
SB4-16	175	22.3	1600	740	60	0.43	–	–	27	12	0.630	0	26.2	0.01506	0.70850	0.70837
SB4-17	178	24.8	1400	950	30	–	–	–	47	32	0.650	0.3	25.2	–	–	–
SB4-18	185	32.1	1700	930	80	–	–	–	21	12	0.610	0.3	25.1	–	–	–
SB4-19	190	19.9	1100	590	70	–	–	–	16	8	0.620	1.1	25.7	–	–	–
SB4-20	195	17.2	340	810	80	0.28	–	–	4.3	10	0.640	3.7	25.6	0.00841	0.70851	0.70844
SB4-21	210	3.1	60	310	110	0.10	–	–	0.55	2.8	0.550	2.5	26	0.00197	0.70786	0.70784
SB4-22	235	2.3	140	300	110	–	–	–	1.3	2.7	0.510	6.8	24.8	–	–	–
SB4-23	240	4.3	190	330	110	–	–	–	1.7	3.0	0.630	2.3	24.7	–	–	–
SB4-24	245	5	170	150	90	–	–	–	1.9	1.7	0.660	–0.9	28.1	–	–	–
SB4-28	400	0.5	30	440	70	0.01	–	–	0.43	6	0.590	1.9	25.1	0.00071	0.70836	0.70836
SB4-29	450	1.8	20	330	70	–	–	–	0.29	4.7	0.630	3	27.1	–	–	–
SB4-30	455	0.8	30	590	80	–	–	–	0.38	7	0.010	2.4	26.5	–	–	–
SB4-31	475	0.6	30	320	70	–	–	–	0.43	4.6	0.640	2.7	25.1	–	–	–
SB4-32	485	0.6	30	400	70	–	–	–	0.43	5.7	0.630	2.2	25.1	–	–	–
SB4-33	495	0.2	30	310	60	0.01	–	–	0.50	5.2	0.670	2.3	28.8	0.00075	0.70817	0.70817
SB4-34	505	2.5	70	880	100	–	–	–	0.70	9	0.590	4.1	24.6	–	–	–
SB4-35	515	2.1	180	2300	90	–	–	–	2.0	26	0.490	2.4	22.6	–	–	–
SB4-36	516	1.2	110	900	90	0.06	–	–	1.2	10	0.580	2.9	24.5	0.00188	0.70852	0.70851

Continued

Table 1 Rb–Sr and Geochemical Characteristics of the Ediacaran–Cambrian Carbonate Rocks of the Tuva–Mongolian Continental Block—cont'd

Sample	Position of Sample *	SA†, %	Content, ppm						Mn/Sr	Fe/Sr	Mg/Ca	$\delta^{13}C$ vPDB, ‰	$\delta^{18}O$ vSMOW, ‰	$^{87}Rb/^{86}Sr$	$^{87}Sr/^{86}Sr$ Measured	$^{87}Sr/^{86}Sr$ Initial
			Mn	Fe	Sr	Rb‡	Mg	Ca								
SB4-37	530	0.9	40	2600	80	–	–	–	0.50	33	0.600	3.50	26.2	–	–	–
SB4-38	535	1	40	2200	70	0.01	–	–	0.57	31	0.560	3.50	26.3	0.00062	0.70846	0.70845
SB4-39	595	3.9	210	4600	80	–	–	–	2.6	58	0.740	1.40	25.2	–	–	–
SB4-40	600	26	430	8400	80	–	–	–	5.4	105	0.640	0.60	24.3	–	–	–
SB4-44	605	1.1	1300	2200	90	–	–	–	14	24	0.750	0.10	27.1	–	–	–
SB4-46	615	12.5	190	6100	70	–	–	–	2.7	87	0.820	2.00	28.7	–	–	–
Hovsgol Group. Burenkhan Deposit																
Section No 1																
MB5-1	5	3.2	60	140	230	–	1600	402,000	0.261	0.61	0.004	0.63	20.55	–	–	–
MB5-2	10	12.7	15	120	200	0.11	1300	412,000	0.075	0.60	0.003	1.75	19.59	0.0016	0.70826	0.70825
MB5-3	15	10.6	30	20	170	0.02	1200	408,000	0.176	0.12	0.003	0.41	21.64	0.0002	0.70829	0.70829
MB5-4	25	1.0	170	50	270	–	1100	331,000	0.630	0.19	0.003	0.73	20.10	–	–	–
Section No 2																
MB5-5	5	2.4	10	130	100	0.03	101,000	227,000	0.100	1.3	0.445	−1.07	23.92	0.0007	0.70839	0.70838
MB5-6	15	10.4	10	160	40	–	124,600	204,000	0.250	4.0	0.611	3.20	25.94	–	–	–
MB5-7	30	12.9	10	260	50	–	108,300	225,000	0.200	5.2	0.481	3.17	27.07	–	–	–
MB5-8	45	17.0	10	50	200	–	1800	395,000	0.050	0.25	0.005	−0.93	20.12	–	–	–
MB5-9	60	17.0	10	60	290	0.08	1500	419,000	0.034	0.21	0.004	−2.43	17.06	0.0007	0.70839	0.70838
Section No 3																
MB5-10	10	20.2	10	80	260	–	1300	392,000	0.038	0.31	0.003	−1.19	16.00			
MB5-11	20	29.9	10	50	320	0.06	1600	399,000	0.031	0.16	0.004	−1.44	17.59	0.0005	0.70820	0.70820
MB5-12	30	30.9	10	130	340	0.05	1900	416,000	0.029	0.38	0.005	−2.65	21.47	0.0004	0.70824	0.70824
MB5-13	35	30.7	10	100	300	–	1900	406,000	0.033	0.33	0.005	−2.75	20.23	–	–	–
MB5-14	40	21.4	10	70	90	0.06	1300	379,000	0.111	0.78	0.003	−2.88	20.06	0.0009	0.70822	0.70821
MB5-16	80	25.8	10	60	250	–	64,500	306,000	0.040	0.24	0.211	−1.47	20.03	–	–	–

Table 1 Rb–Sr and Geochemical Characteristics of the Ediacaran–Cambrian Carbonate Rocks of the Tuva–Mongolian Continental Block—cont'd

Sample	Position of Sample *	SA†, %	Content, ppm						Mn/Sr	Fe/Sr	Mg/Ca	$\delta^{13}C$ vPDB, ‰	$\delta^{18}O$ vSMOW, ‰	$^{87}Rb/^{86}Sr$	$^{87}Sr/^{86}Sr$ Measured	$^{87}Sr/^{86}Sr$ Initial
			Mn	Fe	Sr	Rb‡	Mg	Ca								
Section No 4																
MB5-18	10	29.2	5	50	510	0.05	1800	399,000	0.010	0.10	0.005	0.68	16.90	0.0003	0.70820	0.70820
MB5-19	20	3.0	10	40	300	–	3100	409,000	0.033	0.13	0.008	1.27	18.26	–	–	–
MB5-20	30	2.1	5	40	1550	0.07	11,100	372,000	0.003	0.03	0.030	–0.61	17.64	0.0002	0.70812	0.70812
MB5-21	40	3.0	20	100	80	–	108,300	203,000	0.250	1.3	0.533	–1.32	25.87	–	–	–
MB5-22	45	6.0	5	210	90	–	96,600	230,000	0.056	2.3	0.420	–1.59	29.20	–	–	–
Section No 5																
MB5-30	10	25.4	10	270	20	–	116,300	173,000	0.500	13.5	0.672	1.66	25.83	–	–	–
MB5-31	30	2.9	15	100	20	0.01	102,100	196,000	0.750	5.0	0.521	1.21	28.90	0.0005	0.70821	0.70821
MB5-32	40	3.7	10	80	20	–	98,300	210,000	0.500	4.0	0.468	1.25	26.53	–	–	–
MB5-33	45	2.0	5	150	30	–	99,600	209,000	0.167	5.0	0.477	2.34	27.91	–	–	–
MB5-34	50	1.9	5	100	60	0.04	110,500	244,000	0.083	1.7	0.453	2.66	28.57	0.0004	0.70802	0.70802
MB5-35	55	1.2	15	100	40	–	104,500	211,000	0.375	2.5	0.495	2.68	29.00	–	–	–
MB5-36	60	1.5	5	80	30	0.01	113,000	229,000	0.058	2.7	0.493	2.97	28.94	0.0006	0.70800	0.70800
MB5-37	70	10.2	5	110	30	–	125,800	244,000	0.167	3.7	0.516	2.96	29.28	–	–	–
Hovsgol Group. Hovsgol Deposit																
M9-1/1	0	7.8	30	70	30	–	122,000	212,000	1.0	2.3	0.575	1.7	28.4	–	–	–
M9-2	5	5.7	40	140	130	0.037	114,000	221,000	0.31	1.1	0.516	1.1	19.6	0.00100	0.70839	0.70838
M9-3	10	9.3	60	150	60	–	118,000	204,000	1.0	2.5	0.578	2.7	24.7	–	–	–
M9-3/1	10	10.3	60	160	60	–	118,000	206,000	1.0	2.7	0.573	2.9	25.9	–	–	–
M9-4/1	22	10.3	50	140	50	–	133,000	247,000	1.0	2.8	0.538	–	–	–	–	–
M9-4	22	11.3	50	280	60	–	133,000	251,000	0.83	4.7	0.530	–	–	–	–	–
M9-5	37	7.4	40	230	60	0.040	131,000	248,000	0.67	3.8	0.528	2.5	21.4	0.00216	0.70829	0.70828
M9-6	47	20.8	100	600	70	–	138,000	258,000	1.4	8.6	0.535	3	23.6	–	–	–
M9-7/1	67	28.0	1300	12,800	520	–	38,800	376,000	2.5	25	0.103	2.6	17.4	–	–	–
M9-8	77	45.5	2300	28,900	120	–	112,000	248,000	19	241	0.452	2.6	28.2	–	–	–
M9-9	127	13.8	720	240	130	–	99,000	216,000	5.5	1.8	0.458	–2.4	28.2	–	–	–
M9-10	132	62.6	370	3200	100	–	128,000	264,000	3.7	32	0.485	–4.7	26.9	–	–	–
M9-11	152	12.8	30	200	1400	0.100	5800	351,000	0.021	0.14	0.017	1.8	21.7	0.00016	0.70862	0.70862
M9-12/1	177	14.1	1200	2500	70	–	135,000	249,000	17	36	0.542	0.8	26.1	–	–	–
M9-13	180	3.7	470	1300	60	–	133,000	245,000	8	22	0.543	–0.4	25.9	–	–	–
M9-15	195	3.7	70	240	50	–	131,000	238,000	1.4	4.8	0.550	–0.1	26	–	–	–

Continued

Table 1 Rb–Sr and Geochemical Characteristics of the Ediacaran–Cambrian Carbonate Rocks of the Tuva–Mongolian Continental Block—cont'd

Sample	Position of Sample *	SA†, %	Content, ppm						Mn/Sr	Fe/Sr	Mg/Ca	δ13C vPDB, ‰	δ18O vSMOW, ‰	87Rb/86Sr	87Sr/86Sr Measured	87Sr/86Sr Initial
			Mn	Fe	Sr	Rb‡	Mg	Ca								
M9-16	245	8.4	110	600	50	–	121,000	221,000	2.2	12	0.548	0.4	28	–	–	–
M9-17	285	6.3	90	590	50	0.041	120,000	223,000	1.8	12	0.538	0.4	26.3	0.00203	0.70817	0.70816
M9-18	325	10.3	40	160	50	–	118,000	216,000	0.80	3.2	0.546	0.1	26.1	–	–	–
M9-19	425	3.3	160	600	50	–	105,000	192,000	3.2	12	0.547	0.1	25.4	–	–	–
M9-20	465	22.8	730	3200	90	–	91,000	187,000	8	36	0.487	−0.2	25.6	–	–	–
M9-21	495	15.0	300	2800	80	–	106,500	195,000	3.8	35	0.546	−2.3	26	–	–	–
M9-22	515	14.2	400	2800	100	–	128,000	232,000	4.0	28	0.552	−2.4	26.3	–	–	–
M9-23	575	6.2	100	500	80	–	133,000	236,000	1.3	6.3	0.564	−2.4	26.1	–	–	–
M9-24	675	7.3	40	510	90	0.098	139,000	250,000	0.44	5.7	0.556	−0.7	24.7	0.00292	0.70828	0.70826
M9-25	875	7.8	30	420	80	0.140	134,000	247,000	0.38	5.3	0.543	−2.1	25.7	0.00507	0.70823	0.70819
M9-26	879	98.2	40	5600	5	–	1700	2800	8	1120	0.607	–	–	–	–	–
M9-27	882	9.4	40	170	70	0.105	133,000	237,000	0.57	2.4	0.561	−2.3	26.6	0.00423	0.70825	0.70822
M9-29	884	18.6	170	3500	260	–	46,000	355,000	0.65	13	0.130	−2.4	25.1	–	–	–
M9-30	889	7.5	20	290	70	–	117,000	211,000	0.29	4.1	0.555	−1.9	25.9	–	–	–
M9-31	893	79.6	280	13,000	140	–	127,800	249,000	2.0	93	0.513	−2.3	25.5	–	–	–
M9-32	908	4.9	10	60	140	–	22,100	395,000	0.071	0.43	0.056	−2.5	25.8	–	–	–
M9-33	913	83.7	160	17,000	250	–	33,000	391,000	0.640	68	0.084	−2.9	21.6	–	–	–
M9-34	918	4.8	10	50	180	0.092	29,000	342,000	0.056	0.28	0.085	−3.3	23.2	0.00134	0.70823	0.70822
M9-35	1018	6.6	20	340	240	–	68,000	325,000	0.083	1.4	0.209	−2.4	21.9	–	–	–
M9-36	1031	70.7	20	2200	60	–	30	71,000	0.333	37	0.000	−1.6	24.1	–	–	–
M9-37	1028	3.6	10	140	210	0.075	48,000	349,000	0.048	0.67	0.138	−3.2	22.6	0.00109	0.70830	0.70829
M9-38	1030	11.0	20	240	170	–	124,000	257,000	0.118	1.4	0.482	−2.5	25.7	–	–	–
M9-40	1032	6.3	20	200	240	–	114,000	247,000	0.083	0.83	0.462	−2.7	25.4	–	–	–
M9-41	1034	3.4	20	140	160	–	104,000	292,000	0.125	0.88	0.356	−2.6	23.3	–	–	–
M9-42	1134	2.8	10	90	200	0.082	53,000	342,000	0.050	0.45	0.155	−3.2	20.9	0.00119	0.70825	0.70824
M9-43	1164	3.3	10	120	190	–	6000	390,000	0.053	0.63	0.015	−2.7	21.9	–	–	–
M9-44	1204	2.8	10	90	220	–	43,000	324,000	0.045	0.41	0.133	−2.9	23.4	–	–	–
M9-45	1254	5.0	10	60	340	–	5900	414,000	0.029	0.18	0.014	−2.9	23.6	0.00037	0.70833	0.70833
M9-46	1304	6.7	10	70	570	0.073	7000	413,000	0.018	0.12	0.017	1.7	11.7	–	–	–
M9-47	1354	3.3	10	70	320	–	19,000	369,000	0.031	0.22	0.051	−0.7	21.6	–	–	–

Table 1 Rb–Sr and Geochemical Characteristics of the Ediacaran–Cambrian Carbonate Rocks of the Tuva–Mongolian Continental Block—cont'd

Sample	Position of Sample *	SA†, %	Content, ppm						Mn/Sr	Fe/Sr	Mg/Ca	$\delta^{13}C$ vPDB, ‰	$\delta^{18}O$ vSMOW, ‰	$^{87}Rb/^{86}Sr$	$^{87}Sr/^{86}Sr$ Measured	$^{87}Sr/^{86}Sr$ Initial
			Mn	Fe	Sr	Rb‡	Mg	Ca								
M9-48	1404	6.9	10	180	680	–	5800	403,000	0.015	0.26	0.014	1.8	24.2	–	–	–
M9-49	1424	5.4	10	90	530	–	35,900	370,000	0.019	0.17	0.097	1.8	24.3	–	–	–
M9-51	1454	6.9	10	40	1800	0.117	23,000	363,000	0.006	0.02	0.063	0.3	24.6	0.00017	0.70829	0.70829
M9-53	1504	5.5	10	100	360	–	64,500	322,000	0.028	0.28	0.200	0	24.9	–	–	–
M9-54	1594	2.3	10	40	490	0.281	3200	425,000	0.020	0.08	0.008	−1.9	22.4	0.00157	0.70830	0.70829
M9-55	1674	0.5	5	40	310	–	21,000	356,000	0.016	0.13	0.059	−0.7	21.4	–	–	–
M9-56	1734	5.6	10	90	260	–	33,000	362,000	0.038	0.35	0.091	−0.7	21.7	–	–	–
M9-57	1794	5.0	10	80	270	–	31,000	354,000	0.037	0.30	0.088	−0.7	24	–	–	–
M9-58	1844	0.3	5	30	290	–	6900	381,000	0.017	0.10	0.018	−1.7	24.2	–	–	–
M9-59	1924	3.4	10	40	260	0.068	31,000	361,000	0.038	0.15	0.086	−0.8	22.1	0.00074	0.70837	0.70837
M9-60	2024	0.3	5	20	250	–	4900	346,000	0.020	0.08	0.014	−1.1	23.4	–	–	–
M9-61	2074	3.7	10	50	320	–	2600	405,000	0.031	0.16	0.006	−1.1	16.3	–	–	–
Muren Formation																
Section I																
MM5-31	10	6.4	5	40	2200	–	470	376,000	0.002	0.018	0.001	5.60	23.23	–	–	–
MM5-32	20	10.5	10	400	2400	0.12	320	374,000	0.004	0.167	0.001	6.07	24.41	0.0002	0.70785	0.70785
MM5-33	30	5.8	10	10	2100	–	230	386,000	0.005	0.005	0.001	6.20	24.65	–	–	–
MM5-35	50	8.5	10	20	2100	–	310	383,000	0.005	0.010	0.001	6.03	24.83	–	–	–
MM5-36	60	3.7	5	20	1900	–	430	378,000	0.003	0.011	0.001	5.59	24.05	–	–	–
MM5-37	70	3.2	5	20	2200	0.12	270	367,000	0.002	0.009	0.001	6.26	25.57	0.0002	0.70786	0.70786
MM5-38	80	2.4	5	40	2400	–	310	362,000	0.002	0.017	0.001	5.32	23.45	–	–	–
MM5-39	90	30.8	10	70	2200	0.06	290	394,000	0.005	0.032	0.001	6.23	26.46	0.0001	0.70784	0.70784
MM5-40	102	41.5	40	100	1800	–	29,700	311,000	0.022	0.056	0.095	5.80	25.78	–	–	–
MM5-41	110	4.8	10	0	2300	0.04	370	362,000	0.004	0.000	0.001	6.30	24.73	0.0001	0.70784	0.70784
MM5-42	120	15.3	10	60	2100	0.03	200	347,000	0.005	0.029	0.001	6.30	26.22	0.0001	0.70783	0.70783
MM5-43	130	16.8	5	30	2100	0.09	260	319,000	0.002	0.014	0.001	6.44	26.53	0.0001	0.70783	0.70783
MM5-44	140	40.7	10	270	2300	–	170	366,000	0.004	0.117	0.0005	6.24	26.33	–	–	–
MM5-45	150	4.7	5	70	2100	0.04	230	359,000	0.002	0.033	0.001	6.28	24.54	0.0001	0.70780	0.70780
MM5-46	160	13.3	100	30	1800	0.06	270	355,000	0.056	0.017	0.001	5.79	24.99	0.0001	0.70779	0.70779
MM5-47	170	17.9	5	30	1800	–	250	372,000	0.003	0.017	0.001	5.11	25.16	–	–	–
MM5-48	180	2.9	5	100	2100	0.07	640	409,000	0.002	0.048	0.002	5.85	25.79	0.0001	0.70780	0.70780
MM5-49	190	9.5	10	20	1800	–	490	345,000	0.006	0.011	0.001	5.85	26.05	–	–	–

Continued

Table 1 Rb–Sr and Geochemical Characteristics of the Ediacaran–Cambrian Carbonate Rocks of the Tuva–Mongolian Continental Block—cont'd

Sample	Position of Sample *	SA†, %	Content, ppm						Mn/Sr	Fe/Sr	Mg/Ca	δ13C vPDB, ‰	δ18O vSMOW, ‰	$^{87}Rb/^{86}Sr$	$^{87}Sr/^{86}Sr$ Measured	$^{87}Sr/^{86}Sr$ Initial
			Mn	Fe	Sr	Rb‡	Mg	Ca								
Section II (Agariingol)																
MA-2	10	0.2	80	1900	2700	0.18	2200	328,000	0.030	0.704	0.007	6.99	21.48	0.0002	0.70742	0.70742
MA-4	300	4.5	40	820	2200	0.16	1200	345,000	0.018	0.373	0.003	8.96	23.87	0.0002	0.70743	0.70743
MA-13	600	4.2	60	1120	1900	0.26	2000	346,000	0.032	0.589	0.006	10.54	22.00	0.0004	0.70725	0.70725
MA-14	760	2.8	5	30	2200	0.14	500	366,000	0.002	0.014	0.001	3.80	22.23	0.0002	0.70829	0.70829
MA-15	780	25.6	10	70	2100	0.19	370	244,000	0.005	0.033	0.002	2.52	22.93	0.0003	0.70840	0.70840
MA-16	790	0.1	40	410	2100	–	1600	376,000	0.019	0.195	0.004	10.34	20.29	–	–	–
MA-18	850	3.5	330	80	2800	0.24	2200	377,000	0.118	0.029	0.006	4.21	21.68	0.0003	0.70852	0.70852
MA-19	900	88.7	300	24,200	2200	–	3100	343,000	0.136	11.0	0.009	4.29	21.88	–	–	–
MA-20	950	11.0	50	1060	2100	0.11	2800	361,000	0.024	0.505	0.008	8.06	22.34	0.0001	0.70735	0.70735
Section III																
MM5-1	20	5.8	5	30	2300	0.06	30	378,000	0.002	0.013	0.0001			0.0001	0.70798	0.70798
MM5-2	200	6.2	5	30	2400	0.09	120	390,000	0.002	0.013	0.0003	3.49	25.48	0.0001	0.70796	0.70796
MM5-3	430	6.4	5	50	2700	–	430	381,000	0.002	0.019	0.001	2.29	23.93	–	–	–
MM5-5	442	3.8	5	30	2500	0.08	60	384,000	0.002	0.012	0.0002	2.55	24.28	0.0001	0.70801	0.70801
MM5-6	480	5.8	5	20	2400	–	110	394,000	0.002	0.008	0.0003	3.73	25.35	–	–	–
MM5-7	540	9.8	5	50	2500	0.11	70	376,000	0.002	0.020	0.0002	2.50	23.87	0.0001	0.70800	0.70800
MM5-8	580	5.8	5	10	1900	–	220	352,000	0.003	0.005	0.001	2.91	24.33	–	–	–
MM5-9	610	5.5	5	10	2200	0.04	80	363,000	0.002	0.005	0.0002	3.97	24.23	0.0001	0.70800	0.70800
MM5-10	640	5.8	5	10	2100	–	90	352,000	0.002	0.005	0.0003	2.26	24.34	–	–	–
MM5-10A	650	6.8	5	10	2400	0.06	190	372,000	0.002	0.004	0.001	3.25	24.97	0.0001	0.70803	0.70803
MM5-11	670	5.6	5	10	2200	–	540	336,000	0.002	0.005	0.002	3.76	24.79	–	–	–
MM5-12	700	5.4	5	20	2300	0.09	160	368,000	0.002	0.009	0.0004	2.64	24.06	0.0001	0.70799	0.70799

Table 1 Rb–Sr and Geochemical Characteristics of the Ediacaran–Cambrian Carbonate Rocks of the Tuva–Mongolian Continental Block—cont'd

Sample	Position of Sample *	SA†, %	Content, ppm						Mn/Sr	Fe/Sr	Mg/Ca	$\delta^{13}C$ vPDB, ‰	$\delta^{18}O$ vSMOW, ‰	$^{87}Rb/^{86}Sr$	$^{87}Sr/^{86}Sr$ Measured	$^{87}Sr/^{86}Sr$ Initial
			Mn	Fe	Sr	Rb‡	Mg	Ca								
Araoshei Formation																
8311/1A	31	2.9	15	130	475	0.0559	3700	396,000	0.39	0.05	0.009	–2.7	18.6	0.00034	0.70844	0.70843
8311/1B	43	2.9	15	860	505	0.0518	2900	405,000	2.46	0.04	0.007	–2.8	18.5	0.00030	0.70849	0.70849
8311/7B	140	0.5	5	590	300.5	0.0229	200	418,000	2.81	0.02	0.0005	0.7	24.9	0.00022	0.70853	0.70852
8313/1	214	0.7	40	130	269.9	0.0297	2900	396,000	0.72	0.22	0.007	1.5	24	0.00032	0.70866	0.70865
8314/2	281	21.5	230	4910	330	–	109,100	257,000	15	0.70	0.425	–	–	–	–	–
8314/8	324	8.5	510	3110	260	–	100,500	260,000	12	1.96	0.387	–	–	–	–	–
19/84	370	1.8	30	360	2020.4	0.3194	32,000	340,000	0.23	0.02	0.094	–0.8	24.1	0.00046	0.70873	0.70873
20a/84	413	0.6	40	590	568	0.4124	15,900	385,000	1.59	0.11	0.041	0.1	24.5	0.00210	0.70875	0.70874
21/84	465	0.5	80	510	329	0.5385	22,800	369,000	1.82	0.29	0.062	0.7	23.7	0.00473	0.70884	0.70881
25/84	526	3.1	170	3660	201	0.3114	1300	389,000	23	1.06	0.003	1.7	23.6	0.00448	0.70883	0.70879
27a/84	587	28.6	210	990	320	–	81,800	304,000	3.09	0.66	0.269	–	–	–	–	–
28/84	648	19.4	210	1730	1730	–	126,700	243,000	1.00	0.12	0.521	–0.5	22.1			
43b/83	691	2.1	250	4510	80	–	134,000	220,000	56	3.1	0.609	–	–	–	–	–
43c/83	734	3.7	380	3610	180	–	134,600	225,000	20	2.1	0.598	–	–	–	–	–
45a/83	786	0.4	10	90	748.2	0.2710	9900	385,000	0.17	0.02	0.026	4.2	25.7	0.00105	0.70873	0.70872
48d/83	838	6.0	100	5000	140	–	2000	393,000	36	0.71	0.005	–	–	–	–	–
49/83	890	3.5	30	170	244.8	0.1641	1700	395,000	0.94	0.17	0.004	1.6	23.4	0.00194	0.7086	0.70862
49a/83	951	23.7	100	1330	590	–	36,900	364,000	2.25	0.17	0.101	–	–	–	–	–
50a/83	1012	6.5	80	950	196.2	0.1847	2200	385,000	9	0.73	0.006	1.3	24.6	0.00272	0.70882	0.70879
51c/83	1082	4.4	25	460	687.1	0.1037	129,200	237,000	0.87	0.05	0.545	1.3	25.2	0.00044	0.70845	0.70845
51h/83	1135	32.8	130	1080	300	–	67,000	367,000	3.60	0.43	0.183	–	–	–	–	–
51d/83	1188	55.8	150	1350	750	–	62,300	321,000	1.80	0.20	0.194	–	–	–	–	–

Note: *Dash, not determined.*

**The position of the sample is given in meters from the section base (negative values show the distance from the mixtite bed).*

†SA, Siliciclastic admixture.

‡Rb content was measured by isotopic dilution, and the contents of the other elements, by AAS. In the calculation of the initial $^{87}Sr/^{86}Sr$ ratios, the age of the rocks was taken to be 550 Ma.

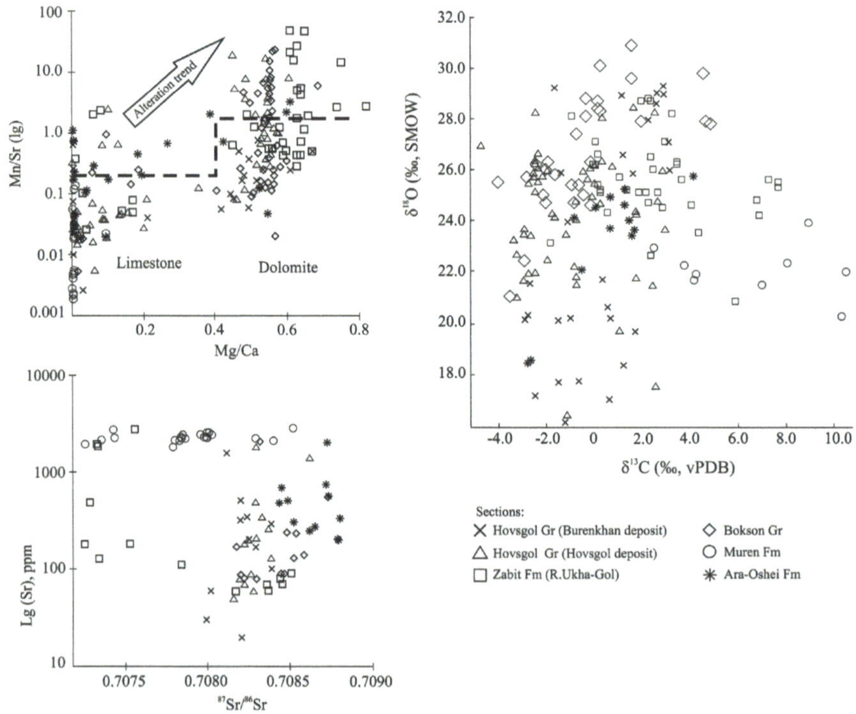

FIGURE 6

Cross-plots of Mn/Sr versus Mg/Ca, Sr contents versus $^{87}Sr/^{86}Sr$ ratio and $\delta^{18}O$ versus $\delta^{13}C$ for Ediacaran–Cambrian Tuva–Mongolian carbonate rocks. Dashed line marks the areas with the least altered carbonates (see explanation in the text).

amounts (more than 29 wt%) are found in beds situated 200 m from the bottom. Higher concentrations of Fe and Mn (1300–7600 ppm and 920–1150 ppm) are observed at this level, The Fe concentrations reach as much as 6400–8700 ppm in the layers near the top of the section. The section also contains carbonates with moderate Fe (440–770 ppm) and Mn (90–660 ppm) and low Sr varying from 30 to 210 ppm (Fig. 6, Table 1) and hence only one sample fits the geochemical criteria for Sr-isotopic chromatography and only six for C-isotopic chemostratigraphy. The $\delta^{18}O_{SMOW}$ value in the Zabit Formation rocks varies from +22‰ to +30.9‰ (Fig. 6), which goes well with sample separation criteria with unaltered isotopic system. But there is positive correlation between Fe/Sr ratio and $\delta^{18}O_{SMOW}$ values, which indicates that some early diagenetic dolomitization took place (Semikhatov et al., 1998; Xiao et al., 2004).

18.4.1.4 Tabinzurta Formation

Carbonate deposits of the Tabinzurta Formation are represented by dolomites and some limestone variations (Mg/Ca = 0.48–0.60; Table 1). Most of the samples contains low amounts of silica-clastic material (1–4 wt%). Mn contents are high in the basal horizon (160 ppm) while in other parts of the section they are quite low (10–40 ppm). Samples can be divided into two groups according to Fe contents: with average (370–640 ppm) and high (1000–6400 ppm) concentrations. Most of the samples of the

Tabinzurta Formation were not suitable for Sr-isotopic composition study. Only two samples in the middle part of the section were selected—K5/197 (Mn/Sr = 0.16, Fe/Sr = 4.6) and K5/198 (Mn/Sr = 0.10, Fe/Sr = 4.3). Higher limiting conditions were accepted for C-isotopic chemostratigraphy and hence, all the samples except two samples were selected. $\delta^{18}O_{SMOW}$ values throughout the section are more than +20‰ (Fig. 6). All these evidence about the absence of significant post sedimentary alterations.

18.4.1.5 Khuzhirtai and Nurgata Formations

According to geochemical data the sediments of these formations are represented mainly by limestone with dolomites (Mg/Ca = 0.40–0.58), and layers of dolomitic limestones (Mg/Ca = 0.06–0.18). The amount of terrigenous material varies from 0.5 to 8.2 wt% (Table 1). Mn contents are low in both the formations (20–35 ppm) and Fe concentration in Khuzhirtai carbonates is twice (1000–1400 ppm) than the Nurgata (540–900 ppm). Sr concentrations vary from 90 to 240 ppm and only one sample of Khuzhirtai Formation has higher value (1200 ppm). All the samples fit the geochemical criteria accepted for Sr-isotopic chemostratigraphy (Table 1). $\delta^{18}O_{SMOW}$ values have been measured for three samples and found varying from +24‰ to +25.4‰.

18.4.1.6 Hovsgol Group

Chemical criteria show that the exposed rocks can be divided into two, namely, lower dolomite (Mg/Ca ~ 0.55) and upper transition from dolomites to limestones (Mg/Ca 0.006–0.05). Dolomites are enriched with Fe (150–3000 ppm at average, reaching 3000 ppm in some samples). They also contain high concentrations of Mn (40–730 ppm) and low concentrations of Sr (30–100 ppm). In the limestones, Mn and Fe contents do not exceed 50–200 ppm and Sr concentrations are high—1800 ppm. $\delta^{18}O_{SMOW}$ values in all the carbonates of this formation vary from +21‰ to +26.6‰ (Table 1; Fig. 6). Due to the high amounts of Fe, only five dolomite samples from basal horizons fit the geochemical criteria. Second part of the section is homogenous. Hence, six samples of limestones were selected for Sr-isotopic composition. Carbonates of post-phosphate part of the Hovsgol Group collected from the Burenkhan deposit have low concentrations of Mn and Fe. Sr concentrations in some samples (MB-21–MB-37) are low (20–90 ppm; Table 1). All the collected samples of the Hovsgol Group were used for C-isotopic chemostratigraphy.

18.4.1.7 Muren Formation

Muren Formation contains limestones (Mg/Ca = 0.0001–0.031) (Table 1, data received from A.B. Kuznetsov). Sr concentrations in carbonate extracted from these rocks are quite high: maximal value is 2700 ppm and minimal is 1900 ppm (Table 1; Fig. 6). Mn, Fe, and Mg contents are low (1.0–40, 10–400, and 30–600 ppm). In few samples, the Fe and Mn are below the detection limit. Amount of insoluble residue represented by quartz and organic matter is low—1–4 wt%. $\delta^{18}O_{SMOW}$ values for all the Muren formation carbonate samples range from +20.2‰ to +26.5‰ and suggests little to no diagenetic transformation. Muren Formation carbonate deposits are unique in geochemical features (Table 1). Geochemical criteria values for carbonate deposits are significantly lower than generally accepted: Mg/Ca $\leq$ 0.0024, Mn/Sr $\leq$ 0.02, and Fe/Sr $\leq$ 0.2. Only five samples from the whole collection do not fit the criteria.

18.4.1.8 Araoshei Formation

The Araoshei Formation sediments were collected from three outcrops located along the Araoshei River. Sediments from all the three sections have similar chemical composition. Rocks here are represented generally by limestones (Mg/Ca ~ 0.01) with rare dolomite layers. The amount of terrigenous material is quite low (<4.5%). Fe and Mn concentrations vary widely (from 90 to 5000 ppm and 5 to 510 ppm, respectively) and show no correlation with the amount of terrigenous material. Sr concentrations also vary widely from 80 to 1540 ppm (Table 1; Fig. 6). Distinctive feature of Araoshei Formation limestone is the amount of Rb. Maximum Rb concentrations (0.53 ppm) are found in samples collected from an exposure located at the upper course of the Araoshei River. Samples from middle course of the Araoshei River have average concentrations of Rb (0.10–0.27 ppm), still high enough for normal marine carbonate rocks. Higher concentration of Rb is probably related to the increased role of basalt material in source areas. Based on the geochemical criteria, nine limestones (with Fe/Sr and Mn/Sr ratios 3.0 and 0.2) were selected for mass spectrometry studies. Additionally, three limestones which do not fit the geochemical criteria were selected for quality control measurements.

18.4.2 CARBONATE ROCKS FROM THE DZABKHAN BLOCK COVER

During the field work, two sections, represented by Ediacaran–Cambrian deposits of the Dzabkhan block (Tsagaan Oloom and Bayan Gol formations, Figs 4 and 5) have been studied. The first section (hereinafter referred as Section N1) is situated along the shore outcrops of Tsagaan Oloom River bank and the second section (hereinafter referred as Section N2) is situated along the Bayan-Gol River bank. Section N1 contains complete sequence of the Tsagaan Oloom Formation. The section N2 exposes the Bayan Gol Formation.

18.4.2.1 Tsagaan Oloom Formation (section 1)

According to geochemical characteristics, the lower part of the section N1 (up to 400 m) is represented by limestones (Mg/Ca < 0.03) that contain low Mn (1–10 ppm), and Fe concentrations (30–80 ppm). Sr concentrations vary from 470 to 2300 ppm. As many as 31 samples fit the geochemical criteria (Mg/Ca ≤ 0.024, Mn/Sr ≤ 0.2, Fe/Sr ≤ 5.0) for further analyses. The upper part of the section shows the occurrences of dolomites (Mg/Ca ~ 0.57) and dolomitic limestones (Mg/Ca 0.05–0.2). They have higher (compared to limestones from the bottom part of the section) Fe and Mn concentrations (from 190 to 5900 ppm and from 20 to 5400 ppm; Table 2; Fig. 7). Sr concentrations vary from 30 to 2000 ppm with the lowest values in sediments overlying the limestones (30–60 ppm). Thus, samples from the upper part were not selected for isotopic analyses.

18.4.2.2 Bayan Gol Formation (section 1)

Carbonates of the Bayan Gol Formation are represented by dolomite limestones (Mg/Ca 0.05–0.2) and limestones (Mg/Ca < 0.03) with higher concentrations of Mn (40–980 ppm) and Fe (490–2800 ppm). Sr concentrations vary from 170 to 870 ppm except for one sample (M9-129 where it is 1500 ppm). Hence, no sample has been selected from this formation at this section.

18.4.2.3 Tsagaan Oloom Formation (section 2)

The lower part of the section (up to 800 m) is represented by dolomites and dolomitic limestones (Mg/Ca from 0.45 to 0.61). These have moderate Mn concentrations (20–170 ppm), high Fe concentrations (70–990 ppm) and low Sr concentrations (30–140 ppm) with the exception of sample M9-145 (Sr concentration 2100 ppm). The data show that these rocks underwent significant diagenetic transformations than the rock of the Section N1. Upper part of the section is also represented by

Table 2 Rb–Sr and Geochemical Characteristics of the Ediacaran–Cambrian Carbonate Rocks of the Dzabkhan Continental Block

Sample	Position of Sample*	SA†, %	Content, ppm						Mn/Sr	Fe/Sr	Mg/Ca	$\delta^{13}C$ vPDB, ‰	$\delta^{18}O$ vSMOW, ‰	$^{87}Rb/^{86}Sr$	$^{87}Sr/^{86}Sr$ Measured	$^{87}Sr/^{86}Sr$ Initial
			Mn	Fe	Sr	Rb‡	Mg	Ca								
Section 1																
Tsagaan Oloom Formation																
M9-74	11	15.9	120	430	680	–	11,000	361,000	0.18	0.63	0.030			–	–	–
M9-75	12	1.5	80	550	1200	–	3000	396,000	0.07	0.46	0.008			–	–	–
M9-78	71	3.4	150	80	1300	0.095	3700	415,000	0.12	0.06	0.009	5.5	23.8	0.00017	0.70715	0.70715
M9-79	94	0.2	1	40	710	–	2100	408,000	0.002	0.06	0.005			–	–	–
M9-80	94	4.8	2	40	670	0.051	2800	410,000	0.002	0.06	0.007			0.00020	0.70718	0.70717
M9-81	97	1.1	1	50	470	–	2400	376,000	0.002	0.11	0.006			–	–	–
M9-82	114	2.3	5	30	560	0.045	2200	426,000	0.009	0.05	0.005	4.8	25.2	0.00022	0.70722	0.70722
M9-83	151	38.5	20	720	1700	–	15,000	374,000	0.012	0.42	0.040			–	–	–
M9-84c	154	0.4	55	1400	1100	–	1600	376,000	0.05	1.3	0.004			–	–	–
M9-86	202	2.1	3	80	1500	0.104	1500	394,000	0.002	0.05	0.004	9.5	24.7	0.00017	0.70741	0.70740
M9-87	219	0.5	5	80	1200	–	1800	372,000	0.004	0.07	0.005			–	–	–
M9-88	242	2.0	10	50	1400	0.108	5100	404,000	0.007	0.04	0.013	9.7	27.8	0.00020	0.70741	0.70741
M9-89	260	1.6	5	70	2300	–	1100	378,000	0.002	0.03	0.003			–	–	–
M9-90	279	0.6	5	50	1000	0.123	1500	406,000	0.005	0.05	0.004	9.9	26.9	0.00029	0.70736	0.70735
M9-91	301	1.0	10	50	1100	–	1100	404,000	0.009	0.05	0.003			–	–	–
M9-92	310	5.2	10	140	1000	0.040	1800	391,000	0.010	0.14	0.005			0.00009	0.70734	0.70733
M9-93	322	0.3	5	40	1600	–	1200	392,000	0.003	0.03	0.003			–	–	–
M9-94	335	0.6	5	50	1600	0.020	1400	393,000	0.003	0.03	0.004	11.9	26.2	0.00003	0.70729	0.70729
M9-95	373	0.5	10	60	1200	–	1100	375,000	0.008	0.05	0.003			–	–	–
M9-96	390	3.0	500	630	60	–	124,000	230,000	8.3	11	0.54			–	–	–
M9-97	391	10.3	200	1600	60	–	119,000	199,000	3.3	27	0.60			–	–	–
M9-98	422	18.9	80	640	50	–	114,000	199,000	1.6	13	0.57			–	–	–
M9-99	424	8.2	30	200	40	–	123,000	207,000	0.75	5.0	0.59	3.9	28.2	–	–	–
M9-100	425	1.1	60	190	30	–	123,000	209,000	2.0	6.3	0.59			–	–	–
M9-101	517	1.4	20	230	40	–	125,000	207,000	0.50	5.8	0.60	3.9	26.5	–	–	–
M9-102	520	1.3	20	230	40	–	120,000	216,000	0.50	5.8	0.556			–	–	–
M9-103	564	18.2	180	480	360	–	103,000	227,000	0.50	1.3	0.45			–	–	–

Continued

Table 2 Rb–Sr and Geochemical Characteristics of the Ediacaran–Cambrian Carbonate Rocks of the Dzabkhan Continental Block—cont'd

Sample	Position of Sample*	SA†, %	Content, ppm						Mn/Sr	Fe/Sr	Mg/Ca	$\delta^{13}C$ vPDB, ‰	$\delta^{18}O$ vSMOW, ‰	$^{87}Rb/^{86}Sr$	$^{87}Sr/^{86}Sr$ Measured	$^{87}Sr/^{86}Sr$ Initial
			Mn	Fe	Sr	Rb‡	Mg	Ca								
M9-104	565	8.3	60	440	2000	–	29,000	337,000	0.03	0.22	0.09	6.0	24.5	–	–	–
M9-105	578	3.9	40	560	50	–	127,000	206,000	0.80	11	0.62			–	–	–
M9-106	591	17.6	40	530	40	–	117,000	204,000	1.0	13	0.57			–	–	–
M9-107	597	6.9	60	810	670	–	32,000	324,000	0.09	1.2	0.099			–	–	–
M9-108	606	10.5	70	530	2000	–	5600	353,000	0.04	0.27	0.016			–	–	–
M9-109	623	11.3	40	520	1600	–	7100	363,000	0.025	0.33	0.020			–	–	–
M9-110	638	10.1	60	600	1100	0.195	33,000	366,000	0.05	0.55	0.090			0.00046	0.70792	0.70791
M9-111	651	10.1	30	240	1700	–	11,000	347,000	0.018	0.14	0.032			–	–	–
M9-112	651	2.7	30	150	380	–	53,000	286,000	0.08	0.39	0.19	6.0	25.9	–	–	–
M9-113	664	11.6	160	5900	40	–	122,000	191,000	4	148	0.64			–	–	–
M9-114	684	9.3	360	4100	50	–	123,000	214,000	7	82	0.57			–	–	–
M9-115	745	2.0	450	2900	60	–	123,000	194,000	8	48	0.63			–	–	–
M9-116	772	14.6	5400	8300	160	–	114,000	195,000	34	52	0.58			–	–	–
M9-117	790	8.4	20	350	2300	–	4000	360,000	0.009	0.15	0.011	0.5	17.5	–	–	–
M9-118	811	10.7	20	550	1100	0.039	7600	344,000	0.018	0.50	0.022			0.00009	0.70848	0.70848
M9-119	835	15.8	60	400	1400	–	4100	340,000	0.04	0.29	0.012			–	–	–
M9-120	863	7.9	40	190	1600	–	2900	401,000	0.025	0.12	0.007			–	–	–
M9-121	894	11.3	20	330	100	–	112,000	204,000	0.20	3.3	0.55			–	–	–
M9-122	901	1.4	10	110	150	–	110,000	240,000	0.07	0.73	0.46	−3.0	23.9	–	–	–
M9-123	933	1.4	20	60	220	–	6500	385,000	0.09	0.27	0.017			–	–	–
M9-124	963	1.7	10	80	230	–	3600	386,000	0.04	0.35	0.009			–	–	–
Bayan Gol Formation																
M9-125	1095	4.6	40	150	170	–	1900	378,000	0.24	0.88	0.005			–	–	–
M9-126	1151	6.4	40	970	480	–	13,800	343,000	0.083	2.0	0.040			–	–	–
M9-126a	1152	3.0	580	2300	400	–	35,800	313,000	1.5	5.8	0.11			–	–	–

Table 2 Rb–Sr and Geochemical Characteristics of the Ediacaran–Cambrian Carbonate Rocks of the Dzabkhan Continental Block—cont'd

Sample	Position of Sample*	SA†, %	Content, ppm						Mn/Sr	Fe/Sr	Mg/Ca	$\delta^{13}C$ vPDB, ‰	$\delta^{18}O$ vSMOW, ‰	$^{87}Rb/^{86}Sr$	$^{87}Sr/^{86}Sr$ Measured	$^{87}Sr/^{86}Sr$ Initial
			Mn	Fe	Sr	Rb‡	Mg	Ca								
M9-127	1153	36.9	700	4800	450	–	6500	229,000	1.6	11	0.028			–	–	–
M9-128	1178	0.6	50	490	270	–	1900	387,000	0.19	1.8	0.005			–	–	–
M9-129	1232	3.7	90	1300	1500	–	4600	355,000	0.06	0.87	0.013			–	–	–
M9-130	1237	61.9	270	2300	410	–	2100	154,000	0.66	5.6	0.014			–	–	–
M9-131	1239	8.4	200	1100	460	–	3000	359,000	0.43	2.4	0.008			–	–	–
M9-132	1239	8.9	280	1100	420	–	3600	331,000	0.67	2.6	0.011			–	–	–
M9-134	1242	10.1	680	1400	680	–	3800	345,000	1.0	2.1	0.011			–	–	–
M9-135	1242	5.9	330	550	670	–	3800	369,000	0.49	0.82	0.010	3.1	16.9	–	–	–
M9-136	1243	6.9	900	3700	870	–	3600	385,000	1.0	4.3	0.009			–	–	–
M9-137	1298	8.2	560	1300	580	–	11,000	376,000	0.97	2.2	0.029	5.1	18.4	–	–	–
M9-138	1391	9.8	260	1100	320	–	2900	349,000	0.81	3.4	0.008			–	–	–
M9-139	1398	9.1	280	880	340	–	2200	360,000	0.82	2.6	0.006	4.1	18.7	–	–	–
M9-140	1433	12.5	980	2800	700	–	3100	354,000	1.4	4.0	0.009			–	–	–
M9-141	1509	1.2	3700	620	300	–	4500	394,000	12	2.1	0.011			–	–	–
Section 2																
Tsagaan Oloom Formation																
M9-142	65	3.9	160	350	430	–	71,000	281,000	0.37	0.81	0.253			–	–	–
M9-143	67	3.6	20	70	470	–	10,700	393,000	0.04	0.15	0.027			–	–	–
M9-144	91	5.3	30	140	590	–	32,800	350,000	0.05	0.24	0.094			–	–	–
M9-145	92	3.6	40	100	2100	–	21,000	379,000	0.019	0.05	0.055			–	–	–
M9-146	110	0.2	30	120	40	–	125,000	225,000	0.75	3.0	0.56			–	–	–
M9-147	129	2.4	40	140	50	–	128,000	231,000	0.80	2.8	0.55			–	–	–
M9-148	162	0.3	30	250	50	–	115,000	218,000	0.60	5.0	0.53			–	–	–
M9-149	218	7.5	20	210	60	–	125,000	235,000	0.33	3.5	0.53			–	–	–
M9-150	223	2.1	30	160	40	–	125,000	222,000	0.75	4.0	0.56			–	–	–
M9-151	249	9.0	50	370	50	–	109,000	214,000	1.0	7	0.51			–	–	–
M9-152	299	1.5	40	830	140	–	120,000	257,000	0.29	5.9	0.47			–	–	–
M9-154	445	0.2	60	150	30	–	131,000	232,000	2.0	5.0	0.56			–	–	–
M9-155	473	3.6	50	980	40	–	123,000	222,000	1.3	25	0.55			–	–	–
M9-156	651	6.7	30	250	30	–	121,000	234,000	1.0	8.3	0.52			–	–	–
M9-157	667	1.0	100	620	50	–	119,000	223,000	2.0	12	0.53			–	–	–
M9-158	715	5.5	170	670	50	–	128,000	232,000	3	13	0.55			–	–	–

Continued

Table 2 Rb–Sr and Geochemical Characteristics of the Ediacaran–Cambrian Carbonate Rocks of the Dzabkhan Continental Block—cont'd

Sample	Position of Sample*	SA†, %	Content, ppm						Mn/Sr	Fe/Sr	Mg/Ca	$\delta^{13}C$ vPDB, ‰	$\delta^{18}O$ vSMOW, ‰	$^{87}Rb/^{86}Sr$	$^{87}Sr/^{86}Sr$ Measured	$^{87}Sr/^{86}Sr$ Initial
			Mn	Fe	Sr	Rb‡	Mg	Ca								
M9-159	729	0.1	70	560	40	–	118,000	213,000	1.8	14	0.55	3.0	29.5	–	–	–
M9-160	780	3.6	40	100	750	0.036	5000	397,000	0.05	0.13	0.013	−0.5	21.0	0.00012	0.70849	0.70849
M9-161	803	9.0	40	320	950	–	20,600	385,000	0.04	0.34	0.054			–	–	–
M9-162	819	4.4	40	270	720	0.064	30,000	353,000	0.06	0.38	0.085			0.00023	0.70849	0.70848
M9-163	819	19.9	30	940	750	–	7800	392,000	0.04	1.3	0.020			–	–	–
M9-164	819	4.4	50	110	1100	0.040	18,000	392,000	0.05	0.10	0.046	−2.4	22.7	0.00009	0.70845	0.70844
M9-165	854	1.5	30	180	590	–	1800	402,000	0.05	0.31	0.004	0.3	20.5	–	–	–
M9-167	780	0.7	50	490	30	–	122,000	221,000	1.7	16	0.552			–	–	–
M9-168	814	1.6	100	840	60	–	118,000	226,000	1.7	14	0.522			–	–	–
M9-169	814	3.9	10	90	610	0.078	2900	402,000	0.016	0.15	0.007			0.00032	0.70844	0.70844
Bayan Gol Formation																
M9-170	941	5.6	30	260	730	–	25,400	372,000	0.041	0.36	0.068			–	–	–
M9-171	949	5.4	20	320	790	0.140	18,200	370,000	0.025	0.41	0.049	−0.3	22.2	0.00042	0.70848	0.70848
M9-172	969	3.5	10	110	1300	–	9200	393,000	0.008	0.08	0.023			–	–	–
M9-173	984	4.2	20	140	1200	0.020	10,400	395,000	0.017	0.12	0.026	−3.9	21.7	0.00004	0.70850	0.70850
M9-174	989	2.8	10	100	1400	–	4800	391,000	0.007	0.07	0.012			–	–	–
M9-175	994	3.6	30	440	880	0.156	15,000	380,000	0.034	0.50	0.039			0.00045	0.70850	0.70849
M9-176	1004	1.8	10	70	700	–	10,600	402,000	0.01	0.10	0.026			–	–	–
M9-177	1039	5.2	40	650	180	–	14,000	377,000	0.22	3.6	0.037			–	–	–
M9-178	1041	1.1	40	480	170	–	37,000	352,000	0.24	2.8	0.105			–	–	–
M9-179	1043	2.1	20	240	350	–	4100	419,000	0.06	0.69	0.010			0.00088	0.70859	0.70858
M9-180	1045	5.3	80	1400	440	0.154	10,000	379,000	0.18	3.2	0.026	−2.4	20.9	–	–	–
M9-181	1048	2.0	30	820	440	–	23,600	395,000	0.07	1.9	0.060			0.00065	0.70857	0.70856

Table 2 Rb–Sr and Geochemical Characteristics of the Ediacaran–Cambrian Carbonate Rocks of the Dzabkhan Continental Block—cont'd

Sample	Position of Sample*	SA†, %	Content, ppm						Mn/Sr	Fe/Sr	Mg/Ca	$\delta^{13}C$ vPDB, ‰	$\delta^{18}O$ vSMOW, ‰	$^{87}Rb/^{86}Sr$	$^{87}Sr/^{86}Sr$ Measured	$^{87}Sr/^{86}Sr$ Initial
			Mn	Fe	Sr	Rb‡	Mg	Ca								
M9-182	1053	0.1	10	200	440	0.099	6400	391,000	0.02	0.45	0.016			–	–	–
M9-183	1058	1.1	10	130	270	–	4600	415,000	0.04	0.48	0.011			0.00024	0.70854	0.70854
M9-184	1063	1.9	10	70	430	0.041	4900	417,000	0.02	0.2	0.012	−2.4	23.3	–	–	–
M9-185	1068	4.9	360	710	670	–	4800	410,000	0.54	1.1	0.012			–	–	–
M9-186	1088	14.8	720	3300	1300	–	5100	347,000	0.55	2.5	0.015			–	–	–
M9-187	1168	12.3	1200	1200	1000	–	3000	472,000	1.2	1.2	0.006			–	–	–
M9-188	1169	66.0	590	5300	250	–	3000	141,000	2.4	21	0.021			–	–	–
M9-189	1173	7.8	270	1600	420	–	4500	423,000	0.64	3.8	0.011			–	–	–
M9-190	1183	3.8	100	630	450	–	2500	526,000	0.22	1.4	0.005	−2.2	20.8	–	–	–
M9-191	1203	0.6	60	160	720	–	2100	416,000	0.08	0.22	0.005			–	–	–
M9-192	1213	2.1	210	1600	320	–	58,300	340,000	0.66	5.0	0.17			–	–	–
M9-193	1223	4.9	1100	660	180	–	1700	393,000	6.1	3.7	0.004			–	–	–

Note: *Dash, not determined.*
**The position of the sample is given in meters from the section base (negative values show the distance from the mixtite bed).*
†SA, Siliciclastic admixture.
‡Rb content was measured by isotopic dilution, and the contents of the other elements, by AAS. In the calculation of the initial $^{87}Sr/^{86}Sr$ ratios, the age of the rocks was taken to be 550 Ma.

limestones and dolomite limestones, however with varying geochemical criteria than the lower part (Mg/Ca from 0.005 to 0.068; Mn: 10–100 ppm; Fe: 70–710 ppm; Sr: 250–1400 ppm).

18.4.2.4 Bayan Gol Formation (section 2)

Bayan Gol Formation in this section is predominated by limestones (Mg/Ca < 0.03). Fe and Mn concentrations vary from 60 to 1200 ppm and from 1600 to 5300 ppm, respectively. It is important to note that maximum concentrations are in samples with lowest solubility. There is a probability that the Fe and Mn in the carbonates are drawn from the marls, as evidenced by the sympathetic correlations between Fe and Mn and amount of insoluble residue. Sr concentrations vary from 250 to 1300 ppm. The geochemical data indicate that most of the dolomites and dolomitic limestones of the lower part of the section have high Mn/Sr and Fe/Sr ratios (Table 2; Fig. 7) suggestive of prevalent significant diagenetic transformation and hence, the are termed to be unsuitable for Sr-isotope analysis. On the other hand, almost all the limestones and dolomite limestones from the upper part of the section fit geochemical criteria that allowed selection of 12 samples for isotopic chemostratigraphy.

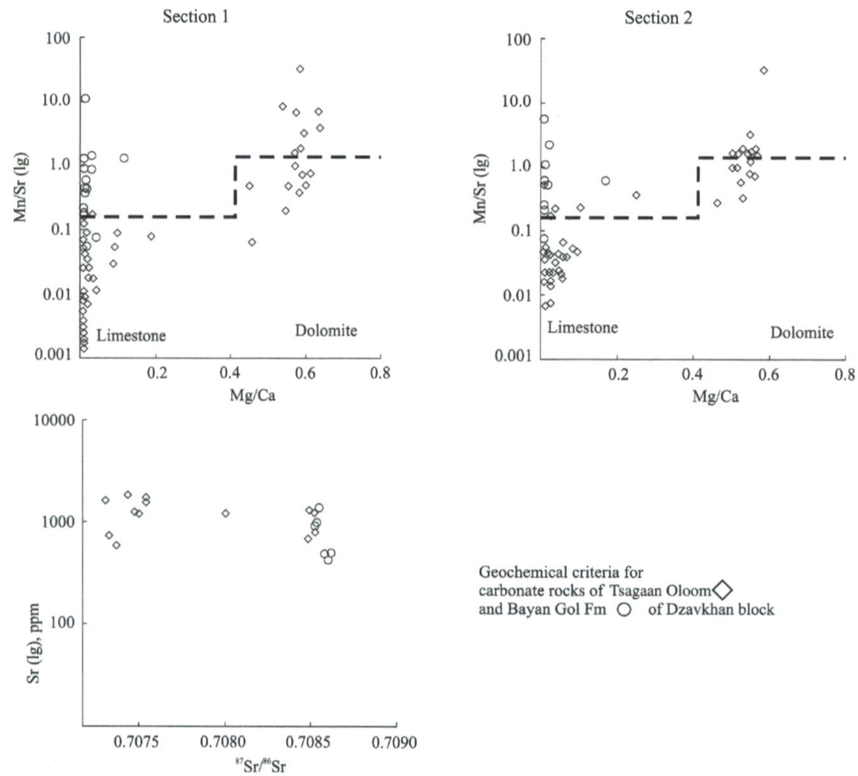

FIGURE 7

Cross-plots of Mn/Sr versus Mg/Ca and Sr contents versus $^{87}Sr/^{86}Sr$ ratio for Ediacaran–Cambrian carbonate rocks of Dzabkhan block. Dashed line marks the areas with the least altered carbonates (see explanation in the text).

18.5 Sr AND C ISOTOPIC COMPOSITION

Study of Sr isotopic composition in the minimally altered carbonate deposits of the Zabit Formation situated in the Ukhagol River show that lower $^{87}Sr/^{86}Sr$ ratios (0.70725–0.70755) are typical for the limestones in the lower part of the section. The upper part of this section, represented by dolomites show a range of $^{87}Sr/^{86}Sr$ ratios from 0.70817 to 0.70852 (Fig. 8; Table 1) (Vishnevskaya and Letnikova, 2013). Initial $^{87}Sr/^{86}Sr$ ratio in the Zabit Formation dolomites of the Urdo-Bokson River and Khoito-Bokson Junction is 0.70817. $\delta^{13}C$ values of the Zabit Formation and Bokson Group deposits (Ukhagol River) in the lower part of the section vary from +2.4‰ to +7.7‰ with an excursion of −1.8‰ (Fig. 8; Table 1). Upper part of the section shows a reduction of $\delta^{13}C$ value with a minimal excursion of 5.2‰. Then there is a peak of +9.5‰. Upper part of the Zabit Formation has $\delta^{13}C$ values varying from +1.9‰ to +3.5‰. $\delta^{13}C$ variation curve for samples from interfluve of rivers Urdo-Bokson and Khoito-Bokson is more simple (Pokrovskii et al., 1999). Near the bottom of the Zabit Formation, maximum $\delta^{13}C$ value of +4.9‰ is observed before showing a negative excursion towards a value of –2.9‰ with three short excurses to zero (+0.3‰).

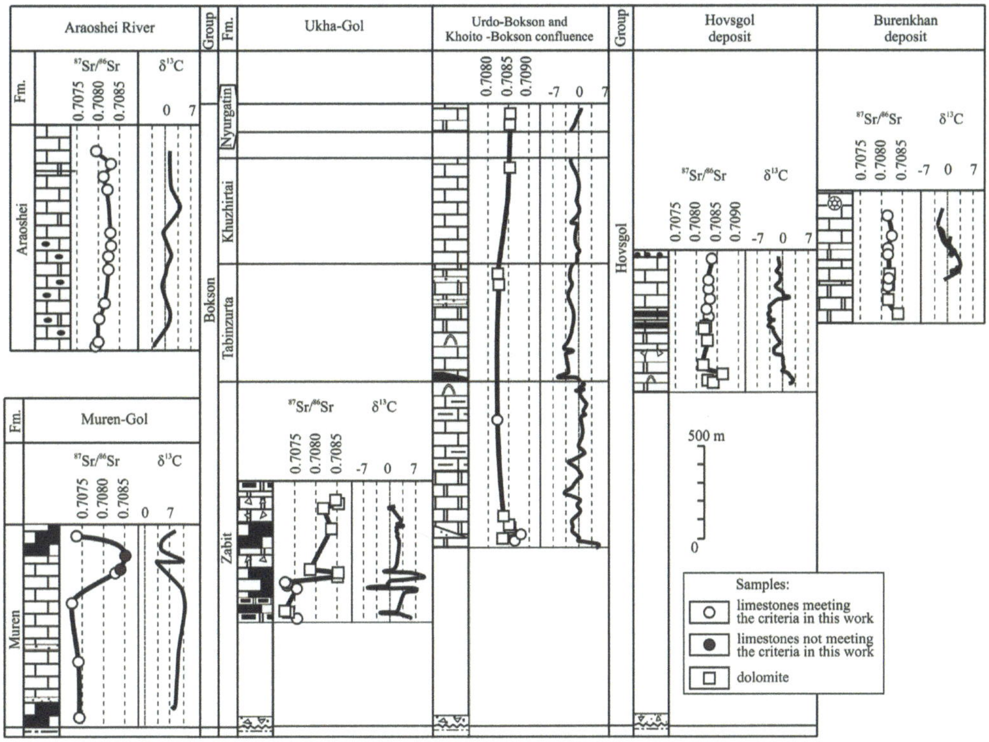

FIGURE 8

Variations in the Sr and C isotope composition of the rocks in the cover of the Tuva–Mongolian block (see Fig. 4 for the legend).

The sediments of the Tabinzurta Formation have moderate negative $\delta^{13}C$ values from −3.5‰ near the bottom of the section (−0.3‰) with some fluctuations C-isotopic values of the Khuzhirtai and Nurgata Formations are close to zero (−1.1‰...+0.2‰) and range to a maximum of −2.0‰ in middle and top parts. Sr-isotopic composition for these formations is monotonous: Tabinzurta Formation dolomites: 0.7082, Khuzhirtai Formation dolomites: 0.7085, Nyurgatin deposits: 0.7083–0.7087.

Initial $^{87}Sr/^{86}Sr$ ratio of the least altered limestones from the bottom part of the Hovsgol Group section (Hovsgol deposit) varies from 0.70822 to 0.70836; in dolomites it reaches 0.70862 (Fig. 6). $\delta^{13}C$ variation curve in the lower part of the section shows positive values (+2.7‰) and fluctuates to negative (from −3.3‰ to −0.7‰) and finally short excursions to positive values (+1.8‰) (Fig. 8; Table 1). The middle and upper parts of the Hovsgol Group section show $^{87}Sr/^{86}Sr$ ratio from 0.70800 to 0.70839 in the second section (Fig. 8; Table 1). $\delta^{13}C$ values in the lower part of the Hovsgol Group section (section N5) vary from +1.2‰ to +2.7‰. In sections N3 and N4 there is an excursion to negative value (−1.5‰). In the middle part of the section (section N2) $\delta^{13}C$ values vary from −2.4‰ to +3.1‰.

The Muren Formation shows $^{87}Sr/^{86}Sr$ ratio of 0.70725–0.70743 in samples from Agaringol section (Vishnevskaya and Letnikova, 2013). Near the terrigenous layers, the $^{87}Sr/^{86}Sr$ ratio increases to 0.70840–0.70888 (Fig. 8; Table 1), because of import of Sr with high isotopic $^{87}Sr/^{86}Sr$ ratio from aluminosilicate rocks. Sr-isotopic composition for samples from sections N1 and N3 was similar and close to 0.7080 (0.70779–0.70803). As the Agaringol section is stratigraphically well constrained, it is taken as a reference for correlation with other sections. In the lower part of Agaringol section, the $\delta^{13}C$ values vary from +7.0‰ to +10.5‰, and decrease with several fluctuations up to +4.0‰ towards top of the section (Fig. 8; Table 1). C-isotopic data for all three studied sections are close to each other and vary from +2.2‰ to +10.5‰ $\delta^{13}C$. The Sr- and C-isotopic compositions of the limestones from Araoshei and Nurgata Formations are similar. Initial $^{87}Sr/^{86}Sr$ ratio in the least altered limestones shows a gradual increase from 0.7084 to 0.7088 towards top. The $\delta^{13}C$ value also shows a gradually increasing trend from −2.8‰ at the bottom and reaches a maximum of 4.2‰.

Thus we conclude that all studied carbonate sediments from Tuva–Mongolian block cover have similar geochemical features. The least altered sediments are from Muren Formation in all of the studied sections. Hovsgol and Bokson Group deposits have various integrity levels. Most complete Sr- and C-isotopic characteristic is observed in the Bokson Group indicating the occurrence of extended carbonate sedimentation event in the range of Tuva–Mongolian block cover.

C-isotopic composition in carbonate deposits of Dzavkhan block rocks varies from −3.9 to anomaly +11.9‰ (Fig. 9). Bottom-up along the section in Tsagaan Oloom Formation, there is a gradual increase of $\delta^{13}C$ values from +5 to +11.9 ‰ and then gradual decrease to −3 ‰. Transition from positive to negative values is in the range of phosphorite layer. $\delta^{13}C$ values in Bayan Gol Formation vary from −3.9 to +5.1 ‰. But for samples with high values (+3.1 to +5.1 ‰) there are incorrect $d^{18}O_{PDB}$ values, about +18‰. It means these samples could be altered and their isotopic compositions are not initial. For the rest rocks, $d^{18}O_{PDB}$ values vary from −10.1 (+20.5 vSMOW) to +1.4 (+29.5 vSMOW) ‰ and are similar in their fluctuations to $\delta^{13}C$ values. Our findings are quite consistent with the data by Brazier et al., (1996).

A total of 20 least altered carbonate deposits from Dzabkhan block cover, each 10 from the N1 and N2 section has been analyzed. Limestones from Tsagaan Oloom Formation have gradual increase of $^{87}Sr/^{86}Sr$ ratio from bottom part of the section N1 to its top: from 0.70715 to 0.70791 (Table 2; Fig. 9). Limestone sample from Bayan Gol Formation has higher $^{87}Sr/^{86}Sr$ ratio: 0.70847. Section N2. Ten limestone and carbonate samples have been analyzed here from. Bottom-up along the section there is a

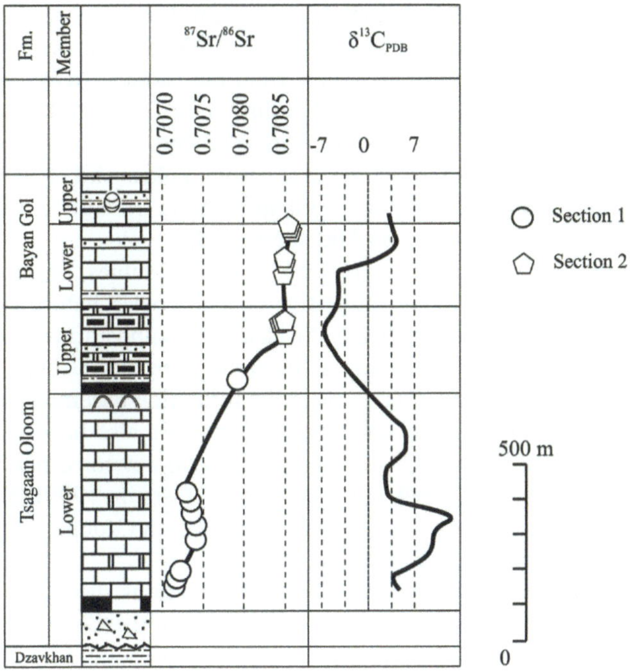

FIGURE 9

Variations in the Sr and C isotope composition of the carbonate rocks of Dzabkhan block (see Fig. 4 for the legend).

gradual increase of $^{87}Sr/^{86}Sr$ ratio from 0.70844 to 0.70858 (Table 2; Fig. 9) in the Bayan Gol Formation N2 section. Published data on isotopic composition and sedimentation time for these deposits (Brazier et al., 1996) show good correlation. C-isotopic composition in carbonate deposits of Dzabkhan block rocks varies from −3.9 to an anomalous value of +11.9‰ (Fig. 9). In the Tsagaan Oloom Formation there is a gradual increase of $\delta^{13}C$ values from +5‰ to +11.9‰ and then a gradual decrease to −3‰. Transition from positive to negative values occurs in the phosphorite layer. $\delta^{13}C$ values in Bayan Gol Formation vary from −3.9‰ to +5.1‰. But for samples with high values (+3.1...+5.1) there are incorrect $\delta^{18}O_{PDB}$ value, about +18‰. It means these samples could have been altered and their isotopic composition are not pristine. For the rest of the rocks, the $\delta^{18}O_{PDB}$ varies from −10.1 (+20.5 vSMOW) ‰ to +1.4 (+29.5 vSMOW) ‰ and is similar in its fluctuations to $\delta^{13}C$ value. Our findings are quite consistent with the data by Brazier et al. (1996).

18.6 DISCUSSION

Comparison of isotopic characteristics of two separate Bokson Group basement strata shows that $^{87}Sr/^{86}Sr$ ratio in dolomites from upper part of the Ukhagol River section and in dolomites from Urdo-Bokson and Khoito-Bokson river junction vary from 0.7081 to 0.7087. Two C-isotopic variation curves have been constructed for the Zabit Formation (Bokson Group). First one was based on the published results of Pokrovskii et al. (1999) for section Urdo-Bokson and Khoito-Bokson river junction (Bokson section) and second was by the authors of this article from the Ukhagol River section (Ukhagol section). At the lower part of the Ukhagol section $\delta^{13}C$ values vary from −5.2‰ to +7.7‰ with many positive and negative excursions. Typical Ediacaran cover sediments have high positive $\delta^{13}C$ values and a gradual decrease (Brazier et al., 1996; Grotzinger et al., 1995; Pokrovskii et al., 2006; etc.). Based on these data, commencement of deposition of Zabit Formation in the Ukhagol River region much earlier than at its stratotype area (Bokson section) could be interpreted.

Sr- and C-isotopic characteristics (0.70725–0.70755; high $\delta^{13}C$ values) of the carbonate deposits of the Zabit and Muren formations are similar, suggestive of their coeval nature and the existence their connectivity in the Tuva–Mongolian terrane shelf. Sr-isotopic compositions of the upper part of the Zabit formation (0.7082–0.7085) and overlaying Bokson and Hovsgol Groups (0.7082–0.7084) are close. They also have similar positive $\delta^{13}C$ values near the formation basement and negative $\delta^{13}C$ values throughout the studied section. Araoshei and Nurgata formation also have similar isotopic characteristics. The published results of Sr- and C-isotopic data thus do not correspond to the results of the Ediacaran–Cambrian Tuva–Mongolian block cover sediments (Osokin, 1999). Though previous studies have assumed coevally deposited nature for Hovsgol and Bokson Groups, our data indicate that the deposition of Bokson Group carbonates commenced earlier.

The Ukhagol section sediments from Zabit and Muren Formations show good correlation with the sediments of Tsagaan Oloom Formation sediments, upper layers of the Charishtag Formation and 1–4 members of Bidzhina Formation from Yenisei Group in terms of C- and Sr-isotopic data and with Yukanda Formation limestone from Yudoma Complex in terms of C-isotopic composition. Hovsgol and upper part of Bokson Group sediments are identical in C- and Sr-isotopic composition with reference to the Bayan Gol Formation. They are also identical in C-isotopic compositional variations present in the Ust-Yudoma and Speckled Formations. Boundary between the Ust-Yudoma and Speckled formations in the Siberian platform is also Ediacaran–Cambrian boundary, and these data go quite well with age determinations of the Hovsgol Group. Sr-isotopic variation curve of the Zabit Formation section at Urdo-Bokson and Khoito-Bokson river junction is similar to that of late Cambrian Khorbusuonka Group deposits of the Olenek uplift (Bowring et al., 1993). Overlying Bokson Group dolomites and limestone are similar in their isotopic characteristics to Yenisei Group deposits from Cambrian part of the section (Letnikova et al., 2011; Ovchinnikova et al., 2011).

The $^{87}Sr/^{86}Sr$ ratio of the Muren Formation limestone varies from 0.7072 to 0.07084. Comparison of these Sr-isotopic data with the Sr-isotopic data published by Halverson et al. (2010) and by Sawaki et al. (2010) shows that the Muren Formation was deposited 600–580 Ma ago. Paucity of Sr-isotopic data on the Bokson Group outcropped at rivers Urdo-Bokson and Khoito-Bokson interfluve ($^{87}Sr/^{86}Sr$ ratio 0.70817–0.70873) evidence that it was formed in the interval from Ediacaran (Zabit Formation) to middle Cambrian (Nyurgatin Formation). These data match well with the previous publications (Kuzmichev, 2004; Belichenko et al., 2003). and based on which—they were formed Deposition of the Hovsgol Group during Ediacaran—middle Cambrian (580–520 Ma ago) is interpreted based on the $^{87}Sr/^{86}Sr$ ratio (0.7080–0.7084).

On the boundary between Proterozoic and Phanerozoic several global glaciation periods have been determined—Marinoan (650–635 Ma (Melezhik et al., 2001; Yoshioka et al., 2003; Chumakov, 2010, 2011)), Gaskiers (580–575 Ma (Melezhik et al., 2001; Halverson et al., 2005; Hoffmann and Li, 2009)) and few others. At the basement of the Bokson Group there is a mixtite horizon genetically related to glaciation

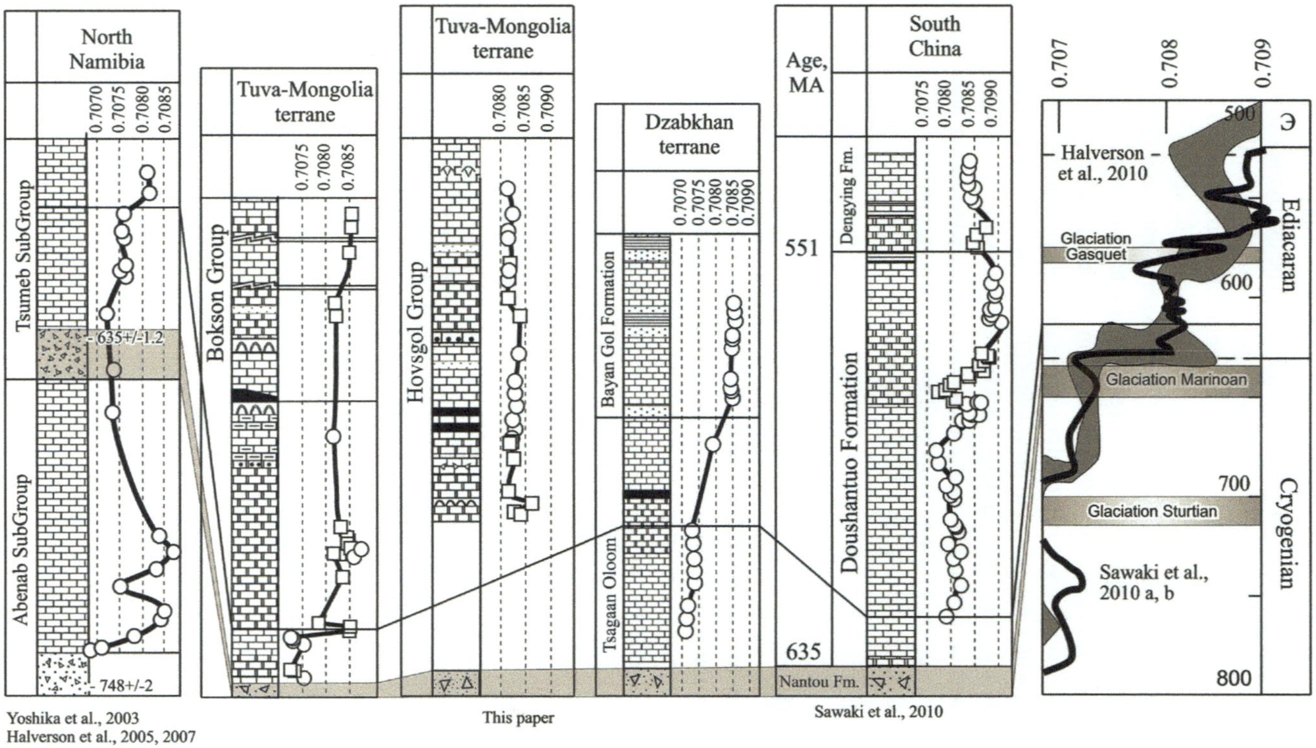

FIGURE 10

Compilation of published Sr isotope data in least-altered carbonate samples from the latest Mesoproterozoic, Neoproterozoic, and Cambrian sections with tillite. Circles, least altered limestone; square, dolomite.

sediments (Osokin and Tyzhinov, 1998). Sedimentation of this formation commenced at 610–600 Ma ago and hence it is logical to correlate this coarse-grained sediments with Marinoan glaciation. At the base of the Hovsgol Group tillite has been reported (Eganov et al., 1999). The overlying carbonate deposits started to accumulate about 560 Ma. Thus, the data provides two insights: first, both tillite horizons can be correlated to Marinoan glaciation and the carbonate sedimentation in the Hovsgol Group started later than in Bokson Group; second, there were two glaciations during Ediacaran in the Tuva–Mongolian block, namely, Marinoan glaciation at the basement of the Bokson Group and Gaskiers glaciation at the basement of the Hovsgol Group. The Bokson and Hovsgol Groups deposits form single carbonate platform and there are no evidences for glaciation in Bokson Group. In the upper part of the Tabinzurta Formation and overlying Khuzhirtai Formation layerwise conformable carbonate (including organogenous) sedimentation is clearly recognized which suggests the prevalence of warm water conditions. Hence, it is safe to conclude that traces of only one glaciation namely, the Marinoan are seen in the Tuva–Mongolian block (Fig. 10) and the sedimentation of the Hovsgol Group carbonate rocks started later than those of the Bokson Group. This interpretation is further supported by the age determinations of the Tsagaan Oloom Formation carbonates by Ovchinnikova et al. (2012)—632 ± 14 Ma.

18.7 CONCLUSION

The results of the study clearly established that chemostratigraphy can be a reliable and accurate tool for correlation even in the poorly biostratigraphically constrained, structurally complex and metamorphosed strata. The study has also helped advancing the existing understanding on glaciation events, and accurately determined the commencement-termination characteristics of sedimentation events. The following age constrains are established by methods of isotope chemostratigraphy. Carbonates of the Tsagaan Oloom Formation were formed during the Ediacaran (590–560 Ma), and the Bayan gol Formation mostly in the early Cambrian (550–520 Ma). Carbonate sedimentation in various parts of the Tuva–Mongolian block was diachronous and started between 600 and 580 Ma and continued into middle Cambrian. Uniformity in geochemical characteristics of the studied rocks and close range of the strontium isotope values suggested the existence of open sea conditions of deposition.

ACKNOWLEDGMENTS

The work was supported by the Russian Foundation of Basic Research (Project nos. 12-05-00569, 12-05-33076 and 14-05-31013) and the Presidium of the Siberian Branch of Russian Academy of Sciences (Project no. 53.68).

REFERENCES

Aktanov, V.I., Doronin, N.A., Posohov, V.F., Sklyarov, E.V., Skopintsev, V.G., 1991. On the structure and age of Gargan blocks (East Sayan). In: Structural Analysis of Crystalline Complexes. IEC SB RAS, Irkutsk, pp. 89–90 (in Russian).

Banner, J.L., Hanson, G.N., 1990. Calculation of simultaneous isotopic and trace element variations during water-rock interaction with applications to carbonate diagenesis. Geochimica et Cosmochimica Acta 54, 3123–3137.

Banner, J.L., Hanson, G.N., Meyers, W.J., 1988. Water-rock interaction history of regionally extensive dolomites of the Burlington–Keokuk Formation (Mississippian): isotopic evidence. In: Shukla, V., Baker, P. (Eds.), Sedimentology and Geochemistry of Dolostones. SEPM Special Publication, vol. 43. SEPM, Tulsa, pp. 97–114.

Belichenko, V.G., Reznitskii, L.Z., Geletii, N.K., Barash, I.G., 2003. Tuva–Mongolia terrane (in the context of microcontinents in the Paleoasian ocean). Russian Geology and Geophysics 44 (6), 554–565.

Boos, R.G. (Ed.), 1991. Paleozoic of Tunka Ridge of East Sayan. Nauka, Novosibirsk (in Russian).

Bowring, S.A., Grotzinger, J.P., Isachsen, C.E., Knoll, A.H., Pelechaty, S.M., Kolosov, P., 1993. Calibrating rates of Early Cambrian evolution. Science 261 (5126), 1293–1298.

Brand, U., Veizer, J., 1980. Chemical diagenesis of a multicomponent carbonate system. – 1. Trace elements. Journal of Sedimentary Petrology 50 (4), 1219–1236.

Brazier, M.D., Shields, G., Kuleshov, V.N., Zhegallo, E.A., 1996. Integrated chemo- and biostratigraphic calibration of early animal evolution: Neoproterozoic–early Cambrian of southwest Mongolia. Geological Magazine 133 (4), 445–485.

Chumakov, N.M., 2011. Late proterozoic African glacial era. Stratigraphy and Geological Correlation 19 (1), 1–20.

Chumakov, N.M., 2010. Precambrian glaciations and associated biospheric events. Stratigraphy and Geological Correlation 18 (5), 467–479.

Dobretsov, N.L. (Ed.), 1988. Geology and Metamorphism of East Sayan. Nauka, Novosibirsk (in Russian).

Eganov, E.A., Rozanov, A.Yu., Zhegallo, E.A., 1999. The Khubsugul phosphorite-bearing basin (KhPB), Mongolia, Russia. In: Shkol'nik, E.L. (Ed.), Nature of Phosphate Grains and Phosphorites in the World's Largest Basins. Dal'nauka, Vladivostok, pp. 32–40 (in Russian).

Gibsher, A.S., Bat-Ireedui, Y., 1990. Late Precambrian–Early Cambrian stratigraphy of the Muren zone of the western Hovsgol area. In: The Late Precambrian and Paleozoic of Siberia: Problems of Regional Stratigraphy. IGiG SO AN SSSR, Novosibirsk, pp. 91–104 (in Russian).

Grotzinger, J.P., Bowring, S.A., Saylor, B.Z., Kaufman, A.J., 1995. Biostratigraphic and geochronologic constraints on early animal evolution. Science 270 (5236), 598–604.

Halverson, G.P., Hoffman, P.F., Schrag, D.P., Maloof, A.C., Rice, A.H.N., 2005. Toward a Neoproterozoic composite carbon-isotope record. GSA Bulletin 117 (9–10), 1181–1207.

Halverson, G.P., Wade, B.P., Hurtgen, M.T., Barovich, K.M., 2010. Neoproterozoic chemostratigraphy. Precambrian Research 182 (4), 337–350.

Hoffmann, P.F., Li, Z.-X., 2009. A palaeogeographic context for Neoproterozoic glaciation. Palaeogeography Palaeoclimatology Palaeoecology 277 (3–4), 158–172.

Il'in, A.V., 1973. The Khubsugul phosphorite-bearing basin. In: Trans. Joint Soviet–Mongolian Research Expedition. Nauka, Moscow. Issue 6 (in Russian).

Jacobsen, S.B., Kaufman, A.J., 1999. The Sr, C and O isotopic evolution of Neoproterozoic seawater. Chemical Geology 161, 37–57.

Khain, E.V., Bibikova, E.V., Degtyarev, K.E., 1999. Paleozoic ocean in the Neoproterozoic and early Paleozoic: new isotope-geochronological data. In: The Geologic Evolution of Proterozoic Structures Paleooceanic Pericratonal and Northern Eurasia, pp. 175–181 Saint-Petersburg (in Russian).

Kheraskova, T.N., Samygin, S.G., 1992. Tectonic conditions for the formation of the Vendian–Middle Cambrian terrigenous-carbonate complex of East Sayan. Geotektonika 6, 18–36.

Khomentovsky, V.V., Gibsher, A.S., 1996. The Neoproterozoic–lower Cambrian in northern Govi-Altay, western Mongolia: regional setting, lithostratigraphy and biostratigraphy. Geological Magazine 133 (1/4), 445–485.

Khomentovsky, V.V., Pak, K.L., Postnikov, A.A., Skopintsev, V.G., 1985. Geologic structure of the Ukha-Gol River basin (East Sayan). In: Khomentovsky, V.V., Terleev, A.A., Bragin, S.S. (Eds.), Stratigraphy of the Late Precambrian and Early Paleozoic of Siberia: Vendian and Riphean. IGiG SO AN SSSR, Novosibirsk, pp. 76–106 (in Russian).

Kovach, V.P., Matukov, D.I., Berezhnaya, N.G., Kotov, A.B., Levitsky, V.I., Barash, I.G., Kozakov, I.K., Levsky, L.K., Sergeev, S.A., 2004. Shrimp zircon age of the Gargan block tonalites—find early Precambrian basement of the Tuvino–Mongolian microcontinent, Central Asia mobile belt. 32nd IGC—Florence. Session: T31.01—Tectonics of Precambrian Mobile Belts, vol. 1. p. 1263.

Kozakov, I.K., Sal'nikova, E.B., Wang, T., Didenko, A.N., Plotkina, Yu.V., Podkovyrov, V.N., 2007. Early Precambrian crystalline complexes of the Central Asian microcontinent: age, sources, tectonic position. Stratigraphy and Geological Correlation 15 (2), 121–140.

Kuzmichev, A.B., Zhuravlev, D.Z., Bibikova, E.V., Kirnozova, T.I., 2000. Upper Riphean (790 Ma) granitoids in the Tuva–Mongolian massif: evidence of early orogenic Baikal. Geology and Geophysics 41 (10), 1379–1383 (in Russian).

Kuzmichev, A.B., 2004. Tectonic History of the Tuva–Mongolian Massif: Early Baikalian, Late Baikalian, and Early Caledonian Stages. PROBEL-2000, Moscow (in Russian).

Kuznetsov, A.B., Semikhatov, M.A., Gorokhov, I.M., 2012. The Sr isotope composition of the world ocean, marginal and inland seas: implications for the Sr isotope stratigraphy. Stratigraphy and Geological Correlation 20 (6), 501–515.

Letnikova, E.F., Geletii, N.K., 2005. Vendian–Cambrian carbonate sequences in sedimentary cover of the Tuva–Mongol microcontinent. Lithology and Mineral Resources 40 (2), 167–177.

Letnikova, E.F., Kuznetsov, A.B., Vishnevskaya, I.A., Terleev, A.A., Konstantinova, G.V., 2011. The geochemical and isotope (Sr, C, O) characteristics of the Vendian–Cambrian carbonate deposits of the Azyr–Tal Ridge (Kuznetsk Alatau): chemostratigraphy and sedimentogenesis environments. Russian Geology and Geophysics 52 (10), 1154–1170.

Levachova, N.M., Kalygin, V.M., Gibsher, A.S., Yff, J., Ryabinin, A.B., Meert, J.G., Malone, S.J., 2010. The origin of the Baydaric microcontinent, Mongolia: constraints from Paleomagnetism and Geochronology. Tectonophysics 485, 306–320.

Li, D., Shields-Zhou, G.A., Ling, H.-F., Thirlwall, M., 2011. Dissolution methods for strontium isotope stratigraphy: guidelines for the use of bulk carbonate and phosphorite rocks. Chemical Geology 290, 133–144.

Macdonald, F.A., Jones, D.S., Schrag, D.P., 2009. Stratigraphic and tectonic implications of a newly discovered glacial diamictite–cap carbonate couplet in southwestern Mongolia. Geology 123–126.

Melezhik, V.A., Gorokhov, I.M., Kuznetsov, A.B., Fallick, A.E., 2001. Chemostratigraphy of Neoproterozoic carbonates: implications for "blind dating". Terra Nova 13 (1), 1–11.

Osokin, P.V., 1999. Central Asian Phosphorite-Bearing Province (Stratigraphy and Phosphorite Potential). Doctoral (Geology Mineral) Dissertation as a Scientific Report (in Russian) (GI SO RAN, Ulan-Ude).

Osokin, P.V., Tyzhinov, A.V., 1998. Precambrian tilloids of the Oka–Khubsugul phosphorite-bearing basin (Eastern Sayany, northwestern Mongolia). Litologiya i Poleznye Iskopaemye 2, 162–176.

Ovchinnikova, G.V., Kuznetsov, A.B., Gorokhov, I.M., Letnikova, E.F., Kaurova, O.K., Gorokhovskii, B.M., 2011. U–Pb age and Sr-chemostratigraphy of limestone from the Sorna formation, Azyr–Tal range, Kuznetsk Alatau. Doklady Earth Sciences 437 (1), 331–334.

Ovchinnikova, G.V., Kuznetsov, A.B., Vasil'eva, I.M., Gorokhov, I.M., Letnikova, E.F., Gorokhovskii, B.M., 2012. U–Pb age and Sr isotope signature of cap limestones from the Neoproterozoic Tsagaan Oloom Formation, Dzabkhan River Basin, Western Mongolia. Stratigraphy and Geological Correlation 20 (6), 516–527.

Podkovyrov, V.N., Semikhatov, M.A., Kuznetsov, A.B., Vinogradov, D.P., Kozlov, V.I., Kislova, I.V., 1998. Carbonate carbon isotopic composition in the Upper Riphean stratotype, the Karatau Group, southern Urals. Stratigrafiya, Geologicheskaya Korrelyatsiya 6 (4), 3–19 (in Russian).

Pokrovskii, B.G., Letnikova, E.F., Samygin, S.G., 1999. Isotopic stratigraphy of the Bokson Group, Vendian–Cambrian of the Sayan Mountains. Stratigrafiya, Geologicheskaya Korrelyatsiya 7 (3), 23–41 (in Russian).

Pokrovskii, B.G., Melezhik, V.A., Bujakaite, M.I., 2006. Carbon, oxygen, strontium, and sulfur isotopic compositions in late Precambrian rocks of the Patom Complex, central Siberia: communication 1. Results, isotope stratigraphy, and dating problems. Lithology and Mineral Resources 5, 450–474.

Ragozina, A.L., Serezhnikova, E.A., Kraushkin, A.V., Dorzhnamzhaa, D., 2009. Vendian–Cambrian biota Dzabkhan structural zone of the West Mongolia. In: Proceedings of the All-Russian Meeting. Paleontological Institute of the Russian Academy of Sciences, Moscow, p. 116.

Sawaki, Y., Ohno, T., Fukushi, Y., Komiya, T., Ishikawa, T., Hirata, T., Maruyama, S., 2008. Sr isotope excursion across the Precambrian–Cambrian boundary in the Three Gorges area, South China. Gondwana Research 14, 134–147.

Sawaki, Y., Ohno, T., Tahata, M., Komiya, T., Hirata, T., Maruyama, S., Windley, B.F., Han, J., Shu, D., Li, Y., 2010. The Ediacaran radiogenic Sr isotope excursion in the Doushantuo Formation in the Three Gorges area, South China. Precambrian Research 176 (1–4), 46–64.

Semikhatov, M.A., Gorokhov, I.M., Kuznetsov, A.B., Mel'nikov, N.N., Podkovyrov, V.N., Kislova, I.V., 1998. The strontium isotopic composition in early Late Riphean seawater: limestones of the Lakhanda Group, the Uchur–Maya region, Siberia. Doklady Earth Sciences 360 (4), 488–492.

Semikhatov, M.A., Serebryakov, S.N., 1967. Vendian and Lower Cambrian of southeastern East Sayan. Izvestiya AN SSSR, Seriya Geologicheskaya, vol. 4, pp. 87–102 (in Russian).

Vishnevskaya, I.A., Letnikova, E.F., 2013. Chemostratigraphy of the Vendian–Cambrian carbonate sedimentary cover of the Tuva–Mongolian microcontinent. Russian Geology and Geophysics 54, 567–586.

Xiao, S., Bao, H., Wang, H., Kaufman, A.J., Zhou, C., Li, G., Yuan, X., Ling, H., 2004. The Neoproterozoic Quruqtagh Group in eastern Chinese Tianshan: evidence for a post-Marinoan glaciation. Precambrian Research 130 (1–4), 1–26.

Yoshioka, H., Asahara, Y., Tojo, B., Kawakami, S., 2003. Systematic variations in C, O, and Sr isotopes and elemental concentrations in Neoproterozoic carbonates in Namibia: implications for glacial to interglacial transition. Precambrian Research 124 (1), 69–85.

CHAPTER

19 CORRELATION OF PHOSPHORITE AND NONPHOSPHORITE CARBONATE SEQUENCES OF THE LOWER ARAVALLI GROUP, NORTHWEST INDIA: IMPLICATIONS ON THE PALEOPROTEROZOIC PALEOENVIRONMENT

Ritesh Purohit
Government College, Sirohi, Rajasthan, India

19.1 INTRODUCTION

Since the earliest reports from the Russian Karelia (Galimov et al., 1968), Finland, and Zimbabwe (Schidlowski et al., 1975, 1976), a vast array of literature is available on the paleoenvironmental evolution of the Paleoproterozoic basins located worldwide and the interest continues unabated (for example, Bekker et al., 2001, 2003a,b, 2004a,b; Bekker and Holland, 2012; Lindsay and Brasier, 2002; Tang et al., 2011; Salmein et al., 2013; Preat et al., 2011; Karhu, 1993; Melezhik et al., 1997, 1999; Yudovich et al., 1991; Baker and Fallick, 1989a,b; Melezhik and Fallick, 1996, 2010; Zagnitko and Lugovaya, 1989; Martin et al., 2103; Maheshwari et al., 2010; Papineau et al., 2013; Purohit et al., 2010). Though these studies have led an insight into the paleoenvironmental evolution of the Paleoproterozoic basins, fewer aspects are still being intensely debated. One among the contentious issue is the coevolution of life in the Earth's early history. The early life came into existence synchronous with the Great Oxidation Event (GOE) at 2.3–2.4 Ga (Bekker et al., 2006; Frie et al., 2009; Guo et al., 2009; Holland, 2006), which in turn was the result of global peak magmatism at ~2.51–2.45 Ga causing higher sea levels (Barley et al., 2005). Sea level fell by ~2.4 Ga after the tectonic quiescence, which was disturbed by continental breakup during ~2.32 Ga (Ahmad et al., 2008; Barley et al., 2005). Oxygen levels rose gradually in the early Paleoproterozoic but more rapidly between 2100 and 2000 Ma (Murakami et al., 2011; Kump et al., 2011).

The GOE is recorded in the sedimentary rocks which have their origin in conjunction with the nutrient bioelements like C, N, Fe, and P. These three elements are intimately involved with the biogeochemical cycles and help to decipher microenvironmental variations. Phosphate bears special

Chemostratigraphy. http://dx.doi.org/10.1016/B978-0-12-419968-2.00019-4

significance in the Paleoproterozoic Aravalli Supergroup rocks because the carbonates host one of the richest Paleoproterozoic phosphate deposits with concentration of about 35% P_2O_5 denoting the world's first major phosphogenic episode (cf. Nelson et al., 2010). Phosphate is hosted by dolomite in some basins which is trapped into the edifices of stromatolitic cyanobacterial colonies (Chauhan, 1979; Roy and Paliwal, 1981). Abundance of P during the deposition of the sediments in some of the basins of the Jhamarkotra Formation of the Lower Aravalli Group led to the cyanobacterial bloom and is most likely responsible for the oxygenation of the Earth's atmosphere (Papineau et al., 2013). P abundance has also affected the N cycle in the region which is reflected in the low heavy N isotope values of the carbonaceous sediments of the Jhamarkotra Formation (Papineau et al., 2009). In addition, this abundant P also led to the variations in the C cycle between the phosphate basins (PBs) and nonphosphate basins (NPBs) of the Lower Aravalli Group. The PB carbonates showed close to zero per mil values of heavy carbon (^{13}C) isotope while the NPB carbonates show moderate to high enrichments in the ^{13}C values (Maheshwari et al., 1999, 2002; 2010; Purohit et al., 2010; Papineau et al., 2013; Sreenivas et al., 2001).

Previous studies indicate that the basins with excess availability of P were restrictive in nature and led to the formation of cyanobacterial blooms whereas the basins that had open bank-shelf depositional environment lacked P and were devoid of the cyanobacterial bloom (Banerjee, 1971; Banerjee et al., 1980; Chauhan, 1979; Choudhuri and Roy, 1986; Roy and Paliwal, 1981). Since both the types of basins deposited carbonates, circulation characteristics and resultant aerobic or eutrophic conditions were interpreted to be the causes of bloom formation and hence the interrelationships between the P and C cycles in both PBs and NPBs need to be examined. The NPBs which lacked organic matter (and hence reduced chances of organic burial) show moderate to high ^{13}C enrichments. Correlation of different carbonate horizons in various subbasins of the Lower Aravalli Group is an archetypal issue since all these basins are proximally located and partially connected with narrow openings paleogeographically (Roy and Paliwal, 1981). Both the PBs and NPBs are considered to be coeval but nonsynchronous (Papineau et al., 2013). The basins were later synchronously deformed and metamorphosed (~1.9 Ga) that had complicated the correlation. Chemostratigraphic studies in conjunction with sequence stratigraphy along selected transect and/or sections can be a useful tool to establish comprehensive correlation between various subbasins. Thus, the scope of this chapter is limited to the correlation among different biogeochemical cycles in the PBs and NPBs with respect to the paleoenvironmental evolution.

19.2 GEOLOGICAL SETTING

The Aravalli Supergroup is the oldest metasedimentary–volcanic ensemble in the northwest India that overlies a gneissic basement, traditionally known as the Banded Gneissic Complex with a profound "erosional unconformity" (Heron, 1953) in between. The erosional unconformity is marked by a metamorphosed white mica paleosol (Roy et al., 1993) which also represents the higher oxidation state due to incipient rise in O levels (Wall et al., 2012). Recent isotopic studies have confirmed a middle-Archean age (c. 3.3 Ga) for the oldest basement gneisses renamed Mewar Gneiss Complex (Roy and Jakhar, 2002, Table 1). Approximately 2.5-Ga-old stabilization age of the Archean basement (Wiedenbeck et al., 1996) confirms the Proterozoic age of the "cover" Aravalli Supergroup (earlier

Table 1 Generalized Stratigraphic Classification of the Shelf Facies Sequence in the Type Area of the Aravalli Supergroup

UPPER ARAVALLI GROUP	Debari Formation	Jharol Formation
	Lakhawalli Formation	(Deep water sequence)
	Kabita Formation	
----------Unconformity----------		
MIDDLE ARAVALLI GROUP	Tidi Formation,	
	Bowa Formation	
	Mochia Formation	
	Udaipur Formation	
----------Unconformity----------		
LOWER ARAVALLI GROUP	Jhamarkotra Formation	Dolomitic carbonate association (locally with a stromatolitic phosphorites) phyllites (with black shale facies rocks)
	Babarmal Formation	Calcitic carbonates and pebbly conglomerates
	Delwara Formation	Metavolcanics with thin beds of metasediments
Mewar Gneissic Complex and granitoids		Archaean basement rocks (Banded Gneissic Complex)

After Roy and Jakhar (2002); revised after Purohit et al. (2012).

considered to be an Archean ensemble Heron, 1953). There is little geochronologic control over the precise age of the cover Aravalli rocks. Lead isotope data from galena associated with barite in the basal Aravalli volcanic sequence not only indicate c. 2075 and 2150 Ma Pb–Pb age for the Aravalli Supergroup (Deb and Thorpe, 2004) but also confirm an approximately 300 Ma hiatus between the opening of the Aravalli rift basins and the stabilization of the basement (Roy and Jakhar, 2002). Recently, Ahmad et al. (2008) using Sm–Nd isotope systematics suggested 2326 ± 321 Ma age for the komatiite lava associated with the basal volcanics. However, the very high error factor deters using the calculated age to fix the maximum age of the Aravalli Supergroup. The minimum (i.e., closing) age for the Aravalli basins is indicated by the c. 1900 Ma Rb–Sr isochron age of the synkinematic migmatitic gneisses (Chaudhary et al., 1984) and the Pb–Pb isochron age of the reconstituted dolomitic carbonates (Sarangi et al., 2006).

In a recent study, Mckenzie et al. (2013) opposed the age of the Aravalli supracrustals (ranging between 2200 and 1900 Ma) and accordingly, the Paleoproterozoic status of the Aravalli Supergroup has been revised to the early Mesoproterozoic (1900–1700 Ma). These authors have determined this age on the basis of detrital zircons from sandstones (earlier described as quartzites) in the Jhamarkotra Formation of the Lower Aravalli Group. The authors also misconceived and inferred a new ^{13}C enrichment anomaly from Jhamarkotra Formation (equivalent to the global Lomagundi-Jatuli Event (LJE), cf. Purohit et al., 2010; Papineau et al., 2013) as a younger event at the onset of the "boring billion." Melezhik et al. (2014) refuted these interpretations considering it as "an unfortunate combination of radiometric dates with the lithological correlation."

In this chapter, the stratigraphic succession of the Aravalli Supergroup established by Roy et al. (1993) and Roy and Jakhar (2002) (Fig. 1, Table 1), and revised thereafter by Purohit et al. (2012) has been followed. The entire ensemble is divisible into Lower, Middle, and Upper Aravalli Groups. The carbonate rocks of the Jhamarkotra Formation and the Babarmal Formation along with the volcanics of the Delwara Formation constitute the Lower Aravalli Group (Purohit et al., 2012). The carbonates were deposited in different shallow subbasins; each characterized by distinct depositional setting. The carbonates of the Lower Aravalli Group were deposited in subtidal to supratidal environments (Banerjee, 1971; Chauhan, 1979; Choudhuri and Roy, 1986; Roy and Paliwal, 1981), while a few were deposited in the bank-shelf depositional environments (Roy and Paliwal, 1981). The major phase of deposition of the carbonates occurred in eperic seas with the development of a stromatolitic horizon near the base (Roy and Paliwal, 1981). Growth characteristics of the stromatolites indicate slow rate of sedimentation under warm and humid tropical climate and mildly aerated depositional environment (Chauhan, 1979; Banerjee et al., 1986; Choudhuri and Roy, 1986). The depositional environment varied considerably within basins due to presence of phosphate. Subbasins which lacked P were devoid of stromatolitic cyanobacteria. On this basis, the Lower Aravalli Group carbonates are divided into PBs and NPBs (Fig. 2).

The PB is represented by the Udaipur Epicontinental Sea and the Jhamarkotra, Sallopat and Jhabua embayments (Banerjee et al., 1980, 1986) in the Lower Aravalli Group of rocks (Fig. 1). The carbonaceous shales which overlie the carbonates have trace element abundances indicative of deposition under a pronounced redoxcline (Papineau et al., 2009). Based on the distribution of the stromatolitic phosphorite horizons, Roy and Paliwal (1981) proposed a complex shoreline configuration of the contemporaneous, virtually landlocked, Udaipur Epicontinental Sea with narrow openings to the Aravalli Epeiric Sea. The Jhamarkotra embayment of the Udaipur Epicontinental Sea was host to the most prolific development of stromatolitic

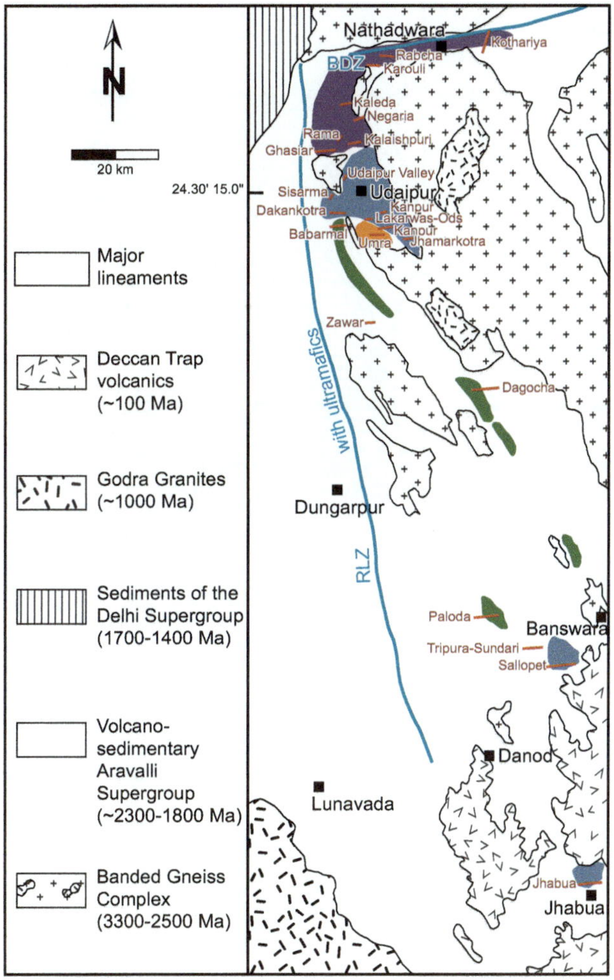

FIGURE 1

Regional geological map modified after Papineau et al. (2013), showing locations of various phosphate and nonphosphate basins of the Lower Aravalli Group from the northwestern Indian shield. The Phosphate Basin (PB) with near-zero $\delta^{13}C_{carb}$ values (blue), the Non-Phosphate Domain (NPD) of the Ghasiar-Rama-Karouli (GRK) bank-shelf with elevated $\delta^{13}C_{carb}$ excursions (purple), and of the Umra isolated basin (orange) and of the calcitic domain at Babarmal-Paloda-Belt (BPB) with moderate $\delta^{13}C_{carb}$ excursions (green). Banas Dislocation Zone (BDZ) and Rakhabdev Lineament Zone (RLZ) are major faults of the Banas dislocation zone and the Rakhabedev lineament zone respectively.

phosphorites along its shorelines (Banerjee, 1971; Chauhan, 1979; Choudhuri and Roy, 1986; Roy and Jakhar, 2002). The rocks of the Aravalli Supergroup show a low to very low grade of green schist facies metamorphism with slight increase in metamorphic grade up to lower amphibolite facies in some parts.

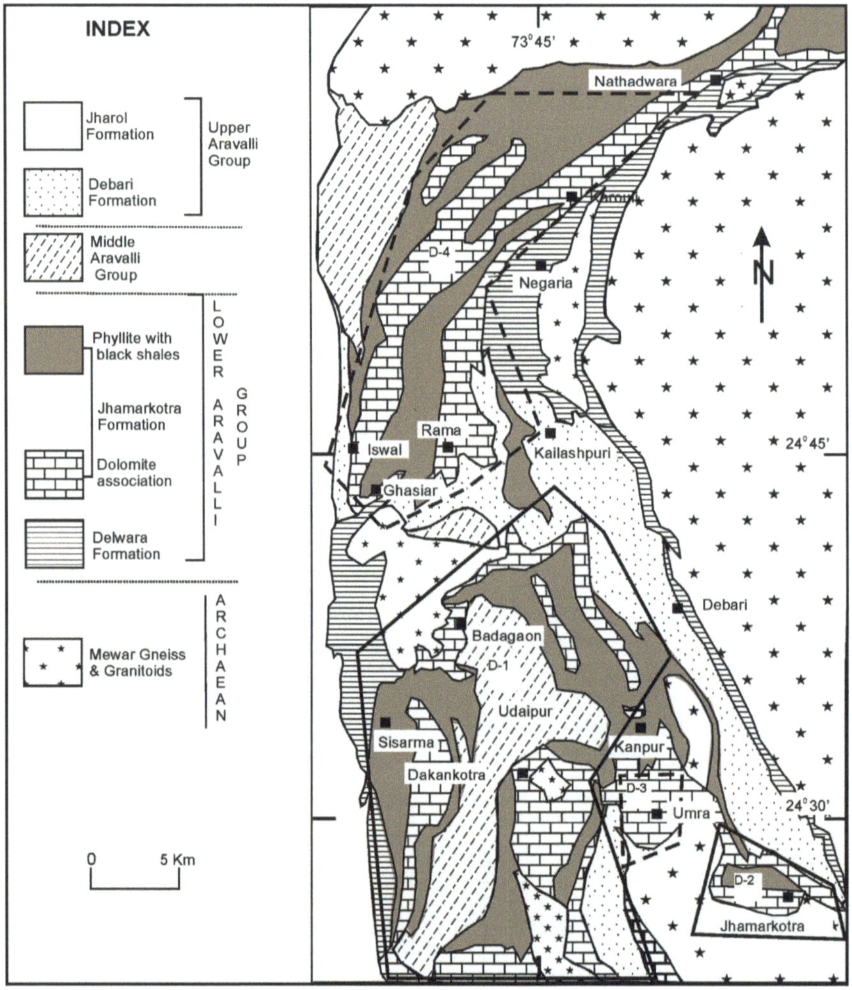

FIGURE 2

Distribution of phosphate basin (PB) and nonphosphate basin (NPB) of the Lower Aravalli Group in the type area near Udaipur city *(after Purohit et al., 2010)*. The map shows proximity between PB (denoted by straight lines) and NPB (denoted by dotted lines).

19.3 PROBLEM OF CORRELATION

Lithostratigraphic subdivisions of the Lower Aravalli Group and correlation between PBs and NPBs are difficult due to the close proximity between subbasins, regional scale complex folding, faulting, and complex deformation. Consequently, a robust lithostratigraphic correlation between different structural subbasins is not straightforward. Consequently, carbonates of both PB and NPB, though having significant differences, were classified and clubbed under the Jhamarkotra Formation of the Lower Aravalli

Group (Roy and Jhakhar, 2002). Recent modification in stratigraphy by Purohit et al. (2012) has however differentiated nonphosphatic calcitic carbonate and associated litho-units as a new formation and named Babarmal Formation in the Lower Aravalli Group (Table 1). It is interesting to take note of the fact that the calcitic carbonates of the Babarmal Formation were earlier correlated as equivalent to the dolomitic Jhamarkotra Formation by Roy and Jakhar (2002) and later younger to the Jhamarkotra Formation by Maheshwari et al. (2002).

Correlation of lithofacies in different subbasins of the Lower Aravalli Group was attempted by Purohit et al. (2010) and the correlation was revised by Purohit et al. (2012) keeping in view the relationship between the various carbonate litho-units (Fig. 3). The revised correlation inferred a hiatus between the carbonates of the PB and NPB though the basins might have opened coeval (cf. Roy and Jakhar, 2002). The NPB carbonates were possibly deposited prior to the PB carbonates (Papineau et al., 2013). However, the issue was not resolved completely and debated later (cf. Mckenzie et al., 2013; Melezhik et al., 2014). Understanding the lateral continuity and sequence stratigraphy are, however, difficult in complexly folded and faulted terrains as there are often misleading repetition of beds. Small-scale sections could, however, be used to characterize different subbasins on the basis of geochemistry. Stable isotope chemostratigraphy for different sections from all the subbasins is used and compared in this study with other known sequences on global scale during the Palaeoproterozoic span (Fig. 4).

19.4 GEOCHEMICAL CHARACTERIZATION OF PBS AND NPBS

19.4.1 CHEMISTRY OF THE PB AND NPB CARBONATES

The PB and NPB carbonates show close similarities and sharp differences in geochemical characters. The carbonates of PB and NPB show close similarity in carbonate composition. Carbonates in both the basins are predominantly dolomitic with exceptional presence of calcitic carbonates in nonphosphatic Babarmal basin (Purohit et al., 2012). The carbonates of PB are predominantly dolomitic with abundant P while the NPB carbonates are calcitic as well as dolomitic with virtual absence of P. Presence of P-bearing cyanobacterial stromatolites in the PB carbonates has led to enrichment in organic carbon in comparison to the NPB carbonates (Papineau et al., 2013). Phosphatic carbonates show higher concentration of Fe and mutual positive correlation (Purohit and Rengarajan, 2007) which suggests that phosphogenesis could be the consequence of Fe-redox pumping of pore water phosphate through a suboxic seafloor (Nelson et al., 2010).

The average carbonate content of the dolomitic carbonates is low in comparison to the calcitic carbonates (Mehta, 2010). Low Mg and Mn contents of the calcitic carbonates point toward low Eh and hyposaline conditions of the coeval seawater at the time of deposition which is a contrasting feature with respect to the dolomitic carbonates which have high Mg and Mn (Purohit et al., 2014). Low Mg values in the calcitic carbonates suggest that the dominantly calcitic content had negligible postdepositional replacement of Ca by Mg whereas the dolomitic carbonates show predominance of the replacement features.

The calcitic carbonate, unlike dolomitic carbonate does not bear phosphate or related organic biomarkers (Purohit et al., 2014). These features depict insignificant diagenetic changes. The low Mg/Ca observed in the calcitic carbonates of the Babarmal subbasin suggests prolonged depositional condition of very low influx of Mg into the basin. Similar trend is also discernible by Sr/Mn index in the carbonates which mirrors permeability of the rocks and a higher state of preservation of the original chemical

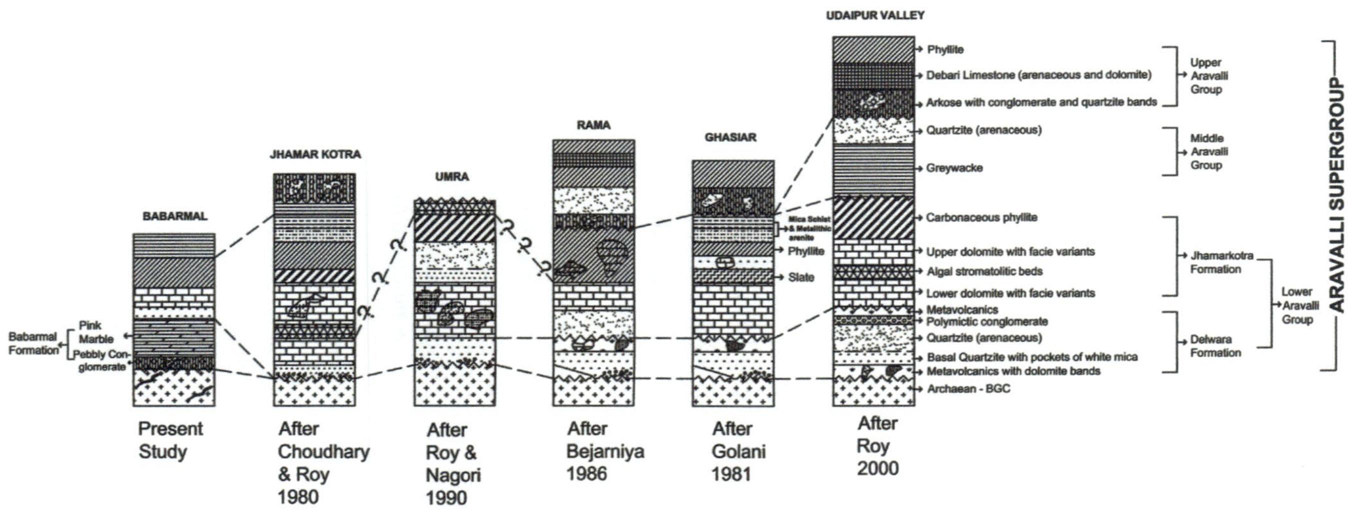

FIGURE 3

Litholog correlation diagram of various lithofacies in different subbasins of the Jhamarkotra Formation, Lower Aravalli Group.

After Purohit et al. (2012).

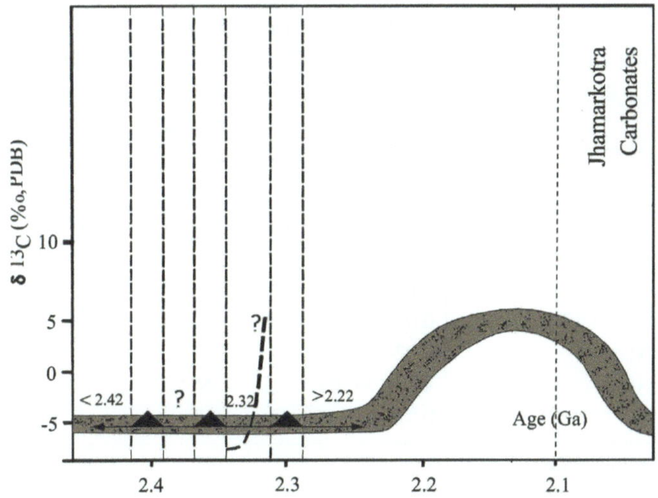

FIGURE 4

The plot showing variations in carbon isotopic composition ($\delta^{13}C$ in PDB) versus age (in Ga), modified after Bekker et al. (2004). The plot shows the position of Lower Aravalli Group carbonates on the Paleoproterozoic time scale.

attributes in the carbonates of the Babarmal basin (Purohit et al., 2012). Mn/Sr values are invariably below 10 ranging from 0.8 to 2.5 which imply that the calcitic carbonate samples are diagenetically unaltered even after metamorphic transformation (Mehta, 2010). High positive Ce anomaly observed in the Babarmal calcitic carbonate suggests its pristine nature (Purohit et al., 2014).

19.4.2 CHEMISTRY OF PB AND NPB BLACK SHALES

Like the carbonates, the black shale facies show diverse geochemical characters in the PB and NPB. Black shales stratigraphically overlie the dolomitic carbonates but are virtually absent in the calcitic carbonates in the Babarmal basin. Black shales show postdepositional alteration through thermal metamorphism and hydrothermal fluid activity related to the regional metamorphism at ~1.8 Ga (Tripathy et al., 2013). Black shales of the PB and NPB show few similarities and some significant differences in geochemical characters (Papineau et al., 2009; Bhu and Singh, 2005). Carbonaceous materials in shales from both the PB and NPB have molybdenum less than 5 ppm (Papineau et al., 2009). These values are less than those expected for carbonaceous shales (few tens of ppm Mo–Scott et al., 2008) deposited after or during the Paleoproterozoic GOE (at ~2.4 Ga). This suggests that the black shales are either not related to the GOE and they were deposited later than the ^{13}C-enriched carbonates. Black shales in the PB are characterized with high Total Organic Carbon (TOC) (Table 2), highest V content (up to 131 ppm), and higher U/Th (Papineau et al., 2009). This suggests association of V with organic matter (cf. Peacor et al., 2000). There are small U enrichments in organic-rich shales of the NPB (Roy, 1995). Mn contents are generally low in both PB and NPB (Bhu and Singh, 2005). Shale-normalized Rare Earth Elements (REE) have flat patterns, some with minor Heavy Rare Earth Elements (HREE) depletion, but without significant Ce or Eu anomalies (Papineau et al., 2009).

Table 2 Range of Values of Various Stable Isotopes and Chemical Ratios of the Black Shale and Carbonate Rocks from the Lower Aravalli Group Phosphate and Nonphosphate Basins

Range of Values	Phosphate Basins		Nonphosphate Basins	
	Black Shales	Carbonates	Black Shales	Carbonates
$\delta^{13}C_{carb}$	NA	−1.5 to +1.5‰	NA	−5.4 to +11.2‰
$\delta^{13}C_{org}$	−18.8 to −29.7‰	−33.1 to −10.1‰	−13.1 to −28.7‰	−27 to −11.9‰
$\delta^{15}N_{WR}$	−4.7 to +26.9‰	0.0 to +3.4‰	+4.4 to +11.6‰	−0.5 to +2.6‰
$\delta^{18}O$	ND	−4 to −17‰	ND	−0.1 to −20.7‰
TOC	0.3–13.8%	0.05–2.17%	0.4–3.6%	0.04–0.36%
TN	ND	0.02–0.16%	ND	0.05–0.1%
P_2O_5	ND	Up to 34.8%	ND	0.01–0.12%
Mg/Ca	NA	0 to 0.6	NA	~0.6

NA, Not applicable; ND, Not determined.
After Purohit et al. (2010) and Papineau et al. (2009, 2013).

19.4.3 STABLE ISOTOPE CHEMISTRY OF THE PB AND NPB

19.4.3.1 Carbonates

Carbonates of the PB exhibit near-zero $\delta^{13}C_{carb}$ values and a wide range of $\delta^{13}C_{org}$ values between −33.3‰ and −10.1‰ (Table 2). These values indicate a complex carbon cycle. On the other hand, those in the NPB show a range of positive $\delta^{13}C_{carb}$ excursions with values up to +11.2‰ (Fig. 3). There are also some other carbonates showing moderately high $\delta^{13}C_{carb}$ values around +6‰ or +3‰ (Table 2). The overall variation in the $\delta^{13}C_{carb}$ excursion suggests smaller carbon cycle perturbations. The $\delta^{15}N$ values in all these rocks vary between −0.7‰ and +3.4‰ (Table 2) which is consistent with the predominance of nitrogen fixation during cyanobacterial blooms in all basin types (Papineau et al., 2013). Low $\delta^{15}N$ in these rocks perhaps occurred as a biological response to high P availability. Increased P availability after the Paleoproterozoic LJE caused cyanobacterial blooms and is considered as a key factor in the oxygenation of the Earth's atmosphere (Papineau et al., 2013).

19.4.3.2 Black Shales

Black shales in both the PB and NPB have distinctive characteristics even though these evolved coeval. Black shales of the PB have trace element concentration (high V and U/Th values) typical of oxygen-deficient environments depict eutrophic conditions during the preceding carbonate deposition. These shales also have a high organic content of up to 14 wt%, generally homogeneous $^{13}C_{org}$ values of around −29‰, C/N_{WR} up to 360 and a wide range of ^{15}N values between −5 and +27‰ (Table 2). The organic matter in the shales is metamorphosed from lower green schist to lower amphibolite facies (Bhu and Singh, 2005). Geochemical data suggest high primary productivity under eutrophic conditions, as well as intense recycling of organic nitrogen through ammonium assimilation and denitrification at the time of sedimentation (Papineau et al., 2009). Black shales from the NPB show trace element concentrations characteristic of oxygenated conditions as compared to the PB. The NPB shales contain less than 3 wt% organic matter with $^{13}C_{org}$ values often between −13 and −18‰, and the $^{15}N_{WR}$ values between +5 and +12‰ (Table 2). Geochemical data of the carbonaceous shales from the NPB also indicate high rates of primary productivity.

19.5 REDOX STATE VARIATIONS IN PBS AND NPBS

Sedimentation in the PB and NPB of the Lower Aravalli Group was a consequence of an early Paleoproterozoic oxidative chemical weathering which was as an aftermath of rifting caused due to magma outpouring (Barley et al., 2005). The global event was recorded in the Aravalli Craton at 2.4–2.3 Ga (Ahmad et al., 2008). This was followed by a rise in the atmospheric oxygen level and changes in the ocean water chemistry leading to the deposition of the carbonates and then the carbonaceous shales. Prevalent changes associated with the redox states during sedimentation were indicated by several geochemical proxies.

Carbonaceous shales in the PB were deposited under semi-restricted and anoxic conditions while in the NPB these were deposited in a relatively higher oxidizing condition (Papineau et al., 2009). Carbonaceous shales from the Umra subbasin of the NPB have noticeable U enrichments that suggest deposition under a reducing environment (Roy, 1995). Papineau et al. (2009) stated that organic-rich shales from the PB have higher V/Sc than shales from NPB which in turn suggests anoxic conditions of deposition for these shales (cf. Tribovillard et al., 2006).

Phosphate accumulation is favored in oxic sediments rather than anoxic sediments (Ingall et al., 1993; Van Cappellen and Ingall, 1996). Phosphate-mediated eutrophic microbial ecosystems develop within the photic zone of the shallow, oxygenated marine realm (Papineau et al., 2013). The accumulation of phosphate in stromatolites has presumably occurred either in the oxygenated uppermost part of the water column or in the shallow oxygen-minimum zone of the PB. Low C and P contents of organic-rich shales from the NPB are consistent with lower U/Th (Papineau et al., 2009) indicating a less pronounced redox gradient in the water column. Intermittent highly oxidized shale layers overlying the dolomites of the NPB suggest increased redox states. These two layers further stand in marked contrast with the reduced white mica paleosols at the base of the Aravalli Supergroup and red ochre beds at the top (Roy and Jakhar, 2002). It is correlated to changes in the redox state with increased oxygen availability. The sharp change in the redox state of the sediments occurring between the lowermost parts and the higher strata of the Lower Aravalli Group indicates increase in oxygen levels.

Redox state variations are also paleogeographic dependent and it becomes relevant because there is a significant difference between the depositional paleogeography of the PB and the NPB. In a recent study, Kah et al. (2012) observed difference in heavy carbon isotope enrichment levels between the epicratonic (open) and the pericratonic (restricted) environments. Similar heavy carbon enrichment patterns are observed in the Aravalli sediments. The prevalence of restricted environments during the deposition of the Aravalli sediments depleted the ^{13}C while the open environments enriched the ^{13}C due to the differences in redox conditions (cf. Kah et al., 2012). In addition to the paleoenvironment, the paleoceanographic factors could also have influenced the redox states. The epeiric seas are typified by restricted circulation patterns that produce water masses with composition different from the open seas (Hiatt and Budd, 2001; Piper, 2001; Algeo and Heckel, 2008).

19.6 DISCUSSION

19.6.1 GLOBAL VERSUS LOCAL SCENARIO

The NPB carbonates display LJE-type high ^{13}C enrichments which have many similar occurrences world over. However, the moderately high ^{13}C enrichment which precedes the high ^{13}C enrichment event in the NPB carbonates of the Lower Aravalli Group is unknown globally. No ^{13}C enrichment is noted in the organically rich PB carbonates. This warrants understanding of the causes of enrichment. Earlier studies by Sreenivas et al. (2001) and Maheshwari et al. (1999, 2002; 2010) correlated the $\delta^{13}C_{carb}$ enrichment in the Lower Aravalli Group with the causes known for enrichment in the Paleoproterozoic carbonates of the Lomagundi Formation of Zimbabwe. Purohit et al. (2010) while studying the Lower Aravalli Group carbonates emphasized the role of the basin-scale depositional environment as the influencing factor for the $\delta^{13}C_{carb}$ variations. These local scale factors build up a coherent story in conjunction with the global models of Paleoproterozoic $\delta^{13}C_{carb}$ trends (cf. Salminen et al., 2013). Papineau et al. (2013) emphasized that the increased P availability might be the dominant cause for cyanobacterial blooms that produced significant quantities of organic matter which possibly was linked to oxygen production related to GOE. These authors highlighted the possibility of global organic burial and stated that the cyanobacterial bloom caused the ^{13}C enrichment which was ironically not evident in the NPB due to open marine basin configuration that lacked phosphogenesis. The abundance of P was the possible cause for the $\delta^{13}C_{carb}$ enrichments which were related to oxygenation. However, the basins which bear P (i.e., PB) do not show enrichments. The reason for lack of evidence of the ^{13}C enrichments in the P-rich basins remained incompletely understood against the known global scenario.

The concept that organic burial can cause increased oxygenation is now widely acknowledged (cf. Bekker et al., 2008; Bekker and Holland, 2012). This is mainly based on the evidence of formation of stoichiometrically equal amounts of oxygen and reduced carbon through the process of oxygenic photosynthesis. However, given cognizance to the statement of Catling and Claire (2005) that "the availability of free O_2 on the Earth's surface during the Paleoproterozoic interval was negligible," this interpretation may not hold true. These authors have disagreed strongly under the premise that a conservative interpretation of the carbon isotope record is not consistent with a secular increase in organic carbon burial rates from the early Archean forward in time through the Phanerozoic. On similar grounds, Kump and Barley (2007) ruled out the possibility of any secular trend indicating a substantial increase in organic carbon burial between 2.5 and 2.4 Ga and argued that any rise of O_2 at this time could have resulted due to a decline in the O_2 sinks. Interestingly, the global scale LJE (~2.06 Ga) which has been linked to permanent increase in oxygenation postdates the 2.4 Ga GOE. Catling and Claire (2005) argued that the high ^{13}C enrichment (perhaps related to oxygenation in the Aravalli carbonates also) took place during the die-out pulse of the GOE while the moderate enrichment could be a preceding weaker pulse.

Hayes and Waldbauer (2006) also disputed the suggestion of any global scale organic burial as the cause of ^{13}C enrichment because of the decoupling between the O_2 levels to the organic carbon burial rates and carbon isotopic signals. These authors argued that if there was an organic carbon burial event that caused both GOE and the subsequent disappearance of the mass-independent fractionation of sulfur isotopes (MIF-S), then the loss of MIF-S is unlikely to precede the carbon-isotopic enrichments. Few other workers (Catling et al., 2001; Hayes and Waldbauer, 2006; Jackson et al., 2007; Falkowski and Isozaki, 2008) argued against any major global scale organic burial as the possible cause of ^{13}C enrichment and suggested that the organic carbon-bearing carbonates were never permanently subducted into the mantle. As an alternative, these workers suggested that the organic carbon was returned to the Earth's surface in various forms where it had another chance to combine with O_2 canceling O_2 gains.

Organic burial should also result in the formation of carbon minerals or organic carbon in the buried rocks. In the case of the Lower Aravalli Group carbonates, evidence of primary or secondary carbon minerals is absent (Purohit et al., 2010). TOC of the carbonates is also low (Table 2). The model of organic burial propagated by Bekker et al. (2001) from the Transvaal Group rocks of Africa stated that the organic burial was related to organic-rich shales which underlie the ^{13}C-enriched carbonates. In the case of the Jhamarkotra Formation of the Lower Aravalli Group, it is the reverse case. Here, the organic-rich shales overlie carbonates that show ^{13}C enrichment. Hence organic burial dynamics do not suit as a cause of the enrichment process in the carbonates of the Lower Aravalli Group.

19.6.2 CORRELATION OF THE PBS AND NPBS

Deposition of carbonate rocks in both the PBs and the NPBs could be broadly coeval but not synchronous (Papineau et al., 2013). Nonphosphatic dolomites were possibly deposited prior to the phosphatic dolomites (Purohit et al., 2012). Two evidence are reported in support of this. One is the field relationship between the NP carbonates and P carbonates along the margins of the Umra basin where the growth direction (younging direction) of stromatolitic columns is markedly away from the nonphosphatic carbonates (Purohit et al., 2010). Second is the study by Purohit et al. (2012) which states that the nonphosphatic calcitic carbonates of Babarmal basin underlie the phosphatic dolomitic carbonates of the PB.

The PB shows significant differences with the NPB on time equivalence. These differences could perhaps be a reflection of interplay of factors that were operational at global as well as local scale. The PB shows abundance of P resulting in the cyanobacterial bloom in the contemporaneous Aravalli Sea. In contrast, the NPB lacked cyanobacterial activities that resulted into the growth of phosphate-bearing stromatolites. The carbonates of the NPB were deposited in open marine bank-shelf conditions which in turn perhaps hindered the cyanobacterial activity. Carbonates of the PB, due to cyanobacterial stromatolites, had higher oxidizing conditions at the time of deposition as compared to the NPB carbonates. However, the abundant cyanobacterial colonies might have contributed significant organic matter which possibly led to the formation of eutrophic conditions (cf. Bekker and Holland, 2012). The transformation of oxic to anoxic conditions in the PB and NPB was most probably synchronous though local scale variations are noticeable between the two. The black shales of PB are nonuraniferous with high TOC and U/Th. These suggest that the paleoenvironmental conditions of black shale deposition in the PB were less oxygenic than the NPB.

Besides the differences in P availability, which consequently led to the differences in redox conditions between the PB and NPB, there were also differences in ^{13}C enrichment between the two. Bimodal ^{13}C enrichment observed in the NPB carbonates did not come into play in PB carbonates with nonspecific reasons. The moderate enrichment perhaps precedes the heavy enrichment and could be identified as an abortive attempt to increase O level that might have led to heavy enrichment. The heavy enrichment has been correlated with the global LJE considering its geochronological relationship (Martin et al., 2013; Bekker and Holland, 2012).

19.7 SUMMARY

The Paleoproterozoic carbonates of the Lower Aravalli Group of northwestern India show congruence with other coeval carbonates worldwide, especially with respect to the high ^{13}C-enrichment event recorded during the ~2.06 Ga LJE. However, the geochemical and stratigraphic characteristics of the Aravalli carbonates are significantly different and can be considered to be unique when compared with the contemporary global examples. The Aravalli carbonates show high abundance of P_2O_5 with prolific cyanobacterial stromatolitic growth with virtual absence of ^{13}C enrichment. Paradoxically, the carbonates which are rich in organic content do not exhibit ^{13}C enrichments while the carbonates which lack source and presence of organic content exhibit moderate to high enrichments of ^{13}C. The NPB carbonates which show moderate enrichment (up to +6‰) probably precedes the high ^{13}C enrichments. This is a unique feature associated with the carbonates of the Lower Aravalli Group and is seldom displayed globally.

Relying on the chemostratigraphic relationship derived from different carbonate ensembles of the Lower Aravalli Group (cf. Purohit et al., 2010, 2012) a stepwise rise in oxygen level is suggested (cf., Murakami, et al., 2011; Kump et al., 2011). The stepwise rise in oxygen level reached its peak at the time of cyanobacterial bloom. Gradual reduction in the oxygen level from the peak ultimately led to the return of anoxic conditions that favored the deposition of black shales. Sharp change in the redox state during the deposition of Lower Aravalli Group of rocks is also indicated by paleosols which are characteristically of white mica at the base and red ochre at the top (Roy and Jakhar, 2002). The occurrences of paleosol are suggestive of transformations in oxygen level. The overall depositional framework of the Lower Aravalli carbonates is indicative of fluctuating O_2 levels from low to high and again back to low during the Paleoproterozoic. Comprehensively, the sequence of events in relation to the O level for the Aravalli rocks can be summarized as follows.

The onset of the Paleoproterozoic Aravalli orogeny at ~2.38 Ga matches with the global GOE dated at 2.4 Ga. The incipient rise in O levels is attributed to the mantle outpouring and the resultant deposition of the basal Aravalli volcanics (Delwara Formation, Table 1) (cf., Barley et al., 2005). This was followed by moderate ^{13}C excursions displayed by the Babarmal calcitic carbonates and Umra dolomitic carbonates in the NPB suggesting increase in O level. Further rise in O_2 levels is noticed with the transformation in oxygenation conditions and is witnessed by increase in ^{13}C enrichment from moderate to high in the bank-shelf NPB carbonates. This is correlatable with the "LJE" at ~2.06 Ga (cf. Bekker and Holland, 2012; Martin et al., 2013). Continued rise in O_2 level coincides with the proliferation of cyanobacterial activity noted in the PBs of the Jhamarkotra Formation. This increase in oxygen levels should have resulted in still higher enrichment levels of ^{13}C, but on the contrary the values of ^{13}C are close to zero. It can be assumed that the cyanobacterial organisms with reasons not known prohibited the ^{13}C enrichment. Prevalence of decoupling between P and C cycles due to excess of availability of former due to increased influx in excess of the riverine input (cf. Aharon, 2005) could be an alternative explanation for this phenomenon.

The carbonates show basin-scale microenvironmental variations which are seldom observed on coeval stratigraphic sections located elsewhere on a global scale. Organic burial as the primary cause of ^{13}C enrichment during the global LJE (Bekker and Holland, 2012 and references therein) cannot be substantiated in the case of the Lower Aravalli carbonates because the black shales (rich in buried organic content) postdate the ^{13}C-enriched carbonates. Conversely, it may be emphasized that the cyanobacterial bloom (which yielded massive organic content) observed in the Lower Aravalli carbonates was a consequence of augmentation of O_2 level due to the ^{13}C enrichment present in the preceding NPB carbonate sequences. It is assumed that the threshold O_2 level yielded from the high ^{13}C-enrichment event (coeval to the LJE), resulted in budding of the cyanobacterial colonies which bloomed under enhanced P levels which in turn was virtually controlled by the Fe-redox pumping. The cyanobacterial bloom coupled with the ^{13}C enrichments are direct evidence of the oxygen overshoot during the late Paleoproterozoic in the Lower Aravalli Group. However, this should be taken up as a hypothesis which requires further testing with more reliable age constraints.

Summarily, it is stated that the stable carbon isotope-based chemostratigraphic studies carried out on different sections of the coeval but nonsynchronous PBs and NPBs have helped to build up a coherent story of paleoenvironmental evolution and basin correlation.

ACKNOWLEDGMENTS

Author thanks the Principal, Government College Sirohi, for providing necessary facility for the work. Thanks are also due to Dr. K. K. Sharma for assistance in designing the figures. Help yielded by Ms Shweta Rajawat in manuscript preparation is also acknowledged.

REFERENCES

Ahmad, T., Dragusanu, C., Tanaka, T., 2008. Provenance of Proterozoic Basal Aravalli mafic volcanic rocks from Rajasthan, Northwestern India: Nd isotopes evidence for enriched mantle reservoirs. Precambrian Research 162, 150–159.

Ahron, P., 2005. Redox stratification and anoxia of the early Precambrian oceans: implications for carbon isotope excursion and oxidation events. Precambrian Research 137, 207–222.

Algeo, T.J., Heckel, P.H., 2008. The late Pennsylvanian Midcontinent sea of North America: a review. Paleogeography, Paleoclimatology, Paleoecology 268, 205–221.

Baker, A.J., Fallick, A.E., 1989a. Evidence from Lewisian limestones for isotopically heavy carbon in two thousand-million-year-old sea water. Nature 337, 352–354.

Baker, A.J., Fallick, A.E., 1989b. Heavy carbon in two-billion-year-old marbles from Lofoten-Vesterålen, Norway: implications for the Precambrian carbon cycle. Geochimica et Cosmochimica Acta 53, 1111–1115.

Banerjee, D.M., 1971. Precambrian stromatolitic phosphorites of Udaipur, Rajasthan, India. Geological Society America Bulletin 82, 2319–2330.

Banerjee, D.M., Basu, P.C., Srivastava, N., 1980. Petrology, mineralogy, geochemistry, and origin of the Precambrian Aravallian phosphorite deposits of Udaipur and Jhabua, India. Economic Geology 75, 1181–1199.

Banerjee, D.M., Schidlowski, M., Arneth, J.D., 1986. Genesis of Upper Proterozoic-Cambrian phosphorite deposits of India: isotopic inferences from carbonate fluorapatite, carbonate and organic carbon. Precambrian Research 33, 239–253.

Barley, M.E., Bekker, A., Krapez, B., 2005. Late Archean to early Proterozoic global tectonics, environmental change and the rise of atmospheric oxygen. Earth and Planetary Science Letters 238, 156–171.

Bekker, A., Karhu, J.A., Eriksson, K.A., Kaufman, A.J., 2003a. Chemostratigraphy of Paleoproterozoic carbonate successions of the Wyoming Craton: tectonic forcing of biogeochemical change? Precambrian Research 120, 279–325.

Bekker, A., Karhu, J.A., Kaufman, A.J., 2006. Carbon isotope record for the onset of the Lomagundi carbon isotope excursion in the Great Lakes area, North America. Precambrian Research 148, 145–180.

Bekker, A., Kaufman, A.J., Karhu, J.A., Beukes, N.J., Swart, Q.D., Coetzee, L.L., Eriksson, K.A., 2001. Chemostratigraphy of the Paleoproterozoic Duitschland Formation, South Africa: implications for coupled climate change and carbon cycling. American Journal of Science 301, 261–285.

Bekker, A., Sial, A.N., Karhu, J.A., Ferreira, V.P., Noce, C.M., Kaufman, A.J., Romano, A.W., Pimentel, M.M., 2003b. Chemostratigraphy of carbonates from the Minas supergroup, Quadrilátero Ferrífero (Iron Quadrangle) Brazil: a stratigraphic record of early proterozoic atmospheric, biogeochemical and climatic change. American Journal of Science 303, 865–904.

Bekker, A., Holmden, C., Beukes, N., Kenig, F., Eglington, B., Patterson, W., 2008. Fractionation between organic and inorganic carbon during the Lomagundi (2.22–2.1 Ga) carbon isotope excursion. Earth and Planetary Science Letters 271, 278–291.

Bekker, A., Holmden, C., Patterson, W., Eglington, B., Coetzee, L., Beukes, N., 2004a. Chemostratigraphy of early Paleoproterozoic carbonates of South Africa. In: Geological Society of America Annual Meeting. http://www.geosociety.org/2004.

Bekker, A., Holland, H.D., 2012. Oxygen overshoot and recovery during the early Paleoproterozoic. Earth and Planetary Science Letters 295-304, 317–318.

Bekker, A., Holland, H.D., Wang, P.-L., Rumble III, D., Stein, H.J., Hannah, J.L., Coetzee, L.L., Beukes, H.J., 2004b. Dating the rise of atmospheric oxygen. Nature 427, 117–120.

Bejarniya, B.R., 1986. Stratigraphy and tectonics of the Precambrian rocks of the Kabita–Lakhawali–Rama triangle, north of Udaipur city, Rajasthan Unpublished (Ph.D. Thesis). University of Rajasthan, Jaipur.

Bhu, H., Singh, K., 2005. Anoxia in proterozoic black shales: a case study near Udaipur, Rajasthan. Indian Journal of Geochemistry 20, 77–90.

Catling, D.C., Claire, M.W., 2005. How Earth's atmosphere evolved to an oxic state: a status report. Earth and Planetary Science Letters 237, 1–20.

Catling, D.C., Zahnle, K.J., McKay, C.P., 2001. Biogenic methane, hydrogen escape, and the irreversible oxidation of early Earth. Science 293, 839–843.

Chaudhary, A.K., Gopalan, K., Sastry, C.A., 1984. Present status of geochronology, of the Precambrian rocks of Rajasthan. Tectonophysics 105, 131–140.

Chauhan, D.S., 1979. Phosphate-bearing stromatolites of the Precambrian phosphorite deposits of Udaipur region, their environmental significance and genesis of phosphorite. Precambrian Research 8, 95–126.

Choudhri, R., Roy, A.B., 1986. Proterozoic and Cambrian Phosphorites-deposits: Jhamarkotra, Rajasthan, India. In: Cook, P.J., Shergold, J.H. (Eds.), Proterozoic and Cambrian Phosphorites. Cambridge University Press, pp. 202–219.

Deb, M., Thorpe, R.A., 2004. Geochronological constraints in the Precambrian geology of Rajasthan and their metallogenic implications. In: Deb, M., Goodfellow, W.D. (Eds.), Sediment-hosted Lead–Zinc Sulphide Deposits. Narosa Publishing House, New Delhi, pp. 246–263.

Falkowski, P.G., Isozaki, Y., 2008. The story of O_2. Science 322, 540–542.

Frei, R., Gaucher, C., Poulton, S.W., Canfield, D.E., 2009. Fluctuations in Precambrian atmospheric oxygenation recorded by chromium isotopes. Nature 461, 250–253.

Galimov, E.M., Kuznetsova, N.G., Prokhorov, V.S., 1968. On the problem of the Earth's ancient atmosphere composition in connection with results of isotope analysis of carbon from the Precambrian carbonates. Geochemistry 11, 1376–1381.

Golani, P.R., 1981. Structural geometry and stratigraphy of the Precambrian rocks around Madar, District Udaipur, Rajasthan Unpublished (Ph.D. Thesis). University of Rajasthan, Jaipur.

Guo, Q., Strauss, H., Kaufman, A.J., Schröder, S., Gutzmer, J., Wing, B., Baker, M.A., Bekker, A., Jin, Q., Kim, S.-T., Farquhar, J., 2009. Reconstructing Earth's surface oxidation across the Archean–Proterozoic transition. Geology 37, 399–402.

Hayes, J.M., Waldbauer, J.R., 2006. The carbon cycle and associated redox processes through time. Philosophical Transactions of the Royal Society B 361, 931–950.

Heron, A.M., 1953. Geology of central Rajputana. Memoir Geological Survey of India 79, 339.

Hiatt, E.E., Budd, D.A., 2001. Sedimentary phosphate formation in warm shallow water: new insights into the paleoceanography of the Permian Phosphoria Sea from analysis of phosphate oxygen isotopes. Sedimentary Geology 145, 119–133.

Holland, H.D., 2006. The oxygenation of the atmosphere and oceans. Philosophical Transactions of the Royal Society B 361, 903–915.

Ingall, E.D., Bustin, R.M., Van Cappellen, P., 1993. Influence of water column anoxia on the burial and preservation of carbon and phosphorus in marine shales. Geochimica et Cosmochimica Acta 57, 303–316.

Jackson, M.G., Hart, S.R., Koppers, A.A.P., Staudigel, H., Konter, J., Blusztajn, J.B., Kurz, M., Russell, J.A., 2007. The return of subducted continental crust in Samoan lavas. Nature 448, 684–687.

Kah, L., Bartley, J.K., Teal, D.A., 2012. Chemostratigraphy of the late Mesoproterozoic Atar Group, Taoudeni Basin, Mauritania: muted isotopic variability, facies correlation, and global isotopic trends. Precambrian Research 203, 82–103.

Karhu, J.A., 1993. Paleoproterozoic evolution of the carbon isotope ratios of sedimentary carbonates in the Fennoscandian Shield. Geological Survey of Finland Bulletin 371, 1–87.

Kump, L.R., Barley, M.E., 2007. Increased subaerial volcanism and the rise of atmospheric oxygen 2.5 billion years ago. Nature 448, 1033–1036.

Kump, L.R., Junium, C., Arthur, M.A., Brasier, A., Fallick, A., Melezhik, V., Lepland, A., Crne, A.E., Luo, G., 2011. Isotopic evidence for massive oxidation of organic matter following the great oxidation event. Science 334, 1694–1696.

Lindsay, J.F., Brasier, M.D., 2002. Did global tectonics drive early biosphere evolution? Carbon isotope record from 2.6 to 1.9 Ga carbonates of Western Australian basins. Precambrian Research 114, 1–34.

Maheshwari, A., Sial, A.N., Chittora, V.K., 1999. High $\delta^{13}C$ paleoproterozoic carbonates from the Aravalli supergroup, Western India. International Geology Review 41, 949–954.

Maheshwari, A., Sial Alcides, N., Gaucher, C., Bossi, J., Bekker, A., Ferreira, V.P., Romano, A.W., 2010. Global nature of the Paleoproterozoic Lomagundi carbon isotope excursion: a review of occurrences in Brazil, India and Uruguay. Precambrian Research 182, 274–299.

Maheshwari, A., Sial, A.N., Chittora, V.K., Bhu, H., 2002. A positive $\delta^{13}C_{carb}$ anomaly in Paleoproterozoic carbonates of Aravalli Craton, Western India – support for a global isotopic excursion. Journal of Asian Earth Sciences 21, 59–67.

Martin, A.P., Condon, D.J., Prave, A.R., Lepland, A., 2013. A review of temporal constrains for the Paleoproterozoic large, positive carbonate carbon isotope excursion (the Lomagundi Jatuli Event). Earth-Science Reviews 127, 242–261.

Mehta, P., 2010. Lithostratigraphic and tectonic studies with emphasis on Barytes Mineralization in Babarmal region, District—Udaipur, Rajasthan Unpublished (Ph.D. Thesis). M.L.S. University, Udaipur, India.

Mckenzie, N.R., Hughes, N.C., Myrow, P.M., Banerjee, D.M., Deb, M., Planavsky, N.J., 2013. New age constraints for the Proterozoic Aravalli–Delhi successions of India and their implications. Precambrian Research 238, 120–128.

Melezhik, V.A., Fallick, A.E., 1996. A widespread positive $\delta^{13}C_{carb}$ anomaly at around 2.33–2.06 Ga on the Fennoscandian Shield: a paradox? Terra Nova 8, 141–157.

Melezhik, V.A., Fallick, A.E., 2010. On the Lomagundi–Jatuli carbon isotopic event: the evidence from the Kalix Greenstone Belt, Sweden. Precambrian Research 179, 165–190.

Melezhik, V.A., Fallick, A.E., Clark, T., 1997. Two billion year old isotopically heavy carbon: evidence from the Labrador Trough, Canada. Canadian Journal of Earth Sciences 34, 271–285.

Melezhik, V.A., Fallick, A.E., Medvedev, P.V., Makarikhin, V.V., 1999. Extreme $^{13}C_{carb}$ enrichment in ca. 2.0 Ga magnesite–stromatolite–dolomite–"red beds" association in a global context: a case for the world-wide signal enhanced by a local environment. Earth-Science Reviews 48, 71–120.

Melezhik, V., Purohit, R., Papineau, D., 2014. Comment on "New age constraints for the Proterozoic Aravalli–Delhi successions of India and their implications" by McKenzie et al. [Precambrian Res. 238 (2013) 120–128]. Precambrian Research 246, 319–320.

Murakami, T., Sreenivas, B., Sharma, D., Sugimori, H., 2011. Quantification of atmospheric oxygen levels during the Paleoproterozoic using paleosol composition iron oxidation kinetics.. Geochimica et Cosmochimica Acta 75, 2982.

Nelson, G.J., Pufahl, P.K., Hiatt, E.E., 2010. Paleoceanographic constraints on Precambrian phosphorite accumulation, Baraga group, Michigan, USA. Sedimentary Geology 226, 9–21.

Papineau, D., Purohit, R., Goldberg, T., Pi, D., Shields, G.A., Bhu, H., Steele, A., Fogel, M.L., 2009. High primary productivity and nitrogen cycling after the Paleoproterozoic phosphogenic event in the Aravalli Supergroup, India. Precambrian Research 171, 37–56.

Papineau, D., Purohit, R., Fogel, M., Shields, G., 2013. High phosphate availability as a possible cause for massive cyanobacterial production of oxygen in the Paleoproterozoic atmosphere. Earth and Planetary Science Letters 362, 225–236.

Peacor, D.R., Coveney, R.M., Zhao, G., 2000. Authigenic illite and organic matter: the principal hosts of vanadium in the Mecca Quarry shale at Velpen, Indiana. Clay Mineralogy 48, 311–316.

Piper, D.Z., 2001. Marine chemistry of the Permian Phosphoria formation and basin, southeast Idaho. Economic Geology 96, 599–620.

Préat, A., Bouton, P., Thiéblemont, D., Prian, J.-P., Ndounze, S.S., Delpomdor, F., 2011. Paleoproterozoic high $\delta^{13}C$ dolomites from the Lastoursville and Franceville basins (SE Gabon): stratigraphic and synsedimentary subsidence implications. Precambrian Research 189, 212–228.

Purohit, R., Bhu, R., Mehta, P., 2014. Contrasting carbonate lithofacies from the paleoproterozoic lower Aravalli group, NW India. Journal of Applied Geochemistry 16 (1), 1–10.

Purohit, R., Sanyal, P., Bhattacharya, S.K., Roy, A.B., 2010. ^{13}C enrichment in Paleoproterozoic carbonates rocks of northwest India: influence of the microenvironment. Gondwana Research 18, 538–546.

Purohit, R., Papineau, D., Kroner, A., Sharma, K.K., Roy, A.B., 2012. Carbon isotope geochemistry and geochronological constraints of the Neoproterozoic sirohi group from northwest India. Precambrian Research 220–221, 80–90.

Purohit, R., Rengarajan, R., 2007. Syn- and Post depositional geochemical alterations in paleoproterozoic Jhamarkotra carbonates from Aravalli Supergroup: implications to Paleoenvironment. Journal of Applied Geochemistry 9 (2), 256–270.

Roy, A.B., Paliwal, B.S., 1981. Evolution of lower proterozoic epicontinental deposits: stromatolites-bearing Aravalli rocks of Udaipur, Rajasthan, India. Precambrian Research 14, 49–74.

Roy, A.B., 1995. Precambrian geology of the Aravalli Mountain and Uranium metallogeny. Exploration and Research for Atomic Minerals 8, 61–71.

Roy, A.B., 2000. Geology of the paleoproterozoic Aravalli supergroup of Rajasthan and northern Gujarat. In: Deb, M. (Ed.), Crustal Evolution and Metallogeny in the Northwestern Indian Shield (A Festschrift for A. Mookherjee). Narosa Publishing House, New Delhi, pp. 87–114.

Roy, A.B., Sharma, B.L., Paliwal, B.S., Chauhan, N.K., Nagori, D.K., Golani, P.R., Bejarniya, B.R., Bhu, H., Sabah, M.A., 1993. Lithostratigraphy and tectonic evolution of Aravalli Supergroup—A Protogeosynclinal sequence. In: Cassyap, S.M., Valdiya, K.S. (Eds.), Rifted Basins and Aulcogens. Gyanodaya Prakasan, Nainital, India, pp. 73–90.

Roy, A.B., Jakhar, S.R., 2002. Geology of Rajasthan (Northwest India) Precambrian to Recent. Scientific Publishers, India. 121.

Roy, A.B., Nagori, D.K., 1990. Influence of differential basement mobility on contrasting structural style in cover rocks: an example from early Precambrian rocks east of Udaipur, Rajasthan. Proceedings of Indian Academy of Sciences (Earth & Planetary Sciences) 99 (2), 291–308.

Salminen, P.E., Karhu, J.A., Melezhik, V.A., 2013. Tracking lateral $\delta^{13}C_{carb}$variation in the paleoproterozoic Pechenga Greenstone Belt, the north eastern Fennoscandian shield. Precambrian Research 228, 177–193. http://dx.doi.org/10.1016/j.precamres.2013.01.005.

Sarangi, S., Gopalan, K., Roy, A.B., Sreenivas, B., Das Sharma, S., 2006. Pb–Pb age of the carbonates of Jhamarkotra formation constraining the age of the Aravalli Supergroup, Rajasthan. Journal Geological Society of India 67, 442–446.

Schidlowski, M., Eichmann, R., Junge, C.E., 1975. Precambrian sedimentary carbonates: carbon and oxygen isotope geochemistry and implications for the terrestrial oxygen budget. Precambrian Research 2, 1–69.

Schidlowski, M., Eichmann, R., Junge, C.E., 1976. Carbon isotope geochemistry of the Precambrian Lomagundi carbonate province, Rhodesia. Geochimica et Cosmochimica Acta 40, 449–455.

Scott, C., Lyons, T.W., Bekker, A., Shen, Y., Poulton, S.W., Chu, X., Anbar, A.D., 2008. Tracing the stepwise oxygenation of the Proterozoic ocean. Nature 452, 456–459.

Sreenivas, B., Das Sharma, S., Kumar, B., Patil, D.J., Roy, A.B., Srinivasan, R., 2001. Positive $\delta^{13}C$ excursion in carbonate and organic fractions from the Paleoproterozoic Aravalli Supergroup, Northwestern India. Precambrian Research 106, 277–290.

Tang, H., Chen, Y., Wu, G., Lai, Y., 2011. Paleoproterozoic positive $\delta^{13}C_{carb}$ excursion in the northeastern Sino-Korean craton: Evidence of the Lomagundi Event. Gondwana Research 19, 471–481.

Tribovillard, N., Algeo, T.J., Lyons, T.W., Riboulleau, A., 2006. Application of trace metals as paleoredox and paleoproductivity proxies. Chemical Geology 232, 12–32.

Tripathy, G.R., Singh, S.K., Bhu, H., 2013. Re–Os isotopes and major trace element geochemistry of carbonaceous shales, Aravalli super group, India: impact of post-depositional processes. Chemical Geology 354, 93–106.

Van Cappellen, P., Ingall, E.D., 1996. Redox stabilization of the atmosphere and oceans by phoshorus-limited marine productivity. Science 271, 493–496.

Wall, H.D., Pandit, M.K., Chauhan, N.K., 2012. Paleosol occurrences along the Archaean–Proterozoic contact in the Aravalli craton, NW India. Precambrian Research 216-219, 120–131.

Wiedenbeck, M., Goswami, J.N., Roy, A.B., 1996. Stabilization of the Aravalli craton of the northwestern India at 2.5 Ga.: an ion-microprobe zircon study. Chemical Geology 129, 325–340.

Yudovich, Y.E., Makarikhin, V.V., Medvedev, P.V., Sukhanov, N.V., 1991. Carbon-isotope anomalies in carbonates of the Karelian Complex. Geochemistry International 28, 56–62.

Zagnitko, V.N., Lugovaya, I.P., 1989. Isotope Geochemistry of Carbonate and Banded Iron Formations of the Ukrainian Shield. NaukovaDumka, Kiev.

Index

Note: Page numbers followed by "f" and "t" indicate figures and tables respectively.

D

O

P

FSC
www.fsc.org
MIX
Papier | Fördert gute Waldnutzung
FSC® C083411

Kolibri 360 GmbH
Ferdinand-Jühlke-Straße 7
99095 Erfurt, Deutschland
produktsicherheit@kolibri360.de